III International Scientific and Practical Conference on Innovations in Engineering and Technology (ISPCIET 2020)

AA007605

IOP Conference Series: Materials Science and Engineering
Volume 939

Veliky Novgorod, Russia
25 – 26 June 2020

Editor:

Sergey B. Sapozhkov

ISBN: 978-1-7138-4774-8
ISSN: 1757-8981

Printed from e-media with permission by:

Curran Associates, Inc.
57 Morehouse Lane
Red Hook, NY 12571

CURRAN ASSOCIATES INC. proceedings.com

Some format issues inherent in the e-media version may also appear in this print version.

This work is licensed under a Creative Commons Attribution 3.0 International Licence.
Licence details: http://creativecommons.org/licenses/by/3.0/.

No changes have been made to the content of these proceedings. There may be changes to pagination and minor adjustments for aesthetics.

Printed with permission by Curran Associates, Inc. (2022)

For permission requests, please contact the Institute of Physics
at the address below.

Institute of Physics
Dirac House, Temple Back
Bristol BS1 6BE UK

Phone: 44 1 17 929 7481
Fax: 44 1 17 920 0979

techtracking@iop.org

Additional copies of this publication are available from:

Curran Associates, Inc.
57 Morehouse Lane
Red Hook, NY 12571 USA
Phone: 845-758-0400
Fax: 845-758-2633
Email: curran@proceedings.com
Web: www.proceedings.com

TABLE OF CONTENTS

IOP Conf. Series: Materials Science and Engineering **939** (2020) 011001 doi:10.1088/1757-899X/939/1/011001

Preface ISPCIET'2020

III International Scientific and Practical Conference on Innovations in Engineering and Technology-2020

For the third year in a row, ISPCIET has become a platform for discussing topical issues in solving problems in the field of modeling and automation of technological processes and productions, materials science and mechanical engineering, materials processing technology, as well as innovative technologies in the development of convergent solutions in engineering and technology. The conference is organized by the Polytechnic Institute of the Yaroslav-the-Wise Novgorod State University.

ISPCIET'2020 was held in person, but the reports of several participants were presented online due to the epidemiological situation caused by COVID-19. The interaction of these participants with the rest of the conference attendees was facilitated through Zoom. Immediately after the report was presented, the listeners were given the opportunity to ask the speaker a question online and take part in the discussion on the topic of the report. They also had the opportunity to contact the authors of the reports after the conference through the organizing committee or by e-mail.

The plenary session was broadcast online for all of the participants and guests of the conference. The conference was attended by 103 people, including 69 online-participants: scientists from universities in Russia, Armenia, Belarus, Bulgaria, Kazakhstan, China, the Netherlands, USA, Uzbekistan, Ukraine. Despite the difference in time zones and a difficult epidemiological situation, there were no setbacks in the conference, the communication of speakers and conference participants went as usual.

The conference organizers express their sincere gratitude to all participants and hope that the Innovations in Engineering and Technology conference will become a permanent platform for discussing innovative technologies in the development of convergent solutions in engineering, technology and cross-industry scientific research.

Following the conference, a collection of scientific articles, prepared in accordance with the international principles of scientific publications and the recommendations of the Committee on Publication Ethics – COPE, was formed.

EDITOR

Sergey B. Sapozhkov

Polytechnic Institute of Novgorod State University – Federal State Budgetary Educational Institution of Higher Education "Yaroslav-the-Wise Novgorod State University (NovSU)

E-mail: ietconf@novsu.ru

Content from this work may be used under the terms of the Creative Commons Attribution 3.0 licence. Any further distribution of this work must maintain attribution to the author(s) and the title of the work, journal citation and DOI.
Published under licence by IOP Publishing Ltd

IOP Conf. Series: Materials Science and Engineering **939** (2020) 011002 doi:10.1088/1757-899X/939/1/011002

Declaration on Peer Review

All papers published in this volume of *IOP Conference Series: Materials Science and Engineering* have been peer reviewed through processes administered by the Editors. Reviews were conducted by expert referees to the professional and scientific standards expected of a proceedings journal published by IOP Publishing.

Details on the procedure for reviewing materials of the III International Scientific and Practical Conference on Innovations in Engineering and Technology-2020 are as follows.

- The type of peer review: *double blind peer review.*
- The management system for the presentation of conference materials:

Call for papers, receipt of the submitted materials, communication with the authors of the papers and articles – via e-mail (e-mail address of the organizing committee: ietconf@novsu.ru).

The conference coordinator and their assistant review applications for participation, clarify details if necessary and enter the application into the register.

Materials received on the conference e-mail box undergo an initial check for compliance with formal requirements and anti-plagiarism check. If necessary, additional materials are requested (separate files of figures, diagrams, clarifications on the translation of certain terms, etc.).

Materials and information on the results of the initial check are transferred to the organizing committee. Further, the materials are submitted for reviewing to one or, if necessary, several reviewers (in a situation when the research belongs to two fields of science or additional expert assessment is required). If the review contains an indication of the need to make corrections, then the materials are returned to the author for revision with a list of comments and questions for further working upon.

Based on the results of reviewing, the Program Committee makes a decision whether to accept or reject the material.

The results of the anti-plagiarism check and copies of reviews are sent to the authors upon request.

In the absence of controversial issues, the materials of the organizing committee remain with the organizing committee coordinator (deputy chairman of the organizing committee).

- The number of applications received is *127.*
- The number of materials sent for consideration is *103.*

Content from this work may be used under the terms of the Creative Commons Attribution 3.0 licence. Any further distribution of this work must maintain attribution to the author(s) and the title of the work, journal citation and DOI.
Published under licence by IOP Publishing Ltd

IOP Conf. Series: Materials Science and Engineering **939** (2020) 011002 doi:10.1088/1757-899X/939/1/011002

• The number of applications accepted is *81*.

• Acceptance rate (number of applications accepted / number of applications received X 100) is 63,77 %.

• The average number of reviews per article is *1,7*.

• The total number of attracted reviewers is *12*.

• Additional information on the peer review process:

If necessary, it is possible to exchange information between authors and reviewers through the organizing committee (in difficult/disputable cases of consideration).

• For inquiries please contact Dr. Oxana Fikhtner, Director of the Center for the Development of Publication Activity of NovSU (Oxana.Fikhtner@novsu.ru).

The use of fuzzy logic for the clean-up systems control for bunkers, containing bulk solids

A M Abramov[1,3], **V Yu Klyukin**[2,4], **V V Kozhikin**[2,5] **and O V Kochneva**[2,6]

[1] Yaroslav-the-Wise Novgorod State University, ul. B. St. Petersburgskaya, 41 173003, Veliky Novgorod, Russia
[2] Peter the Great St.Petersburg Polytechnic University (SPbPU), Polytechnicheskaya, 29 195251, St.Petersburg, Russia

[3] Email: alexandr.abramov@novsu.ru
[4] Email: kljukin@mail.ru
[5] E-mail: tageree@mail.ru
[6] E-mail: kov_mirny@mail.ru

Abstract. The issues of eliminating the bulk solids overhanging in bunkers using pneumatic pulse devices are considered in the following article. An algorithm for automatic control of pneumatic collapse systems based on the use of fuzzy logic methods is proposed. A constructive solution for this algorithm is given, based on the assessment of the level of bulk material on the conveyor installed at the exit of the hopper. The Mamdani algorithm is used as a fuzzy inference algorithm. The fuzzy Logic Toolbox package of the MATLAB computing environment was used for modeling fuzzy systems. The use of mid-maximum and center-of-gravity methods as a defuzzification method is considered. It is demonstrated that the use of fuzzy logic makes it possible to significantly simplify the development of the control system algorithm.

1. Introduction

In various industries, both domestic and foreign, various bunkers are widely used for storing a variety of bulk solids. The volume of such bunkers allows to place from several liters to thousands of cubic meters of bulk solids. Bunkers are used, for example, for storing flour, fertilizers, ore, etc. They can be arranged both indoors and outdoors. Moreover, in very rare cases, they provide the necessary temperature and humidity conditions; basically, these are natural parameters of the environment.

Depending on the material to be loaded, the size of the bunkers and their shape, and the state of the environment, some of the material may be deposited on the walls of the bunkers due to sticking, freezing, etc., or the material particles may affect each other and form an arch [1]. If a funnel is formed or the material freezes (figure 1), it is possible to stop the material supply from the bunker, which can occur relatively quickly and cause the entire process to stop, even on equipment placed in a warm and dry room. It is necessary to ensure uninterrupted supply of bulk material from the bunker and this is ensured by the use of various systems and devices that provide mechanical shock or vibration effects on the walls of the bunker, as well as pneumatic pulse effects on the material itself placed in the hopper.

Content from this work may be used under the terms of the Creative Commons Attribution 3.0 licence. Any further distribution of this work must maintain attribution to the author(s) and the title of the work, journal citation and DOI.
Published under licence by IOP Publishing Ltd

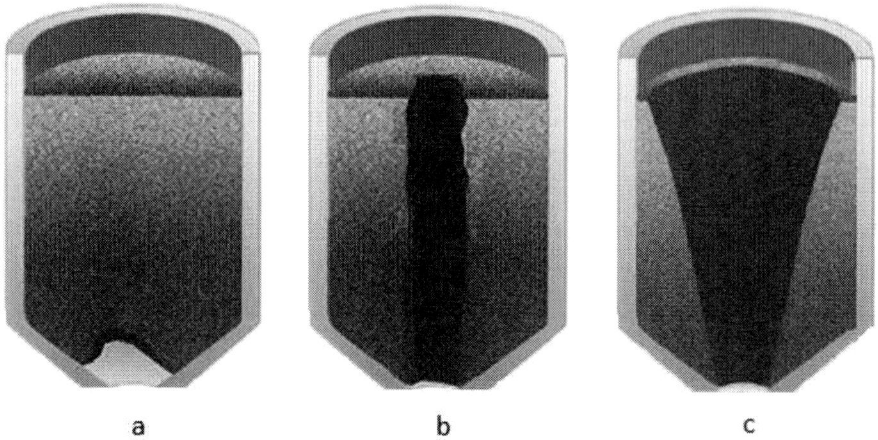

Figure 1. Types of bulk solids stagnation in bunkers (a – arch, b – funnel, c – overhanging).

These systems and devices include: vibrators, magnetic pulse systems, and pneumatic pulse systems [2]. Manual cleaning is also used, which is time-consuming and sometimes dangerous. As practice shows, the use of pneumatic pulse devices is less energy-intensive and more effective in combating overhanging in bunkers. The destruction of the bulk materials formed in the bunker occurs due to the impact of compressed air or nitrogen created by the action of a pneumatic pulse device that generates gas pulses. The pneumatic pulse device is filled with compressed gas, and then the accumulated gas is ejected in a fraction of a second, creating a shock effect on the bulk material.

The pneumatic pulse bunker cleaning system is demonstrated in figure 2.

Figure 2. Pneumatic pulse devices mounted on the bunker.

Usually the bunker is equipped with from one to forty pneumatic pulse devices. The number of devices depends on the shape and size of the bunker, as well as on the characteristics of the material being loaded. The efficiency of the system depends largely on the pneumatic pulse devices place of installation. They must be installed in places where the bulk material freezes. Also, much depends on the power of pneumatic pulse devices. When there is insufficient power, the efficiency of using devices drops sharply.

IOP Conf. Series: Materials Science and Engineering **939** (2020) 012001 doi:10.1088/1757-899X/939/1/012001

2. The processes of pneumatic pulse destruction of material hangings in bunkers and selection of criteria for evaluating the quality of the processes

The process of stagnant areas pneumatic collapse of the material in the bunker occurs due to the impact of a shock gas wave formed by a pneumatic pulse device on the material. The barrel, through which compressed gas escapes from the pneumatic pulse device, is directed to the place of possible stagnation in the bulk material bunker. After a pneumatic pulse shock, the stagnant zone may collapse immediately or after subsequent impacts. After the destruction of stagnant zones, the material begins to flow to the exit of the bunker.

However, the choice of the possible formation place of overhangings and material arches is made by the service personnel on the basis of existing practical experience in the operation of the equipment.

Most quality criteria for emptying bulk material bunkers are based on several variables [3]. Usually, a graded evaluation of such qualitative variables is performed and used to evaluate the process.

Bunkers (especially small sized), as a rule, do not provide sensors for the level of bulk material in the bunker, and if they are provided, they determine the level of material only in certain places, without giving a complete picture of the overhanging formation. Visual inspection usually does not produce results due to the high dust content inside the bunker.

Therefore, the simplest way to judge the formation of overhanging is by the absence of material at the exit of the bunker. Usually, under the exit, a conveyor is installed, where the material that came out of the bunker is fed to the consumer. According to the amount of material on the conveyor, a conclusion is made about the material leaving the bunker. The amount of material on the conveyor is estimated by the level of material on the conveyor belt using a simple sensor, which is a swinging plate resting on the material on the conveyor. If there is no material on the conveyor, the plate goes down and acts on the sensor. The signal from this sensor indicates that there is no material at the outlet of the bunker.

Control of the level of bulk materials in the bunker is carried out using several sensors, for example, ultrasonic [4]. These sensors can be used to assess the level of the substance in the bunker. The accuracy of the estimation depends on the number of sensors per unit area of the bunker section. In this way, you can estimate the amount of substance in the bunker and its distribution, but it is difficult to predict the dynamics of emptying. Therefore, the first criterion for evaluating the formation of material stagnation in the bunker after the material exits the bunker is used more often.

The proposed solution is based on the use of a sensor interacting with the plate to control the level of bulk material, similar to the one described above, but having an analog signal at the output for the entire range of plate movement, or monitoring several specific positions of the plate corresponding to specific levels of bulk material on the conveyor (figure 3).

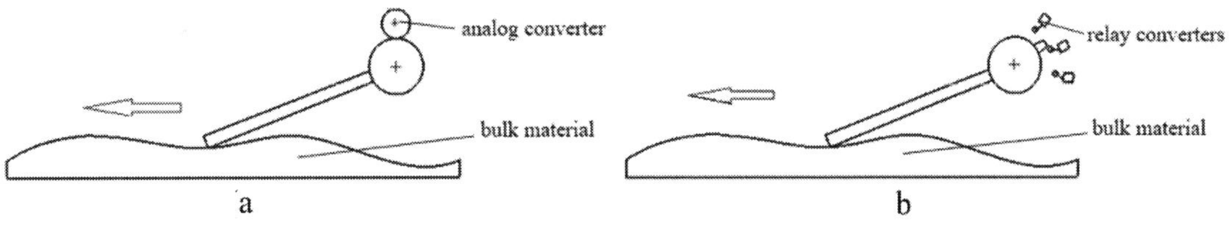

Figure 3. Types of bulk material sensors on the conveyor (a – using a sensor with an analog converter, b – using a sensor with multiple relay-type converters).

Let us consider using an analog converter. The sensor plate changes its angular position depending on the thickness of the material layer on the conveyor, as a result, the sensor outputs an analog signal, the value of which is proportional to the angle of rotation of the plate and is associated with the thickness of the bulk material layer.

When using multiple relay converters, each of them is adjusted to a specific level of material on the conveyor.

3. Control of bunker cleaning systems using fuzzy logic methods

The control algorithm can be built on the basis of the results obtained by mathematical modeling of the processes of formation of hangings or arches. However, process modeling is a complex task that requires solving a large number of equations together. Solving the problem requires a lot of computing time and resources, and the resulting recommendations can only be presented in a certain range or with a certain probability, and not always high [3].

In many ways, obtaining information to control the processes of collapse of vaults or destruction of hangings by mathematical modeling of the process is inefficient. Recommendations for managing the collapse of hangings and vaults obtained during modeling can only be of a recommendatory nature.

Currently, the most commonly used algorithm is that pneumatic pulse devices start cycling simultaneously or sequentially when there is no material at the outlet of the bunker. This algorithm is easy to implement, but leads to a high consumption of compressed gas, because the system starts working only when a hang or arch has already formed. To increase efficiency, it is necessary to monitor the dynamics of changes in material consumption from the bunker and when there is a pronounced trend of reducing material consumption, it is necessary to turn on pneumatic pulse devices [5].

In this case, it is advisable to use fuzzy logic methods. The sensors demonstrated in figure 3 produce a signal proportional to the thickness of the bulk material layer on the conveyor. If there is no material on the conveyor, the sensor will show the minimum value, and if the layer thickness is close to the maximum – the maximum value. To ensure efficient discharge of bulk material, it is necessary to maintain a large volume of material being discharged per unit of time. This mode of operation corresponds to the maximum reading of the sensor. If the thickness of the material layer on the conveyor decreases, the sensor reading will also decrease, and it is necessary to strengthen the operation of pneumatic pulse devices.

Effective management of this system requires monitoring the dynamics of changes in the thickness of the material on the conveyor. For this purpose, you can install a second sensor located at a certain distance from the first one, or you can do it virtually using a time delay. In this way, you can track the system's response to the use of air crash systems.

The signal received from the sensor requires fuzzification, i.e. reduction to fuzziness. For the value of the analog signal from the sensor, a linguistic variable is created that contains terms describing the signal strength. The degree of belonging of the received signal to the specified terms is determined by the set membership functions. Then a fuzzy logical output is made using the Mamdani algorithm based on a pre-formed rule base. The database of fuzzy rules is formed on the basis of expert assessments and can be adjusted when new experimental data is obtained. Next, defuzzification is performed using one of the known methods. The choice of defuzzification method depends on many factors and is selected depending on the customer's needs. After receiving a clear value of the output variable, the pneumatic pulse device generates shock pneumatic signals to eliminate the stagnation of bulk material in the bunker. The system responds to changes in the sensor readings, which ensures uninterrupted unloading of material from the bunker.

Various software packages can be used to create models of fuzzy systems. We use the FuzzyLogicToolbox package of the MATLAB computing environment. The FIS fuzzy output system editor is the main tool for creating and editing fuzzy output systems in graphic mode.

We will develop an algorithm for controlling the pneumatic pulse system for cleaning bunkers for bulk materials based on fuzzy logic. The inputs are the sensor readings and the value of the sensor readings 5 seconds before the current value, which allows tracking the dynamics of changes in the height of the bulk material layer. The values of the sensors are in the range from 0 to 10 V. As an output, we use the duration of the pneumatic pulse and the frequency of their occurrence. However, at the first stage, we will consider the pulse duration constant, since the pulse duration is selected from the condition of emptying the receiver with compressed gas. Therefore, the output variable will be the frequency of pneumatic pulses.

Let's denote the first input variable "Current sensor value", and the second "Previous sensor value". Let's denote the output variable as "Pneumatic pulse frequency". After that, the FIS editor window will look like this (figure 4).

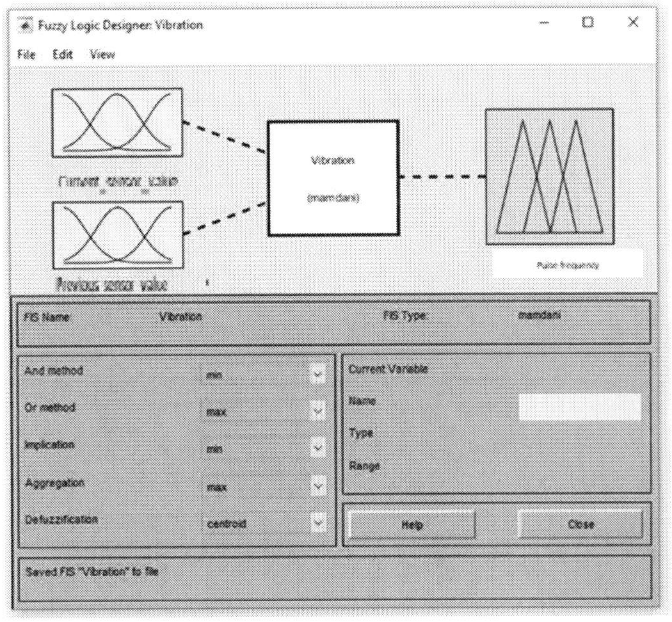

Figure 4. FIS editor window with input and output variables.

Let's denote the terms and their membership functions for each of the variables. For the sensor values, the terms "Low", "Medium" and "High" are set. For the values of the frequency of pneumatic pulses, the terms "Absent", "Low", "Medium", "High" and "Maximum" are set.

For the terms "Low" and "High" of the input variables, select the trapezoidal membership functions with the parameters [0 0 1 4] for the term "Low" and [6 9 10 10] for the term "High". For the term "Medium" of input variables, we define a triangular membership function with parameters [2 5 8].

For the terms of the output variable, we define triangular membership functions with the following parameters: [0 0 6] for the term "Absent", [0 6 12] for the term "Low", [6 12 18] for the term "Average", [12 18 24] for the term "High", and [18 24 24] for the term "Maximum".

Membership functions can be easily adjusted after analyzing the process being implemented.

When creating rules, you should rely on common sense and expert assessments. Let's set the first rule as if the current sensor value is high and the previous sensor value is high, then the pneumatic pulse device does not work (the frequency is zero – the term is "Absent").

Since this system has two input variables, each of which has three terms, it is possible to define nine unique rules. The resulting rules are shown in table 1 and figure 5.

Table 1. Rules for fuzzy inference.

	Current value		
Previous value	Low	Medium	High
Low	Maximum	High	Medium
Medium	High	Medium	Low
HIgh	Medium	Low	Absent

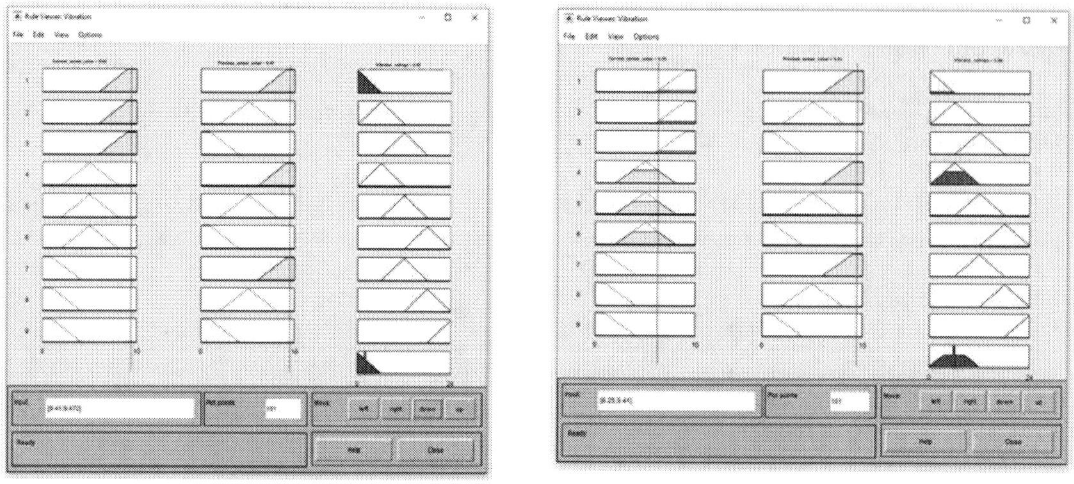

a b

Figure 5. The result of system operation at high (a) and lowered (b) signal levels.

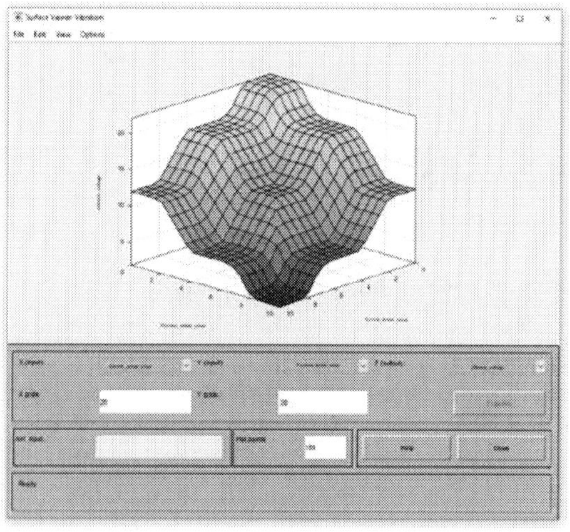

a

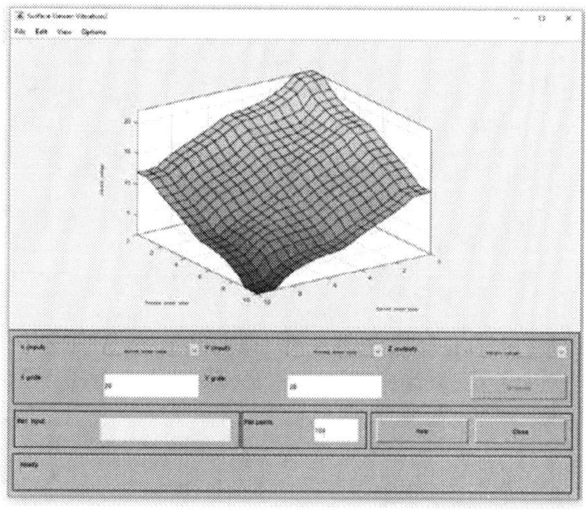

b

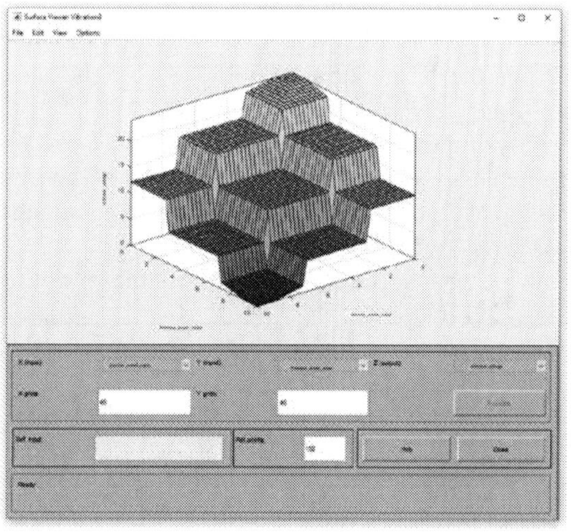

c

Figure 6. Fuzzy output surface for different membership functions (a, b) and when using the "mom" defuzzification method».

After setting all the parameters, we run the modeling. Some of the results obtained are presented in figure 6.

The value of the output variable for the system under consideration, as can be seen from the results, changes unevenly, there are flat sections (figure 6a). By modifying the membership functions of the terms of input variables, you can achieve a smoother change in the output variable (figure 6b). The surface of the fuzzy output is noticeably smoothed, which corresponds to a smoother change in the output variable.

In the surface demonstrated in figures 6a and 6b, the center of gravity method was used as a defuzzification method, which provides smooth control and takes into account all active rules in its operation. However, when using this method, it is not possible to obtain the zero frequency of the pneumatic pulse device.

Using the mid-maximum method allows you to use the entire definition area. The membership functions remain the same (figure 6c), only the defuzzification method changes.

4. Conclusion

The use of fuzzy logic methods allows us to obtain new algorithms for controlling the operation of pneumatic pulse devices. The choice of the defuzzification method and the type of accessory functions depends on the material to be loaded, the shape and size of the bunker, and the customer's requirements. The mid-highs method will provide a great benefit if the pneumatic pulse device rarely works. When the device is operating in a mode that is close to constant, using the center of gravity method will be effective.

Using fuzzy logic allows you to significantly speed up the development of the control system, and also provides tools for fine-tuning the resulting system after conducting pilot tests. The control system determines the desired frequency of operation of pneumatic pulse devices based on the readings of a single level sensor of bulk material on the conveyor installed at the outlet of the bunker. Despite its simplicity, the system can provide high control accuracy and high response speed.

Creating control algorithms based on fuzzy logic is much easier and more demonstrative than traditional methods that use the results of modeling complex mathematical models. Fuzzy logic methods allow getting results that meet the set tasks with significantly less effort.

References

[1] Schulze D 2008 *Powders and Bulk Solids. Behavior, Characterization, Storage and Flow* (Springer: Heidelberg) p 517

[2] Schulze D 2016 *Storage and Flow of Powders and Bulk Solids* (Springer-Verlag,Particle Technology Series) **25** 425–478

[3] Siirde A E, Yurkin S V and Yakutov V V 2006 Prevention of fuel hanging with the help of an air-impact knocking system *Power Technol Eng* **40** 357–361

[4] Sandler U and Tsitolovsky L 2008 *Neural Cell Behavior and Fuzzy Logic* (Springer) p 478

[5] Kruglov V V and Dli M I 2002 *Intellektualnye informatsionnye sistemy: kompyuternaya podderzhka system nechetkoy logiki I nechetkogo vyvoda* [Intelligent information systems: computer support for fuzzy logic and fuzzy inference systems] (Moscow: Fizmatlit) [In Russ.]

Classifications of schematic solutions of the geokhod knife operating body and the interaction surface of the geokhod operating body with bottom rock

V V Aksenov[1,4,6], A B Efremenkov[2], V Yu Sadovets[3,7], D A Pashkov[4,8] and V A Efremenkov[5]

[1] Research Center "Siberian NPO" Ltd, Sovetsky pr., 56, Kemerovo, 650002 Russia
[2] Yaroslav-the-Wise Novgorod State University, ul. B. St. Peterburgskaya, 41 173003, Veliky Novgorod, Russia
[3] T.F. Gorbachev Kuzbass State Technical University, ul. Vesennjaja, 28 650000, Kemerovo, Russia
[4] Institute of Coal of FRC for Coal and Coal Chemistry, Siberian Branch of the RAS, Leningradsky pr., 10, Kemerovo, 650610 Russia
[5] Peter-the-Great St.Petersburg Polytechnic University (SPbPU), Polytechnicheskaya, 29 195251, St.Petersburg, Russia

[6] E-mail: 55vva42@mail.ru
[7] E-mail: vsadovec@yandex.ru
[8] E-mail: pashkov.d.a@inbox.ru

Abstract. When developing geokhods, it is necessary to take into account the helical movement of the machine in the rock mass. In this regard, the substantiation of the technical and structural elements of the geokhod is relevant. The purpose of the work is to classify the schematic solutions of the geokhod knife operating body and the surface of interaction of the geokhod operating body with the bottom rock. As a result, schematic solutions were developed for the knife operating body of the geokhod and the surface of interaction of the geokhod operating body with the bottom rock, their classification is proposed.

1. Introduction

The development of civilization is accompanied by the expansion of infrastructure and the growth of both the population and the number of vehicles, which leads to a reduction in undeveloped territories on the planet's surface and the search for new places to solve modern urban problems – territorial, transport, environmental, energy [1–3].

At the present stage of development of our society, the greatest interest is the underground space [4–7].

For the formation of cavities in the underground space there are many machines, one of them is a geokhod [8–12].

When developing geokhods, it is necessary to take into account the helical movement of the machine in the rock mass [13–16]. In this regard, the substantiation of the technical and structural elements of the geokhod is relevant.

Content from this work may be used under the terms of the Creative Commons Attribution 3.0 licence. Any further distribution of this work must maintain attribution to the author(s) and the title of the work, journal citation and DOI.
Published under licence by IOP Publishing Ltd

2. Research Methods

The operating body (hereinafter – OB) is one of the main elements of the geokhod, affecting the power parameters of the underground apparatus [17, 18]. For destruction of soft rocks with geokhods, the knife-type OB was selected [19, 20].

Given the structural possibility of changing the geometry of the knife, it becomes possible to obtain a number of options for schematic solutions of geokhod knife OB for the destruction of soft rocks, some of which are shown in figures 1 and 2.

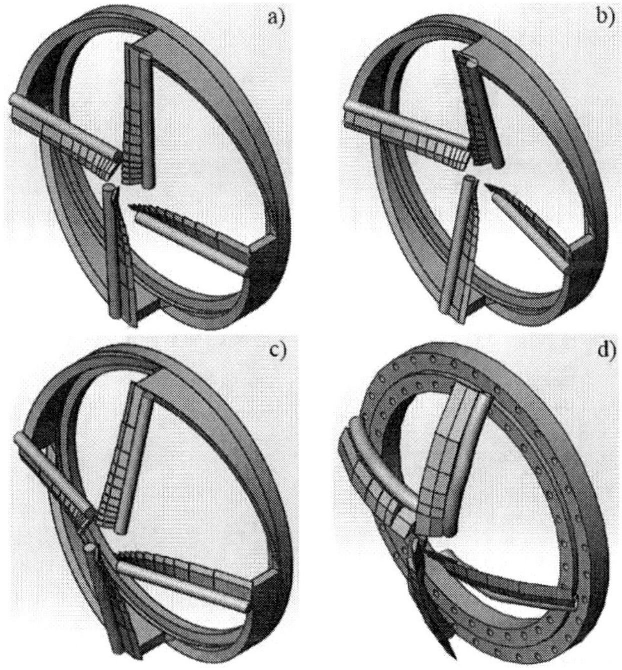

Figure 1. Schematic versions of a four-shaft geokhod knife OB with different knife geometry: a) flat shape; b) inverse cone shape; c) convex cone shape; d) convex ball shape.

Figure 1 shows the options for a four-beam geokhod knife OB, which make various forms of faces. So, the option that forms the flat shape of the face (figure 1, a) is a traditional form when tunneling with the shield method. With this scheme, the knives are perpendicular to the axis of rotation of the geokhod. When the knives are tilted at an angle γ_{OB}, the OB forms a conical shape face. The cone can be either convex (figure 1, b) or concave (inverse) (figure 1, c) relative to the geokhod. These schemes allow keeping the direction relative to the axis of the geokhod, but at the same time reducing the maneuverability of the machine. The geokhod OB, which forms a convex ball face (figure 1, d), in contrast, increases maneuverability.

The geokhod OB can have a different number of shafts. Figure 2 shows the schematic versions of a three-shaft geokhod knife OB for destruction of soft rocks. Schematic versions of the geokhod OB (figure 2, a, b) form the bottom of the conical shape. The geokhod knife OB forming the torus shape face (figure 2, c, d) differ from the spherical shape OB in that they form a spherical shape on the periphery and a cone in the center. The OB with a convex torus shape (figure 2, c) as well as the geokhod OB of a convex ball shape increases the maneuverability of the machine, but closer to the axis of the mine, a rational form of the generatrix appears [21], which leads to a shift of the main stresses in the direction of tension.

Based on the proposed schematic options for a geokhod knife OB (figure 1, 2), a classification of geokhod knife OB is proposed.

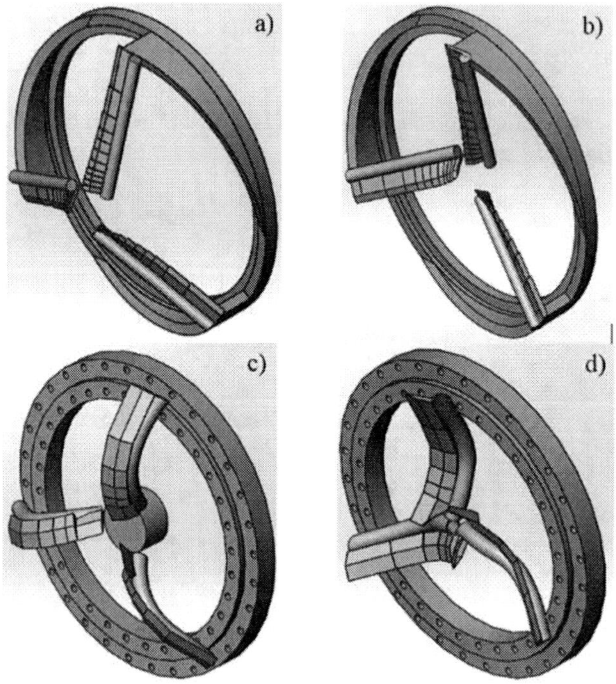

Figure 2. Schematic versions of a three-shaft geokhod knife OB with different knife geometry: a) convex cone shape; b) inverse cone shape; c) convex torus shape; d) concave torus shape.

Features of classification of geokhod knife OB:

1. Location relative to the face surface:
   - convex into mass (figure 1, c, d; figure 2, a, c)
   - direct (figure 1, a)
   - concave (figure 1, b; figure 2, b, d).

2. The shape of the knife:
   - straight (figure 1, a);
   - inclined (figure 1, b, c; figure 2, a, b);
   - radial cone;
   - concave (ball shape) (figure 1, d);
   - circumferential (torus shape) (figure 2, c, d).

3. Number of shafts:
   - two-shaft;
   - three-shaft (figure 2);
   - four-shaft (figure 1);
   - multi-shaft.

4. Presence of helicoidity:
   - straight;
   - helicoid (figure 1, 2).

When the opearting body is working, the interaction surface (hereinafter – IS) of the geokhod OB with the bottom rock is formed.

One of the signs of IS is the shape of the face. Taking into account the schematic solutions of the geokhod knife OB, schematic solutions of various forms of the face were developed (figure 3) with various forms of the generatrix.

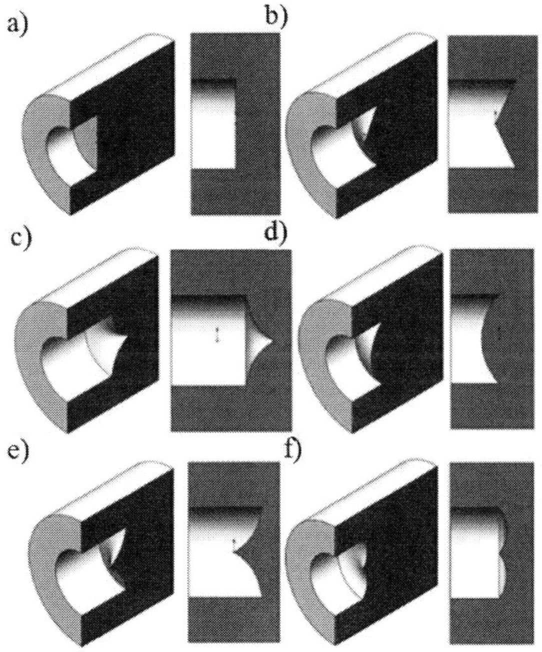

Figure 3. Schematic options for the face shape: a) flat; b) conical; c) with a convex radial-conical generatrix; g) concave; d) with a concave radial-conical generatrix; e) with a circumferential generatrix.

Another sign of interaction surface between the operating body and the face rock is the presence (figure 4, a) or the absence of a bench (figure 4, b).

OB forming a bench include knife, bar, screw, drum OB. Destruction of the face without the formation of a bench is carried out by such OB as faceplate, rotary, planetary, drum.

V.Yu. Beglyakov [21] found out that in the presence of a bench, the total stresses from the action of the operating body lead to a shift in the value of the main stresses in the local zones of action of individual cutters in the direction of tension relative to the same effect with a flat face; the presence of a bench creates the prerequisites for a decrease in the specific energy intensity of fracture in comparison with the energy intensity of fracture in a flat face; the advantages of destruction of the face with a bench are manifested to one degree or another in any direction of cutting forces.

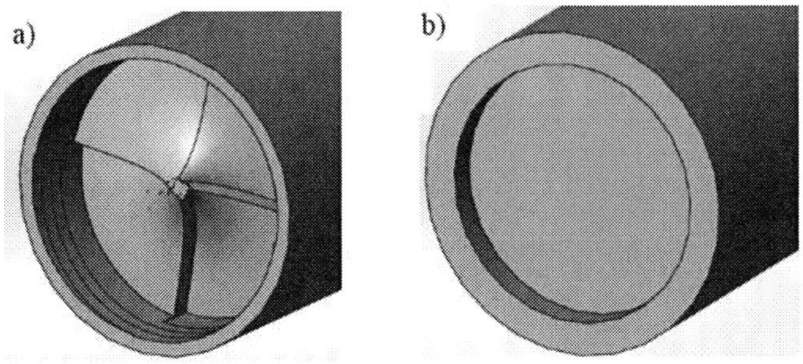

Figure 4. Schematic versions of the interaction surface of geokhod OB with the face rock: a) with a bench formation; b) without a bench formation.

Also the visible sign of IS is the number of the geokhod OB shafts, it is also the number of face benches (figure 5).

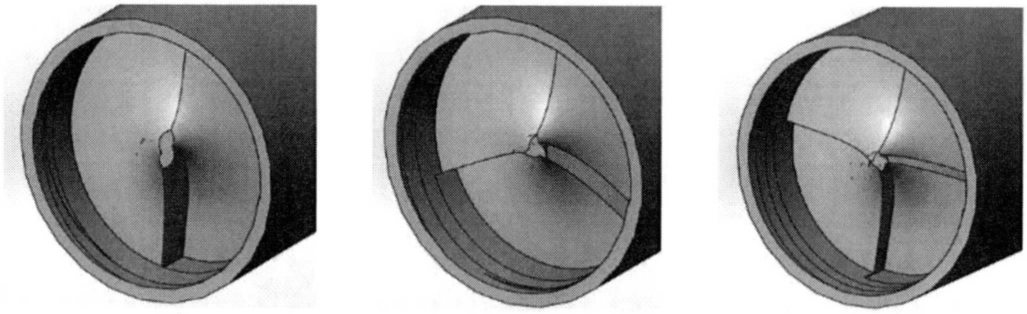

a) b) c)

Figure 5. Schematic versions of the interaction surface of the geokhod OB with the face rock with a different number of benches: a) two-bench; б) three-bench; в) four-bench.

Based on the material presented above, one can classify the surface of the face by visible signs, i.e. signs that give a general characteristic of the interaction surface at a qualitative level, determine the type of face being formed by the presence or absence of certain signs.

Features of face surface classification:

1. Location relative to the face surface:
   - convex into a mass (figure 3, c, f);
   - direct (figure 3, a);
   - concave (figure 3, b, d, e).

2. Generatrix shape:
   - straight (figure 3, a);
   - inclined (figure 3, b);
   - radial cone (figure 3, c, d);
   - concave (ball shape) (figure 3, d);
   - circumferential (torus form) (figure 3, f).

3.Number of benches:
   - two-bench (figure 5, a);
   - three-bench (figure 5, b);
   - four-bench (figure 5, c);
   - multi-bench.

4. Presence of bench:
   - with a bench (figure 4, a);
   - without a bench (figure 4,b).

3. Conclusion

A classification of geokhod knife OB for the destruction of soft rocks is proposed. Based on the schematic solutions of the geokhod knife OB, the schematic solutions of the surface of the interaction of the geokhod operating body with the face rock are developed, and a classification of the schematic solutions of the interaction surface of the geokhod OB with the face of the rock is proposed.

For further research, it is necessary to evaluate each schematic solution of a geokhod knife OB for the destruction of soft rocks for compliance with their requirements.

References

[1] Aksenov V V, Sadovets V Yu, Pashkov D A and Rezanova E V 2018 Granichnyye usloviya opredeleniya kharakternykh tochek nozhevogo ispolnitel'nogo organa geokhoda [Boundary conditions for determining the characteristic points of a knife operating body of a geokhod] *Vestnik of Kuzbass State Technical University* **2(126)** 166–173 [In Russ.]

[2] Aksenov V, Sadovets V and Pashkov D 2018 *E3SWebConf* **41** 03002

[3] Aksenov V V, Efremenkov A B, Sadovets V Yu and Rezanova E V 2010 Formirovaniye strukturnogo portreta geokhodov [Formation of the structural portrait of geokhods] *Vestnik of Kuzbass State Technical University* **1(77)** 35–41 [In Russ.]

[4] Aksenov V V, Khoreshok A A and Beglyakov V Yu 2018 Kontseptsiya sozdaniya perspektivnogo tekhnologicheskogo uklada formirovaniya (osvoyeniya) podzemnogo prostranstva na baze operezhayushchego razvitiya novykh podkhodov v stroitel'noy geotekhnologii i geotekhnike [The concept of creating a promising technological structure for the formation (development) of underground space on the basis of the advanced development of new approaches in construction geotechnology and geotechnics] (Part 1) *Vestnik of Kuzbass State Technical University* **4** 105–113 [In Russ.]

[5] Aksenov V V, Khoreshok A A and Beglyakov V Yu 2018 Kontseptsiya sozdaniya perspektivnogo tekhnologicheskogo uklada formirovaniya (osvoyeniya) podzemnogo prostranstva na baze operezhayushchego razvitiya novykh podkhodov v stroitel'noy geotekhnologii i geotekhnike [The concept of creating a promising technological structure for the formation (development) of underground space on the basis of the advanced development of new approaches in construction geotechnology and geotechnics] (Part 2) *Vestnik of Kuzbass State Technical University* **5** 43–52 [In Russ.]

[6] Aksenov V, Sadovets V and Pashkov D 2017 *E3SWeb of Conferences* **21** 03008

[7] Aksenov V V, Sadovets V Yu, Preis E B and Pashkov D A 2018 Improving the mathematical model for determining the power parameters of the geokhod blade cutting body *Mining Equipment and Electrical Engineering* **5(139)** 16–22

[8] Aksenov V, Sadovets V and Pashkov D 2017 Opredeleniye silovykh parametrov nozhevogo ispolnitel'nogo organa geokhoda dlya razrusheniya porod maloy kreposti [Determination of force parameters of the geokhod knife operating body for destruction of low hardness rock body] *Vestnik of Kuzbass State Technical University* **3 (121)** 116–126 [In Russ.]

[9] Aksenov V V, Horeshok A A, Ananjev K A and Ermakov A N 2014 Razrabotka skhemnykh resheniy ispolnitel'nykh organov geokhodov [Development of schematic solutions of geokhod operating bodies] *Izvestiya vysshikh uchebnykh zavedeniy. Gornyy zhurnal* **3** 73–76 [In Russ.]

[10] Aksenov V V, Efremenkov A B, Sadovets V Yu, Pashkov D A and Efremenkov V A 2019 *IOP Conference Series: Materials Science and Engineering* **656** 012002

[11] Aksenov V V, Efremenkov A B, Sadovets V Yu, Pashkov D A and Efremenkov V A 2019 *IOP Conference Series: Materials Science and Engineering* **656** 012003

[12] Pashkov D A 2019 Obzor metodik opredeleniya energoyemkosti razrusheniya porody [Review of methods for determining the energy capacity of rock destruction] *Young Russia. Collect. of materials of the XI All-Russian Sc. and Pract. Conf. with int. part.* 10310 [In Russ.]

[13] Aksenov V V, Efremenkov A B, Sadovets V Yu and Pashkov D A 2018 *IOP Conference Series: Materials Science and Engineering* **441** 012005

[14] Aksenov V, Sadovets V and Pashkov D 2019 Razrabotka metodiki opredeleniya energoyemkosti razrusheniya gornoy porody nozhevym ispolnitel'nym organom geokhoda [Development of a method for determining the energy intensity of rock destruction by a geokhod knife operating body]. *Gornoye oborudovaniye i elektromekhanika* **2(142)** 30–38 [In Russ.]

[15] Aksenov V V, Khoreshok A A, Beglyakov V Yu and Efremenkov A B 2019 *IOP Conference*

Series: Materials Science and Engineering **656** 012004

[16] Aksenov V V, Khoreshok A A, Beglyakov V Yu and Efremenkov A B 2019 *IOP Conference Series: Materials Science and Engineering* **656** 012005

[17] Sadovets V Yu 2007 *Obosnovaniye konstruktivnykh i silovykh parametrov nozhevykh ispolnitel'nykh organov geokhodov* [Substantiation of the design and power parameters of geokhod knife operating bodies] (Candidate of Technical Science Dissertation) Kemerovo: Kuzbass State Technical University named after T.F. Gorbachev) [In Russ.]

[18] Nesterov V, Aksenov V, Sadovets V, Pashkov D and Beysebayeva Zh 2019 *E3S Web of Conferences* **105** 03024

[19] Nesterov V, Aksenov V, Sadovets V and Pashkov D 2019 *E3S Web of Conferences* **105** 03001

[20] Aksenov V V, Sadovets V Yu and Pashkov D A 2016 Obosnovaniye neobkhodimosti sozdaniya ispolnitel'nogo organa geokhoda dlya razrusheniya porod maloy kreposti [Justification of the need to create an operating body of the geokhod for the destruction of soft rocks] *Vestnik of Kuzbass State Technical University* ***6 (118)*** 8–15 [In Russ.]

[21] Beglyakov V Yu 2012 *Obosnovaniye parametrov poverkhnosti vzaimodeystviya ispolnitel'nogo organa geokhoda s porodoy zaboya* [Substantiation of surface parameters of interaction of the geokhod operating executive body with face rock] (Candidate of Technical Science Dissertation) (Yurga: Kuzbass State Technical University) [In Russ.]

IOP Conf. Series: Materials Science and Engineering **939** (2020) 012003 doi:10.1088/1757-899X/939/1/012003

Determining the interaction surface parameters of the geokhod knife operating body with the face rock

V V Aksenov[1,4,5], **A B Efremenkov**[2,6], **V Yu Sadovets**[3,7], **D A Pashkov**[4,8] **and V A Efremenkov**[5]

[1] Research Center "Siberian NPO" Ltd, Sovetsky pr, 56 650002, Kemerovo, Russia
[2] Yaroslav-the-Wise Novgorod State University, ul St. Petersburgskaya, 41 173003, Veliky Novgorod, Russia
[3] T.F. Gorbachev Kuzbass State Technical University, ul. Vesennyaya, 28 650000, Kemerovo, Russia
[4] Institute of Coal of FRC for Coal and Coal Chemistry, Siberian Branch of the RAS, Leningradsky pr., 10 650610, Kemerovo, Russia
[5] Peter the Great St.Petersburg Polytechnic University, Polytechnicheskaya ul., 29 195251, St.Petersburg, Russia

[5] E-mail: 55vva42@mail.ru
[6] E-mail: abe@novsu.ru
[7] E-mail: vsadovec@yandex.ru
[8] E-mail: pashkov.d.a@inbox.ru

Abstract. One of the main devices of the geokhod having a direct impact on the force characteristics of the underground apparatus is the operating body. During the work of the operating body, the interaction surface of the geokhod operating body with the face rock is formed. The purpose of the research is to determine the parameters of the surface of interaction of the geokhod knife operating body with the face rock.
As a result, it was revealed that the interaction surface parameters of the geokhod knife operating body with the face rock can be divided into two groups: general, which are characteristic for each variant of operating body, and variable, which are characteristic for each variant.

1. Introduction

Throughout the world, the construction of underground facilities for various purposes has become one of the priority areas. Their number in developed countries doubles every 10 years, and in the future we should expect a further increase in the rate of development of the underground space [1–6].

One of the directions for the formation of a cavity in the underground space is geokhod technology, where the basic element is a geokhod [7–11].

When developing technical and constructive solutions for devices and elements of geokhods, it is necessary to take into account the complex, helical movement of the machine to the bottom of the mine. In addition, the developed methods for determining the force parameters of devices and elements of the geokhod, which interact with the environment and among themselves, must take into account the complex nature of the movement of the geokhod [12–16].

Content from this work may be used under the terms of the Creative Commons Attribution 3.0 licence. Any further distribution of this work must maintain attribution to the author(s) and the title of the work, journal citation and DOI.
Published under licence by IOP Publishing Ltd

IOP Conf. Series: Materials Science and Engineering **939** (2020) 012003 doi:10.1088/1757-899X/939/1/012003

One of the main devices of the geokhod that has a direct impact on the force characteristics of the underground apparatus is the operating body [17–21]. During the work of the operating body, the interaction surface of geokhod operating body (hereinafter – OB) with the face rock is formed.

The interaction surface of the operating body with the face rock is a secondary factor and is not taken into account when designing the operating body.

Therefore, the research work aimed at substantiating the parameters of the interaction surface of geokhod operating body with the face rock is relevant.

2. Research methods

When cutting with one straight knife of the geokhod OB, all the power of blocked cutting can be represented as the sum of the three constituent forces (figure 1).

1. Forces to overcome the frontal resistance of the soil by the front edge of the knife P_{cv}, proportional to the cross-sectional area of the slot in front of the front edge of the knife and depending on the cutting angle and soil strength;

2. Forces to overcome the resistance of the soil to fracture in the lateral extensions of the slot P_{side}, proportional to the area of these parts of the slot, which depends on the strength of the soil and does not depend on the cutting angle and cut width;

3. Forces to overcome the resistance of the soil to the cut with side edges of the knife at the bottom of the slot $P_{side.mid}$, proportional to the thickness of the cut, depending on the strength of the soil and not depending on the width of the cut and the angle of cutting.

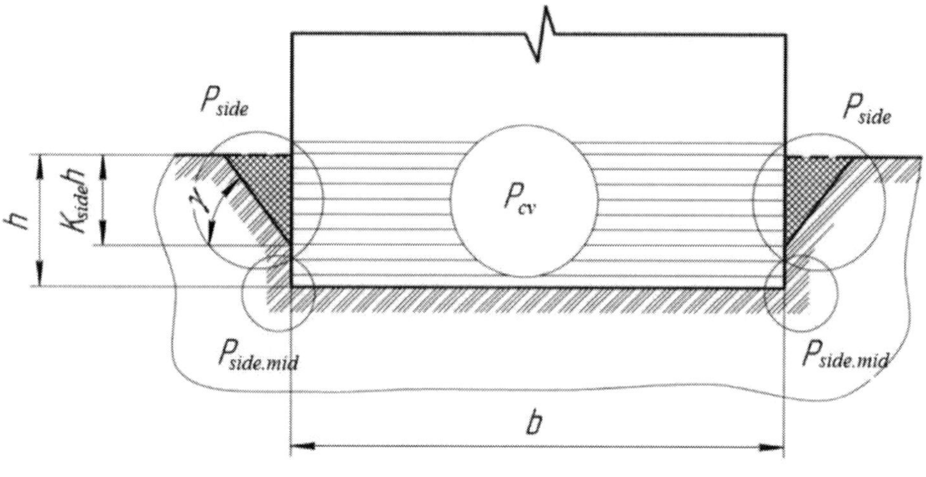

Figure 1. The area of action of the cutting force constituents with a straight sharp knife.

The interaction surfaces (hereinafter – IS) of the knife OB of the geokhod were constructed taking into account the three constituents of the cutting forces shown in figure 1.

Figure 2 shows the helicoid conical surface of the interaction between the OB of the geokhod and the face rock.

From figure 2 it follows that the parameters of the helicoid conical IS of the geokhod knife OB are the radius of the geokhod r_g, which consists of the cutting width b and the generatrix radius r_0, the cutting depth h, as well as the angle of inclination of the radial knife to a plane perpendicular to the axis of geokhod rotation γ_{OB}.

The parameters of the helicoid convex IS of the geokhod knife OB with the face rock (figure 3) include the geokhod radius r_g, which consists of the cutting width b and the generatrix radius r_0, the cutting depth h, as well as the radius of curvature of the cutting edge of the knife $r_{\text{кр.о}}$, which center is located on the axis of mine. Another parameter will be the segment angle Ω.

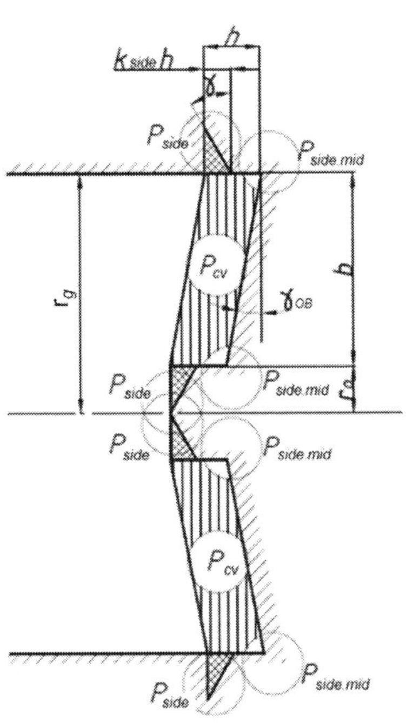

Figure 2. The parameters of the helicoid conical IS of the geokhod knife OB with the face rock.

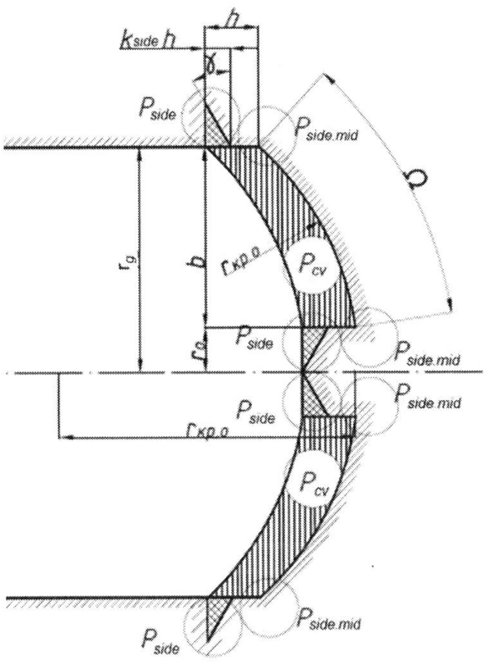

Figure 3. The parameters of the helicoid convex IS of the geokhod knife OB with the face rock.

If the curvature center of the knife cutting edge does not coincide with the axis of mine, the shape of IS of the geokhod knife OB with the face rock will have a convex torus shape (figure 4).

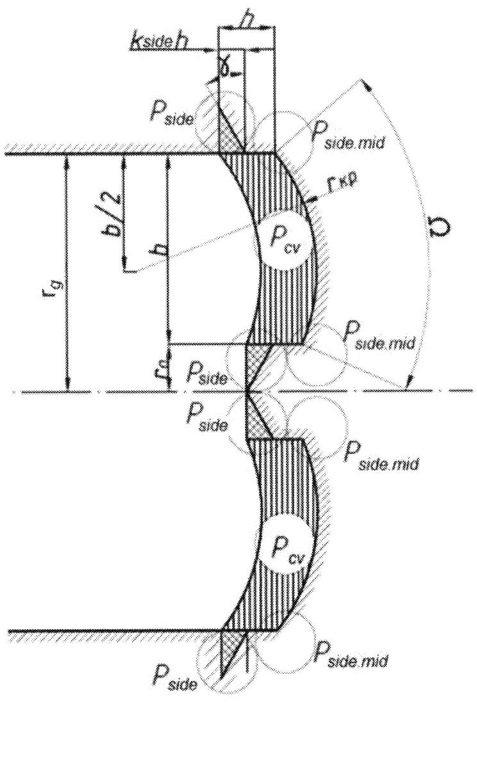

Figure 4. Parameters of helicoid IS with a convex torus shape.

From figure 4 it follows that the parameters of a helicoid IS with a convex torus shape are the geokhod radius r_g, which consists of the cutting width b and the generatrix radius r_0, the cutting depth h, as well as the radius of curvature of the knife edge $r_{кр}$ with a center located at a distance $b/2$ from the mine contour. Another parameter will be the segment angle Ω.

Another chosen variant of the IS of the geokhod knife OB is helicoid with the shape of a concave torus (figure 5). The difference from the helicoid IS with the shape of a convex torus will be the direction of concavity from the face rock.

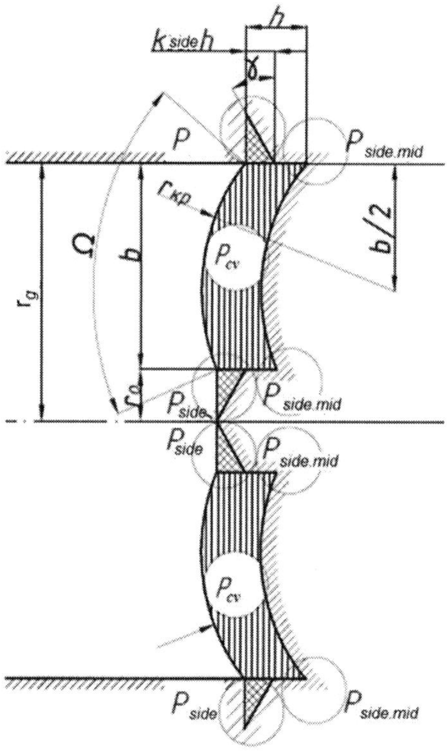

Figure 5. Parameters of helicoid IS with a concave torus shape.

Consequently, the parameters of a helicoid IS with a concave torus shape will be similar to the parameters of a helicoid IS with a convex torus shape; the only difference will be the location of the curvature center of the knife cutting edge. For a helicoid IS with a convex torus shape, the curvature center is located on the worked out space, and for a helicoid IS with a concave torus shape it is located on the side of the rock mass.

3. Conclusion

It should be noted that the parameters of IS of the geokhod knife OB with the face rock can be divided into two groups: general, which are characteristic for each variant of IS, and variable, which are characteristic of each variant.

The general parameters of IS include the geokhod radius r_g, cutting width b, generatrix radius r_0, and cutting depth h. The remaining parameters are variable and characterize each variant of IS separately.

References

[1] Aksenov V, Sadovets V, Pashkov D and Rezanova E V 2018 Granichnyye usloviya opredeleniya kharakternykh tochek nozhevogo ispolnitel'nogo organa geokhoda [Boundary conditions for determining the characteristic points of a knife operating body of a geokhod] *Bulletin of the Kuzbass State Technical University* **2(126)** 166–173 [In Russ.]

[2] Aksenov V, Sadovets V and Pashkov D 2018 Reasoning of the model sizes in modeling the interaction between tool and rock *E3S Web of Conferences* **41** 03002

[3] Aksenov V V, Efremenkov A B, Sadovets V Yu and Rezanova E V 2010 Formirovaniye strukturnogo portreta geokhodov [Formation of the structural portrait of geokhods] *Bulletin of the Kuzbass State Technical University* **1(77)** 35–41 [In Russ.]

[4] Aksenov V V, Horeshok A A and Beglyakov V Yu 2018 Kontseptsiya sozdaniya perspektivnogo tekhnologicheskogo uklada formirovaniya (osvoyeniya) podzemnogo prostranstva na baze operezhayushchego razvitiya novykh podkhodov v stroitel'noy geotekhnologii i geotekhnike [The concept of creating a promising technological structure for the formation (development) of underground space on the basis of the advanced development of new approaches in construction geotechnology and geotechnics] *Bulletin of the Kuzbass State Technical University* Part 1 **4** 105–113 [In Russ.]

[5] Aksenov V V, Horeshok A A and Beglyakov V Yu 2018 Kontseptsiya sozdaniya perspektivnogo tekhnologicheskogo uklada formirovaniya (osvoyeniya) podzemnogo prostranstva na baze operezhayushchego razvitiya novykh podkhodov v stroitel'noy geotekhnologii i geotekhnike [The concept of creating a promising technological structure for the formation (development) of underground space on the basis of the advanced development of new approaches in construction geotechnology and geotechnics] *Bulletin of the Kuzbass State Technical University* Part 2 **5** 43–51 [In Russ.]

[6] Golik V I, Razorenov Yu I and Karginov K G 2017 Mining industry – the basis of the sustainable development of North Ossetia-Alania *Sustainable development of mountain territories* vol 9 **2(32)** 163–171

[7] Aksenov V, Sadovets V and Pashkov D 2017 The influence of parameters on the generatrix of the helicoid form guide of geokhod bar working body *E3S Web of Conferences: The Second International Innovative Mining Symposium* **21** 03008

[8] Aksenov V V, Sadovets V Yu, Preis Ye V and Pashkov D A 2018 Sovershenstvovaniye matematicheskoy modeli opredeleniya silovykh parametrov nozhevogo ispolnitel'nogo organa geokhoda [Improving the mathematical model for determining the force parameters of a geokhod knife operating body] *Mining Equipment and Electromechanics* **5(139)** 16–22 [In Russ.]

[9] Aksenov V, Sadovets V and Pashkov D 2017 Opredeleniye silovykh parametrov nozhevogo ispolnitel'nogo organa geokhoda dlya razrusheniya porod maloy kreposti [Determination of force parameters of the geokhod knife operating body for destruction of low hardness rock] *Bulletin of the Kuzbass State Technical University* **3(121)** 116–126 [In Russ.]

[10] Aksenov V V, Horeshok A A, Ananjev K A and Ermakov A N 2014 Razrabotka skhemnykh resheniy ispolnitel'nykh organov geokhodov [Development of schematic solutions of geokhod operating bodies] *Izvestiya vysshikh uchebnykh zavedeniy. Gornyy zhurnal* **3** 73–76 [In Russ.]

[11] Aksenov V V, Efremenkov A B, Sadovets V Yu, Pashkov D A and Efremenkov V A 2019 *IOP Conf. Ser.: Mater. Sci. Eng.* **656** 012002

[12] Aksenov V V, Efremenkov A B, Sadovets V Yu, Pashkov D A and Efremenkov V A 2019 *IOP Conf. Ser.: Mater. Sci. Eng.* **656** 012003

[13] Pashkov D A 2019 Obzor metodik opredeleniya energoyemkosti razrusheniya porody [Review of methods for determining the energy capacity of rock destruction] *Rossiya molodaya. Sbornik materialov XI Vserossiyskoy nauchno-prakticheskoy konferentsii s mezhdunarodnym uchastiyem* 10310 [In Russ.]

[14] Aksenov V V, Efremenkov A B, Sadovets V Yu and Pashkov D A 2018 *IOP Conf. Ser.: Mater. Sci. Eng.* **441** 012005

[15] Aksenov V, Sadovets V and Pashkov D 2019 Razrabotka metodiki opredeleniya energoyemkosti razrusheniya gornoy porody nozhevym ispolnitel'nym organom geokhoda [Development of a method for determining the energy intensity of rock destruction by a

geokhod knife operating body] *Mining Equipment and Electromechanics* **2(142)** 30–38 [In Russ.]

[16] Aksenov V V, Khoreshok A A, Beglyakov V Yu and Efremenkov A B 2019 *IOP Conf. Ser.: Mater. Sci. Eng.* **656** 012004

[17] Aksenov V V, Khoreshok A A, Beglyakov V U and Efremenkov A B 2019 *IOP Conf. Ser.: Mater. Sci. Eng.* **656** 012005

[18] Sadovets V Yu 2007 *Obosnovaniye konstruktivnykh i silovykh parametrov nozhevykh ispolnitel'nykh organov geokhodov* [Substantiation of the design and power parameters of geokhod knife operating bodies] Candidate of Technical Science Dissertation. (Kemerovo: Kuzbass State Technical University named after T.F. Gorbachev) [In Russ.]

[19] Nesterov V, Aksenov V, Sadovets V, Pashkov D and Beysebayeva Zh 2019 Determination of the energy capacity of face rock breaking by the geokhod's knife operating element and its dependence on the external propeller's pitch *E3S Web of Conferences IVth International Innovative Mining Symposium* **105** 03024

[20] Nesterov V, Aksenov V, Sadovets V and Pashkov D 2019 Solution for the location of rock cutting elements relative to the rotation center of geokhod *E3S Web of Conferences IVth International Innovative Mining Symposium* **105** 03001

[21] Aksenov V V, Sadovets V Yu and Pashkov D A 2016 Obosnovaniye neobkhodimosti sozdaniya ispolnitel'nogo organa geokhoda dlya razrusheniya porod maloy kreposti [Justification of the need to create an operating body of the geokhod for the destruction of soft rocks] *Bulletin of the Kuzbass State Technical University* **6(118)** 8–15 [In Russ.]

IOP Conf. Series: Materials Science and Engineering **939** (2020) 012004 doi:10.1088/1757-899X/939/1/012004

Influence of the knife shape on the operating body cutting force

V V Aksenov[1,4,6], **A B Efremenkov**[2], **V Yu Sadovets**[3,7], **D A Pashkov**[4,8] **and V A Efremenkov**[5]

[1] Research Center "Siberian NPO" Ltd., Sovetsky pr., 56, Kemerovo, 650002 Russia
[2] Yaroslav-the-Wise Novgorod State University, ul. St.Petersburgskaya, 41 173003 Veliky Novgorod, Russia
[3] T.F. Gorbachev Kuzbass State Technical University, ul. Vesennyaya, 28 650000, Kemerovo Russia
[4] Institute of Coal of FRC of Coal and Coal Chemistry, Siberian Branch of the RAS, Leningradskiy pr., 10 650610 Kemerovo, Russia
[5] Peter-the-Great St.Petersburg Polytechnic University (SPbPU), Polytechnicheskaya, 29 195251, St.Petersburg, Russia

[6] E-mail: 55vva42@mail.ru
[7] E-mail: vsadovec@yandex.ru
[8] E-mail:pashkov.d.a@inbox.ru

Abstract. The basic part of any mining machine is the operating body used for mine working. The parameters of the mining machine are particularly affected by the geometric and force parameters of their operating bodies. The knife shape is one of the geometric indicators of the mining machine cutting part. The existing methodology for calculating a geokhod cutting body does not take this indicator into account. The purpose of the work is to determine the influence of operating body knife shape on the strength of the blocked cutting. The research shows that the knife shape affects only the forces for overcoming the frontal resistance of the soil by the front edge of the knife. It is possible to determine the influence of the cutting edge shape of the operating body on the force of blocked cutting. As a result of the study, it was revealed that when the knife is deepened to the entire cutting depth and when it destroys the face corresponding to its shape, the knife shape does not influence the strength of the blocked cutting.

1. Introduction

The growth in the volume and scale of the effective development of the underground urban space is observed today throughout the world. It is associated with the ever-increasing concentration of the urban population and the continuous increase in the number of vehicles that give rise to almost all of the most acute contemporary urban problems – territorial, transport, environmental, and energy. As noted in research, the depth of development of the underground space is 20–50 m [1–5].

Kuzbass at a depth of 20–50 m is characterized by such soils as loams and clays. The depth of their occurrence ranges up to 95 m [6]. These soils belong to soft rocks according to the scale of M. Protodyakonov [7–9].

Content from this work may be used under the terms of the Creative Commons Attribution 3.0 licence. Any further distribution of this work must maintain attribution to the author(s) and the title of the work, journal citation and DOI.
Published under licence by IOP Publishing Ltd

IOP Conf. Series: Materials Science and Engineering **939** (2020) 012004 doi:10.1088/1757-899X/939/1/012004

A promising trend for solving these issues is the use of geokhod technology for the cavity formation in the underground space, where the geokhod is the basic element [10–15].

The basic part of any mining machine is the operating body used for mine working. The parameters of the mining machine are particularly influenced by the geometric and force parameters of their operating bodies [16–18].

Therefore, the research aimed at substantiating the parameters of the geokhod operating body for the destruction of soft rocks is relevant.

The research purpose is to determine the influence of the knife shape of operating body on the force of blocked cutting.

2. Research Methods

Taking into account the structural possibility of changing the knife geometry, it becomes possible to obtain a number of options for schematic solutions of geokhod cutting bodies used for the destruction of soft rocks, some of which are presented in figure 1.

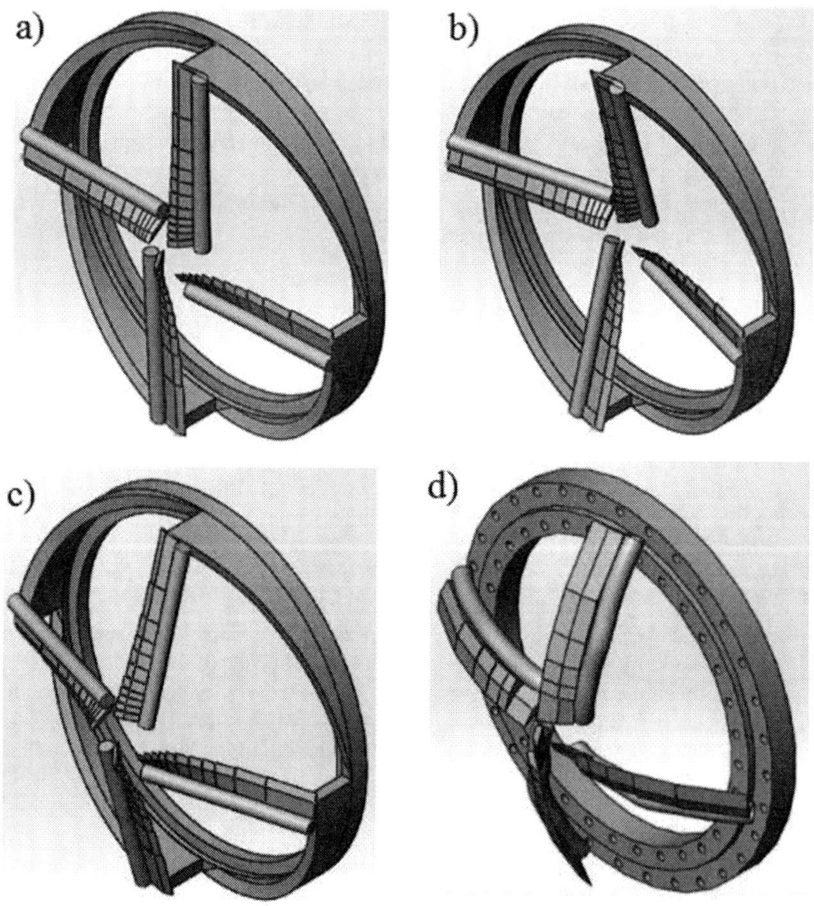

Figure 1. Schematic versions of a four-beam knife of geokhod operating body with different knife geometry: a) a flat shape; b) an inverse cone shape; c) convex cone shape; d) a convex ball shape.

The methodology for determining the force parameters of the geokhod knife operating body is based on the methodology for determining the cutting forces proposed by Yu.A. Vetrov [19–21]. The methods of Yu. A. Vetrov are based on the laws of the process of cutting soil with simple knives that separate the chips. Under a simple knife, a rectangular-shaped knife is understood (figure 2).

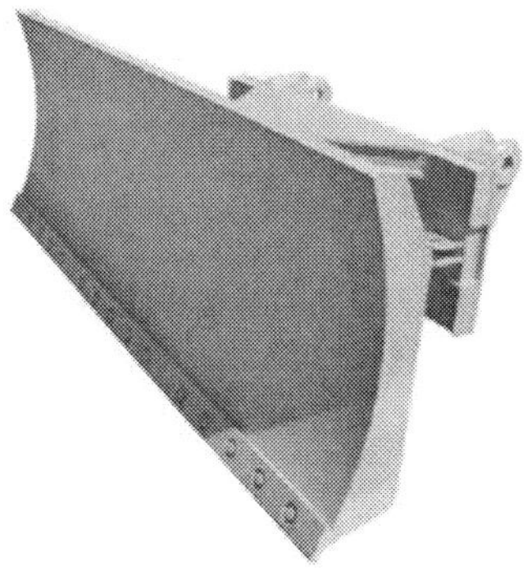

Figure 2. A rectangular-shaped knife.

When cutting with one straight knife of the geokhod operating body, all the power of blocked cutting can be represented as the sum of the three constituent forces (figure 3).

1. Forces to overcome the frontal resistance of the soil with the front edge of the knife P_{sv}, proportional to the cross-sectional area of the slot in front of the front edge of the knife and depending on the cutting angle and soil strength;
2. Forces to overcome the resistance of the soil to fracture in the lateral extensions of the P_{side} slot, proportional to the area of these parts of the slot, depending on the strength of the soil and not depending on the cutting angle and width of the cut;
3. Forces to overcome the resistance of the soil to the cut side edges of the knife at the bottom of the slot $P_{side\ mid}$ proportional to the thickness of the cut, depending on the strength of the soil and not depending on the width of the cut and the angle of cutting.

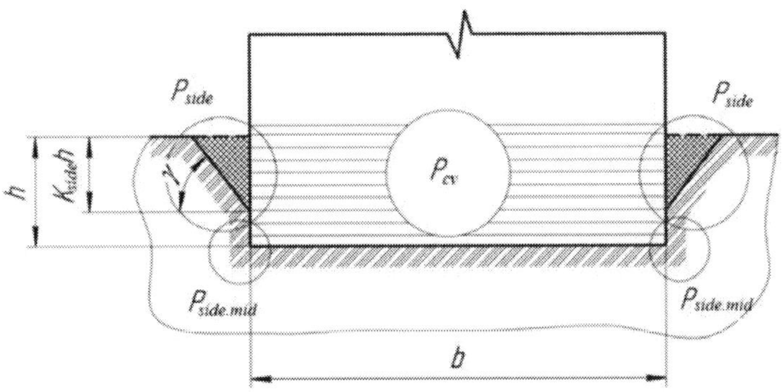

Figure 3. The action areas of a sharp straight knife cutting force constituents.

To determine the influence of the shape of the knife of the cutting body on the force of blocked cutting, we assume that the recess of the knife occurs throughout the depth of cutting and the knife destroys the face corresponding to its shape.

When cutting with a spherical shape knife of the geokhod operating body, all the power of blocked cutting can be represented as the sum of the three constituent forces (figure 4).

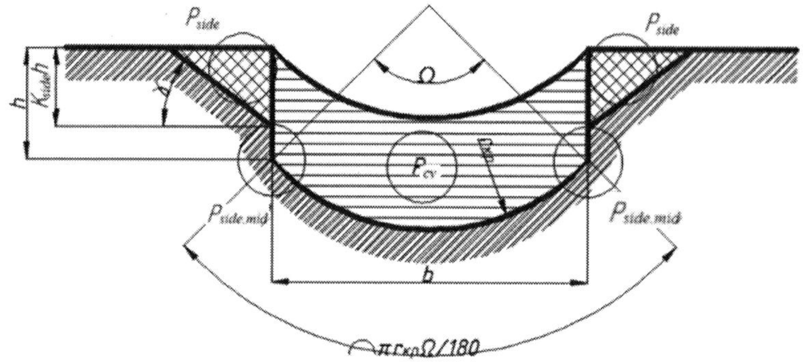

Figure 4. The action areas of a sharp spherical knife cutting force constituents.

Figures 3 and 4 show that the shape of the knife only affects the forces to overcome the frontal resistance of the knife (in figure 3 and 4 indicated as P_{sv}).

The entire force of blocked cutting with a straight sharp knife [19–21]:

$$P_{mid} = \varphi m_{cv} bh + 2m_{side} h^2 + 2m_{side.mid} h, \tag{1}$$

where $\varphi \ m_{cv} bh$ - forces to overcome frontal resistance to the knife (in figure 3 indicated as P_{cv}), N;

$2m_{side} h^2$ – soil fracture force in lateral slit extensions (P_{side} in figure 3), N;

$2m_{side.mid} h$ – side shear forces ($P_{side.mid}$ in figure 3), N;

φ – coefficient taking into account the influence of the cutting angle;

m_{cv} – specific cutting force to overcome ground resistance with the front face at a cutting angle of 45^0, Pa;

b – knife width, m;

h – cutting depth, m;

m_{side} – coefficient characterizing the strength of soil destruction in the lateral parts of the slot, Pa;

$m_{side.mid}$ – coefficient characterizing the specific shear strength of one of the side ribs of the knife, N/m.

Since $\varphi \ m_{cv} bh$ environmental parameters multiplied by the cut area (S) in the forehead of the knife (figure 3), then with the spherical shape of the knife, it is necessary to determine the cut area ($S_{sp.sh}$) in the forehead of the knife shown in figure 4.

As seen from figure 4, the figure has a complex spherical shape; to find the area, it is necessary to divide the figure into elementary figures (figure 5).

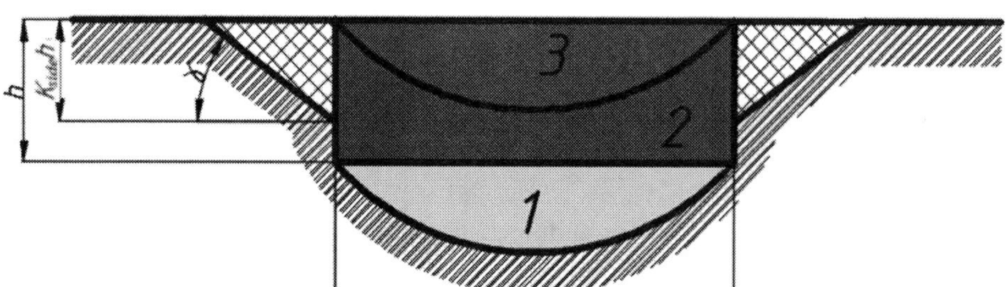

Figure 5. Elementary figures forming a cut with a spherical shape knife.

From figure 5, figures 1 and 3 are circular segments, figures 2 and 3 jointly form a rectangle with sides b and h. Figures 1 and 3 are the same and have the same area, since figure 3 is figure 1 in the previous cut.

In view of the foregoing, the cut area in the forehead of spherical shape knife will be equal to

$$S_{sp.sh} = bh, \tag{2}$$

where b – cut width, m;

h – cutting depth, m.

Since the knife width and the cut width are equal, the forces to overcome the frontal resistance to the knife P_{cv} will be the same for both a straight knife and a spherical knife with equal values of b and h.

Therefore, the entire force of blocked cutting will not depend on the spherical shape of the knife.

When cutting with one knife when the knife is tilted at an angle γ_{ob}, the entire force of the blocked cutting can also be represented as the sum of the three constituent forces (figure 6).

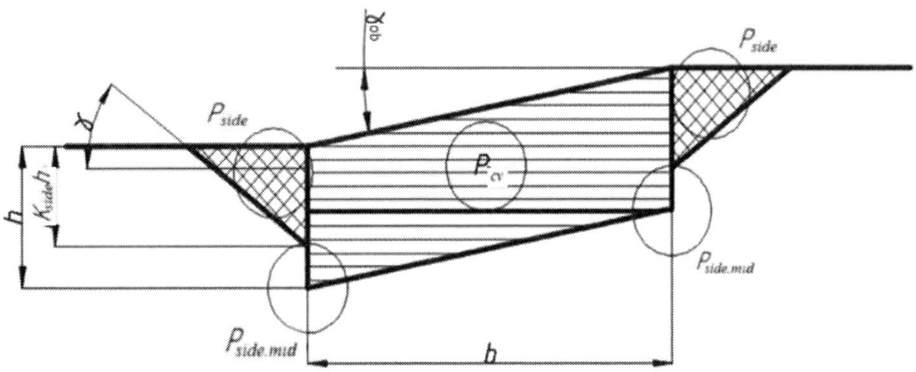

Figure 6. The action areas of a sharp knife cutting force constituents when the knife is tilted at an angle γ_{ob}.

Since $\varphi\ m_{cв}bh$ are environmental parameters multiplied by the cut area in the forehead of the knife (figure 3), when the knife is tilted at an angle γ_{ob}, it is necessary to determine the cut area in the forehead of the knife depicted in figure 6.

From figure 6, the cut area in the forehead of the knife will be equal to the area of the parallelogram.

In view of the foregoing, the cut area in the forehead of the knife when the knife is tilted at an angle γ_{ob} will be equal to

$$S_{tilt} = bh, \tag{3}$$

where b – parallelogram height or cut width, m;

h – a parallelogram side or cutting depth, m.

Hence, the forces to overcome the frontal resistance to the knife will be the same for a straight-shaped knife, as well as for a knife inclined at an angle γ_{ob} with equal values of b and h.

Therefore, the force of blocked cutting will not depend on the shape of the knife.

3. Conclusion

1. The influence of the shape of the knife of the knife operating body on the strength of the blocked cutting is determined.

2. When the knife is deepened to the entire depth of cutting and destroys the face corresponding to its shape, the shape of the knife does not affect the strength of the blocked cutting.

It is necessary to determine the influence of the shape of the knife of the knife operating body on the blocked cutting force when the shape of the knife and the shape of the face do not match, in the notching situation.

References

[1] Shashkin A G, Zentsov V N and Ulitskiy V M 2018 Razvitiye podzemnogo prostranstva megapolisa [Underground Space Development in Megalopolis] *Zhilishchnoye stroitel'stvo* [Housing Construction] *9* 30–36 [In Russ.]

[2] Aksenov V, Sadovets V and Pashkov D 2018 Reasoning of the model sizes in modeling the interaction between tool and rock *E3S Web of Conferences* **41** 03002

[3] Aksenov V V, Yefremenkov A B, Sadovets V Yu and Rezanova Ye V 2010 Formirovaniye strukturnogo portreta geokhodov [Formation of the structural portrait of geokhods *Bulletin of the Kuzbass State Technical University* **1(77)** 35–41 [In Russ.]

[4] Aksenov V V, Horeshok A A and Beglyakov V Yu 2018 Kontseptsiya sozdaniya perspektivnogo tekhnologicheskogo uklada formirovaniya (osvoyeniya) podzemnogo prostranstva na baze operezhayushchego razvitiya novykh podkhodov v stroitel'noy geotekhnologii i geotekhnike [The concept of creating a promising technological structure for the formation (development) of underground space on the basis of the advanced development of new approaches in construction geotechnology and geotechnics *Bulletin of the Kuzbass State Technical University* part 1 **4** 105–113 [In Russ.]

[5] Aksenov V V, Horeshok A A and Beglyakov V Yu 2018 Kontseptsiya sozdaniya perspektivnogo tekhnologicheskogo uklada formirovaniya (osvoyeniya) podzemnogo prostranstva na baze operezhayushchego razvitiya novykh podkhodov v stroitel'noy geotekhnologii i geotekhnike [The concept of creating a promising technological structure for the formation (development) of underground space on the basis of the advanced development of new approaches in construction geotechnology and geotechnics] *Bulletin of the Kuzbass State Technical University* part 2 **5** 43–51[In Russ.]

[6] Aksenov V, Sadovets V and Pashkov D 2017 The influence of parameters on the generatrix of the helicoid form guide of geokhod bar working body *E3S Web of Conferences The Second International Innovative Mining Symposium* **21** 03008

[7] Shtumpf G G 1994 Mekhanicheskiye svoystva gornykh porod Kuznetskogo ugol'nogo basseyna i zakonomernosti ikh izmeneniya [Mechanical properties of rocks of the Kuznetsk coal basin and the laws of their change] *Fizicheskiye problemy razrabotki poleznykh iskopayemykh* **4** 43–50 [In Russ.]

[8] Aksenov V, Sadovets V and Pashkov D 2017 Opredeleniye silovykh parametrov nozhevogo ispolnitel'nogo organa geokhoda dlya razrusheniya porod maloy kreposti [Determination of force parameters of the geokhod knife operating body for destruction of low hardness rock] *Bulletin of the Kuzbass State Technical University* **3 (121)** 116–126 [In Russ.]

[9] Aksenov V V, Horeshok A A, Ananjev K A and Ermakov A N 2014). Razrabotka skhemnykh resheniy ispolnitel'nykh organov geokhodov [Development of schematic solutions of geokhod operating bodies] *Izvestiya vysshikh uchebnykh zavedeniy. Gornyy* **3** 73–76 [In Russ.]

[10] Aksenov V V, Efremenkov A B, Sadovets V Yu, Pashkov D A and Efremenkov V A 2019 Impact of the number of blades of the geokhod cutting body on the energy intensity of the rock destruction *IOP Conf. Ser.: Mater. Sci. Eng.* **656** 012002

[11] Aksenov V V, Efremenkov A B, Sadovets V Yu, Pashkov D A and Efremenkov V A 2019 Impact of the inclination angle of a blade of the geokhod cutting body on the energy intensity of rock destruction *IOP Conf. Ser.: Mater. Sci. Eng.* **656** 012003

[12] Begljakov V Y, Timofeev V Y and Dokhnenko M V 2014 *Appl. Mech. Mater.* **682** 282

[13] Beglyakov V Yu 2012 *Obosnovaniye parametrov poverkhnosti vzaimodeystviya ispolnitel'nogo organa geokhoda s porodoy zaboya* [Substantiation of surface parameters of interaction of the geokhod operating executive body with face rock] Dissertatsiya na soiskaniye uchenoy stepeni kandidata tekhnicheskikh nauk [Candidate of Technical Science Dissertation] (Yugra: Kuzbass State Technical University) [In Russ.]
[14] Aksenov V, Sadovets V and Pashkov D 2019 Razrabotka metodiki opredeleniya energoyemkosti razrusheniya gornoy porody nozhevym ispolnitel'nym organom geokhoda [Development of a method for determining the energy intensity of rock destruction by a geokhod knife operating body] *Gornoye oborudovaniye i elektromekhanika* **2(142)** 30–38 [In Russ.]
[15] Blaschuk M Yu, Dronov A A and Miheev D A 2015 Geokhod propel effort mathematical model *Applied Mechanics and Materials* vol 770 391–396
[16] Blashchuk M Y, Kazantsev A A and Chernukhin R V 2014 *Appl. Mech. Mater.* **682** 418
[17] Sadovets V Yu 2007 *Obosnovaniye konstruktivnykh i silovykh parametrov nozhevykh ispolnitel'nykh organov geokhodov* [Substantiation of the design and power parameters of geokhod knife operating bodies] Dissertatsiya na soiskaniye uchenoy stepeni kandidata tekhnicheskikh nauk [Candidate of Technical Science Dissertation] (Kemerovo: Kuzbass State Technical University named after T.F. Gorbachev) [In Russ.]
[18] Nesterov V, Aksenov V, Sadovets V, Pashkov D and Beysebayeva Zh 2019 Determination of the energy capacity of face rock breaking by the geokhod knife operating element and its dependence on the external propeller's pitch *E3S Web of Conferences IVth International Innovative Mining Symposium* 03024
[19] Nesterov V, Aksenov V, Sadovets V and Pashkov D 2019 Solution for the location of rock cutting elements relative to the rotation center of geokhod *E3S Web of Conferences IVth International Innovative Mining Symposium* 03001
[20] Vetrov Yu A 1971 *Rezaniye gruntov zemleroynymi mashinami* [Soil cutting by earth-moving machinery] (Moscow: Mashinostroyeniye) [In Russ.]
[21] Vetrov Yu A 1985 *Raschet sil rezaniya i kopaniya gruntov* [Calculation of soil cutting and digging forces] (Kiyev: Izd-vo Kiyev. un-ta) [In Russ.]
[22] Vetrov Yu A and Baladinskiy V L 1980 *Mashiny dlya spetsial'nykh zemlyanykh rabot* [Machines for special earthworks] (Kiyev: Izd-vo Kiyev. un-ta) [In Russ.]

IOP Conf. Series: Materials Science and Engineering **939** (2020) 012005 doi:10.1088/1757-899X/939/1/012005

Development of a methodology for modeling complex shaped geokhod operating body in *SolidWorks*

V V Aksenov[1,4,7], A B Efremenkov[2,8], V Yu Sadovets[3,9], D A Pashkov[4,10], R V Sadovets[5], E V Rezanova[3,11] and V A Efremenkov[6]

[1] Research Center "Siberian NPO" Ltd, Sovetsky pr., 56, Kemerovo, 650002 Russia
[2] Yaroslav-the-Wise Novgorod State University, ul. St.Petersburgskaya, 41 173003, Veliky Novgorod, Russia
[3] T.F.Gorbachev Kuzbass State Technical University, ul. Vesennyaya, 28 650000 Kemerovo, Russia
[4] Institute of Coal of FRC for Coal and Coal Chemistry, Siberian Branch of the RAS, Leningradsky pr., 10 650610, Kemerovo, Russia
[5] Secondary School No 14, ul. Druzhby, 7 650024, Kemerovo, Russia
[6] Peter the Great St.Petersburg Polytechnic University, Polytechnicheskaya, 29 195251, St.Petersburg, Russia

[7] E-mail: 55vva42@mail.ru
[8] E-mail: abe@novsu.ru
[9] E-mail: vsadovec@yandex.ru
[10] E-mail: pashkov.d.a@inbox.ru
[11] E-mail: elen-rezanova@yandex.ru

Abstract. Modeling complex, curved surfaces of manufactured parts is an elaborate design task. The article discusses the development of a methodology for modeling a knife with a complex helicoid surface and a geokhod knife operating body in SolidWorks. The development of the modeling methodology presented in the article allows making constructive solutions not only for the knife-type operating body of any size, but for other types of geokhod operating bodies. In addition, the solid model of the operating body knife allows assessing the stress-strain state of the operating body elements.

1. Introduction

Geokhods represent the new class of mining machines, combining the performance of technological mining cycle operations on a single platform [1–5]. The technology of cavity formation in the underground space using a geokhod is called a geokhod technology [6–8].

The main operation for the formation of a cavity in a rock mass is the destruction of the rock mass in the bottom of the mine. The destruction operation is carried out by the operating body [9–12].

At shallow depths of cavity formation in the rock mass, rocks have low physical and mechanical properties [3] and belong to the category of soft rocks according to the scale developed by M.M. Protodyakonov. In geokhod technology, knife-type operating bodies are used to destroy such rocks [13]. One of the technical solutions of the knife operating body of the geokhod is shown in figure 1.

Content from this work may be used under the terms of the Creative Commons Attribution 3.0 licence. Any further distribution of this work must maintain attribution to the author(s) and the title of the work, journal citation and DOI.
Published under licence by IOP Publishing Ltd

Operating bodies of geokhods have structural and technological features [14, 15], the main of which is the dependence of the geometrical parameters of the knife operating body on the diameter of the head section and the geometrical parameters of the external mover of the geokhod [16].

To develop options for constructive solutions of a knife operating body, as well as to assess the stress-strain state of the knife and the elements of the operating module pairing with each other and with the head section of the geokhod, it is necessary to develop a method for constructing a solid model of a helicoid operating body used for face destruction [17].

Figure 1. Technical solutions of the geokhod knife operating body.

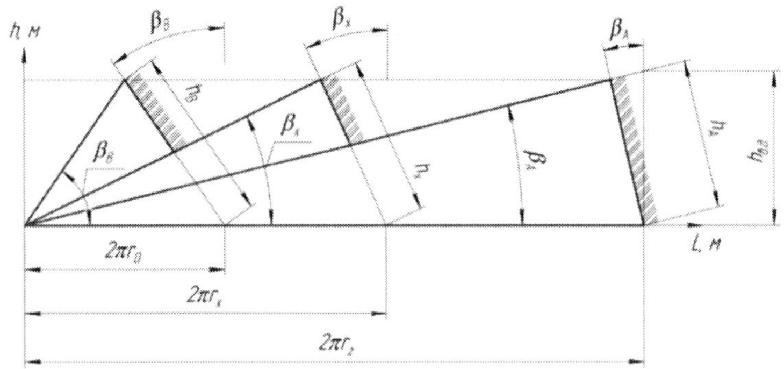

Figure 2. The direction of movement of the points of the operating body knife depending on their location on the knife along the radius.

Purpose of work is to develop a methodology for creating a solid model of a knife operating body for face destruction.

2. Methodology

Any point of the knife located at a distance x (figure 1) from the axis of rotation of the geokhod moves to the bottom at an angle of [18]

$$\beta_x = arctg\frac{h_B}{2\pi x} \quad (1)$$

For one revolution of the geokhod each point of the knife of geokhod operating body must move to the bottom of the mine by an amount equal to the step of the external mover of the geohod $h_{в.д}$. Moreover, the length of the path for each point in the process of moving will depend on the radius of the location on the knife of the operating body. Knife points located closer to the axis of rotation of the geokhod move to the bottom at a larger angle (point B at an angle β_B) than points located on the periphery of the knife (point A at an angle β_A) (figure 2). The size of the cut-off layer for each point of the knife of the geokhod operating organ will depend on the angle of movement of the point to the bottom of the mine.

The point located on the periphery of the knife moves to the bottom at an angle β_A that should be equal to the angle of inclination of the external mover of the geokhod.

We will carry out the development of a method for constructing a solid model of a knife operating body for a geokhod diameter of 1.8 m. The values of angles β_x calculated by expression (1), depending on the location on the radius of the knife, are presented in table 1.

Table 1. The values of the angles β_x, depending on the location on the radius of the knife.

No	0	1	2	3	4	5	6	7	8	9	10
β_x , degrees	90	37.36	23.41	16.81	13.06	10.67	9.01	7.79	6.65	5.54	5.05
r_x , mm	0	90	180	270	360	450	540	630	720	810	900

1. We select the plane in front in which the cross section of the knife of the operating body will be built (figure 3). In the selected plane, we construct the cross section of the knife (figure 4).

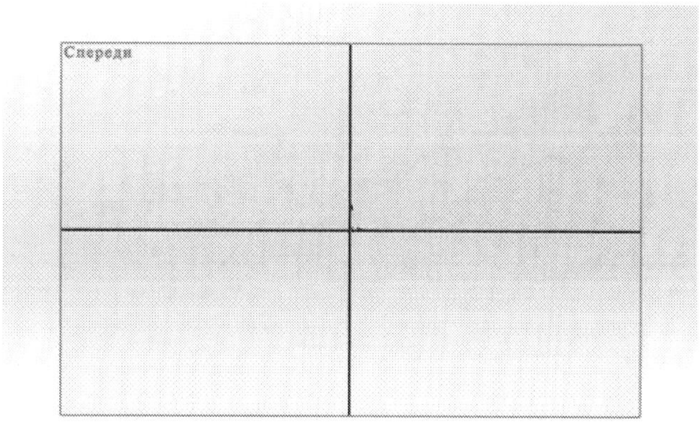

Figure 3. Base plane.

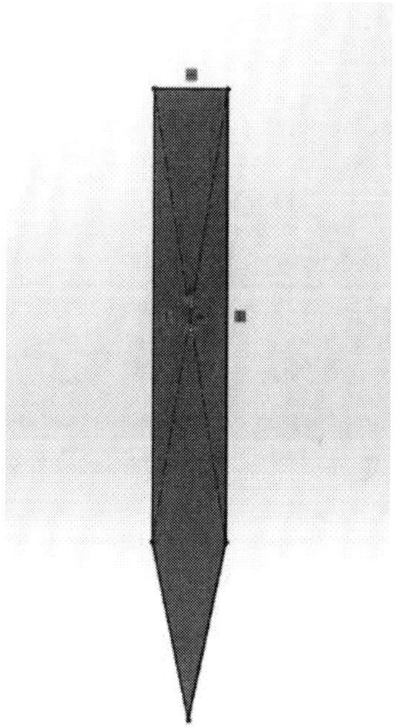

Figure 4. Knife cross section.

2. Perpendicular to the constructed sketch, we build 10 additional planes parallel to each other at a distance of 90 mm (as a component for the plane we select the plane in front and the points on the segment) (figure 5).

3. In each of the additional planes constructed, we build exactly the same sketch of the knife cross section (figure 6) as in the first construction plane (the center of the sketches will be the segment characterizing the length of the knife). After constructing the sketches in additional planes, we turn each sketch by the calculated angle (table 1) relative to the first sketch.

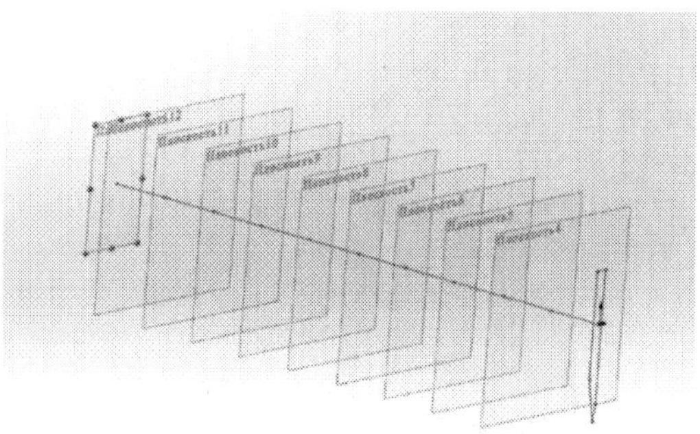

Figure 5. Additional planes.

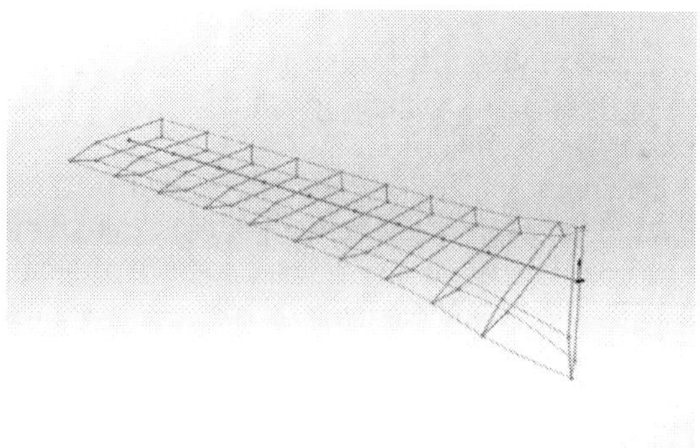

Figure 6. Copying the cross section of the knife into each plane.

4. Using the boss arrangement method, we build a solid model of the knife of geokhod operating body on the base (figure 7).

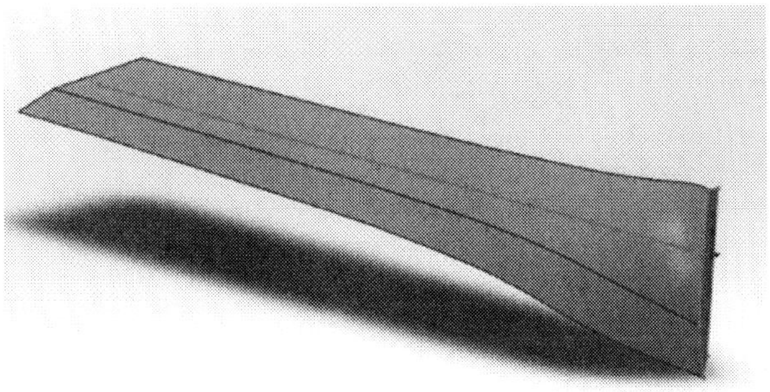

Figure 7. Solid model of the knife.

5. We make a circular array 90 degrees to the previously constructed solid model of the knife (figure 8). The operating body may have a number of knife shafts. We will construct a four-shaft model of the geokhod knife operating body.

Figure 8. Four shafts of operating body.

Figure 9. Simulating central part.

6. We build a circle in the center of the connection of the four knives on top and squeeze in both directions. Then, in the constructed cylinder, we cut out the cavity of a given radius (figure 9).

7. From the center of the part we build up the center line of a given length. We add a new plane parallel to the plane at the top and the end point of the additional centerline. We build two circles in this plane and make a cutout to the side of the part (figure 10).

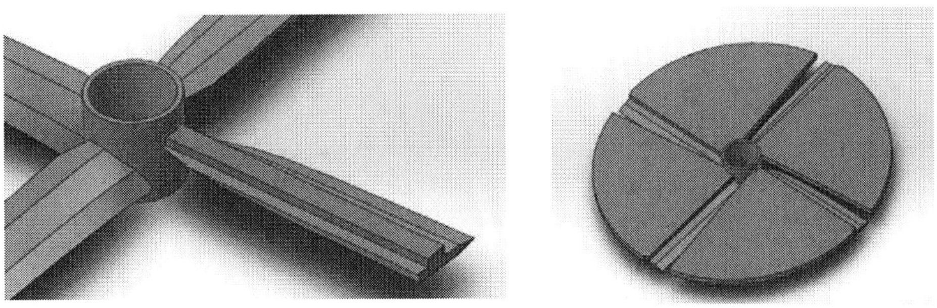

a) b)

Figure 10. Aperture simulation.

8. We draw the center line from the center of coordinates at a given distance. We build a rectangle, positioning it so that the auxiliary line that we built is located in the center of the rectangle and extrude it to a given distance (figure 10, a). We make a circular array relative to the origin.

9. Using the tool "Arc Center" from the center of coordinates we build two circles of a given radius. In the center of the rectangle (point 8) we connect both ends of the arcs. We extrude out for a given distance. We make a circular array of this part (figure 10, b).

10. We add a circle with a radius equal to the length of the knife. The solid model of the knife operating body of the geokhod is ready (figure 11).

Figure 11. Model of the helicoid geokhod operating body.

3. Conclusion

The developed technique for constructing a solid model of the helicoid shape of a knife of geokhod operating body allows

– simulating the stress-strain state of the elements of geokhod knife operating body, as well as the fastening elements of the operating body to the geokhod bulk;

– substantiating the design parameters of both the elements of geokhod knife operating body and the elements of their fastening to each other and to the head section of the geokhod;

– developing new constructive solutions for various types of operating bodies that meet the requirements for geokhod operating bodies.

References

[1] Aksenov V V, Horeshok A A and Beglyakov V Yu 2018 Kontseptsiya sozdaniya perspektivnogo tekhnologicheskogo uklada formirovaniya (osvoyeniya) podzemnogo prostranstva na baze operezhayushchego razvitiya novykh podkhodov v stroitel'noy geotekhnologii i geotekhnike [The concept of creating a promising technological structure for the formation (development) of underground space on the basis of the advanced development of new approaches in construction geotechnology and geotechnics] *Bulletin of the Kuzbass State Technical University* Part 1 **4** 105–113 [In Russ.]

[2] Aksenov V V, Horeshok A A and Beglyakov V Yu 2018 Kontseptsiya sozdaniya perspektivnogo tekhnologicheskogo uklada formirovaniya (osvoyeniya) podzemnogo prostranstva na baze operezhayushchego razvitiya novykh podkhodov v stroitel'noy geotekhnologii i geotekhnike [The concept of creating a promising technological structure for the formation (development) of underground space on the basis of the advanced development of new approaches in construction geotechnology and geotechnics] *Bulletin of the Kuzbass State Technical University* Part 2 **5** 43–51 [In Russ.]

[3] Sadovets V Yu, Beglyakov V Yu and Pashkov D A 2016 *Fiziko-mekhanicheskiye svoystva gornykh porod maloy kreposti* [Physico-mechanical properties of rocks of low strength]

Perspektivy innovatsionnogo razvitiya ugol'nykh regionov Rossii: Proc. V Int. Sc. and Pr. Conf. 142–147 [In Russ.]
[4] Aksenov V, Sadovets V and Pashkov D 2017 The influence of parameters on the generatrix of the helicoid form guide of geokhod bar working body *E3S Web of Conferences: The Second International Innovative Mining Symposium* **21** 03008
[5] Aksenov V, Sadovets V and Pashkov D 2018 Reasoning of the model sizes in modeling the interaction between tool and rock *E3S Web of Conferences* **41** 03002
[6] Aksenov V V and Sadovets V Yu 2006 Sintez tekhnicheskikh resheniy nozhevogo ispolnitel'nogo modulya geokhoda [Synthesis of technical solutions for a knife executive module of a geokhod] *Bulletin of KuzSTU* **6** 33–37 [In Russ.]
[7] Aksenov V V, Efremenkov A B, Sadovets V Yu and Rezanova Ye V 2010 Sozdaniye innovatsionnogo in-strumentariya dlya formirovaniya podzemnogo prostranstva [Creation of innovative tools for the formation of underground space] *Bulletin of KuzSTU* **1(77)** 42–47 [In Russ.]
[8] Aksenov V V, Efremenkov A B, Sadovets V Yu, Pashkov D A and Efremenkov V A 2019 *IOP Conf. Ser.: Mater. Sci. Eng.* **656** 012002
[9] Gorbunov V F, Aksenov V V and Sadovets V Yu 2006 *Strukturnaya matritsa gornoprokhodcheskikh sistem* [Structural matrix of mining systems] Sluzheniye delu (Kemerovo: KuzGTU) 77–84 [In Russ.]
[10] Aksenov V V, Efremenkov A B, Sadovets V Yu, Pashkov D A and Efremenkov V A 2019 *IOP Conf. Ser.: Mater. Sci. Eng.* **656** 012003
[11] Aksenov V V and Sadovets V Yu 2007 Modelirovaniye osobennostey dvizheniya geokhoda [Modeling the features of the movement of the geokhod] *Bulletin of KuzSTU* **1** 20–22 [In Russ.]
[12] Nesterov V, Aksenov V, Sadovets V, Pashkov D and Beysebayeva Zh 2019 *E3S Web of Conferences* **105** 03024
[13] Nesterov V, Aksenov V, Sadovets V and Pashkov D 2019 *E3S WebofConferences* **105** 03001
[14] Sadovets V Yu and Pashkov D A 2016 Opredeleniye funktsii linii kontakta nozha ispolnitel'nogo organa geokhoda s poverkhnost'yu zaboya [Determination of the function of the contact line of a knife of geokhod operating body with a bottom surface] *Tekhnologii i materialy* **4** 9–14 [In Russ.]
[15] Aksenov V, Sadovets V and Pashkov D 2017 *E3SWeb of Conferences* **21** 03008
[16] Aksenov V, Sadovets V, Rezanova E and Pashkov D 2017 *E3SWeb of Conferences* **15** 03015
[17] Aksenov V, Sadovets V and Pashkov D 2018 *E3SWeb of Conferences* **41** 03002
[18] Aksenov V V, Efremenkov A B, Sadovets V Yu and Pashkov D A 2018 *IOP Conf. Ser.: Mater. Sci. Eng.* **441** 012002

Areas of research on the construction of tunneling underground machines of the Geokhod class

V V Aksenov[1,2,6], **A A Khoreshok**[3], **A B Efremenkov**[4], **V Yu Beglyakov**[5,7], **A V Koperchuk**[5,8], **M Yu Blaschuk**[5,9], **V Yu Sadovets**[3,10] **and D A Pashkov**[2,11]

[1] Scientific Research Center OOO "Sibirskoye SPA", Sovetskiy pr., 56 650099, Kemerovo, Russia
[2] Institute of Coal of FRC of Coal and Coal Chemistry, Siberian Branch of the RAS, Leningradskiy pr., 10 650610, Kemerovo, Russia
[3] T.F. Gorbachev Kuzbass State Technical University, Vesennyaya ul., 28 650000, Kemerovo, Russia
[4] Yaroslav-the-Wise Novgorod State University, B. St. Peterburgskaya ul., 41 173003, Veliky Novgorod, Russia
[5] Yurginsky Institute of Technology, branch of Tomsk Polytechnic University, Leningradskaya ul., 26 652052, Yurga, Russia

[6] E-mail: 55vva42@mail.ru
[7] E-mail: begljakov@rambler.ru
[8] E-mail: avkop@mail.ru
[9] E-mail: mby.tpu@gmail.com
[10] E-mail: vsadovec@yandex.ru
[11] E-mail: pashkov.d.a@inbox.ru

Abstract. The article presents the features that characterize the complexity and novelty of the main functional systems of underground machines of the "Geokhod" class as an object of design and production. The authors discuss the problems that developers encountered at all stages of creating test models and prototypes of a new class of underground machines "Geokhod". The lack of special scientific and methodological support is a deterrent to the creation of a new class of underground vehicles. Taking into account the peculiarities of work, interaction with the geo-environment of tunneling underground machines of the "Geokhod" class, the need arose to create a scientific specialty that will be associated with the design and construction of tunneling underground equipment. Thus, by analogy with the aircraft industry, the support for the creation of underground machines interacting with the geo-environment should be assigned to a new scientific profile and the scientific specialty "Construction of underground machines that interact with the geo-environment". A new scientific direction is considered as a key element of an advanced technological structure for the development of underground space on the basis of advanced development and new approaches in construction geotechnology and geotechnics.

1. Introduction

Currently, with the active participation of OOO "Sibirskoye SPA", a Concept for creating an advanced industrial structure based on the rapid development of key engineering components has been formed and is being implemented. One of the approaches of the Concept is the thesis [1]:

Content from this work may be used under the terms of the Creative Commons Attribution 3.0 licence. Any further distribution of this work must maintain attribution to the author(s) and the title of the work, journal citation and DOI.
Published under licence by IOP Publishing Ltd

"It is difficult to enter the industrial structure formed by global corporations with standard products, but you can set a goal to replace this structure to the maximum extent. If we propose more economical technologies, we can significantly expand our participation in the international division of labor, as well as secure a priority for ourselves in creating more advanced mechanisms and machines".

The team of OOO "Sibirskoye SPA" Research Center is working on the creation of Geokhod – a new class of tunneling underground machines (UM) moving in the bowels of the Earth using an edge rock mass (geo-environment) [2–6]. One example of a "Geokhod" class UM is shown in figure 1. In contrast to existing approaches, the authors consider underground mining **as a process of movement of a solid body (underground vehicle) in a solid medium (geo-environment)** [4–19].

a) b)

Figure 1. The prototype UM model "401" with a diameter of 3.2 m: a) on the assembly slipway, b) bottom-hole area in tests.

2. Main part

UM of the "Geokhod" class so far **have no analogues among existing mining systems** (MS).

Distinctive features of the "Geokhod" class UM and the advantages arising from them are given in [4, 6, 9, 10]; here we indicate only the main ones:

– The presence of an external (contour) musculoskeletal system, including new functional and structural elements that interact with the geo-environment.

– The use of the rock massif to form traction and pressure forces (using the geo-environment), while the massif is a kinematic link, is meshed with the marginal musculoskeletal system.

– Possibility of horizontal and vertical maneuvering with a turning radius commensurate with the diameter of the output.

– Versatility in tilt angles in a wide range.

– Exclusion of the weight of the machine from the process of forming traction and pressure forces.

– To create on the executive body of the destruction of the slaughter of UM sufficient pressure without artificial weight gain UM.

As part of the "Geokhod", **new functional systems** have appeared that are absent in serial MS:

– marginal elements: external mover (EM), counter-rotation elements (CRE) [8, 14–19];

– executive parts (EP) of contour elements – EP EM and EP CRE [8, 14–19];

– unit (module) for pairing sections [20];

– a rotating shell [12, 21];

– "steering" system [7, 8, 15–19].

Other "Geokhod" systems have undergone changes that practically exclude the borrowing of ready-made aggregates from serial tunneling equipment.

Signs characterizing the complexity and novelty of the main functional systems of the UM class "Geokhod" as an object of design and production [2, 3, 21–26]:

Common to all systems and user agents in general:

– lack of analogues among existing MS;

– there are no methods of calculation and design;

– the combination of rotation (movement) and the performance of the functions of temporary support;

– placement of working mechanisms in a confined space;

– placement of working mechanisms on a rotating supporting structure;

– the need to transfer energy from a non-rotating stabilizing section to consumers located on the rotating section;

– the need to manufacture parts and assemblies of large sizes with high accuracy;

– the need to ensure the possibility of modification in mining;

– ensuring the performance of all operations of the driving cycle with their combination in time;

– providing the possibility of horizontal and vertical maneuvering;

– performance at any angles of inclination of the production;

– the possibility (feasibility) of reducing metal consumption due to the exclusion of weight from the number of useful and necessary parameters;

Specific to individual systems:

– peripheral placement of power elements of the transmission with the possibility of multi-threaded power transfer;

– significant torque and axial force on the housing and interface module;

– the complex nature of the movement of destructive elements of the AI to the face;

– a different approach to rock destruction in the central and peripheral parts of the face;

– the need to coordinate the parameters of the EM and the main EP, due to the rigid kinematic connection of their movements;

– the formation of EP EM and EP CRE of the deep channel with a complex profile.

When creating experimental models and prototypes of a new class of UM "Geokhod" [2, 3] **at the design stage**, developers were faced with the problem of the lack of methodology for determining the parameters of the main elements and systems of the UM interacting with the geo-environment during its movement: the main face [9, 11]; UM case [12]; marginal elements (external mover, anti-rotation elements and their EP) [8, 14–19], etc.

At the production stage, they faced the problem of the lack of a methodology for ensuring the accuracy and manufacturability of specific UM systems [21–25]. The problem of the scientific and methodological "deficit" was also present **at the testing stage** [27, 28].

The lack of **special scientific and methodological support** is a limiting factor [9, 23–30] on the way to creating a new class of UM.

New scientific and methodological problems arose at all stages of creating a new generation of UM: **design, manufacturing and testing.**

And if there are separate guidelines for **designing** mining machines, textbooks and educational disciplines, then **production** is based on the general technology of mechanical engineering, and **tests** are based on operating experience and bench tests of individual elements. While in the aircraft industry there are scientific and engineering specialties that are "tailored" to each of the stages of creating a new aircraft: there are research institutes and design institutes that accompany the process of creating new equipment at each stage, there is a "test pilot" profession, testing grounds and full-scale laboratory facilities (wind tunnels, etc.).

Papers [31–35] describe the concept of creating an advanced technological structure for the formation (development) of underground space on the basis of the advanced development of new approaches in construction geotechnology and geotechnics. Figure 2 presents the structure of the proposed technological structure of the development (formation) of the underground space.

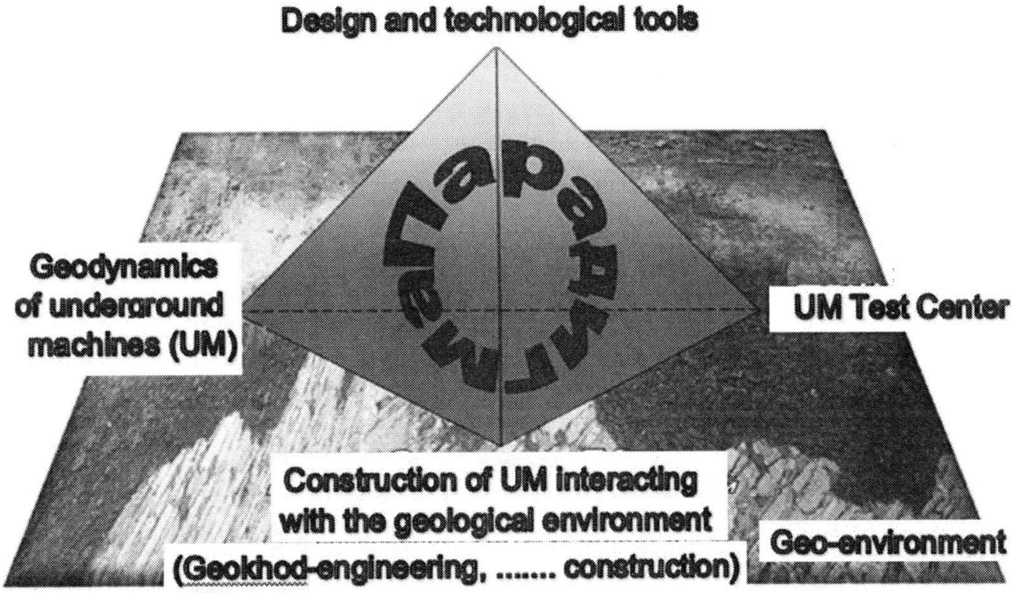

Figure 2. The structure (Paradigm) of an advanced technological structure for the development (formation) of the underground space on the basis of the advanced development of new approaches in construction geotechnology and geotechnics.

All the key structural elements of the technological structure have their own special purpose, are closely interconnected and together form a technical and economic paradigm [36]. **The construction of UM interacting with the geo-environment** is the main structural element of the technological structure of the development (formation) of the underground space. When creating a promising technological structure, the experience and advantages of the through technological structure created in the aircraft industry were fully taken into account [31, 35].

One of the key structural elements of the technological structure in the aircraft industry is aircraft construction (helicopter construction, rocket construction).

The term **"Aircraft industry"** has two closely related meanings:

1) Scientific discipline related to the design and construction of aircraft.

2) Aircraft manufacturing industry.

It should be specially noted that aircraft construction at one time initiated the need for industrialization of the country, and also became one of the main consumers of the created components of mechanical engineering: materials; bearings, mechanisms, etc.

The name of the science supporting aircraft industry is 05.07.02 "Design, construction and production of aircrafts".

From the specialty formula 05.07.02: "A distinctive feature of the specialty is that its main content is a systematic study of the design, construction and production of engineering objects taking into account economics and automation, an analysis of the experience in creating samples of rocket and space and aviation equipment, and conversion use of the abovementioned methods and works. The importance of solving the scientific and technical problems of this specialty lies in improving the theoretical, methodological, experimental and production base, which can improve the quality (reliability, maintainability, carrying capacity, etc.) and reduce the cost of developing, manufacturing and operating aircraft (reduce metal consumption, power consumption, etc.). This area of science is complex, covering the identification of physical, chemical, mechanical and other laws with the aim of using in practice the most efficient and economical design and technological processes that require the least time, human and material resources. "

In mining, there is a science "Mining machines" described as "a field of science and technology, including research, development and operation of machinery and equipment (units and apparatuses) of the main and auxiliary industries in the mining industry and engaged in the study of relationships and patterns in order to create new and improvement of existing mining machines and equipment and their elements, which have increased productivity (efficiency), durability (reliability), safety and environmental friendliness."

Taking into account the peculiarities of work, interaction with the geo-environment of tunneling UMs of the "Geokhod" class, the best approach to the formulation of research areas for the scientific specialty "Construction of UMs, interacting with the geo-environment", in our opinion, is shown by the specialty 05.07.02 "Design, construction and aircraft manufacturing. "

We propose the following research areas for the scientific specialty "Construction of UMs interacting with the geo-environment":

– Development of design and construction methods, mathematical and software-algorithmic support for choosing the optimal look and parameters, layout and structural power scheme, units and systems of UM.

– Development of assessment methods and research of the organization and management of design and design work of design bureaus with a high level of application of CALS-technologies.

– Development of search methods for optimal design and technological solutions in the early stages of designing a UM.

– Research and analysis of ways to intensify the design and modernization of existing UMs, taking into account the experience gained.

– Creation and development of fundamentally new design solutions for the implementation of nodes, systems and user agents as a whole. Study of their characteristics and assessment of application prospects.

– The study of the dynamics (changes) in the reliability of systems and UM in the process of the life cycle in order to form an optimal plan for their development.

– Study of the impact on the technical characteristics of systems and the structural design of the UM body of geodynamic processes in the designed structure.

Development of methods of model and mathematical support for solving (research) of functional problems:

- selection of the optimal composition of UM;
- selection of the trajectory of movement, etc.

– Development of methods, models and software for making optimal decisions in order to study design problems under given restrictions, taking into account their compromise nature, risk and distinguishability of compared product options (processes).

– A study of the economic feasibility of creating UMs with a multi-purpose purpose (for example, civil and military), as well as the effectiveness of using existing products and ground complexes for the same purposes.

– Technological preparation for the production of UMs and their systems, including:

- constructive and technological solutions that allow for advanced preparation of production;
- manufacturability of structures;
- directive technological materials for the production of new UM structures, their systems and assemblies;
- systems and means of automated production preparation;
- other methods and means of development and implementation of technological processes of production.

– Technological processes, special equipment for the manufacture of UM parts, including technology:

- manufacturing of cast parts;
- manufacturing of parts by pressure treatment (forging, stamping, etc.);
- manufacture of parts using beam energy beams and other physical and physico-chemical methods;
- manufacture of parts from liquid, powder or fibrous materials; - manufacturing of parts from non-metallic materials, including parts of thermal protection;
- coating;
- manufacture of parts from composite materials; - manufacture of parts by galvanic plastic;
- thermal, thermomechanical and chemical-thermal treatment of parts;
- metal cutting;
- manufacturing of parts by combined and complex methods, including in flexible production systems.

– Technological processes, special and specialized equipment for assembly, installation and testing, repair of UM, their systems and units, including technology and tools:

- nodal, modular and general assembly;
- the formation of detachable and one-piece joints during the assembly of UM using bolted joints, welding, soldering, riveting, gluing and their combinations, including using robotic systems;
- cleaning the internal and external surfaces of assembly units and UM as a whole;
- installation and testing of on-board systems in workshop conditions;
- performance of newly developed compounds specific for the production of UM.

– Technological processes for monitoring, testing and metrological support in the production of UM, their systems and units, including technology and tools:

- control of geometric parameters; surface quality control;
- functional control and testing (pneumohydraulic, vibroacoustic, electrophysical, force impacts);
- control of physical and technical parameters;
- non-destructive testing of parts, assemblies, products and integral joints, as well as other types of control;
- metrological support of controlled quality indicators of production facilities.

– Technological processes of designing, programming and information support in the production of UM, including technology and tools:

- computer-aided design of technological processes and their management;
- mathematical modeling of technological processes;
- dimensional linking of aggregates;
- control of the shape, size and relative position of the surfaces of the units;
- manufacturing of technological surfaces, equipment and parts, including machines with CNC;
- control of technology and programs for the manufacture of parts and assembly units.

– Organization and economics of manufacturing UM, including:

- the production structure of the enterprise and the features of the organization of technological processes in pilot, single, small-scale and mass production;
- enlarged methods for calculating material consumption, labor and cost of products;
- methods for evaluating the manufacturability of products;
- methods for calculating the technical and economic efficiency of technological processes, equipment and organizational and technical measures.
- Study of the process of introducing scientific, technological and technical achievements in the design and construction of UM in other areas of science, engineering and technology.

In the direction of creating and practicing fundamentally new design solutions for the implementation of nodes, systems and UM as a whole, the first studies were performed and the results obtained [9, 10, 19, 26]. But these results, as experience shows, can generally be regarded as obtaining "intelligence" data for making decisions on the choice of areas for further research and development of experimental samples of UM.

With regard to the creation of a new class of user agents, priority tasks are:

– Creation (improvement) of the general layout of the "Geokhod" class UM.
– Creation (improvement) of UM body.
– Creation (improvement) of UM transmissions.
– Creation (improvement) of external movers for UM.
– Creation (improvement) of counter-rotation elements for UM.
– Creation (improvement) of executive bodies for destruction of the face of UM.
– Creation (improvement) of the paramount executive bodies of the UM.
– Creation (improvement) of UM controls along the route of development.
– Creation (improvement) of UM loading modules.
– Creation (improvement) of transport systems for UM.
– Creation (improvement) of power plants for UM.
– Creation (improvement) of UM starting devices.
– Creation (improvement) of UM mounting hardware modules.

3. Conclusion

The need to develop the provisions of a new scientific direction and scientific specialty "Construction of underground machines interacting with the geo-environment" is based on the distinctive features of the "Geokhod" class UM with the ensuing advantages, as well as highlighted features that characterize the complexity and novelty of the main functional systems UM. By analogy with the aircraft industry, the authors of the article have proposed research areas for the scientific specialty "Construction of underground machines that interact with the geological environment."

Currently, work is underway to create new elements and systems of the "Geokhod" class UM. The existing scientific and methodological support for the development and construction of mining machines basically does not meet the requirements and features of the new class of "Geokhod" UM. For most systems and elements of a new class of UMs for which there are no calculation methods and a reasonable choice of rational parameters. The lack of scientific and methodological support hinders the creation of new elements and systems of UM "Geokhod", which dictates the need to expand the areas of research areas of the passport of the specialty "Mining machines".

References

[1] *Kontseptsiya sozdaniya perspektivnogo promyshlennogo uklada na baze operezhayushchego razvitiya klyu-chevykh mashinostroitel'nykh komponentov* [The concept of creating an advanced industrial structure on the basis of the accelerated development of key engineering components] http://pro-kompas.info/strategy/ (Access date: 05.13.2020)

[2] Aksenov V V, Beglyakov V Y, Kazantsev A A, Valter A V and Efremenkov A B 2016 Opyt uchastiya v proekte po organizacii vysokotekhnologichnogo proizvodstva [Experience of participation in the project on the organization of high-tech production] *Mining equipment and electrical engineering* **8** 8–15

[3] Aksenov V V, Beglyakov V Y, Blashchuk M Yu, Efremenkov A B, Kazantsev A A, Khoreshok A A and Valter A V 2016 Geokhod: zadachi, harakteristiki, perspektivy [Geokhod: objectives, characteristics, prospects] *Mining equipment and electrical engineering* **8** 3–8

[4] Aksenov V V 2004 *Nauchnyye osnovy geovinchesternoy tekhnologii provedeniya gornykh vyrabotok i sozdaniya vintopovorotnykh agregatov* [Scientific foundations of geovinchester technology for mine workings and the creation of screw-rotary assemblies] Abstract of a dissertation for the degree of Doctor of Technical Sciences (Kemerovo: Institute of Coal and Coal Chemistry of the Siberian Branch of the Russian Academy of Sciences) [In Russ.]

[5] Aksenov V V and Efremenkov A B 2008 Geovinchesternaya tekhnologiya i geokhody – innovatsionnyy podkhod k osvoyeniyu podzemnogo prostranstva [Geovinchester technology

and Geokhods – an innovative approach to the development of underground space] *Bulletin of Higher Educational Institutions. Mountain Journal* **4** 19–28 [In Russ.]
[6] Aksenov V V, Efremenkov A B, Sadovets V Yu and Rezanova E V 2010 Sozdaniye innovatsionnogo instrumentariya dlya formirovaniya podzemnogo prostranstva [Creation of innovative tools for the formation of the underground space] *Bulletin of Kuzbass State Technical University* **1(77)** 42–47 [In Russ.]
[7] Aksenov V V and Sadovets V Yu 2007 Modelirovaniye osobennostey dvizheniya geokhoda [Modeling of Geokhod movement features] *Bulletin of Kuzbass State Technical University* **1** 20–22 [In Russ.]
[8] Aksenov V V, Kostinets I K and Beglyakov V Yu 2013 Osobennosti raboty vneshnego dvizhitelya geokhoda [Features of the external propulsion of Geokhod] *Mining Information and Analytical Bulletin* **6** 419–425 [In Russ.]
[9] Aksenov V V, Efremenkov A B, Sadovets V Yu, Pashkov D A and Efremenkov V A 2019 *IOP Conference Series: Materials Science and Engineering* **656** 012002
[10] Beglyakov V Yu and Aksenov V V 2012 *Poverkhnost' zaboya pri prokhodke gornoy vyrabotki geokhodom* [The surface of the face during the mining of a mine by Geokhod] (Saarbrücken, Germany: LAP LAMBERT Academic Publ.) p 139 [In Russ.]
[11] Nesterov V, Aksenov V, Sadovets V and Pashkov D 2019 Solution for the location of rock cutting elements relative to the rotation center of Geokhod *E3S Web of Conferences IVth International Innovative Mining Symposium* **105** 03001
[12] Aksenov V V, Beglyakov V Yu and Kapustin A N 2014 Analiz nesushchikh konstruktsiy (korpusov) izvestnykh tekhnicheskikh sistem primenimykh v kachestve korpusa (nositelya) geokhoda [Analysis of load-bearing structures (cases) of known technical systems applicable as a body (carrier) of Geokhod] *Vestnik KuzSTU* **6(106)** 34–36 [In Russ.]
[13] Aksenov V V, Khoreshok A A and Begljakov V Y 2013 Justification of creation of an external propulsor for multipurpose shield-type heading machine – Geokhod *Applied Mechanics and Materials* **379** 20–23
[14] Beglyakov V Yu, Aksenov V V, Kostinets I K, Efremenkov A B and Khoreshok A A 2018 *IOP Conference Series: Materials Science and Engineering* **441** 012008
[15] Beglyakov V Yu, Aksenov V V, Kostinets I K and Khoreshok A A 2018 Opredeleniye sil vzaimodeystviya osnovnykh sistem geokhoda s geosredoy i mezhdu soboy [Determination of the forces of interaction of the main systems of Geokhod with the geo-environment and among themselves] *Mining Sciences and Technologies* **4** 23–30 [In Russ.]
[16] Beglyakov V U, Aksenov V V, Kazantsev A A and Kostinets I K 2017 Razrabotka zakonturnoj oporno-dvizhitel'noj sistemy geohodov [Development of outside the contour support-propulsion system of geokhod] *Vestnik KuzSTU* **6(124)** 175–181 [In Russ.]
[17] Aksenov V V, Beglyakov V Yu, Kazantsev A A, Kostinets I K and Koperchuk A V 2016 Klassifikatsiya geometricheskikh parametrov vneshnego dvizhitelya geokhoda [Classification of the geometrical parameters of the external mover of Geokhod] *Mining Equipment and Electromechanics* **8** 33–39 [In Russ.]
[18] Beglyakov V Yu, Aksenov V V, Kostinets I K and Khoreshok A A 2018 Skhemy nagruzheniya pri modelirovanii protsessa vzaimodeystviya vneshnego dvizhitelya geokhoda prikonturnym massivom porod [Loading schemes for modeling the process of interaction of an external mover of a geohod by a contour rock mass] *Mining Sciences and Technologies* **3** 3–10 [In Russ.]
[19] Kostinets I K 2018 *Obosnovaniye parametrov opornoy poverkhnosti vneshnego dvizhitelya geokhoda* [Justification of the parameters of the supporting surface of the external mover of the Geokhod] Dissertation for the degree of Candidate of Technical Sciences (Kemerovo: T.F. Gorbachev Kuzbass State Technical University) [In Russ.]
[20] Dronov A A, Beglyakov V Yu, Aksenov V V, Efremenkov A B and Efremenkov V A 2019 *IOP Conference Series: Materials Science and Engineering* **656** 012015

[21] Aksenov V V, Walter A V and Beglyakov V Yu 2014 Obespecheniye geometricheskoy tochnosti obolochki pri sborke sektsiy geokhoda [Ensuring the geometric accuracy of the shell when assembling sections of Geokhod] *Metal Processing (technology, equipment, tools)* **4** 19–28 [In Russ.]

[22] Walter A V, Nozirzoda Sh S and Ivkin A N 2016 Issledovaniye tochnosti izgotovleniya krupnogabaritnykh korpusov vrashcheniya opytnogo obraztsa geokhoda [The study of the accuracy of manufacturing large-sized rotation cases of a prototype Geokhod] *Concept* **11** 2026–2030 [In Russ.]

[23] Walter A V 2016 Faktory, obuslavlivayushchiye pogreshnosti poverkhnostey vrashcheniya krupnogabaritnykh korpusov geokhoda [Factors causing errors in the surfaces of rotation of large-sized bodies of the Geokhod] *Mining Equipment and Electromechanics* **8 (126)** 19–25

[24] Walter A V, Borovikov I F, Chernukhin R V and Nozirzoda S S 2016 Research on geometric errors of intermediate unit shell of the Geokhod *Urgent Problems of Modern Mechanical Engineering* 12017

[25] Walter A V, Chernukhin R V and Kapustin A N 2016 Otkloneniya formy obolochki stabiliziruyushchey sektsii geokhoda [Deviations of the shell shape of the stabilizing section of Geokhod] *Technologies and Materials* **1** 4–7 [In Russ.]

[26] Blaschuk M Yu 2012 *Obosnovaniye parametrov transmissii geokhoda s gidroprivodom* [Substantiation of transmission parameters of the Geokhod with a hydraulic drive] Dissertation for the degree of Candidate of Technical Sciences (Yurga: Kuzbass State Technical University) [In Russ.]

[27] Koperchuk A V, Kazantsev A A, Beglyakov V Yu and Filonov V V 2015 Obosnovaniye neobkhodimosti razrabotki startovogo ustroystva geokhoda [Justification of the need to develop a launch device for Geokhod] *Technologies and Materials* **1** 29–3 0[In Russ.]

[28] Aksenov V V, Khoreshok A A, Adamkov A V and Ermakov A N 2015 Obosnovaniye neobkhodimosti razrabotki novoy tekhnologii stroitel'stva podzemnykh vyrabotok [Justification of the need to develop a new technology for the construction of underground workings] *Bulletin of Kuzbass State Technical University* **4(110)** 21–26 [In Russ.]

[29] Aksenov V V, Beglyakov V Yu and Ganovichev S I 2015 Obosnovaniye neobkhodimosti razrabotki uni-fitsirovannoy matematicheskoy modeli geokhoda [Substantiation of the need to develop a unified mathematical model of the Geokhod] *Technologies and Materials* **3** 9–13 [In Russ.]

[30] Aksenov V V, Beglyakov V Y, Kazantsev A A and Doroshenko I V 2016 *IOP Conference Series: Materials Science and Engineering* **127** 012031

[31] Aksenov V V, Khoreshok A A, Beglyakov V Yu and Efremenkov A B 2019 *IOP Conference Series: Materials Science and Engineering* **656** 012004

[32] Aksenov V V, Khoreshok A A, Beglyakov V Yu and Efremenkov A B 2019 *IOP Conference Series: Materials Science and Engineering* **656** 012005

[33] Golik V I, Gabaraev O Z, Maslennikov S A, Khasheva Z M and Shulgaty L P 2016 The provision of development conversion perspectives into undeground one for russian iron ore deposits development *The Social Sciences* vol 11 **18** 4348–4351

[34] Aksenov V V, Khoreshok A A and Beglyakov V Yu 2018 Kontseptsiya sozdaniya perspektivnogo tekhnologicheskogo uklada formirovaniya (osvoyeniya) podzemnogo prostranstva na baze operezhayushchego razvitiya novykh podkhodov v stroitel'noy geotekhnologii i geotekhnike [The concept of creating a promising technological structure for the formation (development) of underground space on the basis of the advanced development of new approaches in construction geotechnology and geotechnics] (Part 1) *Vestnik of Kuzbass State Technical University* **4(128)** 105–113 [In Russ.]

[35] Aksenov V V, Khoreshok A A and Beglyakov V Yu 2018 Kontseptsiya sozdaniya perspektivnogo tekhnologicheskogo uklada formirovaniya (osvoyeniya) podzemnogo prostranstva na baze operezhayushchego razvitiya novykh podkhodov v stroitel'noy geotekhnologii i geotekhnike

[The concept of creating a promising technological structure for the formation (development) of underground space on the basis of the advanced development of new approaches in construction geotechnology and geotechnics] (Part 2) *Vestnik of Kuzbass State Technical University* **5(129)** 43–52 [In Russ.]
[36] Lvov D S and Glazyev S Yu 1986 Teoreticheskiye i prikladnyye aspekty upravleniya NTP [Theoretical and applied aspects of NTP management] *Economics and Mathematical Methods* **5** 793

Research areas of the scientific specialty "Geodynamics of underground machines"

V V Aksenov[1,2,6], **S V Magazov**[1,7], **A A Khoreshok**[3], **A B Efremenkov**[4], **V Yu Beglyakov**[5,8], **V Yu Sadovets**[3,9] **and D A Pashkov**[2,10]

[1] Scientific Research Center LLC "Siberian Research and Production Association", Sovetskiy pr., 56 650099, Kemerovo, Russia
[2] Institute of Coal of FRC for Coal and Coal Chemistry, Siberian Branch of the RAS, Leningradskiy pr., 10 650610, Kemerovo, Russia
[3] T.F. Gorbachev Kuzbass State Technical University, Vesennyaya ul., 28 650000, Kemerovo, Russia
[4] Yaroslav-the-Wise Novgorod State University, ul. B. St. Petersburgskaya, 41 173003, Veliky Novgorod, Russia
[5] Yurginsky Institute of Technology, Branch of Tomsk Polytechnic University, Leningradskaya ul., 26 652052, Yurga, Russia

[6] E-mail: 55vva42@mail.ru
[7] E-mail: s.mag@mail.ru
[8] E-mail: begljakov@rambler.ru
[9] E-mail: vsadovec@yandex.ru
[10] E-mail: pashkov.d.a@inbox.ru

Abstract. At all stages of creating experimental models and prototypes of a new class of underground machines "Geokhod", developers faced problems. At the design stage the problem was the lack of methods for determining the parameters of the main elements and systems of underground machines that interact with the geo-environment during their movement. At the production stage, they faced the problem of the lack of a methodology to ensure the accuracy and manufacturability of specific systems of underground machines. In addition, the problem of scientific and methodological "deficit" was present at the test stage. The development of systems of a new class of underground machines is constrained by the lack of special scientific and methodological support. When creating a new class of underground devices moving in the bowels of the Earth and using the geo-environment, there is an urgent need to study the effects of the geo-environment on bodies moving in the geo-environment. In the aircraft industry there is a fundamental science that studies the forces arising on the surface of a solid body moving in the air - aerodynamics of aircrafts. Thus, by analogy with aircraft construction, studies of the effects of the geo-environment on bodies moving in the geo-environment should be assigned to a new scientific direction and the scientific specialty "Geodynamics of underground machines". The article proposes a draft passport of the specialty and a field of research in the new direction "Geodynamics of underground machines". This scientific direction is one of the basic elements of an emerging advanced technological structure for the development of the underground space, based on new approaches in construction geotechnology and geotechnics.

Content from this work may be used under the terms of the Creative Commons Attribution 3.0 licence. Any further distribution of this work must maintain attribution to the author(s) and the title of the work, journal citation and DOI.
Published under licence by IOP Publishing Ltd

1. Introduction

To develop and implement new approaches and projects in any engineering industry, the Concept of creating an advanced industrial structure based on the rapid development of key engineering components has been formed and is being implemented. OOO "Sibirskoye SPA" took an active part in developing the provisions of this concept, one of which is the thesis [1]:

"It is difficult to enter the industrial structure formed by global corporations with standard products, but you can set a goal to replace this structure to the maximum extent. If we propose more economical technologies, we can significantly expand our participation in the international division of labor, as well as secure a priority for ourselves in creating more sophisticated mechanisms and machines."

One of the R&D directions pursued by the team of OOO "Sibirskoye SPA" Research Center is the creation of a new class of tunneling underground machines (UM) named "Geokhod" that move in the bowels of the Earth using an edge rock mass (geo-environment) [2–5]. The sample of the new "Geokhod" class is presented in figure 1. When developing the UM, the authors rely on a new approach to the tunnel work and consider it **as a process of movement of a solid body (underground apparatus) in a solid medium (geo-environment)** [4–6].

Figure 1. The prototype UM model "401" with a diameter of 3.2 m.

Distinctive features of the "Geokhod" class UM [7–10]:

– The principle of operation is based on the use of a contour massif of rocks for the formation of traction and pressure forces (use of a geo-environment).

– The presence of an external (marginal) musculoskeletal system, including new functional and structural elements that interact with the geo-environment.

– General functional layout.

– Qualitatively new functionality: horizontal and vertical maneuvering, versatility in the angles of inclination of the mine in a wide range.

In this case, **the near-mass array (geo-environment)** is a kinematic link, is meshed with the UM running system and is used [4, 7–10]:

– as a supporting element involved in the creation of traction of the UM and the formation of pressure on the executive body;

– for the perception of reactive efforts when driving a driving unit (underground unit);

– to perform basic technological operations, including securing the mine with a permanent support.

Using the geo-environment allows you to:

– Exclude the weight of the machine from the process of forming traction and pressure forces.

– To carry out the movement of UM in any direction of the bowels of the Earth.

– To create on the executive body of the destruction of the slaughter of UM sufficient pressure without artificial weight gain for UM.

2. Main part

At the stage of designing experimental samples and prototypes of a new class of UM "Geokhod" [2, 3], the developers faced the problem of the lack of a methodology for determining the parameters of elements and UM systems that interact with the geo-environment during its movement. The main such systems are: executive bodies (EB) of the main face [9–11]; UM case [11]; external mover, counter-rotation elements and their EB (edge elements) [8,12–18], etc. At this stage of development, it is possible to modernize the existing calculation methods presented in the guidelines, textbooks and educational disciplines on mining machines, to the requirements of "Geokhod" class UM.

The problem of the lack of a methodology to ensure manufacturing accuracy and manufacturability of original UM elements and devices arose **at the stage of production** of a new "Geokhod" class UM [19–23]. At this stage of the development of experimental and prototypes, production is based solely on the general principles and approaches of mechanical engineering technology.

In addition, **at the test stage**, the problem of the lack of scientific and methodological support for testing new equipment [24, 25] was especially acute. In this case, the developers are based on the experience of operating mining machines and bench tests of individual elements.

The indicated problems arising during the creation of a new "Geokhod" class UM create an urgent need to develop and justify special scientific and methodological support [9, 21, 24–27].

At the same time, in the aircraft industry, in the development of experimental aircraft models and prototypes, there are scientific and engineering schools and specialties, "sharpened" for each of the stages of creating a new aircraft. To accompany the process of creating new technology, at each stage we can see research and design institutes, test sites, and full-scale laboratory facilities (wind tunnels, there is the profession of a test pilot, etc.).

To form a new scientific structure, the papers [28–31] developed and presented the concept of creating an advanced technological structure for the formation (development) of underground space on the basis of the accelerated development of new approaches in construction geotechnology and geotechnics. The structure of the proposed technological structure for the development (formation) of the underground space is shown in figure 2.

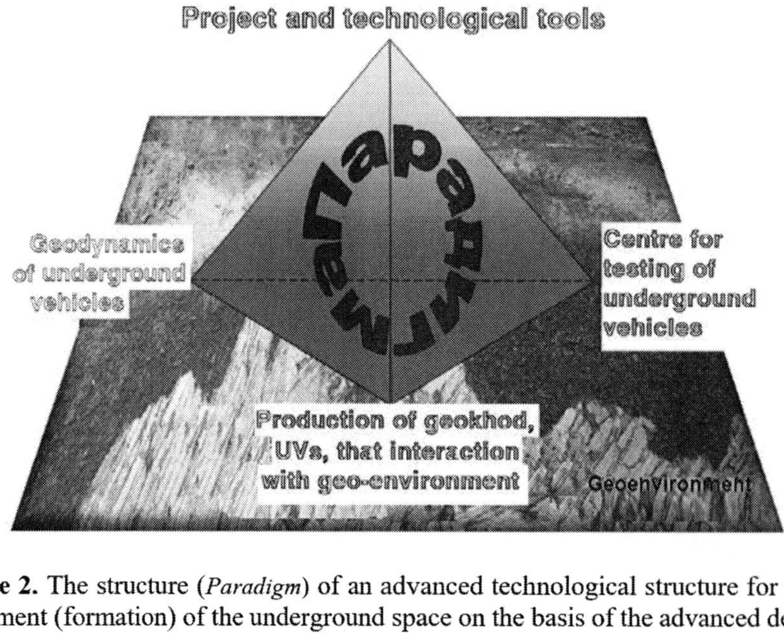

Figure 2. The structure (*Paradigm*) of an advanced technological structure for the development (formation) of the underground space on the basis of the advanced development of new approaches in construction geotechnology and geotechnics.

IOP Conf. Series: Materials Science and Engineering **939** (2020) 012007 doi:10.1088/1757-899X/939/1/012007

The key structural elements of the technological structure located on the tops of the tetrahedron have their special purpose, they are closely interconnected and form the techno-economic paradigm [32]. **Geodynamics of UM** is one of the main structural elements of the technological structure of the development (formation) of the underground space. When developing the provisions of the new technological structure, the experience and advantages of the end-to-end technological structure created in the aircraft industry [29, 31] were fully taken into account.

One of the key structural elements of the technological structure in the aircraft industry is fundamental science, which studies the forces arising on the surface of a solid body moving in the air – **Aerodynamics of aircraft** [33–38].

The very same role of fundamental science in the new technological structure of the development (formation) of the underground space should be played by the new scientific direction **"Geodynamics of UM"**. But, at present, there is no science with that name!

To formulate a definition of a new term, we turn to the established terminology in aircraft construction and shipbuilding. For convenience, we will present the existing formulations of the variants of terms in binding to the medium in the form of a table.

Table 1. Terminology.

Term	Definition	Notes
Aerodynamics (from classical Greek ἀηρ – air and δύναμις – power)	– A science that studies the movement of air masses and ***their effect on bodies in a stream***. – The science of air movement and ***its effect on the bodies streamlined by it*** [38].	It originated in connection with the first attempts to create aircraft. Aerodynamics is the theoretical basis of aviation and rocket and space technology [37]. The main task of aerodynamics [33] is the choice of the rational form of aircraft in order to obtain the specified flight performance, as well as determine the aerodynamic loads and heat fluxes acting on the surface of the aircraft for strength calculations.
Hydrodynamics (from classical Greek ὕδωρ «water» + dynamics)	– A branch of continuum physics that studies the motion of ideal and real liquids and gas and ***their force interaction with solids.*** – The science of the movement of fluids under the influence of external forces and the mechanical ***interaction between a fluid and bodies in contact with it during relative motion***; is part of hydromechanics. – Section of hydromechanics, in which the motion of incompressible liquids and their ***interaction with solids*** or interfaces with other liquids (gas) are studied.	As a science began to emerge in the 16th-17th centuries from experimental studies of the resistance of a mobile medium to the movement of a rigid body (Leonardo da Vinci, Galileo Galilei, E. Toricelli, I. Newton, etc.). In the 18th -19th centuries the science obtained analytical, empirical and methodological foundations with the participation of the most prominent researchers of that time (L. Euler, J.L. Dalamber, John, Jacob and Daniel Bernoulli, O. Reynolds and others). It formed the basis of Aerodynamics [33, 34]. The main task of hydrodynamics is the selection of rational geometric parameters of hydraulic units with the aim of obtaining the given characteristics [39-41].
Geodynamics	– A branch of geology that studies the nature of deep-seated forces and processes that arise as a result of the planetary evolution of the Earth, and which determine the movement of matter inside the planet.	Geodynamics as science began to isolate itself from other Earth sciences in the 1950s. Objects of the study of geodynamics are not accessible to direct study and they can be judged by indirect signs, theoretical constructions and the results of their manifestation on the surface of the Earth [42].

Term	Definition	Notes
	– The science of the deep forces and processes arising as a result of the evolution of the Earth as a planet and determining the movement of the masses of matter and energy inside the Earth and in its outer hard shells. Objects of geodynamic research are inaccessible to direct study and they can be judged by indirect signs, theoretical constructions and the results of their manifestation on the Earth's surface.	

From the above definitions, it becomes clear that in relation to the air and water environment, the corresponding sciences (aerodynamics and hydrodynamics) study not only the movement of the environment itself, but also **its effect on bodies located in the environment under consideration.**

As for the term geodynamics, such a scientific direction – **the study of the effects of the geo-environment on bodies located (moving) in the geo-environment – does not exist at the moment.**

This situation is a consequence of the fact that in underground conditions external movers (tracked, wheeled, wheel-rail or spreader-walking) were used and used to move the tunneling apparatus, which showed themselves well when working on the earth's surface, and to create the traction force of the UM and pressure effort on the executive body in no way involved the surrounding geo-environment, but only the solid surface of the mine at the contact of the geo-environment and the air, or with the shield method of penetration - a powerful permanent lining. For this class of mining machines, the corresponding scientific and methodological support was created, but which cannot provide the design of new UM.

When creating a new class of UM moving in the bowels of the Earth using the geo-environment [7–18], there was **an urgent need to study the effects of the geo-environment on bodies that are (moving) in the geo-environment**.

The solution to this problem by analogy, for example, with aircraft construction should be assigned to the **"Geodynamics of UM".**

At the initial stage, for the preliminary formulation of the passport of the specialty "Geodynamics of underground machines", the passport of the specialty 05.07.01 "Aerodynamics and aircraft heat transfer processes" was adopted as an analogue.

3. Conclusion

The authors of the article propose the following:

The specialty formula for the scientific specialty "Geodynamics of UM":

Geodynamics and heat transfer processes of underground machines (UM) is a field of science that studies the laws of interaction between UM and their parts with various geo-environment, including the development of theoretical and experimental methods for studying stationary, unsteady force and thermal effects of these geo-environments on UM.

Research areas of the scientific specialty "Geodynamics of UM":

– Geo-environment, geodynamic parameters. Formulations (terms), parameters, systematization.

– Theoretical and experimental studies of the interaction of UM and their elements with various geo-environments.

– Computational and experimental studies of the geodynamic parameters (characteristics) of UM and their elements.

– Development of methods for determining geodynamic parameters, including algorithms and software for CAD UM.

– Studies of the influence of various geo-environments on the geodynamic characteristics of UM.

– The geodynamic calculation of the UM motion controls.

– The geodynamic characteristics of UM and surface heating under conditions of interaction with the geo-environment.

– Studies of the impact of UM on the environment (deformation, displacement).

– Development of tools and methods for experimental geodynamic and thermal studies (methods, installations, full-scale underground studies).

– Experimental and theoretical studies of the force, thermal and physico-chemical interaction of various geo-environments with the surfaces of structural elements from various structural materials.

If we compare the air environment (aerodynamics of the aircraft) and the geomedium (geodynamics of the UM), then the geo-environment is much more diverse and multifaceted in its properties and manifestations in comparison with the air environment.

The indicators of physical and mechanical properties of rocky and non-rocky soils (rocks), and there are only twenty of them by name [43], differ quite significantly among themselves.

With regard to the creation of a new class of user agents, the priority ones in terms of solving the problems of geodynamics of user agents are:

– Development of geodynamic forms of UM elements.

– Development of geodynamic forms of supporting elements of the geo-environment

– Definitions of geodynamic resistance, power, strength parameters of UM elements interacting with the geo-environment.

In the direction of developing the geodynamic forms of UM elements, the first studies were performed and the results were obtained [9–11, 14–18].

The development and construction of the "Geokhod" class UM allows solving the problems that arise during the construction of underground structures of various directions in the bowels of the Earth. The lack of scientific and methodological support for the development of UM elements and systems interacting with the geological environment inhibits the creation of new machines. Such a basis should be a new scientific direction "Geodynamics of underground machines." Obtaining and systematizing new knowledge in the presented research areas is a priority.

References

[1] http://pro-kompas.info/strategy/

[2] Aksenov V V, Beglyakov V Y, Kazantsev A A, Valter A V and Efremenkov A B 2016 Opyt uchastiya v proekte po organizacii vysokotekhnologichnogo proizvodstva [Experience of participation in the project on the organization of high-tech production] *Mining equipment and electrical engineering* **8(126)** 8–15

[3] Aksenov V V, Beglyakov V Y, Blashchuk M Yu, Efremenkov A B, Kazantsev A A, Khoreshok A A and Valter A V 2016 Geokhod: zadachi, harakteristiki, perspektivy [Geokhod: objectives, characteristics, prospects] *Mining equipment and electrical engineering* **8 (126)** 3–8

[4] Beglyakov V Yu and Aksenov V V 2012 *Poverkhnost' zaboya pri prokhodke gornoy vyrabotki geokhodom* [The surface of the face during the mining of a mine by Geokhod] (Saarbrücken, Germany: LAP LAMBERT) p 139 [In Russ.]

[5] Aksenov V V and Efremenkov A B 2008 Geovinchesternaya tekhnologiya i geokhody – innovatsionnyy podkhod k osvoyeniyu podzemnogo prostranstva [Geovinchester technology and Geokhods – an innovative approach to the development of underground space] *Bulletin of Higher Educational Institutions* **4** 19–28 [In Russ.]

[6] Beglyakov V Yu 2012 *Obosnovaniye parametrov poverkhnosti vzaimodeystviya ispolnitel'nogo organa geokhoda s porodoy zaboya* [Justification of the surface parameters of the interaction of the executive body of the geohod with the rock of the face] Diss. for the degree of Candıo of Technical Sciences (Yurga: Kuzbass State Technical University) [In Russ.]

[7] Aksenov V V and Sadovets V Yu 2007 Modelirovaniye osobennostey dvizheniya geokhoda [Modeling of Geokhod movement features] *Bulletin of Kuzbass State Technical University* **1**

20–22 [In Russ.]

[8] Aksenov V V, Kostinets I K and Beglyakov V U 2013 Osobennosti raboty vneshnego dvizhitelya geohoda [Features of the external propeller of a geokhod] *Mining Information and Analytical Bulletin* 419–425

[9] Sadovets V Yu 2007 *Obosnovaniye konstruktivnykh i silovykh parametrov nozhevykh ispolnitel'nykh or-ganov geokhodov* [Substantiation of design and power parameters of knife executive bodies of Geokhods] Dissertation for the degree of Cand. of Technical Sciences (Kemerovo: T.F. Gorbachev Kuzbass State Technical University) [In Russ.]

[10] Ananyev K A 2016 *Sozdaniye ispolnitel'nogo organa geokhoda dlya razrusheniya porod sredney kreposti* [Creation of the executive body of the geohod for the destruction of rocks of medium strength] Dissertation for the degree of Cand. of Technical Sciences (Kemerovo: T.F. Gorbachev Kuzbass State Technical University) [In Russ.]

[11] Sadovets V Yu and Rezanova E V 2011 Obosnovaniye konstruktivnykh i tekhnicheskikh resheniy geokhodov [Substantiation of constructive and technical solutions for geological ducts] [in Russ] *Innovations in Mechanical Engineering. Proc. 2nd Int. Scient.c and Pract. Conf.* ed by V Yu Blumenstein 176–183

[12] Nesterov V I and Ananyev K A 2015 Opredeleniye tochki vkhoda reztsa barabannogo ispolnitel'nogo organa geokhoda v kontakt s porodoy [Determination of the entry point of the cutter of the drum executive organ of the geohod into contact with the rock] *Mining Equipment and Electromechanics* **7 (116)** 45–48

[13] Nesterov V I and Ermakov A N 2015 Sravneniye iznosa rezhushchego instrumenta na zakonturnykh ispolni-tel'nykh organakh geokhodov [Comparison of wear of cutting tools on the circulating executive bodies of geokhodom] *Mining Equipment and Electromechanics* **7 (116)** 41–45

[14] Ermakov A N 2016 *Obosnovaniye parametrov zakonturnnykh ispolnitel'nykh organov geokhodov dlya razrusheniya porod sredney kreposti* [Justification of parameters of contouring executive bodies of geokhods for the destruction of medium-strength rocks] Dissertation for the degree of Cand. of Technical Sciences (Kemerovo: T.F. Gorbachev Kuzbass State Technical University) [In Russ.]

[15] Ermakov A N 2016 Otsenka koeffitsiyenta variatsii krutyashchego momenta na zakonturnykh ispolnitel'-nykh organakh geokhoda [Assessment of the coefficient of variation of torque on the contour executive bodies of the geokhod] *Mining Equipment and Electromechanics* **8 (126)** 25–29

[16] Beglyakov V U, Aksenov V V, Kazantsev A A and Kostinets I K 2017 Razrabotka zakonturnoj oporno-dvizhitel'noj sistemy geohodov [Development of outside the contour support-propulsion system of geokhod] *Vestnik KuzSTU* **6(124)** 175–181 [In Russ.]

[17] Aksenov V V, Beglyakov V Yu, Kazantsev A A, Kostinets I K and Koperchuk A V 2016 Klassifikatsiya geometricheskikh parametrov vneshnego dvizhitelya geokhoda [Classification of the geometrical parameters of the external mover of Geokhod] *Mining Equipment and Electromechanics* **8** 33–39 [In Russ.]

[18] Walter A V and Lagunov S E 2015 Opredeleniye pripuska na poverkhnosti vrashcheniya sbornykh korpusnykh izdeliy geokhoda [Determination of the allowance on the surface of rotation of prefabricated hull products of the Geokhod] *Actual Problems in Mechanical Engineering* **2** 152–157

[19] Valter A V, Aksenov V V, Beglyakov V U and Chazov P A 2015 Opredelenie pogreshnosti raspolozheniya sektorov stabiliziruyushchej sekcii geohoda na osnove dannyh koordinatnogo kontrolya [Determination of the error of the location of the sectors of the stabilizing section of the geohod based on the coordinate control] *Metal processing (technology, equipment, tools)* **4 (69)** 31–42

[20] Walter A V, Nozirzoda Sh S and Ivkin A N 2016 Issledovaniye tochnosti izgotovleniya krupnogabaritnykh korpusov vrashcheniya opytnogo obraztsa geokhoda [The study of the

accuracy of manufacturing large-sized rotation cases of a prototype Geokhod] *Concept* **11** 2026–2030 [In Russ.]

[21] Walter A V 2016 Faktory, obuslavlivayushchiye pogreshnosti poverkhnostey vrashcheniya krupnogabaritnykh korpusov geokhoda [Factors causing errors in the surfaces of rotation of large-sized bodies of the Geokhod] *Mining Equipment and Electromechanics* **8 (126)** 19–25

[22] Walter A V, Borovikov I F, Chernukhin R V and Nozirzoda S S 2016 Research on geometric errors of intermediate unit shell of the Geokhod *Urgent Problems of Modern Mechanical Engineering* 12017

[23] Walter A V, Chernukhin R V and Kapustin A N 2016 Otkloneniya formy obolochki stabiliziruyushchey sektsii geokhoda [Deviations of the shell shape of the stabilizing section of Geokhod] *Technologies and Materials* **1** 4–7 [In Russ.]

[24] Koperchuk A V, Kazantsev A A, Beglyakov V Yu and Filonov V V 2015 Obosnovaniye neobkhodimosti razrabotki startovogo ustroystva geokhoda [Justification of the need to develop a launch device for Geokhod] *Technologies and Materials* **1** 29–30 [In Russ.]

[25] Sadovets V Yu and Koperchuk A V 2017 Razrabotka konstruktivnykh resheniy ustroystv dlya vozvedeniya kre-pi v usloviyakh geokhodnoy tekhnologii [Development of constructive solutions for the construction of crepes in the conditions of geo-mining technology] [in Russ] *Mining Information and Analytical Bulletin* **1** 327–335 [In Russ.]

[26] Koperchuk A V 2019 Startovoye ustroystvo geokhoda. Trebovaniya k konstruktsii, rezul'taty ispytaniy [Starting device of the Geokhod. Design requirements, test results] *Innovative Technologies in Mechanical Engineering:* Proc. X Int. Scient. and Pract. Conf. 107–110 [In Russ.]

[27] Koperchuk A V and Kazantsev A A 2015 Obosnovaniye neobkhodimosti razrabotki startovogo ustroystva geokhoda [Justification of the need to develop a start-up device for the Geokhod] *Innovative Technologies and Economics in Mechanical Engineering* ed by D A Chinakhov (Yurga: YTI Publ) 439–440 [In Russ.]

[28] Aksenov V V, Khoreshok A A, Beglyakov V Yu and Efremenkov A B 2019 *IOP Conference Series: Materials Science and Engineering* **656** 012004

[29] Aksenov V V, Khoreshok A A, Beglyakov V Yu and Efremenkov A B 2019 *IOP Conference Series: Materials Science and Engineering* **656** 012005

[30] Aksenov V V, Khoreshok A A and Beglyakov V Yu 2018 Kontseptsiya sozdaniya perspektivnogo tekhnologicheskogo uklada formirovaniya (osvoyeniya) podzemnogo prostranstva na baze operezhayushchego razvitiya novykh podkhodov v stroitel'noy geotekhnologii i geotekhnike [The concept of creating a promising technological structure for the formation (development) of underground space on the basis of the advanced development of new approaches in construction geotechnology and geotechnics] (Part 1) *Vestnik of Kuzbass State Technical University* **4** 105–113 [In Russ.]

[31] Aksenov V V, Khoreshok A A and Beglyakov V Yu 2018 Kontseptsiya sozdaniya perspektivnogo tekhnologicheskogo uklada formirovaniya (osvoyeniya) podzemnogo prostranstva na baze operezhayushchego razvitiya novykh podkhodov v stroitel'noy geotekhnologii i geotekhnike [The concept of creating a promising technological structure for the formation (development) of underground space on the basis of the advanced development of new approaches in construction geotechnology and geotechnics] (Part 2) *Vestnik of Kuzbass State Technical University* **5** 43–52 [In Russ.]

[32] Lvov D S and Glazyev S Yu 1986 Teoreticheskiye i prikladnyye aspekty upravleniya NTP [Theoretical and applied aspects of NTP management] *Economics and Mathematical Methods* **5** 793

[33] Baydakov V B and Klumov A S 1979 *Aerodinamika i dinamika polota letatel'nykh apparatov* [Aerodynamics and flight dynamics of aircraft] (Ripol Classic) [In Russ.]

[34] Pyshnov V 2013 *Aerodinamika samoleta* [Aerodynamics of the aircraft] (Ripol Classic) [In Russ.]

[35] Petrov K P 1998 *Aerodinamika tel prosteyshikh form* [Aerodynamics of bodies of the simplest

forms] (Moscow: Factorial) [In Russ.]
[36] Ostoslavsky I V 1957 *Aerodinamika samoleta* [Aerodynamics of an airplane] (Moscow: Oborongiz) [In Russ.]
[37] Krasnov N F 2010 *Aerodinamika* [Aerodynamics] (URSS) [In Russ.]
[38] Sedov G A 1961 *Lotchiku o prakticheskoy aerodinamike* [Practical aerodynamics for pilots][in Russ] (Ripol Classic) [In Russ.]
[39] Lavrentiev M A 1973 *Problemy gidrodinamiki i ikh matematicheskiye modeli* [Problems of hydrodynamics and their mathematical models] (Ripol Classic) [In Russ.]
[40] Khoreshok A A, Kuznetsov V V, Ananyev K A, Aksenov V V, Blashchuk M Yu and Beglyakov V Yu 2014 *Gidromekhanika* [Hydromechanics] (Tomsk) [In Russ.]
[41] Altshul A D 1987 *Gidravlika i aerodinamika: uchebnik dlya vuzov* [Hydraulics and aerodynamics: a textbook for high schools] (Moscow: Stroyizdat) p 414 [In Russ.]
[42] Hain V E and Lomize M G 2005 *Geotektonika s osnovami geodinamiki* [Geotectonics with the basics of geodynamics] (Moscow: KDU) [In Russ.]
[43] Rzhevskij V V and Novik G Y 1964 *Osnovyj fiziki gornyh porod* [Fundamentals of rock physics] (Moscow: Science Publ) [In Russ.]

Mathematical modelling of the process of crest formation by the working body of the cultivator – dump type

D N Algazin[1,4], A A Kem[2,5], A P Shevchenko[3,6], M A Begunov[3,7] and V S Koval[3,8]

[1] Yaroslav-the-Wise Novgorod State University, ul. B. Sankt-Peterburgskaya, 41 173003, Veliky Novgorod, Rissia
[2] Omsk Agrarian Scientific Center, pr. Koroleva, 26 644012, Omsk, Russia
[3] Tarsky branch of FSBEI HE Omsk State Agricultural University, ul. Tyumenskaya, 18 646532, Tara, Omsk Region, Russia

[4] E-mail: dna@novsu.ru
[5] E-mail: 55asc@bk.ru
[6] E-mail: ap.shevchenko@omgau.org
[7] E-mail: maksim-begunov@mail.ru
[8] E-mail: koval_v.s@mail.ru

Abstract. The results of theoretical studies of the cultivator - dump type ridger are presented. The analysis of the forces acting on the working body is carried out. An expression is obtained for determining the second consumption of soil mass transported by a working body. The rational parameters of the working body are determined.

1. Introduction

When cultivating row crops, such as corn, mechanical interrow cultivation is widely used [1], as one of the most effective and environmental methods of controlling weed vegetation [2].

The complexity of this method lies in the need for the simultaneous cutting and destruction of at least 95% of weeds in the immediate vicinity of cultivated plants and hilling (creation of a ridge) to ensure their reliable nutrition [3].

2. Model

The process of creating ridges is complex and multifactorial and it is directly affected by the following parameters: the speed of the machine-tractor aggregate (MTA), the size, shape and material of the working body, the mechanical composition of the soil, its physical properties, moisture, etc. [4].

We proposed a scheme of the working body of a ridge-forming cultivator of the cultivator-dump type, which provides simultaneous high-quality cutting of weeds and hilling of plants (ridge formation) in accordance with the agrotechnical requirements for row spacing to ensure high-quality and efficient implementation of the technological process of ridge formation during row-spacing processing (figure 1) [5].

Content from this work may be used under the terms of the Creative Commons Attribution 3.0 licence. Any further distribution of this work must maintain attribution to the author(s) and the title of the work, journal citation and DOI.
Published under licence by IOP Publishing Ltd

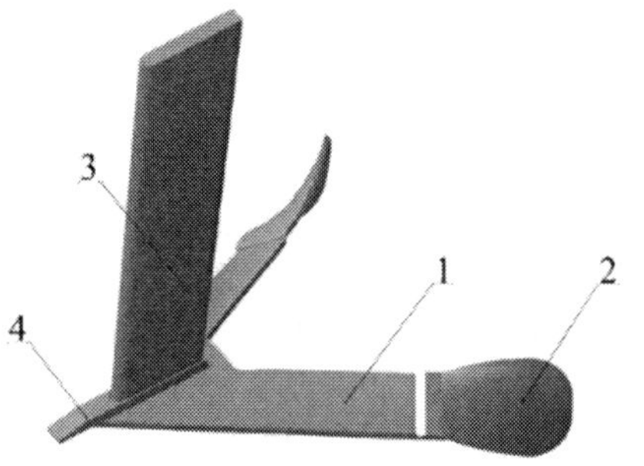

Figure 1. General view of the working body of the cultivator-dump type ridger: 1 – lancet paw; 2 – dump; 3 – rack; 4 – chisel.

The principle of operation of the proposed working body is as follows: the working body is installed in the row-spacing and installed at a predetermined depth of processing, then it starts to move under the action of the MTA traction, the soil in the central part of the row-spacings is cut by the lancet paw 1 and chisel 4, while partially shifted by the lancet paw to the periphery of the row-spacings. Dumps fixed at the ends of paws 2 cut the top layer of soil and form a ridge in the immediate vicinity of cultivated plants, hilling plants.

Qualitative indicators of the execution of operations depend on the design parameters of the working body (figure 2) and the rational choice of its operating modes, such as: v_s- MTA speed, m/s; a – depth of tillage, m; b, d, c – geometric parameters of the paws, m; r is the radius of the screw, which is the generatrix of the blade, m; h is the width of the blade, m; φ is the elevation angle of the helix (screw) forming the blade, deg; Ψ is the angle of elevation of the axis of the screw to the surface of the soil forming the blade, deg; L is the length of the helix forming the blade, m; σ is the angle of the circular segment bounded by the blade, deg.

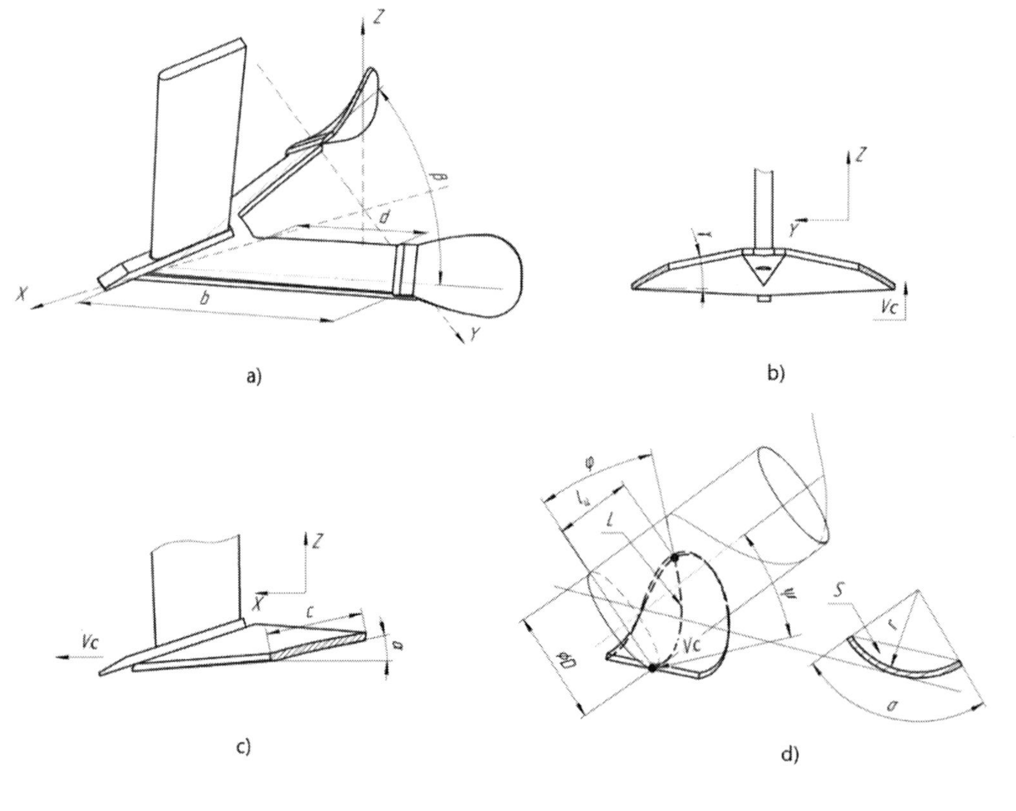

Figure 2. The experimental design of working body of the cultivator-dump type ridger: a – general view; b – rear view; c – in-side view; d – dump scheme.

In order to create a mathematical model of the interaction of the working body with the soil, it is necessary to take into account the forces (figure 3) acting on it [6]: F_I – inertia force acting on the soil layer, N; F_d – force to counteract soil deformation, N; G is the weight of the soil layer (gravity), N; N is the force of action of the working body on the soil layer, N; F_{fr} – friction force, N. Soil deformation resistance of the soil F_d is the vector sum:

$$\vec{F}_d = \vec{F}_{Fk} + \vec{F}_{Fd} + \vec{F}_{Fg} \tag{1}$$

where F_{Fk} – resistance to soil chipping, H;
F_{Fd} – soil anti-shear strength, H;
F_{Fg} – soil anti-compression force, H.

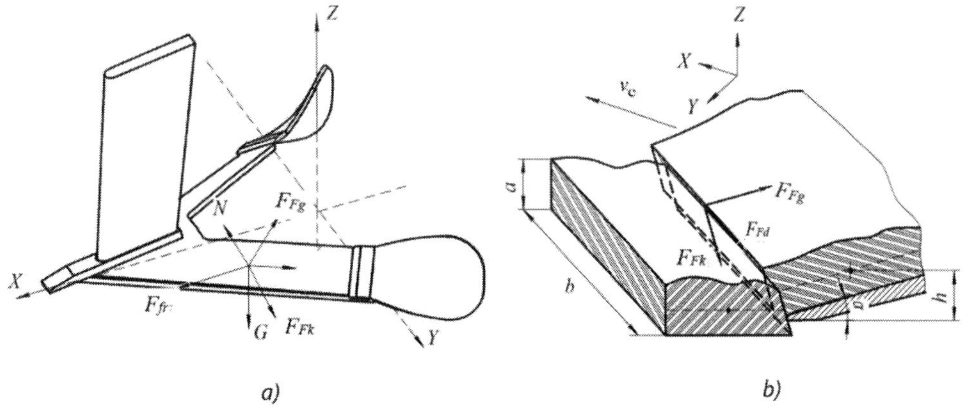

Figure 3. Scheme of forces acting on the working body.

We obtain an expression for determining the weight of the soil moved by the working body, going to the formation of the ridge, per unit time (mass flow) taking into account the physical properties of the soil and the analytical expressions of the forces acting on the working body:

$$M_r = \xi\left(k \cdot \frac{r^2}{\sin 2\varphi}\left(\frac{\pi}{180}\sigma - \sin\sigma\right) \cdot \cos\beta + f \cdot \left(c \cdot \sin\alpha + \frac{a^3}{a - c \cdot \sin\alpha}\right) \cdot b \cdot \cos\frac{\beta}{2}\right) \cdot v_c \cdot \frac{\gamma_v}{g} -$$
$$-\xi \cdot 0{,}05 \cdot \rho \cdot S_S \cdot \sqrt{\frac{r \cdot g \cdot \cos\psi \cdot \cos\varphi - L(f \cdot \sin\psi + 1)}{1 - 2 \cdot r}} \tag{2}$$

where k – soil losses that occur when the cultivating part of the working body moves from the central part of the aisle to the periphery;
ξ – soil loss for shedding and scatter, taken equal 0,6;
f – coefficient of friction;
γ_v – bulk soil weight, g/cm^3;
β – paw opening angle, deg.
S_S – cross-sectional area of the soil carried by the dump, m^2;
ρ – soil density, kg/m^3.

The physical meaning of the indicator of the mass flow rate of soil M_G, moved by the working body, for the formation of the ridge in accordance with agrotechnical requirements is that it shows the weight of the soil, which is moved by the working body to form the ridge at a speed of v_s for a time t equal to one second. Let us move from mass flow to specific volume:

$$V_r = \frac{M_r}{v_c \cdot \rho} \tag{3}$$

The obtained dependences 2 and 3 allow us to determine the ranges of technological and geometric parameters of the working body, at which a ridge is formed with the parameters in accordance with agricultural requirements.

3. Results and discussion

Based on the mathematical model, we will construct the response surfaces (figure 4–6):

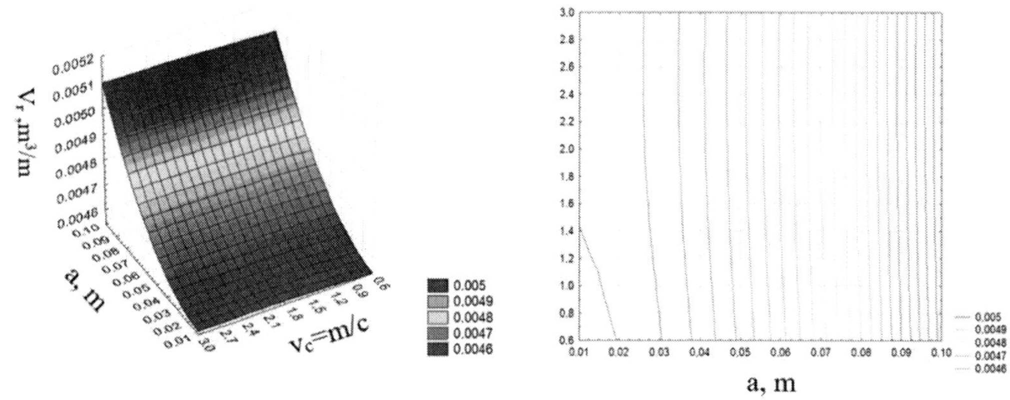

Figure 4. Response surface: $V_r = f(\alpha, v_c)$, r =0,3 m; Ψ=10°; φ=66°; h = 0,09 m; L = 0,15 m.

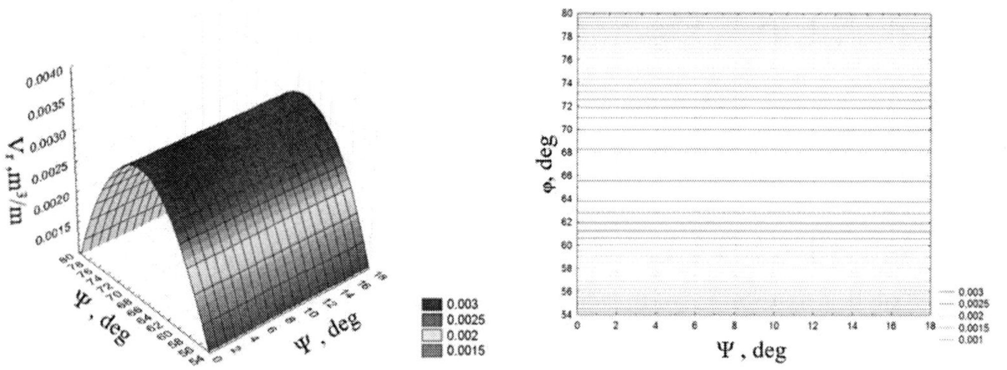

Figure 5. Response surface: $V_r = f(\varphi, \Psi)$, a =0,6 m; r =0,3 m; v_c = 3m/c; h = 0,09m; L = 0,15 m.

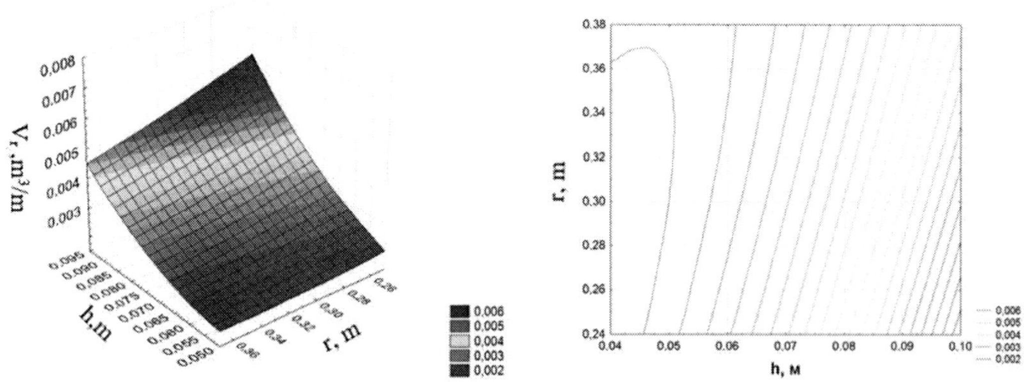

Figure 6. Response surface: $V_r = f(h,r)$, a =0,6 m; Ψ=10° ; φ=66°; v_c = 3m/c; L = 0,15 m.

We find rational ranges of the values of the working body parameters from the obtained dependencies and the response surfaces constructed from them:

Technological:

– soil cultivation depth: a – has the greatest influence on the volume of ridge formation, the rational range of the parameter is within 4–10 cm.

Construction:

– the screw radius r, the rational range of the parameter is within 0.2–0.3 m;

– the width of the cross section of the blade h has a directly proportional effect on the volume of the moved soil and providing its optimal value at values h = 0.07– 0.09 m;

– the screw elevation angle φ, has rational values in the range 60°–70°;

The remaining parameters, such as: MTA speed v_s, elevation angle of the screw axis Ψ and the length of the blade generatrix L do not significantly affect the ridge formation.

4. Conclusion

Analytical dependences are obtained, which make it possible to determine the influence of structural and technological parameters on the formation of the ridge during the processing of row-spacing by the working body of the cultivator-dump type. This made it possible to choose rational ranges of their values: screw radius r in the range 0.2–0.3 m, blade width h 0.07– 0.09 m, screw elevation angle φ 60–70°.

The studies conducted in the field experiment in the soil channel and directly in the field established the rational shape and size of the working body of the cultivator-dump type. The ridge volume V_{gr} corresponding to the agrotechnical requirements was achieved with a soil cultivation depth a of 6 cm, a screw radius r of 0.3 m, a blade cross section h of 0.08 m and a screw angle of elevation φ of 60°.

The created mathematical model of the interaction of the working body with the soil and modeling of the process of crest formation will allow the creation of new tools for inter-row cultivation of row crops.

References

[1] Abbaspour-Gilandeh Y, Fazel Niari Z, Shahgoli G and Bavafa M 2011 Design and Construction of a High Speed Inter-Row Cultivator *Applied Mechanics and Materials* 110–116 DOI: 10.4028/www.scientific.net/AMM.110-116.4914

[2] Perez-Ruiz M, Carballido J, Agüera J and Rodríguez-Lizana A 2013 Development and Evaluation of a Combined Cultivator and Band Sprayer with a Row-Centering RTK-GPS Guidance System *Sensors* **13** 3313–3330

[3] Werf H M G, Klooster J J, Schans D A, Boone F R and Veen B W 2008 The Effect of Inter-row Cultivation on Yield of Weed-free Maize *J. of Agronomy and Crop Science* **166(4)** 249–258

[4] Abbaspour-Gilandeh Y, Fazeli M, Roshanianfard A, Hernández-Hernández M, Gallardo-Bernal I and Hernández-Hernández J 2020 Prediction of Draft Force of a Chisel Cultivator Using Artificial Neural Networks and Its Comparison with Regression Model *Agronomy* vol 10 451 DOI: https://doi.org/10.3390/agronomy10040451

[5] Orlov A N, Algazin D N and Krasilnikov E V Vliyaniye parametrov grebneobrazovatelya kul'tivatorno-otval'nogo tipa na obrazovaniye grebney [Influence of the parameters of the combing machine of the cultivator-dump type on the formation of ridges] Omsk Scientific Bulletin **1(117)** 124–127 [In Russ.]

[6] Sineokov G N and Panov I M 1977 Teoriya i raschet pochvoobrabatyvayushchikh mashin [Theory and calculation of tillage machines] (Moscow: Engineering) p 328 [In Russ.]

IOP Conf. Series: Materials Science and Engineering **939** (2020) 012009 doi:10.1088/1757-899X/939/1/012009

Zirconia crystals for tribotechnical applications doped by rare earth elements

V V Alisin[1]

[1] Institute of Machines Science named after A.A.Blagonravov of the Russian Academy of Sciences, Russia

[2] E-mail: vva-imash@yandex.ru

Abstract. The article discusses the impact of rare-earth microadditives on the mechanical properties of nanostructured zirconia crystals stabilized with yttrium oxide as applied to the friction units of devices. The mechanical properties of the surface layers were studied by the nondestructive method of kinetic microindentation. Compressive strength was determined by the standard destructive compression test method. Comparative tribological tests were carried out with dry sliding friction of the test samples on steel according to the disk-finger scheme. The effect of the ionic radius of the alloying element on the compressive strength is investigated. It was established that the doping of zirconia crystals is ambiguous, because it can improve and worsen the tribological properties of crystals. On the basis of the study, a comparison is made of the mechanical properties of crystals and ceramics used in the supports of the axis of precision instruments and areas that are promising for the use of zirconia crystals in instrumentation are identified.

1. Introduction

Increasing the service life and reliability of the mechanical modules of devices and microelectronic mechanical systems is mainly achieved by improving the tribological and strength properties of materials for moving mates of parts due to the use of supports containing ceramics and crystals. One of the most important tasks of modern materials science is the creation of new structural non-metallic materials in which strength, wear resistance, high fracture toughness and chemical inertness are combined with high resistance in an oxidizing medium over a wide temperature range. These materials include partially stabilized zirconia dioxide (PSZ). At present, partially stabilized tribological zirconia dioxide is obtained mainly by ceramic technology [1]. Nanostructured powders are used to improve the mechanical properties of ceramics [2–4]. In the article [5], it is noted that the morphology of the starting powders is a key factor for optimizing the mechanical properties of the material. Various technologies for compacting and sintering ceramic materials are being improved: self-propagating high-temperature synthesis [6, 7], sintering in a nitrogen medium [8], gelcasting method combined with pressureless sintering, which ensured excellent mechanical properties (bending strength 890 MPa, fracture toughness 10,2 MPa • m 0.5, Vickers hardness 13.2 GPa and the lowest porosity) [9]. The low-temperature degradation and mechanical properties of zirconium ceramic synthesized by a solid-state reaction are studied [10]. The prospects of using partially stabilized zirconia dioxide crystals (PSZ crystals) obtained from a melt by crystallization in a cold container for friction nodes have been established [11]. Materials based on zirconium dioxide obtained in practice are single or

Content from this work may be used under the terms of the Creative Commons Attribution 3.0 licence. Any further distribution of this work must maintain attribution to the author(s) and the title of the work, journal citation and DOI.
Published under licence by IOP Publishing Ltd

multiphase solid solutions of zirconium oxide with one or more oxides of alkaline earth, rare earth and some other elements. As a rule, such materials are metastable at room temperature, but nevertheless, under certain conditions, they have a fairly high degree of stability and do not undergo phase transformations for a long time under repeated temperature influences. When synthesized from a melt, a growing PSZ single crystal has an initial cubic structure, and phase transformations occur in it during cooling in the solid phase. The transition of the cubic phase to the tetragonal phase is accompanied by the formation of a domain structure in crystals with sizes of 10–100 nm. PSZ crystals have a unique combination of mechanical properties [12], however, they have significant anisotropy along the crystal axes [13]. There is a search for ways to improve the tribological and strength properties of crystals [14]. The structure and transport properties of PSC crystals [15, 16], spectral-luminescent properties [17, 18] are studied. Doping with rare-earth elements has spread to improve the various properties of PSC crystals [15–18], however, the mechanical properties of doped PSC crystals have been studied very little. With respect to tribological applications, interest is mainly in hardness, fracture toughness, elastic modulus, and compressive strength [1]. The mechanical treatment of the surfaces of the samples during their manufacture leads to a change in surface properties. The most reliable information on the mechanical properties of the surface layer of the material, namely, plasticity and elasticity, can be obtained by the method of testing for kinetic microhardness.

The purpose of this work is to study the influence of rare earth microadditives on the tribological and strength properties of nanostructured partially stabilized zirconia crystals (PSZ crystals).

Materials and equipment. Samples from PSZ crystals of the composition ZrO_2-3 mol% Y2O3c with impurities of transition and rare earth elements were made in the form of rectangular fingers measuring 5x5x8 mm. Typical crystal sizes were 10–20 mm in cross section and 40–60 mm in length at growth rates of 15 mm/h. The crystals were colored depending on the type of impurity used. Crystal faces are predominantly shiny; a dull surface was present in the form of small sections on separate faces. A quantitative chemical electronic microanalysis was carried out at the facility «CAMEBAX SX-50» for an adequate interpretation of the results of studies of samples from PSZ crystals.

Tests to determine the mechanical properties of PSZ crystals were performed on a MNT_Z_AE_000 kinetic microhardness meter manufactured by CSM Instruments (Switzerland) in accordance with the ISO/DIS 14577 standard with a Vickers tetrahedral indenter according to the method [19]. Tribological tests were performed on the installation of UMT-1 according to the disk-finger scheme according to the method [20].

2. Results and Discussion

Samples were selected with the same content of a stabilizing additive of yttrium oxide for testing. A typical surface morphology of the growth faces of crystals of various chemical composition is shown in figure 1.

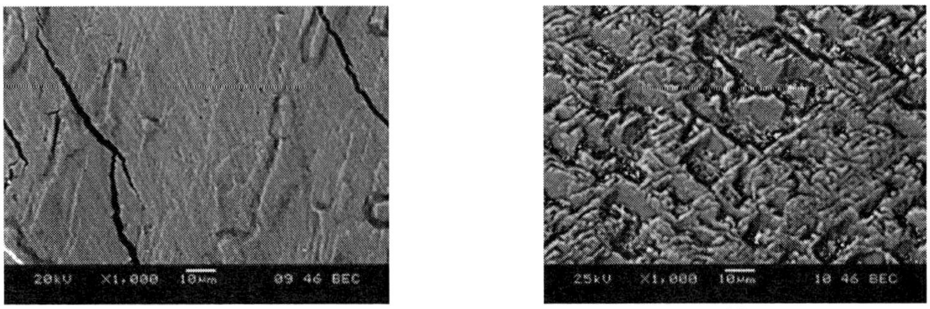

a) b)

Figure 1. The growth surface of the crystal PSZ with the following composition: a) 97.2 mol % ZrO_2+1.0 mol % Y_2O_3+1.8 mol % CeO_2; b) – 97.2 mol % ZrO_2 +2.0 mol % Y_2O_3+0.8 mol % CeO_2.

The presence of a monoclinic phase in samples with a concentration of 1 mol% Y_2O_3 + 1.8 mol% CeO_2 (Nd_2O_3) is accompanied by the formation of microcracks on the surface of PSZZcrystals (figure 1a). With an increase in the concentration of yttrium oxide, there are no microcracks, and the surface morphology has the form of a "parquet" microstructure typical of the tetragonal phase composition of PSZ crystals (figure 1b); therefore, samples with a stabilizing additive content of 3 mol% were chosen for experimentsY_2O_3.

The most important characteristics of the mechanical properties of the surface layers of rubbing parts are hardness, modulus of elasticity and the coefficient of irreversible energy loss, taking into account the energy costs of plastic deformation and cracking. The irreversible loss coefficient K_p is determined by the formula:

$$K_p = \eta_{IT} = W_{plast} / W_{gen}$$

where η_{IT}– elastic component of the indentation work;

$$\eta_{IT} = W_{elast.} / W_{gen.} x100,$$

where W_{elast} – is reverse work with elastic deformation of indentation (energy of elastic restoration of the indent after unloading), N·m; W_{plast} – work during plastic deformation of indentation (energy absorbed in the loading – unloading cycle, N·m; W_{gen} – general (full) mechanical work of indentation, H·m, $W_{gen} = W_{elast} + W_{plast}$.

In fact, the coefficient K_p reflects the irreversible energy loss during indenter indentation, which consists of the actual energy losses due to plastic deformations and the energy losses spent on brittle fracture (crack formation). Samples were tested to determine the strength characteristics of PSZ crystals of the composition ZrO_2 – 3 mol% Y_2O_3 with impurities of transition and rare earth elements. Fingerprints were deposited on the surface of samples of PSZ crystals of arbitrary crystallographic orientation. The number of prints (number of diagrams) for each surface is from 3 to 10 for different samples. These figures are not always sufficient for a correct statistical analysis because of the possible variation in the shape of the deformation diagrams for a particular sample, especially for friction surfaces. Such a spread should lead to an ambiguity in the values of microhardness over the surface of the sample. In this regard, in some experiments there was a need to increase the number of prints on the surface of the test material. Table 1 shows the data on the composition of the crystals of PSZ and test results.

Table 1. Indentation test results (average values).

№	Composition	HV, GPa	E, GPa	Wgen, μм J	Wpl, μм J	K_p
1	ZrO_2-2.8 mol %Y_2O_3	16.74	213	0.71	0.38	0.535
2	ZrO_2-3.7 mol %Y_2O_3	17.17	217	0.73	0.37	0.506
3	ZrO_2-2.8 mol %Y_2O_3 +0.1weight % CeO_2	17.07	171	0.81	0.39	0.479
4	ZrO_2-3.7 mol %Y_2O_3 +0.1weight % CeO_2	17.95	204	0.72	0.36	0.499
5	ZrO_2-2.8 mol %Y_2O_3 +0.6 weight % CeO_2	16.67	209	0.71	0.37	0.521
6	ZrO_2-3.7 mol %Y_2O_3 +0.6 weight % CeO_2	17.16	204	0.73	0.37	0.506
7	ZrO2-2.8 mol %Y2O3 +1 weight % CeO2;	17.29	207	0.73	0.38	0.516
8	ZrO_2-3.7 mol %Y_2O_3 +1 weight % CeO_2;	17.80	200	0.73	0.36	0.492
9	ZrO_2-2.8 mol %Y_2O_3 +0.1 weight % Nd_2O_3	16.66	196	0.73	0.37	0.508

10	ZrO_2-3.7 mol %Y_2O_3+0.1 weight % Nd_2O_3	17.73	203	0.72	0.36	0.502
11	ZrO_2-3.7 mol %Y_2O_3+0.6 weight % Nd_2O_3	17.84	186	0.74	0.35	0.472
12	ZrO_2-3.7 mol %Y_2O_3+0.9 weight % Nd_2O_3	17.46	187	0.74	0.35	0.480
13	ZrO_2-2.8 mol %Y_2O_3+0.6 weight % Nd_2O_3	17.74	208	0.72	0.36	0.503
14	ZrO_2-2.8 mol %Y_2O_3+0.9 weight % Nd_2O_3	16.76	193	0.74	0.38	0.511
15	ZrO_2-2.8 mol %Y_2O_3+0.1 weight % CoO	18.55	236	0.67	0.33	0.50
17	ZrO_2-2.8 mol %Y_2O_3+0.5 weight % CoO	16.76	214	0.72	0.37	0.51
18	ZrO_2-3.7 mol %Y_2O_3+0.1 weight % CoO	16.19	210	0.73	0.38	0.52

We use the radial type of cracks formed in the vicinity of the indentation to assess the crack resistance. During the test, a force of the order of 8 N is applied to the Vickers indenter, keep under a load of 15 seconds, then the average length of the half-diagonals of the microprint a_{av} and the average length of cracks in the corners of the imprint of l_{av}, measured no later than 2 minutes after unloading. The value of the critical stress intensity factor Kc was determined by the formula with a calibration coefficient (table 2):

$$\text{Kc} = 0{,}015 \times \left(\frac{a}{l}\right)^{1/2} \times \left(\frac{E_{IT}}{H_{IT}}\right)^{2/3} \times \left(\frac{F}{c^{3/2}}\right), \tag{4}$$

where F is the maximum load, N; a is the length of the semi-diagonal microprint, microns; l is the length of the radial crack, microns; c is the length of the crack counted from the center of the indent, microns; E_{IT} is the modulus of elasticity, GPa; H_{IT} is the hardness, GPa.

$$A_{av} = (a_1 + a_2)/2 = (d_1 + d_2)/4,$$

where a_1 *u* a_2 *are the* lengths of the half-diagonal microprint, microns; d_1 *and* d_1 *are* the lengths of the diagonals of the print, microns; c = $a_{av}+l_{av}$, microns.

All surfaces have roughness. The characteristic of a single roughness is the angle between the tangent to the roughness surface at the base and the surface plane. For the roughest machining surfaces, this angle does not exceed 2–3 degrees, i.e. unevenness is compressed in the contact; the ultimate compressive strength of crystals according to [13] is given in the table 2.

Table 2. Fracture and compression strength test results.

Additional supplement	Weight, %	Color of amples	Ionic radius, nm	Strength σ, MPa	Kc MPa·m^{0,5}
Cu	0,3	greenish blue	0.096	2535	11.1
Mn	0,3	dark grey	0.091	2340	8.5
Co	0,3	violet	0.082	2531	8.2
Er	0,3	pink	0.104	2461	7.7
Nd	0,5	lilac	0.115	1752	7.5
Pr	0,5	yellow	0.1	2094	6.5
Ce	0,6	orange red	0.118	1702	9.9

Since the deformation of the crystal lattice is most strongly affected by the ionic radius of the implanted element, it is of interest to determine the dependence of the compressive strength on the ionic radius (figure 2). It is shown that alloying with rare-earth elements with a higher ionic radius leads to a decrease in the compressive strength of the crystal. Thus, it can be considered that alloying with additional impurities can both improve the mechanical characteristics and worsen. The stability of zirconia doped with microimpurities is based on the local strengthening of the Zr-O chemical bond as a result of the redistribution of the electron density near the impurity atom ion. The role of oxygen vacancies, the interaction of defects, the electronic structure of impurity ions and their valency, which affect the adhesion strength of molecular bonds in the areas of actual contact during external friction is confirmed in [21].

The results of tribological tests showed (table 3) a significant effect of microalloying on the coefficient of friction (f) and the wear rate (I) of the PSZ crystals.

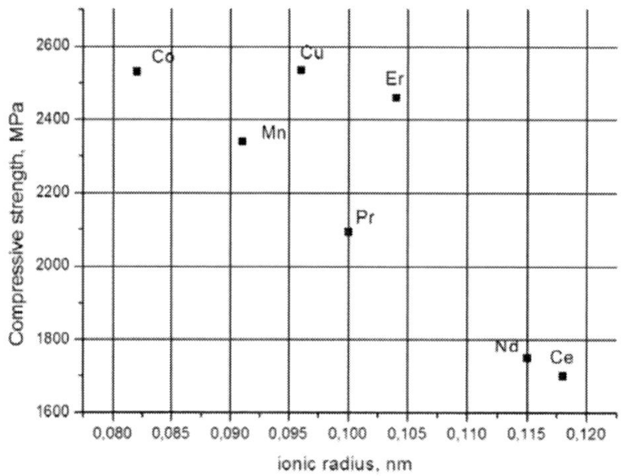

Figure 2. Dependence of the compressive strength of a PSZ crystal on the ionic radius of an alloying element.

Table 3. Tribological properties of PSZ crystals with microadditives of rare earth elements.

№ crystal	Alloying elements	f	I	note
1	CeO_2+Nd_2O_3	0.19	$2.25\cdot10^{-8}$	
2	3.3% mol Y_2 O_3	0.23	$2.52\cdot10^{-9}$	Annealing, 2000C^0
5	CeO_2+Er_2O_3	0.12	$1.22\cdot10^{-9}$	
6	Er_2O_3	0.25	$1.92\cdot10^{-9}$	
7	CeO_2+Er_2O_3	0.21	$1.17\cdot10^{-9}$	
8	CeO_2	0.21	$1.39\cdot10^{-9}$	
9	Pr	0.20	$1.29\cdot10^{-9}$	

The results indicate that the microalloying of PSZ crystals with rare-earth elements is an effective technological method for increasing the wear resistance of PSZ crystals. By introducing the CeO_2 + Er_2O_3 microadditive, it is possible to increase the wear resistance by about two times, namely, the wear rate of PSZ crystals after annealing in vacuum 3.12 10^{-9} decreases to 1.17×10^{-9}. Table 4 shows a comparison of the mechanical characteristics of ceramic materials that are promising for use in mechanical friction units of mechanical modules of devices and microelectronic mechanical systems (MEMS).

Table 4. Characteristic characteristics of the mechanical properties of ceramic materials used in precision instruments.

Material	E, GPa	Strength, GPa	K_{IC}, MPa m$^{1/2}$	Cross-breaking strength, MPa
ZrO_2+3molY_2O_3	353	15	10	924
SiC	430	29	4	100–440
Si_3N_4	300	15	5–8	800
Al_2O_3	380	22	2.3–3.5	450–700

3. Conclusion

It has been experimentally established that nanostructured zirconia crystals are promising for solving microtribology problems, especially in microelectronics and mechanical modules of devices (axle supports "on stones"). Improving the characteristics of microelectronic mechanical systems (MEMS) of the future is largely determined by higher values of the characteristics of the mechanical and tribological properties of the new structural material. The experiments suggest that the only alternative to silicon crystals are nanostructured PSZ crystals. A comparison of the properties of silicon crystals with PSZ crystals shows that:

a) The elastic modulus of silicon (190 GPa) is close to steel, nickel and is much higher than that of quartz, as well as many alkaline borosilicate lead-containing glasses, but it is almost two times less than the elastic modulus of ChSC crystals (350–400 GPa).

b) The microhardness of silicon (8.5 GPa) is close to quartz and significantly higher than glass (5.3 GPa), but is almost 1.5 times inferior to the microhardness of a PSZ crystal (10–15 GPa).

c) The characteristics of viscosity and elasticity of crystals of PSZ are significantly higher than that of any other ceramic materials, as well as silicon crystals, which are brittle material. Due to the fragility of crystals in MEMS manufacturing technology, it is necessary that all mechanical processes, such as cutting, turning and polishing, be minimized or eliminated. The machining results in brittle edges and surface defects that can lead to chips and/or create internal stresses with subsequent consequences for breakage.

d) Silicon crystals do not withstand bending tests, and PSZ crystals have the greatest bending strength of ceramic materials.

References

[1] Drozdov Yu N, Yudin E G and Belov A I 2010 *Applied Tribology* ed Yu N Drozdov (Moscow: Eco-Press) p 604
[2] Peng Z, Luo X, Xie Z and Yang M 2020 Sintering behavior and mechanical properties of spark plasma sintering SiO2-MgO ceramics *Ceramics International* **46(3)** 2585–91
[3] Song J et al 2018 *Ceramics International* **44(5)** 5371–77
[4] Lapteva V G, Alisin V V, Kuksenova L I, Sherbakov Y I and Khrennikova I A 2011 *Journal of Machinery Manufacture and Reliability* **40**(2) 135–9
[5] Vorotilo S, Levashov E A, Petrzhik M I and Kovalev D Yu 2019 *Ceramics International* **45** A2 1503–12
[6] Wang X, Liu C, Li J, Qiao L and Bai Y 2018 *Ceramics International* **44(18)** 22686–91
[7] Wang Y, Hou Q, Tian L, Chen P and Xiao J 2020 *Ceramics International* **46(7)** 8725–9
[8] Fregeac A, Ansart F, Selezneff S and Claude 2019 *Ceramics International* **45** B17 23740–9
[9] Liu P, Li Z, Xiao P, Luo H and Jiang T 2018 *Ceramics International* **44(2)** 1394–403
[10] Jing Q *et al.* 2019 *Ceramics International* **45** A18 24212–20
[11] Borik M A *et al.* 2015 *Journal of the European Ceramic Society* **35(6)** 1889–94

[12] Borik M A *et al.* 2019 *Journal of Alloys and Compounds* **792** 1255–60
[13] Moskvitin G V, Mel'shanov A F and Pugachev M S 2008 *Journal of Machinery Manufacture And Reliability* **37(5)** 452
[14] Borik M A *et al.* 2013 *Physics of the Solid State* **55(8)** 1690–6
[15] Agarkov D A *et al.* 2018 *Russian Journal of Electrochemistry* **54(6)** 459–63
[16] Agarkov D A *et al.* 2019 *Kimya Problemleri* **17(2)** 235–45
[17] Borik M A *et al.* 2015 *Optics and Spectroscopy* **118** 6 918–23
[18] Borik M A *et al.* 2016 *Physics of the Solid State* **58** 7 1308–13
[19] Alisin V V et al. 2015 *Inorganic Materials* **51(6)** 548–52
[20] Kuksenova L I, Lapteva V G, Kolmakov A G and Rybakova L M 2001 *Friction and Wear Test Methods* (Moscow: Intermetnengineering) p 152
[21] Borik M A *et al.* 2017 *Physics of the Solid State* **59(10)** 1934–9

Mechanics of inspection robot traveling over the electrical line with constant velocity

M R Bahrami[1]

[1] Innopolis University, Innopolis 420500, Russia

E-mail: mo.bahrami@innopolis.ru

Abstract. This article aims to the construction of mathematical models for a situation in which the electrical line stretched and supported by two isolators, one horizontal and the other vertical and robot inspector moves with a constant speed on it. The expansion theorem is used to construct the equation of motion of the system and has been solved by two methods: expansion method and Galerkin method. As a result of even constant speed of motion of robot inspector along the electrical line, vibrations have been observed. These vibrations can cause damage to the line or the robot structure as well. Then, the effect of the resistance force, wind force, has been studied. As shown in results, resistance force acts as a damper to the system and causes a decrease in the amplitude of vibrations. The practical results are useful for engineers to select proper parameters and speed for designing and constructing the robot inspectors.

1. Introduction

Nowadays electrical energy has an inseparable role in human life. Inspection and maintenance of electrical transmission lines are necessary and done regularly according to the standards of each country [1–2], since any problem during the transmission of the electrical energy lines to cities, industries, etc. may affect human life.

Usually, inspection and maintenance of electrical transmission lines are done by human forces. By this, human forces are located in a potential hazard environment. Therefore, to reduce this risk, different types of robot inspectors are designed and used in some regions [1–10]. Each of these robots is capable of doing different tasks. One of these robots shown in figure 1 [11].

Executive and mathematical modeling experience shows that even with a constant speed movement of the robot over the line, vibrations occur [3–10, 12–16]. These vibrations can cause damage as well to the electrical line as to the robot. Therefore, it is needed to conduct mathematical modeling for different situations that may robot faces [11–16].

This article, which is the extended work of [12–15], deals with the construction of mathematical models for a situation in which the electrical line stretched and supported by two isolators, one horizontal and the other vertical and robot inspector moves with a constant speed on it. It follows by studying the effect of wind force (resistance force).

Content from this work may be used under the terms of the Creative Commons Attribution 3.0 licence. Any further distribution of this work must maintain attribution to the author(s) and the title of the work, journal citation and DOI.
Published under licence by IOP Publishing Ltd

IOP Conf. Series: Materials Science and Engineering **939** (2020) 012010 doi:10.1088/1757-899X/939/1/012010

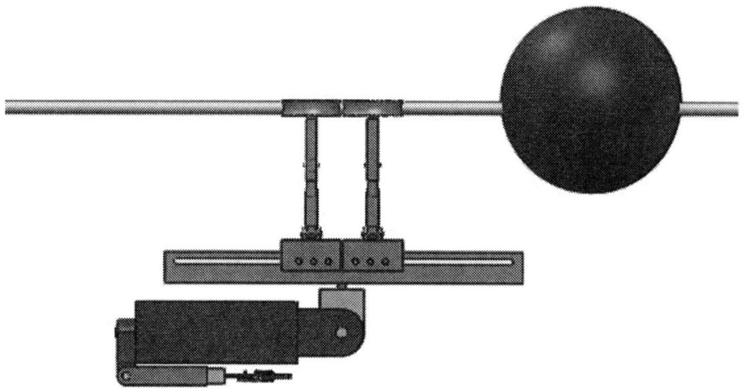

Figure 1. Electrical transmission inspection robot scheme [11].

2. The fluctuation of the stretched string with a moving point load

This section deals with the dynamics of the conductor with a moving load is considered (figure 2). We consider a case in which the electrical line stretched and supported by two isolators, one horizontal and the other vertical (figure 2a). The horizontal one acts as simple pin support and the other is a massless roller that can move vertically. As illustrated in figure 2b, here, we consider the conductor is a taut string while the robot inspector is considered as a moving point load.

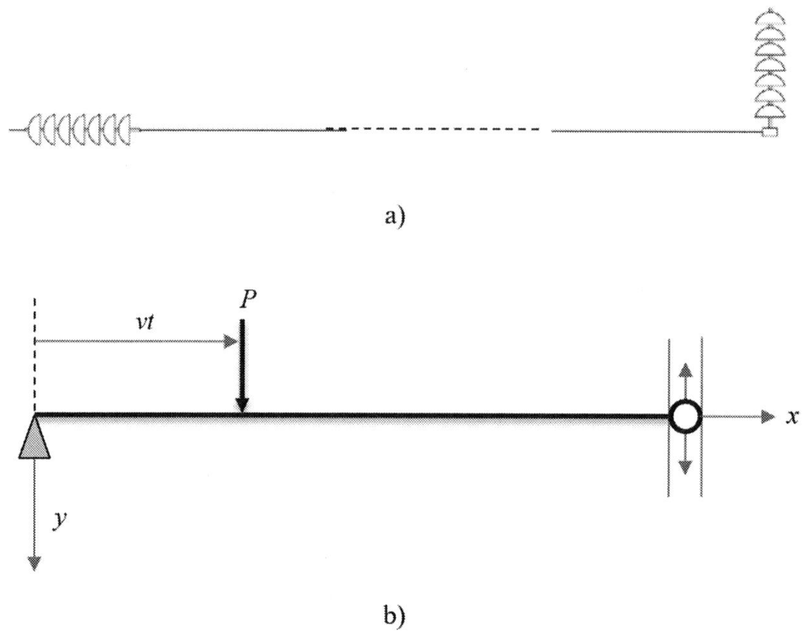

Figure 2. (a) Electrical line hangs between two supports, and (b) concentrated load moves over the stretched string.

2.1. The expansion method

We start to construct the problem statement for the dynamic deflection of a string using the expansion method. The kinetic energy (K) and the potential energy (U) of the string are:

$$K = \frac{1}{2}\int_0^l \rho \dot{u}^2 \mathrm{d}x,$$
$$U = \frac{1}{2}\int_0^l T u'^2 \mathrm{d}x. \tag{1}$$

where T is the string tension force, ρ is the density, l is the string length. Differentiation with relative to the coordinate x and time t is shown by prime and dot.

Utilizing the Lagrangian $L = K - U$, when a system is subject to generalized forces $q(x, t)$, one can introduce them in extension Hamilton's principle through the virtual work expression $\delta A = q(x, t)\delta u$. Extension Hamilton's principle can be written as $\delta \int_0^{t_f} (L + A)\mathrm{d}t = 0$.

The equation of motion can be obtained as explained in [17]:

$$\int_0^l \rho \dot{u}\delta u \big|_0^{t_f} \mathrm{d}x - \int_0^{t_f} Tu'\delta u \big|_0^l \mathrm{d}t + \int_0^{t_f}\int_0^l (Tu'' - \rho \ddot{u})\delta u \, \mathrm{d}x\mathrm{d}t + q(x,t) = 0. \tag{2}$$

Since the variation δu at the initial and final times are zero, then the first term in equation (2) is always zero. Equating the third term in equation (2) to zero leads to the dynamic equation for the string as:

$$\rho \ddot{u} - Tu'' = q(x,t). \tag{3}$$

Setting the second term in equation (2) to zero gives us the boundary conditions of the problem, i.e.:

$$x = 0:\ u = 0,$$
$$x = l:\ Tu' = 0, \tag{4}$$

Initial conditions are $t = 0:\ u = 0, \dot{u} = 0$.

The loading point (Constant P) moves with a constant speed of v. Therefore, the force concentrated at a point x will be $q(x,t) = P\,\delta(x - vt)$ (δ is the Dirac delta function).

The equation of motion for the systems described above can be written in the general form

$$\ddot{u} - c^2 u'' = p\delta(x - vt),$$
$$c = \sqrt{\frac{T}{\rho}},\ p = \frac{P}{\rho}. \tag{5}$$

The solution to the problem (5) can be constructed by the method of eigenfunctions [18] knowing the eigenvalues and eigenfunctions are given by $\omega_n = \lambda_n c, \varphi_n = \sin \lambda_n x,\ \ \lambda_n = \frac{n\pi}{l}$ as:

$$u(x, t) = \sum_{n=1}^{\infty} u_n(t)\varphi_n(x), \tag{6}$$

where the unknown modal coordinate is designated by $u_n(t)$.

Recalling the properties of eigenfunctions

$$\int_0^l \varphi_n \varphi_k dx = \delta_{kn} \equiv \begin{cases} 1, & k = n \\ 0, & k \neq n \end{cases} \Rightarrow u_n = \int_0^l u\varphi_n dx, \tag{7}$$

and multiplying both sides of equation (5) by φ_n and integrating, we obtain ordinary differential equations (ODE) for u_n:

$$\ddot{u}_n + \omega_n^2 u_n = f_n(t),$$
$$f_n(t) = \frac{\int_0^l \varphi_n(x) q(x,t)\, \mathrm{d}x}{\int_0^l \varphi_n^2(x)\, \mathrm{d}x} = \frac{2}{l} p \ \sin(\lambda_n vt). \tag{8}$$

One can write the solution of equation (8) as:

$$u_n(t) = C_{1n} \cos(\omega_n \mathrm{t}) + C_{2n} \sin(\omega_n \mathrm{t}) + \frac{2pl}{n^2\pi^2(c^2 - v^2)} \sin(\lambda_n vt), \tag{9}$$

where C_{1j} and C_{2j} are arbitrary integration constants. Therefore, the solution for the string is obtained substituting equation (9) in equation (6) as:

$$u(x,t) = \sum \left[C_{1n} \cos(\omega_n \mathrm{t}) + C_{2n} \sin(\omega_n \mathrm{t}) + \frac{2pl}{n^2\pi^2(c^2 - v^2)} \sin(\lambda_n vt) \right] \sin(\lambda_n x). \tag{10}$$

Under zero initial conditions, one can obtain $C_{1n} = 0$ and $C_{1n} = -\frac{2plv}{cn^2\pi^2(c^2 - v^2)}$.

Substituting integration constants in equation (10), we obtain the solution of problem (5):

$$u(x,t) = \frac{2pl}{\pi^2(c^2 - v^2)} \sum \frac{1}{n^2} \left[\sin(\lambda_n vt) - \frac{v}{c} \sin(\lambda_n c\, \mathrm{t}) \right] \sin(\lambda_n x). \tag{11}$$

The deflection of the conductor solving equation (11) is shown in figure 3. Parameters for calculations are: T=10 kN, P=1 kN, v=2 m/s, ρ= 5 kg/m, l= 200 m. One can see from figure 3 that even the robot moves with a constant speed over the line, oscillations appear in the vertical plane.

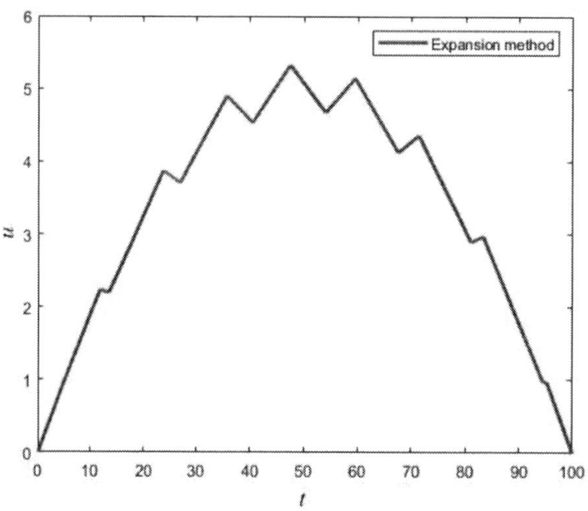

Figure 3. Deflection of the electrical line with a moving load on it.

2.2. *Galerkin method*

In this section, we construct an approximate solution of equation (3) in the matrix form:

$$u(x,t) = \sum_{n=1}^{\infty} u_n(t)\varphi_n(x) = U^T(t)\Phi(x), \tag{12}$$

IOP Conf. Series: Materials Science and Engineering **939** (2020) 012010 doi:10.1088/1757-899X/939/1/012010

where $\mathbf{\Phi}(x)$ is known as comparison functions which are orthogonal and orthonormal.

The approximate solution (12) will also satisfy all the boundary conditions, but in general, will not satisfy (3) identically since there remains an error defined by

$$e(x,t) = \ddot{\mathbf{U}}^T\mathbf{\Phi} - c^2\mathbf{U}^T\mathbf{\Phi}'', \tag{13}$$

which one can force it to have a zero projection on the chosen functions $\mathbf{\Phi}(x)$.

$$\int_0^l e(x,t)\varphi_n \, \mathrm{d}x = 0. \tag{14}$$

Substituting the expression of the error from equation (13) in equation (14), and rewriting, we have:

$$\begin{aligned} &\mathbf{M}\ddot{\mathbf{U}} + \mathbf{K}\mathbf{U} = \mathbf{f}, \\ &\mathbf{M} = \int_0^l \mathbf{\Phi}\mathbf{\Phi}^T \, \mathrm{d}x, \\ &\mathbf{K} = c^2\int_0^l \mathbf{\Phi}\mathbf{\Phi}''^T \, \mathrm{d}x, \\ &\mathbf{f} = p\mathbf{\Phi}. \end{aligned} \tag{15}$$

The equation (15) can be transited to a discrete model of a wire with a moving load to an ODE system (linear, with constant coefficients). System (15) is represented in the form of first-order equations:

$$\begin{cases} \dot{\mathbf{U}} = \mathbf{Z} \\ \dot{\mathbf{Z}} = \mathbf{M}^{-1}(\mathbf{f} - \mathbf{K}\mathbf{U}) \end{cases}. \tag{16}$$

The result is shown in figure 4. Parameters are the same as the previous section.

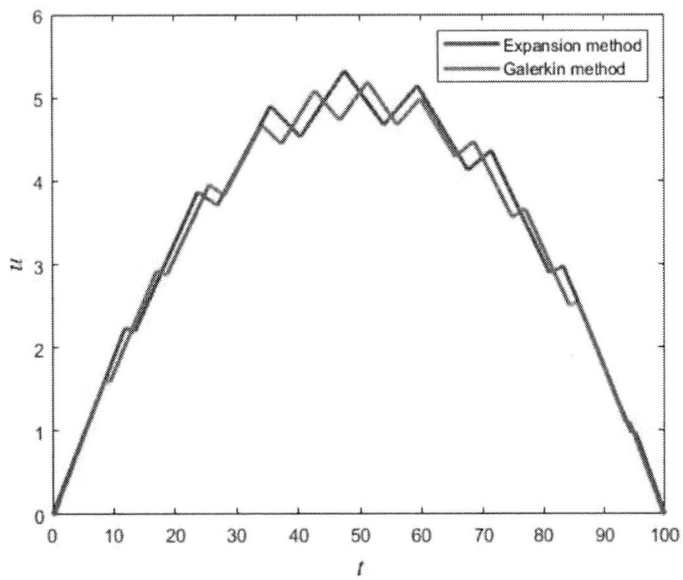

Figure 4. Deflection of the conductor presented by two methods.

3. Effect of wind force on the vibration of a stretched string with a moving load on it

In this section, we consider the resistance force from the wind on the electrical line. To take into account the distributed forces of external resistance in the string equation (3), the new form will be obtained [17–20]:

$$\rho \ddot{u} - Tu'' + b\dot{u} = q(x,t) \tag{17}$$

To solve the equation (17) using the Galerkin method, the following change can be made in the equation (15):

$$\begin{aligned} &\mathbf{M}\ddot{\mathbf{U}} + \mathbf{K}\mathbf{U} = \mathbf{f}_{new}, \\ &\mathbf{M} = \int_0^l \mathbf{\Phi}\mathbf{\Phi}^T \,\mathrm{d}x, \\ &\mathbf{K} = c^2 \int_0^l \mathbf{\Phi}\mathbf{\Phi}''^T \,\mathrm{d}x, \\ &\mathbf{f}_{new} = p\mathbf{\Phi} - b\dot{\mathbf{U}}. \end{aligned} \tag{18}$$

One can represent the system (19) in the form of first-order equations to solve:

$$\begin{cases} \dot{\mathbf{U}} = \mathbf{Z} \\ \dot{\mathbf{Z}} = \mathbf{M}^{-1}(\mathbf{f} - \mathbf{K}\mathbf{U} - b\dot{\mathbf{U}}) \end{cases} . \tag{19}$$

The results are shown in figure 5. Graphs, shown in figure 5a, have parabolic shape since the quasistatic deflection (deflection obtained by setting all time-functions in equation (3) zero) is also included. To see a better difference in the effect of wind force, we omitted this deflection from the results and show just the difference Δ in figure 5b. Parameters are the same as the previous section including b=0.5 Ns/m.

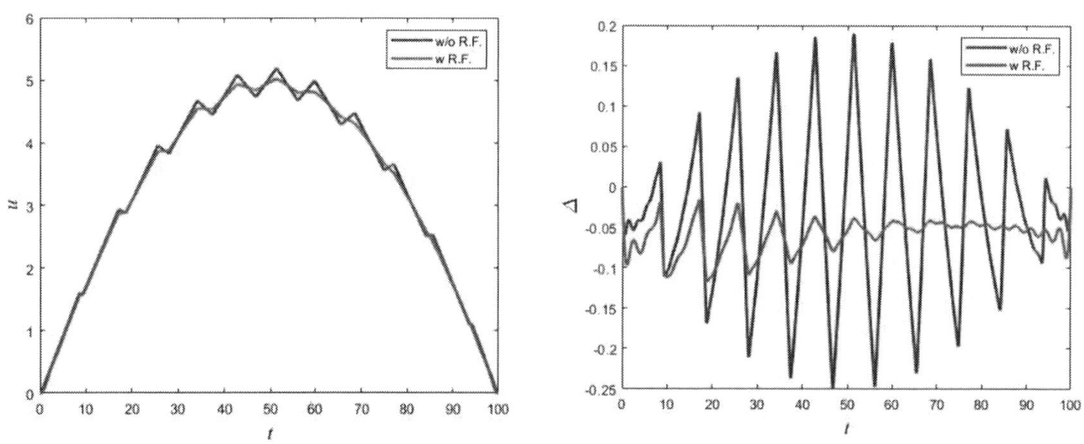

Figure 5. (*a*) General and (*b*) dynamic deflection of the conductor when resistance force considered.

4. Conclusion

The dynamical behavior of the electrical transmission lines robot inspector while moving over the line with a constant speed has been studied. To construct the equation of motion of the system by two methods: the expansion method and Galerkin method have been used. As a result of mathematical modeling, it is shown that motion of the robot even with a constant speed along the line vibration appears in the vertical plane. Then, the effect of the resistance force, wind force, has been investigated. The practical results are useful for engineers to select proper parameters and speed for designing and constructing the robot inspectors.

References

[1] Aggarwal R K, Johns A T, Jayasinghe J A S B and Su W 2000 An overview of the condition monitoring of overhead lines *Elect. Pow. Sys. Res.* **53** 15–22
[2] Dong G, Chen X and Wang B 2012 Inspecting transmission lines with an unmanned fixed-wing aircraft *Appl. Rob. for the Pow. Ind. (CARPI)* 173–174 DOI: 10.1109/CARPI.2012.6473355
[3] Toussaint K, Pouliot N and Montambault S 2009 Transmission line maintenance robots capable of crossing obstacles: state-of-the-art review and challenges ahead *J. of Field Rob.* **26** 477–99
[4] Sawada J, Kusumoto K, Munakata T, Maikawa Y and Ishikawa Y A 1991 Mobile robot for inspection of power transmission lines *IEEE Trans. on Power Del.* **6** 309–15
[5] Higuchi M, Maeda Y, Tsutani S and Hagihara S 1991 Development of a Mobile Inspection Robot for Power Transmission Lines *J. of Robotic Soc.* **9** 57–63
[6] Tsujimura T and Morimitsu T 1997 Dynamics of mobile legs suspended from wire *Rob. and Aut. Sys.* **20** 85–98
[7] Montambault S and Pouliot N 2007 Design and validation of a mobile robot for power line inspection and maintenance *6th Int. Conf. on Field and Ser. Rob. (FSR)* 1–10
[8] Barbosa C F and Nallin F E 2014 Corrosion detection robot for energized power lines *Appl. Rob. for the Pow. Ind. (CARPI)* 1–6 DOI: 10.1109/CARPI.2014.7030059
[9] Alhassan A B, Zhang X, Shen H, Jian G, Xu H and Hamza K 2019 Investigation of Aerodynamic Stability of a Lightweight Dual-Arm Power Transmission Line Inspection Robot under the Influence of Wind *Math. Prob. in Eng.* 1–16
[10] Miller R, Abbasi F, and Mohammadpour J 2017 Power line robotic device for overhead line inspection and maintenance *Ind. Rob.* **44** 75–84 DOI: 10.1108/IR-06-2016-0165
[11] Bahrami M R 2016 A novel design of electrical transmission line inspection machine *Lect. Notes in Mech. Eng.: Adv. in Mech. Eng.* 67–73
[12] Bahrami M R 2018 Mechanics of diagnostic machine on electrical transmission lines conductors *MATEC Web of conferences* **224** 02021 DOI: 10.1051/matecconf/201822402021
[13] Eliseev V V and Bahrami M R 2016 Strength suspension of inspector robot on the electrical transmission line wire *Bulletin of engineering* (Moscow) **6** 19–22
[14] Bahrami M R and Abed S A 2019 Mechanics of robot inspector on electrical transmission lines conductors: performance analysis of dynamic vibration absorber *Vibroengineering PROCEDIA* **25** 60–64 DOI:10.21595/vp.2019.20807
[15] Bahrami M R and Abed S A 2020 Mechanical challenges of electrical transmission lines inspection robot *IOP Conf. Ser.: Mater. Sci. Eng.* **709** 022099
[16] Bahrami M R 2020 Mechanics of robot inspector on electrical transmission lines conductors with unequal heights supports *Vibroengineering PROCEDIA* **30** 20–25
[17] Hagedorn P and DasGupta A 2007 *Vibrations and waves in continuous mechanical systems* (Chichester: Wiley)
[18] Eliseev V V 2006 *Mechanics of deformation of the rigid body.* (St. Petersburg: Polytech Uni Pub)
[19] Tikhonov A N and Samarsky A A 1979 *Equations of mathematical physics.* (Moscow: Science)
[20] Oden J T and Reddy J N 2012 *Variational methods in theoretical mechanics.* (Springer Science and Business Media)

IOP Conf. Series: Materials Science and Engineering **939** (2020) 012011 doi:10.1088/1757-899X/939/1/012011

Deformation-force effects of shape memory of TiNi alloys under conditions of incomplete martensitic transformation

V N Belyakov[1]

[1] Yaroslav-the-Wise Novgorod State University, ul. B. St. Petersburgskaya, 41 173003 Veliky Novgorod, Russia

E-mail: val-belyakov@mail.ru

Abstract. The article is devoted to studying the influence of preliminary and plastic processing, applied stress, degree of phase transformation on the deformation effects manifested in conditions of incomplete martensitic transformation. The laws of generation and relaxation of reactive stresses were studied depending on the stiffness of the reaction, the magnitude of the preliminary deformation and the selected interval of heat changes. It is shown that the accumulation and return of deformation, the effect of generation and relaxation of reactive stresses in a narrow temperature range occurs only in structurally inhomogeneous objects, determined with the statistical spread of characteristic temperatures.

1. Introduction

Materials with the effect of shape memory, due to their unique properties, are widely used in many fields of technology [1, 2]. Most often, titanium nickelide and alloys based on it are used as thermosensitive elements. In a series of materials with reversible martensitic transformations, TiNi-based alloys occupy a special place due to the unique combination of mechanical and physical properties and high values of the shape memory effect. However, the presence of a temperature hysteresis of 40-60 K complicates the task of using a material with shape memory effect in devices that maintain a given temperature regime in a narrow range of 5–15 K (relay sensors, instrument containers, thermostats, etc.) [3]. It is not possible to significantly reduce the conversion hysteresis due to a change in the chemical composition of the alloy or in plastic and heat treatment [4, 5]. The only possible way to perform regulatory functions with thermoset elements of titanium nickelide in a narrow temperature range is to realize the effects of shape memory, generation and relaxation of reactive stresses under conditions of incomplete phase transformation. The urgency of the problem is also determined with the fact that the properties of shape memory during heat changes in the interval of incomplete martensitic transformation (IMT) are practically not studied. There are no systematic studies on the influence of the structural state of the material, loading conditions, and the interval of heat changes on the deformation-force effects of shape memory. Without these data, it is impossible to design and create workable mechanisms and devices operating in a narrow temperature range.

2. Objects and research methods

As objects of study, TiNi alloys with a content of 49.5, 50.0, 50.2, 50.5 at % Ni were chosen, in which, depending on the heat and plastic processing, various phase transition chains and, accordingly, various

Content from this work may be used under the terms of the Creative Commons Attribution 3.0 licence. Any further distribution of this work must maintain attribution to the author(s) and the title of the work, journal citation and DOI.
Published under licence by IOP Publishing Ltd

loop shapes of hysteresis can be observed. The samples were made of wire with a diameter of 2 mm and cantilevered beams of polycrystalline materials subjected to the following types of pretreatment.

1) Quenching from 1073 K in water.

2) Cold rolling with compression strain ε = 15–20% and subsequent annealing at a temperature of 723K for 2 hours.

3) Drawing at a temperature of 823-873K with a reduction ratio of 10–12% and subsequent annealing, T_{OT} = 723K, τ_{OT} = 2 hours.

The experimental studies were performed on two specially designed installations, which allow measuring the deformation-force parameters of selected objects in a given temperature range under conditions of constantly applied stress, constrained deformation with finite and infinite stiffness of the opposing system. X-ray diffraction studies of the phase composition of alloys at various temperatures were carried out on a DRON-2.0 X-ray diffractometer using an attachment for heating (cooling) the test sample.

3. Experimental results and discussion

Thermomechanical loops in the incomplete MT interval were formed on the complete hysteresis curve obtained after 10–15 preliminary thermal cycles. At the stage of direct MT at the temperature of M_K^H (the temperature of the end of the incomplete martensitic transformation), cooling was replaced with heating to the temperature of A_K^H (the temperature of the end of the incomplete austenitic transformation) and thermal cycling was performed in this interval $\Delta T_H = A_K^H - M_K^H$. In a similar way, small loops were formed from the heating stage. The experiments showed (figure 1) that regardless of the location of the heat transfers interval relative to the characteristic temperatures and its magnitude, the thermal deformation line always closes, forming a hysteresis loop (small loop). That part of the deformation that has accumulated during the cooling process is completely restored during the heating process.

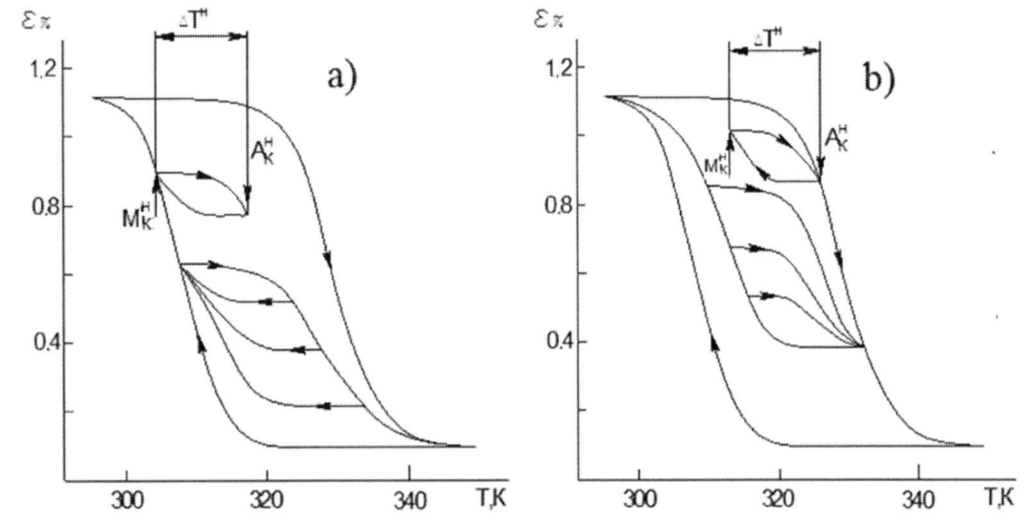

Figure 1. The accumulation and return of deformation in a Ti-50 at% Ni alloy under a constant voltage σ = 35 MPa during heat transfer under conditions of IMT from the stage of cooling (a) and heating (b).

It was established that the nature of the change in deformation during heat changes in a limited temperature range is determined with the structural state of the material, the shape of its full hysteresis figure (figure 2).

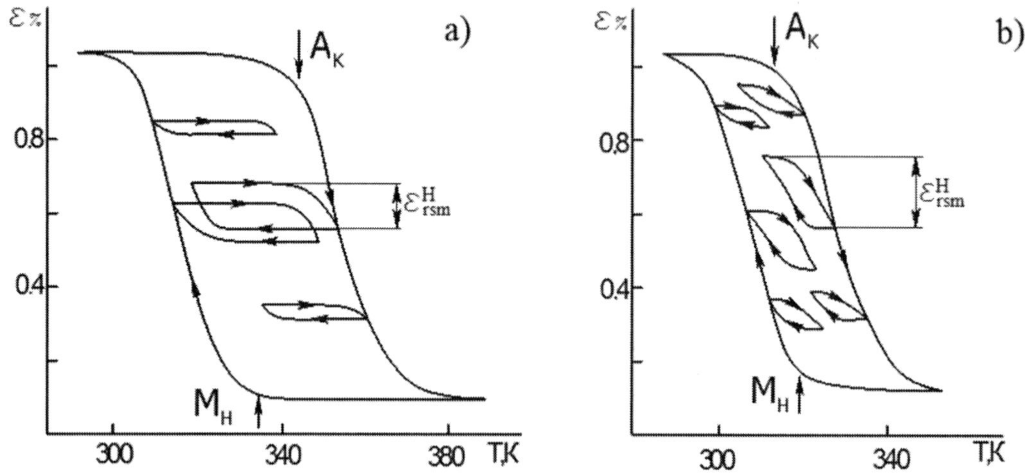

Figure 2. Hysteresis loops of incomplete phase transformation during heat changes under constant voltage σ = 30 MPa for alloys. a) Ti-49.5 at% Ni after quenching from 1073 K; b) Ti-50.2 at% Ni after rolling (ε = 15%) and annealing (T_{OT} = 723K, τ_{OT} = 2 hours).

In quenched alloys with a homogeneous structure and a wide hysteresis loop (ratio of characteristic temperatures (A_K>M_H), the shape of the thermodeformation curves of the incomplete transformation does not depend on the temperatures of the onset of heat changes (formation conditions) and it is characterized with a wide hysteresis of termination. Accumulation and return of deformation does not occur until the interval of temperature change exceeds the width of the total hysteresis (figure 2a). An increase in the heterogeneity of the structure of the alloys, associated with an increased nickel content C_{Ni} = 50.2–50.5 at% and plastic processing, leads to a change in the shape of the full hysteresis loop (A_K<M_H) and affects the kinetics of accumulation and return of deformation in the temperature range of the IMT. The shape of the small hysteresis loops changes depending on the specified heat exchange interval, and deformation is observed even with a slight change in temperature (figure 2b).

X-ray structural studies showed that in the quenched alloys with a nickel content of 49.5–50.2 at%, only the B2→B19' martensitic reaction occurs, and in the plastic treated alloys Ti-50.5 at% Ni, along with the B2→B19' reaction, phase reaction occurs according to the B2→B19' scheme, characterized by a narrow transformation hysteresis. As a result, it can be noted that the structural heterogeneity of the alloy and the presence of volumes with different intervals of martensitic transformations make it possible to accumulate and return strain at sufficiently small temperature changes, which was also noted in other studies [6].

The study of the influence of the heat change interval on the magnitude of the deformation of the small loop (the value of the reversible deformation that accumulated during cooling and fully recovered during heating $\varepsilon_{оп\phi}{}^{H}$) shows the following result. The maximum value of the reversible deformation is achieved if the initial temperatures of the heat change interval $M_K{}^H$ or $A_K{}^H$ are located in the middle of the M_H–M_K intervals or A_H–A_K, that is with the degree of phase transformation η = 0,4–0,6 (figure 2)

The degree of direct ηM and reverse ηA martensitic transformation is determined with the relation

$$\eta_M = \frac{M_H - M_K^H}{M_H - M_K} \text{ or } \eta_A = \frac{A_K^H - A_H}{A_K - A_H} \quad (1)$$

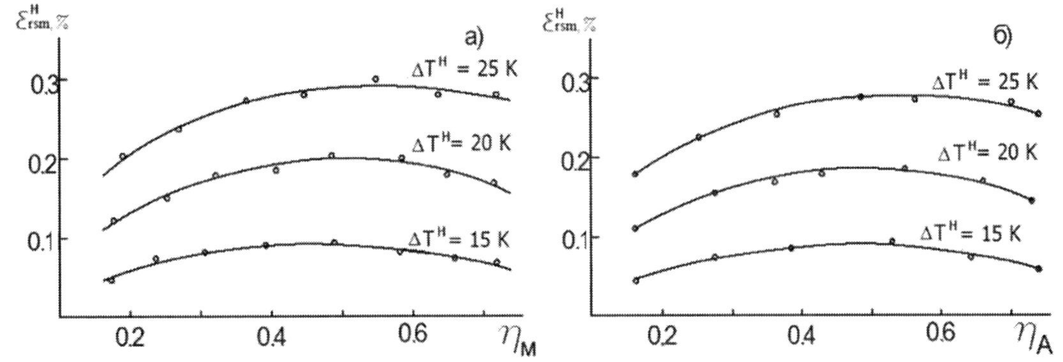

Figure 3. Change in the reversible deformation $\varepsilon_{rsm}{}^{H}$ in the Ti-50 at% Ni alloy under a stress of σ = 35 MPa as a function of the degree of phase transformation η during heat changes from cooling stage (a) and heating (b) in a given temperature range ΔT_H.

It was established that the kinetics of changes in reactive stresses is also determined with the structure of the material, the shape of its complete hysteresis loop [7, 8]. The effect of stress generation and relaxation during heat changes in a narrow temperature range not exceeding the width of the total hysteresis is observed only in structurally inhomogeneous alloys (figure 4).

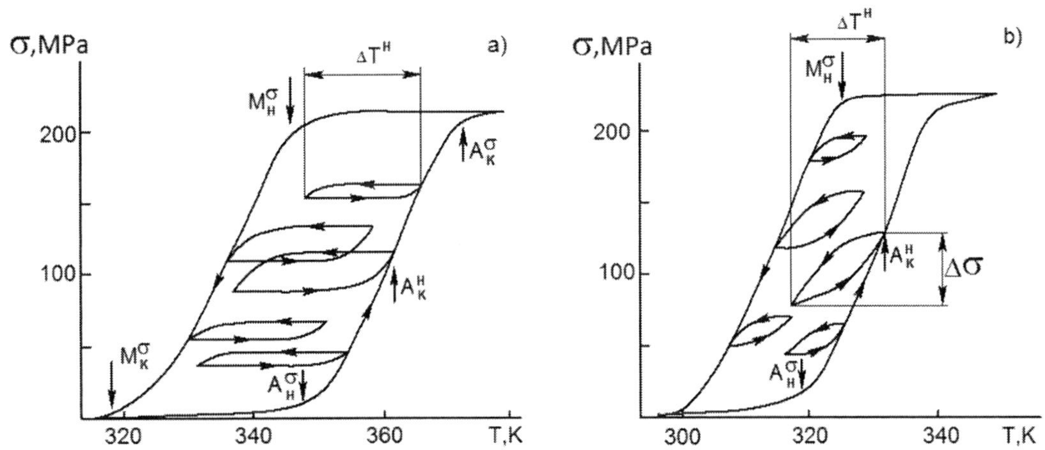

Figure 4. The effects of the generation and relaxation of reactive stresses in the conditions of IMT with infinite stiffness of reaction (K→8, ε_{pr} = 1.2%) in alloys: a) Ti-49.5 at% Ni (quenching from 1073 K in water); b) Ti-50.2 at% Ni (rolling εоб = 15%, T_{OT} = 723K, τ_{OT} = 2 hours).

The influence of the heat change interval, the degree of preliminary deformation, the stiffness of the counteraction on the level of reactive stresses $\sigma_r{}^{H}$ generated when heated to the temperature of the $A_K{}^{H}$, and the voltage swing $\Delta\sigma$ of the small hystere,sis loop are studied. It was found that the stress range $\Delta\sigma$ increases linearly with increasing maximum reactive stresses $\sigma_r{}^{max}$ generated in the alloy upon heating above AKσ temperature (figure 5).

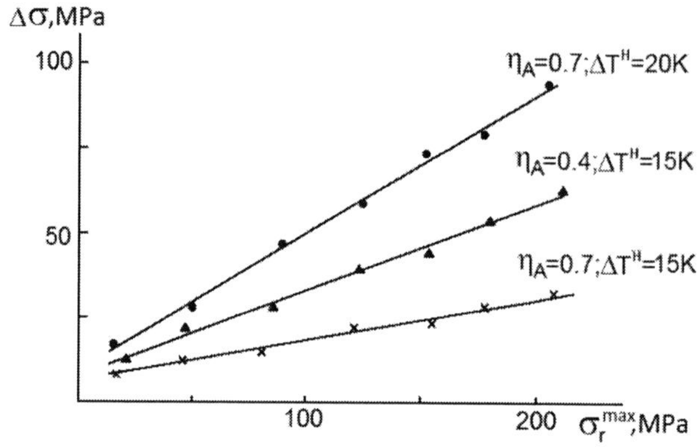

Figure 5. The dependence of the magnitude of the stress range $\Delta\sigma$ of the small loop on the maximum level of generated stresses σ_r^{max} in the Ti alloy – 50.2 at% Ni.

Studies have shown that the proportionality coefficient (relative stress range) $\alpha\Delta\sigma = \Delta\sigma / \sigma_r^{max}$ depends on the cycling interval and the degree of reverse phase transformation (Fig. 6). The level of reactive stress σ_r^{H} is also determined with the maximum stresses σ_r^{max} and can be found from the relation.

$$\sigma_r^{H} = \eta_A * \sigma_r^{max} \quad (2)$$

where η_A is the degree of austenitic transformation.

It was established that an increase in the statistical dispersion of characteristic temperatures leads to a transformation of the complete hysteresis loop (the slope of the hysteresis figure increases, the temperature Mn exceeds the temperature of the A_K) and to a narrowing of the incomplete conversion hysteresis loop. In structurally heterogeneous materials, a change in deformation or stress occurs virtually at any temperature change.

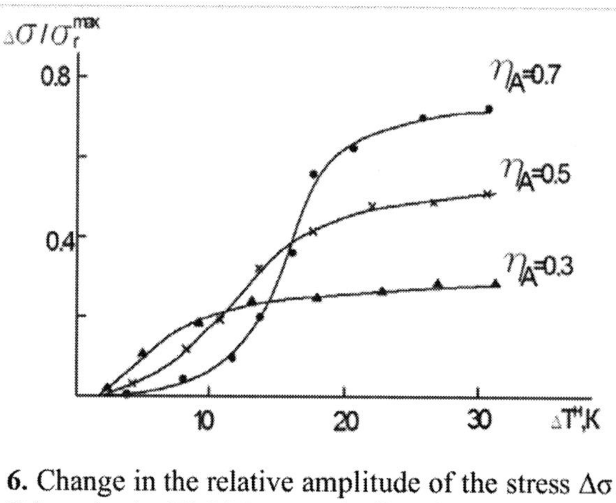

Figure 6. Change in the relative amplitude of the stress $\Delta\sigma / \sigma_r^{max}$ of the small loop in the Ti-50.2 at% Ni alloy from the cycling interval ΔT_H at specified values of the degree of inverse MP (ηA).

4. Conclusion

Studies have shown that the thermomechanical hysteresis of an incomplete transformation depends on the shape of the full hysteresis figure. In alloys with a wide hysteresis loop, accumulation or return of deformation in the temperature range less than the width of the total hysteresis is not observed. In alloys

IOP Conf. Series: Materials Science and Engineering **939** (2020) 012011 doi:10.1088/1757-899X/939/1/012011

with a narrow hysteresis of phase transformations, thermomechanical small loops are formed in a small temperature range (5–10 K). It has been established that the structural heterogeneity of the alloy, with the presence of volumes with different intervals of martensitic transformations, makes it possible to manifest the effects of shape memory, generation and relaxation of reactive stresses at sufficiently small temperature changes. The regularities of the change in the shape of small hysteresis loops depending on the location of the heat change interval relative to the characteristic temperatures are revealed. The maximum value of the deformation range or stresses of the small hysteresis loop is achieved during heat changes with an initial degree of phase transformation lying in the range 0.4–0.6. It was found that with increasing stress, the magnitude of the reversible deformation $\varepsilon_{\text{опф}}^{\text{н}}$ during heat changes in the temperature range not exceeding the width of the total hysteresis increases slightly, and it is determined largely with the initial temperature and thermal cycling interval.

It was found that the magnitude of the thermosilic hysteresis stresses increases linearly with increasing maximum level of reactive stresses generated in the alloy, and the proportionality coefficient (relative stress range) is determined with the heat change interval, the degree of phase transformation, and the alloy structure.

References

[1] Jiang P C, Zheng Y F, Tong Y X, Chen F, Tian B, Li L, Gunderov D V and Valiev R Z 2014 Transformation hysteresis and shape memory effect ofanultrafine-grained TiNiNb shape memory alloy *Intermetallics* **54** 133–135

[2] Gusev D E, Kollerov M Yu and Sharonov A V 2015 Investigation of reactive stresses generated with implants from titanium nickelide alloys *Titanium* **1** 42–49

[3] Tong Y, Chen F, Guo B, Tian B, Li L, Zheng Y F, Gunderov D V and Valiev R Z 2013 Superplasticity and its stability of an ultrafine-grained Ti49.2Ni50.8 shape memory alloy processed by equal channel angular presing *Materials Science and Engineering:* A **587** 61–64

[4] Kim H Y and Miyazaki S 2016 Several issues in the development of Ti-Nb-based shape memory alloys *Shape Memory and Superelasticity* vol 2 **4** 380–390

[5] Belyaev S P, Volkov A E and Evard M E 2017 Modeling irreversible deformation and fracture of titanium nickelide during thermal cycling *Deformation and fracture of materials* **4** 65–73

[6] Andronov I N, Demina M Yu, Kormshchikova Z I and Matveeva O A 2017 Influence of incomplete thermal cycles on the performance of a coil spring of titanium nickelide *Deformation and destruction of materials* **8** 51–58

[7] Isaenkova M G, Perlovich Yu A, Fesenko V A and Zaripova M M 2019 Orientational dependences of the functional properties of alloys with the effect of shape memory and superelasticity *Chelyabinsk Journal of Physics and Mathematics* vol 4 **2** 221–240

[8] Mashikhin A E, Kazarina S A and Movchan A A 2016 Generation of reactive stresses in the experiment on the oriented transformation of a shape memory alloy in a non-stressed state *Bulletin of the Tambov University* **3** 1152–1155

History review of aircraft performance improvement

A Bobkov[1,2] and M Krivenko[1]

[1] Komsomolsk-na-Amure State University, Lenin Prospect, 27, Komsomolsk-on-Amur, 681013, Russia

[2] E-mail: bobkov822@yandex.ru

Abstract. This paper is a review of trends in performance improvement of all-purpose military remotely piloted aircrafts in the last years. The review results are intended for use by designers of such aircrafts. Performance improvement of a machine, aircrafts in particular, is aimed at cost-effective optimization. The fundamental of a design specification for a new all-purpose aircraft is drafting requirements to its military characteristics. It is required to determine if an aircraft requires new or improvement of existing functions that influence the aircraft's weight taking into account novel technologies to appear soon. No look-ahead analytic base available makes it difficult to determine needs and capacities of the future engineering period. This paper focuses on an analysis of the past period to forecast the aircraft engineering future. A history chart is built to show the design improvements of all-purpose military remotely piloted aircrafts starting from 2005.

1. Introduction

A study of aircraft performance allows finding principles to be used for selecting a right aircraft design concept when developing a design specification [1–4]. A new design of a military remotely piloted aircraft ("aircraft") must consider many interconnected requirements, among them are: target load, flying distance, course speed, take-off and landing characteristics [5-12]. Each factor has qualitative constituents achievable by increased aircraft's size and take-off weight, specific wing load, improved aircraft's aerodynamic and weight characteristics and engine capacity [10, 13].

The review of aircraft performance evolution is the review of design innovations improving the aircraft's performance and eventually its functions [14].

2. Aircraft Performance Improvements

Selection of specific performance measures of an aircraft design depends on its purpose, operation conditions, manufacture and maintenance costs and other factors [2-4]. Let us look at typical measures.

2.1. Transport Performance

The aircraft's transport performance defines the ability to transport a target load to a point. It can be evaluated by delivery parameters such as consumed power, capacity, flying duration and distance [15, 17–19]. In terms of "transportation-costs", the process efficiency must be studied: hourly capacity $m_{tlw} \cdot V$ or flying capacity $m_{tlw} \cdot L$ where m_{tlw} is a target load weight, kg; V is course speed, m/s; L is flying distance, m. Then the transport performance TP can be found by formula [10] taking into account consumed power:

Content from this work may be used under the terms of the Creative Commons Attribution 3.0 licence. Any further distribution of this work must maintain attribution to the author(s) and the title of the work, journal citation and DOI.
Published under licence by IOP Publishing Ltd

$$TP = m_{tlw} \cdot V^2 \cdot L \,. \tag{1}$$

Therefore, the squared target load delivery speed is most important in improving the transport performance.

2.2. Aircraft Engineering Level

Engineering level of an aircraft is evaluated by comparing its performance with previous designs. As high engineering level can be achieved not only by innovations but also by an increased weight, then the compared aircrafts must be of a similar type and same weight. Engineering level (EL) increase is a full differential of function of several variables dEL of the aircraft design and transportation process:

$$dEL = \frac{\partial EL}{\partial W} dW + \frac{\partial EL}{\partial m_{tlw}} dm_{tlw} + \frac{\partial EL}{\partial V} dV + \frac{\partial EL}{\partial L} dL + \frac{\partial EL}{\partial m_0} dm_0 \ , \tag{2}$$

where W is effective performance probability, unsized [20]; m_0 is the aircraft maximum take-off weight, kg.

3. Example of All-Purpose Aircraft Performance Improvement History

The example uses a known all-purpose reconnaissance and strike MQ-1B Predator 2005 [16], ref. figure 1. Predator has a traditional aerodynamic design with a one-beam body and aft propeller and with a reverse V-tail and Rotax engine.

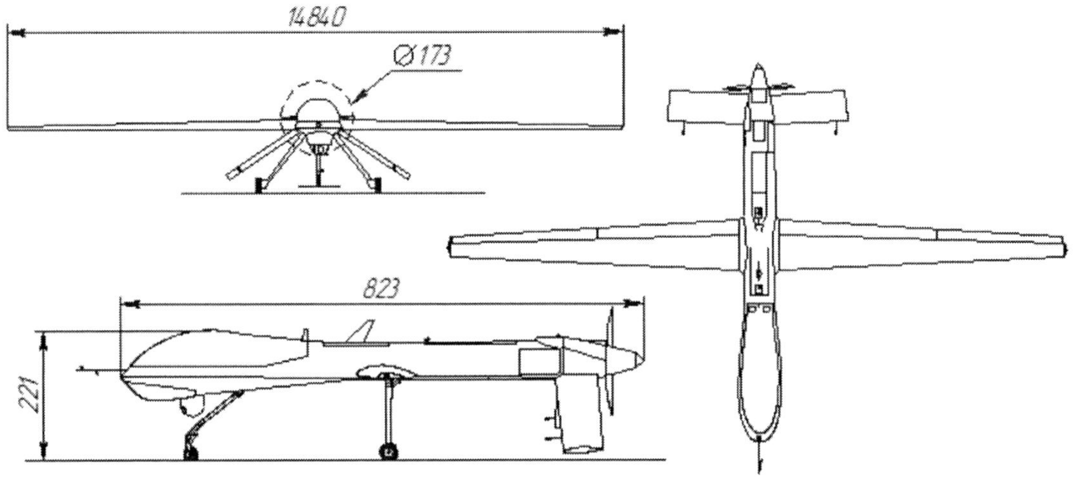

Figure 1. Reconnaissance and Strike MQ-1B Predator.

Target load (radiotronics and missiles) is placed in the fuselage and on external slings. This aircraft is a classic MALE-aircraft (Medium Altitude Long Endurance), a prototype of most of other developed all-purpose aircrafts. Table 1 contains transport functions of four MALE-aircrafts with the design trends from 2005 to 2015.

Table 1. Aircraft Transport Functions.

Parameter	MQ-1B Predator	Heron 1	Hermes 900	Dozor-600
Year	2005	2007	2011	2015
Maximum take-off weight, kg	1020	1250	970	720
Maximum payload, kg	204	250	300	120
Course speed, kph	130	130	130	130
Flying distance, km	1100	1000	1500	3700
Engine	Rotax 914 F	Rotax 914 F	Rotax 914 F	Rotax 914 F

Figure 2 illustrates a history chart of the aircraft's main transport efficiency elements represented as: L/L_r is flying distance; $m_0/(m_0)_r$ is maximum take-off weight; m_{tlw}/m_0 is specific take-off weight where "r" refers to MQ-1B Predator. Figure 2 illustrates a trend chart of increased flying distance with the same course speed. Maximum take-off weight and target load weight show an unclear reduction tendency. The course speed is the same for all aircrafts due to gasoline reciprocating engine Rotax 914 the screw jack ram is based on. The screw jack ram is not likely to be changed. For low flying speeds, such screw jack ram design is more efficient comparing to gas-turbine engines: screw or reactive. Figure 3 illustrates a change of the integral parameter of the aircraft transport efficiency TP that is calculated by formula (1). Aircraft transport performance has been improved by 2.8 times in 10 years of the design upgrading.

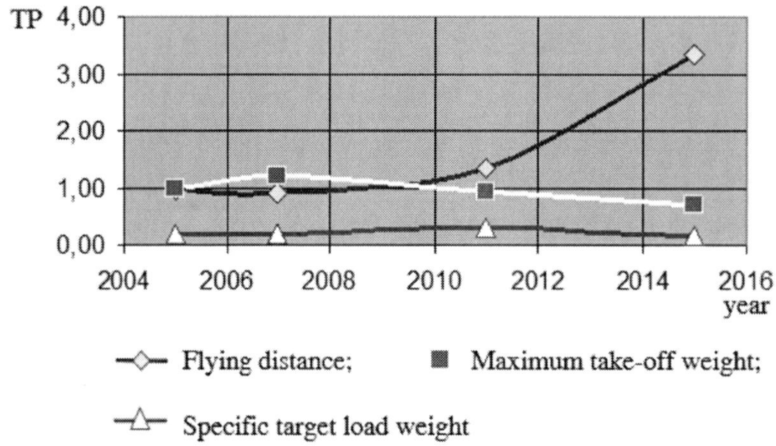

Figure 2. Historic All-Purpose Aircraft Performance Trends.

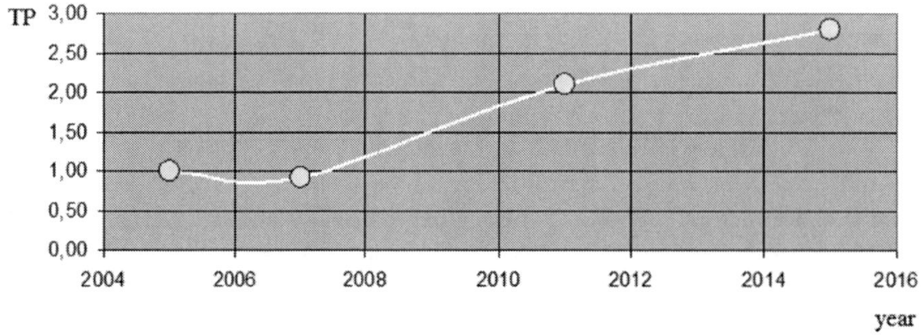

Figure 3. All-Purpose Aircraft Transport Performance Trends.

4. Conclusion

The study of all-purpose aircraft performance improvement trends with the study of performance changing trends shows that the new military aircrafts have not demonstrated a brainstorm design. Yet, engineering based on the aircraft development trends is quite promising.

References

[1] Yeager C M, Mishin V F and Lisacek N K 1983 *Aircraft Design: Textbook for universities* (Moscow: Engineering) p 616
[2] Pankov S E *The Role of military technologies in the development of the armament system of the Armed Forces of the Russian Federation: Department of advanced interspecies research and special projects of the Ministry of defense of the Russian Federation* (http://federalbook.ru/files/OPK/Soderjanie /OPK-10/III/Pankov.pdf)
[3] Simonavicius D P, Hot D N and Filippov V V 2017 Innovative activity in the Armed Forces of the Russian Federation *Proceedings of the all-army scientific and practical conference* (Saint Petersburg: Military Academy of communications) pp 270–272
[4] Tseplyaeva T P and Morozova O V 2009 Stages of development of unmanned aerial vehicles *Open information and computer integrated technologies* (Kharkov: National aerospace University) pp 10–16
[5] Fetisov V S, Neugodnikova L M, Adamovskiy V V and Krasnoperov R A 2014 *Unmanned Aviation: Terminology, Classification, Current State* (Ufa: Foton) p 217
[6] Grebenikov A G, Myalitsa A K and Parfenyuk V V 2008 *General Views and Characteristics of Unmanned Aircrafts: Guide Book* (Kharkov: National Aerospace University "Kharkov Aviation Institution") p 377
[7] Preobrazhenskiy N 2014 World market of UAVs *Military-industrial courier* (http://vpk.name/news/104319_mirovoi_ryinok_bespilotnikov.html) p 521
[8] Yuferev S Prospects for the development of Russian UAVs *ARMY. MAN. INFO* (http://armyman.info/stati/15140- perspektivy-razvitiya-rossiyskih-bpla.html)
[9] Kondratev A 2011 Prospects of development and application of unmanned and robotic means of armed struggle in the armed forces of leading foreign countries *Zarubezhnoe voennoe obozrenie* **5** 14–21
[10] Petrash V Ya and Turkin I K 2011 Review of Perspectives of Unarmed Aircraft Improvement Trends *Trudy MAI* **49** 17
[11] Vavrinyuk S A, Antonov A N and Sheludko M S 2017 Innovative activity in the Armed Forces of the Russian Federation *Proceedings of the all-army scientific and practical conference* (Saint Petersburg: Military Academy of communications) pp 73–76
[12] Boitsov A I and Yakovitsky S A 2017 Concept of combined use of manned and unmanned aerial vehicles in the US army *Innovative activity in the Armed Forces of the Russian Federation. Proceedings of the all-army scientific and practical conference* (Saint Petersburg: Military Academy of communications) pp 61–65
[13] Golubev I S and Levochkin S B 2008 *Aircraft Quality and Competitive Ability* (Moscow: MAI) p 223
[14] Myshkin L V 2006 *Forecasting the development of aviation technology: theory and practice* (Moscow: Fizmatlit) p 304
[15] Pripadchev A D, Sultanov N Z, Shatalova T N and Tikhonova O A 2009 *Economic Evaluation of Passenger Airplanes: Text Book* (Orenburg: State Educational Institution Orenburg State University) p 121
[16] General Atomics MQ-1 Predator *Warrior Lodge Media* https://warriorlodge.com/pages/general-atomics-mq-1-predator
[17] Petrash V L, Kovalenko A I and Kuchumov D V 2006 *Project evaluation of the extension of the far border of the zone of application for saving the mass of AIRCRAFT* (Moscow: MAI) p 96
[18] Sheinin V M 1962 *Weight and transport efficiency of passenger aircraft* (Moscow: Oborongiz

publ.) p 363
[19] Antonov O K and Tolmachev V I 1968 Transport plane today and tomorrow *Aviation and cosmonautics* pp 18–22
[20] Wentzel E S 1969 *Probability theory. Fourth edition, stereotypical* (Moscow: Nauka) p 576

Feasibility review of substitution a high-lift device by a telescopic wing with self-similarity of reynolds

A Bobkov[1,2] **and T Mitashova**[1]

[1] Komsomolsk-na-Amure State University, Lenin Prospect, 27 681013 Komsomolsk-on-Amur, Russia

[2] E-mail: bobkov822@yandex.ru

Abstract. This Paper describes results of the feasibility review of a telescopic wing application in unmanned aerial systems with self-similarity of Reynolds Re. Low Re is achieved through low-velocity flow of the lifting surface and small dimensions of an aircraft. Such combination is typical for a mini remotely piloted system according to UVS International classification. At low Reynolds the wing lift and high-lift device performance reduce. While in the high-lift mode, it is possible to adjust the wing area rather than the lift ratio. This can be achieved by a telescopic wing consisting of fixed and retractable sections. A telescopic wing is based on Morphing design with simultaneous wing geometry adapting to current flight parameters in one, two or three planes. The wing is simple by the design and can function as a high-lift device or aileron depending on whether the sections move out at the same time. The area of a telescopic wing with its sections moving out in the high-lift mode needs to be 2.0...2.18 times greater the initial area for Reynolds variation $(2.7...4)\cdot 10^5$.

1. Introduction

Lifting surface lift of an aircraft depends on few factors. The main factors are the boundary layer flow (represented by Reynolds Re) and wing geometry. Ideally, the latter must give a non-separating flow across the surface with maximum static pressure drop per square. Only one effective wing geometry / incidence angle variant correspond to each combination of aerofoil parameters: velocity, temperature, viscosity, density, pressure, ram air non-uniformity. To improve the wing performance, the wing geometry must change according to the instant ram air. This requirement is met by Morphing Wing. “Morphing” means transformation. The morphing design utilizes a bird-like wing when the wing changes its shape to match the flying conditions [1]. Morphing wing of an aircraft can also change its geometry: area, proportion and coning angle or cross section [2–5]. A well-known example of Morphing Wing in today’s aviation is a wing with variable sweep. It consists of fixed and retractable parts and adapts to flight speeds. It is commonly used in military aircrafts.

2. Morphing Wing Types

In the past ten years, people are more focusing on Morphing Wing because of novel smart materials, effective drives and actuators, and design concepts allowing 3D transformations of a wing. Figure 1 and figure 2 show how the concept options can be implemented by changing the wing geometry in horizontal and vertical planes [6].

Content from this work may be used under the terms of the Creative Commons Attribution 3.0 licence. Any further distribution of this work must maintain attribution to the author(s) and the title of the work, journal citation and DOI.
Published under licence by IOP Publishing Ltd

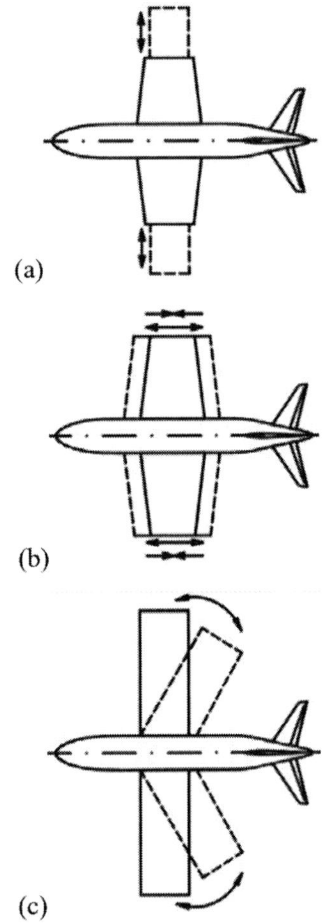

Figure 1. Horizontal Changes to Wing Geometry: (a) Variable Span; (b) Chord Morphing; (c) Sweep Morphing.

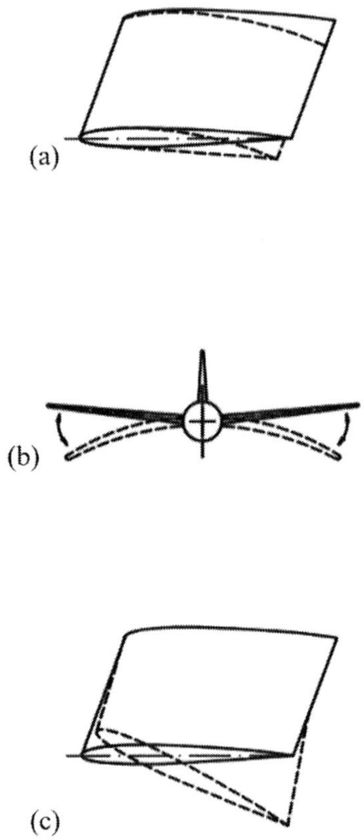

Figure 2. Vertical Changes to Wing Geometry: (a) Chord-Wise Morphing; (b) Span-Wise Morphing; (c) Twist Morphing.

Morphing Wing design requires a non-standard approach. Let us see a telescopic wing where the linear surface geometry is changed by retractable sections, ref. to figure 1a. The wing consists of a main fixed section and one or more retractable sections. Retractable sections allow changing the wing span and chord which growth reduces the lift-induced drag from the non-maneuverable aircraft aerodynamics point of view [7, 8]. In addition, the retractable sections make the wing area greater and the lift eventually increases.

3. Aerodynamic Characteristics of Small-Size Aircrafts

Today, remotely piloted small-size aircrafts are more widely used [9]. A designer who determines a small-size aircraft design and parameters must know the difference between wing aerodynamic characteristics of a small-size aircraft and full-size aircraft. For example, the paper [10] shows that a smaller size and takeoff weight result in low Reynolds for the wing chord Re_b that influences the lifting surface flow. When Reynolds drops to $Re_b \leq 4 \cdot 10^5$, the boundary layer flow becomes non-self-similar, laminar or transient laminar-turbulent [11, 12]. The layer loses its kinematic energy when the flow cannot overcome positive static pressure. The resulting laminar separation bubble effect on the lift surface affects the wing's ability to produce lift. The separation area can be over 5% of the wing chord [10] and proportionally reduce the lift. Thus, the aerodynamic specific is that a small-size aircraft wing's ability to generate lift per square represented by the lift ratio is less than for a full-size aircraft. This must always be considered when studying a small-size aircraft wing's ability to generate the required lift in takeoff and landing regimes.

4. Evaluation of Telescopic Wing Applicability in Small-Size Aircrafts for Takeoff and Landing Regimes

The studied Y_a lift control capabilities of an aircraft show that the lift varies under three factors and two of them directly or indirectly depend on the wing geometry. This appears from the lift empiric formula for the wing's subsonic flow:

$$Y_a = c_{ya} \cdot q_\infty \cdot S_w \tag{1}$$

- where c_{ya} is wing's lift ratio;
- q_∞ is ram air, Pa;
- S_w is wing area, m^2.

Coefficient c_{ya} is the wing's ability to transform ram air q_∞ into static pressured drop per wing's square. The better the wing shape, the higher coefficient c_{ya} is and therefore the smaller wing area S can be for a given aircraft weight.

For a constant wing geometry in plane i.e. S_w=const., it is possible to maintain the lift Y_a equal to or exceeding the aircraft weight in stationary flight regime by changing: either the wing lift ratio or ram air q_∞ (flaying speed). The takeoff-landing regime causes q_∞ to drop. It is possible to compensate it with S_w=const. by increasing the coefficient c_{ya} accordingly using the wing incidence angle and airfoil transformation and high-lift devices i.e. wing chord extending slats and flaps. Such control of the lift Y_a is power-consuming due to considerably increased drag yet it is common in full-size aircrafts in terms of the high coefficient c_{ay} [13–15].

For a small-size aircraft, the aerodynamic specific is the flow regime under low Re_b due to the small size and slow flying speed v. This causes the coefficient c_{ay} drop comparing to a full-size aircraft. The conventional approach to controlling the lift in takeoff and landing regime by high-lift devices must be adjusted for small-size aircrafts because a small-size aircraft wing chord extension (which is minor in absolute terms) gives minor lift increase in a small-size aircraft. This suggests an alternative to be studied: controlling an aircraft's Y_a by changing the area S_w of a telescopic wing with retractable sections. S_w was evaluated for a telescopic wing model with plane geometry parameters illustrated in figure 3.

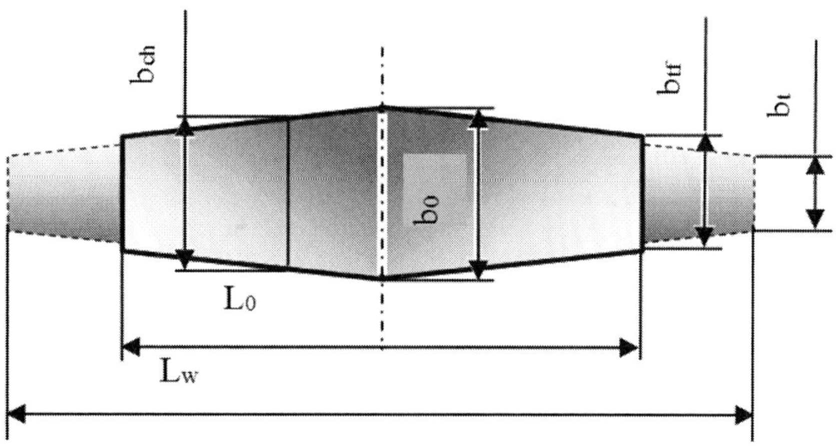

Figure 3. Telescopic Wing Geometry in Plane.

The evaluation applied the following parameters: L_0 – fixed section span, m; L_w – wing span with retracted sections, m; b_0 – root chord, m; b_{tf} – tip chord of fixed section, m; b_t – tip chord with retracted sections, m; b_{ch} – mean chord with retracted sections, m. The tapered wing area S_w is limited by the fixed and retractable sections and is the sum of the section areas:

$$S_w = 2 \cdot \left(\left(\frac{b_0 + b_{tf}}{2} \cdot \frac{L_0}{2} \right) + \left(\frac{b_{tf} + b_t}{2} \cdot \frac{L_w - L_0}{2} \right) \right) \tag{2}$$

A hypothetic small-size aircraft was studied for the following parameters [12]: the wing span L_0=1.5 m; chord b_{ch}=0.21, wing area S_w=0.315 m^2, airfoil TsAGI 731, wing's lift ratio in cruising regime c_{ya}=1.2. With chord b_{ch} and speed v=30 m/s Reynolds was Re_b= 3.9 10^5 and the aircraft weight was 18.35 kg.

For Reynolds Re_b= 2.7 10^5...3.9 10^5 typical for small-size aircraft takeoff-landing regime was applied to evaluate the compensation of the lift ratio c_{ya} drop 10% and 20% by increasing its area of the telescopic wing. The study results are illustrated in figure 4. Comparing to the initial variant, with non-retracted sections and given Reynolds Re_b range the required wing area increase was $\bar{\Delta}S_w$ $\bar{\Delta}S_w$=0...1.00 for the lift ratio drop by 10% to 0.9·c_{ya} and ΔS_w=0...1.18 for the lift ratio drop to 0.8·c_{ya}.

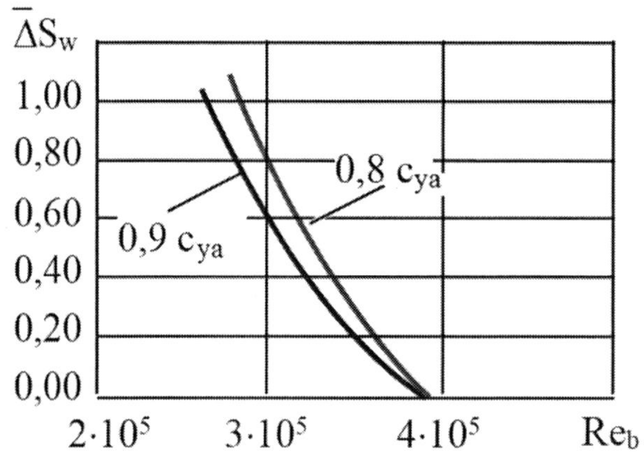

Figure 4. Relation between Wing Area Increase $\bar{\Delta}S_w$ and Reynolds Re_b.

5. Conclusion

The evaluation results are theoretical prediction. Field studies are required to evaluate aerodynamic characteristics of a telescopic wing with self-similarity of Reynolds. This will allow obtaining objective and trusted data on effectiveness of lift control by changing the wing of a small-size aircraft.

References

[1] Djojodihardjo H 2018 An assessment of a linear aerodynamic modeling of a generic flapping wing ornithopter *Int. J. Astronaut Aeronautical Eng.* **3** 017

[2] Martinez J M, Scopelliti D, Bil C, Carrese R and Marzocca P 2017 Design, analysis and experimental testing of a morphing wing *25th AIAA/AHS Adaptive Structures Conf.* DOI: 10.2514/6.2017-0059

[3] McGowan R, Washburn A E, L G Horta, R G Bryant, D E Cox, E J Siochi, S L Padula and N M Holloway 2002 *Recent results from NASA's morphing project proceedings of the SPIE 9th Annual Symposium on Smart Structure and Materials* (San Diego: California) DOI: 10.1117/12.475056

[4] Barbarino S, Bilgen O, Ajaj Rafic M, Friswell Michael I and Inman Daniel J 2011 A review of morphing aircraft *J. Intell. Master. Syst. Struct.* vol 22 823–877

[5] Neal D A, Good M G, Johnston C O, Robertshaw H H, Mason W H and Inman D J 2004 Design and wind-tunnel analysis of a fully adaptive aircraft configuration *45th AIAA/ASME/ASCE/AHS/ASC Structures, Structural Dynamics and Materials Conf.* (Palm Springs, California) pp 2004–1727

[6] Sofla A Y N, Meguid S A, Tan K T and Yeo W K 2010 Shape morphing of aircraft wing: Status and challenges *Materials and Design* **31** 1284–1292

[7] Krivokhatko I S and Sukhov V V 2014 Aerodynamic performance definition of UAV with complex planform wing at low Reynolds numder *Aviatsionno-kosmicheskaia tekhnika i tekhnologiia* **5(102)** 79–85

[8] Krivokhatko I S and Sukhov V V 2013 Experimental investigation of aerodynamic performance of a small uav with a telescopic wing *Actual Problems of Unmanned Air Vehicles Developments Proc. (APUAVD), IEEE 2nd Int. Conf.* (Kyiv) p 17

[9] Iliushko V N, Mitrakhovich M M, Samkov A V, Silkov V I, Solovev A V and Strelnikov V I 2009 *Bespilotnye Letatelnye apparaty: metodiki priblizhennykh raschetov osnovnykh parametrov i kharakteristik* (Kiev: TSNII VVT VS Ukrainy) p 302

[10] Pelletier A and Mueller T J 1999 Aerodynamic force / moment measurement at very low *Reynolds Proc. of the 46th annual conf. of the Canadian Aeronautics and Space Institute* (Montreal) pp 59–68

[11] Schmitz F W 1963 *Aerodinamika malykh skorostei* (Moscow: DOSAAF) p 60

[12] Voronich I V, Kolchev S A, Panchuk D V, Pesetskii V A, Silkin A A, Tkachenko V V and Nguen Tkhan Tung 2019 On the peculiarities of aerodynamics of a small-sized aircraft of a normalschem *Trudy MAI* p 109 DOI: 10.34759/trd-2019-109-8

[13] Bragin N N, Bolsunovskii A L, Buzoveria N P, Gubanova M A, Skomorokhov S I and G VKhoziainova 2013 Research to improve the aerodynamics of the takeoff and landingmechanization of the wing of a passenger plane *Uchenye zapiski TsAGI* **44(4)** 1–14

[14] Grechikhin L I, Laptsevich A A and Kuts N G 2012 Aerodynamics of high-lift wing *Nauchnyi vestnik UVAU GA(I)* vol 4 8–20

[15] Gubskii V V 2013 The use of adaptive wing mechanization in a light transport aircraft *Trudy MAI-68*

Simulating the pyrolysis process of sludge-lignin with the production of active carbons

N I Bogdanovich[1], K B Vorontsov[1], B C Labudin[1,3] and K A Varenik[2,4]

[1] North (Arctic) Federal University named after M.V. Lomonosov, the Northern Dvina Embankment 17 163002 Arkhangelsk, Russia
[2] Yaroslav-the-Wise Novgorod State University, ul. B. St. Petersburgskaya, 41 173003 Veliky Novgorod, Russia

[3] E-mail: sevned@mail.ru
[4] E-mail: vkirillv89@mail.ru

Abstract Using the method of the planned experiment, mathematical models that establish a relationship between the technological parameters of the pyrolysis process of sludge-lignin with the characteristics of a porous surface of the synthesized active carbons are obtained. The nature of the influence of temperature and duration of the pyrolysis, the dosage of activating agent on the yield of carbon, the volume and specific surface of the pores are determined. The optimal conditions for obtaining adsorbents from sludge-lignin are established; the specific BET surface area was about 2000 m^2/g.

1. Introduction

Active carbons (AC) are used in various fields of human activity: medicine and pharmaceuticals, treated water production and wastewater treatment, electronics, food, chemical and other industries. Along with wood, technical lignins formed at wood processing and pulp and paper industries (PPI) can be used as raw materials for AC production: hydrolysis [1–3] and sulfate lignin [3–7], lignosulfonates [8, 9], black liquor [10, 11].

The raw material for the synthesis of carbon adsorbents can also be lignin sludge, a precipitate formed in the processes of coagulation treatment of lignin-containing wastewater and water treatment [12]. Pyrolysis of sludge-lignin is the most effective way of processing it to produce a marketable product – activated carbons [8], it will contribute to a wider implementation of local coagulation systems for wastewater treatment of pulp and paper industry and, as a result, will improve the environmental situation.

An urgent task in the production of activated carbons is the synthesis of adsorbents with desired properties and characteristics, which should include in the first place the parameters that determine the porous structure of adsorbents: volume and specific surface. This problem can be solved by developing mathematical models that connect the characteristics of activated carbon (AC) with the conditions for obtaining it. The modeling of the pyrolysis process of technical lignins can be carried out using the methods of the planned experiment [2, 13]; similar work has not been carried out in relation to pyrolysis of sludge-lignin.

The purpose of the work is to obtain mathematical models that establish the dependence of the characteristics of activated carbons from sludge-lignin with the pyrolysis conditions by the method of

Content from this work may be used under the terms of the Creative Commons Attribution 3.0 licence. Any further distribution of this work must maintain attribution to the author(s) and the title of the work, journal citation and DOI.
Published under licence by IOP Publishing Ltd

the planned experiment, as well as a comparative analysis of the characteristics of AC from sludge-lignin with the characteristics of adsorbents from other types of technical lignins.

2. Research objects and research methods

Sludge-lignin was obtained in laboratory conditions by treating lignin-containing wastewater with aluminum oxychloride, based on the optimal conditions and operating parameters established by us earlier [14], dried to an air-dry state, and subjected to pyrolysis. Pyrolysis was carried out by thermochemical activation, the advantage of which is the production of carbons with high characteristics of porous surface [15]. Sodium hydroxide was used as an activating agent, as an effective reagent for the pyrolysis of carbon materials having an disordered structure [3].

Previous studies have established that an experimental study of the pyrolysis process can be carried out using a three-factor rotatable central compositional plan of the second order [1, 2]. This plan was used by us to simulate the pyrolysis process of sludge-lignin. The regression equation describing the experimental data in accordance with the specified plan has the following form:

$$y = b_0 + b_1X_1 + b_2X_2 + b_3X_3 + b_{12}X_1X_2 + b_{13}X_1X_3 + b_{23}X_2X_3 + b_{11}X_1^2 + b_{22}X_2^2 + b_{33}X_3^2$$

For the experiment, the most significant thermochemical activation factors were selected: X_1 is the pyrolysis temperature (T, °C), X_2 is the duration of the pyrolysis (τ, min), and X_3 is the dosage of the activating agent (D,%). The following characteristics were used as output parameters: coal yield (Y, %); micropore volume calculated according to the Dubinin-Radushkevich equation of micropore volumetric filling theory (V_{DR}, cm^3/g); micropore volume (V_{BET}, cm^3/g), total pore volume (VΣ, cm^3/g) and the specific BET surface (S_{BET}, m^2/g) calculated according to the Brunauer-Emmett-Teller (BET) equation of the theory of polymolecular adsorption; Mesopore volume determined by the Barrett-Joyner-Halenda method (V_{BJH}, cm^3/g); specific surface of meso- and macropores determined by the t-plot method (S_M, m^2/g).

The characteristics of the porous structure of active carbons were determined by low-temperature nitrogen adsorption (at a temperature of 77 K) using the ASAP 2020MP automated system (Micromeritics).

Based on the results of preliminary studies, we determined the levels of variation of the experimental factors (table 1) and, in accordance with the chosen plan, compiled a planning matrix (table 2).

Table 1. Factor variation levels.

Factors	Factor levels				
	–1.682	–1	0	1	1.682
Pyrolysis temperature	650	670	700	730	750
Pyrolysis duration	30	42	60	78	90
Sodium hydroxide dosage	120	144	180	216	240

Table 2. Experiment planning matrix.

Experiment №	Coded notations			Actual values		
	X_1	X_2	X_3	X_1	X_2	X_3
1	–1	–1	–1	670	42	144
2	1	–1	–1	730	42	144
3	–1	1	–1	670	78	144
4	1	1	–1	730	78	144
5	–1	–1	1	670	42	216

6	1	–1	1	730	42	216
7	–1	1	1	670	78	216
8	1	1	1	730	78	216
9	– α	0	0	650	60	180
10	α	0	0	750	60	180
11	0	– α	0	700	30	180
12	0	α	0	700	90	180
13	0	0	– α	700	60	120
14	0	0	α	700	60	240
15	0	0	0	700	60	180
16	0	0	0	700	60	180
17	0	0	0	700	60	180
18	0	0	0	700	60	180
19	0	0	0	700	60	180
20	0	0	0	700	60	180

3. Results and discussion

We conducted experimental studies, the obtained data we used to calculate the coefficients of the regression equations and the development of statistical models. Models were checked according to the Fisher criterion: all models adequately describe the processes under study, since for all the equations obtained, the calculated value of the Fisher criterion does not exceed the tabulated one (at a significance level of 0.05). Using the equations of mathematical models, we constructed response surfaces that clearly demonstrate the influence of experimental factors on the output characteristics.

Mathematical models:

$$Y = 19.8 - 2.0X_3 - 3.5X_1X_3 - 3.1X_1^2 + 2.6X_3^2 \quad (1)$$

$$V_{DR} = 0.31 + 0.13X_3 - 0.06X_1X_2 + 0.06X_1X_3 + 0.02X_2^2 \quad (2)$$

$$V_{BET} = 0.36 + 0.15X_3 - 0.06X_1X_2 + 0.07X_1X_3 \quad (3)$$

$$V_{BJH} = 0.08 + 0.07X_3 + 0.03X_1X_3 + 0.05X_3^2 \quad (4)$$

$$V_{\Sigma} = 0.45 + 0.23X_3 - 0.08X_1X_2 + 0.1X_1X_3 + 0.08X_3^2 \quad (5)$$

$$S_{BET} = 788 + 374X_3 - 138X_1X_2 + 162X_1X_3 + 41X_2X_3 - 88X_1^2 + 69X_2^2 + 86X_3^2 \quad (6)$$

$$S_M = 145 - 45X_2 + 119X_3 - 55X_1X_2 + 87X_1X_3 + 63X_2^2 + 38X_3^2 \quad (7)$$

The dosage of sodium hydroxide negatively affects the AC output; the mutual influence of the pyrolysis temperature and alkali dosage is also negative; the coefficient at X_1^2 is negative, and at X_3^2 is positive, this means that the response surface will be convex with respect to the pyrolysis temperature and concave with respect to the dosage of sodium hydroxide (1). This generally agrees with the data of previous studies [1], but the amount of coal output from sludge-lignin is somewhat lower and amounted to 20–30%, which is associated with the presence of a larger amount of inorganic substances (aluminum compounds) in the composition of the initial sludge-lignin.

The mathematical models obtained for the micropore volume calculated by the Dubinin-Radushkevich (2) and BET (3) equations are almost identical; the presence of a quadratic coefficient in equation (2) gives some curvature of the response surface relative to factor *X2* — the duration of the pyrolysis (figure 1b). The dependences of these parameters on the pyrolysis conditions are of a similar nature: the factor that has the most significant positive effect is the dosage of the activating agent; at temperatures below 700°C, with an increase in the duration of pyrolysis, the micropore volume increases, and at higher temperatures it decreases (figure 1a, figure 2a). At a constant pyrolysis temperature (figure 1b, figure 2b) with increasing τ, the micropore volume decreases slightly. In general, the micropore volume of the obtained AC samples reaches 0.8 cm^3/g.

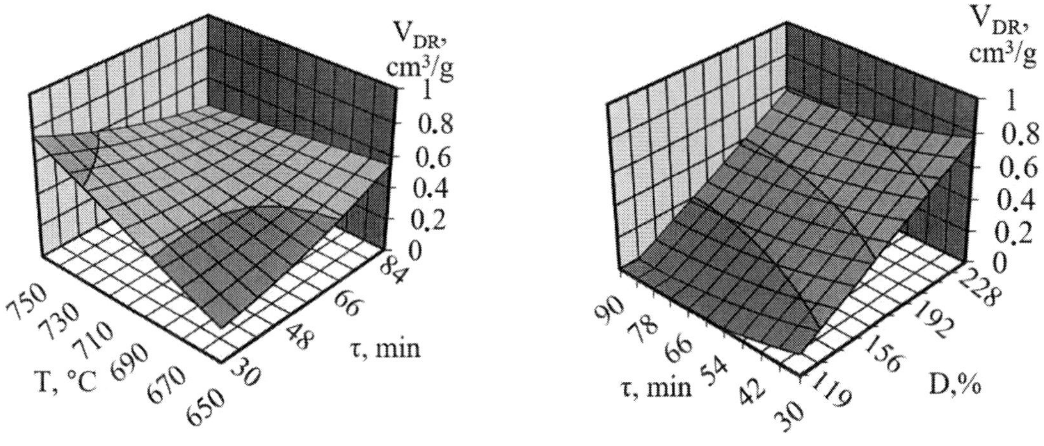

a) b)

Figure 1. V_{DR} versus AC production conditions: at a constant dosage of NaOH (a) and at a constant pyrolysis temperature (b).

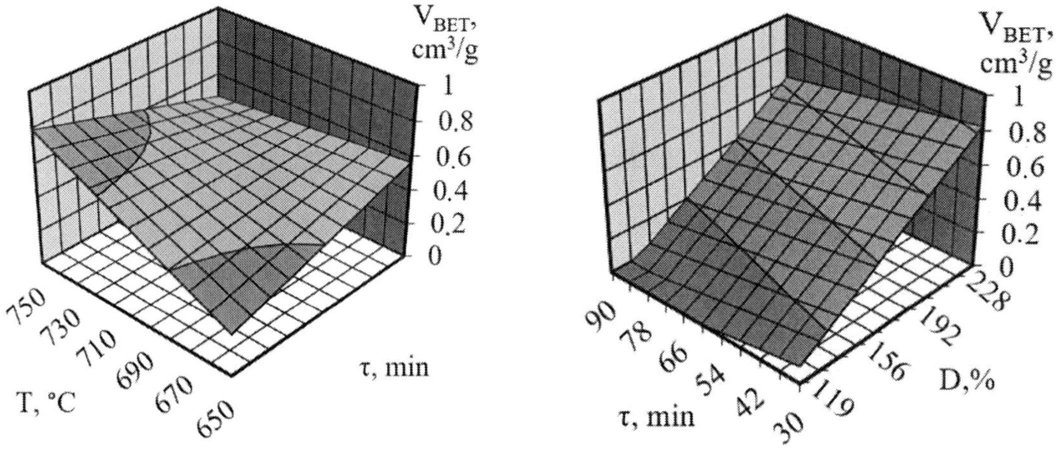

a) b)

Figure 2. The dependence of V_{BET} on the conditions for obtaining AC: at a constant dosage of NaOH (a) and at a constant pyrolysis temperature (b).

The dosage of sodium hydroxide has a positive effect on the mesopore volume at values of the latter 160% and higher; in this interval, the response surface has a significant degree of curvature (figure 3b), which is due to the quadratic coefficient b_{33} in equation (4). The duration of the pyrolysis process does not affect V_{BJH}; with an increase in the pyrolysis temperature, the volume of mesopores increases linearly in the entire temperature range (figure 3a). The value of V_{BJH} reaches 0.2–0.3 cm^3/g.

The dependences constructed for the total pore volume (figure 4) and micropore volume (figure 1, figure 2), in general, have a similar character. This is due to the fact that a significant proportion of the pore volume (up to 80%) falls on micropores. Thus, the adsorbents we synthesized from sludge-lignin have mainly a microporous surface structure.

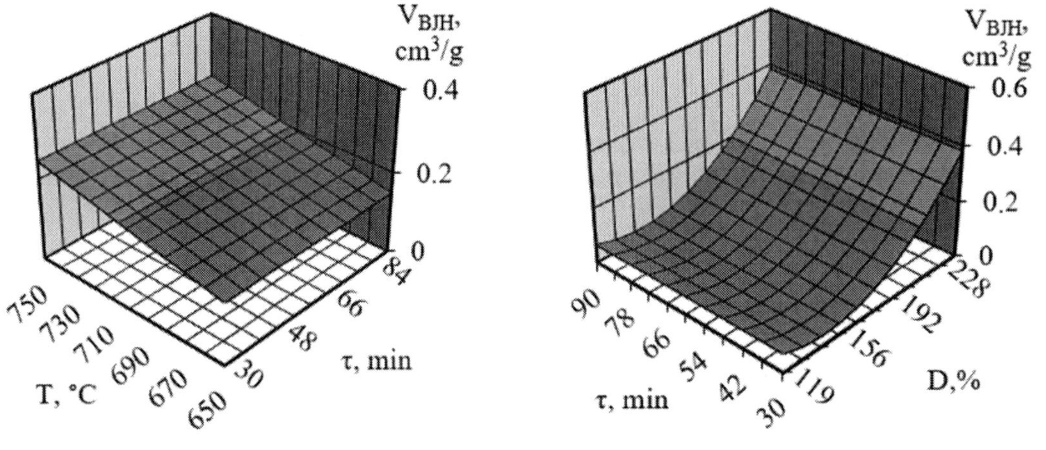

a) b)

Figure 3. The dependence of V_{BJH} on the conditions for obtaining AC: at a constant dosage of NaOH (a) and at a constant pyrolysis temperature (b).

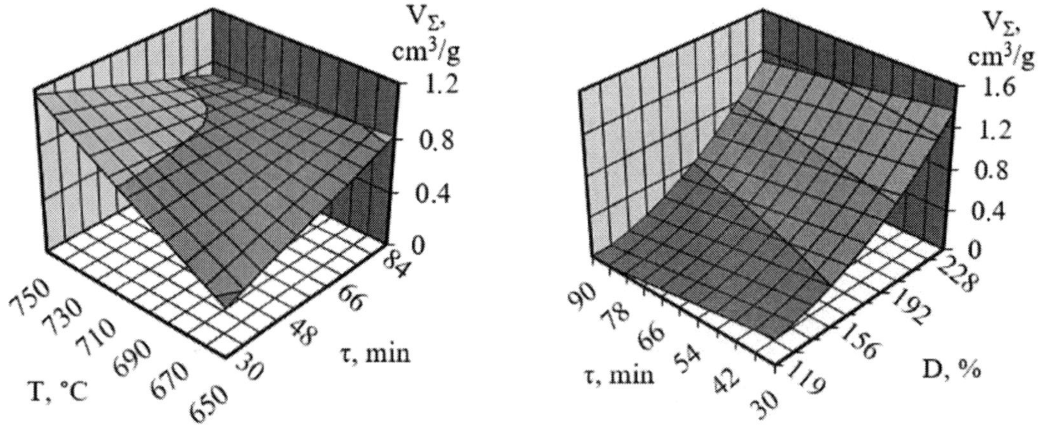

a) b)

Figure 4. The dependence of V_{Σ} on the conditions for obtaining AC: at a constant dosage of NaOH (a) and at a constant pyrolysis temperature (b).

The specific surface of S_{BET} has a significant degree of curvature due to the presence of quadratic effects coefficients in the equation, the dosage of sodium hydroxide has a major and positive effect, therefore, the mutual effects of the pyrolysis temperature and alkali dosage and the duration of pyrolysis and alkali dosage also have a positive effect, and the mutual influence of temperature and the duration of the pyrolysis process is negative (6).

From the presented data (figure 5), it is possible to establish the optimal conditions for the pyrolysis of sludge-lignin, based on the fact that the obtained samples of activated carbon have a developed microporous structure: temperature – 700 °C, duration – 80 min, dosage of sodium hydroxide – 200%. The specific surface area was about 2000 m^2/g.

The specific surface of meso- and macropores under optimal conditions was about 500 m^2/g (figure 6). The size of this parameter is positively affected by the dosage of alkali and negatively by the duration of the pyrolysis process (7).

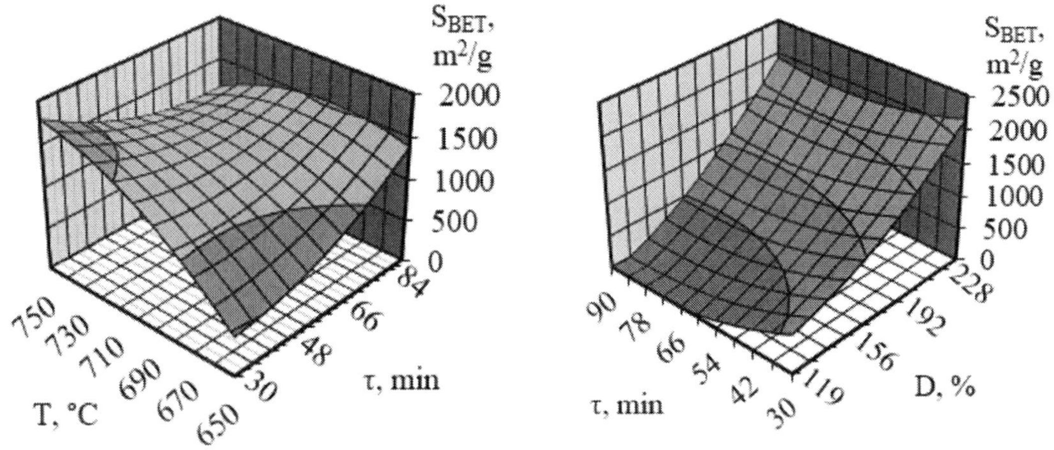

a) b)

Figure 5. The dependence of S_{BET} on the conditions for obtaining AC: at a constant dosage of NaOH (a) and at a constant pyrolysis temperature (b).

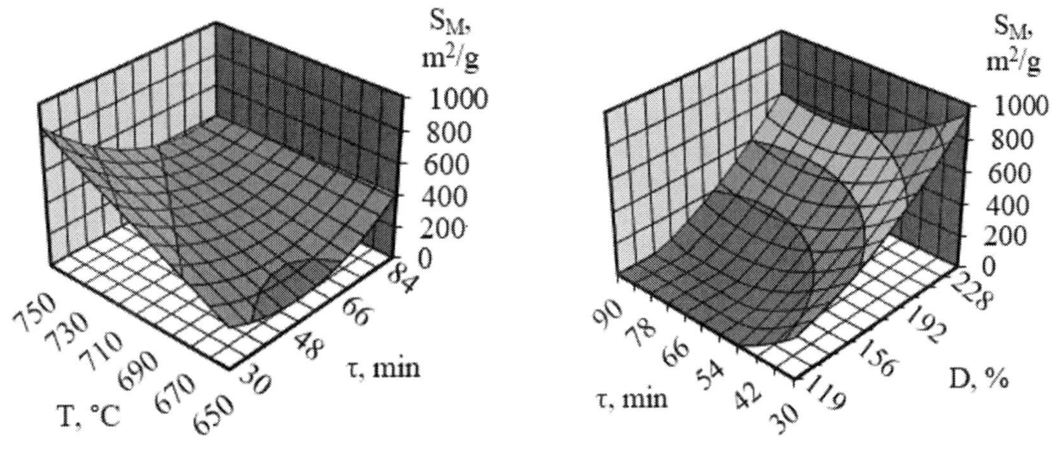

a) b)

Figure 6. The dependence of S_M on the conditions for obtaining AC: at a constant dosage of NaOH (a) and at a constant pyrolysis temperature (b).

Table 3. Active carbons of thermochemical activation with sodium hydroxide.

Raw material	V_{BET}, cm^3/g	S_{BET}, m^2/g
Sludge-lignin	0.8	2000
Sludge-lignin [12]	–	98–112
Hydrolytic lignin [16]	–	1300
Sulphate lignin [5]	1.2	2400
Sulphate lignin [7]	0.8–1.08	1900–2430
Sulphate lignin [17]	–	800–2000
Sulphate lignin [18]	0.75	1440

Considering the generally accepted characteristics of adsorbents – the volume and the specific BET surface area, it can be noted that the samples of coal from sludge-lignin synthesized by us are practically inferior, and in some cases, superior to ACs obtained from various types of technical lignins by the method of thermochemical activation with sodium hydroxide (table 3).

4. Conclusion

For the implementation of specific tasks requires activated carbons with certain characteristics. First of all, they include the characteristics of the porous surface – volume and specific surface. By changing the technological parameters of the pyrolysis process, one can obtain adsorbents with desired properties and characteristics.

Our mathematical models link the conditions for the synthesis of AC from sludge-lignin with their characteristics and allow one to simulate the pyrolysis process, while receiving adsorbents with pre-predicted properties.

It was found that the dosage of the activating agent (sodium hydroxide) has the most significant effect on all the output parameters that we studied: in all models, except the AC output equation, this factor is positive. Thus, increasing the dosage of alkali, it is possible to obtain coals with a highly developed porous structure, however, their output will decrease. The temperature and duration of pyrolysis have a smaller effect on the characteristics of activated carbons from sludge-lignin. At the same time, a simultaneous increase in these factors leads to a slight decrease in V_{BET} and S_{BET} and an increase in S_M. Using the effect of the mutual influence of T and τ, we can to some extent change the ratio of the parameters of micropores and mesopores.

Based on the ratio of the volumes of micropores and mesopores and the specific surfaces of S_{BET} and S_M, it was concluded that the structure of the obtained AC is mainly microporous.

According to the nature of the influence of experimental factors on the S_{BET} value, the optimal operating parameters of the sludge-lignin pyrolysis were established: T – 700 °C, τ – 80 min, D – 200 %. Active carbons obtained under optimal conditions have high porous surface characteristics (S_{BET} = 2000 m^2/g) and, in general, are not inferior in their properties to adsorbents synthesized from other types of technical lignins by thermochemical activation with sodium hydroxide.

References

[1] Beletskaya M G and Bogdanovich N I 2013 The formation of adsorption properties of nanoporous materials by thermochemical activation *Khimija Rastitel'nogo Syr'ja* **3** 77–82 [In Russ.]

[2] Romanenko K A, Bogdanovich N I and Kanarskiy A V 2017 Obtaining of Activated Carbons by Pyrolysis of Hidrolytic Lignin *Lesnoy zhurnal* **4(358)** 162–171 [In Russ.]

[3] Carrott P J M *et al.* 2007 Lignin–from natural adsorbent to activated carbon: a review *Bioresource technology* vol 98 **12** 2301–2312

[4] Khezami L *et al.* 2005 Production and characterisation of activated carbon from wood components in powder: Cellulose, lignin, xylan *Powder Technology* vol 157 **1–3** 48–56

[5] Torné-Fernández V *et al.* 2009 Statistical optimization of the synthesis of highly microporous carbons by chemical activation of Kraft lignin with NaOH *Journal of Chemical & Engineering Data* vol 54 **8** 2216–2221

[6] Bedmohata M A *et al.* 2015 Adsorption capacity of activated carbon prepared by chemical activation of lignin for the removal of methylene blue dye *International Journal of Advanced Research in Chemical Science (IJARCS)* vol 2 **8** 1–13

[7] Kwiatkowski M, Fierro V and Celzard A 2017 Numerical studies of the effects of process conditions on the development of the porous structure of adsorbents prepared by chemical activation of lignin with alkali hydroxides *Journal of colloid and interface science* vol 486 277–286

[8] Bogdanovich N I 1998 Pyrolysis of technical lignins *Lesnoy zhurnal* **2–3** 120–132 [In Russ.]

[9] Vaysman Y I, Glushankova I S, Shirinkina E S and Davletova S F 2018 Method for processing

lignin-containing wastes from the paper industry to produce sorbents for wastewater treatment *Teoreticheskaya i prikladnaya ekologiya* **3** 93–99 [In Russ.]

[10] Zhao X-Y *et al.* 2010 High surface area activated carbon prepared from black liquor in the presence of high alkali metal content *Journal of chemical engineering of Japan* vol 43 **12** 1029–1034

[11] Zhang J *et al.* 2011 Spherical microporous/mesoporous activated carbon from pulping black liquor *Journal of Chemical Technology & Biotechnology* vol 86 **9** 1177–1183

[12] Amriani F *et al.* 2015 Activated carbon from lignin-based black liquor coagulated by polyaluminium chloride Procedia Chemistry vol 16 134–140

[13] Nieto-Delgado C and Rangel-Mendez J R 2011 Production of activated carbon from organic by-products from the alcoholic beverage industry: Surface area and hardness optimization by using the response surface methodology *Industrial crops and products* vol 34 **3** 1528–1537

[14] Sedova E L, Vorontsov K B and Burkova S A 2019 Influence of coagulation treatment on the efficiency of lignin containing wastewater purification *Lesnoy zhurnal* **4** 159–167 [In Russ.]

[15] Chistyakov A V and Tsodikov M V 2018 Methods for preparing carbon sorbents from lignin *Russian Journal of Applied Chemistry* vol 91 **7** 1090–1105 [In Russ.]

[16] Ponomarev N and Sillanpää M 2019 Combined chemical-templated activation of hydrolytic lignin for producing porous carbon *Industrial crops and products* vol 135 30–38

[17] Hayashi J *et al.* 2000 Preparation of activated carbon from lignin by chemical activation *Carbon* vol 38 **13** 1873–1878

[18] Hu S and Hsieh Y L 2013 Ultrafine microporous and mesoporous activated carbon fibers from alkali lignin *Journal of Materials Chemistry* A vol 1 **37** 11279–11288

Methodology for static tuning of the HEV fuel flow measuring system

S T Bozhkov[1], I Milenov[1], R V Petrov[2], V S Leontiev[2] and P Bozhkov[3]

[1] Faculty of Machinery and Construction Technologies in Transport, Faculty of Telecommunication and Electrical Equipment in Transport, Todor Kableshkov University of Transport, Sofia, Bulgaria
[2] Yaroslav-the-Wise Novgorod State University, 41, ul. B. St.Peterburgskaya, 173003, Velikiy Novgorod, Russia
[3] Olimex Ltd, Plovdiv, Bulgaria

[4] E-mail: roman.petrov@novsu.ru

Abstract. The modern automobiles are subject of more stringent requirements [1] corresponding to power, torque, fuel economy and ecology legislations, which is led to designing of new power sources and constructions, such as the hybrid electric vehicles (HEV). They are moving by the energy [2], which is ensured by the internal combustion engine (ICE) and the battery. The main factor in this area is the HEV fuel system, which is controlled by the electronic control unit (ECU) [1]. The electronic control of the Spark Ignition Engines (SI engines), as well as the Direct Ignition Engines (DI engines) is based on the certain sensors signals, program maps and management algorithms. The result in this electronic control is the management of the fuel injectors. The management of the fuel injectors consists in the start of injection, injection duration, number of injection events, injection advance, injection pressure etc. Moreover, the fuel consumption and fuel efficiency are the main factors, which are determining the HEV advances. The learning and measuring the HEV fuel consumption, as well as, the conventional automobiles, is the ground for achievement of quality results in the education of the automotive engineers, as well as for obtaining of scientific researching for developments and innovations. Significant meaning in this concept has the real, live and practical performance with the help of testing equipment and test-benches. This paper renders the methodology of static tuning the Fuel Flow Measuring System EFMS100 on the test bench SAV-1 with the support of controller Matrix MIAC MI0245 and Flowcode 7 softwar.

1. Introduction

The hybrid electric vehicles or simply hybrid vehicles use both electric motors and an SI engine for delivering the propulsion power [3]; these vehicles have lower emissions compared to a similarly sized conventional vehicle, resulting in less environmental pollution. The ICE used in a HEV is, of course, downsized compared to an equivalent vehicle engines. The SI engine in combination with the electric motor and an energy storage unit battery provide an extended range for HEV and bring down pollution. The HEV serves as a compromise for the environmental pollution problem and the limited range capability of today's purely electric vehicle (EV). The HEV energy efficiency is the main factor for its advantage and evaluating. This efficiency depends directly from the HEV fuel consumption. It is very important to measure the fuel consumption in correct manner and with suitable equipment to

Content from this work may be used under the terms of the Creative Commons Attribution 3.0 licence. Any further distribution of this work must maintain attribution to the author(s) and the title of the work, journal citation and DOI.
Published under licence by IOP Publishing Ltd

IOP Conf. Series: Materials Science and Engineering **939** (2020) 012015 doi:10.1088/1757-899X/939/1/012015

obtain the correct results. Meanwhile, the fuel consumption metering equipment must be calibrated and adapted to the current tests.

Also, to ensure optimal operation of the HEV engine, it is necessary to use highly sensitive and accurate sensors of various types. One of these types of sensors is current sensors. The current sensors in HEV are used to monitor current in various units of the unit, the range of measured currents can reach 500 A. One of the possibility for solving this problem can be current sensors based on the magnetoelectric (ME) effect, made in the form of a clip, with subsequent signal processing on the head unit of the car. A sensitive element of such a sensor are materials based on multiferroic layered structures. The ME current sensor can operate in different modes, in resonant and non-resonant. It is assumed that the use of ME current sensors in HEV will optimize the operation of the car.

2. Structure

The fuel consumption metering equipment is specialized set of fuel sensors, metering units and display units. I accordance with the fuel to use there is different kind of sensors and metering techniques. In this case the EFMS100 [4] metering system is used. It consists from two sets of fuel sensors, metering units and display units (figure 1).

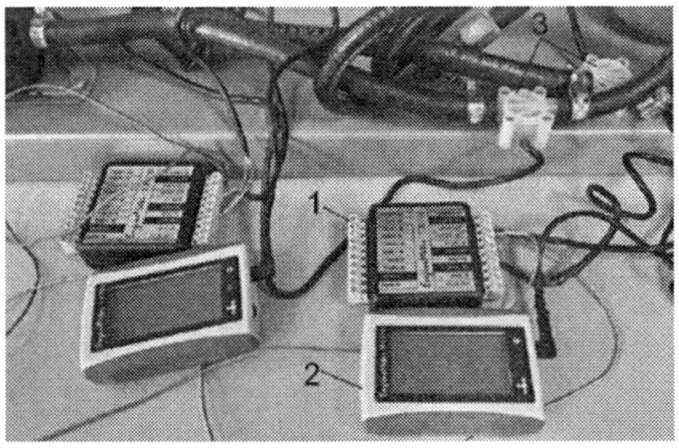

Figure 1. EFMS fuel flow metering system [4]: 1 – metering unit; 2 – display unit; 3 – fuel flow sensor.

The fuel metering system must be calibrated according to the amount of the fuel flow in the current application. In the case of metering the HEV fuel consumption it is very useful to apply test benches, which can adjust and calibrate the equipment according to the flow range. Such test bench is the test bench SAV-1 [1] for automated management of the automotive gasoline fuel injectors SAV-1 (figure 2). The test-bench can be programmed and managed by the Flowcode 7 software [5], which ensures adjustment to the real work mode of the fuel injectors in the modern automobiles and HEVs.

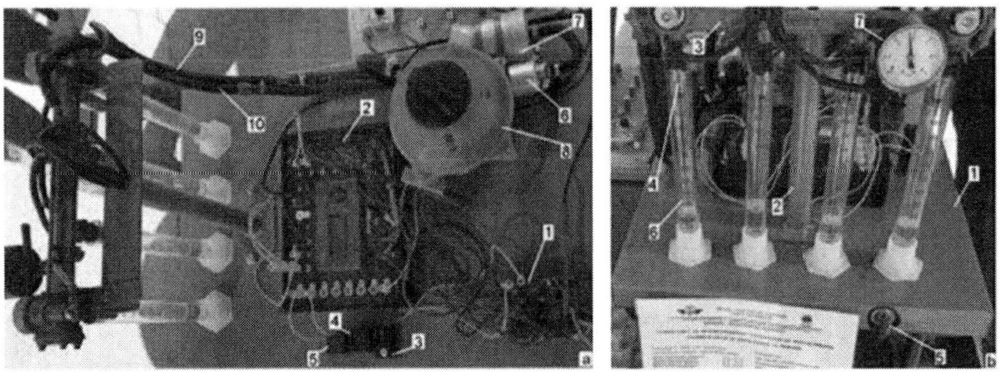

Figure 2. Common view of the test bench SAV-1 for automated management of the automotive gasoline fuel injectors: a) 1 – control panel; 2 – controller MI0245; 3 – fuses; 4 – fuel pump relay; 5 – main relay; 6 – fuel pump; 7 – fuel filter; 8 – fuel tank; 9 – pump line; 10 – return line b) 1 – base plate; 2 – fuel rail bar; 3 – fuel rail; 4 – fuel injector; 5 – ignition key; 6 – metering glass; 7 – manometer.

3. Methodology

The EFMS100 equipment is connected to the test bench SAV-1. The fuel sensors 3 (figure 1) are connected to the pump line 9 and return line 10 (figure 2a). The fuel to use is gasoline, which is in the fuel tank 9. The gasoline is pumped by the fuel pump 6 and is pressurized in the fuel rail 3 (figure 2b). The gasoline pressure is maintained by the fuel pressure regulator in limits of 3-4 bar. The gauge 7 measures the pressure. The controller MI0245 2 (figure 2a) manages the fuel injectors 4 (figure 2b). The Flowcode 7 software develops the management program.

The management program determines the opening and closing time of the injectors, i.e. the injection duration and injection advance.

The metering glasses 6 are used to measure the fuel sprayed from the injectors. The metering glasses are an A class according to the DIN 12680. The quantity of sprayed fuel in the metering glasses is compared to the quantity registered by the EFMS100 system. The whole equipment is powered form the 12V battery or by an adaptor from the electric set.

The two fuel sensors are metering the fuel flow in the pump line and the return line. The measured values of the sensors are registered in the metering units and displayed on the display units. The difference between measured values is the real fuel quantity which is essential for the calibration. The static tuning, or calibration is carried on by the values from the pump line fuel sensor. In this case the test bench is set to pump mode, i.e. the regulator does not permit the fuel to the return line. The dynamic tuning is carried on by the values from the two sensors, i.e. by the differential value of the sensors. The results from this tuning is object of another researching.

The static calibration is performed in the four modes, which have main automotive application. These modes are:

1) Continuous – all injectors are activated and produce continuous spraying;
2) Simultaneous – all injectors are pulse activated and produce pulse spaying;
3) Semi-sequential – the injectors are pulse activated in groups and produce group pulse spraying;
4) Sequential – the injectors are individual pulse activated and produce individual pulse spraying.

The pulse has three parameters:

1) ON_{time} – the time in which the injector is opened;
2) OFF_{time} – the time in which the injector is closed;
3) Duty cycle D:

$$D = \frac{ON_{time}}{ON_{time}+OFF_{time}}.100,\% \qquad (1)$$

The timing of the ON_{time} and OFF_{time} are varying from 5 to 45 ms at the step of 5 ms, and D is varying from 10 to 90%. There are the following combinations, which are represented on the table 1. These values are corresponding to the real injection automotive processes.

Table 1. Timing combinations.

ON_{time}, ms	OFF_{time}, ms	D, %
5	45	10
10	40	20
15	35	30
20	30	40
25	25	50
30	20	60
35	15	70
40	10	80
45	5	90

The quantity of fuel sprayed in the metering glasses is set to be 100 ml. After that, the test bench SAV-1 is paused and is performing the comparison between the sprayed and registered fuel quantity. Then is performed the EFMS100 adjustment if required. The adjustment is made by the EFMS100 menu, which gave the access to the parameter *PVR* – pulse/volume ratio. This parameter is calculated the by the formulae:

$$PVR = \frac{FPC}{Q_f}, \tag{2}$$

where *FPC* is the fuel flow pulse count, which is the number of revolutions of the flow sensor turbine to the measured fuel quantity; Q_f. – the measured fuel quantity.

4. Results

After the performed experiments is obtained the following results, which are displayed in graphic diagrams. The fig.3 displays the fuel metering results during the Continuous mode. In this case the fuel injectors are spraying continuously. The characteristic shows equal dependence between metering glasses and calibrated EFMS100 system. The deviation of the experiment 3 is because of fuel flow fluctuations, which has chance character.

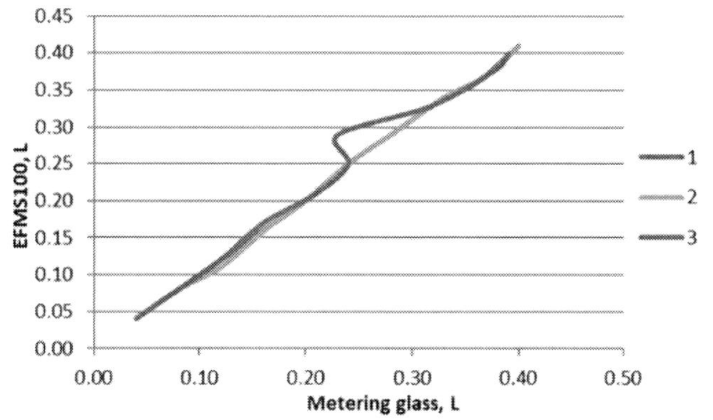

Figure 3. Fuel flow metering characteristic at Continuous mode:
1, 2, 3 – number of experiments.

At the next figure 4 is displayed the fuel flow characteristic during the Simultaneous mode. At this mode, the injectors work at varying the ON_{time} and OFF_{time} timing as mentioned above.

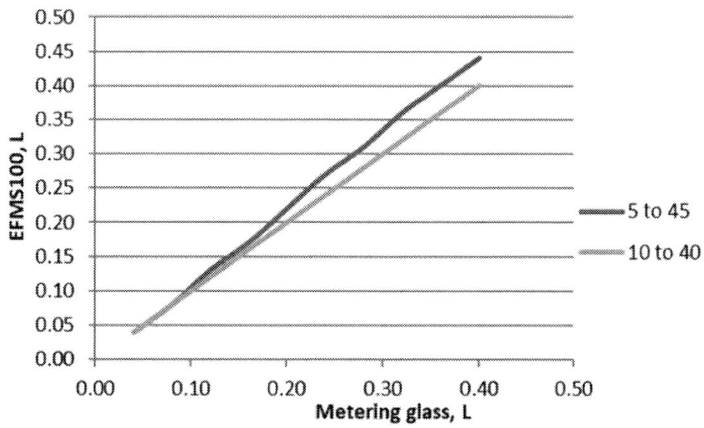

Figure 4. Fuel flow metering characteristic at Simultaneous mode.

The remain combinations are the same with the second combination 10 to 40 ms.

The Semi-sequential mode is displayed on the figure 5. The timing of the ON_{time} and OFF_{time} is the same as previous Simultaneous mode. All the combination shows equal characteristic and are layered one above other.

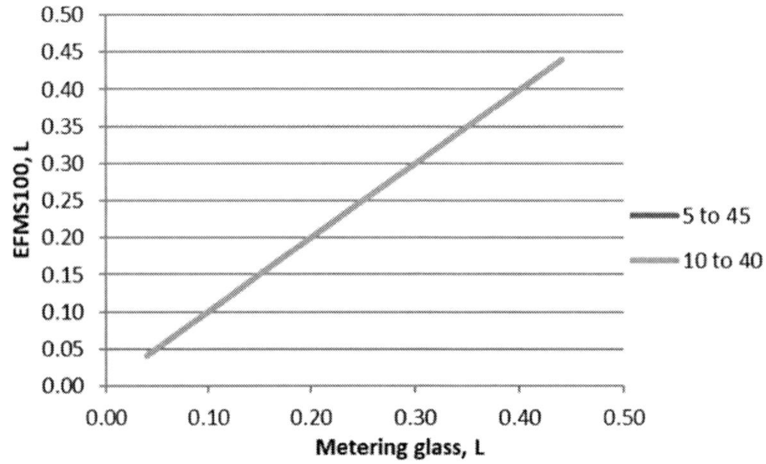

Figure 5. Fuel flow metering characteristic at Semi-sequential mode.

The last and most modern automotive injection sequention is the Sequential mode. Its characteristic is shown on the figure 6 and the timing is the same.

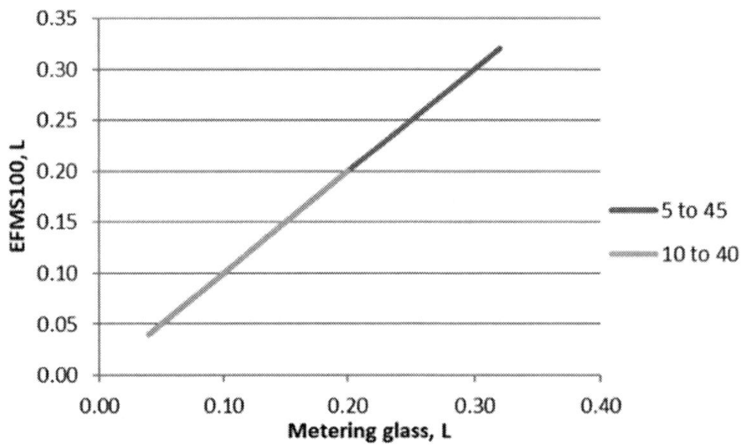

Figure 6. Fuel flow metering characteristic at Sequential mode.

As can be seen, all the fuel flow characteristic has approximate dependence, which is essential to the static tuning, i.e. static calibration of the calibrated system.

5. Calibration

During the experiments at the previous point can be seen that the main difference is rendered at the little quantities of injected fuel, i.e. at the ON_{time} timing of 5 to 10 ms. To achieve the comparative characteristic of the calibration or calibration characteristic it is important to compare the *FPC* values with the fuel quantity values during the four modes. The values for the Continuous and Simultaneous mode is represented in the table 2.

Table 2. Comparative value for the Continuous and Simultaneous mode.

Continuous		Simultaneous			
fuel, L	FPC	ON5ms, OFF45ms		ON10ms, OFF40ms	
		fuel, L	FPC	fuel, L	FPC
0.04	636	0.04	1411	0.04	975
0.08	1153	0.08	2721	0.08	1855
0.12	1770	0.12	4590	0.12	2729
0.16	2545	0.16	6008	0.16	3633
0.20	3134	0.20	7663	0.20	4502
0.24	3781	0.24	9323	0.24	5475
0.28	4425	0.28	10866	0.28	6361
0.33	5079	0.32	12352	0.32	7337
0.36	5656	0.36	13822	0.36	8172
0.40	6152	0.40	15168	0.40	9063

The values in the table and the remain values are represented on the figure 7. As was mentioned above the range of little fuel quantities is critical for the adjustment and calibration.

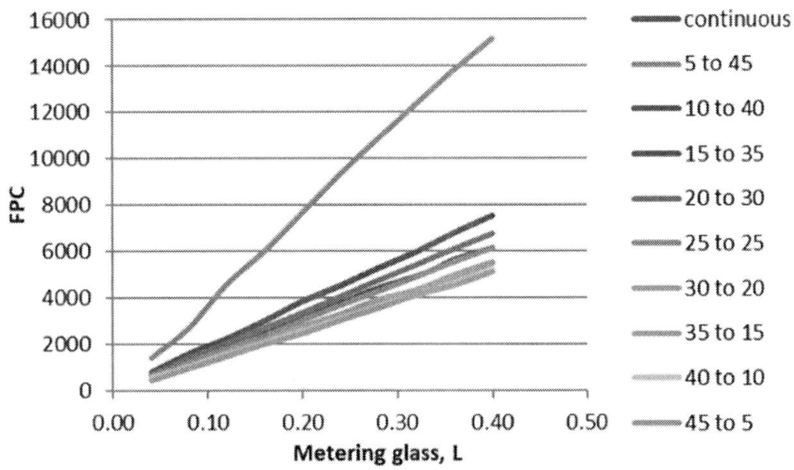

Figure 7. Calibration characteristic at Continuous and Simultaneous mode.

The values for the Semi-sequential and Sequential mode is represented in the table 3.

Table 3: Comparative value for the Semi-sequential and Sequential mode.

Semi-sequential				Sequential			
ON5ms, OFF45ms		ON10ms, OFF40ms		ON5ms, OFF45ms		ON10ms, OFF40ms	
fuel, L	FPC	fuel, L	FPC	fuel, L	FPC	fuel, L	FPC
0.04	2055	0.04	1340	0.04	2383	0.04	1783
0.08	4163	0.08	2604	0.10	6811	0.08	3783
0.18	9241	0.18	5626	0.20	13041	0.12	5895
0.30	15736	0.31	9808	0.28	18619	0.16	7970
0.20	7663	0.20	4502	0.32	22867	0.20	9914

The values in the table and the remain values are represented on the figure 8.

As it is shown on the figure 7 and figure 8, there is critical area in the range of little fuel quantities, i.e. at the little ON_{time} values. This area is more likely to be the reason for incorrect fuel flow measurements. So, the static calibration must be performed in this area.

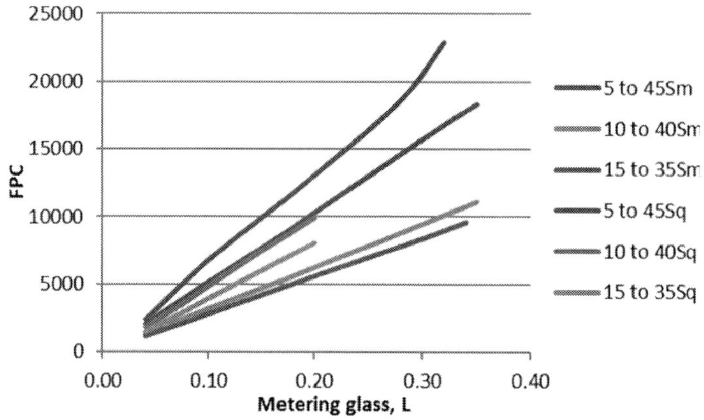

Figure 8. Calibration characteristic at Semi-sequential and Sequential mode.

The correct calibration consists the following:

1) Setting up the timing to the minimal ON_{time} values (for example 5 ms).
2) Defining the control fuel quantity Q_f (for example 200 ml).
3) Performing the experiment.
4) Comparing the values of sprayed fuel quantity Q_f by the metering glasses and by the display of calibrating system.
5) If there is difference between the readings the parameter PVR must be update, according to below.
6) Reading the values of sprayed fuel quantity Q_f by the metering glasses and the values of the parameter *FPC*.
7) Calculating the PVR parameter by the (2).
8) Setting up the updated PVR value in the menu of the calibrated system.
9) Repeating the experiment and comparing the readings.

After the correct calibration, the fuel metering system can be attached to the real HEV and can be made experiments to determine its fuel consumption during different driving modes in different areas. This will give the accurate evaluation of the HEV fuel efficiency.

6. Conclusion

The experimental calibration of the fuel metering system for the HEV fuel consumption is needful and important factor for the correct fuel metering.

Static tuning or calibration must made in the little fuel flow quantity modes to achieve the accurate tuning of calibrated system. Thus, the fuel consumption during idle mode of the HEV will be accurately define, moreover the idle mode is very common during the city traffic conditions.

The use of ME current sensors for HEV will optimize engine performance, as well as increase vehicle safety.

Acknowledgement

The article is related to the implementation of the project "Modeling and design of position sensors based on multiferroic layered structures" under Contract № KP-06-Russia/20 28.09.2019, Todor

Kableshkov University of Transport – Sofia, of the Bulgarian side. For the Russian side the reported study was funded by RFBR and NSFB according to the research project № 19-58-18001.

References

[1] Bozhkov Slavcho, Kirilov Boris, Valev Teodor, Pircheva Evelina and Bozhkov Penko 2018 Test Bench SAV-1 for Automated Management of the Automotive Gasoline Fuel Injectors *Mechanics Transport Communication* vol 16 **3/3** 1676

[2] Milenov Ivan, Bozhkov Slavcho and Tonkov Georgi 2018 Layout and electric propulsion features of hybrid electric vehicles *Mechanics Transport Communications* vol 16 **3/2** 1716

[3] Bozhkov S, Mutafchiev M, Milenov I, Bozhkov P, Bichurin M I and Petrov R V 2019 Method for determination of the hybrid electric vehicle energy efficiency in urban transportation *Vestnik NovSU* **4(116)** 4–8 DOI: 10.34680/2076-8052.2019

[4] http://www.enginemeter.com/site/

[5] http://www.matrixtsl.com

Digital design and technological innovation in the small machine building sector

V F Bulavin[1,2], **T G Bulavina**[1,3], **C A Stepanov**[1,4], **V V Yahrichev**[1,5] **and N S Grigoriev**[1,6]

[1] Vologda state University, Vologda, Russia

[2]E-mail: bulavin35@mail.ru
[3]E-mail: tamara.bulavina.53@mail.ru
[4]E-mail: alex.stepanov@mail.ru
[5]E-mail: yahrichev@yandex.ru
[6]E-mail: grigorevns@vogu35.ru

Abstract. Integration of CAD products is the governing factor of technical refitting and enhanced competitive advantage of small machine-building enterprises. The digital support associated with PLM and MES platforms makes it possible to state the beginning of a transfer of enterprises to a new type of production. The primary components of the design and engineering stage are as follows: digital design, automated process preparation, and structural engineering. In this process, the part's virtual copy is the central element of the navigation through the process preparation for production. The digital form of documentation in the PLM platform defines the requirements to start production. Electronic versions of documents are an attribute of project management organization in the team, a condition for rapid modification of the object, and a factor of coordination of work of the whole chain: from the project manager and single operators to stakeholders.

1. Introduction

Convergent use of CAD/CAM/CAPP/CAE ideologies is in its early days, but some of the process components have been widely used by enterprises of all sizes. The accumulated experience and growing expertise allow designing and executing both relatively simple and complex high-technology projects at the level of small enterprises.

2. Design support of production

Digital design and engineering support of production consists of 3D design, automated process preparation, and engineering support of calculations on the basis of CAE platforms. The first stage consists of creation and adjustment of digital copies of all the parts and subassemblies on the basis of a CAD platform. This process is accompanied by the development of respective working drawings depicting the required projections, sections, and cross sections [3]. The associative link between electronic prototypes and types of drawings helps to quickly record all the modifications of the digital version in the technical documentation.

The engineering stage involves specification of the layout, detalization of kinematical connections and composition of the future product based on the principle of unification and the maximum involvement of standard subassemblies.

Content from this work may be used under the terms of the Creative Commons Attribution 3.0 licence. Any further distribution of this work must maintain attribution to the author(s) and the title of the work, journal citation and DOI.
Published under licence by IOP Publishing Ltd

It is also the time to define physicochemical properties of the materials to be used, refine the appearance and ergonomic parameters. Completion of this stage results in the object's design documentation package with subsequent control of the drawings using an expert-level application. The functions of this module include finding faults and flaws in the drawings, check of standards and dimension chains before the documentation may be used in production [4].

The parametrization technique allows obtaining parts of a similar configuration using a single prototype. Forming elements are added to design and technology databases for use in future projects. See an 3D-modeled electronic copy of the "Base" part of an optoelectronic device in figure 1. The software code allows viewing the virtual layout of a designed object in 3 dimensions and various projections, giving it a photorealistic image in accordance with the assigned material to tentatively assess engineering, ergonomic, and design solutions (figure 2).

Figure 1. An electronic copy of the "Base" part.

Electronic copies of assembly units and the whole object help to reveal and eliminate collisions, rule out design flaws, and set tolerances on the basis of the principle of interchangeable parts of a structure. In small machine-building enterprises, 3D design reflects the policy of digital formats in the economy and serves as one of the foundation stones of evolution of production of a new type.

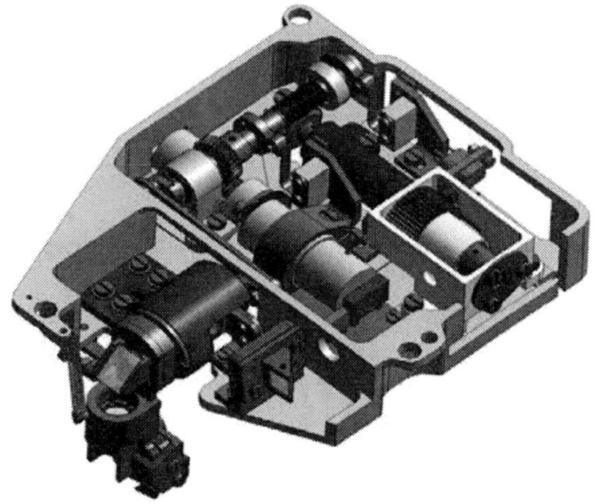

Figure 2. An electronic model of the optoelectronic device's module.

An exploded parts list is drawn up for the purpose of visualization of the structure, assembly and disassembly of the object's modules. The CAD platform's software allows exploding the structure into modules and present the result in an exploded view (figure 3). The result must be quickly editable and graphic. At the last stage of the engineering part, the electronic list is drawn up [5].

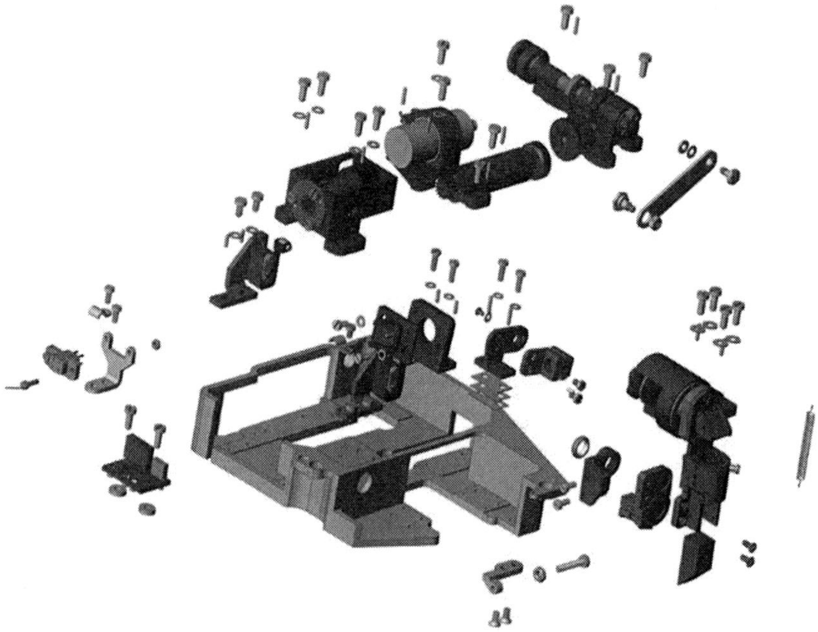

Figure 3. An exploded view of the optoelectronic device's module.

3. Technological support of production

The next stage, i.e., process preparation for production, involves automated design of the part production technologies on the basis of a CAPP platform: operations of machine processing and assembly of units and the whole structure. Developers ought to fulfill the part and structure manufacturability conditions [6, 7]. In the process of operation, the CAPP platform interacts with standard, group, and single operating procedures. These are the principles followed to modernize the existing and develop new part manufacturing technologies taking into account the possibilities offered by the available equipment, assign instruments, choose materials and shapes of workpieces, calculate labor indicators, and determine required qualifications of the operators.

Multi-purpose, structured and constantly growing data banks ensure automated support of the process preparation of production [6, 7]. This stage involves design of new technological processes, determination of additive technologies, modes of machine processing and heat treatment, the order of assembly of subassemblies and the structure in whole.

The software for metal-working machines and centers with a digital control code is developed in specialized applications. Interaction with a CAM platform implies choosing the tool assigned in the CAPP databases, finding the optimal trajectory of the cutting tool, as well as simulating the forming process of the given object. Functional possibilities of such a platform allow estimating the time spent on each machine operation and calculating the total metal working time [4, 5, 7].

Machine processing simulation with the cutting tool trajectory imitation taking into account movements of all the actuating and ancillary elements, as well as of the tooling, ensures there are no errors and collisions in the control software. The back-end setup provides a finished part processing software product for a given digitally controled machine column. Simulation of the cutting tool trajectory for a turning operation within the process of production of the "Shaft" part in the CAM space is given in figure 4.

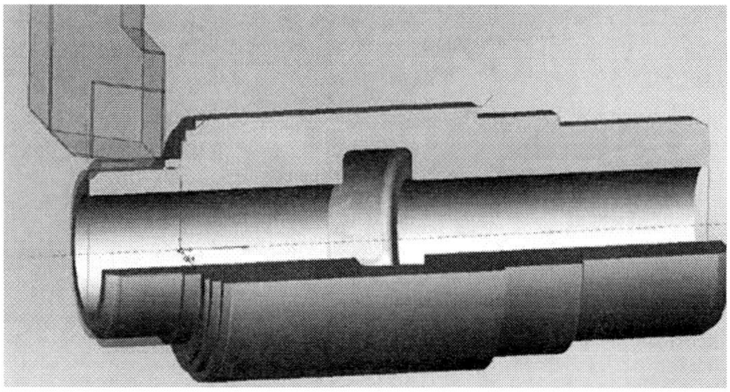

Figure 4. The "Shaft" part processing simulation.

4. Engineering support of design

Engineering analysis of parts and structures is based on studying the object's stress–strain state by means of computer simulation. This allows predicting reliability of the given parts and the structure in whole both in the standard operating mode and in extreme conditions. Possible types of analysis include the following: calculation for statically loaded parts and subassemblies, modal diagnostics, and thermal deformations [8, 9]. The result of simulation of the water flow in the event of unstable UV treatment tank filling is given in figure 5.

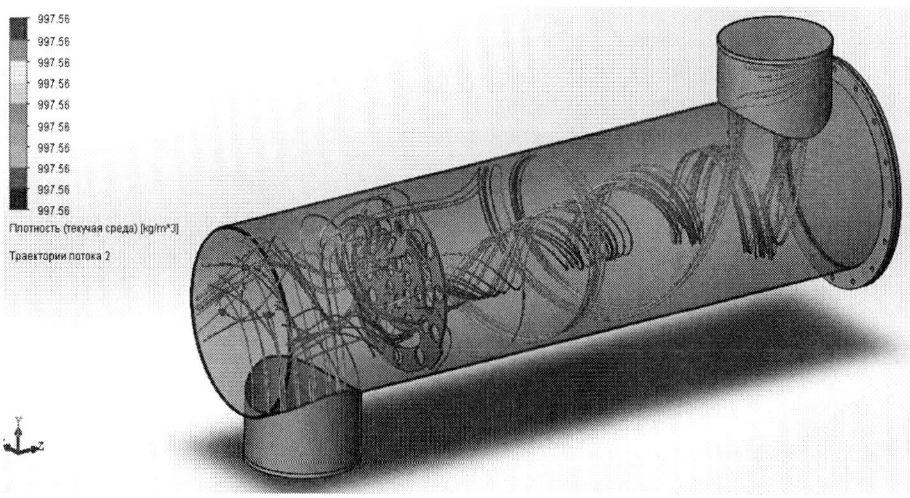

Figure 5. Unsteady water treatment tank filling flow movement visualization.

The calculation results in a graphic report depicting the object's shape after the study, distribution of velocities, pressure, forces, and displacements supplemented with a color chart with parameter values. The color legend helps to define the zones of the maximum values of the study parameters and draw conclusions as to strength or other characteristics of a structure [8, 9].

The result of experimental static loading of the "Body" part is given in figure 6. It helps to reveal dangerous areas, sections and the maximum deformations of the structure under consideration, as well as the factor of safety. The color legend helps to visualize simulation results and allows drawing a conclusion as to the part's operability [10, 11].

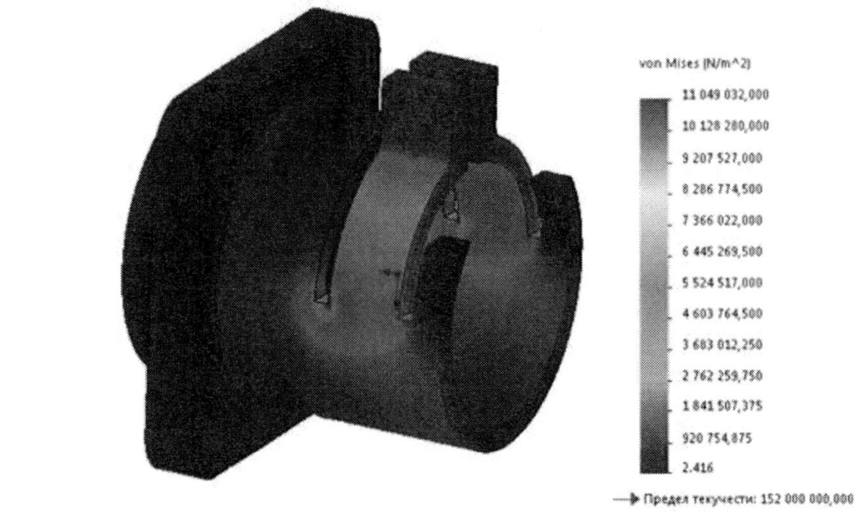

Figure 6. The distribution of normal stresses in the "Body" part.

5. Design support in virtual reality technology

Use of virtual and augmented reality (VR/AR) technologies—a source of new technological possibilities—is an innovative sphere of digital design. Being in the virtual reality and interacting with the 3D object, the designer gains an objective impression of the project's maturity, can check dimensions of the subassemblies of interest, identify faults and flaws of the layout. The VR technology allows evaluating serviceability and usability of the product, as well as whether it is comfortable. A visualization fragment of a crane cabin imported to a VR system from a CAD platform is given in figure 7.

Figure 7. VR view from the crane cabin.

The use of inverse kinematics in VR/AR systems is accompanied by full control of position of all human body parts and use of tracking systems to reach a higher level of preparation and management of manufacturing processes.

The VR technology helps to largely dispense with the stages of building a full-scale mockup and improve safety of manufacturing process when developing a new product. It enables perfecting and controlling all the technological operations involved in the processes of production and assembly of a product.

The interaction of digital machine-building products is reflected in the PLM platform, which forms the product tree. The structure contains models, drawings, technology descriptions, accompanying documents, and supporting information. In the created project's data bank, the product is saved as a virtual prototype linked to all the parts, subassemblies, and technologies accompanying the development process.

Digital design and support allow modifying structure elements with product configuration management options. These are operated by the PLM platform, where the product tree is formed as an object structure linked to all the components of the structure and the respective production processes. Implementation of a project in the common information space of technological and engineering services is one of the conditions of a digital enterprise. The digital document package formed in the PLM platform constitutes the first stage of production and the condition of future modifications of the product, defines the project group's work activity management rules [12], and serves as the means of coordination of all the links of the manufacturing chain. This offers powerful possibilities for industrial enterprises to implement innovative technologies and new structural and organizational practice principles [12].

The report [13, p. 8] describes a new approach to measuring the real impact of the digital economy, which, according to the authors, "reflects the value of the accompanying digitalization effects" and "the ensuing direct and indirect value." Digital economy management requires coordination of interests of all the social and public groups interested in the development of the parties, aggregation of organizational, labor, and financial resources, as well as joined effort of businesses [1,2,13]. The comprehensive approach to the transition of production to the digital space is premised on the integration of both technical and organizational and economic, social, and psychological aspects.

6. Conclusions

Penetration of digital technologies into enterprise activities is reflected by introduction and active use of CAD/CAM/CAPP/CAE, ERP, and PLM systems. The use of high-technology software platforms is a characteristic feature of the modern pattern of development of enterprises of a new type, results in higher labor productivity, faster introduction of new types of products, and improved consumer properties of such products. A stronger role of digital technologies is required not only in the production sphere, but also in the marketing policies of enterprises to improve their effectiveness and potential. Digital design reflects the technical part of the problem of transition of small enterprises to virtual production on the basis of innovative solutions and development of a manufacturers' information space.

The digital field of small enterprises is a constituent part of the global information space, and all the innovative technologies are reflected in it. The digitalization policy manifests itself on all the levels of the person–society–material production interaction.

References

[1] Sharonov A, Atnashev M et al 2017 Digital production. Methods, ecosystems, technologies *Working report of the Corporate Training Department of the Moscow School of Management SKOLKOVO* http://assets.fea.ru/uploads/fea/news/2017/11_november/17/tsifrovoe_proizvodstvo_112017.pdf (Accessed 29 Nov 2017)

[2] Schuh G, Anderl R et al 2018 Industrie 4.0 Maturity Index – Managing the Digital Transformation of Companies *The National Academy of Science and Engineering of*

Germany https://www.acatech.de/wpcontent/uploads/2018/03/acatech_STUDIE_Maturity_Index_eng_WEB.pdf (Accessed 03 Mar 2018)

[3] Bulavin V, Bulavina T and Yahrichev V 2017 Validation of CAD products in small enterprises of engineering sector *Fundamental and applied problems of engineering and technology* **5** 64–72

[4] Bulavin V, Yahrichev V and Stepanov A 2019 Politics of digital technologies in small machine-building enterprises *News of higher educational institutions Engineering* **9** 35–45 doi: 10.18698/0536-1044-2019-9-35-45

[5] Bulavin V and Yahrichev V 2018 Digital technologies in small business of machine-building industry *CAD and Graphics* **6** 52–55

[6] Bulavin V, Bulavina T, Yakhrichev V and Stepanov A 2020 Digital format of instrument manufacturing preparation Part I Design and development phase *News of higher educational institutions Instrument making* **3(63)** 157–166 doi: 10.17586/0021-3454-2020-63-3-157-166

[7] Bulavin V, Bulavina T, Yakhrichev V and Stepanov A 2020 Digital format of instrument manufacturing preparation Part II Technological stage *News of higher educational institutions Instrument making* **3(63)** 167–172 doi: 10.17586/0021-3454-2020-63-3-167-172

[8] Alyamovskiy A 2010 *Engineering calculations in Solid Works Simulation* (Moscow: DMK-Press publ)

[9] Zamriy A 2007 *Practical training course CAD/CAE ARM.* WinMachine System (Moscow: DMK–Press publ)

[10] Bulavin V, Bulavina T and Yakhrichev V 2018 Engineering analysis and new technologies in the finite element method *Fundamental and Applied Problems of Engineering and Technology* **2(328)** 109–120

[11] Bulavin V, Bulavina T, Yahrichev V, Stepanov A and Frolov A 2019 Digital support of production small business preparation in engineering *Journal of Physics: Conference Series,* **1399** 033045

[12] Bulavin V, Yahrichev V and Glazkov B 2018 PLM-strategy in small-scale production of machine-building industry. News of higher educational institutions Engineering **8** 37–49 doi: 10.18698/0536-1044-2018-8-37-49

[13] Xu W, Cooper A et al 2017 Concomitant digitalization effect. Measuring the real impact of the digital economy *Presentation by Huawei and Oxford Economics* https://www.huawei.com/minisite/gci/en/digital-spillover/index.html (Accessed 5 Sep 2017)

IOP Conf. Series: Materials Science and Engineering **939** (2020) 012017 doi:10.1088/1757-899X/939/1/012017

Simulation of operation of a sequential hybrid drive of a haul truck with a traction battery and a bilateral DC-to-DC converter

N V Buzunov[1,4], R D Pirozhkov[1,5], A B Kartashov[1,6], D M Dubinkin[2,7] and A B Efremenkov[3,8]

[1] Department of Wheeled Vehicles SM-10, Bauman Moscow State Technical University, 2nd Baumanskaya St., 5105005, Moscow, Russia
[2] Department of Metal Cutting Machines and Tools, Kuzbass State Technical University, Vesennyaya St., 28 650000, Kemerovo, Russia
[3] Yaroslav-the-Wise Novgorod State University, B. St. Peterburgskaya str., 41 173003, Veliky Novgorod, Russia

[4] E-mail: buzunovnv@bmstu.ru
[5] E-mail: pirozhkovrd@student.bmstu.ru
[6] E-mail: kartashov@bmstu.ru
[7] E-mail: ddm.tm@kuzstu.ru
[8] E-mail: abe@novsu.ru

Abstract. In this article we consider one of the possible options for organizing the traction drive of a haul truck. The used scheme of the drive organization is given, and its performance is estimated by means of the drive simulation. The consideration is given to DC-to-DC converters in drives of hybrid traction systems. The conclusion is made about their influence on the efficiency of the hybrid traction drive.

1. Introduction

There are three main schemes for organizing the traction system of vehicles with combined power plants [1–4]: sequential, parallel and mixed. The circuits may include an internal combustion engine (ICE), generator, batteries, capacitors, traction motor (TM), rectifiers and converters. The role of the internal combustion engine and generator can replace fuel cells.

All electrical devices in the sequential scheme are connected to a common direct current bus (DC bus). The internal combustion engine is not connected to the wheels mechanically and can be switched off while driving; in this case the vehicle is driven by battery power. The internal combustion engine operates at a constant speed corresponding to the most optimal mode. If the power required for movement is less than that developed by the generator, the excess is used to recharge the batteries. If there is a lack of power developed by the generator, the batteries come into operation. The advantages of a sequential scheme include greater freedom of layout, maximum energy efficiency of recovery, and the disadvantages – significant dimensions and weight relative to other schemes, energy losses during its transformations.

Content from this work may be used under the terms of the Creative Commons Attribution 3.0 licence. Any further distribution of this work must maintain attribution to the author(s) and the title of the work, journal citation and DOI.
Published under licence by IOP Publishing Ltd

In a parallel scheme, the internal combustion engine and traction motor have a rigid mechanical connection with the wheels. In the vast majority of cases, the internal combustion engine operates in optimal mode for a significant part of the time. Traction motor is activated in situations when the engine does not provide the required power. If the internal combustion engine provides an excess of power, the traction motor is switched to generator mode. The advantages of this scheme are smaller sizes and weight compared to the sequential scheme and less energy loss; among the disadvantages there are the need to use complex mechanical transmissions (an electric car combined with a transmission element) and greater complexity of the layout compared to the sequential scheme.

The mixed scheme uses features of parallel and sequential schemes: the internal combustion engine is connected mechanically to two electrical machines by means of a transmission with two degrees of freedom. The transmission gearbox is excluded in this case. In terms of advantages and disadvantages, such a scheme is close to a parallel one and is sometimes considered a variation of it.

2. Modeling of a hybrid sequential traction drive

The sequential scheme is suitable for use in heavy machinery due to its simplicity and layout capabilities. In this article we consider the sequential scheme, presented in the figure 1.

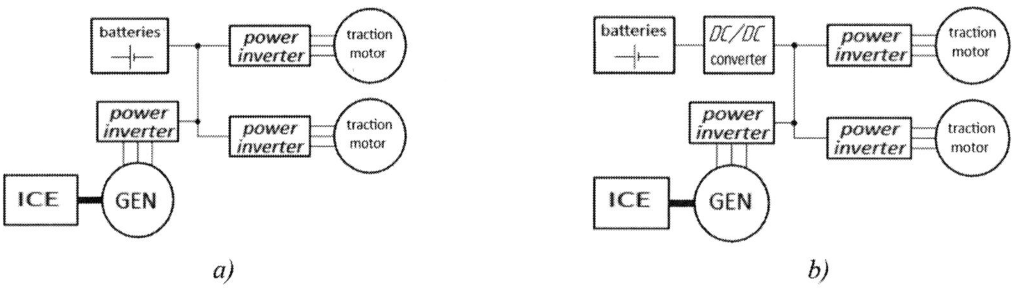

Figure 1. Conditional traction drive scheme: *a*) without DC-to-DC converter; *b*) with DC-to-DC converter.

As previously noted, the generator, traction motor and batteries are connected to a common bus. This solution has the following limitations – when using a common bus with multiple devices connected in parallel, it is not possible to control the power flows between them. They are determined by the internal resistance of these devices, in general, depending on a variety of parameters. In order to provide the traction motor with energy in full, it is necessary to match the exact generator voltage and the batteries voltage, while the algorithms for controlling the generator voltage become more complex.

In all types of batteries, together with a decrease in the charge level, there is a decrease in voltage. In this regard, the generator voltage must also be reduced as the batteries are discharged to ensure that the generator and batteries work together on a common load. Together with a decrease in the voltage on the common DC bus, the power of the traction motor will fall, which may make it impossible to continue driving when working in difficult conditions.

This problem is overcome by using bilateral DC converters (DC-to-DC converters). Their use allows combining devices with different rated voltages and controlling the electrical power flow in the system [5].

There are many different implementations of DC-to-DC converters. The efficiency of existing schemes exceeds 90% [6–9]. For some converters, the efficiency can be 96–98% with a power of 100 kW [10].

2.1. Prototype

The appliance of the above scheme is planned for a promising haul truck truck with a 4x4 wheel arrangement, the total weight of the haul truck is 50 tons, the curb weight is 25 tons. The following requirements are applied to the truck traction drive based on the simulation of the working cycle [11–14]:

– nominal rating power, kW, not less than 220;
– peak output, kW, not less than 350;
– maximum traction motor rotations per minute, rpm 3335;
– peak frequency torque, N·m 743.

Based on the requirements, the following devices are selected: traction motor – as part of the front and rear axles, internal combustion engine, generator and battery pack. EM-PMI540-T1500-2400-DUAL electric motors are used as traction motor. Table 1 shows the characteristics of electric motors. The LSM280A-HV electric motor is used as a generator, the characteristics of the generator are shown in table 2. Microvast 6C+2B lithium-titanate (LpTO) batteries are used as a battery pack, the battery characteristics are shown in table 3.

Table 1. Traction motor characteristics.

EM-PMI540-T1500-2400-DUAL		
Coolant temperature +40°C		
Continuous torque, N·m	Long-standing power, kW	Rated current, A
1510	380	522
Coolant temperature +40°C/+65°C		
Rotation speed, rpm	Maximum rotations per minute, rpm	Peak torque, N·m
2400	4000	2135

Table 2. Generator characteristic.

LSM280A-HV		
Parameters	Theoretical value (600 B_{DC}, 45°C)	Theoretical values (750 B_{DC}, 45°C)
Continuous mechanical power	225 kW	281 kW
Continuous torque	1852 N·m	1852 N·m
Rated rotation speed range	0–3400 rpm	
Maximum sustained current	350 A_{DC}	
Highest efficiency	0.948	

Table 3. Battery bank.

Microvast 6C+2B		
Parameters		Quantity
Nominal voltage/Capacity		556,6 V/140 Ah (78 kWh)
Operating voltage		400 V ~ 677.6 V
Module configuration		7p66s/1C, 7p44s/1V
System configuration		264s14p
Cell type		LpTO
Charging at a temperature of 25°C	Maximum DC	480 A
	Maximum peak current	500 A < 10 sec
Discharging at a temperature of 25°C	Maximum DC	365A (20% ~ 100% SOC)
	Maximum peak current	500 A < 10 c

2.2. Simulated mathematical model

The mathematical model of the described hybrid traction system, in versions without DC-to-DC converter and in combination with DC-to-DC converter, was obtained in the MATLAB Simulink programming environment. The block schematic diagram of the model is shown in the figure 2.

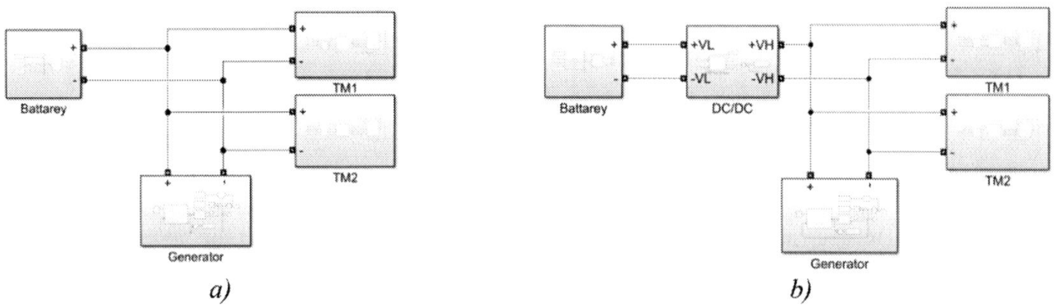

Figure 2. Block schematic diagram of the mathematical model of the traction system in the Simulink programming environment: *a)* without DC-to-DC converter; *b)* in combination with DC-to-DC converter.

The presented model consists of blocks: two traction motors, generator, DC-to-DC converter and battery. The subsystems are connected to a common DC bus.

The traction motor blocks contain mathematical models of an inverter, vector traction motor controller, and a model of a synchronous motor with permanent magnets. The generator block contains a mathematical model of a controlled DC power source. The battery block subsystem contains a mathematical model of a lithium-ion (LiIon) battery; the model takes into account the processes of the battery charging and discharging.

2.3. Modeling exercises

To assess the need for DC-to-DC converter in the scheme used, the following experiments are performed.

The performance of the traction drive is evaluated when the engines are running in constant power mode for 20 seconds. In the simulation, the generator recoil current is controlled at the level of no more than the allowed generator current in accordance with table 2. Modeling exercises are performed for cases where the drive operates without a battery, with a low battery level, and with a high battery level. As a result of the modeling exercises, power flows in the drive are estimated.

2.3.1. Battery-free operation. The first experiment involves running a generator on a traction drive without using a battery and a DC-to-DC converter. The graphical interpretation of the simulation results (figure 3) is: the generator current, the total current of the traction motor, the voltage on the common DC bus, and the mechanical power developed by the traction motor.

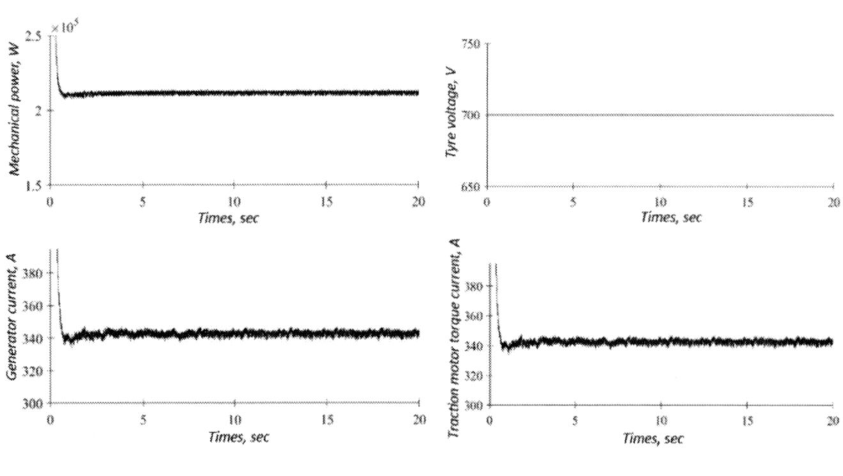

Figure 3. Test parameters when the drive is operated without a battery.

The load on the shafts of the electric motors is 1000 N·m, the speed of the electric motors is 1000 rpm, and the power developed by the two engines is 210 kW.

2.3.2. Drive operation without using DC-to-DC converter. In these modeling exercises without a DC-to-DC converter, the operation of the drive is considered, in which the battery and generator voltages are equalized. In all experiments, the motor shaft speed is maintained at 1000 rpm.

In the current experiment, the load on the shafts of electric motors is 1550 N·m, the power developed by two engines is 327.5 kW, the initial battery charge level is 90%, the voltage on the DC bus was maintained at 578 V in order to ensure long-term permissible recoil currents of the battery and generator.

A graphical interpretation of the experiment results is shown in figure 4.

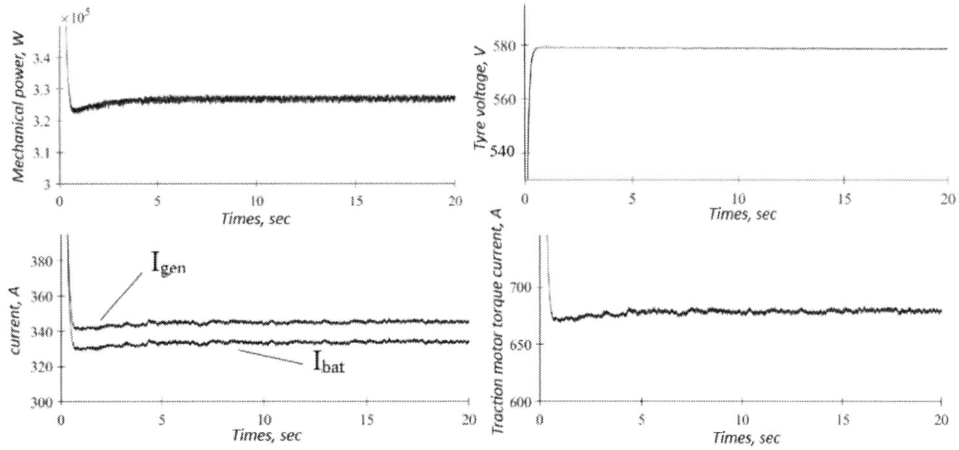

Figure 4. Drive parameters without DC-to-DC converter, battery charge level – 90%.

The third experiment started with a battery charge of 30%. At the same time, the overcome moment of resistance on each electric motor was 1460 N·m, the power developed by the two engines was equal to 307.5 kW, and the voltage on the common DC bus decreased to about 550 V. The graphical interpretation of the experiment results is shown in figure 5.

In the next experiment, the generator voltage is controlled to maintain the battery charge current of no more than 30 A. The graphical interpretation of the experiment results is shown in figure 6.

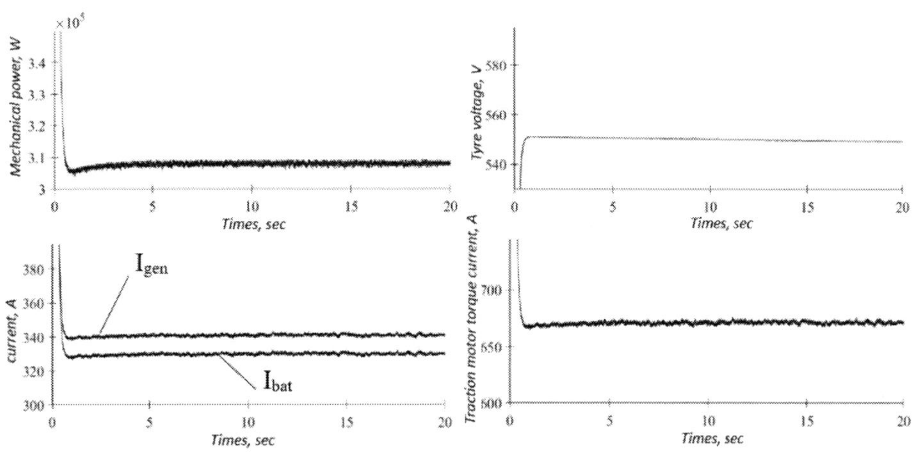

Figure 5. Drive parameters without DC-to-DC converter, battery charge level – 30%.

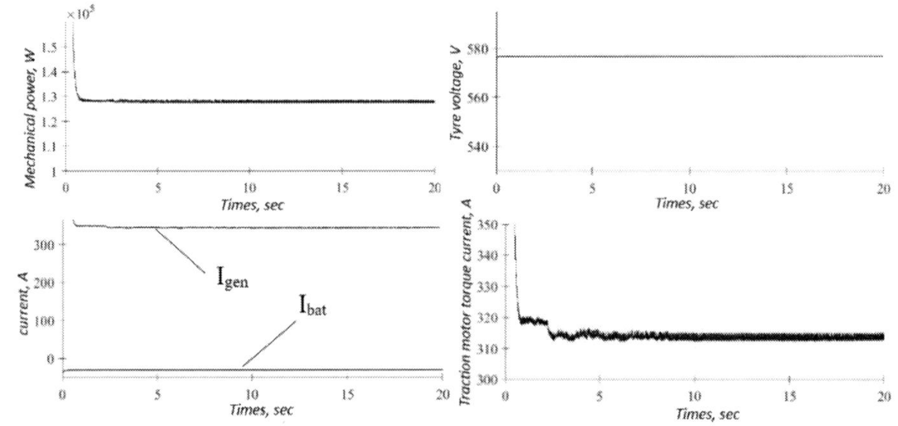

Figure 6. Experimental parameters when the drive is operating without voltage equalization, the initial battery charge level – 10%.

The load on the shafts of electric motors is 600 N·m, the power developed by two engines is 127.5 kW, the battery charge level at the initial time is 10%. In this case, the voltage on the common DC bus was 577 V.

2.3.3. Drive operation while using DC-to-DC converter. Figure 7 shows the simulation results when using a DC-to-DC converter running on a battery with an initial charge level of 10%. The battery charge current is controlled by a DC-to-DC converter and is 30 A regardless of the common DC bus voltage (700 V). At the same time, the overcome moment of resistance on each electric motor was 850 N·m, and the power developed by the two engines was equal to 180.2 kW.

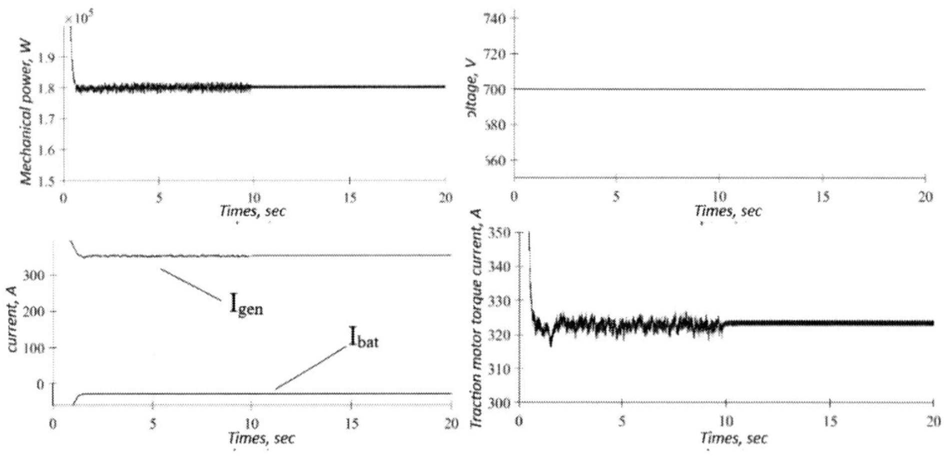

Figure 7. Experimental parameters when the drive is operating with DC-to-DC converter in the mode of the battery charge, the initial battery charge level – 10%.

The last experiment involved the joint operation of a DC-to-DC converter and generator on a common load with a battery charge of 90%. The results are shown in figure 8. The voltage on the common DC bus was also 700 V, the resistance moment on each electric motor was 1730 N·m, and the developed total power was 365.8 kW.

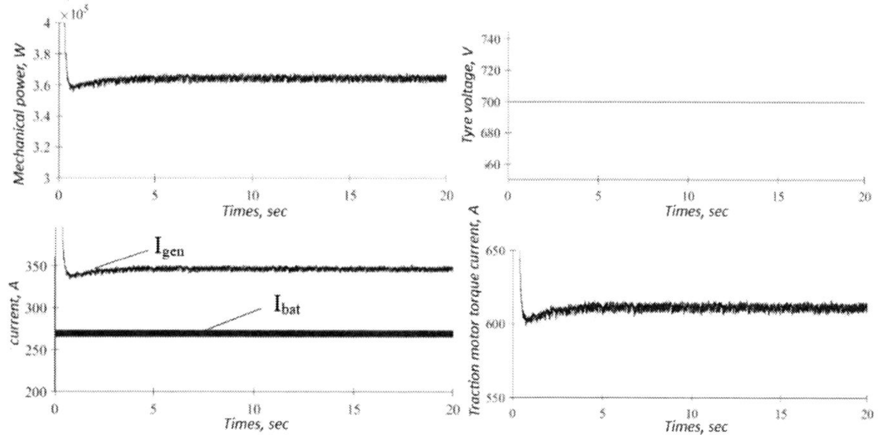

Figure 8. Drive parameters operating with DC-to-DC converter in the mode of the battery charge, the initial battery charge level – 90%.

3. Conclusion

Based on the results of previously conducted modeling exercises [15], the dependence of the torque of one of the drive engines on the driving time was formed for the vehicle in question (figure 9).

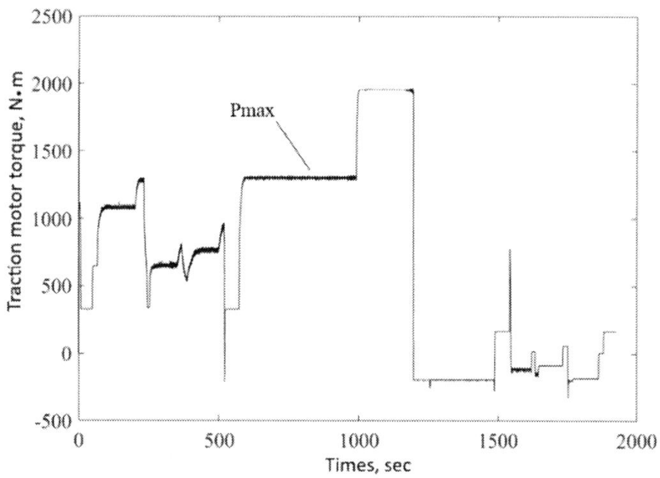

Figure 9. Dependence of the drive motor torque on driving time.

In the Pmax section, the engine speed was 78.9 radian per second. The total power of the two engines is 207.86 kW. Table 4 shows a comparison of the capacities that can be obtained under different modes of operation of onboard power sources of a hybrid vehicle, both with the use of a DC-to-DC converter and without the use of this device.

The following conclusions can be drawn from the presented data:

– using DC-to-DC converter allows to get more power at both high and low battery levels;

– this device allows to disconnect the discharged battery from the common DC bus without mechanical switches, while the generator will provide the necessary power for driving;

– when the battery is low, the DC-to-DC converter can provide a low current charge, while the generator can provide the power needed to move at a slightly lower speed compared to a typical driving cycle.

Table 4. Onboard power source capacities.

Operation mode	Maximum continuous power, kW
DC-to-DC converter and traction motor generator, battery charging – 90%	365.8
Generator and traction motor battery (initial battery charge level – 90 %)	327.5
Generator and traction motor battery (initial battery charge level – 30 %)	307.5
Traction motor generator (battery-free)	210
Standardized power	207.86
DC-to-DC converter at the battery charge (charging rate – 30 A, initial battery charge level – 10%) and traction motor generator	180.2
Traction motor generator and battery charge (charging rate – 30 A, initial battery charge level – 10%)	127.5

4. Confirmation

This study was performed at the Bauman Moscow State Technical University financially supported by the Russian Ministry of Education and Science (Agreement about federally funded grants provided in the form of subsidies No. 14.577.21.0287 (Unique identifier of the work: RFMEFI57718X0287).

References

[1] Yaroslavcev M V 2016 *Energoeffektivnyj tyagovyj privod gorodskogo bezrel'sovogo transporta* [Energy efficient traction drive for urban vehicles]: diss. …PhD (Eng) 05.09.03 (Novosibirsk State Technical University) p 157 [In Russ.]

[2] Siang F T and Chee W T 2013 A review of energy sources and energy management system in electric vehicles *Renewable and Sustainable Energy Reviews* **20** 82–102

[3] Lajunen A, Sainio P, Laurila L, Pippuri-Mäkeläinen J and Tammi K 2018 Overview of powertrain electrification and future scenarios for non-road mobile machinery *Energies* **11(5)** 1184

[4] Gao Y, Ehsani M and Miller J M 2005 Hybrid Electric Vehicle: Overview and State of the Art *Proceedings of the IEEE International Symposium on Industrial Electronics* (Dubrovnik, Croatia) 307–316 DOI: 10.1109/ISIE.2005.1528929

[5] Al Sakka M, Van Mierlo J and Gualous H 2011 DC/DC Converters for Electric Vehicles *Electric Vehicles – Modelling and Simulations* pp. 309–332

[6] Solero L, Lidozzi A and Pomilio J A 2005 Design of multiple-input power converter for hybrid vehicles *IEEE Transactions on Power Electronics* vol 20 **5** 1007–1016 DOI: 10.1109/TPEL.2005.854020

[7] Destraz B, Louvrier Y and Rufer A 2006 High Efficient Interleaved Multi-channel dc/dc Converter Dedicated to Mobile Applications *Conference Record of the 2006 IEEE Industry Applications Conference Forty-First IAS Annual Meeting* (FL: Tampa) pp. 2518–2523 DOI: 10.1109/IAS.2006.256893

[8] Ehsani M, Gao Y and Miller J M 2007 Hybrid Electric Vehicles: Architecture and Motor Drives *Proceedings of the IEEE* vol 95 **4** 719–728 DOI: 10.1109/JPROC.2007.892492

[9] Elsied M *et al.* 2015 Efficient Power-Electronic Converters for Electric Vehicle Applications *2015 IEEE Vehicle Power and Propulsion Conference (VPPC)* (Montreal, QC) 1–6 DOI: 10.1109/VPPC.2015.7352941

[10] Eckardt B, Hofmann A, Zeltner S and Maerz M 2006 Automotive Powertrain DC/DC Converter with 25kW/dm3 by using SiC Diodes *Proc. 4th Int. Conf on Integrated Power Systems (CIPS 2006)* pp. 227–232

[11] Buzunov N V, Kotiev G O and Padalkin B V 2019 *IOP Conf. Ser.: Mater. Sci. Eng.* **534** 012021

[12] Kotiev G O, Miroshnichenko A V, Staduhin A A and Kositsyn B B 2019 *IOP Conf. Series: Mater.*

Sci. Eng. **534** 012004
[13] Gorelov V A, Butarovich D O, Staduhin A A and Skotnikov G I 2019 *IOP Conf. Series: Mater. Sci. Eng.* **534** 012013
[14] Volskaia V N , Zhileykin M M and Zakharov A Y 2018 *IOP Conf. Series: Mater. Sci. Eng.* **315** 012028
[15] Kartashov A B and Skotnikov G I 2020 *IOP Conf. Series: Mater. Sci. Eng.* **819** 012010

IOP Conf. Series: Materials Science and Engineering **939** (2020) 012018 doi:10.1088/1757-899X/939/1/012018

Hybrid technology for the synthesis of ultrafine carbon powder material in an amorphous state

V A Chaika[1,4], V V Savin[1], I S Zherebcov[1], M E Boiko[2] and A M Boiko[3]

[1] Immanuel-Kant Baltic-Federal University, Nevsky st., 14, 23616 Kaliningrad, Russia
[2] Ioffe Physical-Technical Institute of RAS, 26 Polytekhnicheskaya, 194021 Saint Petersburg, Russia
[3] Herzen Pedagogical University, embankment of the Moika River, 48, 191186 Saint Petersburg, Russia

[4] E-mail: v_chayka8@rambler.ru

Abstract. The results of using a hybrid technology for the synthesis of ultrafine carbon powder material in an amorphous state are considered. Different modes of hybrid technology are proposed, including mechanochemical synthesis in a ball mill and the process of self-propagating sustainable combustion in a natural atmosphere to produce carbon powder materials (CPM) in an amorphous state. Using X-ray diffraction methods: diffractometric, radiographic, and small-angle X-ray scattering, heterogeneities of the CPM structure were studied. Areas of structural heterogeneity having a flat shape and containing up to 10 graphene-like layers are revealed.

1. Introduction
There processing, abrupt changes in the lattice period and graphite properties are noted [4]. Intense changes in the macro- and microstructure of the material being processed alternate with relaxation processes, resulting structural inhomogeneities and defects.

The aim of this work was to obtain an amorphous state in carbon powder materials using the method of mechanochemical synthesis (MCS) during long-term processing in a planetary ball mill of MPG-7 graphite.

2. Results
MHS was carried out in a PM-100 planetary ball mill (manufactured by RetschGmbH) at 600 rpm in a steel can with steel balls in the atmosphere of residual air. The feedstock was particles of graphite MPG-7 fractions of less than 1 mm. The ratio of the mass of the balls to the mass of the charge charge corresponded to a ratio of 19: 1.

During the synthesis, 4 processing modes were used: 1) four times for 6 hours with stops for 18 hours; 2) six times for 4 hours with stops for 8 hours; 3) 12 times for 2 hours with stops at least 3 hours; 4) 24 times for 1 hour with stops for 0.5 hours.

After MCS, when glasses are opened in natural atmospheric conditions, the process of oxidation (burning) of the powder material occurs. The process is realized by the local occurrence of one or several centers (foci) of combustion and a propagating stable combustion front according to the type of self-propagating high-temperature synthesis (SHS). The most intense course of the combustion

Content from this work may be used under the terms of the Creative Commons Attribution 3.0 licence. Any further distribution of this work must maintain attribution to the author(s) and the title of the work, journal citation and DOI.
Published under licence by IOP Publishing Ltd

process (according to the time of occurrence of the foci, their number, and the velocity of the combustion front) is observed in the powder material after MCS under conditions No. 3 and No. 4. Under conditions No. 1 and No. 2, combustion passed into the process of decaying decay of one or more local ignition centers.

Figure 1 shows a fragment (frame) of a video of the propagating stable combustion front of the MHS powder obtained in mode N3 in PM100 with a steel headset.

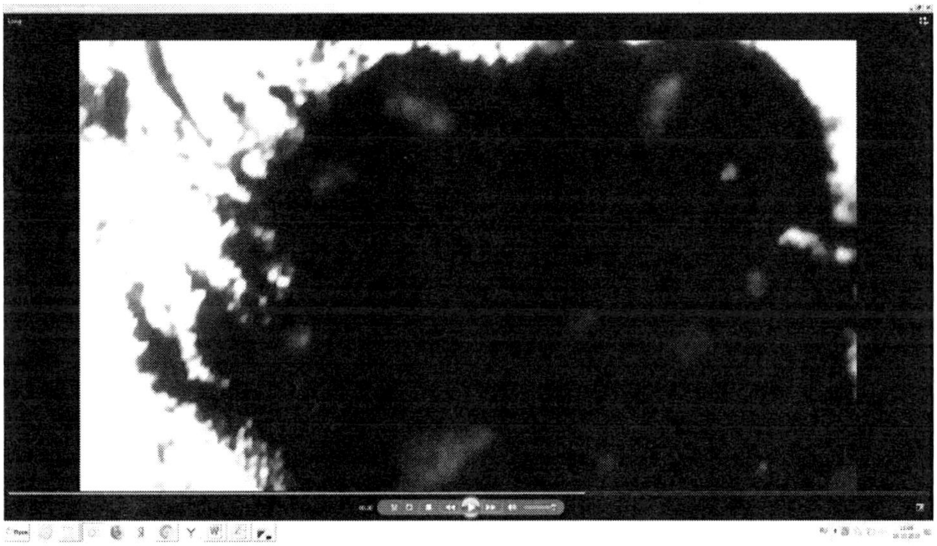

Figure 1. Video image of powder decay after 24 hours of processing with stops every 2 hours in PM100 with a steel headset.

In the work, powder carbon materials obtained by the MCS method and subjected to combustion according to the SHS type were studied in detail, i.e. obtained by hybrid synthesis technology (MCS + SHS).

The resulting material is characterized by high dispersion. Figure 2 shows the results of metallographic studies of the shape and structure of the powder material synthesized by the hybrid (MCS + SHS) method. As can be seen from the data presented, all particles have a "fragmentation" shape. After hybrid processing, white and brown particles are present in the array of powders. It was experimentally established that powder material obtained according to regime N3 is prone to agglomeration.

To clarify the chemical composition of the powder material, detail the morphology and microstructure of the synthesized powder particles, scanning electron microscopy (SEM) studies were carried out on a JSM-6390LV instrument (JEOL, Japan) with an OxfordINCAEnergy energy dispersive analyzer (Oxford Instrument, England). The research results are shown in figurer 3.

The bulk (up to 90%) of the synthesized powder material has a particle size of less than 10 microns. According to the chemical composition data, there are practically no impurities in the obtained material, except oxygen.

As shown by X-ray studies (DRON-3M diffractometer, Cu-K_α radiation) in all powder samples, the diffraction maxima of graphite in the amorphous type are weakened and broadened (figure 4). It is noteworthy that the center of gravity of the main peak is shifted nonlinearly.

Estimation of the CSR size (D) and crystallite size along the crystallographic axis "a" (La) of the test material is presented in table 1. According to the data in table 1, ultrafine powder is obtained under processing conditions N2 and N3.

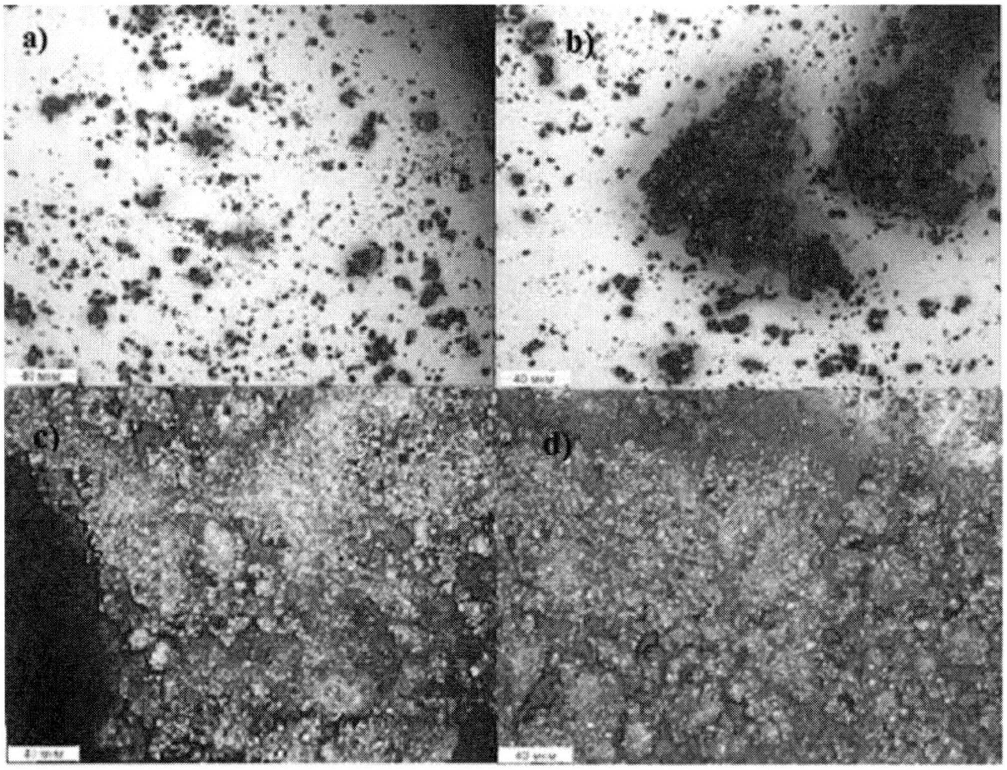

Figure 2. Optical image of the powder after 24 hours of processing in PM100 with a steel headset under processing conditions: N1 – (d); N2 – (c); N3 – (b) and N4 – (a).

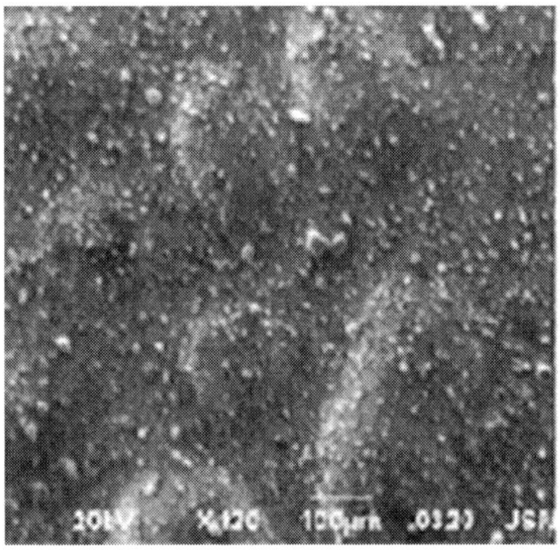

Processing Parameter: All elements analyzed

Spectrum	In stat	C	O	Al	Fe
Spectrum 1	Yes	95.85	4.10	0.01	0.03
Spectrum 2	Yes	94.45	5.51	0.01	0.03
Max		95.85	5.51	0.01	0.03
Min		94.45	4.10	0.01	0.03

All results in atomic %

Figure 3. SEM image of the powder after 24 hours of processing with stops every 1 hour (mode N4) in the PM100 with a steel headset.

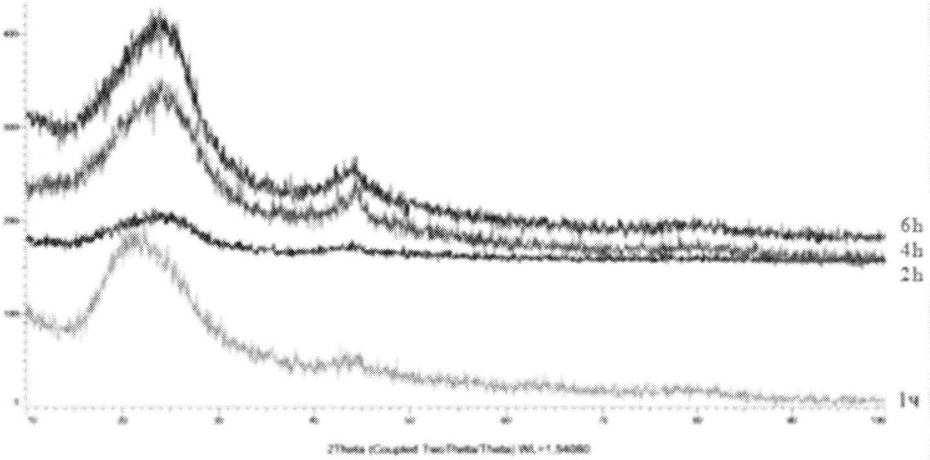

Figure 4. X-ray powder diffraction patterns MCS.

Table 1. Estimation of the size of the CSR material after MCS synthesized by hybrid (MCS + SHS) technology.

	Processing Modes			
Parameters	N4	N3	N2	N1
D, nm	96.4	6.63	6.65	27.5
L_a, nm	4.7	1.37	1.89	4.56

Based on the location of the (002) line, the average interatomic distance R1 in the material was estimated (see table 2).

Table 2. The concentration of heavy metals in the diets of feeding bulls of the black-and-white breed, mg.

	Processing Modes			
Parameters	N4	N3	N2	N1
2θ, degree	26.06	23.99	26.66	25.33
d/n, Å	3.415	3.705	3.339	3.512
H	150	122	30	120
FWHM, degree	2.76	8.2	3.36	6.15
R_1, Å	4.2	4.557	4.107	4.32

When comparing the data of tables 1 and 2, we can conclude that the bulk of the material is a flat flakes with a thickness of about 3–10 atomic layers.

Figure 5 shows the radiographic image (MCS + SHS) of the powder material (modes No. 1 and No. 2). The image was obtained on a unique scientific facility (USI) – a scientific and educational multifunctional complex for preparing and conducting synchrotron studies "SynchrotronLike" (X-ray source ExcilliumMetalJet D2 manufacturer – Excillum AB, Swedishcorporation). The image shows transparent flakes of processed graphite, consisting of many small particles. Transparency indicates a small number of atomic layers in the particles.

IOP Conf. Series: Materials Science and Engineering **939** (2020) 012018 doi:10.1088/1757-899X/939/1/012018

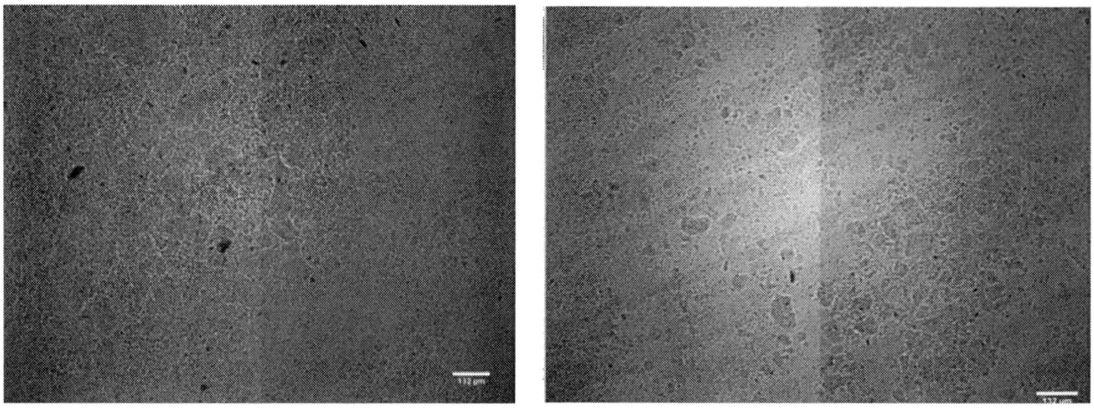

a) *b)*

Figure 5. Radiographic image of the material after a long MCS in PM100 with a steel headset with stops every 4 (a) and 6 (b) hours.

To clarify the grain size, small-angle X-ray scattering studies were performed [6] and FWHM = 0.102 was obtained for the material after the hybrid technology in regime N3 (see figure 6). The values of the Porod modulus (from left to right) were determined 2.7 – 2.4, which indicates that the shape of the particles is more gravitated to the shape of the membranes than to the shape of the particles of dense powder. The gyration radius (R giration) of graphite particles is determined in the range of 22– 34 nm, which is much larger than the CSR sizes and is explained by the flat shape of the particles.

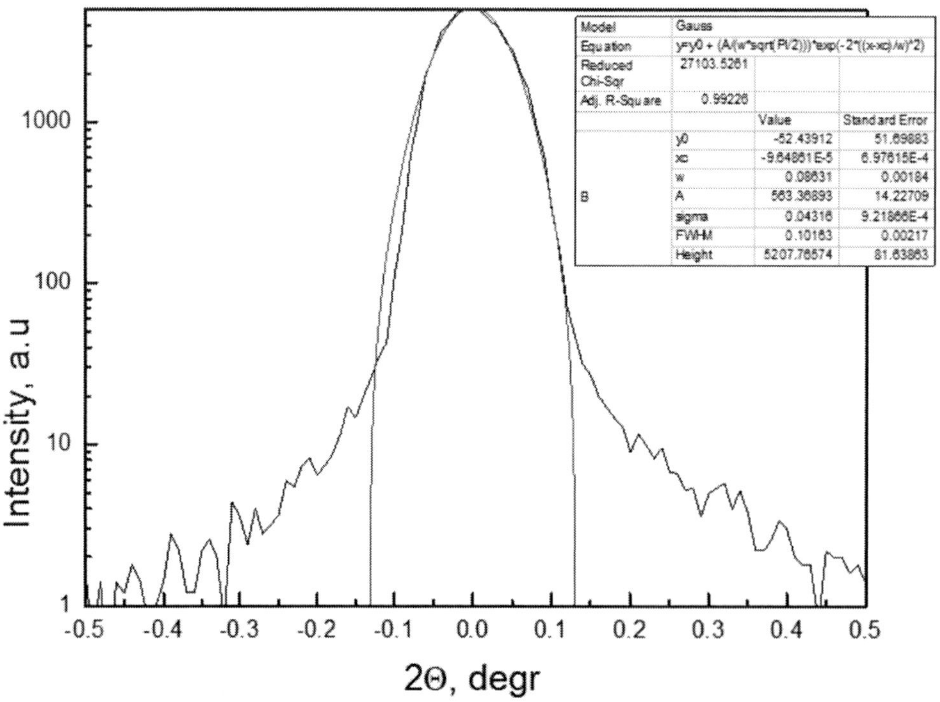

Figure 6. An example of a small angle X-ray scattering curve for a powder material obtained by the hybrid (MCS + SHS) technology according to mode N3.

3. Conclusion

1. For the synthesis of carbon materials in an amorphous state, a hybrid technology has been applied, including mechanochemical synthesis (MCS) in a PM100 ball planetary mill with a steel headset and the process of self-propagating sustainable combustion in a natural atmosphere (SHS).
2. The modes of hybrid technology for the synthesis of powder ultrafine carbon powder materials in an X-ray amorphous state are proposed.
3. X-ray methods were used to determine the flat shape of material particles with a thickness of the order of 3-10 atomic layers.

Acknowledgments

The study was supported by a subsidy allocated for the implementation of the Competitiveness Program of the Immanuel Kant Baltic Federal University.

References

[1] Opalev SV and Belenkov E A 2004 An experimental study of changes in the structure of graphite during mechanical grinding *Physical chemistry and technology of inorganic materials. Bulletin of the Chelyabinsk Scientific Center* **3(24)** 27–30
[2] Nikonova R M, Pozdeeva N S and Ladyanov V I 2011 The deformation behavior of copper during mechanical activation with carbon *Chemic. physics and mesoscopy* 1(13) 88–93
[3] Belenkov E A and Greshnyakov V A 2016 Structure, properties and possible mechanisms of the formation of diamond-like phases *Solid State Physics* **58**(10) 2145–2154
[4] Borunova A B 2015 The effect of the dose of mechanical activation on the defective structure of artificial graphite *Colloid. Journal* **77** 134–143
[5] Chaika V A, Savin V V, Savina L A, Osadchy A V and Zherebcov I S 2018 Structural-phase changes of graphite during mechanochemical treatment *Key Engineering Materials* **777** 205–209
[6] RU patent. N2548601 C1: Method for X-ray spectral determination of the composition and size of nanoparticles in a sample. dated 11/20/2013

IOP Conf. Series: Materials Science and Engineering **939** (2020) 012019 doi:10.1088/1757-899X/939/1/012019

The effect of coagulation on the processes of growth and changes in the numerical concentration of carbon dioxide crystals

M M Danilov[1,4], O S Opicina[1], I G Arendatelev[2,5] and I M Zawadzki[3]

[1] Saint-Petersburg National Research University of Information Technologies, Mechanics and Optics, Kronverksky pr., 49 197101 Saint Petersburg, Russia
[2] Yaroslav-the-Wise Novgorod State University, ul. B. St. Petersburgskaya, 41 173003 Veliky Novgorod, Russia.
[3] Novgorod Chemical-Industrial College, ul. Kochetova, 28 173025 Veliky Novgorod, Russia

[4]E-mail: dmm@trumgame.ru
[5]E-mail: Igor.Arendatelev@ novsu.ru

Abstract. The process of phase transition to the solid state of carbon dioxide vapors contained in an expanding gas mixture stream consists of the formation of crystallization centers and subsequent growth of the crystals formed, due to an increase in the degree of supercooling of the stream. Crystals nucleate in certain sections of the flow part of the expanders and grow as they move in the flow. Therefore, crystals of carbon dioxide of various sizes can be found in each of certain sections. As a result of their interaction, coagulation, expressed as the adhesion of smaller crystals to larger ones, can occur. The influence of the coagulation process on the change in the number of smaller crystals and on the growth of larger ones is considered. Equations are given on the basis of which the calculations of the studied processes are made.

1. Introduction

A mathematical model of the process of formation of carbon dioxide crystals in the volume of an expanding gas stream [1] allows determining, among other things, the numerical concentration of the formed carbon dioxide nucleus crystals and their size, as well as the size of crystals growing as the flow moves along the flow part of the expansion device [2]. Although this model is designed to evaluate the quantitative and dispersed composition of crystalline carbon dioxide, it does not take into account coagulation, i.e. adhesion of one crystal to another as a result of their collision with each other [3]. In the above-mentioned research works, it was shown that in the presence of particles of various sizes in the stream, collisions and coalescence of these particles occur due to differences in their velocities. Two methods for calculating coagulation are presented: the Euler method (coagulation of predetermined fractions) and the Lagrange method (coagulation of individual particles).

For the stated problem, we apply the Euler method. Carbon dioxide crystals formed and growing in an expanding flow have a size of less than 1 μm [4], therefore, they can be attributed to the class of highly dispersed aerosols [5]. Such aerosols display coagulation due to the Brownian motion of particles, i.e. thermal coagulation. The calculation of coagulation by the Euler method involves

Content from this work may be used under the terms of the Creative Commons Attribution 3.0 licence. Any further distribution of this work must maintain attribution to the author(s) and the title of the work, journal citation and DOI.
Published under licence by IOP Publishing Ltd

the one-dimensional motion of all particles, which, when they collide, necessarily stick together [6]. Based on this method, one should solve the problem of creating a mathematical apparatus for describing the process of coagulation of carbon dioxide crystals and verify the practical effect of coagulation on the numerical concentration of crystals and their size growth.

2. Results and discussion

Coagulation of carbon dioxide crystals. In each of the subsequent sections of the flowing part of the expansion device (expander), there may be several groups of carbon dioxide crystals having different sizes. These are both embryonic crystals formed in this section and crystals formed in the previous section and grown in this section, as well as crystals formed and grown in previous sections. Since the equilibrium shape of the carbon dioxide crystal is a cube, it is assumed that the crystals growing both by increasing the degree of supercooling of the flow and by coagulation will have a cubic shape.

If we consider a crystal that moves relative to other motionless crystals of the same size, then its contact with the motionless crystals will occur when the distance between the centers of these crystals is equal to a (a is the size of the face of the cubic crystal). The center of the moving crystal, each side face of which (relative to the direction of movement) touched the motionless crystals, will be in the center of the parallelepiped with a base area of $4a^2$ (square area with side $2a$). The volume of such a parallelepiped per time unit can be expressed as

$$V = 4a^2 \cdot \bar{c_c},$$

where $\bar{c_c} = \sqrt{3kT/m_c}$ – average value of the relative velocity of Brownian motion of crystals in a moving stream [7];

k – Boltzmann constant;

T – flow temperature;

m_c – crystal mass.

Number of contacts

$$Z = V \cdot N,$$

where N is a number of crystals per volume unit.

Considering that all crystals in the flow move with an average speed $\bar{c_c}$, the average speed of motion relative to other crystals should be used

$$\bar{c}_{reg.} = \bar{c_c} \cdot \sqrt{2}.$$

In order to reduce the recording of patterns of the coagulation process, we introduce the coagulation constant

$$K = 4a^2 \cdot \bar{c_c} \cdot \sqrt{2}.$$

In each contact, two crystals participate. In order not to take into account one contact twice, the actual coagulation constant must be reduced by half

$$K_r = \frac{K}{2} = 2a^2 \cdot \bar{c_c} \cdot \sqrt{2}.$$

Then, the number of contacts per time unit

$$Z_r = 2a^2 \cdot \bar{c_c} \cdot \sqrt{2} \cdot N = K_r \cdot N.$$

If we multiply the number of crystals per volume unit by the number of contacts per time unit, we obtain the number of crystals, which decreased as a result of the coagulation process in a volume unit per unit time

$$\frac{dN}{d\tau} = -N \cdot Z_r = -K_r \cdot N^2.$$

Let us turn to the real picture of the formation of carbon dioxide in the flow part of the expander, when in the same section there are several groups of crystals of different sizes (with different facet sizes a). In this case, the growth of larger crystals can occur due to coagulation with smaller crystals, the number of which will decrease (up to the disappearance of whole groups of crystals).

We denote the size of small and large crystals as a_s and a_l, respectively. A small moving crystal comes into contact with large motionless (as an assumption) crystals within a parallelepiped with a volume per time unit

$$V = (a_s + a_l)^2 \cdot \bar{c}_{c_s}.$$

Given the speed of large crystals, it is possible to calculate the average Brownian motion velocity of crystals of various sizes

$$\bar{c}_{c_{avg}} = (\bar{c}_{c_s} + \bar{c}_{c_l})/2$$

The actual coagulation constant can be represented as

$$K_r = (a_s + a_l)^2 \cdot \frac{\bar{c}_{c_{avg}}}{2}$$

The number of contacts of small crystals with large crystals (the number of small crystals whose centers lie inside the box) per time unit

$$Z_r = K_r \cdot N_s.$$

Multiplying the number of large crystals per volume unit by the number of collisions of small crystals with them per time unit, we obtain the number of small crystals that disappear during coagulation with large crystals in volume unit per time unit

$$\frac{dN_s}{d\tau} = -N_l \cdot Z_r = -K_r \cdot N_s \cdot N_l. \quad (1)$$

Conditions for the implementation of the process of coagulation of carbon dioxide crystals. Two consequences can be considered the result of coagulation: a reduction in the number of small crystals (or their complete disappearance) and growth of large crystals due to this. If we talk about small crystals, then coagulation can be considered significant when the number of vanishing small crystals dN_s is comparable with their number N_s. When the number of dN_s is several orders less than the number of N_s, then this group of small crystals will not significantly decrease.

It should be noted that in the flow part of the expander, nucleating crystals are formed with a cube facet size a equal to 3 a_m and 2 a_m (a_m is the parameter of a molecular crystal of carbon dioxide) [4]. The number of crystals per volume unit formed in different sections of the flowing part varies from $1 \cdot 10^4$ (crystals with the parameter a = 3 a_m) to $1 \cdot 10^{18}$ (crystals with the parameter $a = 2a_m$). This group of formed crystals will always be smaller in size than the groups of already grown crystals that appeared earlier; therefore, the number of such crystals per unit volume can be denoted as N_s.

Now let us trace the growth of larger crystals due to the addition of small crystals to them. We will consider the growth of a cubic crystal with parameter a. It is assumed that the crystal grown as a result of coagulation will also have a cubic shape (equilibrium with the gas phase). Then a crystal with parameter $2a$ will consist of 8 crystals with parameter a (other 7 of the same crystals should join the crystal with parameter a), a crystal with parameter $3a$ will consist of 27 crystals, etc. If we denote the parameter of the grown crystal na, where n is a positive integer larger than 1, then the number of crystals with the parameter a, which should join it, can be expressed as $n^3 - 1$ (coagulation of crystals of the same size).

In fact, the size of the attached crystals will be smaller than a. If the parameter of small crystals is f times smaller than a, i.e. a/f, then to increase the crystal with parameter a to size na, it will be necessary to attach small crystals to it in the amount of $(n^3 - 1) \cdot f^3$.

IOP Conf. Series: Materials Science and Engineering **939** (2020) 012019 doi:10.1088/1757-899X/939/1/012019

For example, the number of crystals with parameter *a*/2, which must be attached to the crystal with parameter *a* in order to increase it to parameter $2a$, will be $(2^3 - 1) \cdot 2^3 = 56$.

Given the number of large N_l crystals, we can calculate the number of small crystals necessary for their growth

$$dN_s = (n^3 - 1) \cdot f^3 \cdot N_l. \quad (2)$$

If the value of dN_s determined by equation (2) is larger than the value of dN_s expressed from equation (1), then the growth of large crystals as a result of coagulation will not occur.

The results of the calculation of the coagulation process for crystals of carbon dioxide. As an example, carbon dioxide crystals were formed and grown during the expansion of a gas mixture containing 10% CO2 in a centripetal turboexpander from a pressure of 200 kPa to a pressure of 110 kPa [4]. The values of the main parameters of the crystal formation process necessary for calculating coagulation are shown in table 1. Several explanations should be given to this table:

- in section 6 of the flow part of the turboexpander, for the first time a significant number of embryonic crystals appear;
- after section 11, the process of crystal formation and growth is completed;
- the subscript "*s*" refers to embryonic crystals formed in the considered sections;
- the subscript "*l1*" refers to crystals formed in previous sections and grown in the sections under consideration;
- the subscript "*l2*" refers to crystals formed two sections before the considered one;
- subscripts "*l3*", "*l4*", "*l5*" refer to crystals formed three, four and five sections before the considered, respectively.

Table 1. Design parameters of carbon dioxide crystals in various sections of the flow part.

No	Parameters, sizes	Sections					
		6	7	8	9	10	11
1	$a_s \cdot 10^9, m$	1.701	1.700	1.132	1.131	1.131	1.131
2	$a_{l_1} \cdot 10^9, m$	–	6.848	7.262	5.821	6.084	6.210
3	$a_{l_2} \cdot 10^9, m$	–	–	18.620	20.360	18.282	19.233
4	$a_{l_3} \cdot 10^9, m$	–	–	–	38.971	43.230	40.894
5	$a_{l_4} \cdot 10^9, m$	–	–	–	–	68.702	75.411
6	$a_{l_5} \cdot 10^9, m$	–	–	–	–	–	106.404
7	$N_s, 1/m^3$	$2.33 \cdot 10^5$	$0.44 \cdot 10^5$	$4.55 \cdot 10^{17}$	$1.99 \cdot 10^{17}$	$0.79 \cdot 10^{17}$	$0.73 \cdot 10^{17}$
8	$N_{l_1}, 1/m^3$	–	$2.33 \cdot 10^5$	$0.44 \cdot 10^5$	$4.55 \cdot 10^{17}$	$1.99 \cdot 10^{17}$	$0.79 \cdot 10^{17}$
9	$N_{l_2}, 1/m^3$	–	–	$2.33 \cdot 10^5$	$0.44 \cdot 10^5$	$4.55 \cdot 10^{17}$	$1.99 \cdot 10^{17}$
10	$N_{l_3}, 1/m^3$	–	–	–	$2.33 \cdot 10^5$	$0.44 \cdot 10^5$	$4.55 \cdot 10^{17}$
11	$N_{l_4}, 1/m^3$	–	–	–	–	$2.33 \cdot 10^5$	$0.44 \cdot 10^5$
12	$N_{l_5}, 1/m^3$	–	–	–	–	–	$2.33 \cdot 10^5$
13	$m_s \cdot 10^{24}, kg$	7.891	7.887	2.278	2.334	2.332	2.332
14	$m_{l_1} \cdot 10^{22}, kg$	–	5.160	6.164	3.179	3.627	3.866
15	$m_{l_2} \cdot 10^{20}, kg$	–	–	1.038	1.360	0.986	1.148

16	$m_{l_3} \cdot 10^{19}, kg$	–	–	–	0.954	1.303	1.104
17	$m_{l_4} \cdot 10^{19}, kg$	–	–	–	–	5.232	6.922
18	$m_{l_5} \cdot 10^{18}, kg$	–	–	–	–	–	1.944
19	$\bar{c}_{c_s}, \frac{m}{s}$	29.523	29.297	53.383	52.977	52.607	52.535
20	$\bar{c}_{c_{l_1}}, \frac{m}{s}$	–	3.622	3.286	4.540	4.218	4.081
21	$\bar{c}_{c_{l_2}}, \frac{m}{s}$	–	–	0.800	0.694	0.809	0.749
22	$\bar{c}_{c_{l_3}}, \frac{m}{s}$	–	–	–	0.262	0.223	0.242
23	$\bar{c}_{c_{l_4}}, \frac{m}{s}$	–	–	–	–	0.111	0.096
24	$\bar{c}_{c_{l_5}}, \frac{m}{s}$	–	–	–	–	–	0.058

Table 2 shows the calculated indicators of the number of smaller crystals that should disappear as a result of coagulation with larger crystals, the crystals designated as “*l1*” will be smaller relative to the crystals “*l2*”, the crystals “*l2*” will be smaller relative to the crystals “*l3* ", etc. These indicators should be compared with the number of smaller crystals themselves, presented in table 1.

For example, in section 7, when coagulating with crystals “*l1*” in a volume unit, $5.12 \cdot 10^{-10}$ crystals “*s*” should disappear, while in this section in a volume unit there are $0.44 \cdot 10^5$ such crystals, t. e. 14 orders more. Therefore, in this section, a significant reduction of smaller crystals as a result of coagulation will not occur. In the same section 7, the number of crystals “*s*”, which must join the crystals “*l1*” to increase their parameter a by a factor of two, is determined by equation (2) and amounts to $1.066 \cdot 10^8$ per volume unit, which is more than the number of these crystals themselves (more than three orders), and even more so than the number of small crystals that can be reduced as a result of coagulation (more than 17 orders).

Table 2. The interaction parameters of groups of carbon dioxide crystals.

No	Crystal Groups	Parameters, Sizes	Sections				
			7	8	9	10	11
1	$s - l1$	$\Delta N_s/\Delta\tau, \frac{1}{m^3 \cdot s}$	$6.20 \cdot 10^{-6}$	$2.01 \cdot 10^7$	$6.28 \cdot 10^{19}$	$1.16 \cdot 10^{19}$	$4.40 \cdot 10^{18}$
		$\Delta N_s, \frac{1}{m^3}$	$5.12 \cdot 10^{-10}$	$1.76 \cdot 10^3$	$5.92 \cdot 10^{15}$	$1.18 \cdot 10^{15}$	$5.05 \cdot 10^{14}$
2	$s - l2$	$\Delta N_s/\Delta\tau, \frac{1}{m^3 \cdot s}$	–	$5.58 \cdot 10^8$	$5.46 \cdot 10^7$	$1.81 \cdot 10^{20}$	$3.18 \cdot 10^{19}$
		$\Delta N_s, \frac{1}{m^3}$	–	$4.91 \cdot 10^4$	$5.15 \cdot 10^3$	$1.85 \cdot 10^{16}$	$3.66 \cdot 10^{15}$
3	$s - l3$	$\Delta N_s/\Delta\tau, \frac{1}{m^3 \cdot s}$	–	–	$9.90 \cdot 10^8$	$9.10 \cdot 10^7$	$7.73 \cdot 10^{20}$
		$\Delta N_s, \frac{1}{m^3}$	–	–	$9.34 \cdot 10^4$	$9.28 \cdot 10^3$	$8.88 \cdot 10^{16}$

4	$s-l4$	$\Delta N_s/\Delta\tau, \frac{1}{m^3\cdot s}$	–	–	–	$1.18\cdot10^{9}$	$2.49\cdot10^{8}$
		$\Delta N_s, \frac{1}{m^3}$	–	–	–	$1.21\cdot10^{5}$	$2.86\cdot10^{4}$
5	$s-l5$	$\Delta N_s/\Delta\tau, \frac{1}{m^3\cdot s}$	–	–	–	–	$2.58\cdot10^{9}$
		$\Delta N_s, \frac{1}{m^3}$	–	–	–	–	$2.97\cdot10^{5}$
6	$l1-l2$	$\Delta N_{l_1}/\Delta\tau, \frac{1}{m^3\cdot s}$	–	$7.06\cdot10^{-6}$	$1.81\cdot10^{7}$	$6.74\cdot10^{19}$	$1.23\cdot10^{19}$
		$\Delta N_{l_1}, \frac{1}{m^3}$	–	$6.19\cdot10^{-10}$	$1.70\cdot10^{3}$	$6.88\cdot10^{15}$	$1.41\cdot10^{15}$
7	$l1-l3$	$\Delta N_{l_1}/\Delta\tau, \frac{1}{m^3\cdot s}$	–	–	$2.55\cdot10^{8}$	$2.38\cdot10^{7}$	$8.62\cdot10^{19}$
		$\Delta N_{l_1}, \frac{1}{m^3}$	–	–	$2.40\cdot10^{4}$	$2.42\cdot10^{3}$	$9.90\cdot10^{15}$
8	$l1-l4$	$\Delta N_{l_1}/\Delta\tau, \frac{1}{m^3\cdot s}$	–	–	–	$2.80\cdot10^{8}$	$2.44\cdot10^{7}$
		$\Delta N_{l_1}, \frac{1}{m^3}$	–	–	–	$2.86\cdot10^{4}$	$2.80\cdot10^{3}$
9	$l1-l5$	$\Delta N_{l_1}/\Delta\tau, \frac{1}{m^3\cdot s}$	–	–	–	–	$2.41\cdot10^{8}$
		$\Delta N_{l_1}, \frac{1}{m^3}$	–	–	–	–	$2.77\cdot10^{4}$
10	$l2-l3$	$\Delta N_{l_2}/\Delta\tau, \frac{1}{m^3\cdot s}$	–	–	$8.67\cdot10^{-6}$	$1.97\cdot10^{7}$	$8.09\cdot10^{19}$
		$\Delta N_{l_2}, \frac{1}{m^3}$	–	–	$8.18\cdot10^{-10}$	$2.00\cdot10^{3}$	$9.30\cdot10^{15}$
11	$l2-l4$	$\Delta N_{l_2}/\Delta\tau, \frac{1}{m^3\cdot s}$	–	–	–	$1.84\cdot10^{8}$	$1.67\cdot10^{7}$
		$\Delta N_{l_2}, \frac{1}{m^3}$	–	–	–	$1.88\cdot10^{4}$	$1.91\cdot10^{3}$
12	$l2-l5$	$\Delta N_{l_2}/\Delta\tau, \frac{1}{m^3\cdot s}$	–	–	–	–	$1.47\cdot10^{8}$
		$\Delta N_{l_2}, \frac{1}{m^3}$	–	–	–	–	$1.69\cdot10^{4}$
13	$l3-l4$	$\Delta N_{l_3}/\Delta\tau, \frac{1}{m^3\cdot s}$	–	–	–	$1.08\cdot10^{-5}$	$2.30\cdot10^{7}$
		$\Delta N_{l_3}, \frac{1}{m^3}$	–	–	–	$1.10\cdot10^{-9}$	$2.64\cdot10^{3}$
14	$l3-l5$	$\Delta N_{l_3}/\Delta\tau, \frac{1}{m^3\cdot s}$	–	–	–	–	$1.72\cdot10^{8}$
		$\Delta N_{l_3}, \frac{1}{m^3}$	–	–	–	–	$1.97\cdot10^{4}$

15	*l*4 – *l*5	$\Delta N_{l_4}/\Delta\tau, \frac{1}{m^3 \cdot s}$	–	–	–	–	$1.31 \cdot 10^{-5}$
		$\Delta N_{l_4}, \frac{1}{m^3}$	–	–	–	–	$1.51 \cdot 10^{-9}$

3. Conclusion

Based on the calculation results of the coagulation process of carbon dioxide crystals formed during the expansion of the gas mixture with the initial parameters: pressure 200 kPa and a concentration of carbon dioxide of 10%, the following conclusions can be drawn.

The most significant possible reduction of smaller crystals will occur in section 10 during the interaction of the "*s* – *l*2" crystal groups, where $1.85 \cdot 10^{16}$ can disappear from the total number (per volume unit) of "*s*" crystals equal to $7.9 \cdot 10^{16}$, that is 23%.

But to increase at least 2 times the size a of crystals "*l*2", the number of crystals "*s*" (per unit volume) equal to$1.35 \cdot 10^{22}$, which is almost 6 orders larger than their number in this section, will be required. Therefore, even this interaction of groups of crystals will not lead to the growth of larger crystals. It can be assumed that the number of crystals "*s*" that is capable of coagulation with crystals "*l*2", but which is not enough to increase the size of crystals "*l*2", will participate in the formation of crystals "*l*1" in the following section (along with other crystals "*s*" in the considered section).

In all the indicated sections of the flow part, the interaction of all groups of crystals will not lead to the growth of larger crystals due to the addition of smaller crystals to them, and the number of small crystals capable of coagulation is insignificant.

Further studies will be devoted to calculations of the process of coagulation of carbon dioxide crystals formed during the expansion of gas mixtures with different initial parameters and in other expansion machines, which will allow evaluating the effect of coagulation under conditions of different rates of crystal formation and growth.

References

[1] Danilov M M 2007 Modelirovanie protsessa vymorazhivaniia dioksida ugleroda v ob'eme rasshiriaiushchegosia gazovogo potoka [Modeling of the process of freezing carbon dioxide in the volume of expanding gas flow] *Izvestiia SPbGUNiPT: Mezhvuz. sb. nauch. tr.* **1** 6–8 [In Russ.]

[2] Danilov M M and Smirnov A S 2014 Osnovnye osobennosti obrazovaniia tverdoi fazy dioksida [Main features of formation of solid phase of dioxide] *Vestnik Mezhdunarodnoi akademii kholoda* **2** pp 37–40 [In Russ.]

[3] Tishin A P and Khairutdinov R I 1971 K raschetu koaguliatsii chastits kondensata v soplakh Lavalia [To the calculation of particle coagulation condensation in Laval nozzles] *Izvestiia AN SSSR. Mekhanika zhidkosti i gaza* **5** 18–185 [In Russ.]

[4] Danilov M M 2015 Analiz protsessa kristalloobrazovaniia dioksida ugleroda v protochnoi chasti tsentrostremitel'nogo turbodetandera [Analysis of the process of carbon dioxide crystallization in the flow part of a centripetal turbo expander] *Nauchnyi zhurnal NIU ITMO Seriia: Kholodil'naia tekhnika i konditsionirovanie* **4** 18–23 [In Russ.]

[5] Sapozhkov S B, Arendateleva S I and Ushakova O V 2019 *IOP Conf. Ser.: Mater. Sci. Eng.* **656** 012046

[6] Li X Y, Brandenburg A, Haugen N E L and Svensson G 2017 Eulerian and Lagrangian approaches to multidimensional condensation and collection *J. Adv Model. Earth Syst* vol 9 **2** 1116–37

[7] Galdin V D and Kondrat'ev N V 2003 Rezul'taty rascheta protsessa koaguliatsii tverdogo dioksida ugleroda pri rasshirenii dymovykh gazov v turbodetandere [Results of calculating the process of solid carbon dioxide coagulation during the expansion of flue gases in a turbodetander] *Omskii nauchnyi vestnik* (Omsk: OmGTU) **21** 80–83

IOP Conf. Series: Materials Science and Engineering **939** (2020) 012020 doi:10.1088/1757-899X/939/1/012020

Mathematical modeling of seed dissemination with a colter drill across the planting acreage

A V Evchenko[1], A P Shevchenko[1], M A Begunov[1], V S Koval[1] and D N Algazin[2]

[1] State Agrarian University named after P.A. Stolypin, ul. Tyumenskaya, 18 646532 Tara, Omskaya oblast', Russia
[2] Yaroslav-the-Wise Novgorod State University, ul. B. St. Petersburgskaya, 41 173003 Veliky Novgorod, Russia

[3] E-mail: maksim-begunov@mail.ru

Abstract. The article is devoted to the development of a mathematical model of seed dissemination process with a colter drill equipped with a discharge channel across the feeding area. This mathematical model makes it possible to determine the optimal technological and constructive parameters of the colter drill of the planter.

1. Introduction

Regularly spaced seed dissemination across the feeding area is one of the important conditions for high-quality grain sowing [1]. The evidence from practice shows that modern seeding machines do not fully meet the requirements for regularly spaced seed dissemination across the feeding area, which ultimately affects the yield of the crop [2]. A colter drill with a discharge channel, see figure 1, capable of eliminating the above-mentioned drawbacks has been developed to solve this problem [3].

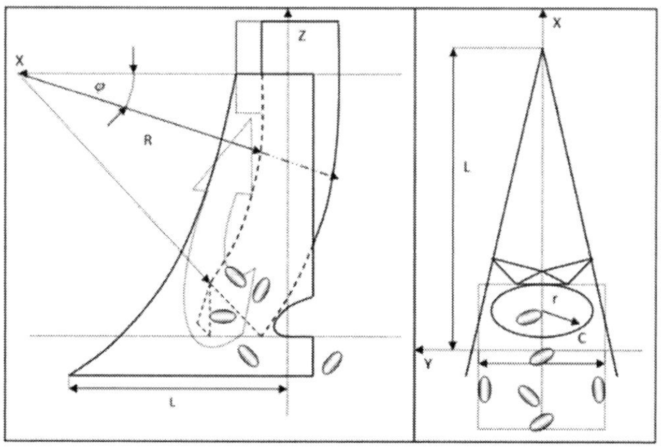

Left view **Top view**

Figure 1. The work scheme of a colter drill with a discharge channel: L – the length of the subdrill space; C-the width of the subdrill space; r – the radius of the drill seed tube; R – the radius of the bend of the drill seed tube; φ – the angle of the bend of the drill seed tube.

Content from this work may be used under the terms of the Creative Commons Attribution 3.0 licence. Any further distribution of this work must maintain attribution to the author(s) and the title of the work, journal citation and DOI.
Published under licence by IOP Publishing Ltd

IOP Conf. Series: Materials Science and Engineering **939** (2020) 012020 doi:10.1088/1757-899X/939/1/012020

2. Model

To determine the rational constructive and technological parameters of the colter drill, it is necessary to develop a mathematical model that will reveal the correlation between the uniformity of seed material distribution and the constructive and technological parameters of the colter drill [4].

Since the seeds are randomly released, the density of dissemination can be calculated using the Poisson formula [5]:

$$P(t) = ae^{-at}, \tag{1}$$

where a is the frequency of seed release.

The mathematical expectation of one seed's release time is equal to:

$$\int_0^{\infty} t\,P(t)dt = \int_0^{\infty} t\,ae^{-at}dt = \frac{1}{a} \tag{2}$$

To ensure the desired average seeding rate (λ=500 PCs/m2), select the parameter value from the following ratio:

$$a = 2 \cdot \mathrm{B} \cdot \lambda \cdot V, \tag{3}$$

where 2B – the width of the furrow opener;

V – the speed of the colter planter;

λ – seeding rate.

In this case

$$a = 0.07 \cdot 500 \cdot 2.5 \approx 87.5 \text{seed/sec}.$$

To generate the point of initial location of the grain when it hits the drill seed tube and its initial speed, it was decided to use the built-in function of generating random values of the Visual Basic translator.

The following ratio was taken for the initial speed:

$$(V_x, V_y, V_z) = (V_x^0, V_y^0, V_z^0) + (E_x, E_y, E_z), \tag{4}$$

where $\vec{\mathrm{E}} = (E_x, E_y, E_z)$– a random vector distributed according to the normal law with a probability density.

(V_x^0, V_y^0, V_z^0) – some vector determined by the design of the drill seed tube.

Since the mathematical expectation of a random vector $\vec{E}$ is equal to the zero vector, $\vec{V}_0$ is the most likely seed release speed.

To generate a seed throw point inside the drill seed tube's rack, the following ratio was taken:

$$\vec{Q} = (Q_x, Q_y, Q_z) + (\mathrm{E}_x, \mathrm{E}_y, \mathrm{E}_z), \tag{5}$$

where $\vec{Q}$ (0, 0, E_z) – the point where the seed is most likely to appear;

$\vec{\mathrm{E}} = (\mathrm{E}_x, \mathrm{E}_y, \mathrm{E}_z)$ – a random vector distributed according to the normal law, with the E_z coordinate determined by the velocity of the air flow.

$$-\frac{d}{2} \le E_x \le \frac{d}{2}, \tag{6}$$

$$-\sqrt{\frac{d^2}{4} - E_x^2} \le E_y \le \sqrt{\frac{d^2}{4} - E_x^2}, \tag{7}$$

IOP Conf. Series: Materials Science and Engineering **939** (2020) 012020 doi:10.1088/1757-899X/939/1/012020

$$0 \leq E_z \leq p \cdot V_z, \tag{8}$$

where $p = \frac{1}{a}$ – density of grain flow.

That is, the coordinates of the grain are formed within the section of the drill seed tube up to the colter drill rack.

The seed flow is formed taking into account the flow density according to the step scheme using the formula:

$$E_z^i = i \cdot p \cdot V_z + E_z. \tag{9}$$

The analyzed connections and dependences of the seed movement make it possible to identify the general flight pattern of the seeds presented in figure 2 in the "drill seed tube-reflector-soil" system after colliding with each other, with the walls of the drill seed tube, with the surface of the reflector, with the soil and to identify their location coordinates in the area of the soil.

According to the above method, it is possible to write a program in Visual Basic for a computer of the JBMPC type. Using this program, calculations were performed to determine the optimal parameters of the discharge channel and the reflector of the colter drill.

Simulating the process of seed dissemination across the feeding area, we obtain the coordinates of the landing places of grain seeds [1]. After analysis and statistical processing of these data, it is possible to assess the quality of seed dissemination across the feeding area [6].

For more convenient calculations the following values were accepted as constant for all variants: seeding rate Ne=5mln. per 1 ha or (λ=500 PCs. / m2), the speed of the colter drill V=2.5 m/s, the average speed of the seed material in the drill seed tube Vc=20 m/s, the width of the colter nozzle b = 70 mm, the bending angle of the drill seed tube φ=45°, the internal size of the drill seed tube – a radius of 14 mm, the bending radius of the drill seed tube R=100 mm.

3. Results and Discussion

The regularity of seed dissemination across the feeding area can be considered ideal if the value of the indicator of uneven dissemination is equal to one [7].

The following variable parameters were taken: the diameter of the discharge channel d; the angle of inclination of the reflector to the horizon α; the angle between the faces of the reflector β; the height of the reflector H, the width of the reflector B, the coordinates of points describing the surface of the reflector:

$$\bar{O} = (O_x, O_y),\ \bar{M}_1 = (M_{1y}, M_{1z}),\ \bar{M}_2 = (M_{2y}, M_{2z}),\ \bar{P}_1 = (P_{1y}, P_{1z}),\ \bar{P}_2 = (P_{2y}, P_{2z}), \bar{F} = (F_x, F_z).$$

By changing the values of the diameter of the discharge channel and the constructive parameters of the reflector, it becomes possible to determine the distances between seeds, when they are fixed along the profile of the seed groove (along the X and Y axes).

Analyzing the results of the carried out calculations it can be noted that when the diameter of the discharge channel d=28 mm, the reflector height H=20 mm, width of reflector B=20 mm, the angle of inclination of the reflector to the horizon α=1050; angle between the faces of the reflector β=1350; and for given numerical values of coordinates of points describing the surface of the reflector seeds

$$\bar{O} = (39, -61),\ \bar{M}_1 = (10, -61),\ \bar{M}_2 = (-10, -61),\ \bar{P}_1 = (-15, -81),\ \bar{P}_2 = (14, -81), \bar{F} = (30, -79)$$

the maximum regularity of seed dissemination both along and across the seed groove is achieved.

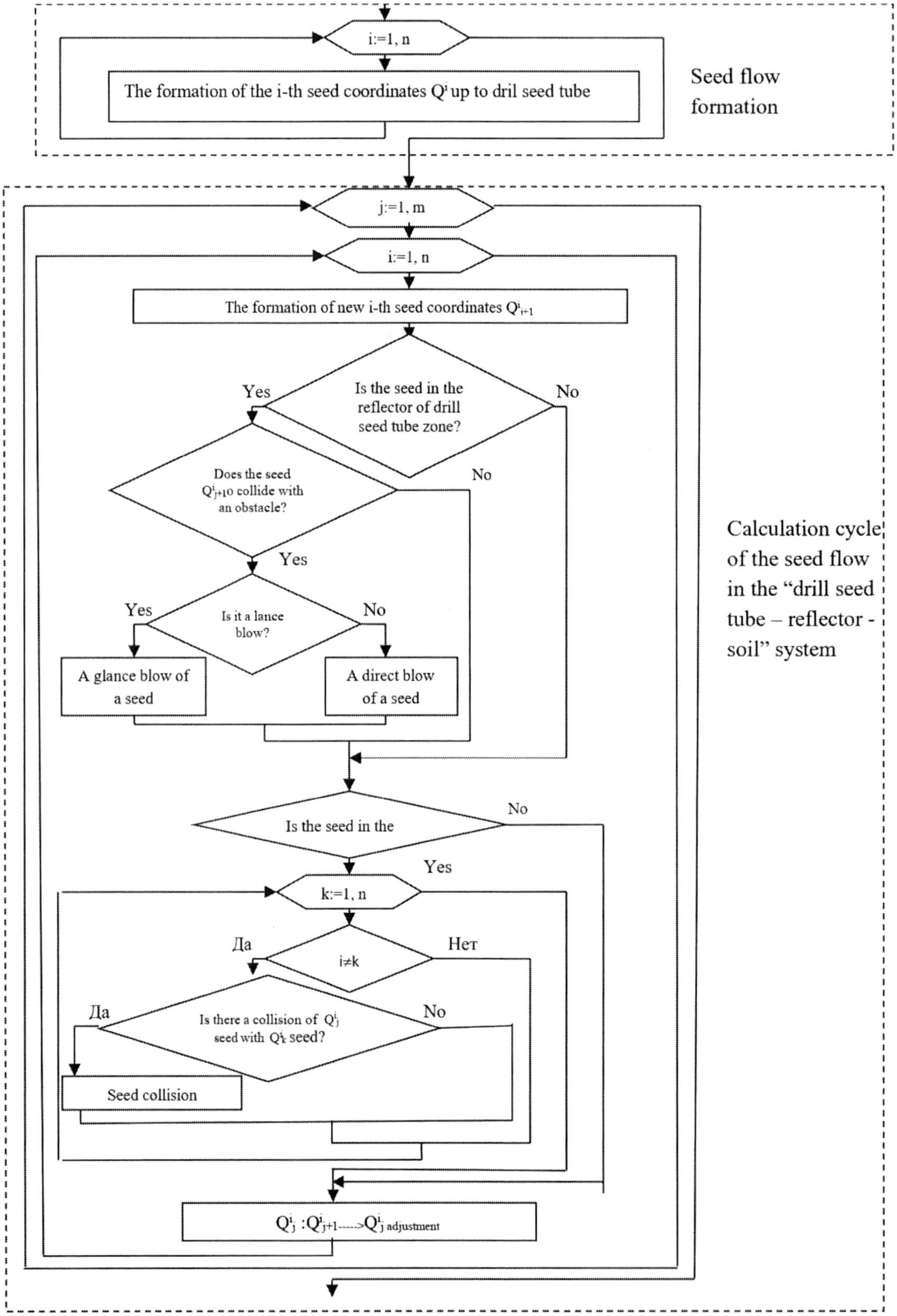

Figure 2. The scheme of seed movement formation.

Figure 3 and figure 4 show the theoretical dependence of the regularity of the seed material distribution along the length and width of the seed groove on the diameter of the discharge channel d mm, angle α degrees, with fixed values of the width of the reflector, the height of the reflector H, angle β.

The peak values of the response surface in the diagram show the areas that correspond to the values of rational parameters of the discharge channel and the seed reflector [9].

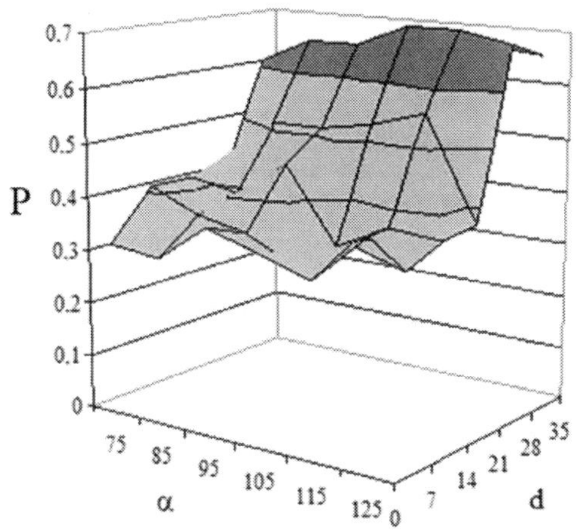

Figure 3. Dependence $P = f(\alpha, d)$ along the seed groove, H = 20 mm; B = 20 m; β = 135°.

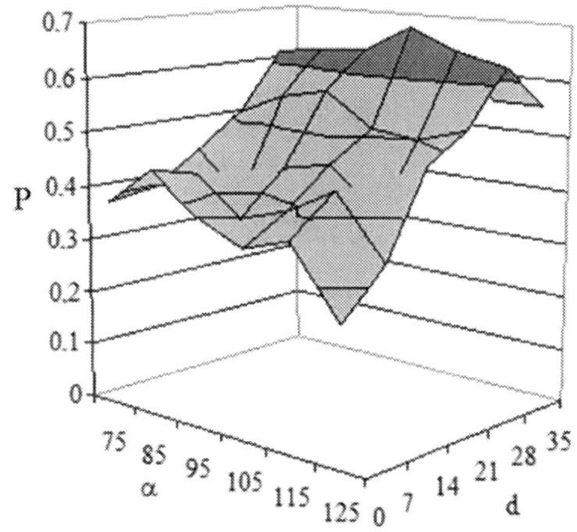

Figure 4. Dependence $P=f(\alpha, d)$ across he seed groove, H = 20 mm; B = 20 mm; β = 135°.

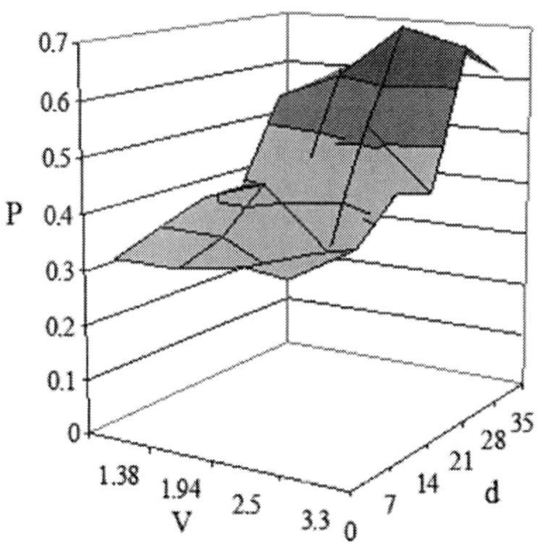

Figure 5. The response surface that characterizes the regularity of seed dissemination along the length of the seed groove, depending on the speed of the machine V m/s and the discharge channel parameters where B=20 mm, H=20 mm, α=105°, β=135°.

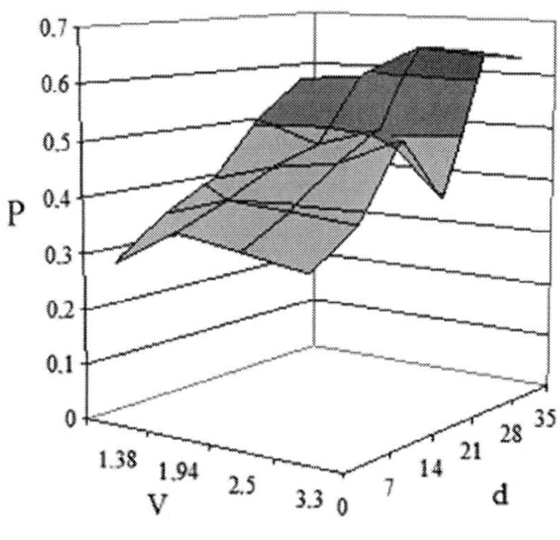

Figure 6. The response surface that characterizes the regularity seed dissemination across the width of the seed groove, depending on the speed of the macine V m/s and the discharge channel parameters where B=20 mm, H=20 mm, α=105°, β=135°.

Figure 5 and figure 6 show the results of theoretical calculations of the regularity of seed dissemination along the length and width of the seed groove depending on the speed of the macine V and the discharge channel parameters d with fixed values of the inclination angle α of the reflector, the angle between the faces β of the reflector, height of the reflector B and width and reflector H.

4. Conclusion

The developed mathematical model makes it possible to reveal the correlation between the regularity of seed material distribution across the feeding area and the constructive and technological parameters of the colter drill. Thus, based on the optimal value of the seed dissemination across the feeding area, using this mathematical model, it is possible to determine the rational constructive and technological parameters of the olter drill.

Based on the research results, it can be noted that when the diameter of the discharge channel d = 26–28 mm, the reflector height H = 20 mm, width of the reflector B = 20 mm, the angle of inclination of the reflector to the horizon α = 102–1070; the angle between the faces of the reflector β = 1350 and the speed of the machine's movement is in the range V = 2...3 m/s, maximum regularity of seed dissemination both along the length and width of the seed groove is achieved.

References

[1] Yevchenko A V, Chernyakov A V, Begunov M A, Koval' V S and Pavlyuchenko K V 2017 Metodika i rezul'taty laboratornykh issledovaniy soshnika po raspredeleniyu semyan po ploshchadi pitaniya [Technique and Results of Laboratory Investigations of the Coulter for Seed Distribution by Area of Nutrition] *Elektronnyy nauchno-metodicheskiy zhurnal Omskogo GAU* **1(8)** 2 [In Russ.]

[2] Shevchenko A P and Begunov M A 2014 Eksperimental'noye issledovaniye dvukhstrochnogo kilevidnogo soshnika dlya poseva semyan l'na [Experimental study of the operation of a two-row keeled opener for sowing flax seeds] *Omskiy nauchnyy vestnik* **1** 152–155 [In Russ.]

[3] Demchuk E V, Sabiev U K, Mylo V V, Soyunov A S and Koval V S 2018 *J. Phys.: Conf. Ser.* **1059** 012009

[4] Soyunov A S *at al.* 2019 *IOP Conf. Ser.: Mater. Sci. Eng.* **582** 012037

[5] Koval V, Chernyakov A, Shevchenko A, Begunov M and Demchuk E 2019 *IOP Conf. Ser.: Mater. Sci. Eng.* **582** 012040

[6] Redreev G V 2016 *IOP Conf. Ser.: Mater. Sci. Eng.* **142** 12085

[7] Shevchenko A P, Bankrutenko A V, Koval V S, Begunov M A and Demchuk E V 2018 *J. Phys.: Conf. Ser.* **1059** 012011

Stability evaluation of cutting edges in operation of composite materials for instrumental purposes based on a metal matrix

A A Fedotov[1,3], **M N Safonova**[1] **and A S Syromyatnikova**[2]

[1] North-Eastern Federal University, Yakutsk, Russia
[2] V.P. Larionov Institute of the Physical-Technical Problems of the North of the Siberian Branch of the RAS, Yakutsk, Russia

[3] E-mail: fedot_andrey@mail.ru

Abstract. The authors analyzed the existing methods for assessing the stability of cutting grains in the article. The most optimal method for assessing the performance of abrasive grains was selected based on the results of the analysis. This method was used to calculate the concentration of diamond grains on the working surfaces of samples of a composite material made on the basis of bronze. The results of the studies showed that when testing the samples for friction and wear, the number of active abrasive grains and the relative solids content of natural diamond grinding powders show the stability of the cutting edges.

1. Introduction

The performance evaluation of grinding tools is made by the value of the specific diamond consumption (by weighing or counting the number of grains dropped), by measuring the linear and mass wear of the tool, productivity and roughness of the treated surface, which is a very time-consuming and costly procedure. However, it is known that grinding performance and processing quality significantly depend on the stability of the tool cutting properties, that is, the stability of the number of active diamond grains during its operation. Tool stability is an indicator characterizing the constancy of the diamond abrasive disk's cutting properties. Therefore, when developing abrasive composite materials, it is especially important to determine the change in the number of active grains during friction and wear.

Most researchers determine the number of active (cutting) grains by various experimental methods. When deriving an equation for calculating the number of grains on the sintering tool surface, one of the main characteristics is their shape. The actual grain shape differs from the indicated models and depends on the brand of diamonds.

All the described methods for calculating the number of grains are not universal; they are developed for special cases of research, taking into account the features of the objects producing [1–8]. Existing calculation methods make it possible to calculate the initial volumetric concentration of active grains in a binder, however, they do not allow to determine its change during the operation of the tool.

The content of grains in the working layer is an important characteristic of a diamond tool, significantly affecting its cutting properties, durability and service life, as well as the forces and temperature that occur during cutting.

Content from this work may be used under the terms of the Creative Commons Attribution 3.0 licence. Any further distribution of this work must maintain attribution to the author(s) and the title of the work, journal citation and DOI.
Published under licence by IOP Publishing Ltd

2. Methods

The object of the study is a standard binder – tin bronze M2-01 (20 mass% tin, 80% copper) with additives from 0–4 mass% of ultrafine natural diamond.

Samples (cylindrical, with a height and diameter of 1 cm) were made by sintering in vacuum (0.1*10^{-3} Pa) in an SNVE furnace. Joint sintering ensured the constancy of the process conditions for a given batch of samples. Empirically determined that the optimal temperature is 700 ℃. The sintering time was 60 minutes.

The calculation and experimental method used to determine the number of active abrasive grains with a statistically uniform distribution in the volume of composite material that was developed in [9] is used. When developing the method, we proceeded from the assumption that abrasive grains are a collection of particles of arbitrary shape and different dispersion, distributed in the bulk of the material statistically uniformly with a random spatial orientation.

When modeling grinding processes, it is important to use the shape of abrasive grains as the base model. In the scientific literature there are various views on the choice of such models: in the form of a cone with a rounded top, a ball, an ellipsoid of rotation, a cylinder, a cube. Given these differences, various models were compared and regulated according to the degree of approximation to the experimental data [10].

In a number of cases, the grain located on the plane is located so that the dimensions visible in the plan and taken as the length and width are smaller than the third size, conventionally called the height and visible on the second projection. Therefore, to select a geometric model of the shape of a real diamond grain, it is necessary to most accurately determine the linear dimensions of the grain from volume measurements in two projections.

The grain size characteristics necessary for simulating the control sieve screening procedures were determined from two projections obtained using a SEM XL-20 scanning electron microscope (Philips) in the secondary electron mode. Measurement of the length and width of the projection of grains, their height by SEM photographs (figure 1).

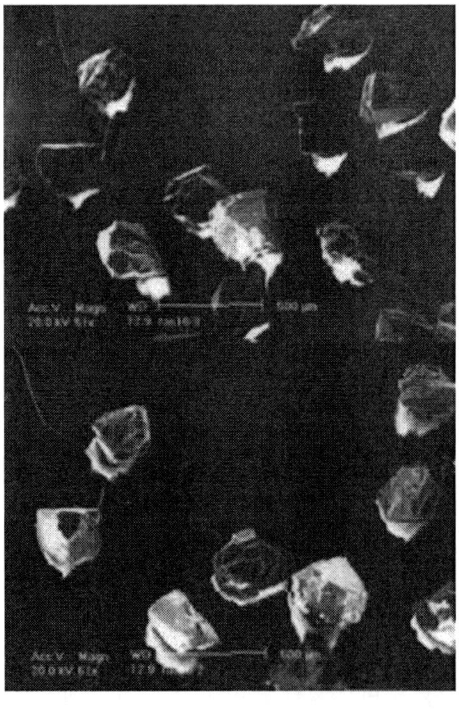

Figure 1. SEM images of grains of diamond grinding powders for modeling grains in two projections.

Method for determining the relative solids content in biphasic alloys.

Along with methods for calculating the quantity and size distribution of microparticles, there is a point analysis method that does not require geometric modeling of particles and the derivation of the distribution law, and which can be used to establish the volume fraction of the second phase. The method of point analysis is based on the fact that the proportion of points randomly printed on the micrograph that fall on the images of the phase under study is equal to the volume fraction (or area fraction) of this phase. In order for this principle to be fulfilled, points must be applied randomly to the micrograph. The volume fraction of phase α is given by the fraction of points that fell on the areas occupied by the studied phase.

In practice, it is usually inconvenient to deal with a random set of points and you can use an ordered point system. Figure 2 shows the application of the method of point analysis to determine the volume fraction of the solid phase (diamond particles) in the volume of the composite based on tin bronze (M1) and the study of their changes during friction and wear.

It is assumed that the microstructural components are distributed randomly. This assumption is mainly satisfied for most ordinary microstructures, but it must be remembered in cases where a completely regular arrangement of structural components is observed.

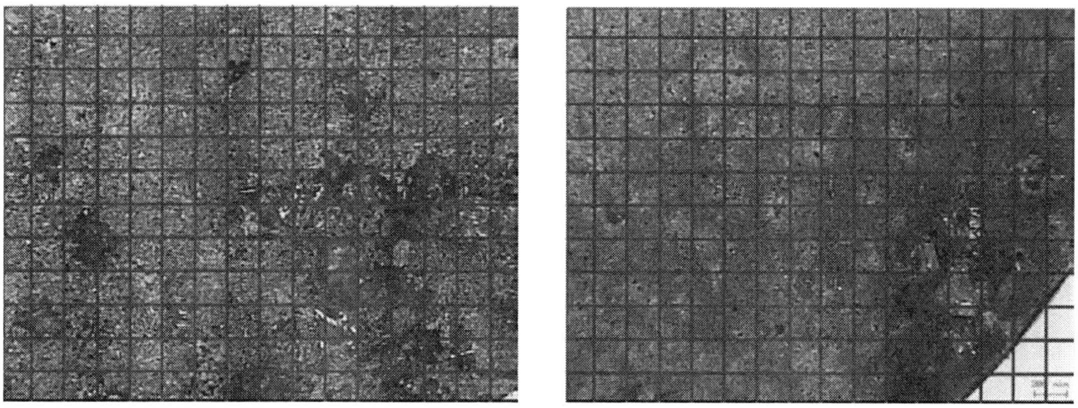

Figure 2. The friction surface of composite materials based on M1 - grinding powders from natural diamonds 315/250 microns.

It is very important that the distance between the nodal points in a rectangular grid is large compared to the dimensions of the studied structural components. If this distance is very small, then the condition for the chaotic arrangement of the components is not satisfied, and, therefore, the error in determining the phase volume fraction can significantly increase and become larger than the expected value. In this case, the calculation of the volume fraction of the phase will strongly depend on the location of the point system with respect to the individual particles of the second phase.

3. Results and discussion

As is known, the shape coefficient *Kф* of an individual grain is defined as the ratio of the length of its projection to its width. The isometricity of grains (in percent), following the methodology of the standard [10], was determined by the formula

$$u = \frac{u_1}{n} \cdot 100,$$

where u_1 is the number of isometric grains, n is the number of measured grains. The number of isometric grains u_1 was calculated by calculating the shape coefficient. Grain is considered isometric, if the shape coefficient *Kф* does not exceed 1.3. The obtained values of the grain isometricity are shown in table 1.

Table 1. Analysis of the adequacy of the choice of the basic model of diamond grains.

Grinding powder	Grain, μm		Geometric model of grain				Isometricity, %
	By GOST	Rectangular box	ellipsoid	spheroid	cube	octahedron	
ND I	50/40	0.18	0.11	0.12	0.123	0.13	26.3
ND II	80/63	0.13	0.14	0.17	0.12	0.14	41.81
ND III	125/100	0.21	0.19	0.23	0.156	0.17	49.3
ND IV	315/250	0.25	0.19	0.27	0.19	0.18	39.5

A visual examination of SEM – images of diamond grinding powders grains showed that the synthetic diamonds grains have a rounded shape, and natural diamonds grains are mainly lamellar and needle-shaped, there are crystals with a clear cut and sharp edges. The variational range of values is wider, the frequency is lower, and its modal value lies in the range of values of the shape factor greater than 1.3 for natural diamond powders. The consequence of this is a significant difference in the isometricity of the grains.

The method developed in [9] was used to calculate the concentration of diamond grains on the working surfaces of samples based on the M1 metal binder with 6% by weight of natural diamond powders with a grain size of 315/250 μm in the initial state and after testing on a friction machine (figure 3) . The principle of operation of the machine is the abrasion of a pair of test samples pressed against each other by force P.

The calculation of the number of particles of the *k-th* size group N_k was carried out according to the formula obtained using the probability distribution F_i for cubic particles [9]:

$$N_k = 1/H_k\,(2{,}433\,n_k - 0{,}971\,n_{k-1} - 0{,}270\,n_{k-2} - 0{,}097\,n_{k-3} - 0{,}044\,n_{k-4} - 0{,}029\,n_{k-5} - 0{,}019\,n_{k-6} - 0{,}012\,n_{k-7} - 0{,}007\,n_{k-8} - 0{,}007\,n_{k-9}), \quad (1)$$

wher $H_k = 1{,}5\,h$ – average cube height if h is the cube edge (h was defined as the half-sum of the sides of the rectangle described around the projection of a random grain section [11]).

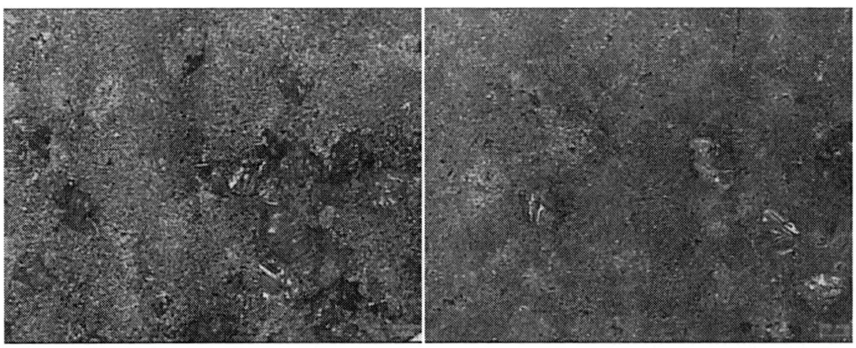

Figure 3. Friction surfaces of composite materials M1 – grinding powders from natural diamonds 315/250 microns.

The relative change in the number of active grains in the binder during friction and wear $C = (n_0 - n_e / n_0) \times 100\%$ characterizes the stability of the number of active grains in the binder.

Figure 4 shows the size distribution of diamond particles in the metal matrix of the studied tin bronze composite materials with 6% by weight of natural diamond grinding powders with a grain size of 315/250 microns.

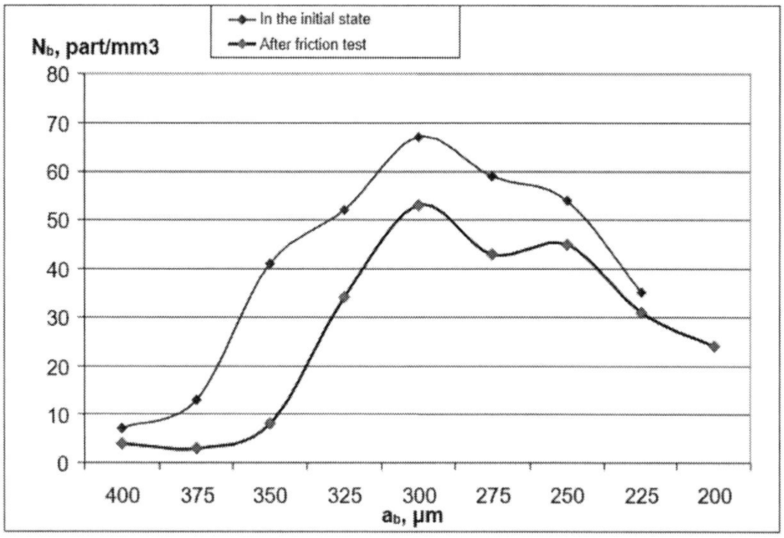

Figure 4. The size distribution of diamond particles in the composite on the working surface in the initial state and after a wear test with 6% by weight of natural diamond grinding powders with a grain size of 315/250 microns.

The volume fraction p_α of phase α is determined from the expression

$$p_\alpha = n_\alpha / n \ , \qquad (2)$$

where n_α – число the number of points located on the phase α, и n – total number of points plotted on the microstructure image.

The standard deviation of the volume fraction depending on the number of points used in the analysis:

$$\sigma_{p\alpha} = [p_\alpha (1 - p_\alpha) / n]^{1/2}. \qquad (3)$$

This expression is also valid for the analysis of microstructures containing more than two phases.

The results of determining the volume fraction of the solid phase (diamond particles) based on a metal binder (M1) are presented in table 2.

Table 2. Determination of the volume fraction of the solid phase (diamond particles) in the volume of the composite based on a metal binder (M1).

	n	n_α	p_α	$\sigma_{p\alpha}$
Sample	Total number of points	The number of points located on phase α	Volume fraction of phase α	The standard deviation of the volume fraction
Sample with 315/250 before testing	59	15	0.294237	0.056688367
Sample with 315/250 after testing	36	8	0.21223	0.069289952

The method of point analysis, which does not require geometric modeling of particles and the derivation of the distribution law, which is used to establish the volume fraction of the solid phase, can be used as an indirect express method for determining the health of composite materials with superhard materials powders fillers.

4. Conclusions

The results of the studies showed that when testing the samples for friction and wear, the number of active abrasive grains and the relative solids content of grinding powders from natural diamonds with a grain size of 315/250 μm show the stability of the cutting edges in the self-sharpening mode in a metal matrix based on tin bronze.

References

[1] Novikov N V and Bogatyreva G P 2008 *Superhard Materials* **2** 3
[2] Petasyuk G A and Sirota Yu V 2012 *Superhard Materials* **6** 70
[3] Petasyuk G A 2010 *Superhard Materials* **2** 80
[4] Dexiang W, Pequi G, Lei Z and Zhenjie Z 2012 *Applied Mechanics and Materials* **474** 229
[5] List E, Vollstaedt H and Frenzel J 2008 *5th Zhengzhou Int. Superhard Materials and Related Products Conf.*
[6] Engels A 2003 *Industrial diamond review* **63(2)** 39
[7] Lavrinenko V I, Bogatyreva G P, Ilnitskaya G D, Petasyuk G A and Smokvina V V 2011 *Rock-cutting and metalworking tools – equipment and technology for its manufacture and use* **15** 484
[8] Zhuo Z, Xia S, Bai Q and Zhou B 2018 *Journal of Materials Science* **53-4** 2844
[9] Safonova M N, Syromyatnikova A S and Shitz E Yu 2007 *Friction and wear* **5** 84
[10] Safonova M N, Petasyuk G A and Syromyatnikova A S 2013 *Computer-analytical methods for diagnosing the operational characteristics of diamond powders and composite materials based on them* (Novosibirsk: Publ. House of the Siberian Branch of the Russian Academy of Sciences)
[11] GOST 9206-80 *Diamond powders. Technical specifications*

IOP Conf. Series: Materials Science and Engineering **939** (2020) 012022 doi:10.1088/1757-899X/939/1/012022

Physical and chemical cleaning of the surface of power metal optics for the purpose of increase in beam firmness

S Filin[1], **V Rogalin**[2], **I Kaplunov**[3,5] **and M Satayev**[4]

[1] Plekhanov National University of Economics, Stremyannyj per. 36 117997, Moscow, Russia
[2] Institute of Electrophysics and Electric Power, Russian Academy of Sciences, Dvorzovaya Naberezhnaya 18 191186, St. Petersburg, Russia
[3] Tver State University, Zhelyabova Str. 33 170100, Tver, Russia,
[4] SILKWAY International University, Tokayev Str., 27a 160011, Shymkent, Kazakhstan

[5]E-mail: kaplunov.ia@tversu.ru

Abstract. Features of the choice of solvents for the physical and chemical cleaning of power optical elements in accordance with the parametric solubility theory were investigated. When cleaning the surface from the model contaminant with halogenated hydrocarbons, visually observed white film of alkali and alkaline earth metal salts that are not soluble by this class of solvents and iridescent bands from the interaction of hydrocarbons with the metal remain on the surface. All this greatly impairs the optical properties of the surface. It was shown that, when using solvents, it is necessary to inhibit the interaction of halogenated hydrocarbons with mirrors by stabilizing solvents or by selecting the regimes for carrying out the physicochemical purification process, or by thereof combination.

1. Introduction

In industry, in science and technology, powerful CO_2 lasers emitting at a wavelength $\lambda = 10.6$ µm are used [1]. In the middle infrared region, CO and HF (DF) lasers are used [2]. Reflectors are usually cooled mirrors [3–9] made of copper, aluminium, beryllium, molybdenum, silver, etc. They have high thermal conductivity and a reflection coefficient in the operating range, which makes it possible to use mirrors without additional interference coatings that reduce the radiation strength. Cooling reduces the effect of thermomechanical distortions in the shape of mirrors, which require regular cleaning to restore performance.

Optical durability of power optics is the main requirement for laser systems operating at high radiation intensities. Optical stability in the continuous mode of radiation is due to a combination of absorptivity and thermomechanical properties of the material [10]. In the pulsed mode, plasma of low-threshold optical disruption of the gas, initiated by evaporation of the ions of the material from the surface, can occur near the surface, which can sometimes be damaged with the appearance of irreversible changes [3–5].

In the framework of the theory of thermal optical breakdown of solid transparent dielectric inclusions, the reason for the appearance of luminous spots on the metal under the action of laser radiation, which are distinguished by their brightness on the general background of the irradiated zone,

Content from this work may be used under the terms of the Creative Commons Attribution 3.0 licence. Any further distribution of this work must maintain attribution to the author(s) and the title of the work, journal citation and DOI.
Published under licence by IOP Publishing Ltd

is explained. Inclusions are local initiators of the development of optical breakdown of air near the surface [5, 11–14]. On these formations, the reflection of light is markedly reduced, and the temperature of their melting and evaporation considerably exceeds the corresponding values of the metal. The effect is enhanced by the interaction of the laser pulse with surface electromagnetic waves [15], since up to 100% of the radiation can be absorbed in this case.

In the process of using mirrors with a gradual increase in power, appears the effect of "laser cleaning". [15–18]. Due to the evaporation and burning out of the absorbing centres, the optical stability threshold increases. However, the selection of the regime is very complicated, since "laser cleaning" can be accompanied by oxidation of the metal.

The parameters of technological contaminations on the surface of metallic mirrors formed during optical processing and during operation and decreasing the optical stability of the mirror are revealed in this work. The efficiency of solvents for cleaning optics was also investigated.

The degree of impurities contamination of the surface of the irradiated optical element affects the occurrence of irreversible changes in the material, in particular, the absorption capacity, and, consequently, the thresholds of gas breakdown at the surface. It was noted in [11–14] that under the action of the first pulse on the unpurified surface, the threshold intensity is always several times lower than under the action of "subsequent" pulses. In [12], when the metal surface was irradiated consecutively with a series of pulses from a CO_2 laser, the energy expenditures for the formation of the plasma rose with increasing in the number of pulses to a certain value, at which they remained constant, i.e. the surface was cleaned in the presence of an air-breakdown plasma by removing the defects which absorb radiation. It was shown in [13] that the initial absorbing power for copper and aluminum targets, was 10–12 % and 6 %, respectively, but after laser cleaning it was reduced to 2–3 %, which is typical for pure unpolished samples.

2. Materials and methods

In reality, metals are covered with a sorbet layer of dielectric, mostly organic, contaminants. These may be areas of the surface covered with a continuous thin film or single ("island") film formations.

Since purification is a reduction of surface contamination to a certain level (surface modification), it is interesting to investigate the plasma formation threshold caused by this contamination as a function of the contamination radius (provided that it is small in comparison with the depth of the optical absorption).

Analysis the increase of the temperature of dielectric impurities on the basis of the theory [19] makes it possible to determine the initial heating of the absorbing contaminants by formula

$$\Delta T = \left[\frac{\omega \chi'' E_0^2}{c_t}\right] t, \tag{1}$$

where ω is the circular frequency of light; $\chi'' = \frac{\varepsilon'' i}{4\pi}$ is the imaginary part of the susceptibility; c_t is the heat capacity; E_0 is the field intensity; t is the time of exposure.

The depth of optical absorption is the reciprocal of the absorption coefficient

$$\mu = \frac{8\pi^2 \chi''}{\lambda n}, \tag{2}$$

where λ is the wavelength in vacuum; n is the refractive index.

For short radiation pulses of duration t_p, it is necessary to use the equation for ΔT, in which $t = t_p$. In many cases, t is determined by the time t_c of heat transfer from the inclusions. For stationary heating

$$t = \frac{t_0 = a^2}{D_\tau}, \tag{3}$$

where D_τ is the thermal diffusion coefficient.

The value D_τ for many solids at room temperature is of the order of ~ 10^{-2} cm^{2}c^{-1}, as a result, t_c ~1 μs for the particles with a radius of ~1 μm. If the particle radius is ≤ 0.1 μm, and t ≥ 10^{-8} s, the increase of the inclusion temperature can be recorded

$$\Delta T_{SS} = \left[\frac{\omega \chi'' E_0^2 a^2}{D_t c_t}\right] t \tag{4}$$

It follows from equation (4) that the field strength required to achieve a critical temperature for inclusions of ~0.01 μm should be ~2.1·10^6 Vcm^{-1} (for λ = 10.6 μm), which is comparable with the plasma formation of a pure material (for example, of glass). Consequently, inclusions with a radius of ~0.01 μm from the point of view of lowering the threshold are less dangerous compared to larger ones. Inclusions ≥ 1 μm can be easily detected and removed. The most dangerous from the viewpoint of reducing the threshold of plasma formation are inclusions with a radius of ~0.1–1.0 μm located in the surface layer and having an optical absorption depth of 10^{-3} ÷ 10^{-4} cm. The critical radius (r_{cr}) of the spherical inclusion, less than which its absorption cross section begins to decrease, is at the wavelength λ = 10.6 μm for dielectrics. The critical radius (r_{cr}) of the spherical inclusion, less than which its absorption cross section begins to decrease, is at the wavelength λ = 10.6 μm for dielectrics

$$r_{cr} = \mu^{-1}, \tag{5}$$

where μ is the radiation absorption coefficient.

Equation (5) defines a minimum limit, indicating below what size on the treated surface after physical and chemical cleaning inclusions may be present. Thus, for λ = 0.63 μm, the maximum impurity size below which it is impractical to clean the surface is ~ 0.1 μm. The presented evaluations have largely a qualitative nature, since under pulse actions the index (D_T) can increase to 10^6 times [3].

Structural defects (grain boundaries, dislocations, vacancies, etc.) contribute to a decrease in the plasma formation threshold by more than 2 orders of magnitude [14].

Surface inclusions, having a much larger (compared to bulk) absorption, along with a decrease in the plasma gas formation threshold near the metal, are also responsible for the destruction thresholds caused by their thermal explosion [11].

The choice of solvents for the physical and chemical cleaning of power cells in accordance with the parametric solubility theory was carried out as follows. An analysis of the process contaminants present on the surface of the metal showed that their bulk (up to 85 % by weight) mainly includes rosin pitch (with a rosin content of up to 70 % by weight), used to manufacture a polisher for processing optics. The pitch is a polymerization type compound whose solubility parameter (δ) is not determined from the heat of evaporation or other quantities. Therefore, it is determined experimentally by the nature of solubility in solvents. For rosin, δ = 18 J$^{1/2}$cm$^{-3/2}$ [20].

3. Result and discussion

At the first stage, a kinetic analysis of the physical and chemical process of dissolving the rosin pitch was carried out in solvents with an indicator (δ) close to rosin. The choice of solvents was carried out from various classes of selectivity according to Snyder [21, 22]. Then we carried out a kinetic analysis of the physical and chemical process of dissolving the rosin pitch in these solvents, for which 1 g of resin was placed in a container and dissolved in solvents. The amount of resin dissolved after each cycle of contact was determined on an EF-3MA electronic fluorometer (table 1).

The data in Table 1 show that acetone and diethyl ketone have the maximum solubility with respect to rosin pitch (after 4–5 cycles of dissolution, the rosin is not detected, completely passing into solution). Somewhat worse, its dissolution occurs in diethyl carbonate, xylol, methylene dichloride, 2–butanol. The bulk of the rosin pitch dissolves after 6–7 cycles (only insignificant amount of visually observed suspension remains undissolved).

The results of the experiment (table 1): rosin pitch is most effectively dissolved in solvents belonging to the class "VI" of polar selectivity according to Snyder. Therefore, in order to obtain the optimum

IOP Conf. Series: Materials Science and Engineering **939** (2020) 012022 doi:10.1088/1757-899X/939/1/012022

solvent for this pitch, similar studies of the process of its dissolution in solvents of the class "VI" of polar selectivity (table 2) were carried out.

According to tables 1 and 2, among solvents with a class "VI" of polar selectivity, acetone and methylethyl ketone have a maximum solubility with respect to rosin pitch. It can be assumed that for rosin pitch, the index (δ) is 19.8 $J^{1/2}cm^{-3/2}$, which corresponds to the solubility parameter of acetone, in which it dissolves most effectively.

Table 1. Amount of model solvent.

Solvent	Amount of model solvent, mg						
	Dissolution cycles						
	1	2	3	4	5	6	7
1.	D	51.5	6.7	6.2		ND	
2.	D	82.3	32.4	11.6	10.8	9.0	8.6 S
3.	D	70.3	50.4	38.5	27.1	14.9	7.3 S
4.	-	-	D	165	37.3	11.3	10.0 S
5.	-	-	D	65.4	33.2	10.6	8.7 S
6.	-	-	D	83.1	49.9	21.3	17.4S
7.	D	63.8	10.1	6.9	6.1	ND	
8.	D	65.9	17.8	11.9	7.2	7.1	ND
9.	D	83.4	41.3	21.2	12.4	11.5	10.9
10.	D	70.1	40.3	25.4	15.2	12.9	8.8
11.	D	79.9	40.4	30.3	21.5	14.7	10.6
12.	-	-	D	63.2	31.4	20.3	14.7 S
13.	-	-	D	23.2	10.0	8.6	6.3 S
14.	-	-	D	161	36.1	10.9	9.8 S
15.	-	-	D	75.0	40.7	22.6	11.2 S
16.	D	69.4	39.2	22,4	14.8	9.9	7.8 S
17.	-	-	D	153	40.9	31.1	23.5 S
18.	-	200	79.9	51.4	35.9	23.7	16.9 S
19.	D	62.0	6.7	4.9	4.3	3.3	3.4 S

Notes. D > 200 mg. S is suspension. ND – not detected.

1. Acetone*, the solubility parameter (δ) = 19.8 $J^{1/2}cm^{-3/2}$; class of polar selectivity is "VI". 2. 1-Butanol*, δ = 23.2; class of polar selectivity is "II". 3. 2-Butanol*, δ = 22.1; class of polar selectivity is "II". 4. Petrol **; class of polar selectivity is "0". 5. Freon-114B2 ****, δ = 15.09; class of polar selectivity is "V". 6. Freon-12 ***, δ = 12.5; class of polar selectivity is "V". 7. Diethyl ketone*, δ = 18.0; class of polar selectivity is "VI". 8. Diethyl carbonate*, δ = 18; class of polar selectivity is "VI". 9. m-Cresol *, δ = 22.7; class of polar selectivity is "VIII". 10. Xylene*, δ = 18.0; class of polar selectivity is "VII". 11. Methylene chloride***, δ = 17.0; class of polar selectivity is "V". 12. Nitromethane***, δ = 25.2; class of polar selectivity is "VIII". 13. 2-Propanol***, δ = 23.6; class of polar selectivity is "II". 14. Petroleum ether**; class of polar selectivity is "0". 15. Freon-113***, δ = 14.8; class of polar selectivity is "V". 16. Furan*, δ = 18.7; class of polar selectivity is "I". 17. Cyclohexane*, δ = 16.8; class of polar selectivity is "0". 18. CCl_4*, δ = 17.8; class of polar selectivity is "V". 19. Ethanol**, δ = 26.6; class of polar selectivity is "II".

Solvents: *with an index (δ) close to rosin; **are used for cleaning widely. Their value (δ) is difficult to determine due to a large range of boiling point (≤ 40 °C) and because of a large range of densities; ***are used for cleaning in the optical and microelectronic industries; ****does not contain chlorine, which destroys ozone, has a low boiling point – 47.35 °C, refers to halogen-substituted aliphatic hydrocarbons considered to be the most effective in cleaning of optics, has a smaller specific gravity – 2.3 g/cm^3 and volatility, as well as has greater resistance to thermal heating and UV radiation in the presence of water and air.

Table 2. Amount of resin in the solvent.

Solvent	Amount of resin in the solvent, mg						
	Dissolution cycles						
	1	2	3	4	5	6	7
1.	>200	58.4	9.1	7.0	1.1		
2.	>200	55.3	8.9	6.8	3.2	CD	
3.	>200	72.6	12.4	9.3	6.5	CD	
4.	>200	62.0	11.9	9.8	6.1	CD	
5.	>200	83.5	15.4	10.0	9.1	4.3	CD

Note. 1. Methyl ethyl ketone (δ = 19.0). Cyclohexanone (δ = 19.7). 3. Diethyl phthalate (δ = 20.5). 4. Ethyl formate (δ = 19.7). 5. 1,4-Dioxane (δ = 19.7).
CD is complete dissolution.

In addition to it, there are other contaminants on the surface of the metal optics. The main mass of contaminants (up to 10 % by weight) are various kinds of oils, greases and fats of mineral and animal origin. Such impurities are most effectively removed by chlorine-substituted hydrocarbons [22]. When choosing the optimum solvent for such impurities, we mainly used solvents with a class "V" of polar selectivity.

As a model contamination, we used spindle and diffusion (VM-5) oils and grease from human hands. These impurities (total amount of 100 mg, taken in equal proportions) were applied to copper mirrors (Mob grade copper), previously purified by acetone. Data on the amount of pollutant that passed into the solvent after each cycle of their contact were determined on the EF-3MA fluorometer (table 3).

Table 3. Number of model pollutant.

Solvent	Number of model pollutant, mg						
	Dissolution cycles						
	1	2	3	4	5	6	7
1.	12.8	10.4	9.3	8.2	7.5		
2.	24.1	18.1	11.3	9.2	8.0	6.5	3.2
3.	25.0	18.9	12.2	9.6	8.1	6.9	4.0
4.	41.0	25.1	14.9	10.9	8.1	CD	
5.	80.0	20.0	CD				
6.	45.0	29.4	16.2	8.5	CD		
7.	51.0	25.1	16.9	7.0	CD		
8.	23.2	14.1	10.6	7.8	6.9	4.9	2.0E
9.	21.1	15.8	13.9	10.9	7.9	5.1	1.5E
10.	38.5	22.1	17.2	13.9	8.5	CD	
11.	76.1	23.9	CD				
12.	65.9	30.9	3.2	CD			
13.	20.0	15.1	13.1	10.5	7.9	5.3	1.0E

Note. *Solvents are used in the optical and microelectronic industries.
1. Acetone*. 2. 1-Butanol*. 3. 2-Butanol*. 4. Gasoline*. 5. Freon-114B2*. 6. Freon-12. 7. Methylene chloride. 8. Nitromethane*. 9. 2-Propanol*. 10. Petroleum ether*. 11. Freon-113 12. CCl_4. 13. Ethanol*.
E is an emulsion.
CD is complete dissolution.

As it was shown in our experiments, the most effective model pollutant is dissolved by halogenated hydrocarbons: after 2–3 cycles it dissolves completely. However, the cleaning of optics by halogen-substituted hydrocarbon solvents was accompanied by a number of negative aspects: on the surface to be cleaned, a white coating was visually observed, and also appeared iridescent stains, that worsened the optical characteristics of the metal.

From the practice of optical processing of nonferrous metals we know that this coating consists of salts of alkaline and alkaline earth metals, relatively poorly soluble in organic solvents, especially in hydrocarbons and their halogenated derivatives, somewhat better – in alcohols, amides, sulfoxides, ketones, nitro derivatives and water [23]. This assumption was tested on solvents of these classes with different solubility parameters after removal of rosin pitch and oil contaminants from the surface. The dissolution cycle time was 5 minutes; the presence of impurities (white coating) on the surface was evaluated visually after each cycle of processing of the optical element. The data on the presence of inorganic salts on the surface of the metal after each contact with the solvent are shown in table 4.

Table 4. The presence of inorganic salts on the surface.

<table>
<tr><th rowspan="2">Solvent</th><th colspan="5">Dissolution cycles</th><th rowspan="2">Note</th></tr>
<tr><th>1</th><th>2</th><th>3</th><th>4</th><th>5</th></tr>
<tr><td>1.</td><td></td><td></td><td></td><td colspan="2">NWC</td><td>IS</td></tr>
<tr><td>2.</td><td></td><td></td><td colspan="3">NWC</td><td></td></tr>
<tr><td>3.</td><td></td><td></td><td colspan="3">NWC</td><td></td></tr>
<tr><td>4.</td><td colspan="5">NWC</td><td>IS</td></tr>
<tr><td>5.</td><td></td><td></td><td></td><td></td><td>WC</td><td>IS</td></tr>
<tr><td>6.</td><td></td><td></td><td></td><td></td><td>WC</td><td></td></tr>
<tr><td>7.</td><td></td><td></td><td colspan="3">NWC</td><td>IS</td></tr>
<tr><td>8.</td><td colspan="5">NWC</td><td></td></tr>
</table>

Note. NWC – no white coating. WC – white coating. IS – iridescent stains. 1. Acetone. 2. 1-Butanol. 3. 2-Butanol. 4. Water (δ = 47.2). 5. Dimethylformamide (δ = 24.4). 6. Nitromethane. 7. 2-Propanol. 8. Ethanol.

4. Conclusions

Water and aliphatic alcohols most effectively dissolve inorganic salts that are present on the surface. After 1–2 cycles, the white coating from the surface disappears.

Along with the effective removal of the model contaminant, halogenated hydrocarbons interact with the metal of the mirror, causing a degradation of the optical characteristics. In connection with this, when using these solvents, it is necessary to inhibit the process of interaction of halogenated hydrocarbons with the surface of mirrors by stabilizing the solvents or by selecting the regimes for carrying out the physicochemical purification process, or by thereof combination.

Acknowledgments

This study was performed using the resources of the Centre for Collective Use of Tver State University and supported by the Ministry of Science and Higher Education of the Russian Foundation within a government contract for scientific work.

References

[1] Mesyac G A, Osipov V V and Tarasenko V F 1991 *Impul'snye gazovye lazery* (Moscow: Nauka) [In Russ.]
[2] Bulaev V D, Gusev V S, Kazantsev S Yu, Kononov I G, Lisenko S L, Morozov Yu B, Poznishev A N and Firsov K N 2010 *Kvant. electron.* **40(7)** 615
[3] Rogalin V E 2013 Materialovedenie **9** 34 [In Russ.]
[4] Rogalin V E 2013 *Izv. vuzov. Mater. Ehlektron. tekhn.* **2** 11 [In Russ.]
[5] Rogalin V E 2012 *Bulletin of the Russian Academy of Sciences: Physics* **76(11)** 1205

[6] Rogalin V E, Ashkinazi E E, Popovich A F, Ralchenko V G, Konov V I, Aranchii S M, Ruzin M V and Uspenskii S A 2012 *Russian Microelectronics* **41(8)** 464
[7] Kaplunov I A, Rogalin V E, Filin S A 2015 *Non-ferrous Metals* **1** 29
[8] Filin S A, Rogalin V E, Kaplunov I A and Zingerman K M 2017 *AIP Conference Proceedings* **1915** 040016
[9] Filin S A, Rogalin V E and Kaplunov I A 2019 *IOP Conf. Ser.: Mater. Sci. Eng.* **656** 012018
[10] Shmakov V A 2004 *Silovaya optika* (Moscow: Nauka) [In Russ.]
[11] Ageev V P, Burdin S G, Goncharov I N, Konov V I, Minaev I M and Chapliev N I 1983 *Itogi nauki i tekhniki. Ser.: radiotekhnika* 31 (Moscow: VINITI) [In Russ.]
[12] Kovalev V I, Morozov V V and Faizullov F C 1975 *Soviet J. of Quant. Electronics* **4(10)** 1208
[13] Arhipov Yu V *at al.* 986 *Soviet J. of Quant. Electronics* **16(1)** 63
[14] Makshantsev B I, Rovinskii R E and Rogalin V E 1985 *Soviet J. of Quant. Electronics* **15(1)** 10
[15] Gandel'man G M and Kondratenko P S *1983 Pis'ma v ZHEHTF* **38** 246 [In Russ.]
[16] Vejko V P, Smirnov V N, Chirkov A M and Shahno E A 2013 *Lazernaya ochistka v mashinostroenii i priborostroenii* (St. Petersburg: NIU ITMO) [In Russ.]
[17] Vejko V P and Shahno E A 2001 *Izv. RAN. Ser. fiz.* **65(4)** 584 [In Russ.]
[18] Kaplunov I A, Rogalin V E and Filin S F 2015 *Nonferrous Metals* **1** 29
[19] Hopper R W and Uhlman D R 1970 *Appl. Phys.* **41(10)** 4023
[20] Drinberg S A and Icko Eh F 1986 *Rastvoriteli dlya lakokrasochnyh materialov* (Leningrad: Himiya) [In Russ.]
[21] Vavruh I 1984 *Vliyanie rastvoritelej i ih vybor* (Har'kov: Torg.-prom. palata USSR) [In Russ.]
[22] Rejnol'ds V V 1967 *Fizicheskaya himiya neftyanyh rastvoritelej* (Leningrad: Himiya)
[23] Fialkov Yu Ya, Tarasevich Yu A and Zhitomirskij A N 1973 *Fizicheskaya himiya nevodnyh rastvorov* (Leningrad: Himiya) [In Russ.]

Magnetoelectric and magnetostriction characteristics of symmetric three layered structures of nickel - lead zirconate titanate – nickel and permendure – lead zirconate titanate – permendure

D A Filippov[1,6], V M Laletsin[2], N N Poddubnaya[2], A Galichyan[3], Jitao Zhang[4] and Gopalan Srinivasan[5]

[1] Yaroslav-the-Wise Novgorod State University, B. St. Peterbursgskaya, 41 173003, Velikiy Novgorod, Russia
[2] Institute of Technical Acoustics National Academy of Sciences of Belarus, Lyudnikava Praspekt, 13, BY-210023, Vitebsk, Belarus
[3] Institute of Mechanics, National Academy of Sciences of Armenia, 24B Baghramyan Ave., 0019 Yerevan, Armenia
[4] Zhengzhou University of Light Industry, Zhengzhou 450002, China
[5] Physics Department, Oakland University, Rochester, Michigan 48309, USA

[6] E-mail: Dmitry.Filippov@novsu.ru

Abstract. The magnetoelectric and magnetostriction characteristics of the symmetric three-layered structures are presented. The composite structures were formed like sandwich structures with a piezoelectric layer between two identical magnetostrictive layers using epoxy adhesive. An analytical expression for magnetoelectric voltage coefficient in terms of parameters for magnetostrictive and piezoelectric subsystems is presented. On the basis of this expression determination of magnetostriction characteristics via magnetoelectric response was offered. This procedure was tested for these three layered structures using the field dependences of the magnetoelectric response. The magnetostrictive characteristics obtained are in good agreement with results, obtained by other methods.

1. Introduction

Composite multiferroics have shown a good magnetoelectric coupling at the room temperature [1, 2]. The magnitude of magnetoelectric (ME) effect in composite multiferroics is several orders of magnitude higher than in single phase multiferroic materials in which the magnetoelectric interaction is due to a change in the spin-orbit interaction of the electron under the action of an external electric field [3]. Since the spin-orbit interaction is a relativistic effect, this leads to small ME coupling. In composite multiferroics, the ME effect is due to the mechanical interaction between magnetostrictive and piezoelectric subsystems or, so called strain induced mechanism [4]. The interconnection of the magnetic and electrical characteristics of multiferroics allows one to create on their basis devices with dual control of both electric and magnetic fields. These are inductors, tunable by both electric and magnetic fields [5, 6], dual electric and magnetic field tunable high-frequency devices [7, 8], gyrators capable of direct conversion of current to voltage [9, 10], The origin of ME coupling in composites arises

Content from this work may be used under the terms of the Creative Commons Attribution 3.0 licence. Any further distribution of this work must maintain attribution to the author(s) and the title of the work, journal citation and DOI.
Published under licence by IOP Publishing Ltd

IOP Conf. Series: Materials Science and Engineering **939** (2020) 012023 doi:10.1088/1757-899X/939/1/012023

due to the elastic interaction between the magnetostriction and piezoelectric subsystems; therefore, the magnitude of the ME effect directly depends on the magnetostriction and piezoelectric parameters of the subsystems. Recently in series of works, theoretically shown and experimentally confirmed that the magnitude of the magnetic resonance line shift under the action of an electric field is proportional to the product of the piezoelectric module d by the magnitude of the saturation magnetostriction λ_S [11, 12], power conversion efficiency in gyrator is proportional to the piezomagnetic module $q = d\lambda/dH$ [9, 10], the change in the inductance of a coil under the action of an electric field is also proportional to the product $d{\cdot}\lambda$ [13]. Thus, the piezoelectric module and magnetostriction constant are the most important parameters that determine the possibility of using composite multiferroics in straintronics. Methods for measuring a piezoelectric module are well known and fairly accurate. For example, the piezoelectric constant d_{31} is determined quite strong using the resonance - antiresonance method [14]. The magnetostriction of ferromagnetic materials is usually measured using a strain gauge, glued to the surface of the sample. The measurement accuracy in this case is ~ 10^{-6} in the field range up to 5 kOe [15]. This measurement method is not entirely convenient, since it requires gluing a strain gauge and magnetic material, which introduces an additional measurement error. It leads to the fact that this method does not allow measuring the magnetostriction of films whose thickness is commensurate with the thickness of the strain gauge. Besides, due to the smallness of the effect, it does not allow one to measure magnetostriction in the region of small fields. The low-field region is of particular interest because it exhibits a non-linear effect, on the basis of which sensitive magnetic field sensors can be created [15, 16]. Since the ME effect is directly related to the appearance of an electric field in a piezoelectric due to deformations caused by magnetostriction of a magnet in a magnetic field, it is of interest to use the ME response to determine magnetostriction. In earlier works [17, 18] this method was developed to determine the magnetostriction of bulk composites. In this paper, we have expanded this method for determining magnetostriction via the magnetoelectric response for layered composites.

2. Theory: magnetoelectric response and magnetostriction

Magnetoelectric response in composite multiferroics is caused by mechanical coupling between the magnetostrictive and piezoelectric phases by strain. In the case of the direct ME effect mechanical deformations occurs by magnetic field in the magnetostrictive phase and are transferred to the piezoelectric phase by mechanical interaction at the interface, that leads to change the electric polarization and produces electric voltage between electrodes of the sample. In the case of converse ME effect an electric field produces mechanical stresses in the piezoelectric phase that are transmitted through the interface into the magnetostrictive phase, which leads to a change in the magnetization of the sample. This change leads to occurrence of voltage on the coil wound on the sample [19]. The ME response for the direct ME effect is characterized by ME voltage coefficient (MEVC) α_E, which is determined by ratio of the average value electric field $<E>$ to the ac magnetic field H, i.e.

$$\alpha_E = <E>/H = (U/{}^p t)/H \qquad (1)$$

where U is the electric voltage, and ${}^p t$ is a piezoelectric layer thickness.

For the ME effect for three layered symmetric structures the theory predicts the following expression for the low frequency MEVC

$$\alpha_E = \frac{{}^pY\,{}^pd_{31}({}^mq_{11}+{}^mq_{12})}{{}^p\varepsilon_3\varepsilon_0\left(1-K_p{}^2\left(\frac{{}^mY\,{}^mt}{{}^mY\,{}^mt+{}^pY\,{}^pt}\right)\right)}\cdot\frac{{}^mY\,{}^mt}{{}^mY\,{}^mt+{}^pY\,{}^pt}, \qquad (2)$$

where ${}^pY$, ${}^mY$are the Young's modules of the piezoelectric and magnetostrictive phases; ${}^pd_{31}$, ${}^mq_{11}$, ${}^mq_{12}$ are the piezoelectric and piezomagnetic coefficients; ${}^pt$, ${}^mt$ are the thickness of the piezoelectric and the two magnetostrictive layers; ${}^{p}\varepsilon_3$ is permittivity, ε_0 is dielectrically constant and $K_p{}^2 = \frac{{}^{p}Y({}^{p}d_{31})^2}{{}^{p}\varepsilon_3}$ is the squared coefficient of electromechanical coupling.

Using the fact that $K_p{}^2 << 1$, we can rewrite Equation (2) in the first approximation in the form

$$\alpha_E = \frac{{}^{p}Y\,{}^{p}d_{31}({}^{m}q_{11} + {}^{m}q_{12})}{{}^{p}\varepsilon_3\varepsilon_0} \cdot \frac{{}^{m}Y\,{}^{m}t}{{}^{m}Y\,{}^{m}t + {}^{p}Y\,{}^{p}t}. \tag{3}$$

The piezomagnetic coefficients ${}^{m}q_{11}$and ${}^{m}q_{12}$ in Equation (3) are defined as follows:

$$ {}^{m}q_{11}(H_{bias}) = \left.\frac{\partial\lambda_{\parallel}}{\partial H_1}\right|_{H=H_{bias}}, \tag{4}$$

$$ {}^{m}q_{12}(H_{bias}) = \left.\frac{\partial\lambda_{\perp}}{\partial H_1}\right|_{H=H_{bias}}, \tag{5}$$

where $\lambda_{\parallel}$ is a longitudinal magnetostriction, $\lambda_{\perp}$ is transverse magnetostriction, H_{bias} is bias magnetic field. In Equation (3) only terms ${}^{m}q_{11}$ and ${}^{m}q_{12}$ have a strong dependence on magnetic field. The terms, including ${}^{m}Y$, have a weak dependence on magnetic field. The remaining terms in expression (3) for the MEVC are independent of the bias field. Thus, the field dependence of the MEVC is determined by the dependence of the piezomagnetic modules ${}^{m}q_{11}$ and ${}^{m}q_{12}$ on the bias field. This allows us to determine the magnetostriction from the field dependence of MEVC $\alpha_E(H_{bias})$ using the relationship:

$$\lambda_{\parallel}(H_{bias}) + \lambda_{\perp}(H_{bias}) = \int_0^{H_{bias}} ({}^{m}q_{11} + {}^{m}q_{12})dH = \frac{1}{C}\int_0^{H_{bias}} \alpha_E(H)\,dH, \tag{6}$$

where $C = \frac{{}^{p}Y\,{}^{p}d_{31}}{{}^{p}\varepsilon_3\varepsilon_0} \cdot \frac{{}^{m}Y\,{}^{m}t}{{}^{m}Y\,{}^{m}t + {}^{p}Y\,{}^{p}t}$ is a constant coefficient for given structures. At the first approximation we can use the fact $\lambda_{\perp} = -\nu\lambda_{\parallel}$. Using this relation we get finally for the magnetostriction follows expression:

$$\lambda_{\parallel}(H_{bias}) = \frac{1}{C(1-\nu)}\int_0^{H_{bias}} \alpha_E(H)\,dH. \tag{7}$$

Saturation magnetostriction coefficient λ_s, according to Equation (7) will be determined as:

$$\lambda_{s\parallel} = \frac{1}{C(1-\nu)}\int_0^{\infty} \alpha_E(H)\,dH. \tag{8}$$

Using this determination we get the expression for magnetostriction coefficient in the following form:

$$\lambda_{\parallel}(H_{bias})/\lambda_{s\parallel} = \int_0^{H_{bias}} \alpha_E(H)\,dH / \int_0^{\infty} \alpha_E(H)\,dH. \tag{9}$$

Expressions (8) and (9) allow us to determine the magnetostriction of a magnetic phase by the field dependence of the MEVC for the composite structures without any addition devices. It is very important for the thin magnetic films on the piezoelectric substrate in the region of low bias magnetic field. For weak magnetic field the value of magnetostriction is very small; therefore there are many difficulties for its determination using resistance strain gauge. The method for the determination of the magnetostriction using ME response has a very high sensitivity, more than strain gage factor, therefore one can use for the small values of the magnetic field.

3. Details of experiment

The ME effect in structures was studied by measuring the voltage arising on samples in ac magnetic field H_{ac}=*1 Oe* and in bias field, the strength of which H_{bias} changed in the range of H_{bias}= *0 – 2000 Oe*. The samples were three-layer Nickel – Lead Zirconate Titanate (PZT) – Nickel and Permendure

(Pe) – Lead Zirconate Titanate – Permendure structures in the form of a parallelepiped, obtained by the gluing using an epoxy. The length of the samples was *20 mm*, the width was *4.5 mm*, the thickness of the PZT plate was ${}^{p}t = 0.3\ mm$, the thickness of magnetic plates was ${}^{m}t = 0.15\ \ mm$of each.

Simultaneously were prepared three-layer Nickel-Quartz-Nickel structures in the form of a parallelepiped, obtained by the method of gluing using epoxy and electrolytic deposition. The length of the samples was *20 mm*, the width was *4.5 mm*, the thickness of the quartz plate was ${}^{p}t = 0.5\ \ mm$. For electrolytic deposition structures, the thickness of nickel plates was ${}^{m}t = 0.25\ \ mm$of each; therefore, these structures can be considered as bulk samples. The thickness of the nickel layer for structures obtained by galvanic deposition was equal ${}^{m}t = 10\ \ \mu m$, therefore, these structures were the thin film structures.

The frequency dependence of MEVC has the resonance character. At the low frequencies MEVC does not dependence on the frequency. To measure the field dependency of MEVC ac magnetic field with frequency *f=1 kHz* was used. The ME coefficient is given by the expression:

$$\alpha_E = U / ({}^{p}t \cdot H_{ac}) \tag{10}$$

where u is the induced voltage between the plates of the sample, ${}^{p}t$ is the thickness of the piezoelectric layer, H_{ac} is the value of the applied ac magnetic field.

4. Results and discussion

The measured low frequency MEVC for Ni – PZT – Ni structures and Pe – PZT – Pe are presented in figure 1a and figure 1b respectively.

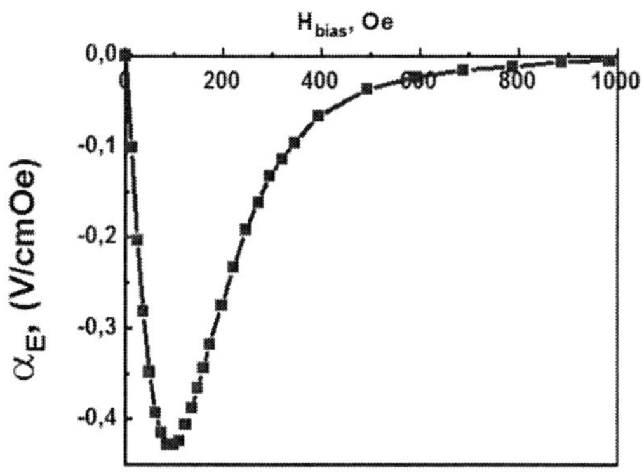

a) Ni – PZT – Ni

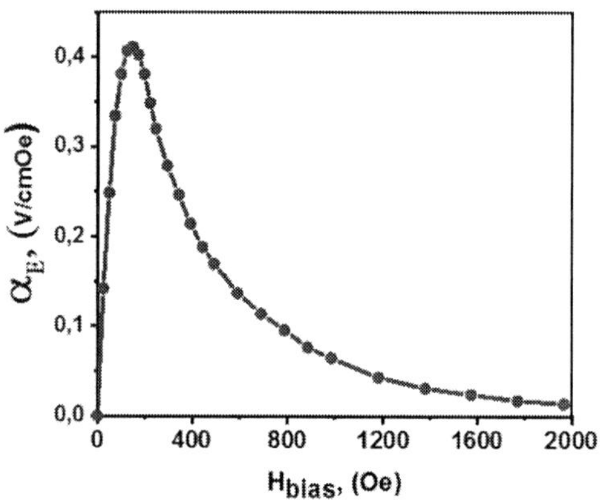

b) Pe – PZT – Pe

Figure 1. Field dependencies of MEVC for three-layered symmetric structures.

As can be seen from figure 1, both field dependences have a typical characteristic for low frequency MEVC. The maximum value of MEVC for Ni – PZT – Ni structure is $\alpha_{E,max}$=0.43 V/cm Oe and observed at the bias magnetic field H_{bias}~100 Oe, while maximum value for Pe – PZT – Pe structure observed at the bias magnetic field H_{bias}~150 Oe and its value is $\alpha_{E,max}$=0.41 V/cmOe. Magnetostriction characteristics for these structures were calculated using Equation 8. For the calculations following values were used for material parameters: ${}^{p}Y$= 67 GPa, ${}^{p}t$=0.3 mm, ${}^{p}d_{31}$= –175 pC/N, $\varepsilon_0=8.85\cdot10^{-12}$F/m, ${}^{m}Y_{Ni}$= 210 GPa, ${}^{m}Y_{Pe}$=207 GPa, ${}^{m}t$=2*0.25 mm. The calculated magnetostriction curves are presented in figure 2.

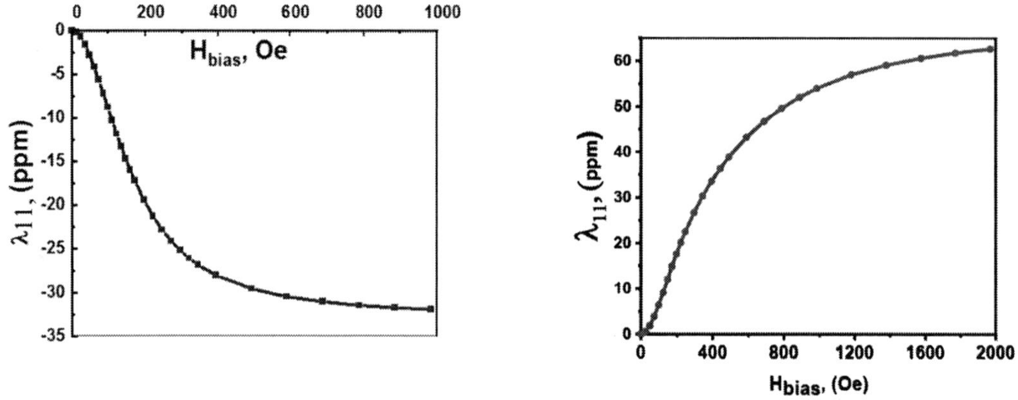

a) Ni – PZT – Ni b) Pe – PZT – Pe

Figure 2. Field dependence of magnetostriction for three-layered symmetric structures.

As can see from figure 2, presented curves are in good agree with results obtained by using resistance strain gauge [22]. Thus, the procedure proposed here provides one with a unique method for determination of magnetostriction characteristic magnetic phase in the layered multiferroic.

We also utilized this procedure for the determination magnetostriction of thin films, fabricated by galvanic deposition of nickel on quartz piezoelectric substrates. The samples were three-layer Nickel-Quartz-Nickel structures in the form of a parallelepiped, obtained by the method of gluing with using epoxy and electrolytic deposition. The length of the samples was *20 mm*, the width was *4.5 mm*, the thickness of the quartz plate was $^{p}t = 0.5\ \ mm$. The thickness of nickel plates the gluing structures was equal $^{m}t = 0.25\ \ mm$; therefore, these structures can be considered as bulk samples. The thickness of the nickel layer for structures obtained by galvanic deposition was $^{m}t = 10\ \ \mu m$, therefore, they were the thin film structures. The field dependency of MEVC for the Ni – Q – Ni structures obtained by epoxy glue and electrolytic deposition is presented in figure 3.

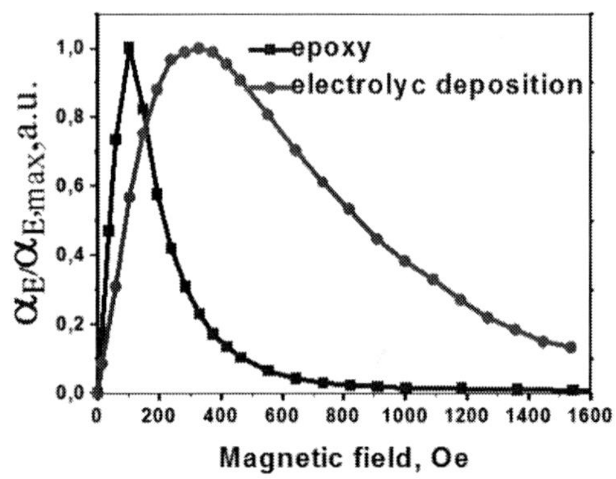

Figure 3. Field dependency of MEVC Ni-Q-Ni three layered structures: black – epoxy gluing structure, red – electrolytic deposition structure.

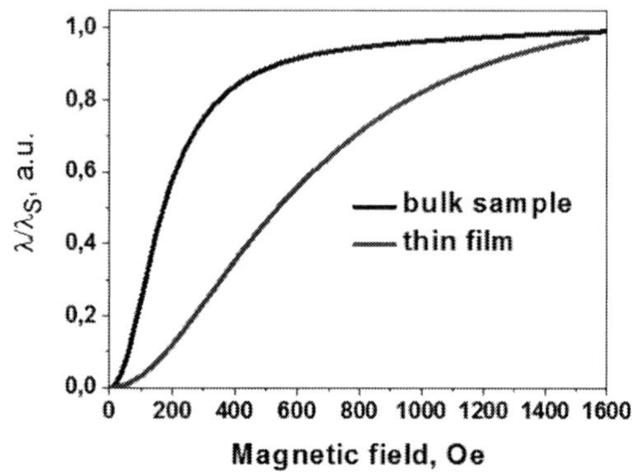

Figure 4. Magnetostriction of Nickel: black – bulk sample, red – Ni thin film on substrate.

As can be seen from figure 3 MEVC has a maximum for both samples, but for epoxy glue structures this maximum occurs when the value of bias field is H_{ac}=*150 Oe*, while for electrolytic deposition structures this maximum is observed at the value of bias field is equal to H_{ac}=*400 Oe*. This fact demonstrates the difference of field dependence of magnetostriction between bulk sample and thin film structures. The field dependencies of magnetostriction obtained by using Equation (9) are presented in Fig. 4. It shows that the magnetostriction for bulk sample has saturation in magnetic field near 500 Oe. These results are in good agreement with those reported in previous work [23]. For the electrolytic deposition sample the magnetostriction tends to saturation at fields above 1500 Oe.

5. Conclusion

The study of the field dependence of MEVC in composite magnetostrictive – piezoelectric structures allows one to estimate the magnetostriction and its field dependence reveal the features expected for ferromagnetic thin films. The magnetostriction coefficient of nickel thin films tends to saturate at the value of magnetic field more than three times higher than the values of saturation magnetic field for bulk samples.

Acknowledgment
The work was jointly supported by the Russian Foundation for Basic Research and Belarussian Foundation for Basic Research (Russian project no. 18-52-00021 and Belarussian project no. F18R-300). The research at Oakland University was supported by grants from the National Science Foundation (DMR-1808892, ECCS-1923732).

References

[1] Ramesh R and Spaldin N A, 2007 *Nature Materials* **6** 21–29
[2] Pyatakov A P and Zvezdin A K 2012 *Physics Uspekhi* **55** 557–581
[3] Bichurin M I and Filippov D A 1997 *Ferroelectrics* **204** 225–232
[4] Bukharaev A A, Zvezdin A K, Pyatakov A P and Fetisov Y K 2018 *Physics Uspekhi* **61** 1175–1212
[5] Lou J, Reed D, Liu M and Sun N X 2009 *Appl. Phys. Let.* **94** 112508
[6] Jitao Zhang, Dongyu Chen, Filippov D A, Kang Li, Qingfang Zhang, Liying Jiang, Weiwei Zhu, Lingzhi Cao and Gopalan Srinivasan 2018 *Appl. Phys. Lett.* **113** 113502
[7] Sun N X and Gopalan Srinivasan 2012 *SPIN* **2** 1240004
[8] Sadovnikov A V, Grachev A A, Beginin E N, Sheshukova S E, Sharaevskii Yu P and Nikitov S A 2017 *Phys. Rev. Applied* **7** 014013
[9] Zhuang X, Leung C M, Li J and Viehland D 2017 *Appl.Phys. Lett.* **111** 103902
[10] Jitao Zhang, Dongyuhen Kang Li, Filippov D A, Bingfeng Ge, Qingfang Zhang, Xinxin Hang, Lingzhi Cao and Gopalan Srinivasan 2019 *AIP Advances* **9** 035137
[11] Bichurin M I, Kornev I A, Petrov V M, Tatarenko A S, Kiliba Yu V and Srinivasan G 2001 *Phys. Rev.* B **64** 094409
[12] Peng Zhou *at al.* 2019 *Phys. Rev. Materials* **3** 044403
[13] Jitao Zhang *at al.* 2019 *J. Phys. D*, **52**, 165001
[14] Mazon W P 1948 *Phys. Rev.* **74**, 1134–1147
[15] Burdin D A, Chashin D V, Ekonomov N A, Fetisov L Y, Fetisov Y K, Srinivasan G and Sreenivasulu G 2014 *JMMM* **358** 98–104
[16] Filippov D A, Laletin V M, Firsova T O 2014 *Physics of the Solid State* **56** 980–984
[17] Filippov D A, Laletin V M, Poddubnaya N N, Shvartsman V V, Lupascu D C, Zhang J and Srinivasan G 2019 *Technical Physics Letters* **45** 1152–1154
[18] Filippov D A, Laletsin V M, Srinivasan Gopalan, Zhang Jitao, Tikhonov A A and Galichyan T A 2019 *IOP Conf. Series: Materials Science and Engineering* **656** 012019
[19] Filippov D A, Galkina T A and Srinivasan G 2010 *Technical Physics Letters* **36** 984–986
[20] Filippov D A, Laletsin U, Srinivasan G 2007 *J. Appl. Phys.* **102** 093901
[21] Loyau V, Morin V, Chaplier G, LoBue M and Mazaleyrat F 2015 *J App. Phys.* **117** 184102
[22] Hörner E, Krykanov I M, Chashin D V, Fetisov Y K, Fetisov L Y and Shamonin 2012 *M Int. J. Mater. Res* **103** 1345–1350
[23] Burdin D A, Chashin D V, Ekonomov N A, Fetisov L Y, Fetisov Y K, Srinivasan G and Sreenivasulu G 2014 *JMMM* **358-359** 98–104

IOP Conf. Series: Materials Science and Engineering **939** (2020) 012024 doi:10.1088/1757-899X/939/1/012024

Processing of steel galvanizing waste using the method of melt immersion

V I Gel[1] **and S B Sapozhkov**[1,2]

[1] Yaroslav-the-Wise Novgorod State University, ul. B. St. Petersburgskaya, 41 173003, Veliky Novgorod, Russia

[2] E-mail: ssb@novsu.ru

Abstract. The article describes the methods of steel galvanizing waste processing by immersion in a zinc melt, furnaces and devices for processing hard zink and drosses into usable products for zink pots, the methods for zinc extraction from waste and their use in industry. It is demonstrated that the hot-dip galvanizing wastes is a serious resource in reducing the consumption of primary zinc, what is more their use does not entail deterioration in the quality of the hot-dip zinc coating. However, not all technologies and types of equipment for processing zinc-containing waste are equally effective. It is appropriate to use vibro-deburring closed machines for processing dross (liming), and furnace with a "roof" heated for processing hard zinc, and in the latter case, according to the technology of fractional loading of reagent – aluminum scrap – processing floating aluminum-iron intermetallic compounds carnallite-cryolite fluxing agent.

1. Introduction

In the last ten years, Russia has been actively developing the methods of galvanizing steel products by immersion in liquid zinc, including power line pylons, pipes, building bars, hardware items. Galvanizing of individual large products is carried out mainly in baths up to 13 meters long, 1.6 meters wide and up to 3 meters deep. At the same time, the average output of waste is: 7.5 kg /t of hard zinc, dross (liming) – up to 3.75 kg/t and zinc dust – up to 3–4 kg/t (for example, when flues blowing-down) of zinc items.

The calculations show that, for example, when the production volume is 30,000 t/year of galvanized products, they form: 225 t of hard zinc and 112.5 t of dross, 30 t (averaged) of zinc dust, which contains a total of about 290 t of zinc.

For example, it is proposed to use hard zinc milled in a ball mill for thermodiffusion zinc coating of small products. It is also possible to prepare synthetic saturated mixture of dross for the same purpose. The dust produced by blowing flues contains only 0.1% of iron and 0.35% of lead and has various direct use options. It is proposed to melt the hard zinc in a gas-heated crucible with loading in three stages and with exposure at a temperature of up to 610 °C for several hours after each loading. The "Korner" distilling furnace for hard zinc regenerates 85–88% of the bonded zinc with a purity of 99.9%. The furnace brings hard zinc to the evaporation temperature of zinc. Then the zinc settles in the cooling retort.

According to the description of another method, the waste is loaded into the melt of an equimolar mixture of sodium and potassium chlorides with the addition of sodium or aluminum fluorides at a temperature of 740–790 °C. Then it is carried 15–30 minutes, then the metal zinc is removed, the

Content from this work may be used under the terms of the Creative Commons Attribution 3.0 licence. Any further distribution of this work must maintain attribution to the author(s) and the title of the work, journal citation and DOI.
Published under licence by IOP Publishing Ltd

operation is repeated 3–5 times. Then the aluminum is added, 5–10% of molten salts weight, galvalume melt is taken away, and the oxides and intermetallic compounds with aluminum and zinc sediments are extracted from molten salts. After that, the operation is repeated [2].

The liming is processed by sieving in a rotating drum screen with the release of metal globules and their return to the hot-dip galvanizing bath. This allows reducing the volume of zinc purchases, and mechanizing the processing process.

The dross (liming) after each galvanizing operation is removed to the end of the galvanizing bath, where it is settled. The spread of dross on the surface of the melt is excluded due to its restriction by a metal partition. Heating of zinc dross pieces is performed at the expense of the temperature of molten zinc – 440–460 °C in the process bath for 20–30 minutes. The reduction of zinc losses with production waste is provided from 85% to 76–77% and a decrease in the yield of dross is from 6.1–6.2 kg/t to 5.6–5.7 kg/t of galvanized products.

In [4], an equipment analysis for processing zinc-containing waste is given. The equipment is patented in twenty countries and is operated in twenty-eight countries. The most common methods of processing zinc-containing waste involve placing them in cylindrical drums or retorts of various designs, heated to a temperature higher than the melting point of zinc. The drums are then rotated to intensify the separation of molten zinc from the sludge.

2. Research and development of new aggregates

One of the first was designed "Carousel melting furnace", consisting of a powerful fixed furnace body, inside which rotates a non-removable metal melting drum, heated from the outside. Through the center of the drum passes a metal pipe that acts as bearing shafts, the ends of which go through the holes in the opposite walls of the furnace body and rotate in special supports. Inside the melting drum, between the main shaft and the walls of the drum, there are blades that perform the functions of a conveyor screw. The loading and unloading devices are made in the form of holes in the opposite ends of the melting drum [5].

The disadvantages of the furnace include the complexity of the design and the use of special heat-resistant materials in the construction of the furnace, as well as environmental pollution by waste flue gases, metal- and chlorine-containing dust, which requires the use of gas cleaning equipment.

In a tilted rotary furnace (figure 1), which is widely used worldwide for melting aluminum scrap, metalized zinc, lead, and copper wastes, the technology for melting zinc dross, there was tested the melting technique of zinc dross at the Taganrog Iron and Steel Factory. The furnace temperature is 500–520 °C, the waste oil consumption is 25–28 l/hour. The furnace charge of dross is 5 t, with a zinc content of 60%. There is collected 2940 kg of zinc and 2050 kg of slag with a residual zinc content of 2.92%. The operating time of the furnace is 2 hours [3].

Figure 1. Rotary kiln for processing of metalized waste of Al, Pb, Zn.

The article [6] describes the experience of processing hot-dip galvanizing, hard zinc and dross wastes in a closed vibration machine (which does not emit dust during operation). The technological scheme for processing zinc liming is demonatrated at the figure 2.

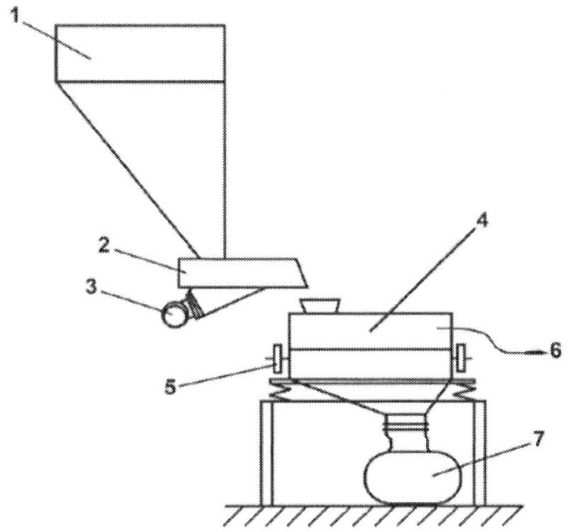

Figure 2. Technological scheme of zinc ash processing in the vibration machine.

The frequency of circular vibrations is 25 Hz, the amplitude is 5 mm, the drive power is 5.0 kWe, the processing time is 45–50 min (500 kg of dross). At the same time, the speed of particles of bulk loading reaches 1.0 m/s, acceleration is up to 150 m/sec^2, impact force is up to 30 N, contact pressure is up to 15 kN, and with an increase in the residence time of metal crowns, even grinding of their surfaces occurs while removing the oxychloride part.

The yield of globule metal from dross (fractions – minus 4 mm) containing 50% zinc, is 160 kg, fraction 4.0-2.0 – 150 kg, fraction minus 2 mm – 190 kg. The yield of zinc in the remelting of the globule metal is 96.5%, in the remelting of the fraction minus 4.0-2.0 – 62.5%, and in the fraction minus 2 mm – 46%. The globule metal was returned to the hot-dip galvanizing bath, and the 4.0–2.0 fraction, containing less than 0.01% chlorine, was used to produce zinc oxide.

Figure 3 shows a photo of a working vibration machine with a loading capacity of 500 kg. During 45–60 minutes of machine operation, the complete separation of the globule metal from the liming is achieved.

Figure 3. Vibro deburring machine separating metal globules from dross.

The fraction minus 2 mm was studied for the preparation of a sulfate-chloride salt, applicable for correcting the composition of galvanizing baths, where the chlorine ion is a necessary component of the electrolyte, improving the adhesion of the coating, increasing the electrical conductivity of solutions, contributing to the uniform dissolution of anodes. The low cost of the solvent, the possibility of separating lead and zinc already at the leaching stage without the introduction of cementation treatment has determined the choice in favor of sulfuric acid leaching. The resulting salt corresponds to the standard for zinc sulfuric acid 7-water [7].

Hard zinc was processed in a furnace with a capacity of 2 t lump hard zink with vaulted electric heaters [10]. The choice of arch heating is due to the nature of *Al-Fe-Zn* intermetallic compounds' liquation to the surface of the melt bath and the possibility of their overheating when mixed with carnallite-cryolite fluxes in order to reduce the amount of liquid zinc in them and turn them into a loose condition [8].

Currently, in order to save energy, a gas furnace with a removable arch and melt mixing with the help of vertical mechanical pump is offered. The diagram of the hard zinc processing furnace is demonstrated at the figure 4.

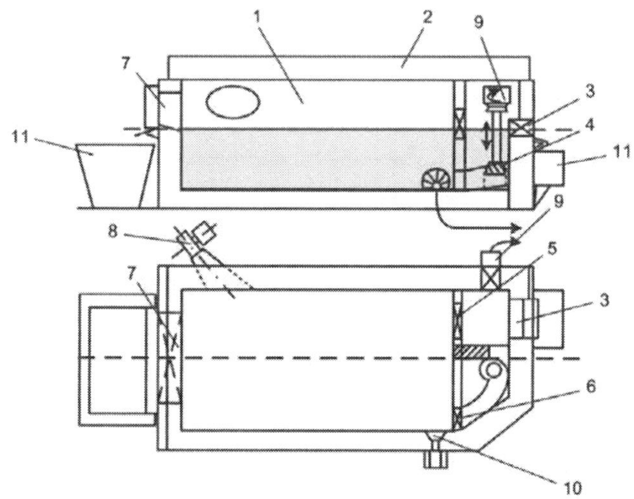

Figure 4. Diagram furnace for processing hard zinc. 1 – furnace cavity; 2 – removable arch; 3 – window for slag removal; 4 – impeller pump; 5 – window for receipt of the melt in the mixing zone; 6 – window for feeding the melt into the furnace cavity; 7 – window for loading reagents; 8 – burner; 9 – smoke removal; 10 – leaks for the metal release; 11 – slag tanks.

The temperature of primary heating is determined by the composition of the initial hard zinc and peritectic lines in the diagram of the metals state. The temperature reduction in steps and a discrete aluminum feed together with the flux provide a minimum output of "drained" flux agents.

The time of processing the melt with aluminum scrap is 16–20 hours, depending on the iron content in the hard zinc. The ratio of aluminum lump scrap with fractional introduction of aluminum into the melt is 1.2:1, where: 1.2 – is the amount of aluminum scrap or chips, 1.0 – is the amount of iron in the load, the flux agent consumption is 2–4% adding to the hard zinc weight. As a result, the content of iron in secondary zinc should not exceed 0.05%, the content of aluminum – 1.2%, which on average amounted to 5.0–5.5% of the zinc consumption for coating and did not worsen its quality. The slag from the melt surface was sent to a vibration machine for processing.

Furnaces installation for processing hard zinc and vibration machines for dross allows to get the maximum zinc return in the form of commercial products, and it is possible to cross-use the resulting zinc intermediates.

3. Conclusion

Hot-dip galvanizing wastes are a significant resource in reducing primary zinc consumption. Many technologies exist and are used for processing zinc-containing waste, both by mechanical grinding and sieving, and by melting.

Not all offered and using waste processing technologies meet the requirements for emissions of harmful substances into the environment and complex processing for full use in the national economy.

Most sufficiently meets these requirements the processing technology of drossb (liming) vibro deburring closed machine, allowing to obtain high quality products: metal globules to return to the hot-dip galvanizing bath, the fraction minus 4 to plus 2 mm – for the production of zinc oxides, the fraction minus 2 mm – for the production of sulfate-chloride salts to stabilize the process in electroplating technology.

Processing of hard zinc occurs most fully in furnaces with arch heating using the technology of fractional loading of the reagent-aluminum scrap - with processing of floating aluminum-iron intermetallic compounds carnallite-cryolite fluxing agent.

References

[1] Alekseev A A 2005 Metallurgicheskaya pererabotka cinkosoderzhashchih othodov proizvodstva [Metallurgical processing of zinc-containing waste products] *Vestnik Inzhekona. Seriya: Tekhnicheskie nauki* **3(8)** 125–128 [In Russ.]

[2] Kazantsev G F, Barbin N M, Moiseev G K and Vatolin N A 2000 *Sposob pererabotki othodov cinka* [Method for zinc waste processing] Patent RU 2147322 [In Russ.]

[3] Tarasov A V, Gel V I, Sheveleva L M 1989 Pererabotka othodov goryachego cinkovaniya [Processing of hot-dip galvanizing waste] Stal' **6** 57–58 [In Russ.]

[4] Alekseev A A 2007 *Organizaciya effektivnogo tekhnologicheskogo processa goryachego cinkovaniya tekhnologij na sudostroitel'nyh predpriyatiyah putem vnedreniya resursosberegayushchih tekhnologij* [Organization of effective technological process of hot-dip galvanizing of technologies at shipbuilding enterprises by implementing resource-saving technologies] (Sankt-Peterburg) p 129 [In Russ.]

[5] Gel V I 2013 *Rotornaya pech* [Rotary kiln] Patent RU 134620 [In Russ.]

[6] Gel V I and Samsonov A I 1991 Kompleksnaya tekhnologiya pererabotki cink- i hlorsoderzhashchih othodov [Complex technology of zinc-and chlorine-containing waste processing] *Voprosy ekologii i resursosberezheniya v pererabotke othodov cvetnoj metallurgii: Sb. nauch. tr.* ed A N Petrun'ko *at al.* 30–33 [In Russ.]

[7] Gel V I, Malyk E F, Makarenko E A and Prohorov S N 1989 *a.s. SSSR* № 1493855 [In Russ.]

[8] Tarasov A B, Besser A D and Maltsev V I 2004 Metallurgicheskaya pererabotka vtorichnogo cinkovogo syr'ya [Metallurgical processing of secondary zinc raw materials] ed A V Tarasov (Moscow: Gincvetmet) p 219 [In Russ.]

Improving the design of a disintegrator for deep utilization of enrichment tailings

V I Golik[1,2,5], **Y I Razoryonov**[3,6] **and A B Efremenkov**[4,7]

[1] Geophysical Institute of Vladikavkaz Scientific Centre of the Russian Academy of Sciences, Markov Str., 93a 362002, Vladikavkaz, Russia
[2] North-Caucasus State Technological University, ul. Kosmonavta Nikolaeva, 44 362021, Vladikavkaz, Russia
[3] South Russian State Technical University, ul. Prosvesheniya, 132 346428, Novocherkassk, Russia
[4] Yaroslav-the-Wise Novgorod State University, B. St. Peterburgskaya str., 41173003, Veliky Novgorod, Russia

[5]E-mail: v.i.golik@mail.ru
[6]E-mail: yiri1963@mail.ru
[7]E-mail: abe@novsu.ru

Abstract. The article discusses the directions of improving the design of disintegrators for activating materials and leaching metals when disposing of solid mineral raw materials using mechanochemical technology. The results of experimental extraction of metals from non-ferrous and ferrous ores, as well as coal are presented. Disintegrator designs differing in the use of physicochemical methods of increasing activity are recommended.

1. Introduction

In mining, the relevance of the full use of raw materials extracted from the bowels of the earth is increasing. The efficiency of the functioning of a mining enterprise is characterized by material, energy and other costs for the production of commercial products, provided that the environment is reliably protected from the negative technological impact.

One of the requirements for mining is the completeness of the use of mineral resources, including the fate of mineral resources extracted to the earth's surface [1–4].

It becomes obvious that when determining the profit from the extraction of minerals, those resources that have not been extracted from the bowels of the earth, or have been extracted and not used, for example, ore processing tailings, must be taken into account.

Over time, the relevance of the development of technologies and processes increases, as a result of which inactive stocks of metal-containing raw materials are involved in production [5–8].

One of the directions of realization of the idea of deep extraction of metals from substandard raw materials is the mechanochemical activation of the processes of leaching of metals in a disintegrator [9–12].

The industrial use of the disintegrator was carried out in the 80s at the filling complex of a uranium mine in Northern Kazakhstan with a capacity of 100,000 m^3 of voids per year. The DU-65 unit was equipped with 4- and 3-row rotors and 250 kW motors and was located in a building with a base area of

Content from this work may be used under the terms of the Creative Commons Attribution 3.0 licence. Any further distribution of this work must maintain attribution to the author(s) and the title of the work, journal citation and DOI.
Published under licence by IOP Publishing Ltd

5×7 m on three levels.

For 10 years, the unit has met the needs of a medium-sized mine in binders for the preparation of hardening mixtures in a system of development with chambers with backfilling of mined-out space.

Processing in a disintegrator provided an increase in the strength of the mixture by 30% more than when processing in a ball mill. The output of 55% of the active class 0.076 mm upon activation of the granular tailings of the Karaganda metallurgical plant made it possible to reduce the cement consumption from 120–150 kg/m^3 to 30 kg/m^3. The equivalent of 1 kg of M-400 commercial cement was 4 kg of activated tailings.

The purpose of this work is to identify ways to optimize the design of the disintegrator to improve the performance of metal recovery from the tailings.

2. Results

Mechanochemical technology is based on the phenomenon of change in the state of matter in a disintegrator when processing at a speed of more than 250 m/s theoretically substantiated and experimentally proved by J. Hint (figure 1).

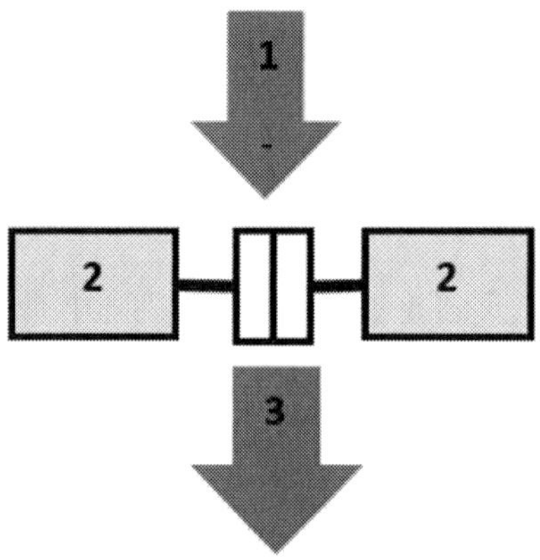

Figure 1. The scheme of the mechanical activation of mineral raw materials: 1 – source raw material; 2 – motors with rotors and working body; 3 – activated raw materials.

The mechanism of development of physicochemical processes in solids is based on the phenomenon of that the larger the surface of the substance involved in the process, the better the processing rates.

Since then, the basic design of disintegrators has undergone improvements.

2. Materials and methods

The phenomenon of change in the state of a substance in a disintegrator during processing at a high speed has been investigated. The quantitative values of the parameters of the extraction of metals into the solution were determined experimentally during leaching of metal enrichment tailings in a laboratory disintegrator. The recovery rates from the tailings by disintegrator leaching are compared with agitational leaching. The results obtained are used to predict the development of the sounded technology for extracting metals from tailings during their disposal.

To improve the processing performance in the disintegrator, the design of the disintegrator is modernized by introducing additional effects on the treated substance.

In order to reduce the adhesion of particles of enrichment tailings in a disintegrator, during processing, they are subjected to vibration and tossing with fluctuations from 30 to 1500 Hz with an amplitude of horizontal oscillations of 2–50 mm and an amplitude of vertical tosses of up to 30 mm (figure 2).

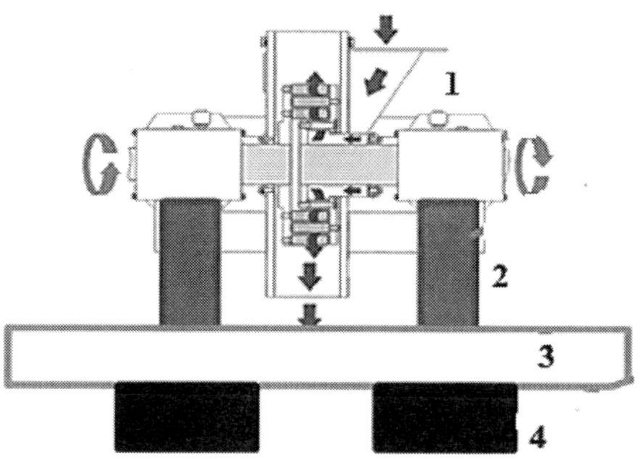

Figure 2. Mechanical activation scheme with vibration: 1 – disintegrator; 2 – vibration exciters; 3 – base; 4 – anti-vibration mountings.

The tails entering the disintegrator also perform a forward motion with tossing. Contacting with the working surfaces of the disintegrator, they are deprived of the opportunity to stick together, which increases the speed of the process and reduces energy consumption to obtain the desired result.

The efficiency of activation of materials is increased by combining mechanical activation with temperature exposure (figure 3).

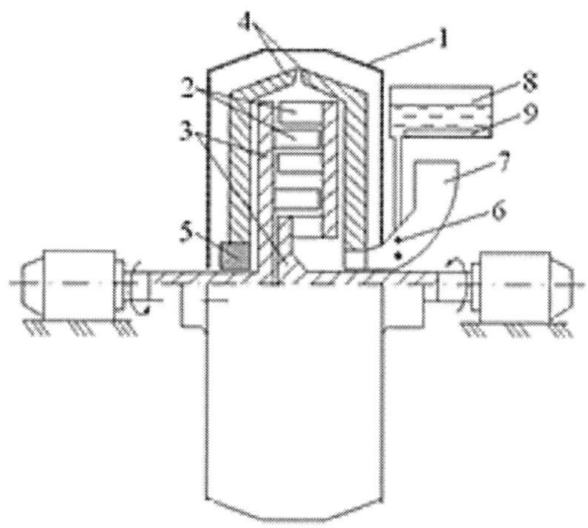

Figure 3. Mechanical activation scheme with temperature exposure: 1 – case; 2 – fingers; 3 – discs; 4 – the gap between the inner and outer discs; 5 – control mechanism; 6 – nozzles; 7 – inlet pipe; 8 – container with liquid; 9 – heating element.

Using a heating element, the temperature of the liquid is brought to 90–95 °C. The tailings mix with the liquid sprayed by the nozzles and enter the disintegrator chamber. The pulp passes through the rows of fingers and is thrown onto the outer discs. Particles larger than the required size are additionally crushed due to the rotation of the discs and collision with newly arriving particles.

One of the options for increasing the efficiency of materials processing is the combination of mechanical activation with electroacoustic action. Electroacoustic emitters are fixed along the perimeter of the working chamber, and the disintegrator is equipped with an electronic generator (figure 4).

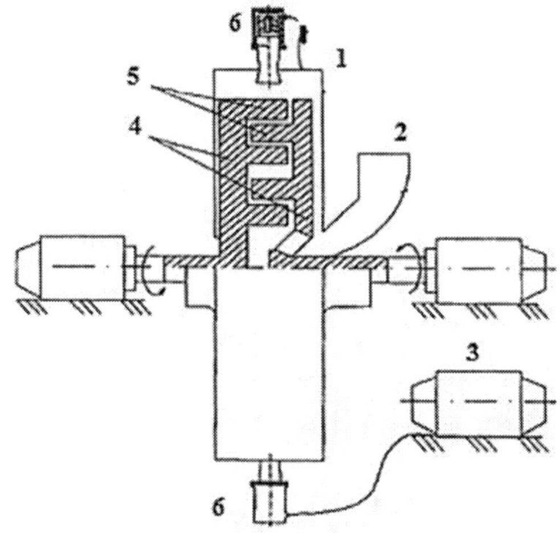

Figure 4. Mechanical activation scheme with electroacoustic effect: 1 – case; 2 – inlet pipe; 3 – electronic generator; 4 – disks; 5 – fingers; 6 – electroacoustic emitter.

In the case of combined mechanochemical activation of metal-containing raw materials, the leaching reagent is fed into the working body of the disintegrator, and the extraction of metals into the solution occurs simultaneously with the destruction of crystals, when the solution is pressed into the resulting cracks. When combining mechanical activation with chemical enrichment, the tailings in a mixture with sulfur are pretreated with sulfuric and nitric acid (figure 5).

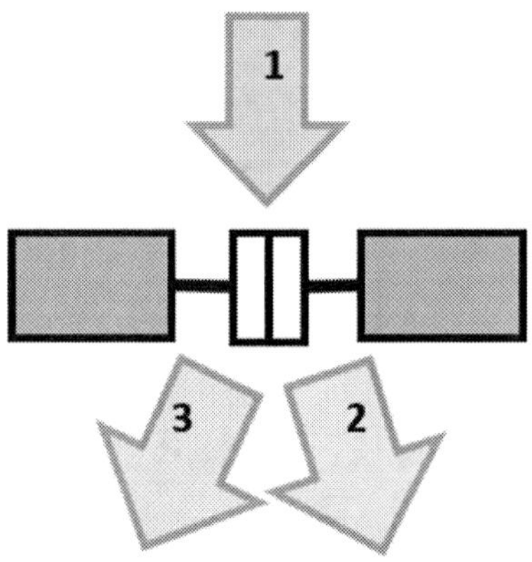

Figure 5. Scheme of the chemical component of the activation of the leaching process: 1 – concentration tailings; 2 – metal-containing solution; 3 – leaching tailings.

When the tailings are exposed to chemical reagents, pulp slurry is prepared from them with a solid to liquid ratio of 1: 2. The tailings are ground in a mixture with elemental sulfur (12% by weight) to a particle size of 0.01 mm. The treatment of the pulp with reagents is carried out at a ratio of the latter by mass of 2: 1. The pH is adjusted to 1 with an increase to 3 within 2 hours. In the leached mass, oxidizing agents are formed, which convert non-oxidized refractory metal minerals into easily openable forms.

After leaching in a disintegrator, the tailings are stacked, washed with water and leached first with a solution of sulfuric acid with a concentration of 10-30 g/l, and then with a solution of sodium sulfidotrioxosulfate with a concentration of 10-20 g/l.

The combination of leaching processes in the disintegrator and in the stack allows the tails activated in the disintegrator to release metals, increasing their recovery in comparison with both methods separately without combining.

The possibility of mechanochemical leaching has been proven experimentally when processing tailings of enrichment of Sadon polymetals, coal from the Russian Donbass and ferruginous quartzites KMA in a DU-11 disintegrator designed by the Center for Applied Mechanochemistry "Gefest". It has been determined that the technology under consideration makes it possible to extract metals from 50 to 80% of the original content in the tailings, and with a decrease in the residual content to the background level, it can be obtained by increasing the processing cycles.

After activation, the tailings can be used in the composition of the hardening mixture not only as inert fillers, but also as binding components, since an increase in the activity of the components by 20-25% increases the strength of concrete products up to 1 MPa.

The concept of deep utilization of waste from processing metal ores meets the principles of interaction between a Human and the biosphere and is especially relevant for solving the problems of diversification of mining in market conditions [13–15].

The results of the study are an element of the system of works of Russian scientists in the direction of resource conservation in mining [16–21].

4. Conclusion

Mechanochemical activation of metal-containing substandard raw materials during deep utilization of tailings makes it possible to extract metals and increase the activity of solid minerals to a state where they exhibit the desired properties.

The design of disintegrators is being improved through the use of new physical and technical processes, solving the problem of increasing the completeness of the use of bowels of the earth, including returning to production raw materials containing metal components inaccessible to traditional enrichment technologies.

References

[1] Golik V I and Komashchenko V I 2017 Othody obogashcheniya zhelezistyh kvarcitov kak syr'e dlya doizvlecheniya metallov i ispol'zovaniya v kachestve zakladochnyh smesej [Waste of enrichment of ferruginous quartzites as a raw material for additional extraction of metals and use as filling mixtures] *Gornyi Zhurnal* **3** 43-47

[2] Gavrishev S E, Kornilov S N, Pytalev I A and Gaponova I V 2017 Povyshenie ekonomicheskoj effektivnosti gornodobyvayushchih predpriyatij za schet vovlecheniya v ekspluataciyu tekhnogennyh georesursov [Increasing the economic efficiency of mining enterprises due to the involvement of technogenic georesources in operation] *Gornyi Zhurnal* **12** 46–51

[3] Golik V I, Komashchenko V I, Stradanchenko S G and Maslennikov S A 2012 Povyshenie polnoty ispol'zovaniya nedr putem glubokoj utilizacii othodov obogashcheniya uglya [Increasing the completeness of subsoil use by deep utilization of coal preparation waste] *Gornyi Zhurnal* **9** 91–95

[4] Golik V I, Razorenov Yu I, Zakharov E I and Abramkin N I 2018 Osvoenie zabalansovyh zapasov metallicheskih rud [Development of off-balance reserves of metal ores] *Bulletin of the Tula State University. Earth Sciences* **3** 158–170

[5] Bunin I Zh, Ryazantseva M V, Samusev A L and Khabarova I A 2017 Teoriya i praktika primeneniya, kombinirovannyh fiziko-himicheskih i energeticheskih vozdejstvij na geomaterialy i vodnye suspenzii [Theory and practice of application, combined physicochemical and energetic effects on geomaterials and water suspensions] *Mining Journal* **11** 65–73

[6] Dmitrak Yu V and Kamnev E N 2016 JSC "Vedushchij proektno-izyskatel'skij i nauchno-issledovatel'skij institut promyshlennoj tekhnologii" – Put' dlinoj v 65 let ["Leading design, survey and research institute of industrial technology" – A 65-year path] *Gornyi Zhurnal* 2016 **3** 6–12

[7] Chanturia V A 2017 Nauchnoe obosnovanie i razrabotka innovacionnyh processov kompleksnoj pererabotki mineral'nogo syr'ya [Scientific substantiation and development of innovative processes of complex processing of mineral raw materials] *Mining journal* **11** 98–105

[8] Voropanova L A 2018 Teoriya i praktika ekstrakcionnyh processov izvlecheniya cvetnyh metallov iz vodnyh rastvorov [Theory and practice of extraction processes of non-ferrous metals extraction from aqueous solutions] (Vladikavkaz: Litera) p 647

[9] Golik V I, Razorenov Yu I and Karginov K G 2017 Osnova ustojchivogo razvitiya RSO-Alaniya – gornodobyvayushchaya otrasl' [The basis of sustainable development in North Ossetia-Alania is the mining industry] *Sustainable Development of Mountain Territories* vol 9 **2(32)** 163–171

[10] Golik V I and Efremenkov A B 2016 *IOP Conf. Ser.: Mater. Sci. Eng.* 125 012038

[11] Golik V I, Razorenov Yu I, Brigida V S and Burdzieva O G 2020 Mekhanohimicheskaya tekhnologiya dobychi metallov iz hvostov obogashcheniya [Mechanochemical technology of metal extraction from ore dressing tailings] *Bulletin of the Tomsk Polytechnic University. Engineering georesources* vol 331 **6** 175–183

[12] Golik V I, Gabaraev O Z, Maslennikov S A, Khasheva Z M and Shulgaty L P 2016 The provision of development conversion perspectives into undeground one for Russian iron ore deposits development *The Social Sciences* (Pakistan) vol 11 **18** 4348–4351

[13] Komashchenko V I, Golik V I, Belin V A and Gaponenko A L 2014 Povyshenie effektivnosti vzryvnoj otbojki na osnove novyh sposobov iniciirovaniya skvazhinnyh zaryadov na kar'erah [Increasing the efficiency of blasting blasting on the basis of new methods of initiation of borehole charges in open pits] *Mining information and analytical bulletin (scientific and technical journal)* **9** 293–304

[14] Dmitrak Yu V and Shishkanov K A 2010 Razrabotka veroyatnostnoj kinematicheskoj modeli melyushchih tel v pomol'noj kamere vibracionnoj mel'nicy [Development of a probabilistic kinematic model of grinding bodies in the grinding chamber of a vibration mill] *Mining information and analytical bulletin (scientific and technical journal)* **12** 302–308

[15] Bosikov I I, Klyuev R V and Kelekhsaev V B 2017 Razrabotka pokazatelej dlya ocenki effektivnosti funkcionirovaniya prirodno-promyshlennoj sistemy v gorno-obogatitel'nom komplekse s ispol'zovaniem metoda analiticheskoj ierarhii [Development of indicators for assessing the efficiency of the functioning of the natural-industrial system in the mining and processing complex using the method of analytical hierarchy] *International Conference on Industrial Design, Application and Production (ICIEAM)* (Russia: Saint Petersburg)

[16] Kaplunov D R 2015 Problema ispol'zovaniya vozobnovlyaemyh istochnikov energii v hode razrabotki mestorozhdenij tverdyh poleznyh iskopaemyh [The problem of using renewable energy sources during the development of solid mineral deposits] *Physical and technical problems of the development of minerals* **1** 88–96

[17] Burmistrov K V and Osintsev N A 2020 Principy ustojchivogo razvitiya gornotekhnicheskih sistem v perekhodnye periody [Principles of sustainable development of mining systems in transition periods] *Bulletin of the Tomsk Polytechnic University. Engineering georesources* vol 331 **4** 179–195

[18] Zaalishvili V B, Melkov D A, Dzeranov B V, Morozov F S and Tuaev G E 2018 Integrated instrumental monitoring of hazardous geological processes under the Kazbek volcanic center *International Journal of GEOMATE* 15(47) 158–163

[19] Meskhi B, Pleshko M, Buligin Yu, Alekseenko L and Molev M 2017 *IOP Conf. Ser.: Earth Environ. Sci.* 012217

[20] Rilnikova M V, Vladimirov D Ya, Pytalev I A and Popova T M 2017 Robotizirovannye

geotekhnologii kak put' povysheniya effektivnosti i ekologizacii osvoeniya nedr [Robotic geotechnology as a way to improve the efficiency and greening the development of mineral resources] *Physical and technical problems of the development of minerals* 1 92–101

[21] Kachurin N M, Stas G V, Korchagina T V and Zmeev M V2017 Geomechanical and aerogasdynamic consequences of undermining the territories of mountain allotments in the mines of Eastern Donbass [Geomekhanicheskie i aerogazodinamicheskie posledstviya podrabotki territorij gornyh otvodov shaht Vostochnogo Donbassa] *Bulletin of the Tula State University: Ser. Earth Sciences* **1** 170–182

Development of a compact plasma torch design for welding and surfacing

M E Goshkoderya[1,2], N A Kondratev[1], V R Nikitina[1] and N A Zhugarev[1]

[1] Peter the Great St. Petersburg Polytechnic University, ul. Polytekhnicheskaya, 29, 195251 Saint Petersburg, Russia

[2] E-mail: Gosmike@yandex.ru

Abstract. The article discusses the development of the design of the plasma torch used for welding and surfacing, the relevance of the plasma type of welding, the stages of work necessary to develop this design, modeling the flow of protective and working gas.

1. Introduction

Relevance of work: plasma technologies have long established themselves in welding operations [1, 2], however, today there is a decline in research activity and the introduction of new developments in this field. Plasma torch is one of the fundamental components of plasma welding; therefore, the development of new torches will allow increasing the efficiency of indicators such as productivity, convenience and quality of welding.

Plasma torches with direct polarity current are used everywhere and are the determining components of any plasma process [3]. One of the successful examples of the application of plasma processes is plasma surfacing, which is used to create special and protective coatings [4]. It is also possible to use plasma surfacing in additive manufacturing.

A plasma-forming gas flows around the electrode inside the burner. The ignition of the pilot arc usually occurs due to a high voltage pulse. As a result of the formation of an electric arc, the gas is heated to a temperature at which its ionization occurs. Moreover, around the cooled surface of the tip, it remains cold and non-conductive. A narrowing of the plasma torch occurs due to it. In addition to these factors, the diameter of the tip through which the arc exits also affects the quality of the formed surfacing and protective coatings.

Plasma torches, which are used for manual welding, should also be as convenient as possible for the welder. Based on this, the design of a compact plasma torch will be proposed in this study, which can be used for both manual and automatic welding.

Purpose of work: development of a compact plasma torch design for welding and surfacing using low and medium current values.

Task of work: development of a plasma torch is a multi-component task, which includes both the study and development of design documentation, followed by modeling of the selected solutions, and the manufacture of prototypes, and conducting real experiments. At this stage, it is necessary to develop the design of the plasma torch, reflected in the drawings and modeling the flow of protective and working gas. Due to the modeling of structural elements where gas is used, it becomes possible to improve the design without going to the stage of prototype manufacturing and conducting real experiments.

Content from this work may be used under the terms of the Creative Commons Attribution 3.0 licence. Any further distribution of this work must maintain attribution to the author(s) and the title of the work, journal citation and DOI.
Published under licence by IOP Publishing Ltd

IOP Conf. Series: Materials Science and Engineering **939** (2020) 012026 doi:10.1088/1757-899X/939/1/012026

2. Analysis of existing designs of plasma torches for welding and surfacing

Plasmatrons for welding and surfacing, using inert gases as a protective and working gas, are now quite rare. In Russia, at the moment, almost no research is being carried out in the development of new plasma torches for welding, despite the fact that plasma welding has a high productivity, as well as the heat-affected zone and the welding seam is smaller compared to other manual welding methods.

The following models were considered as similar plasmatron designs:

A plasmatron is considered in [5], figure 1, which has small mass and size characteristics and can be used as an ignition device, or in plasma processing of materials in domestic conditions.

The following differences can be distinguished when comparing this design with the proposed one: since the plasmatron [5] is offered for domestic use, it did not provide for the protection of the weld using shielding gas, which affects the obtained characteristics of the weld. Air is used as a working gas. The cathode is a copper electrode. The possibility of water cooling of the anode is not provided, which reduces its service life. There is no possibility of quick electrode replacement.

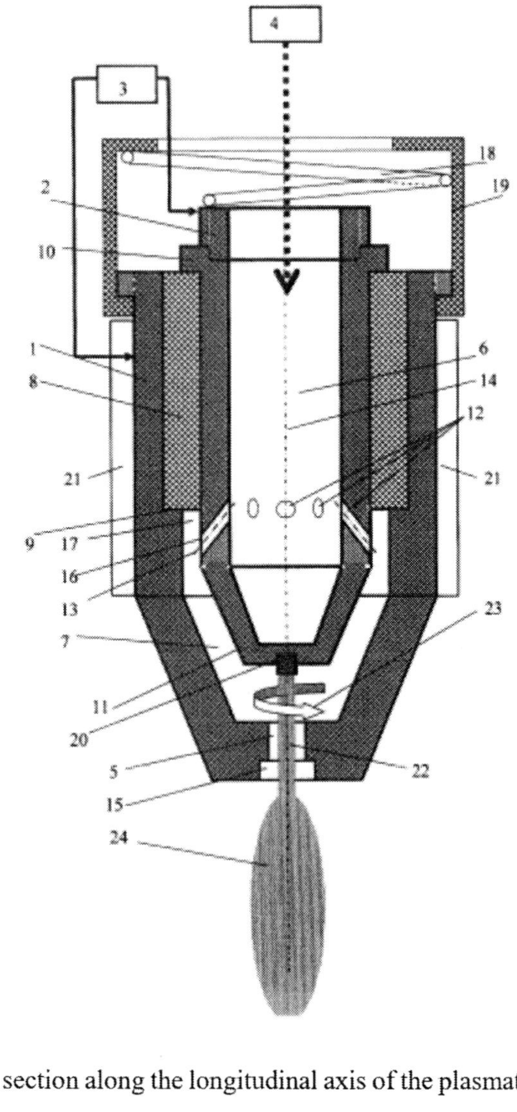

Figure 1. A section along the longitudinal axis of the plasmatron device [5].

The objective of the invention [6] was the creation of a small plasma torch for manual welding with direct current, figure 2, creating a stable electric arc burning between the electrode and the workpiece.

In the design [6] there is no possibility of using water cooling of the anode. The use of a copper electrode is assumed as a cathode. Also, the type of method for creating plasma acts as a difference; in this work, a plasmatron with the possibility of direct and indirect connection is presented. There is no possibility of quick electrode replacement.

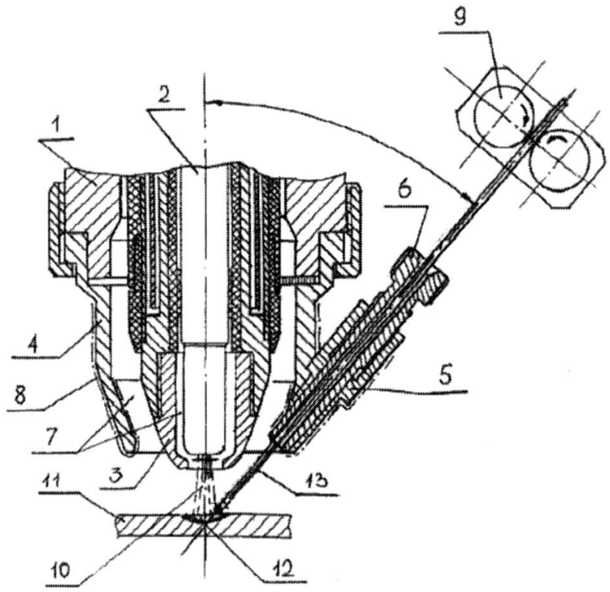

Figure 2. Section along the longitudinal axis of the plasmatron device [6].

The design of the plasmatron, figure 3, having reduced overall dimensions, with the existing water-cooling system, in contrast to the aforementioned, is presented in [7]. The main difference of the proposed device from the design [7] is the significant simplicity of the design. The design lacks the ability to quickly replace the electrode. The plasma torch [7] also uses a copper electrode as a cathode.

3. Requirements for the developed design of the plasma torch

The plasmatron is structurally quite a complex device. The plasma torch is such a complex design due to the content of elements operating under high temperature loads, requiring intensive cooling during welding / surfacing, as well as the influence of the shape of the plasma-forming unit on the protective properties.

The plasma torch must meet the following criteria:

- Elements subject to greater thermal effects should have maximum thermal conductivity;
- Presence of a sufficient level of structural cooling achieved by water and gas;
- Plasma forming unit should have the maximum possible volume with a constant surface area;
- Electrode must be firmly fixed and centered using the electrode holder;
- Lack of double arcing;
- Operation of the structure at low (microplasma welding) and medium currents;
- Ability to replace the electrode quickly and easily;
- Compliance with certain overall dimensions;
- Use for welding and surfacing.

IOP Conf. Series: Materials Science and Engineering **939** (2020) 012026 doi:10.1088/1757-899X/939/1/012026

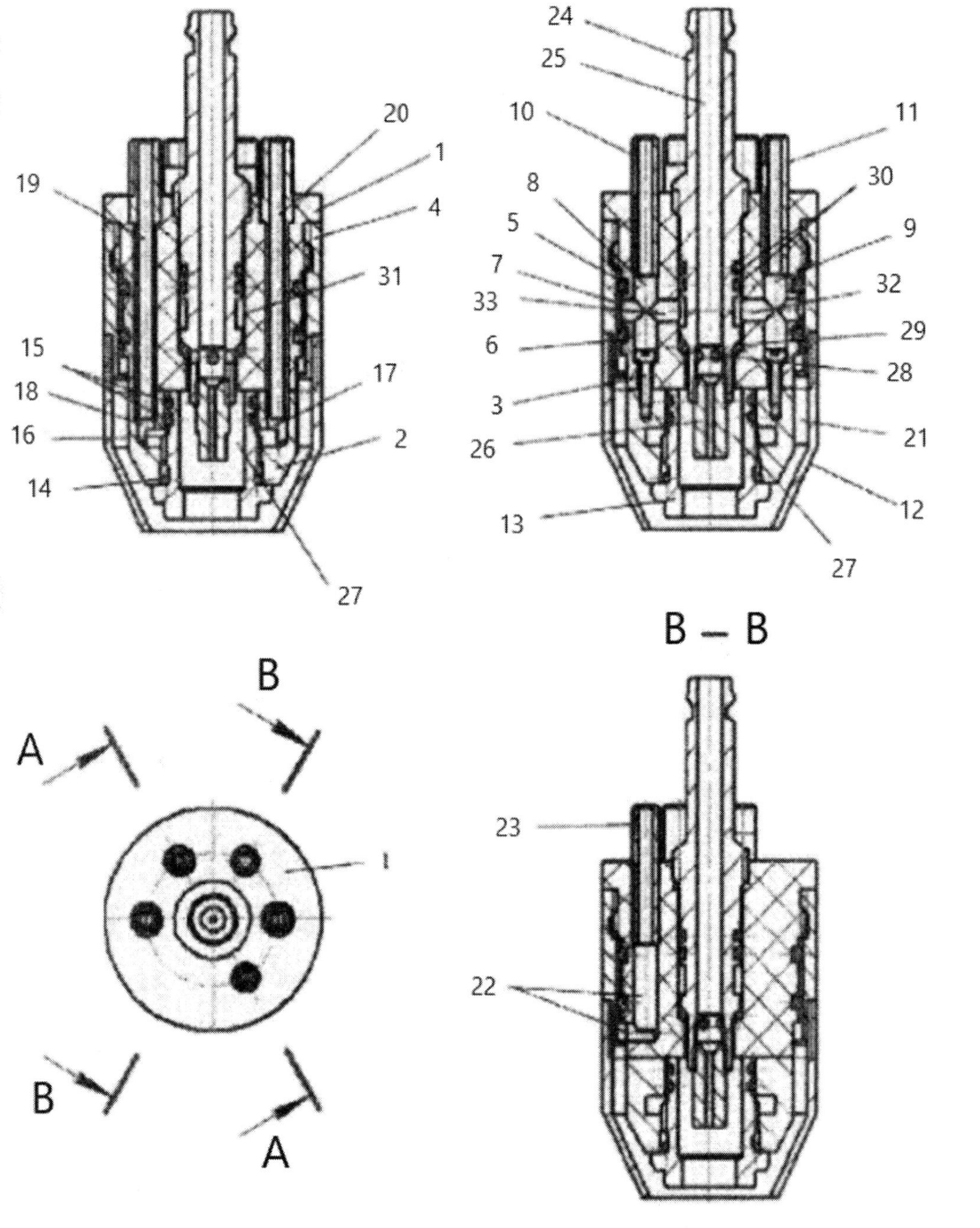

Figure 3. Drawings of the plasma torch assembly [7].

4. Developed plasma torch design

Initially, a sketch of the plasmatron was developed, according to which the design was developed, figure 4. The requirements that the plasmatron must meet were taken into account, and, as the main task, the possibility of reducing the overall structure was considered.

IOP Conf. Series: Materials Science and Engineering **939** (2020) 012026 doi:10.1088/1757-899X/939/1/012026

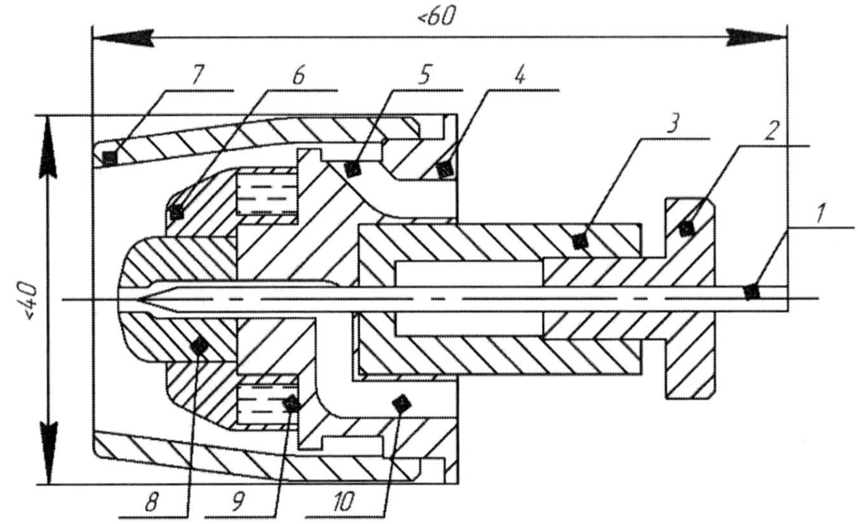

Figure 4. Sketch of the developed plasma torch with maximum overall dimensions.

The plasma torch consists of a tungsten electrode 1, which is tightly fixed due to the threaded connection of the clamping screw 2 and the electrode holder assembly 3. In the nozzle channel 4, in which the channels for the protective gas 5, working gas 10 and water cooling 9 are located, the electrode holder assembly 3 is fixed. An adapter for tip 6 and a protective nozzle 7 are attached to the nozzle channel 4. A plasma is formed in the region of the tip 8.

5. Modeling of structural units of gas outflow in a plasmatron

After developing the sketch and solid-state structural elements, the task arose to understand how the protective and working gas will behave in the channels designed for them. "Ansys" engineering analysis and numerical simulation software was used for this task.

Shielding gas outflow modeling was performed to improve the geometry of the gas channel, figure 5. As a result, it was possible to achieve a more uniform distribution of the gas flowing around the structural units that need to be cooled, figure 6. Also, thanks to the shielding gas, better weld protection during welding and surfacing is achieved.

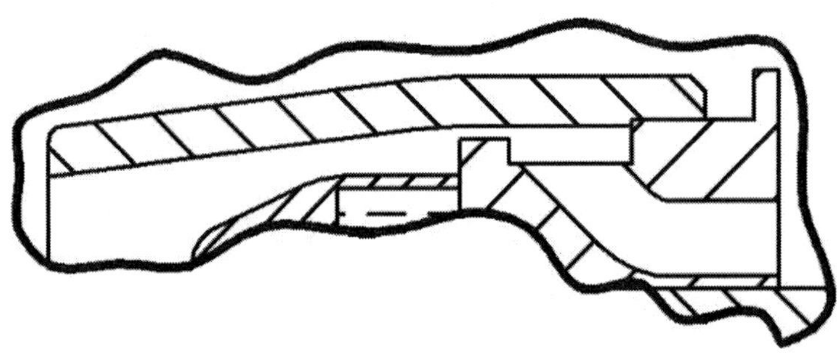

Figure 5. Simulated shielding gas supply unit.

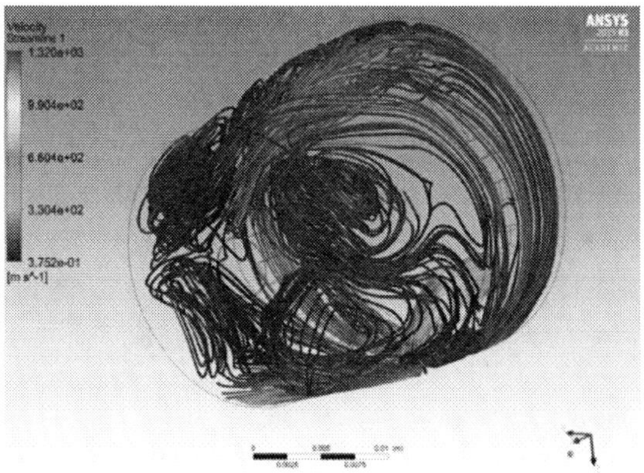

a)

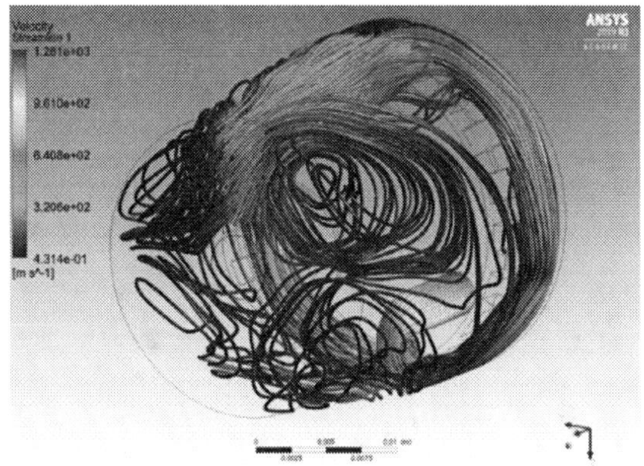

b)

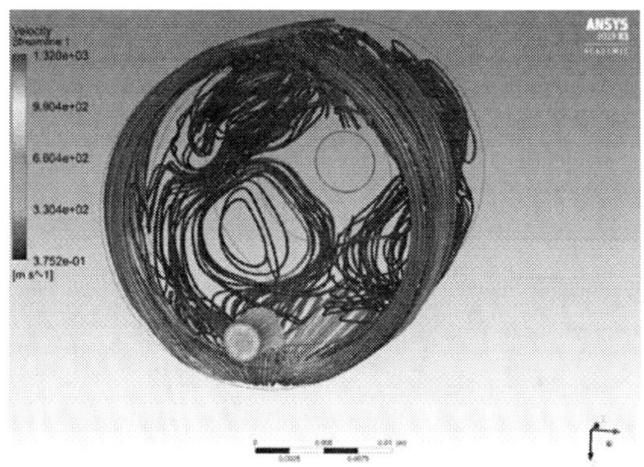

c)

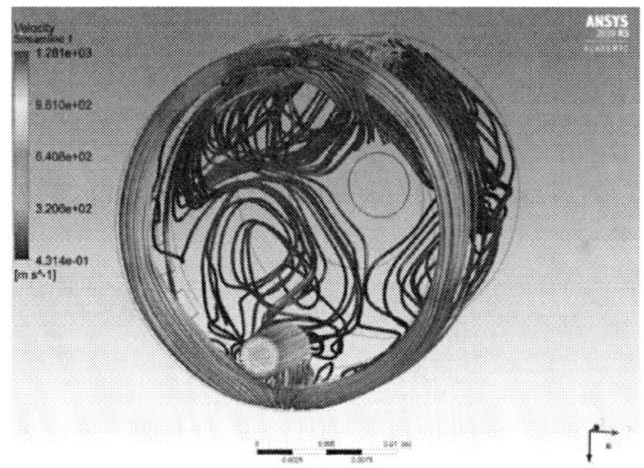

d)

Figure 6. Channel simulation results for shielding gas *a*), *c*) – the first sample is the most successful; *b*), *d*) – the second sample.

As a result of modeling the working gas unit, figure 7, various results were obtained, figure 8. It was proved that when the gas moves through the channels, turbulences are formed in the region of its collision with the electrode, which should be minimized, since the flow rate decreases. In the vicinity of the gas outlet from the plasma torch, it is possible to use various tip designs, however, we need the gas flow to be laminar, since we use the plasmatron for welding and surfacing, and not for cutting. In this case, the correct tip geometry necessary for the formation of a plasma flow should be preserved.

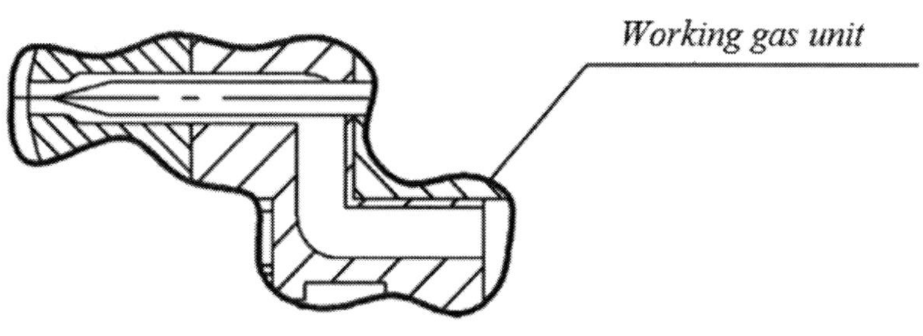

Figure 7. Unit for modeling the flow of working gas.

IOP Conf. Series: Materials Science and Engineering **939** (2020) 012026 doi:10.1088/1757-899X/939/1/012026

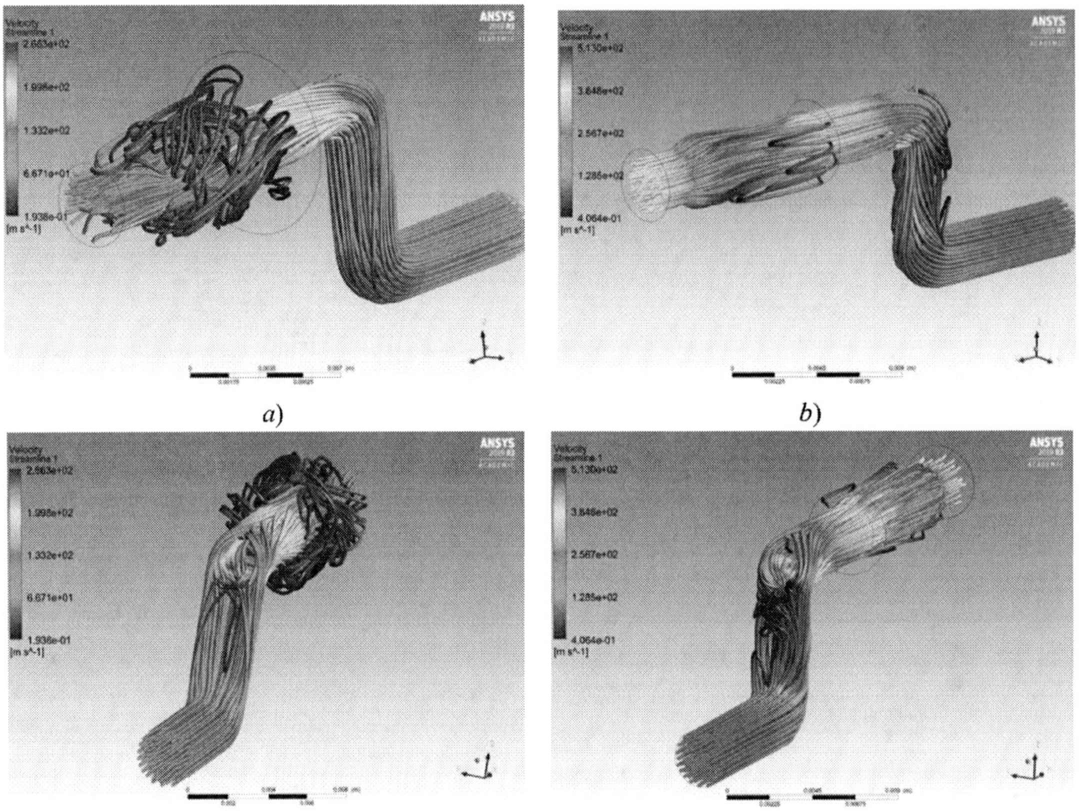

c) *d)*

Figure 8. Channel simulation results for working gas *a*) – the first sample; *b*) – second sample.

6. Results

The developed 3D model of the plasma torch was designed using the "Compass 3D" program. It becomes possible to conduct surface and solid modeling of 3D objects using this software. Figure 9 shows a 3D model of the assembly of a plasma torch in isometric projection.

This model is noteworthy for its geometric characteristics, due to the reduction of some elements, as well as combining certain parts together; it was possible to achieve the minimum dimensions of the structure. This allows reducing the basic cost of the plasma torch, while maintaining quality at a high level, since it is supposed to use argon as a protective and working gas in this design. Also, the level of comfort during the work process increases by reducing the total weight acting on the welder, when using this model for manual welding. Water cooling was provided in the tip design to avoid overheating of components.

Figure 9. 3D model of a plasma torch.

IOP Conf. Series: Materials Science and Engineering **939** (2020) 012026 doi:10.1088/1757-899X/939/1/012026

7. Conclusion

As a result of the work, the design of a compact plasma torch for welding and surfacing was developed. This plasmatron has a number of promising advantages for further modernization. Data were also obtained on the influence of the geometric characteristics of the channels for the protective and working gas. It is possible to achieve a more uniform distribution with the right direction of the shielding gas channel. Also, it is possible to get a laminar flow without violating the geometry of the tip when changing the channel for the working gas. The next stage in the development of a plasma torch will include the manufacture of prototypes of plasma torches and laboratory welding experiments.

References

[1] Shchitsyn Iu D 2004 *Plasma technology for welding* (Perm: Perm National Technic University) Part 1 p. 73

[2] Redchits A V and Ovchinnikov V V 2004 *Improving the energy efficiency of plasma welding* Svarochnoe proizvodstvo **8** 21–23

[3] Sosnin N A, Ermakov S A and Topolansyi P A 2013 *Plasma technology* (St. Petersburg) p 405

[4] Das B, Yadaiah N, Ozah R, Chowdhury S, Mondal A K and Muralidhar M A 2017 Perspective Review on Estimation of Keyhole Profile during *Materials today: proceedings* **5(2)** 6345–6350

[5] Gerasimov V A and Dostovalov V A RF patent № 2004138676/22 27.05.2005 *Plasma torch* Patent of Russia № 45888 28.12.2004

[6] Bochkarev V P, Nazaruk V K, Gorbach V D and Chikerda S I RF patent № 2005128353/02 10.03.2008 *Welding electric arc plasma torch* Patent of Russia № 2318639 12.09.2005

[7] Terentyev S A RF patent № 2019107987 08.10.2019 *Plasma torch* Patent of Russia № 2702512 20.03.2019

IOP Conf. Series: Materials Science and Engineering **939** (2020) 012027 doi:10.1088/1757-899X/939/1/012027

Development of intelligent algorithms for the continuous diagnostics and condition monitoring subsystem of the equipment as part of the process control system of a stainless steel pipe production enterprise

E S Grishin[1]

[1] Immanuel Kant Baltic Federal University, ul. A Nevskogo, 14 236041 Kaliningrad, Russia

E-mail: negrshn@mail.ru

Abstract. This article describes the process of development and implementation of an integrated APCS and continuous diagnostics and equipment condition monitoring subsystems using intelligent machine-learning-based algorithms for a stainless steel pipe manufacturing enterprise with combining local automatic control systems into a single information technology the system. Briefly, the author describes the technological equipment and the production cycle, its features, as well as its change as a result of the implementation of APCS. The software tools and their structure, the interaction of elements in the system, the main tasks solved with APCS and their impact on the quality of finished products are described. A feasibility study and justification of the application of PCS for this production are carried out.

1. Introduction

The LLC Techno Tube enterprise specializes in the production of welded stainless pipes for a wide range of consumers. The production of stainless welded pipes includes the use of welding equipment. Regardless of which welding method is used (electric welding, laser welding, plasma welding, and electron beam welding), the essence of the process is the same: a stainless steel sheet (strip is a flat billet in the form of a roll) is rolled using rolls of a rolling mill, and its edges are connected using a weld. After this, the seam is cleaned, sanded and passes all the necessary tests proving its strength. At the last stage, the pipe is calibrated and is cut into pieces of the desired length. According to GOST 11068–81, (Electric-welded pipes from corrosion-resistant steel. Technical conditions) [1] the permissible diameter of welded stainless steel pipes is from 8 to 102 mm and the permissible thickness of their walls is from 0.8 to 4 mm. In order to control the quality of products, to control the compliance of the actual production process with flow charts and schemes for various types of materials and to confirm product parameters to consumers, the company plans to create a process control system with a subsystem of intelligent continuous diagnostics and equipment condition monitoring.

2. Development of APCS enterprise

The company has five rolling mills and a stainless steel strip (flat billet) preparation section (cutting machine from material rolls), and it is also planned to install a mill for the production of pipes with

Content from this work may be used under the terms of the Creative Commons Attribution 3.0 licence. Any further distribution of this work must maintain attribution to the author(s) and the title of the work, journal citation and DOI.
Published under licence by IOP Publishing Ltd

increased requirements for the chemical and nuclear industries. It is planned to include mills with a laser welding process and a mounted mill in APCS. In total there are up to five rolling mills and a preparation section for billets.

2.1. Organization of the technological process before the implementation of the developed APCS
At present, pipe production lines are a system of local automated control systems on microcontrollers of various manufacturers (Siemens [2], Omron [3]), synchronized with a moveable work piece. The Italian-made equipment is used.

The disadvantages of the existing production system include the following:

1) The parameters of the technological process are not recorded, it is impossible to establish the relation between the batch number of the manufactured products and the parameters of its production life cycle.

2) It is impossible to control the compliance of the parameters of the technological process of production with the technological map from the database for various types and grades of steel.

3) Maintenance of the equipment and polishing of the rolls, elimination of the defects that occur, are performed on the basis of operating time.

2.2. Organization of the technological process after the implementation of the developed APCS
After the implementation of APCS, the following tasks will be solved and the organization and control of the production process will change:

1) For various grades of steel, significant parameters of the technological process will be controlled in accordance with the database [4] containing the required parameters of the technological process depending on the grade of raw material and its geometric steam meters.

2) All significant process parameters will be recorded in the database, the batch number will be associated with a set of technological parameters.

3) Product quality will be continuously monitored with nondestructive testing methods.

4) Production process will be visualized, equipment operation control will be carried out continuously with access for persons controlling the production process.

5) Continuous monitoring of the state of the equipment will be carried out on the basis of the analysis of trends in the change (departure) of technological parameters from operating characteristics. Accordingly, a logical conclusion will be formed about the need for technical maintenance or the acquisition of spare parts for equipment items that may fail [5]. Knowledge will be accumulated on troubleshooting equipment in the database, which will reduce equipment downtime due to failures. A model will be trained that will significantly accelerate the analysis of trends in technological parameters from operating characteristics (analysis of the dynamics of the time series).

This will allow service personnel to carry out maintenance and equipment repairs according to its actual condition. Studies [6] showed that the effectiveness of the service strategy as estimated at 30% of the cost of the total machines and can reduce by 10–15% the output of substandard products that are detected with non-destructive testing methods for quality control.

The scientific novelty in this work is the machine learning algorithm and additional model training "using examples" using classical methods of mathematical statistics to determine the dynamics of a time series and methods of logical predictive inference in the process of continuous equipment diagnostics.

3. Development of a subsystem of intelligent continuous diagnostics and monitoring of equipment status

The subsystem of intelligent continuous diagnostics and monitoring of equipment status is based on the production model of knowledge representation of expert intelligent systems with direct logical inference [7]. In this model, knowledge is represented in the form of rules: "If (condition [and condition ... and condition]), then (conclusion)". A system with direct inference performs processing from conditions to conclusion. To adapt to the diagnostic tasks and ease of use of the diagnostic subsystem by operators of production lines, the rules have been expanded to the form: "If (condition [and condition ... and condition]), then (defect). Recommended: (action). " The rules are formed in advance by a specialist when implementing APCS with analyzing technical documentation, statistical information on the maintenance and repair of diagnosed equipment, as well as formalizing the knowledge of service personnel. When rules are entered into the system, the rules are automatically checked for consistency. General view of the production model is:

$$N =< S; L; A \rightarrow B; Q; R >, \tag{1}$$

where N – product model name;

S – description of the class of situations;

L – product activation condition;

A – core products;

Q – inference of the production rule;

R – recommendations to the operator.

The development of equipment defects during the operation phase is mainly associated with the physical aging of the equipment, which is caused with the processes of wear and corrosion and mechanical damage under various conditions, and occurs smoothly. The development of each defect is accompanied with some changes in the characteristics of the part or unit of equipment (temperature, vibration, and others). Analyzing the dynamics of changes in characteristics, grouping changes in characteristics corresponding to certain defects, a diagnostic model of the equipment fleet of this enterprise was built into the diagnostic rules.

An analysis of the trend in the movement of the value (dynamics) of the observed characteristics is based on the analysis of the absolute and average absolute increase in the series of dynamics, as well as the sum of the absolute chain increments of levels. Absolute growth shows the rate of change of the series and is defined as the difference between the current and base levels. For the base level, the level of the previous period is taken.

In view of the large number of measurements and the corresponding increase in the complexity of the calculations, the author proposed to reduce the data with quantizing the analyzed time interval into equal segments according to formula (2).

$$n = \frac{q}{k}, \tag{2}$$

where n – quantization step;

q – sampling depth;

k – regulatory coefficient.

The quantization step is essentially a sampling step of depth q. The control coefficient allows you to adjust the overall complexity of the algorithm and the calculation error in particular. To obtain an error in the calculation of the algorithm of less than 1%, the coefficient k = 26280 was selected empirically.

Comparisons of the analysis of the complete set of technological data and the reduced set showed that the reduction of the data did not significantly affect the result of the analysis of the dynamics of the observed characteristics. In figure 1, blue is the average absolute increase in the initial series, red is the average relative increase in the reduced series, for example, 30 different series are shown levels.

The computational complexity of the algorithm for determining the dynamics of the initial series is directly proportional to the number of observed characteristics (A) and sample depth (B), and is determined by formula 3, the same is true for the reduced sequence, but inversely proportional to the value *n* of formula (2) and is determined by formula 4 .

$$T = A * B + C, \tag{3}$$

where C – some constant.

$$T = \frac{A * B}{n} + C, \tag{4}$$

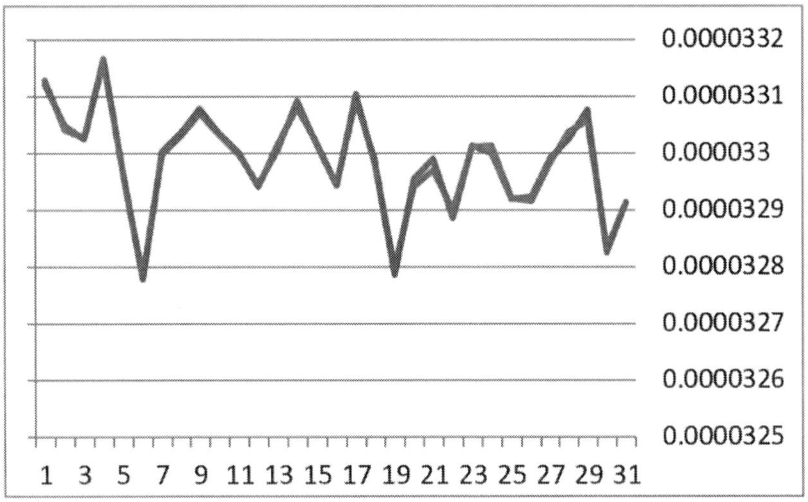

Figure 1. Average Absolute Growth Levels.

Thus, the theoretical computational complexity of the algorithm for 10^3 pieces of equipment with 10^2 observed characteristics for each with a maximum sampling depth will be approximately 10^{13} for the initial series, for the reduced series it is approximately 10^9. This method of data reduction allows you to comfortably use the diagnostic system; however, the complexity of the algorithm is still linear.

The task of determining the dynamics of a series is highly specialized, and to accelerate its solution, you can use the machine-learning model. In this situation, the "learning by examples" model of the ML.NET development tool was used. The training examples for this model are generated during the first six months of diagnosis. From the moment the diagnostic subsystem is introduced into production, the model begins to learn the data from the analysis of the dynamics of the observed characteristics obtained with the method described above. The training data consists of time series, which, in turn, are divided into three types: those with a tendency to increase, with a tendency to decline, and static (taking into account possible fluctuations in the measurement values). The ML.NET development tool machine learning model contains implementations of various algorithms: linear (average perceptron, stochastic double coordinate wise lift, Broyden - Fletcher - Goldfarb - Shanno algorithm, character-by-symbol stochastic gradient descent), decision trees (gradient boosting, fast tree, fast forest, generalized additive model), metaalgorithms and others [8]. Empirically, while training the model on various input data, the algorithm of averaged perceptron was chosen, because it reaches the established or absolute accuracy faster than the others do. This is a linear classification algorithm based on an artificial neural network (a classification problem with more than two linearly separable classes is solved with including several neurons in the output (computational) layer of the perceptron, which makes its predictions with finding a separating hyperplane [9]. The algorithm was built on the basis of the perceptron model proposed by Frank Rosenblatt [10], who, together with his colleagues, proved its convergence [11].

Data for primary education is formed by a package containing tuples, of the form: "row, type of dynamics". The number of tuples for each type of row dynamics is balanced. A package is considered formed if one of the following conditions is met: its size has reached 10 megabytes; a month has passed of the continuous formation of the package. Upon completion of the formation of the data package for primary training, it is transferred to the machine-learning model and the process of its training is launched. The duration of model training for a given volume of training data is up to 60 minutes, it depends on the computational performance and the performance of the computer memory subsystem.

IOP Conf. Series: Materials Science and Engineering **939** (2020) 012027 doi:10.1088/1757-899X/939/1/012027

After the training process is completed, an arbitrary sample is formed from the previously obtained data, with a volume of 1000 tuples for each type of dynamics of the series, and an analysis of the error in the calculation of the model for each type of dynamics is performed. If the error for any type of dynamics exceeds 1%, then the process of additional training of the model for this type of dynamics is performed. Training data is grouped in a package of up to 5 megabytes, additional model training takes up to 20 minutes. After that, a second analysis of the error is performed and, if necessary, the process is repeated. A generalized scheme of the model-learning algorithm is presented in figure 2.

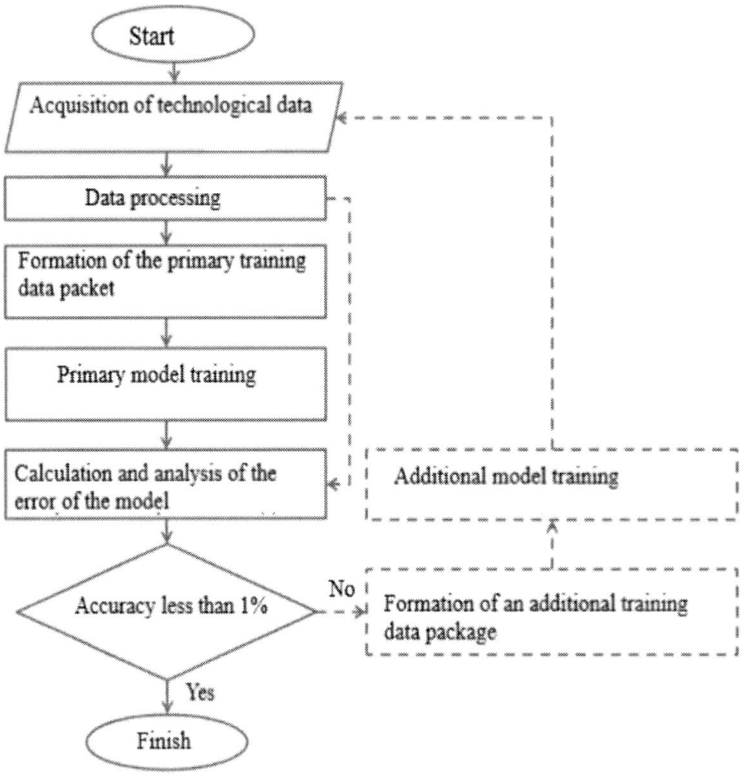

Figure 2. Generalized algorithm diagram.

After training the model to the established accuracy, the statistical algorithm for analyzing the trend of value movement is replaced with a machine learning model, but it is not completely withdrawn from work and continues to be used in the training data preparation module, preparing data tuples of the form: "row, type of dynamics". To reduce the computational load, the time series analysis interval is reduced. Every day, to verify the model in an automated mode with a certain time interval, a repeated analysis of accuracy is performed. If for some type of dynamics the machine-learning model ceases to fit into the established error, then the additional training procedure described above is carried out for this type.

The average speed of the model was analyzed on 10^2 samples of time series of $18 \cdot 10^3$ levels in each and it is less than 250 milliseconds (figure 3).

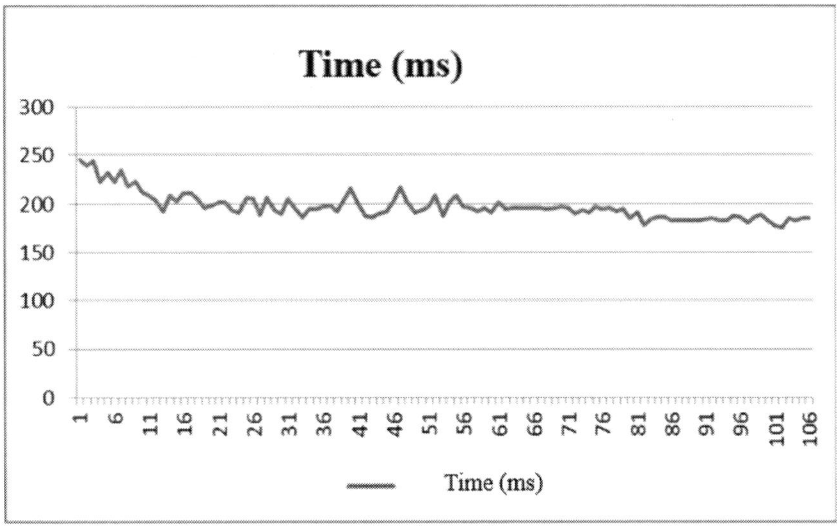

Figure 3. The speed of the algorithm.

In addition, when diagnosing equipment, its residual life at a given time is taken into account according to the formula (5)

$$t = \frac{n-k}{T}, \quad (5)$$

where t – the residual life, hours;

n – the threshold value of the characteristic;

k – the current value of the characteristic;

T – the average absolute increase in the characteristic value per hour.

Taking into account the above factors, equipment is clustered into states according to the highest degree of wear of its component parts and residual resource. There are three levels:

• Initial – if the value of the observed characteristic is between the lower and upper warning settings and the residual resource of each node is more than 6 months, it is necessary to consider the condition of the equipment as stated or normal.

• Medium – if the value of the observed characteristic is between the lower and upper warning settings and the residual resource of each node is more than 3 months, it is necessary to consider the condition of the equipment as transient;

• Pre-emergency – if the value of the observed characteristic is between the lower and upper warning settings and the residual resource of each node is more than 1 month, it is necessary to the condition of the equipment pre-emergency.

For the convenience of using the diagnostic subsystem by the operator, the author suggested dividing the equipment nodes according to the degree of criticality into five levels:

1) Low;

2) Below average;

3) Medium;

4) Above average;

5) High.

For each criticality level, a time interval is established that determines the sampling depth obtained with a statistical study of the service intervals of various equipment and their components:

• 60 months;

• 36 months;

• 12 months;

• 6 months;
• 3 months.

4. Results

Sample reduction allows the use of a statistical algorithm for analyzing the trend of the observed characteristics with linear complexity quite long and efficiently until the time when the processed data becomes enough to train the machine-learning model. In turn, constant monitoring of the model error does not allow errors in the diagnosis. In the presented work, it was obtained an intelligent algorithm that functions at a constant speed and fits within one second. This allows doing continuous diagnostics of equipment and promptly notifying maintenance personnel of the occurrence of defects.

5. Conclusion

The introduction of intelligent algorithms of the continuous diagnostics subsystem and monitoring the status of equipment as part of APCS at the automation facility will give the following positive results:

1) It will allow doing continuous quality control of products based on the analysis of the measured technological parameters and the actual condition of the equipment.

2) It will allow monitoring the state of the fleet of industrial equipment used and optimizing the costs of its maintenance.

3) It will allow the management staff of the enterprise to control the progress of the process and its significant parameters.

4) It will allow management of technical maintenance and regulation of industrial equipment in actual condition.

5) It will allow management of spare parts, tools and accessories based on knowledge of the actual condition of the equipment.

Acknowledgments

The research was supported by the Ministry of Science and Higher Education of the Russian Federation, unique identification number RFMEFI57817X0252.

References

[1] GOST 11068–81 *Electric-welded pipes from corrosion-resistant steel. Technical conditions*
[2] *Siemens LTD* (Official website: https://www.siemens-pro.ru/components/s7-1500.htm)
[3] *Guzzetti SPA* (Official site: http://www.guzzetti.com/en/azienda/) retrieved May 11, 2020
[4] *A typical industry-specific technological database for the automated generation of sets of technological documentation for all types of technological conversions using licensed software products from Intermech* 2019 URL: http://lab18.ipu.ru/projects/conf2006/2/22.htm (Retrieved December 13, 2019)
[5] GOST 34. 601. 90. *Automated systems. The stages of creation*
[6] Stephan H and Karl F 2000 Knowing plant – Decision supporting and planning for engineering design *Intelligent Systems in Design and Manufacturing III. Proc. SPIE* pp 376–384
[7] Popov E V 1987 *Expert systems* (Moscow: Nauka)
[8] *Guide to ML.NET* URL: https://docs.microsoft.com/ru-ru/dotnet/machine-learning/how-to-choose-an-ml-net-algorithm (Retrieved May 11, 2020)
[9] *Description of the averaged perceptron algorithm* URL: https://docs.microsoft.com/en-us/dotnet/api/microsoft.ml.trainers.averagedperceptrontrainer?view=ml-dotnet (Retrieved May 11, 2020)
[10] *Frank Rosenblatt* URL: https://ru.wikipedia.org/wiki/Rosenblatt,Frank (Retrieved May 11, 2020)
[11] Perceptron URL: https://ru.wikipedia.org/wiki/Perceptron

IOP Conf. Series: Materials Science and Engineering **939** (2020) 012028 doi:10.1088/1757-899X/939/1/012028

Characteristics of droplet transfer of electrode metal during MMA depending on the chemical composition of the material of the rod of the coated electrode

D P Il'yaschenko[1,2,5], **Iu I Chebotarev**[3,6] **and S B Sapozhkov**[4,7]

[1] Yurga Technological Institute branch of Tomsk Polytechnic University, ul. Leningradskaya, 26, 652050 Kemerovo region, Yurga, Russia
[2] Institute of strength physics and materials science SB RAS, Tomsk, Russia
[3] Irkutsk National Research Technical University
[4] Yaroslav-the-Wise Novgorod State University, ul. B. St. Petersburgskaya, 41 173003 Veliky Novgorod, Russia

[5]E-mail: mita8@rambler.ru
[6]E-mail: chebotarev-irk@mail.ru
[7]E-mail: ssb@novsu.ru

Abstract. This article presents the results of experimental studies of the effect of the coated electrode material for MMA welding on the droplet transfer parameters of electrode metal. It has been established that the use of coated electrodes based on copper (OZCH-4) and aluminum (Castolin EC 4001) alloys, compared to traditionally used metal rod based electrodes (Sabaros me20|10, LB-52U), affects the mass transfer characteristics of droplets electrode metal , in the form of an increase in the average duration of arc burning at intervals of melting of the electrode and the period of formation of drops of electrode metal up to 4 times.

1. Introduction

Over the past three years, a stable level of production and consumption of coated electrodes has been observed in Russia. In 2019, 99 296.7 tons of coated electrodes for MMA were produced in Russia, which is 3.4% more than the previous year's production volume [1].

It was established in [2–6, 9] that the grade of the coated electrode for MMA welding with coated electrodes intended for welding steels, namely the chemical composition of the metal rod and coating, have a significant effect on the parameters of the transferred droplets of the electrode metal during welding, and, accordingly, on received performance characteristics of welded joints. However, data on the influence of the chemical composition of the electrode rod, which includes non-ferrous metals, on the characteristics of droplet transfer during MMA are absent.

Objective: to determine the effect of the chemical composition of the coated electrode rod for MMA on the stability of melting and transfer of electrode metal into the weld pool, as the main indicators of heat and mass transfer of the welding process, affecting the operational and strength properties of the formed permanent joints.

Content from this work may be used under the terms of the Creative Commons Attribution 3.0 licence. Any further distribution of this work must maintain attribution to the author(s) and the title of the work, journal citation and DOI.
Published under licence by IOP Publishing Ltd

It was established in [2, 7] that the main parameters for assessing the stability of heat and mass transfer during arc welding with a consumable electrode are:

– the duration of the short circuit of the arc gap,, $\tau_{к.з.}$, ms;
– the duration of the cycle (the period of formation and transfer of the droplet), $T_{к.з.}$, ms;
– values of currents: maximum – Imax and minimum I_{min}, A;
– current slew rate: V increase Isb, and V decrease Isb, A / s.

2. Experimental part

A comparative assessment of the stability of the welding process and the droplet transfer parameters of the electrode metal was carried out with building up rollers on the surface of the plates of the material for welding which they are intended.

The following ones were selected as coated electrodes for the experiment: Castolin EC 4001, OZCh-4, Sabaros me20 | 10, LB-52U (table 1).

Surfacing was carried out using a classic power source for MMA welding – welding rectifier VD-306.

Oscillograms of the current in the welding circuit and the voltage between the electrode and the product recorded changes in the energy characteristics of the surfacing mode (figure 1) and are presented in table 1.

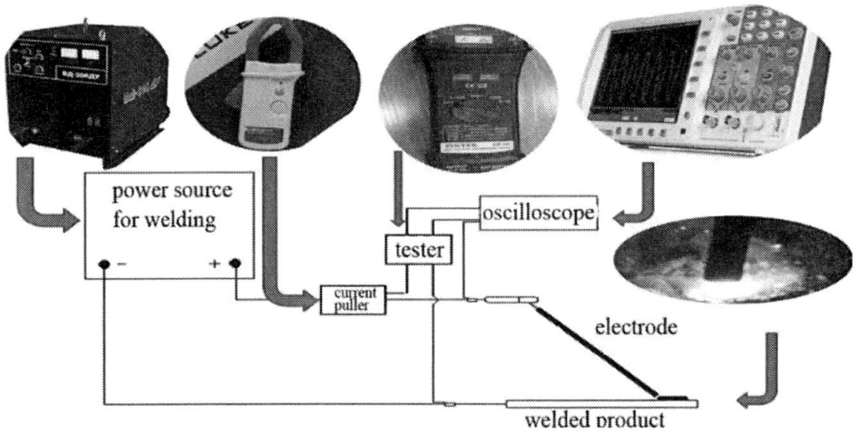

Figure 1. Experimental procedure to determine stability criteria for MMA welding [4].

Table 1. Welding conditions for different electrodes and power sources.

Electrode	Material to be welded	Average values of parameters (oscilloscope AKIP -4122/1V)
Castolin EC 4001	Aluminum	Current 90.7±2,7 A; Voltage 20.7±0.6 B
OZCh-4	Cast iron	Current 90±2.7 A; Voltage 23.5±0.6 B
Sabaros me20|10	Austenitic steel	Current 86.7±2.5 A; Voltage 24.5±0.6 B
LB-52U	Low alloy steel	Current 89±2.7 A; Voltage 20.8±0.6 B

Studying the stability of arc burning (figure 2) of the electrodes under consideration (table 1) was carried out according to the method described in [2, 4] based on a comparison of current and voltage waveforms (figure 3) obtained using: an AKIP-4122/22 digital storage oscilloscope; differential probe "Pintek Electronics" DP-50 "; "Fluke i1010" clamp meters and the "OWON_Oscilloscope" program.

IOP Conf. Series: Materials Science and Engineering **939** (2020) 012028 doi:10.1088/1757-899X/939/1/012028

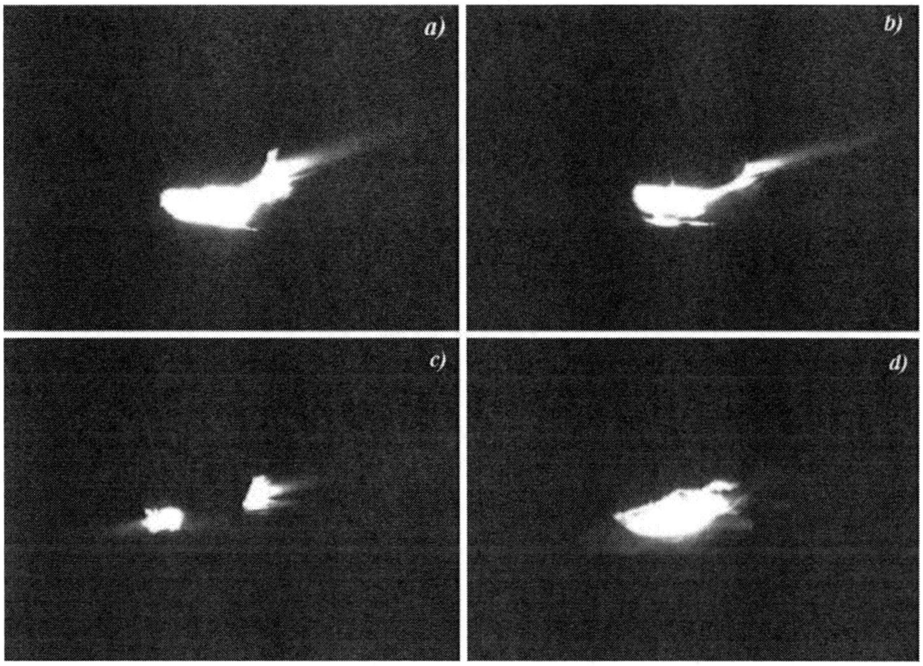

Figure 2. Kinograms of the process of transferring drops of electrode metal during manual arc welding: *a* – the beginning of the formation of a drop on the surface of the coated electrode; *b* – the growth of a drop of liquid metal on the surface of the electrode; *c* – drop transition from the electrode surface to the weld pool with the closure of the arc gap; *d* – the process of the onset of nucleation of the subsequent drop.

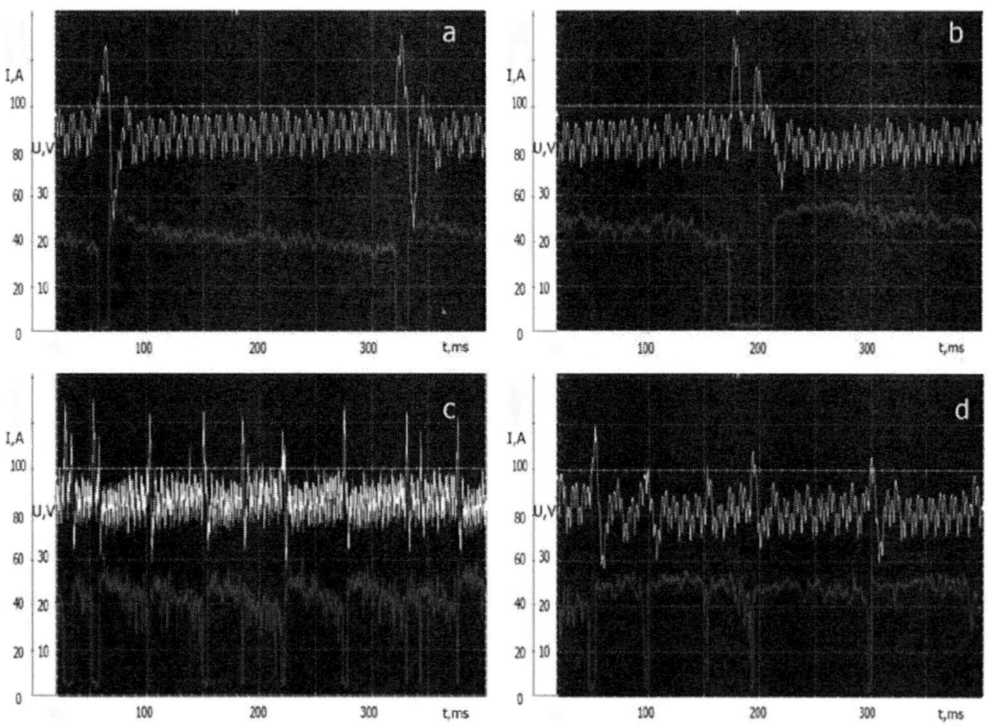

Figure 2. Current and voltage oscillograms in MMA welding with coated electrodes: a) Castolin EC 4001; b) ОЗЧ-4; c) Sabaros me20|10; d) LB-52U.

Table 2. Statistical data on parameters of electrode metals droplet transfer.

Parameter	Electrode			
	Castolin EC 4001	OZCh-4	Sabaros me20|10	LB-52U
Arc gap period shortcut τ_{sc}, MS± mean root square deviation of period shortcut, $\sigma_{\tau sc}$, ms	34.5±3.1	61.3±9.4	13.3±6.6	13.14±4.13
Period of cycle T_{sc}, MS± mean root square deviation of period of cycle σ_{Tsc}, ms	280±177	750±295	148.7±69.03	162±57
I_{min}, A	49.3±4.6	58.28±1.8	59.7±3	60±5.5
I_{max}, A	127.1±3.8	127±6.5	123±4.7	123.6±6

The use of coated electrodes based on copper (OZCh-4) and aluminum (Castolin EC 4001) alloys for MMA compared to traditionally used metal rod based electrodes (Sabaros me20 | 10, LB-52U) affects the mass transfer characteristics of electrode metal drops. Namely, it increases in the average duration of arc burning at intervals of melting of the electrode and the period of formation of drops of electrode metal up to 4 times.

The experimental patterns established above (table 2) can be explained with the different balance of forces acting on the drop, the welding process using these types of electrodes: Castolin EC 4001, OZCH-4 is large-drop, because a drop of molten electrode metal is in a dense shell formed with an oxide film of aluminum and copper, respectively. Due to the fact that this film has high mechanical strength, a considerable force is required to detach the droplet from the electrode. G.D. Nikiforov [8] believes that as the shell grows, the film stretches and cracks form in it that should lead to its destruction, however, in the presence of oxygen in the atmosphere, the metal oxidizes at the points of rupture and the continuity of the oxide film is restored. The rupture of the film and the transition of the droplet into the weld pool is also facilitated with the pressure of the gases generated in the droplet.

3. Conclusion

The use of coated electrodes based on copper (OZCH-4) and aluminum alloy (Castolin EC 4001) alloys for MMA compared to traditionally used metal rod electrodes (Sabaros me20 | 10, LB-52U) affects the mass transfer characteristics of electrode metal droplets. Namely, there is an increase in the average duration of arc burning at the intervals of electrode melting and the period of formation of drops of electrode metal up to 4 times.

References

[1] *The market for welding electrodes in Russia. Current situation and forecast 2020-2024* 2020 https://alto-group.ru/otchot/rossija/333-rynok-svarochnyh-jelektrodov-v-rossii-tekuschaja-situacija-i-prognoz-2020-2024-gg.html (Accessed 06.04.2020) [In Russ.]

[2] Saraev Y N, Chinakhov D A, Ilyashchenko D I, Kiselev A S and Gordynets A S 2017 Investigation of the stability of melting and electrode metal transfer in consumable electrode arc welding using power sources with different dynamic characteristics *Welding International* **31(10)** 784–790

[3] Vaz C T, Bracarense A Q, Felizardo I and Pereira Pessoa E C 2012 Impermeable low hydrogen covered electrodes: Weld metal, slag, and fumes evaluation *J. Mater. Res. Technol.* **1(2)** 64–

70 doi: 10.1016/S2238-7854(12)70012–1
[4] Il'yaschenko D, Chinakhov D, Makarov S, Danilov V, Galinsky A and Verkhoturova E 2020 Effect of ultra-disperse powder in electrode coating on properties of welds in MMA welding *Engineering Solid Mechanics* **8(1)** 41–48
[5] Mazel A G and Dedjuh R I 1978 O stabil'nosti processa ruchnoj dugovoj svarki modulirovannym tokom [On the stability of the process of manual arc welding with modulated current] *Svarochnoe proizvodstvo* **12** 11–15 [In Russ.]
[6] Erokhin A A 1973 *Basics of fusion welding. Physicochemical laws* (Moscow: Engineering) p 448
[7] Lankin Yu N 2011 Pokazateli stabil'nosti processa dugovoj svarki plavjashhimsja jelektrodom [Stability indices of the process of arc welding with consumable electrode] Avtomat. svarka **1** 7–15 [in Russ.]
[8] Nikiforov G D 1972 *Metallurgy of fusion welding of aluminum alloys Engineering* p 264
[9] Sapozhkov S B and Burakova E M 2018 *IOP Conf. Ser.: Mater. Sci. Eng.* **441** 012042

IOP Conf. Series: Materials Science and Engineering **939** (2020) 012029 doi:10.1088/1757-899X/939/1/012029

Method for predicting the composition of the protective coating in MMA

D P Il'yaschenko[1,2,6], **E A Zernin**[3,7], **S B Sapozhkov**[4,8] **and L G Loskutov**[5,9]

[1] Yurga Technological Institute branch of Tomsk Polytechnic University, ul. Leningradskaya, 26, 652050, Kemerovo region, Yurga, Russia
[2] Institute of strength physics and materials science SB RAS, Tomsk, Russia
[3] Kuzbass State Technical University (KuzSTU), Kemerovo, Russia
[4] Yaroslav-the-Wise Novgorod State University, ul. B. St. Petersburgskaya, 41 173003 Veliky Novgorod, Russia
[5] JSC "Azot" Cargo building 1, 650021, Kemerovo, Kuzbass, Russia

[6] E-mail: mita8@rambler.ru
[7] E-mail: yuti_sp@bk.ru
[8] E-mail: ssb@novsu.ru
[9] E-mail: svarkamett@mail.ru

Abstract. This article presents the results of experimental studies to determine the optimal coating composition for MMA. The optimization is based on the idea of multi-factor planning, which allows us to disseminate the obtained results not only on welding processes, but also related technologies.

1. Introduction

Splashing of the electrode metal represents one of the weaknesses of consumable electrode arc welding. As a result of this process, a great number of molten metal splashes (droplets) with a variety of sizes are ejected from an arc zone and initiate physical and chemical reactions with surface layers of a metal to be welded, which change its structure and phase state. When welding high-strength alloying steels under a droplet there is an annealed structure, seen as a stress riser, worsening operation properties of a construction. For instance, a protective surface coating is deposited to prevent splashing of molten metal when carrying out assembly and welding operations, /1/. Figure 1 demonstrates the effect of a protective coating used in MMA [1, 2, 9].

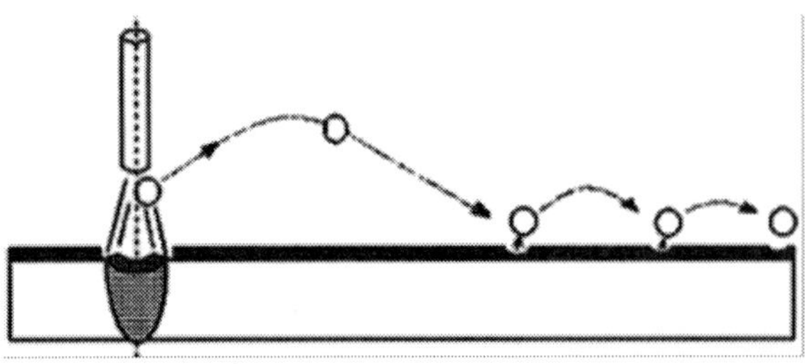

Figure 1. A droplet trajectory on the coated surface.

Content from this work may be used under the terms of the Creative Commons Attribution 3.0 licence. Any further distribution of this work must maintain attribution to the author(s) and the title of the work, journal citation and DOI.
Published under licence by IOP Publishing Ltd

Research and analysis of the coatings used showed [3, 4] that there is no methodology for predicting the composition of the coating taking into account technological and rheological properties. The technique is especially relevant when developing coatings containing more than two components, since each component has its own functional purpose.

2. The main part

When developing and testing the methodology for predicting the composition of the coating, the known coating was taken as an analogue [5], having the following composition:

1) a binder – sulphite-alcohol stillage (LWC);
3) functional additive (to increase the viscosity of the composition) – Glycerin;
2) stabilizing substance (preventing the spread of putrefactive bacteria) – Furacilin.

The first and second components are introduced into the coating composition in the form of a solution, the third – in the form of a powder.

Optimization of the composition of the coating was carried out according to the following technological and rheological indicators:

1) a criterion indicator of protective properties (the number of hard-to-remove sprays) – β (%);
2) conditional viscosity – η_y (c).

Before starting the optimization of the coating composition, it was found that the parameters of the MMA mode affect the amount of hard to remove spatter. The research results showed that the maximum adhesion of the spray to the surface of the product is observed at a current strength of I_{CB}=100–120A and arc voltage Ud = 25V (coated electrodes with a diameter of 3 mm) [6], which were taken as the basis for optimizing the composition protective coating.

Each component was assigned the designation: binder (LWC) – X; furatsilin – Y; glycyrin – Z.

At the initial stage, the boundaries of changes in the *X*, *Y*, and *Z* components, as well as the change step (the results are shown in table 1) were experimentally determined. The number of components is given in grams per liter of water. The decisive factor in choosing the boundaries of the change in components was the correspondence of the behavior during shear of the coating, similar to the Shvedov – Bingham body [7]. The shear tests of the coatings were carried out using the setup described in [7].

Table 1. Determining the boundaries of component changes.

Component Change Boundaries	Parameter Change Step
$60<X<120$	$\Delta X=30$
$0<Y<6$	$\Delta Y=1$
$60<Z<160$	$\Delta Z=20$

When planning studies, the idea of factor planning was used [8]. The main condition for factor planning is to ensure the orthogonality of the used “Latin (magic) square” (factor experiment).

As a result, three values were considered for component *X*, six values for components *Y* and *Z*. Variants of factors are indicated with indices 1, 2, 3, ... (X_1, X_2, X_3, Y_1, Y_2, ..., Y_6, Z_1, Z_2, ..., Z_6). Variation of factors is given in table 2.

Table 2. Variation of factors.

	Y_1	Y_2	Y_3	Y_4	Y_5	Y_6
Z_1	X_1	X_2	X_3	X_1	X_2	X_3
Z_2	X_3	X_1	X_2	X_3	X_1	X_2
Z_3	X_2	X_3	X_1	X_2	X_3	X_1
Z_4	X_1	X_2	X_3	X_1	X_2	X_3
Z_5	X_3	X_1	X_2	X_3	X_1	X_2
Z_6	X_2	X_3	X_1	X_2	X_3	X_1

In order to exclude the influence of systematic errors caused with external conditions, using a table of evenly distributed random numbers, the experiments were randomized. Next, experiments were conducted on the effect of components on the protective and viscous characteristics of the protective coating.

To find the optimal composition of the protective coating, the dimensionless function Φ was determined from the expression:

$$\Phi=\beta_D \cdot \eta_D, \tag{1}$$

where β_D is the dimensionless value of the protective property (difficult to remove spray); η_D – dimensionless value of conditional viscosity.

The dimensionless quantity for hard to remove sprays is found from the expression:

$$\beta_D = \frac{\beta_i}{\beta_c}, \tag{2}$$

where β_i is the actual value of the number of hard-to-remove sprays in the i-th experiment; β_C is the systemic value of the amount of hard to remove spatter (taken for welding without coating equal to 2.1% [4]).

Dimensionless value for conditional viscosity of protective coatings:

$$\eta_D = \frac{\eta_i}{\eta_c}, \tag{3}$$

where η_i is the actual value of the conditional viscosity of the protective coating during the i-th experiment; η_C is the systemic value of the conditional viscosity (we take for water equal to 10 by [7]).

To ensure reliable protection of the surface from splashes of molten metal during welding, parameter β should tend to a minimum. The ease of application of the protective coating is due to its low viscosity; therefore, the parameter η should also tend to a minimum.

Therefore, the optimal composition of the protective coating will correspond to the minimum value of the dimensionless function Φ (figure 2):

$$\Phi=\beta_D \cdot \eta_D \rightarrow \min$$

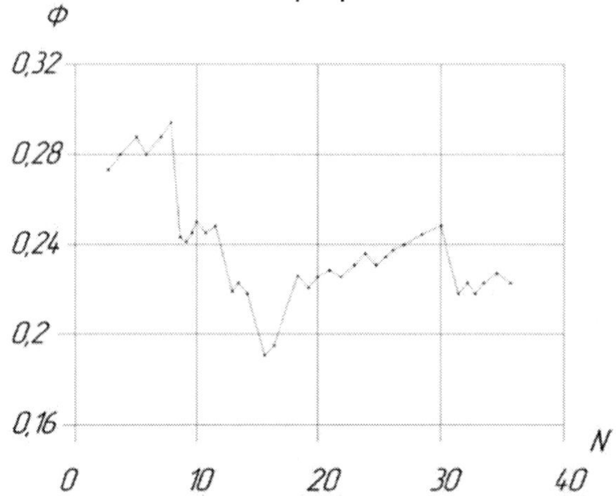

Figure 2. Determination of the optimal composition of the protective coating: Φ – dimensionless function, N – experiment number.

Based on a multivariate experiment on the influence of various components on the technological and rheological properties of the coating, the coating composition was obtained with combining factors X_2,

Y_4 and Z_3. The minimum value of the dimensionless function Φ was obtained at $N = 16$ (figure 2) (according to the order of implementation, the thirty-third experiment).

As a result, we obtained coating compositions that take into account technological and rheological properties (in grams per liter of water): coating 1 (optimized) – binder (LWC) – 80–100; Furacilin – 1–3; Glycerin – 90–110; coating 2 (corresponds to experiment number 5, figure 1) – binder (CBF) – 60–70; Furacilin – 1–3; Glycerin – 70–80.

Based on the results of further studies, it was established that the nature of the flow curve of coating 1 corresponds to the curve of the flow of the Shvedov – Bingham body, and the protective properties of the coating are necessary and sufficient, that cannot be said about coating 2.

For the obtained coating compositions, a study was carried out to determine the amount of spraying according to the method [4], the coating scheme for the parts to be welded is presented in figure 3.

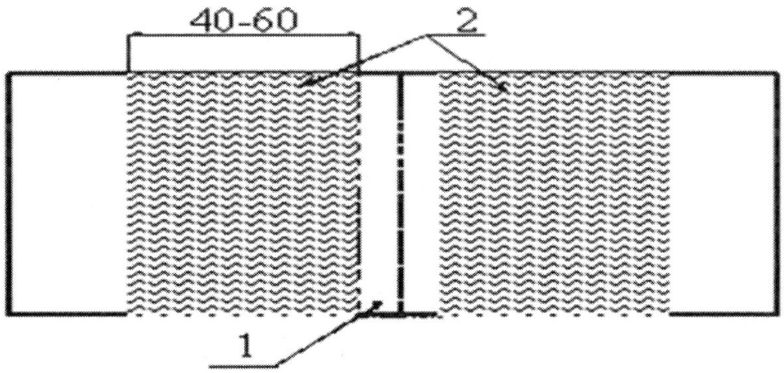

Figure 3. The scheme of applying a protective coating for MMA: 1 – weld; 2 – protective coating.

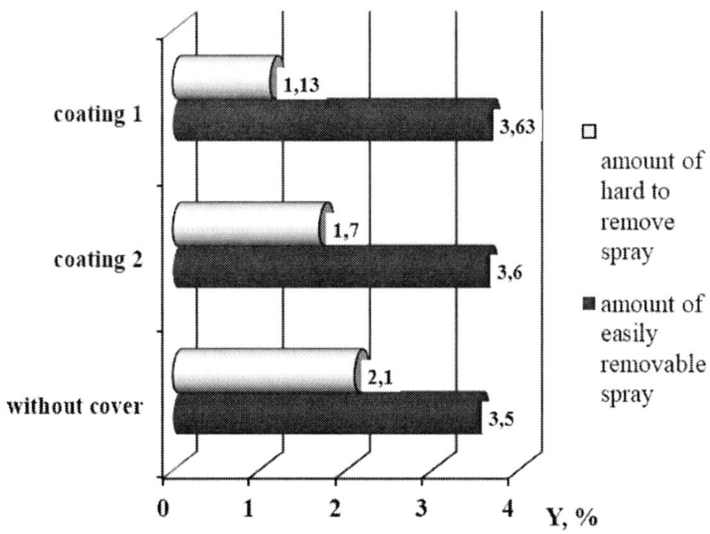

Figure 4. The effectiveness of the use of a protective coating with MMA coated electrodes UONI 13/55 (diameter 3 mm; I_{CB}=110 ± 4A; Ud = 25 ± 2V).

Analysis of the data presented in figure 4 shows a 47% reduction in the number of hard-to-remove droplets on the surface to be welded when using coating 1, in comparison with welding without coating.

3. Conclusion

The developed approach to predicting the coating composition for MMA is the foundation when creating a mathematical model for designing multifunctional coatings for various technological purposes: not only for welding production, but also for metallurgy (non-stick coatings) and other related technologies.

References

[1] Ilyaschenko D P, Chinakhov D A, Verkhoturova E V and Lavrova E V 2020 Stability of MMA welding with protective coatings *Structural integrity and life* vol 20 **1** 33–36

[2] Il'yaschenko D P, Chinakhov D A, Verkhoturova E V and Mamadaliev R A 2020 Effect of Protective Coatings in Welded Constructions on Welding Stability and Electrode Metal Transfer into a Weld Pool in MMA *Solid State Phenomena* vol 303 32–38

[3] Fedko V T 2004 *Technology, theoretical foundations and means of reducing the complexity of welding in carbon dioxide* (Tomsk: Publishing house Tom. University) p 398

[4] Fedko V T and Zernin E A 2003 Rheological properties of coatings to protect the surface of products from splashes of molten metal during welding in CO_2 *Technology of metals* **2** 16–20

[5] Sapozhkov S B, Zernin E A , Kolmogorov D E and Ilyashchenko D P 2006 *Coating composition for protecting the surface of the welded product from sticking of splashes of molten metal during fusion arc welding* Pat 2329885 (RF) V23K 35/36 C09D 1/00 (Applicant and patent holder of Tomsk Polytechnic University) Application 11/07/06 Publ 07/27/2008 bull **21**

[6] Il'yashchenko D P and Sapozhkov S B 2008 Splashing in manual arc coated electrode welding and methods of reducing splashing *Welding International* **22(12)** 874–877

[7] Zernin E A and Fedko V T 2003 Installation for studying the effect of shear rate on coating viscosity Factory Laboratory *Diagnostics of materials* **4** 36–38

[8] Lukyanov S I, Panov A N and Vasilev A E 2014 *Fundamentals of an engineering experiment* (Moscow: IC RIOR: SIC INFRA-M) p 99

[9] Sapozhkov S B and Burakova E M 2018 *IOP Conf. Ser.: Mater. Sci. Eng.* **441** 012042 doi:10.1088/1757-899X/441/1/012042

Prerequisites for obtaining wear-resistant metal coatings with a high-entropy structure by atmospheric plasma spraying, reflow, and hardening

A M Kadyrmetov[1], D A Popov[1,2], E V Snyatkov[1] and A A Plakhotin[1]

[1] G. F. Morozov Voronezh State Forestry University of Forestry and Technologies, Timiryazeva str., 8 394087, Voronezh, Russia

[2] E-mail: qaz.7@mail.ru

Abstract: The prerequisites for the use of atmospheric plasma spraying processes with multiple reflow and hardening of coatings are considered as a method of obtaining a highly entropic structure in them as applied to wear-resistant coatings.

1. Current state of research for multicomponent plasma coatings

In the last decade, the properties of a new class of multicomponent high-entropy alloys (HEA) have been actively studied in metal science. Active research in the field of multicomponent alloys, caused by additional interest due to the discovery of previously poorly studied compositions of alloying elements and substrates in equiatomic concentration, was started in the early 2000s in connection with the tasks of obtaining materials operating under extreme (elevated temperatures and loads) conditions, in particular, wear-resistant and heat-resistant alloys based on iron and nickel [1–2]. Such alloys consist of at least 5 basic elements in concentrations from 5% to 35% at. and have a high entropy of mixing, which can be considered as a measure of the probability of maintaining the state of their system [3]. This gives them advantages in comparison with traditional alloys in terms of physical and mechanical properties, thermal stability, resistance to corrosion and wear, and other properties [4]. The properties of such alloys depend on the chemical composition and method of preparation.

The available results of fundamental and experimental studies demonstrate the unique physical-mechanical properties of HEA, which can significantly expand the scope of modern metal materials [5]. The unique physical-mechanical properties of HEA led to the need for their use as coatings for various functional purposes.

For machine parts operating under friction and wear conditions, the urgent task was to obtain thermostable wear-resistant coatings from HEA. Parts that work under conditions of erosion and wear at high temperatures include parts of gas distribution mechanisms and a cylinder-piston group, internal combustion engines (valves, cylinders), brake parts, details of rocket engines (parts of gas turbines).

The entire spectrum of such materials and their possible properties has not been disclosed [6], while the majority of HEA studies are focused on elucidating the dependences of the properties of alloys on their microstructural parameters, and only a little attention is paid to the complex scientific and technical problem of studying and developing new methods for producing HEA. Among these works should be attributed, first of all, the review [7].

Content from this work may be used under the terms of the Creative Commons Attribution 3.0 licence. Any further distribution of this work must maintain attribution to the author(s) and the title of the work, journal citation and DOI.
Published under licence by IOP Publishing Ltd

IOP Conf. Series: Materials Science and Engineering **939** (2020) 012030 doi:10.1088/1757-899X/939/1/012030

The vast majority of works are devoted to the study of all-metal samples from HEA. In their regard, many researchers obtained excellent results, but the practical application of new materials is hampered by the high cost of obtaining them. A small part of the work is devoted to the study of the production of multicomponent coatings with high entropy (HEA coatings). Such coatings are suitable as functional coatings for machine parts operating under high and extreme loads. This applies primarily to wear-resistant coatings with a thickness of 1 ... 3 mm. In this case, the parts can be made of traditional alloys, and it would be techno-economically feasible to apply HEA coatings on the friction surface using one of the progressive methods.

Until now, such coatings have been obtained mainly by expensive vacuum, ion-plasma methods, and above all, by magnetron sputtering and vacuum-arc spraying of chemical compounds of metals with nonmetals [8], as well as by mechanical and laser alloying [9]. The use of vacuum-free technologies would reduce the cost of obtaining such coatings and increase their productivity. These technologies include gas thermal technology and, above all, plasma spraying in an open atmosphere. However, such methods have not been studied, or insufficiently studied [10]. In this regard, it is of great scientific and practical interest to obtain wear-resistant HEA coatings by combined processes based on atmospheric plasma spraying.

2. Conditions for obtaining multicomponent coatings with high entropy

In operation, HEA coatings must ensure the stability of the specified properties in phase and structural composition over a wide range of conditions and, above all, high adhesion properties that ensure coating retention under high operating loads. Coating processes must be high performance, affordable and economical.

The development of the technology of creating HEA coatings is possible while ensuring uniform mixing and homogenization of the composition when coating is applied to achieve high mixing entropy and to obtain single-phase disordered substitution solid solutions in the limit. Moreover, the entropy of mixing is a necessary but not sufficient criterion for the formation of a single-phase solid solution [7]. For the thermodynamic stability of a solid solution, a high degree of solubility of the constituent elements in each other is necessary. In turn, this requires that any element of the alloy can be replaced by another element with close values of electronegativity and atomic size, while maintaining the total number of constituent elements [1]. The influence of enthalpy of mixing and nonconfigurational contributions of entropy to Gibbs free energy should also be taken into account [7]. Only with a certain elemental composition does a disordered solution form, the mixing entropy of which completely compensates for the influence of both positive and negative enthalpy values [7].

The disorder of elements with different atomic radii in a solid solution due to their different combinations leads to lattice distortions, which affects the structure and properties of alloys. Due to the high density of defects, HEA occupy an intermediate position between crystalline and amorphous alloys [11]. The high density of defects leads to a decrease in the thermal conductivity of the alloy, which is undesirable for antifriction materials of machine parts from the point of view of heat removal from the friction zone.

Due to the fact that the normalized value of the diffusion energy of elements in different matrices is much higher than in pure metals, the following procedure for changing the diffusion coefficients [12] has been established: HEA <stainless steel <pure metal.

The low values of the diffusion coefficient of atoms in a multicomponent alloy determine the strength at high temperature [13], the formation of nanostructures and nanoscale precipitates [14]. This, in turn, can contribute to stability and increase the wear resistance of the alloy.

In highly entropic alloys, mixing leads not only to substitutional solid solutions, but also to other ordered and disordered phases, which contribute to the properties of the HEA. First of all, this relates to the correlation relation between the elastic limit – hardness, and also the relation between the yield strength – ductility [7]. These ratios in comparison with other metal alloys show the best combination of strength and ductility. An increase in ductility can be a prerequisite for reducing the likelihood of friction on the surface during the wear of a microcrack network, and an increase in hardness can be a

IOP Conf. Series: Materials Science and Engineering **939** (2020) 012030 doi:10.1088/1757-899X/939/1/012030

prerequisite for an increase in the wear resistance of HEA. At the same time, highly entropic alloys with a BCC lattice have predominantly high strength and low ductility, while alloys with an FCC lattice have low strength and high ductility. Therefore, for high wear resistance, it is presumably advisable to provide BCC phases.

The task of obtaining a coating with uniform mixing and homogeneity of the alloy components involves the study of unknown mechanisms and patterns of coating formation, as well as the influence of its factors on the structural and phase parameters, physical-mechanical and tribotechnical properties of the coatings, which are the criteria for the HEA coatings.

Factors of the processes of obtaining HEA are parameters of the composition of the components (alloy elements) and their amount in the alloy; external environment, means, modes and frequency rate of thermal, thermo-mechanical and/or mechanical effects on the formed alloy, coordinate distribution and time dynamics of this effect.

Known processes for producing HEA in which these factors are realized are casting processes, mechanical alloying, laser cladding, magnetron sputtering, and self-propagating high-temperature synthesis.

The solution to the problems of obtaining plasma coatings 1–3 mm thick with a uniformly mixed homogeneous composition is similar to the process of casting samples as a kind of melting method.

3. Patterns of HEA formation during casting

Some results of studies of foundry production of ingots are presented in [1, 15–20] and, in particular, in works in which, before melting, preliminary mechanical alloying of components in the form of powder in a high-energy ball mill was used [7]. The main method for producing ingots from HEA is multiple melting in vacuum or in an inert medium to increase the uniformity of the chemical composition and control the cooling rates during crystallization.

It was shown in [17] that an effective factor in controlling the formed structure of an alloy is its crystallization rate, which significantly affects the mechanism of alloy decay. With a fivefold remelting of HEA AlCrFeCoNiCu and low cooling rates of 10 $K \cdot c^{-1}$, a dendritic (BCC) and interdendritic (FCC) phases are formed. These phases undergo spinodal decomposition with the release of the cubic phases A2, B2, LI_2, which differ in elemental composition and morphology.

At a high quenching rate of 10^6-10^7 $K \cdot c^{-1}$, followed by heat treatment, an ultrafine-grained structure without dendrites with the presence of a B2 nanodomain superstructure is formed. Upon subsequent thermal annealing, crystalline multicomponent cubic phases are formed in a small amount [18].

In a study of the evolution of HEA $Al_{0.5}CoCrFeNi$ depending on the aging temperature (350–950 °C) [19], it is noted that the negative enthalpy of mixing of alloy elements with a high concentration of them promotes the formation of intermetallic compounds of nickel with aluminum, that is, ordered structures. An increase in the aging temperature leads to a decrease in the proportion of intermetallic compounds.

An increase in the concentration of certain elements also leads to the formation of intermetallic compounds. Thus, in [20], the formation of CoCr and Ti_2Ni intermetallic compounds in the dendritic phase of HEA $Al_{0.5}CoCrCuFeNiTi_x$ at $x = 1.2$ was found.

4. The effect of plastic deformation and heat treatment on HEA properties

Intense plastic deformation by comprehensive forging of HEA AlCrCuNiFECo results in improved strength characteristics [21]. In comparison with a cast alloy, the brittle-viscous transition decreases from 700–800 °C to 600–700 °C after forging. At temperatures above the temperature of the brittle-viscous transition, the softening of the forged alloy is more noticeable than cast. In this case, a brittle-viscous transition occurs at the same values of tensile strength (350 MPa) for both states. This can positively affect the wear resistance of the alloy with respect to preventing the formation of fatigue cracks on the friction surface of the alloy.

High temperature treatment is accompanied by a transition to a superplastic state. In this case, the cast alloy has a region of local deformation and fracture occurs with a small elongation. Forged alloy is deformed more uniformly without a neck and with a large elongation.

Annealing significantly affects the microstructure and properties of HEA. An increase in the annealing time at 700 °C changes the microstructure of HEA $Al_{05}CoCrCuFeNi$ from elongated dendrites to a polygranular structure [22]. The authors suggest that this is due to the formation of an intermetallic compound — the σ-phase, or solid solutions based on CrFe and CoCr. An increase in temperature to 900 °C leads to the formation of a fine-grained structure without rod-like secretions and elongated dendrites. Annealing at 1100 °C led to the formation of a phase composition similar to the composition of the samples after casting, but with a fine-grained structure.

Due to the weak diffusion in HEA, the cooling of the heated sample during annealing proceeds rapidly, which does not allow completion of the phase transformation with the formation of new phases.

Annealing of samples from HEA $CoCrFeNiTiAl_x$ leads to an increase in Vickers hardness [23]. The authors attribute this to an increase in intermetallic compounds having a higher hardness than traditional alloys, as well as the effect of solid solution hardening. Annealing for this alloy showed very high compressive strength (2.6 GPa) at $x = 0.5$ at room temperature and ductility (13%). A further increase in aluminum content reduces strength and ductility.

5. Prerequisites for obtaining multicomponent coatings with high entropy by processes based on atmospheric plasma spraying, reflow, and electromechanical processing

A review of the methods for producing highly entropic alloys based on the casting method, akin to the plasma spraying method with multiple fusion in an inert medium, allows us to draw some practical conclusions.

The maximum entropy of a multicomponent alloy is determined by the formation of a disordered single-phase substitutional solid solution, which is ensured primarily by the elemental composition. For this, it is necessary to take into account the influence of enthalpy of mixing and non-configurational contributions of entropy to the Gibbs free energy. Only with a certain elemental composition does a disordered solution form, the mixing entropy of which completely compensates for the influence of both positive and negative enthalpy values. For the thermodynamic stability of a solid solution, a high degree of solubility of the constituent elements in each other is necessary so that any element of the alloy can be replaced by another element with close values of electronegativity and atomic size, while maintaining the total number of constituent elements.

Due to the high density of defects, HEA occupy an intermediate position between crystalline and amorphous alloys. The high density of defects leads to a decrease in the thermal conductivity of the alloy, which is undesirable for antifriction materials of machine parts from the point of view of heat removal from the friction zone.

The normalized value of the energy of diffusion of elements in different matrices of HEA is much higher than in pure metals. Low values of the diffusion coefficient of atoms in a multicomponent alloy determine high strength at high temperatures, the formation of nanostructures and nanoscale precipitates. This, in turn, can contribute to stability and increase the wear resistance of the alloy.

The ratio of the elastic limit - hardness and the ratio of yield strength - ductility in comparison with other metal alloys show the best combination of strength and ductility. An increase in ductility can be a prerequisite for reducing the likelihood of friction on the surface during the wear of a microcrack network, and an increase in hardness can be a prerequisite for an increase in the wear resistance of HEA. At the same time, highly entropic alloys with a BCC lattice have predominantly high strength and low ductility, while alloys with an FCC lattice have low strength and high ductility. Therefore, for high wear resistance, it is presumably advisable to provide BCC phases.

An effective factor in the regulation of the formed alloy structure is its crystallization rate, which significantly affects the mechanism of alloy decay. The high quenching rate of 10^6–10^7 $K{\cdot}c^{-1}$ with subsequent heat treatment forms an ultrafine-grained structure without dendrites with the presence of a B2 nanodomain superstructure. Upon subsequent thermal annealing, crystalline multicomponent cubic phases are formed in a small amount. Annealing significantly affects the microstructure and properties of HEA. For example, annealing of samples from HEA $CoCrFeNiTiAl_x$ leads to an increase in Vickers hardness. Due to the weak diffusion in HEA, the cooling of the heated sample during annealing proceeds quickly, which does not allow to complete the phase transformation with the formation of new phases, which can be considered a positive factor in relation to the production of single-phase structures.

The negative enthalpy of mixing of alloy elements with their high concentration promotes the formation of intermetallic compounds, i.e., ordered structures. An increase in the aging temperature leads to a decrease in the proportion of intermetallic compounds. An increase in the concentration of certain elements also leads to the formation of intermetallic compounds.

Intensive plastic deformation by comprehensive forging, for example, HEA AlCrCuNiFECo, leads to an improvement in strength characteristics, leads to a decrease in the temperature of the brittle-viscous transition. This can positively affect the wear resistance of the alloy with respect to preventing the formation of fatigue cracks on the friction surface of the alloy.

6. Conclusion

Known facts allow us to offer methodological recommendations for obtaining wear-resistant wind turbines-coatings by plasma spraying with multiple reflow:

– the use of multiple reflow in an inert environment;

– selection of the elemental composition of the alloy with high solubility of the constituent elements in each other and with close values of electronegativity and atomic size;

– for high wear resistance of the HEA coatings, it is presumably advisable to provide BCC phases;

– for the formation of an ultrafine structure without dendrites, it is necessary to provide a high cooling rate of up to 10^6–10^7 $K{\cdot}c^{-1}$ with subsequent heat treatment, which can increase the hardness of the coating;

– improving the strength characteristics and wear resistance, as well as preventing the formation of fatigue cracks on the friction surface can be obtained using intense plastic deformation. An effective way to do this is to electromechanically harden the coating.

Additional control parameters for producing HEA coatings are plasma spraying, reflow, and electromechanical processing of local and general effects in steady and dynamic modes with modulation of parameters [10, 24].

References

[1] Cantor B *et al.* 2004 *Mater. Sci. Eng.* A **375** 213–218

[2] Otto F *et al.* 2013 *Acta Mater.* **61** 2628–2638

[3] Yeh J W 2006 *Annales de Chimie Science des Material* **31** 633–648

[4] Karpets M V and Maliuchenko O M *et al.* 2014 Properties of a multicomponent high-entropy alloy AlCrFeCoNi doped with copper *Problems of Friction and Wear* **2** 103–111

[5] Zhang Y *et al.* 2014 *Progr. Mater. Scien.* **61** 1–93

[6] Lu Z P *et al.* 2015 *Intermetallics* **66** 67–76

[7] Pogrebnjak A D, Bagdasaryan A A, Yakushchenko I V and Beresnev V M 2014 *Russ. Chem. Rev.* **83 (11)** 1027–1061

[8] Firstov S A *et al.* 2013 Superhard coatings from high-entropy alloys *Science and Innovation* **9 (5)** 32–39

[9] Bin He 2017 The Phase Evolution and Property of FeCoCrNiAlTix High-Entropy Alloying Coatings on Q253 via Laser Cladding *Coatings* **7** 157

[10] Kadyrmetov A M, Popov D A, Vikulin A V, Voronetsky V I and Stegantsev R V 2018 Prospects of obtaining multicomponent coatings by atmospheric plasma spraying *Voronezh Scientific and Technical Bulletin* vol 4 **4(26)** 46–54

[11] Kao Y F, Chen S K and Chen T J 2011 *Alloys Compd.* **509** 1607

[12] Pogrebnjak A D *et al.* 2014 *Mater. Chem. Phys.* **147** 1079

[13] Senkov O N, Wilks G B, Scott J M and Miracle D B 2011 *Intermetallics* **19** 698
[14] Tsai M H *et al.* 2013 *Intermetallics* **32** 329
[15] Salishchev G A, Tikhonovsky M A, Shaisultanov D G, Stepanov N D, Kuznetsov A V, Kolodiy I V, Tortika A S and Senkov O N 2014 *J. Alloys Compd.* **591** 11
[16] Shun T T, Hung C H and Lee C F 2010 *J. Alloys Compd.* **493** 105
[17] Ivchenko M V, Pushin V G and Wanderka N 2014 *J. of Tech. Physics* **84** 57
[18] Singh S, Wanderka N, Kiefer K, Siemensmeyer K and Banhart J 2011 Ultramicroscopy **111** 619
[19] Lin C M, Tsai H L 2011 *Intermetallics* **19** 288
[20] Chen M R, Lin S J, Yeh J W, Chen S K, Huang Y S and Tu C P 2006 *Mater. Trans.* **47** 1395
[21] Kuznetsov A V, Shaysultanov D G, Stepanov N D, Salishchev G A and Senkov O N 2012 *Mater. Sci. Eng.* **533** 107
[22] Ng C, Guo S, Luan J, Shi S and Liu C T 2012 *Intermetallics* **31** 165
[23] Zhang K B, Fu Z Y, Zhang J Y, Wang W M, Lee S W and Niihara K 2011 *Mater. Sci. Eng.* **20** 012009
[24] Chen Y Y, Duval T, Hung U D, Yeh J W and Shih H C 2005 Microstructure and electrochemical properties of high entropy alloys – a comparison with type-304 stainless steel *Corros Sci* **47(9)** 2257–79

IOP Conf. Series: Materials Science and Engineering **939** (2020) 012031 doi:10.1088/1757-899X/939/1/012031

Word Visualization of the morphology of the temperature field of the heat pipe with defects of the active type

V A Karachinov[1,2], A V Petrov[1], A M Abramov[1] and D A Evstigneev[1]

[1] Yaroslav-the-Wise Novgorod State University, ul. B. St. Petersburgskaya, 41 17300 Veliky Novgorod, Russia

[2]E-mail: Polnovo@yandex.ru

Abstract. As part of a computer experiment using the finite element method, a method for visualizing the morphology of the temperature field of a heat pipe with defects of the active type – a heat energy source (ITP) was developed. It is shown that when the starting mode of the pipe ($\tau = 0 \div 120$ C) and the ITP power $q \geq 9\ 10^4$ W/m2, a typical isotherm structure with a convex front is formed. The passage of a rectangular ITP isotherm is accompanied by the formation of a local isothermal contour. Significant variability of the geometric characteristics of isotherms is noted

1. Introduction

Heat pipes (HP) being special cooling devices effectively provide continuous removal of thermal energy from various electronic devices [1–6]. In the serial production of HPs adapted to the product design, the identification of technological defects at the production stage, their analysis and the elimination of the factors that caused the defect become an important problem. Traditionally, a significant role in solving this problem belongs to visualization methods based on experimentally recorded x-ray and thermal contrasts with subsequent digital image processing [7, 8]. Of scientific and practical interest are computer methods for visualizing the morphology of the temperature field of a HP, which do not require full-scale experiments. Such studies are known in the field of HP quality diagnostics, but mainly they relate to visualization methods for passive defects [9–11]. To assess the degree of influence of localized active defects on the diagnosed characteristics of the temperature field of a HP, it is necessary to develop new thermal and mathematical models of HP. This work is devoted to solving problems.

2. Research methodology

2.1. The geometric model

A well-known basic model was chosen in the form of a flat, steel HP with a symmetrical structure (figure 1) [11]. In the center of the hull of the HP was a rectangular heat source pulsed heat flux (HSPHF). The geometric parameters of the HP model were respectively equal: $L \approx 150 \div 500$ mm; $h \approx 20 \div 30$ mm; $\delta \approx 5 \div 8$ mm. As an active defect, a rectangular defect was used, which located on the longitudinal axis of the HP. Its localization, size could be changed during a computer experiment.

Content from this work may be used under the terms of the Creative Commons Attribution 3.0 licence. Any further distribution of this work must maintain attribution to the author(s) and the title of the work, journal citation and DOI.
Published under licence by IOP Publishing Ltd

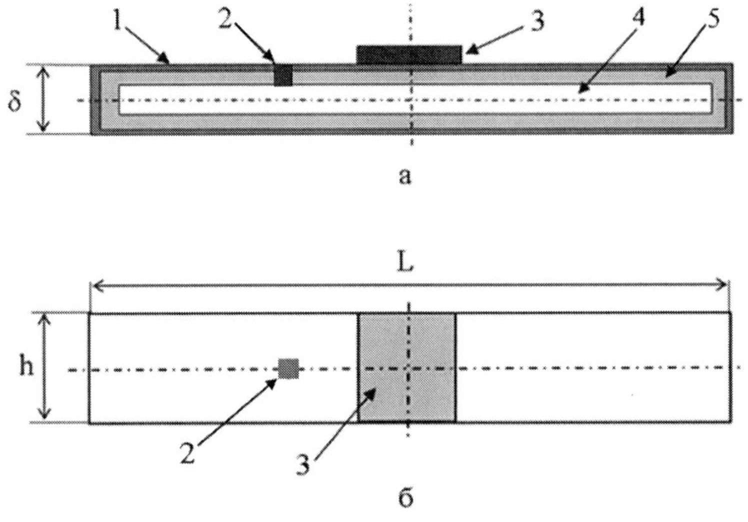

Figure 1. A geometric model of a HP with an active defect in a rectangular case and its structure. *Example*: a – section along the axis; b – top view. 1 – Case; 2 – Defect (HS); 3 – HSPHF; 4 – Steam channel; 5 – Wick [11].

2.2. Features of the thermal and mathematical model

When conducting research, we used the basic thermal and mathematical models of a bounded rod. For an unsteady problem in a plane-parallel formulation the initial heat equation (1), taking into account the formulated limitations, can be written in the form [12, 13]:

$$C\rho\frac{\partial T}{\partial t}=\lambda_x\frac{\partial^2 T}{\partial x^2}+\lambda_y\frac{\partial^2 T}{\partial y^2}+q_0 \quad (1)$$

where: c_P, ρ – specific heat and density of the material; λ_x, λ_y – thermal conductivityи; q_o – heat output per unit volume of heat energy sources; T – temperature; x, y – coordinate.

For the developed basic thermal model, the following initial and boundary conditions were used [12, 13]:

1) at the initial moment of time, for the edges of all the bodies included in the model, a constant temperature was set:

$$T_{i,\tau=0}=T_C=const \quad (2)$$

2) for all the edges of the HSPHF, taking into account the surface isothermality, a condition of the first kind was specified:

$$T=T_p \quad (3)$$

3) for all HSPHF edges (defects), a second-order boundary condition was set in the form of heat flux density (q, W / m^2) [12, 13].

4) a condition of the third kind was set on the edges of the HP model, which describes both convective and radiant heat transfer with the environment:

$$\lambda_p\frac{\partial T}{\partial n}=-\alpha_к(T-T_c)-\beta(T^4-T_c^4) \quad (4)$$

IOP Conf. Series: Materials Science and Engineering **939** (2020) 012031 doi:10.1088/1757-899X/939/1/012031

where: β – value equal to the product of the Stefan-Boltzmann constant (σ_0=5,7 · 10^{-8} W/m^{2}K^4) the emissivity of the surface of the probe material; $\alpha_к$ – convection heat transfer coefficient. Numerical values of the coefficient $\alpha_к$ were obtained from preliminary calculations using the well-known similarity method [12, 13]. The heat flux through the conductive connections of the HP fasteners was not taken into account.

The numerical values of HS heat production (defects) were selected from preliminary experiments taking into account reliably recorded changes in isothermal lines and amounted to $q = 9 \cdot 10^4 \div 3 \cdot 10^5$ W / m^2. As the defect material, steel was used with characteristics similar to that of the HP case, while the condition "isotropy" was specified in the "block" label through the equality of the thermal conductivity coefficients [13]. Computer experiments were performed for the maximum temperatures of the HSPHF T_{HSPHF} = 350 ÷ 400 K and a constant ambient temperature (air) T_a = 300 K. All other characteristics of the computer experiment technology were the same as for the case of HP with passive defects [11]. The system of equations (1–5) was solved by a numerical method (finite element method) on a PC [14].

3. Research results and discussion

It is known that the heat production (power) of HS, taking into account secondary factors, determines the main characteristics of the temperature field of an object containing this source [12, 13]. Due to the lack of knowledge, scientific and practical interest is provided by information on the influence of the thermal power of a defect on the evolution of isothermal.

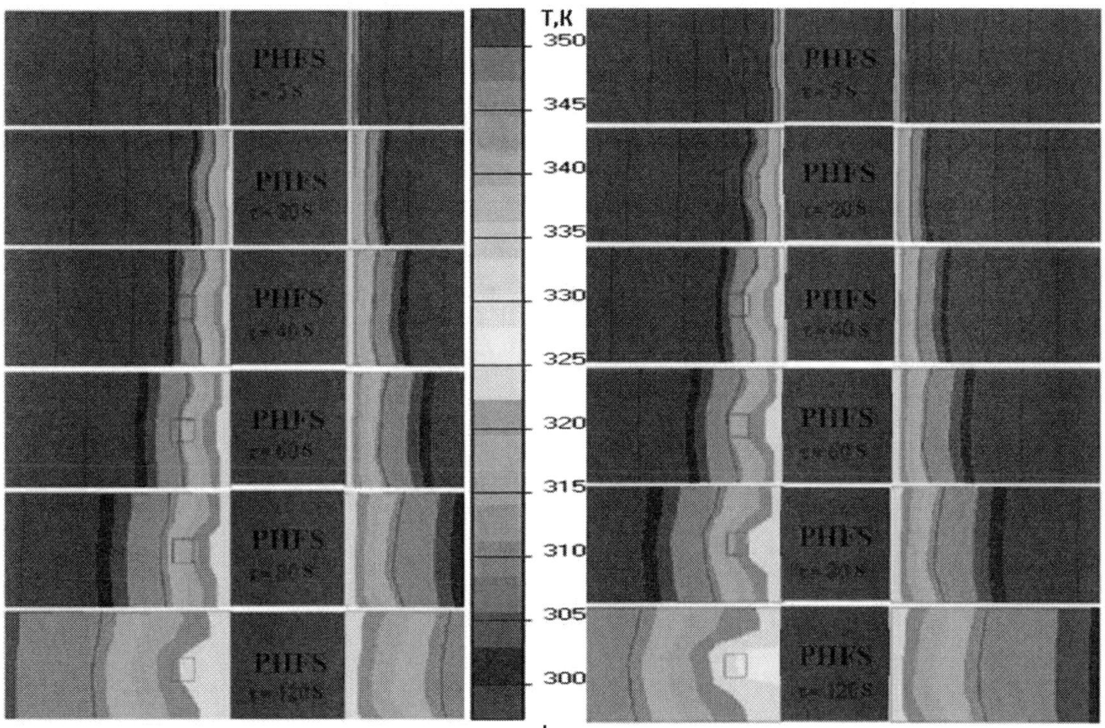

a) b)

Figure 2. Evolution of an isotherm in a steel HP with an active defect (HS) of a square shape (steel, a = 1.22 · 10^{-5} m^2/s, dimensions L_1 x L_2 = 6x6 mm). Example. Modeling, finite element method, T_{PHFS} = 350 K, scale 50K, k_λ = 10. HS power: a – q = 2 · 10^5 W/m^2; b – q = 3 · 10^5 W / m^2.

Analysis of the temperature field of HPs with defects in the following modes: thermograms, isolines, and selected contours, as well as in the case of passive defects, showed the time variability of the shape of the isotherms and the nature of the temperature distribution along the axis. As an example in figure 2. Figure 2 shows the evolution of the selected isothermal line (50 K scale) for two values of thermal power of the investigated rectangular defect. Two features of the morphology of the temperature field are clearly visible. The first is that noticeable changes in the shape of the isotherm from the defective HP region are already goodare observed in the initial period of the transition process (τ = 5s) and these changes gradually increase in time, even beyond the localization of the defect. Computer experiments showed that an increase in the heat output of a rectangular defect only intensifies this process.

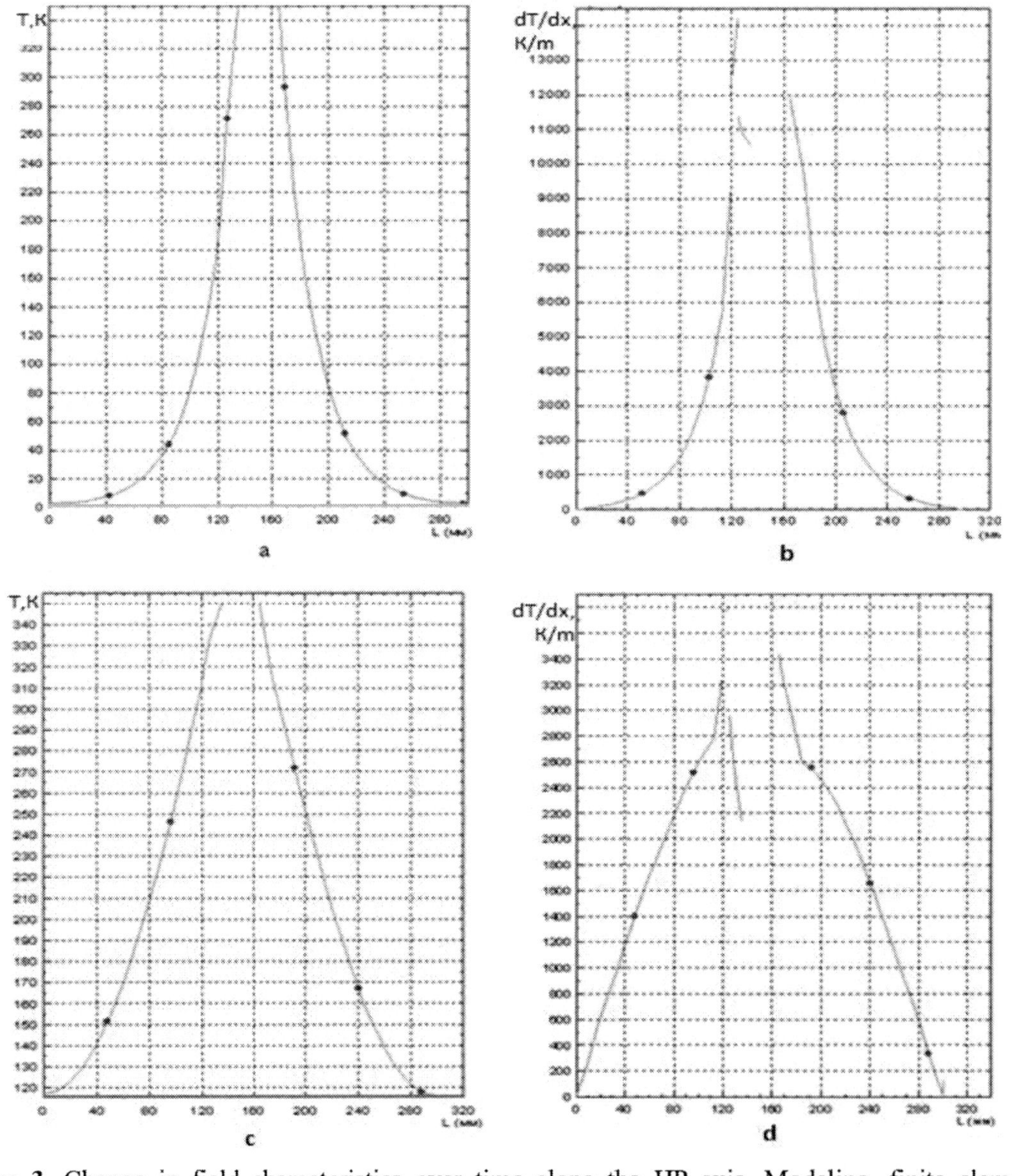

Figure 3. Change in field characteristics over time along the HP axis. Modeling, finite element method (HP with active defect (HS) of square shape (steel, a = 1.22 $\cdot 10^{-5}$ m^2/s, dimensions L_1 x L_2= 6x6 mm)). Example: T_{PHFS} = 350 K, T_C = 300 K. a, b – τ = 5 s; c, d – τ = 40 s.

The second morphological feature is related to the shape of the front of the isothermal line – at the entire stage of evolution from the side of the defective region of the HP, it remains convex. The revealed qualitative features of the morphology of the temperature field were also manifested in changes in the field characteristics (figure 3, figure 4). Thus, studies of the temperature distribution, temperature gradient, and heat flux density along the selected contour along the HP axis at different points in time of the transition process allowed us to fix the formation of local extrema in the defect region and the general asymmetry of the graphs relative to HSPHF, which has been preserved throughout the entire evolutionary period. The presence of a localized active defect in the HP from the standpoint of heat transfer is equivalent to the formation of a system of conductively coupledsources of heat flux in which HSPHF creates background overheating [15]. In this case, overheating of HP temperatures within the localization of a defect is determined on the basis of the principles of superposition and local influence (the scientific foundations of these principles are described in detail in [15]). Therefore, an increase in the thermal power of the defect should lead to an increase in local overheating; therefore, the introduced active defect should become a "generator" of local isothermal circuits. Such an effect for a selected scale of isotherms was discovered during computer experiments when the thermal specific power of a rectangular defect exceeded $P > 3\ 10^5$ W/m^2 (figure 5). It should be noted that the "irregular" shape of the isothermal line of the contour outside the defect is due to the thermal conductivity anisotropy introduced in the HP model (kλ = λX / λY).

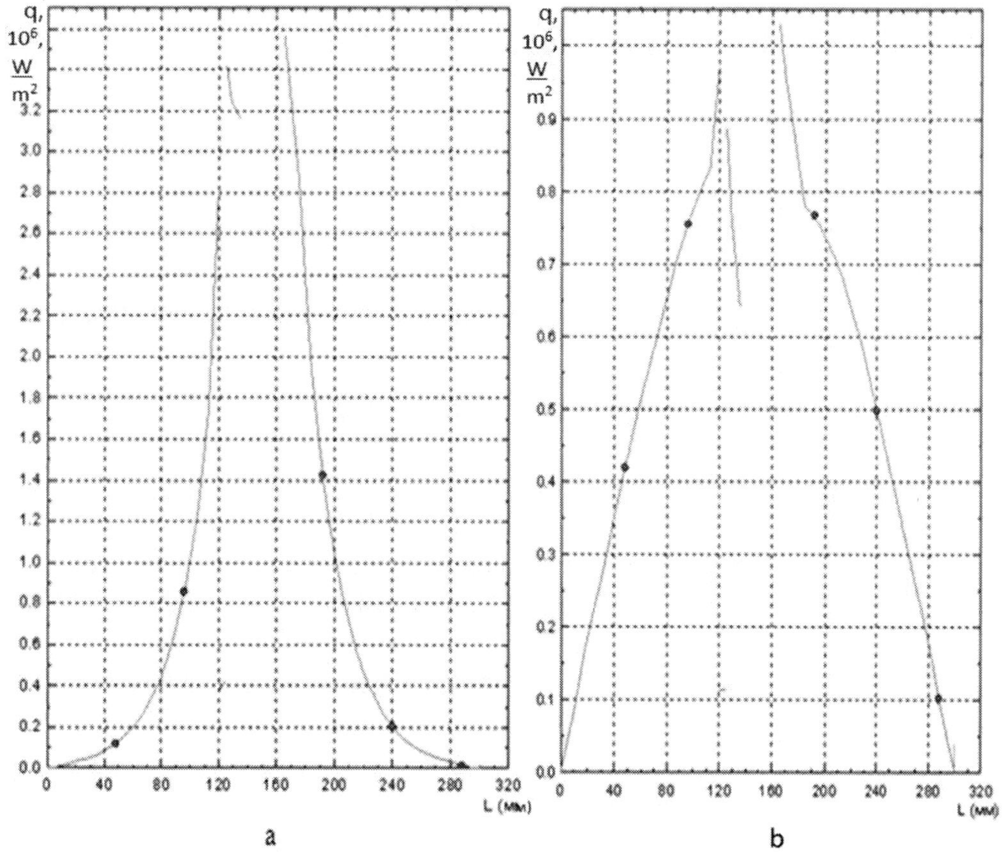

Figure 4. Change in heat flux density over time along the HP axis. Modeling, finite element method (GP with active defect (HS) of a square shape (steel, a = 1.22 · 10^{-5} m^2/s, dimensions L_1 x L_2 = 6x6 mm). Example: T_{PHFS} = 350 K, T_C = 300K. a – τ = 5s; b – τ = 40s.

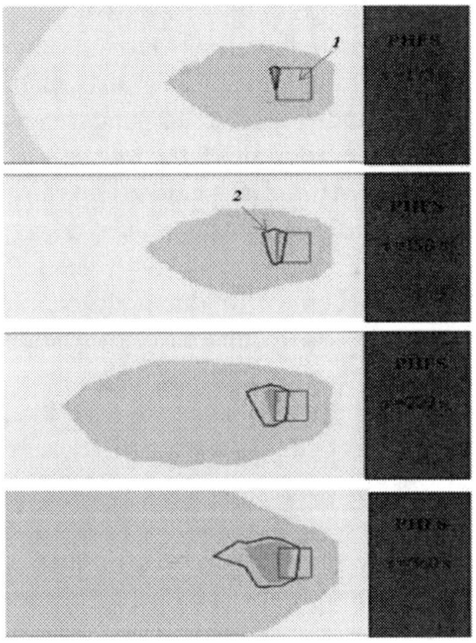

Figure 5. The effect of the formation of a local isothermal circuit with an active defect (HS) of a square shape (steel, a = 1.22 · 10^{-5} m^2/s, dimensions L_1 x L_2 = 6x6 mm) in a steel HP. Example. Modeling, finite element method, T_{PHFS} = 350 K, scale 50K, k_λ = 10, specific power of HS: $q > 3 \cdot 10^5$ W/m²; 1 – defect; 2 – isotherm (contour).

The results of the analysis of the morphological features of the temperature field of a HP with an active defect were also confirmed by conducting targeted computer experiments during which measurements of previously introduced geometric characteristics were carried out:

1) the distance from the center of the HP (HSPHF) to the front of the isothermal line (*L*);
2) the perimeter (length) of the contour of the isothermal line (*P*);
3) the surface area of the HP within the isothermal line (*S*) [8].

Table 1. Time-varying average values of the geometric characteristics of the thermal field of a HP with an active defect (HS) of a square shape (steel, a = 1.22 · 10^{-5} m^2/s, dimensions L_1 x L_2 = 6x6 mm). in the sliding isotherm mode. Computer experiment. Example: TPHFS = 350 K, specific power of ITP, $q = 3 \cdot 10^5$ W/m^2.

	Geometrical characteristics of the heat field					
	The left side of the HP (with a defect)			The right side of the HP (without defect)		
Time, τ c	Front Distance, *L*, pixels	Perimeter, P, pixels	Area, S, pixels	Front Distance, *L*, pixels	Perimeter, P, pixels	Area, S, pixels
5	98.5	559.5	16250	94.2	555.4	15541
20	119.8	607.6	19406	112.5	599.1	18235
40	140.0	742.9	23522	125.1	731.4	21024
60	164.9	734.4	27870	138.8	720.8	23464
80	197.9	796.2	33241	158.3	777.8	26603
120	427.5	1031.0	73100	225.3	855.4	38535
140	418.2	1040.7	70260	287.7	928.1	48338

A typical example is given in table 1. Quantitative estimates were obtained using the well-known software "The Geometer's Sketchpad V4". An analysis of the obtained data showed that the geometric characteristics at the level of average numerical values have different sensitivity with respect to the active defect localized in the HP, but undergo more significant changes compared to the HP containing a passive defect. The use of local values of the distance L also made it possible at the level of a computer experiment to assess the degree of influence of the defect's heat production (figure 6). Considering that the defect in the left block lies on the horizontal axis of the HP, the local change in the shape of the isotherm (convex front, figure 2), as well as ΔL, are pronounced in time. Regardless of the location of the isothermal line in the left defective HP block, ΔL always remains positive and its increase in time is observed according to a law close to parabolic. An increase in the defect's heat production only shifts the time dependence of graph (2) to the region of large ΔL values. Such a character of the dependences is typical for the studied range of variation of the thermal power q.

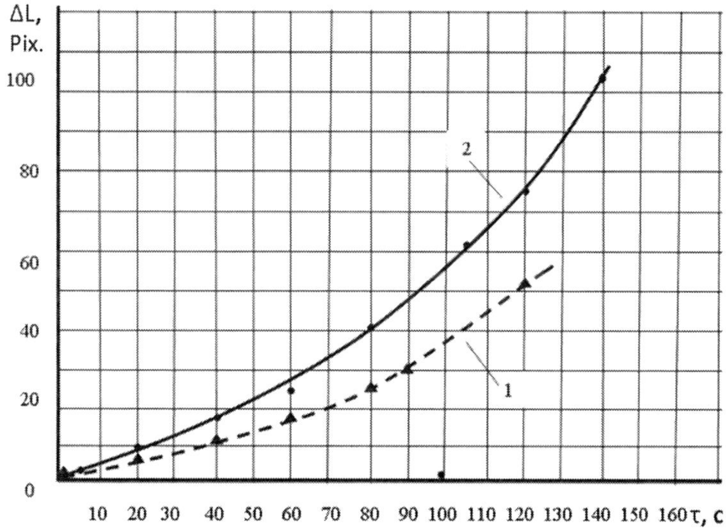

Figure 6. Change in the absolute value of the local distance from the center of the HP to the front of the moving isotherm in time. Computer experiment (HP with an active defect (HS) of square shape (steel, a = 1.22 · 10^{-5} m^2/s, dimensions L_1 x L_2 = 6x6 mm)). Example: T_{PHFS} = 350 K, T_C = 300 K. 1 – q = 2 · 10^5 W/m^2; 2 – q = 3 · 10^5 W/m^2.

4. Conclusion

According to a computer experiment, the presence of a localized active defect, irrespective of shape, causes variability of the shape of the isotherms and the nature of the temperature distribution along the HP axis; moreover, the magnitude of the heat production of the defect has the greatest influence on the morphology of the temperature field; its increase causes the formation of a local isothermal contour.

References

[1] Dan P D and Ray D A 1979 *Heat pipes* (Moscow: Energy) p 272
[2] Dulnev G N and Belyakov A P 1985 *Heat pipes in electronic temperature stabilization systems* (Moscow: Radio and communications) p 96
[3] Karachinov V A 2007 *Protection of RES in extreme conditions. Special cooling devices. Tutorial* (Veliky Novgorod: NovSU) p 176
[4] Lux A L and Matveev A G 2007 Study of high-performance ammonia heat pipes of energy-saving thermal control systems for large-sized structures of the spacecraft *Bulletin of Samara State University: Natural science series* **6(56)** 401–418
[5] Derevyank V A, Nesterov D A, Kosenko V E, Ringer V D, Chebotarev V E, Fatkulin R F and Suntsov B B 2013 Flat heat pipes for heat removal from electronic equipment in spacecraft *Vestnik SibGAU* **6(52)** 111–116
[6] Karachinov V A, Petrov D A and Kiliba Yu V 2019 Analysis of heat pipe quality control methods *Bulletin of NovSU: Ser. Eng.* **2(114)** 14–18

[7] Petrov D A, Karachinov V A, Petrov A V, Ionov A S and Evstigneev D A 2020 *IOP Conf. Ser.: Mater. Sci. Eng.* **734** 012058
[8] Kuznetsov G V and Sandu S F 2004 Mathematical modeling of the operation of high-temperature heat pipes in heat transfer devices of modern power plants *News of universities . Nuclear energy* **2** 102–108
[9] Karachinov V A, Evstigneev D A, Abramov A M and Petrov D A 2019 Diagnostics of thermal pipes with symmetric structure thermal impact method *Machines. Technologies. Materials* vol 13 **2** 83–85
[10] Karachinov V A, Petrov D A and Evstigneev D A 2019 *J. Phys.: Conf. Ser.* **1352** 012026
[11] Ivanovsky M N, Sorokin V P and Yagodkin I V 1978 *Physical foundations of heat pipes* (Moscow: Atomizdat) p 258
[12] Lykov A V 1978 *Heat and mass transfer* (Moscow: Energy) p 480
[13] 2019 *ELCUT. Modeling of two-dimensional fields by the finite element method. User's manual. Research and Production Cooperative "Tor"* (St. Petersburg, 1912– 2019)
[14] Dulnev G N, Parfenov V G and Sigalov A V 1990 *Methods for calculating the thermal regime of devices* (Moscow: Radio and communications) p 312

IOP Conf. Series: Materials Science and Engineering **939** (2020) 012032 doi:10.1088/1757-899X/939/1/012032

Synthesis and application of manganese dioxide nanoparticles as part of anti-corrosion coatings

V E Katnov[1,2], S N Stepn[1], M R Zganshina[1], A A Kayumov[1] and V A Mendelson[1,3]

[1] Kazan National Research Technological University, 68 Karl Marx str., 420111, Kazan, Russian Federation

[2] Kazan Federal University, 18 Kremlyovskaya str., 420111, Kazan, Russian Federation

[3] E-mail: vallery78@mail.ru

Abstract. The paper presents a study of synthesized manganese dioxide nanoparticles effect on the properties of transparent coatings to protect the rolled metal from corrosion. Capacitive-ohmic measurements of the electrochemical cell formed on the studied coatings demonstrate corrosion processes inhibition under the coating with a nanoparticle content of 0.025% and higher.

1. Introduction

Issues related to the production of nanoparticles (NPs) dispersions and composite nanoparticles based on various chemical compounds and substances with given sizes, polydispersity, shape, surface and structure characteristics have been one of the most important areas of physical chemistry in recent decades. However, the correct usage of nanostructures in specific applications is an important aspect. One of the ways to develop research in this area is the design and creation of the so-called organo-inorganic nanocomposites that attract increasing attention due to the fact that they demonstrate special thermal, mechanical, electrical, optical and optoelectronic properties [1–11].

As far as paint and varnish industry is concerned the core idea of nano-objects application is that they should be used not as ordinary fillers in paints and varnishes formulations, but as functional additives that give noticeable improvement to the basic characteristics of coatings [12, 13].

One of the key tasks in paint and varnish technology is metals protection from corrosion. Given a highly developed characteristic of nanoparticles surface, the paintwork components used for corrosion protection endowment can achieve a significantly greater anticorrosive effect with smaller amounts of functional substance than when using micro size components.

The studies presented in this article are aimed at demonstrating the possibility of using manganese dioxide nanoparticles synthesized as stable colored dispersions in organic solvent in transparent coatings with an anticorrosive effect.

2. Methods

The nanoparticles synthesis was carried out according to the method developed by the authors. It calls for the usage of available components and the possibility of adjusting the dispersion medium composition for the specific film-forming base. The obtained sol of MnO2 nanoparticles was characterized by the methods of dynamic light scattering (DLS), optical and electrophoretic

Content from this work may be used under the terms of the Creative Commons Attribution 3.0 licence. Any further distribution of this work must maintain attribution to the author(s) and the title of the work, journal citation and DOI.
Published under licence by IOP Publishing Ltd

spectroscopy. Measurements by DLS and electrophoretic spectroscopy methods were carried out on ZetaPALS and 90 Plus/BI-MAS instruments manufactured by Brookhaven.

The AS-548 varnish analogue was used as a film-forming base. It is a mixture of solutions of acrylic copolymer 5B with epoxy resin E-40 with addition of resylated glyphthalic resin № 90. The nanoparticle and the binder solution were mixed using a laboratory magnetic stirrer.

In order to study the nanoadditive effect on the structured coatings properties, their physicomechanical and protective properties were determined. Coatings were applied with the help of automatic applicator model 1137 manufactured by Sheen Instruments in 1 layer, onto previously cleaned by P100 sandpaper and fat-free steel plates 150×70×1 mm in size. The coating was formed in a drying chamber for 15 min at 170 °C. The coating thickness determined with the help of a TT210 thickness indicator, was 15± 2μm.

In order to study the nanoadditive effect on the structured coatings properties, their physicomechanical and protective properties were determined. Coatings were applied with the help of automatic applicator model 1137 manufactured by Sheen Instruments in 1 layer, onto previously cleaned by P100 sandpaper and fat-free steel plates 150×70×1 mm in size. The coating was formed in a drying chamber for 15 min at 170 °C. The coating thickness determined with the help of a TT210 thickness indicator, was 15± 2μm.

Impact strength was determined according to the national standard GOST 4765-73 on a U-1 device, adhesion was evaluated by the method of grating notches according to GOST 31149-2014-1a, relative hardness was determined according to GOST 5233-89 (ISO 1522-73). Barrier properties were evaluated by capacitive-ohmic method in electrochemical cell "painted metal-electrolyte" under static exposure to a 3% sodium chloride solution, according to the procedure described in [14, 15] using an RCL Fluke PM 6303A automatic immittance meter.

Simultaneously, the coatings' resistance was determined under constant exposure to condensing moisture according to ISO 6270-2 in an artificial climate chamber CH-300 manufactured by Dycometal at a temperature of 40 °C. To determine the anticorrosive effect, we evaluated the behavior of the corrosion potential in the "painted metal-electrolyte" electrochemical cell under the static action of a 3% sodium chloride solution.

3. Results and Discussion

Manganese dioxide nanoparticles were obtained by redox precipitation in a solution of a stabilizing substance in an aqueous-organic medium. During the synthesis, a color change of the reaction mixture was observed, which was recorded by the extinction spectra using a spectrophotometer. Together with optical spectroscopy in the course of nanoparticles synthesis the dispersed characteristics of the reaction mixture were recorded by dynamic light scattering. The characteristic spectra at different time intervals and the size distribution diagram are shown in figures 1 and 2 respectively.

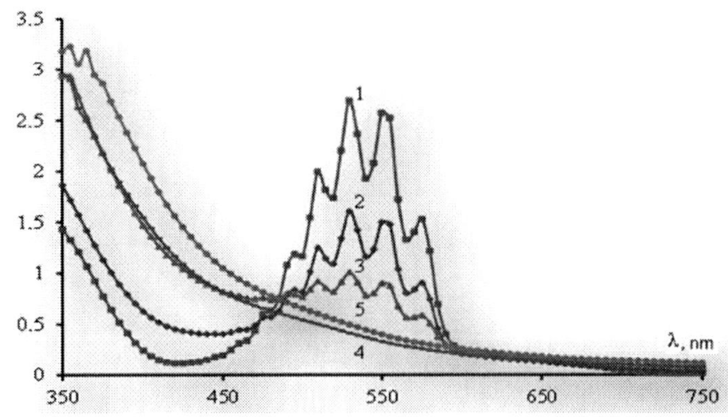

Figure 1. Absorption spectra of the reaction mixture at various time intervals for the synthesis of nanoparticles: 1 – 0 min, 2 – 5 min, 3 – 10 min, 4 – 15 min, 5 – 30 min.

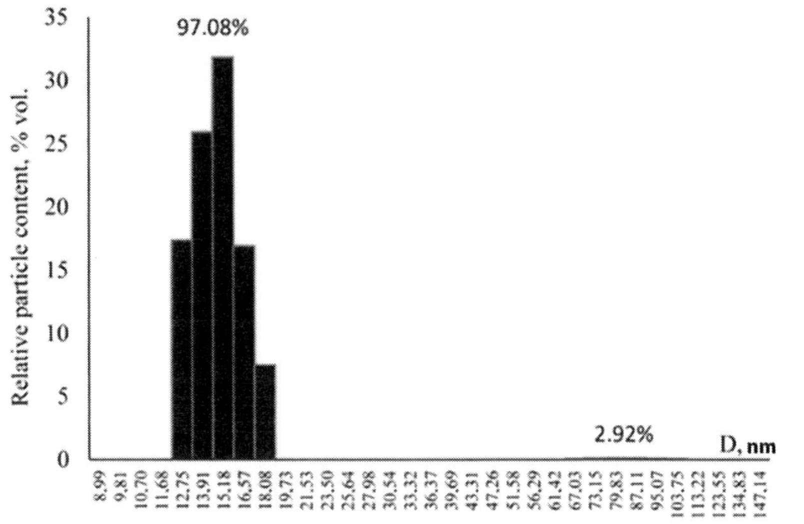

Figure 2. Histogram of the of MnO_2 particles size distribution.

It can be seen from figure 1 that in the course of reaction the extinction peaks disappear in the wavelength range of 500–600 nm, characteristic of the soluble manganese compound. The reaction completion was monitored by the absence of changes in the absorption spectra upon subsequent addition of a reducing agent, and the solution color change from violet to tan was observed.

Analysis of the histogram of the particle size distribution shown in figure 2 showed that dispersion contains two fractions of particles with sizes of 12.75–18.08 nm, the content of which is 97.08% vol., and particles with sizes 73.15–87.11 nm, the content of which is 2.92% vol. The average particle size of the obtained manganese dioxide particles, determined with the help of Brookhaven ZetaPALS nanoparticle analyzer, was 15.5 nm.

The obtained nanoparticles were included in the composition of coatings based on a mixed film former. The results of determining the physicomechanical properties of nanostructured coatings with different contents of the dispersed phase are shown in table 1.

Table 1. Physico-mechanical properties of coatings with different contents of nano-additives.

Characteristics	The nanoparticle content in the coating, %				
	0	0.025	0.05	0.075	0.1
Relative hardness, conv. units	0.77	0.79	0.86	0.87	0.7
Impact strength, cm			50		
Adhesion score			0		

Analysis of the data from table 1 demonstrates high impact strength and adhesion of coatings, and the hardness of the coatings is extremely dependent with a maximum of 0.87 conventional units with a nano-particles content of 0.075%.

Protective coatings of metal substrates that are in contact with corrosive environments during operation must have high barrier properties and the ability to inhibit corrosion processes that can occur on the protected metal surface. The inhibitory ability of manganese-containing pigments has been confirmed by a number of works [16–18]. The change in the barrier properties of coatings was recorded by the kinetics of the electric capacitance (figure 3), determined by the capacitance-ohmic method in the electrochemical cell, and by the appearance of the painted plates (figure 4), tested in a humidity chamber for 120 h.

As a rule, after pouring a 3% solution of sodium chloride into electrochemical cells, an increase in electric capacity occurs within 24 hours due to the swelling of the coating under the action of electrolyte solution, which is obvious for a sample that does not contain NPs. Further kinetics of the electrical capacitance of the coating that does not contain NPs is extremely unstable, and high values of this indicator stand for the possible formation of a microdefect in which direct contact of electrolyte solution with steel substrate takes place. The presence of above mentioned contact should certainly provoke an electric capacitance increase by 1–2 orders of magnitude, however, its decrease is observed due to the filling of the formed defects with corrosion products, leading to a decrease in the diffusion of corrosive substances to the substrate.

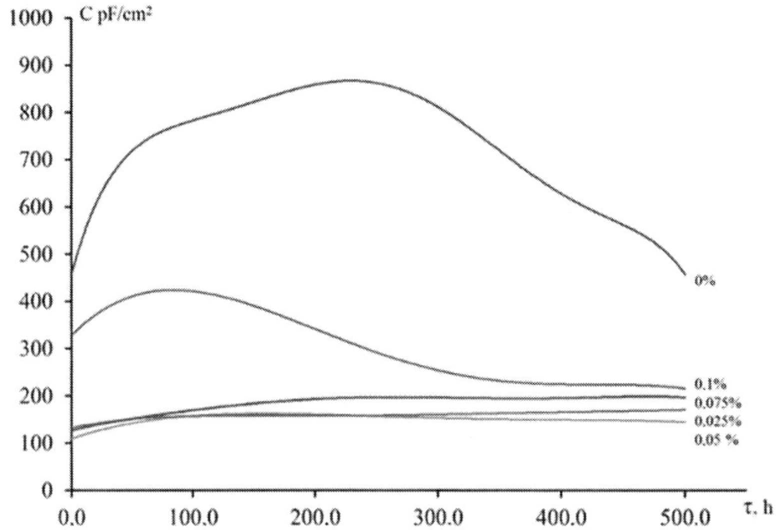

Figure 3. Kinetic dependences of the electric capacitance in the system painted metal – electrolyte with different contents of nanoparticles.

When analysing the kinetic dependences of coatings containing nanoparticles in the range of 0.025–0.075%, a significant improvement in the insulating ability can be noted. That happens supposedly due to the formation of a more perfect structure and increased adhesion. With an increase in NPs content up to 0.1% of the mass, there is an increase in electric capacity and consequently a decrease in barrier properties. As was said above this fact may be related to the degradation of polymer matrix supramolecular structure due to the excess of some critical concentration, because spherical NPs with particle size of 15.5 nm have specific surface area more than 75 m^2/g. Nevertheless, by the end of the test, the electric capacitance value for this sample is stabilized and becomes close to coatings with a lower content of NPs.

Analysis of the samples physical configuration after they have been tested in humidity chamber for 120 hours shows the appearance of proportionate and pinprick-like corrosion lesions on the sample with a coating containing no NPs. Notably, a sample with a coating containing 0.025% of the mass. particles has passed the test successfully.

IOP Conf. Series: Materials Science and Engineering **939** (2020) 012032 doi:10.1088/1757-899X/939/1/012032

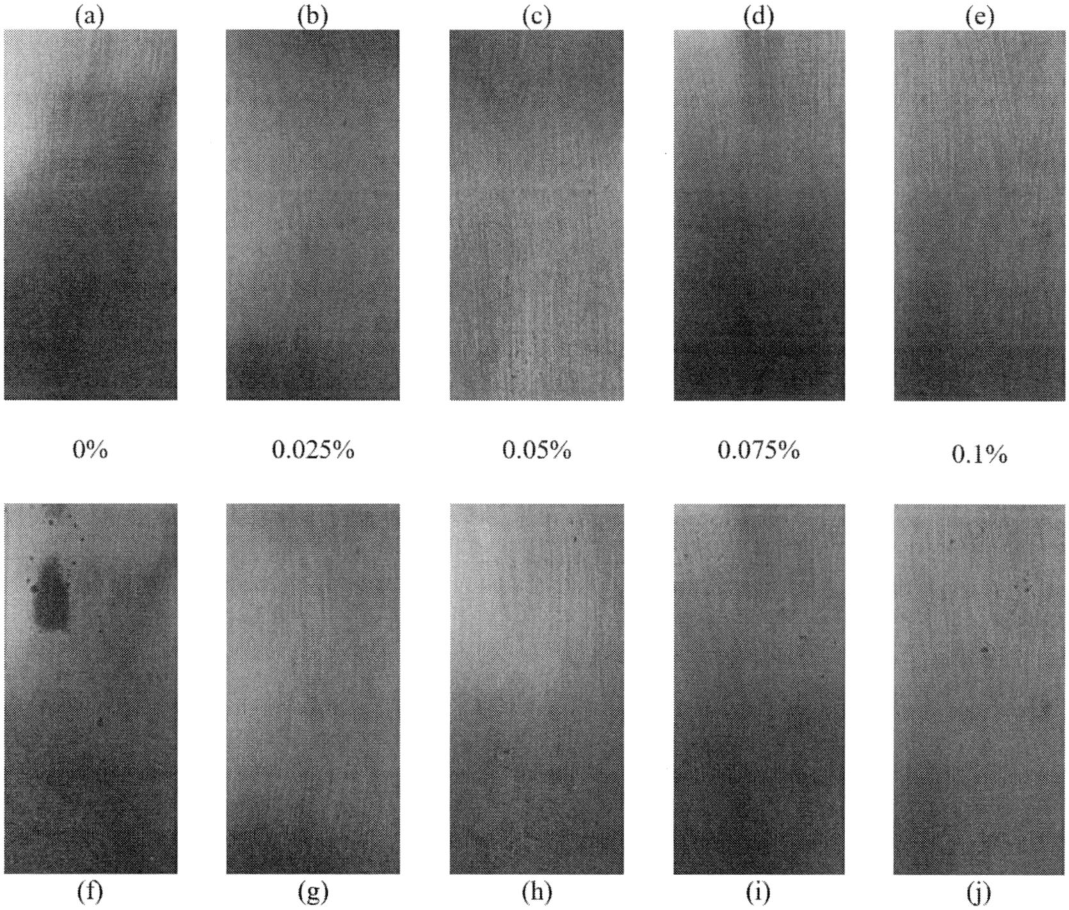

Figure 4. The physical configuration of coatings with different nanoparticle contents before (a), (b), (c), (d), (e) and after (f), (g), (h), (i), (j) tests in humidity chamber at the temperature of 40 ° C for 120 hours.

With further increase of nano-additive concentration in the coating, a strong decrease in resistance to high humidity is not observed, however, after testing, traces of pitting corrosion appear on these samples.

The direct ability of a painted metal to resist corrosion can be determined by measuring the corrosion potential on exposure to electrolyte solution. This indicator monitoring was carried out in electrochemical cell for 500 hours (figure 5) according to the procedure described in [14, 15].

Analysis of the kinetic curves shown in figure 5 makes it possible to conclude that nanostructured coatings in all the studied concentrations of nano-additive contribute to the inhibition of corrosion processes, transferring the metal surface under the coating to passive state.

In this case, the sample covered with the coating containing no nanoparticles shows a downward curve of corrosion potential in the range of –136 – (–437) mV kinetic dependence, thereby indicating active oxidation of iron.

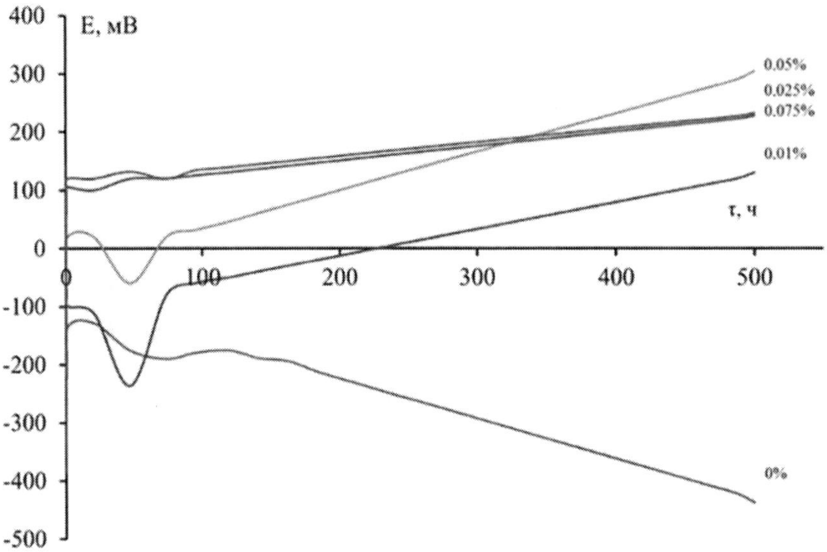

Figure 5. Kinetic dependences of corrosion potential in an electrochemical cell "painted metal – electrolyte" for coatings with different contents of manganese dioxide nanoparticles.

4. Conclusions

In the course of this work, we synthesized MnO_2 NPs in organic solvent medium suitable to be further included in the of organic-diluted coatings composition. Barrier and protective characteristics of polymer composite coatings containing synthesized manganese dioxide nanoparticles were studied. The possibility of the coatings protective properties increase by introducing into their composition manganese dioxide nanoparticles in an amount of 0.025–0.075% mass was revealed.

References

[1] Chau J L H, Hsieh C C, Lin Y M and Li A K 2008 Preparation of Transparent silica – PMMA Nanocomposite Hard Coatings *Progress in Organic Coatings* **62** 436–39 DOI: 10.1016/j.porgcoat.2008.02.005

[2] Gadomsky O N, Altunin K K, Stepin S N, Katnov V E, Rusin A A and Pereskokov E A 2008 Near-field effect in composite nanomaterials with a quasi-zero refractive index *Opt. Comm.* **315** 286–94 DOI: 10.1016/j.optcom.2013.11.035

[3] Grala, Magdalena, Bartczak Z and Różański A 2016 Morphology, thermal and mechanical properties of polypropylene/SiO_2 nanocomposites obtained by reactive blending *J. Polym. Res.* **23** DOI: 10.1007/s10965-015-0914-0

[4] Hakouk K, Oms O, Dolbecq A, Moll H, Marrot J, Evain M, Molton F, Duboc C, Deniard P, Jobic S, Mialane P and Dessapt R 2013 Sulfonium polyoxometalates: a new class of solid-state photochromic hybrid organic-inorganic materials *Inorganic chemistry* **52** 555–57

[5] Kickelbick Guido 2014 Hybrid materials – past, present and future *Hybrid Materials* **1** DOI: 10.2478/hyma-2014-0001

[6] Luo K, Zhou S and Wu L 2009 High refractive index and good mechanical property UV-cured hybrid films containing zirconia nanoparticles *Thin Solid Films* **517** 5974–80 DOI: 10.1016/j.tsf.2009.03.162

[7] Ramesh S and Liew Ch W 2010 Investigation on the effects of addition of SiO_2 nanoparticles on ionic conductivity, FTIR, and thermal properties of nanocomposite PMMA–$LiCF_3SO_3$–SiO_2 *Ionics* **16** 255–62 DOI: 10.1007/s11581-009-0388-3

[8] Rammelt U and Reinhard G 1992 Application of electrochemical impedance spectroscopy (EIS) for characterizing the corrosion-protective performance of organic coatings on metals *Prog. Org. Coat.* **21** 205–26 DOI: 10.1016/0033-0655(92)87005-u

[9] Vakhitov T R, Katnov V E, Grishin P V, Stepin S N and Grigoriev D O 2017 Biofriendly nanocomposite containers with inhibition properties for the protection of metallic surfaces. *Proceedings of the Royal Society A: Mathematical, Physical and Engineering Science* **473** 2016082

[10] Xue D-J, Wang J, Wang Y, Xen S, Guo Y-G and Wan L J 2011 Facile Synthesis of germanium nanocrystals and their application in organic-inorganic hybrid photodetectors *Advanced Materials* **23** 3704–07 DOI: 10.1002/adma.201101436.

[11] Yu Y-Y and Hui-Huan Yu 2013 High refractive index organic–inorganic composites with TiO_2 nanocrystal *Thin Solid Films* **529** 195–99 DOI: 10.1016/j.tsf.2012.05.052.

[12] Stepin S N, Abdullin I Sh, Svetlakova T N, Ziganshina M R and Svetlakov A P 2009 Nanoscale objects in the field of corrosion protection with polymer coatings *Lakokrasochnye Materialy i Ikh Primenenie* **3** 40–44

[13] Stepin S N, Katnov V E, Petrovnina M S and Vakhitov T R 2013 Obtaining and properties of nanoscale dispersed materials and composites based on them *Bulletin of Kazan Technological University* **14** 86–89

[14] Rammelt U and Reinhard G 1992 Application of electrochemical impedance spectroscopy (EIS) for characterizing the corrosion-protective performance of organic coatings on metals *Prog. Org. Coat.* **21** 205–26 DOI: 10.1016/0033-0655(92)87005-u

[15] Stepin S N, Vakhin A V, Sorokov A V and Ziganshina M R 2000 Method to asses anticorrosive properties of pigments and pigmented coatings *Lakokrasochnye Materialy i Ikh Primenenie* **1** 25–8

[16] Ziganshina M R, Peshkova M S, Stepin S N and Katnov A E 2005 The effect of synthesis conditions on the properties of precipitated manganese-based anticorrosive pigments *Lakokrasochnye Materialy i Ikh Primenenie* **10** 34–6

[17] Ziganshina M R and Nurislamova E A 2018 Anti-corrosion properties of coatings with manganese compounds pigmentation IOP Conference Series: Materials Science and Engineering **311** 012017 DOI: 10.1088/1757-899x/311/1/012017

[18] Solodov V A, Ziganshina M R and Bayburina E A 2015 Anticorrosive efficiency of coprecipitated manganese compounds IV *International Journal of Applied Chemistry* **11** 601–9

Theoretical study of the plant residue movement along the needle of a rotating working organ

A A Kem[1], A P Shevchenko[2], M A Begunov[2,4], V S Koval[2] and D N Algazin[3]

[1] FSBSI "Omsk Agricultural Research Center", Korolev prospect, 26 644012, Omskaya oblast', Omsk, Russia
[2] Tara Branch of Federal State Budgetary Educational Institutional of Higher Education "Omsk State Agrarian University named after P.A. Stolypin" (FSBEI HE Omsk SAU), Tyumenskaya Street, 18 646532, Omskaya oblast', Tara, Russia
[3] Yaroslav-the-Wise Novgorod State University, ul. B. St. Petersburgskaya, 41 173003 Veliky Novgorod, Russia

[4] E-mail: maksim-begunov@mail.ru

Abstract. The article is devoted to the identification of dependences to determine the main rational constructive and technological parameters of the proposed rotating working organ.

1. Introduction

Sowing is paid special attention to in the technological system of cultivation of grain crops, in addition to such operations as preparation of seeds [1] and soil, harvesting, The future development of the crop depends mainly on the quality of this operation, which ultimately affects the quantitative and qualitative yield indexes. The technological requirements for the process of tillage and sowing on stubble backgrounds are based on several conditions:

1) high-quality seeding to a depth of 0.06–0.08 m;
2) stable operation of coulters (without soil bulldozing in the inter-coulter space) [2, 3, 4, 6];
3) reduction of tractive resistance [5].

The application perspectiveness of designing a rotating working organ for sowing machines on stubble backgrounds predetermined the need for research to substantiate the parameters of its working organs.

The analysis of works on the existing selection seeders, patent and literature sources shows that they can not perform sowing on stubble backgrounds that dominate crop rotations in the Omsk region. During the seeding process, the soil is loaded with plant residues and the inter-coulter space is clogged.

At the same time, it is possible to eliminate the clogging of the inter-coulter space with plant residues and improve the quality of work of the selection seeder on the stubble background, by installing active needle-shaped rotating hoe wheels between the coulters on the common shaft, which allow shifting plant residues back along the course of the unit's movement. This hypothesis was used as the foundation for designing a modernized selection seeder for sowing on a stubble background on the basis of the SCS-6-10 mock-up model.

Content from this work may be used under the terms of the Creative Commons Attribution 3.0 licence. Any further distribution of this work must maintain attribution to the author(s) and the title of the work, journal citation and DOI.
Published under licence by IOP Publishing Ltd

In order to ensure high-quality seeding, it is necessary to use machines that meet the cutting-edge requirements. For this reason, the design and development of technologically advanced and promising sowing machines and their working organs is one of the main tasks of modern mechanical engineering.

To meet the above-mentioned requirements, it is necessary to theoretically justify the main design and technological parameters of the proposed working organ.

2. Model

The scheme of the proposed model of the coulter group of the selection seeder is shown in figure 1.

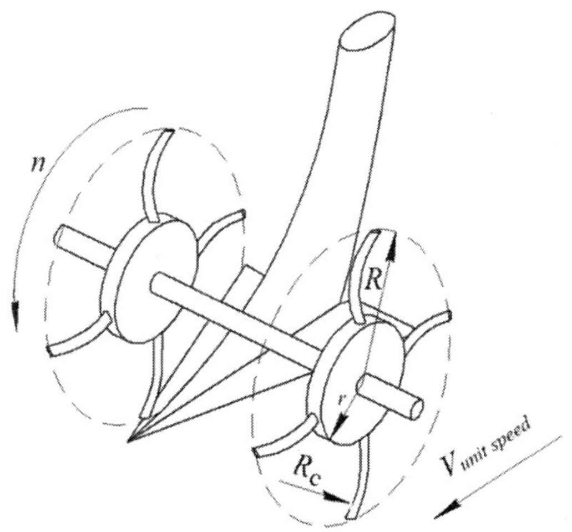

Figure 1. The scheme of the coulter group of the selection seeder with active needle-shaped rotating hoe wheels.

The direction of the circumferential speed of rotation of the needle rotating working organs is opposite to the direction of movement of the coulter, in view of this, the needles of the working organs capture plant residues and shift them in the direction opposite to the movement of the unit. Due to the rotation of the rotating working organs, the plant mass completely moves from the front to the back of the coulter. Intensive cleaning of the coulter tip and the distance between coulters makes it possible to move steadily in the soil and ensure high-quality seeding down to a given depth.

To implement the above-mentioned requirements, it is necessary to justify the most significant design and technological parameters of the proposed working organ. See figure 2:

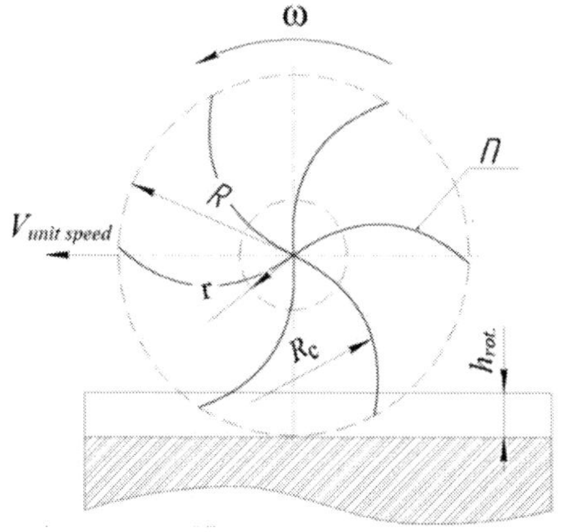

Figure 2. Variable parameters of the working organ.

IOP Conf. Series: Materials Science and Engineering **939** (2020) 012033 doi:10.1088/1757-899X/939/1/012033

Technological:
$V_{unit\ speed}$ – unit speed, m/s;
$h_{rot.}$ – processing depth;
$\lambda \pm$ – the kinematic parameter.
Constructive:
R – the radius of the rotating working organ;
r – the radius of the transition points of a straight needle to a curved needle;
R_c – the needle radius;
ω – rotation frequency of the working organ, rad/s.

The movement of plant residues in the zone of the rotating working organ can be divided into 3 stages:
– movement of plant residues when interacting with the disk needles;
– movement of plant residues on the working part of the needle;
– movement of soil after leaving the surface of the needle.

Let's consider the above-given stages in more detail, provided that the final parameter, which depends on the above-mentioned factors, is the specific volume (*sp.vol.*) moved by the needle working organ.

During the operation of active needle-shaped rotating working organs, partial movement of plant residues occurs, and a common front of soil bulldozing is formed. The unit's speed with plant residues is equal to the speed of the working organ's movement, i.e. $v_{unit\ speed} = v_{\omega}$. The purpose of the proposed coulter is to shift the residues from the front to the back of the working organ's leg. The following conditions must be met:
– working organs must not be involved in tillage;
– plant residues should not significantly change their location relative to the soil. See figure 3. That means that they should not be randomly scattered.

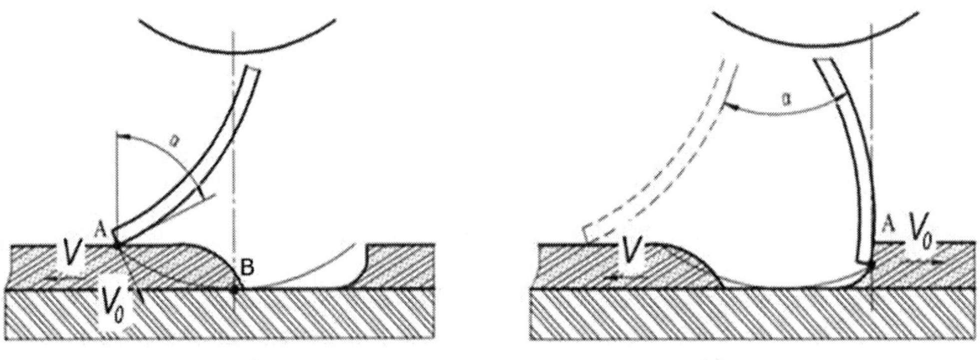

Figure 3. The interaction of plant residues with the needle: a) the moment when the needle touches the plant residues; b) the moment when the plant residues descend from the needle.

To meet the above-given conditions, the following equality must be observed:

$$v_{unit\ speed} = v_{\omega} = v_0 \cdot \cos\alpha;\ m_1 = m_2,$$

where m_1, m_2 – accordingly, the mass of unloaded plant residues and the mass of plant residues moved by the disk needles;
α – the angle formed by the contact surface of the needle and the normal to the soil surface, degrees;
v_0 – circumferential speed of the upper point of the needle (point A).

When working, the rotating disk crushes plant residues with its needle in the soil with its curved part (the part curved forward, or the occipital part). This results in a resistance force (P) proportional to the volume being crushed [5]. The compression resistance is much higher than the resistance to other types of deformation: shear, stretching, etc. Therefore, the basis of the reaction that pushes the needle out of the volume of plant residues, which prevents rotation, is precisely the compression force.

The previously conducted works devoted to the assessment of needle working organs considering the degree of their impact on the soil showed that the quality of tillage depends on the configuration of needles at fixed values of the radius of the working space of the working organ (R) and the radius of curvature of the needle (R_c).

Moving along the needle of the working organ figure 4, the mass of plant residues acquires resultant velocity Vres, which is the result of the geometric addition of the maximum peripheral speed V_p and the maximum speed of movement along the needle of the working organ (further – along the needle) V_r.

The peripheral speed (V_p) is defined as:

$$V_p = \frac{\pi \cdot R}{30} n , \quad (1)$$

where R – the distance from the center of rotation to the tip of the needle of the working organ, m;
n – rotation frequency of the rotating working organ, min^{-1}.

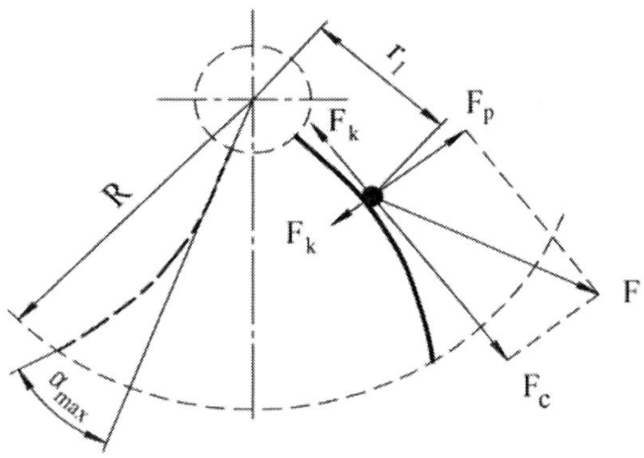

Figure 4. The analysis of forces acting on plant residues interacting with the needle of the working organ.

The value of the movement speed of plant residues along the needle of the working organ is determined by the dependence:

$$V_r = m \cdot \omega^2 \cdot p , \quad (2)$$

where, m – the value the mass of the plant residues moving along the needle, kg;
ω – the value of the angular velocity of the rotating working organ, rad/s;
p – the value of the radius-vector of the center of gravity, particles of plant residues located along the needle of the rotating working organ, m.

The friction force F_{fr} consists of the following values: $\vec{F} = \vec{F_c} + \vec{F_p} = \sqrt{F_c^{\,2} + F_p^{\,2}}$ [2],

where, F_c – the value of the centrifugal force acting along the needle of the working organ, H;
F_p – the value of the force acting perpendicular to the surface of the needle of the working organ H.
F_c – this value is aimed at moving plant residues in the direction along the needle of the disk and is calculated using the following expression:

$$F_c = m \cdot \omega^2 \cdot p \cdot sin\, \alpha , \quad (3)$$

where, α – the value of the inclination angle of the needle of the working organ, degrees.

IOP Conf. Series: Materials Science and Engineering **939** (2020) 012033 doi:10.1088/1757-899X/939/1/012033

F_p – the force that affects the friction force, H:

$$F_p = m \cdot \omega^2 \cdot p \cdot \cos\alpha \, . \tag{4}$$

The value of the Coriolis force for plant residues moving along the needle of the rotating working body is determined by the expression:

$$F_K = m \cdot \omega \frac{dr_1}{dt} \, . \tag{5}$$

The Coriolis force acts perpendicular to the surface of the needle of the working organ and affects the friction force (F_{fr}).

3. Findings and Discussion

Using the formulas considered above we can identify the friction force as:

$$F_{fr} = f \cdot (F_K - F_p) = f(2m \cdot \omega \frac{dr_1}{dt} - m \cdot \omega^2 \cdot p \cdot \cos\alpha) \, , \tag{6}$$

where f – the friction coefficient of plant residues on the material from which the needle of the working organ is made.

The equation of the plant residue movement along the surface of the needle will have the following form:

$$F_c - F_{fr} = m \frac{d^2 \cdot r_1}{dt^2} \, . \tag{7}$$

Substituting the expressions of the force F_c и F_{fr} values in the equation 7, we reduce the mass value m considering that $p \cdot \sin\alpha = r_1$, $p \cdot \cos\alpha = r_1 \cdot tg\,\alpha$. After these mathematical transformations, we obtain a differential equation for the plant residue movement along the surface of the needle of the rotating working organ:

$$m \frac{d^2 \cdot r_1}{dt} + 2f \cdot \omega \frac{dr_1}{dt} - \omega^2 \cdot r_1 - f \cdot \omega^2 \cdot r_1 \cdot \tan\alpha = 0 \, . \tag{8}$$

Based on the analysis of expression 4, we determine the required value of the rational angle of the needle of the working organ. Since the needle of the working organ rotates, the value of the rational angle will be in the range of values from α_{min}=0° to α_{max}=25°, on the basis of which the needle of the rotating working organ will take the form of a curved profile.

The work of the rotating working organ is based on the conditions that the rotation of the disk will occur, provided that the magnitude of the external applied forces (M_p) must exceed the value of the sum of the moments of forces that originate from plant residues, namely the moment of the friction forces of the Mfr and the moment of resistance to peripherical rotation (M_{per}).

$$M_p > M_{fr} + M_{per}. \tag{9}$$

The value of friction force moment is identified as:

$$M_{fr} = f \cdot F_{f.a.} = f \cdot \frac{J \cdot \omega^2}{r_1} \, , \tag{10}$$

where, $\frac{J \cdot \omega^2}{r_1}$ – the component of the centrifugal force;

$F_{f.a.}$ – force acting on plant residues;

J – the value of the inertia moment of plant residues moving on the surface of the needle.

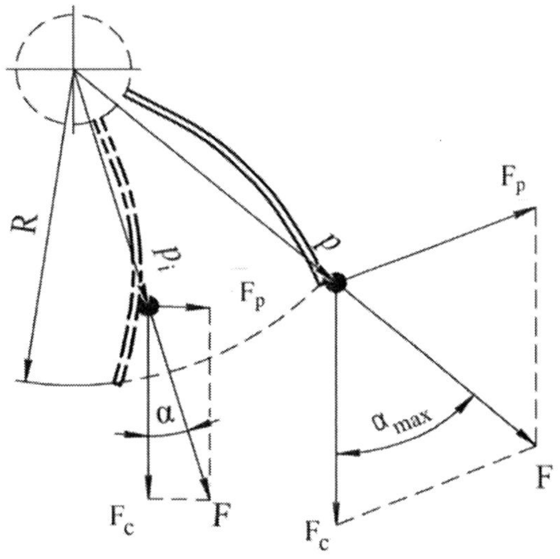

Figure 5. The analysis of the plant residue movement along the needle of the rotating working organ during their descent.

The moment of circumferential effort is identified as:

$$M_{per} = F_{per} \cdot R = \frac{N}{n \cdot z}, \quad (11)$$

where, R – the distance from the rotation center the needle tip of the working organ, m;
F_{per} – the value of circumferential effort, H;
N – the value of the power transmitted to the rotating working organ, H;
n – the rotation frequency of the working organ, min^1;
z – the number of blades, items
Based on the analysis of expression 10 and inequality 9, we get:

$$M_p \geq f \cdot \frac{J \cdot \omega^2}{r_1} + \frac{N}{n \cdot z}. \quad (12)$$

The minimum value of the inertia moment required for the plant residue movement can be determined using the expression:

$$J = \frac{(M_p - \frac{N}{n \cdot z})}{f \cdot \omega^2}. \quad (13)$$

The total value of the projection of all forces on the surface of the needle is defined as:

$$\sum P = mR \cdot \omega^2 \cdot \sin\tau - m \cdot \tan\phi \cdot R \cdot \omega^2 \cdot \cos\tau = mR \cdot \omega^2 \cdot \frac{\cos(\tau - \phi)}{\cos\phi}, \quad (14)$$

where, τ – the value of the angle between the V_r and $V_{per.}$ velocity vectors;
φ – the value of the angle of static friction of plant residues on the material from which the needle is made.

By determining the total force P, it is possible to identify the variable t from the equation 5. The value of the absolute velocity is determined based on the condition – $p \cdot sin a = r_1$, $p \cdot cos a = r_1 \cdot tg a$:

$$V_p = \sqrt{V_{per}^2 - V_r^2 - 2V_{per} \cdot V_r \cdot \cos(\tau - \phi)} \quad (15)$$

Expressing the absolute velocity in terms of the trajectory length (needle length), we obtain an equation of the form:

$$V_p = \sqrt{\omega^2 - R^2 + l \cdot R \cdot \omega^2 \frac{\sin(\tau - \varphi)}{\cos\varphi} + 2\omega \cdot R \frac{\sqrt{l \cdot R \cdot \omega^2 \cdot \sin(\tau - \varphi)}}{\cos\varphi} \cdot \cos\tau}, \quad (16)$$

where, l – the distance covered by the plant residues along the needle of the rotating working organ, m.

The rotating working organ performs two types of movement: forward movement, when moving the unit, and rotational movement, performed by the needles of the working organ. The trajectory of needles with a horizontal axis of rotation is geometrically a cycloid, the change in shape of which is affected by the ratio of the peripheral and translational velocity.

The value of the cycloid step is determined by the following equation:

$$\lambda = \frac{V}{V_a} = \frac{\omega \cdot R}{V_a}, \quad (17)$$

where, ω – the value of the angular velocity of the rotating working organ, rad./s.;
R – the distance from the rotation center to the needle tip of the working organ, m;
V_a – the value of the translational velocity of the unit, m/s.

As a rule, in rotary tillage machines $\lambda>1$ [4], so the absolute trajectory of the working organs is an elongated cycloid. The absolute velocity of the working organ's movement or cutting speed is determined by the expression:

$$V_0 = V_a\sqrt{\lambda^2 \pm 2\lambda \sin\beta + 1}, \quad (18)$$

where, $\beta = \omega \cdot t$ – the value of the rotation angle of the rotating working organ at a time, t.
The "minus" sign in formula (18) refers to the forward rotation, and the "plus" sign refers to the reverse rotation. In the cutting zone, the maximum velocity corresponds to: $V_0 = V_a\sqrt{\lambda^2 + 1}$ in forward rotation, at $V_0 = V_a \cdot (\lambda^2 + 1)$ rotation.

Based on the theoretical calculations and design data, it is possible to determine the performance and operating parameters of the unit.

$$a = d \cdot \xi \quad (19)$$

$$b = \left(\frac{1}{\cos\varphi}\left[R \cdot arccos\left(\frac{r^2+R^2}{2R \cdot r}\right) - arccos\left(\frac{R-r}{R}\right)\right] + \frac{R_{C-}\sqrt{4R^2-r^2-(r^2+R^2)^2}}{2R}\sqrt{h_{dep}(2R - h_{dep})}\right) \cdot \theta \cdot \lambda \quad (20)$$

where: φ – the angle of friction, degrees;
a – the width of the needle grab, m;
b – the length of the needle grab zone, m;
d – the working width of the needle grab, m;
R – the external radius of the working organ, m;
r – the radius of the transition points from the straight surface to the curved surface, m;
R_c – the radius of the needle curvature, m;
$h_{dep.}$ – the depth of processing, m;
ξ, θ – the coefficient of the soil bulldozing;
λ – the kinematic parameter.
The volume of plant residues moved with a single needle is identified as:

$$Q_0 = \frac{1}{3}\pi \cdot a \cdot b \cdot h_{dep} \tag{21}$$

The volume of plant residues moved by the working organ per one rotation is found based on experimental data:

$$Q_0 = \frac{1}{3}\pi \cdot a \cdot b \cdot h_{dep} \cdot \mu \tag{22}$$

where μ – the influence coefficient of the number of needles on the volume of plant residues moved per 1 rotation (determined experimentally):

$$\mu = n_d(30n_d^2 - n_d + 9) \cdot 10^{-4} \tag{23}$$

where n_d – the number of needles on a disc, item.

We obtain:

$$Q_0 = \frac{1}{3}\pi \cdot a \cdot b \cdot h_{dep} \cdot n_d(30n_d^2 - n_d + 9) \cdot 10^{-4} \tag{24}$$

$$Q_{sp.vol} = \frac{Q_0 \cdot \lambda}{L_{unit.}} = \frac{\pi \cdot a \cdot b \cdot h_{dep} \cdot \omega \cdot R \cdot n_d(30n^2 - n + 9)}{3 \cdot 10^4 \cdot V_{dep} \cdot L_{unit}} \tag{25}$$

where: ω – rotation frequency of the working organ, rad./s.

L_{unit} – the distance covered by the unit is calculated as follows $L_{unit} = 1$ m.

Based on the equation (25), let us construct the graphs of the dependence of plant residues movement on the structural and technological parameters. See figures 6–9.

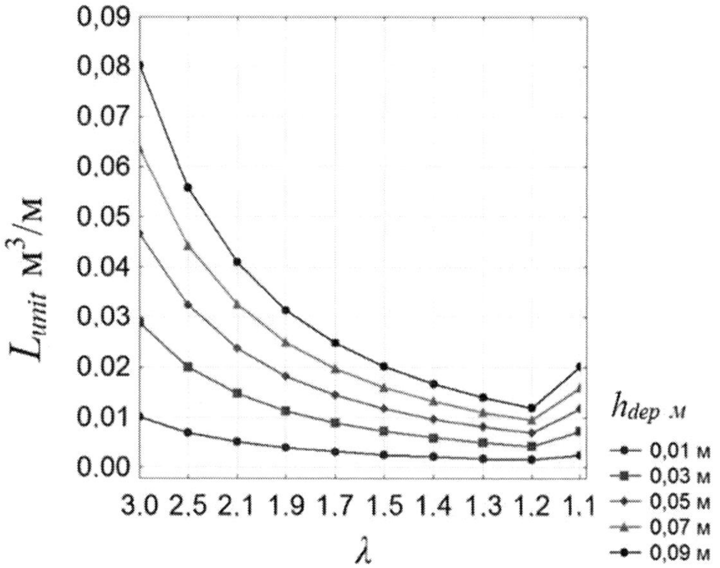

Figure 6. Theoretical dependences of the movement of plant residues $Q_{sp.vol.}$ (m³/m) depending on the kinematic coefficient λ and the depth of processing h_{dep}(m), $Q_{sp.vol.} = f(\lambda, h_{dep})$.

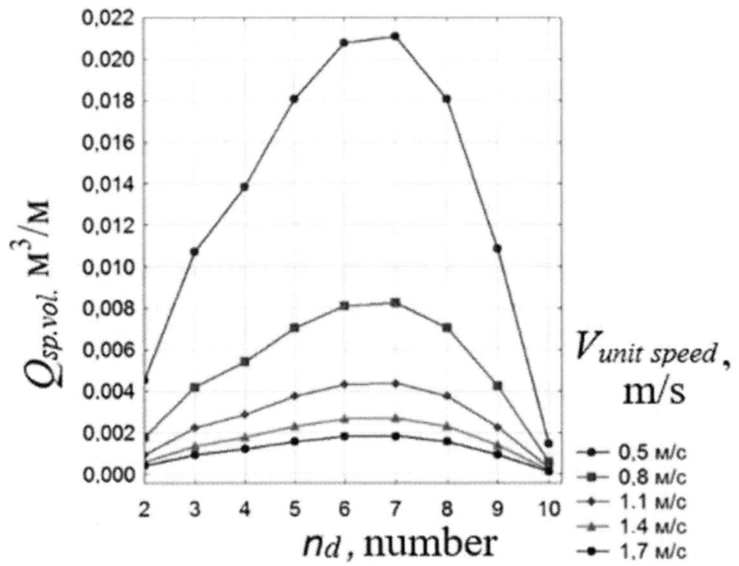

Figure 7. Theoretical dependences of the movement of plant residues $Q_{sp.vol.}$ (m^3/m) depending on the number of needles $n_д$ and the unit's speed $V_{unit\ speed}$ (m/s), $Q_{sp.vol} = f(n_d, V_{unit})$.

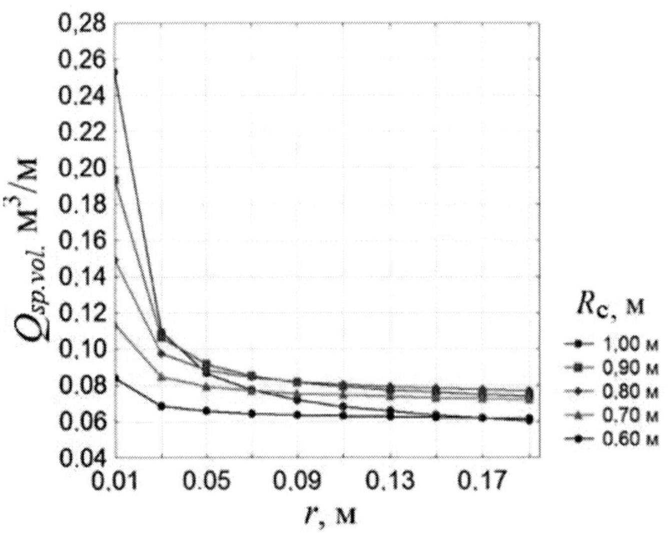

Figure 8. Theoretical dependences of the movement of plant residues $Q_{sp.vol.}$ (m^3/m) depending on the radius of needle curvature $R_д$ and the radius of the transition points of the straight needle surface to the curved needle surface r (m), $Q_{sp.vol.} = f(R_c, r)$.

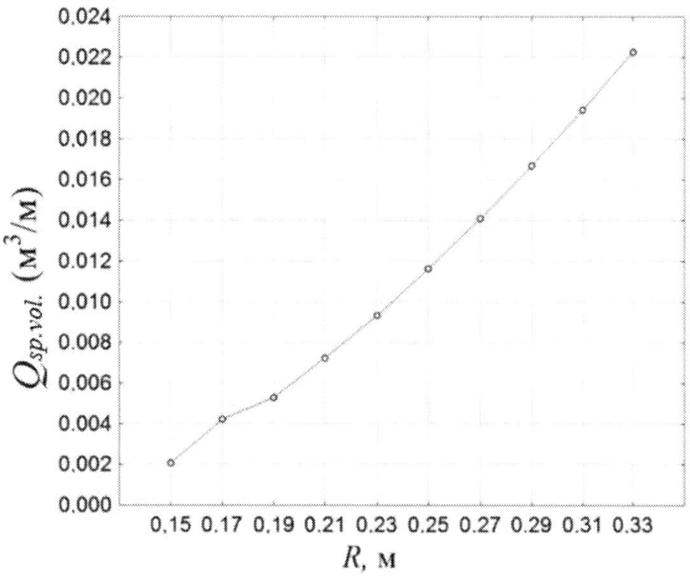

Figure 9. Theoretical dependence of the movement of plant residues $Q_{sp.vol.}$ (m^3/m) depending on the disc radius R, $Q_{sp.vol.} = f(R)$.

4. Conclusion

The obtained dependences make it possible to select the optimal performance parameters of the rotating working organ. The movement of plant residues is mainly influenced by the depth of processing, the number of needles in the working organ, and the kinematic coefficient. The optimal performance parameters include the following ranges: the disk radius of 0.2–0.3 m, the number of needles on the disc equals 2–6 items, the kinematic coefficient is from 1.5 to 3.5. Due to design considerations, we take the radius of the transition points of the straight needle surface into the curved needle surface of 0.1 m, the curvature radius of the needles equals 0.8 m; the needle diameter is 0.015 m.; the overhang distance of rotation axis of the disks forward the axis of the coulters is 40 mm; the relative angular displacement of the needles of adjacent disks in the planes of rotation – half the step of the needles' installation.

References

[1] Koval V, Chernyakov A, Shevchenko A, Begunov M and Demchuk E 2019 *IOP Conf. Ser.: Mater. Sci. Eng.* **582** 012040

[2] Kem A A *at al.* 2019 Obosnovaniye konstruktivno-tekhnologicheskoy skhemy poseva zernovykh kul'tur po neobrabotannomu fonu [Justification of the constructive and technological scheme of grain crops on the raw background] *Sovremennoye nauchnoye znaniye v usloviyakh sistemnykh izmeneniy materialy* pp 110–113 [In Russ.]

[3] Shevchenko A P, Begunov M A, Koval' V S and Dyuba I A 2019 Faktory, vliyayushchiye na kachestvo poseva [Factors affecting quality of crops] *Sovremennoye nauchnoye znaniye v usloviyakh sistemnykh izmeneniy materialy: Tret'yey natsional'noy nauchno-prakticheskoy konferentsii s mezhdunarodnym uchastiyem* p 156–160 [In Russ.]

[4] Shevchenko A P and Domrachev V A 2005 Povysheniye effektivnosti funktsionirovaniya posevnykh mashin putem optimizatsii konstruktivnykh parametrov rabochikh organov [Improving the efficiency of seeding machines by optimizing the constructive parameters of

working organs] Monography (Omsk: Omsk State Agrarian University Publ.) p 120 [In Russ.]

[5] Kem A A and Shevchenko A P 2018 *Mekhanizatsiya polevykh operatsiy v selektsionno-opytnoy rabote* [Mechanization of field operations in selection and experimental work] Monography (Omsk: Omsk State Agrarian Univesity named after P.A. Stolypin) [In Russ.]

[6] Kem A A, Shevchenko A P, Begunov M A and Koval' V S 2019 Eksperimental'nyye issledovaniya vzaimodeystviya rabochikh organov igol'chatogo diska s rastitel'nymi ostatkami [Experimental studies of the needle disc working organs interaction with plant residues] *Bulletin of Omsk State University* **1(33)** 134–141 [In Russ.]

IOP Conf. Series: Materials Science and Engineering **939** (2020) 012034 doi:10.1088/1757-899X/939/1/012034

Use of antivibration gloves when working with a petrol brushcutter Husqvarna 525RX

E A Khlopkov[1,2,3,6], V V Smirnov[4,7], M V Bulateckii[5,8] and Y N Vyunenko[2,9]

[1] Peter the Great St. Petersburg Polytechnic University Russia, Polytechnicheskaya ul., 29 195251, St. Petersburg, Russia
[2] LLC "OPTIMIKST LTD", Peredovikov St., 9 195426, St. Petersburg, Russia
[3] Yaroslav-the-Wise Novgorod State University, ul. B. St. Petersburgskaya, 41 173003, Veliky Novgorod, Russia
[4] North-West Public Health Research Center, 2nd Sovetskaya st., 4 191036, St. Petersburg Russia
[5] Saint-Petersburg State Forest Technical University named after S.M. Kirov Institutsky per., 5 194021, St. Petersburg, Russia

[6] E-mail: hlopkovelisey@mail.ru
[7] E-mail: vvsmirnov00042@rambler.ru
[8] E-mail: djmaksimsunset@mail.ru
[9] E-mail: 6840817@mail.ru

Abstract. With the onset of the season in large cities, extensive work begins on green spaces care. The article demonstrates the effect of using antivibration gloves on the level of acoustic impact on the of the Husqvarna 525RX brushcutter operator hands when mowing the lawn. The indicators are measured using the SVAN 948 vibration and sound velocity analyzer.

1. Introduction

Anti-vibration gloves are recommended for use when the vibration level exceeds the accepted one when working with vibrating tools or electrical equipment [1–3, 4]. The activities of workers in engineering, mining, metallurgy, construction and other industries associated with the devices, having a negative impact on human health due to the generated vibration [4]. Prolonged contact with vibrating tools or electrical equipment and poor protection may cause workers to develop hand-arm vibration syndrome. The main protective equipment and occupational diseases prevention today in almost all these industries are anti-vibration gloves and mittens. They are also relevant in one of the most labor-intensive activities in the green spaces care in large cities – lawn mowing.

The use of petrol brushcutters and lawn mowers in recent years has significantly increased the productivity of gardening professionals. Lawn mowers are usually used in large areas with grass cover. In this case, the surface of the lawn should be fairly smooth. Brushcutters have a number of advantages in comparison with lawn mowers. They are lighter, more maneuverable, and more convenient to work on small areas, sloping surfaces, and bumps. However, the operation of these mechanisms is often accompanied by intense acoustic radiation and vibrations. Protection against high sound pressure is provided by ear protectors or earbuds (earplugs). For the safety of operators' hands, anti-vibration gloves must be used.

Content from this work may be used under the terms of the Creative Commons Attribution 3.0 licence. Any further distribution of this work must maintain attribution to the author(s) and the title of the work, journal citation and DOI.
Published under licence by IOP Publishing Ltd

The purpose of this work is to determine the levels of vibration impact of the brushcutter in the octave spectrum, practical and theoretical assessment of the use of personal protective equipment (hereinafter PPE) when mowing the lawn.

2. Objects and research methods

To conduct an experiment on the applicability of PPE, first of all, the octave spectrum of vibration velocity was determined on the handle of a brushcutter with a petrol engine, Husqvarna 525 (figure 1). Vibration characteristics were measured during lawn mowing (figure 2). The speed of rotation of the brushcutter spindle did not exceed 8500 rpm. The vibration speed sensor was installed on the handle under the right hand. The SVAN 948 noise and vibration analyzer was used to measure surface vibration characteristics. The effectiveness of PPE was determined when using experimental anti-vibration gloves of the NTOT series (the historical name of a series of products designed to protect hands from vibrations) as vibration protection.

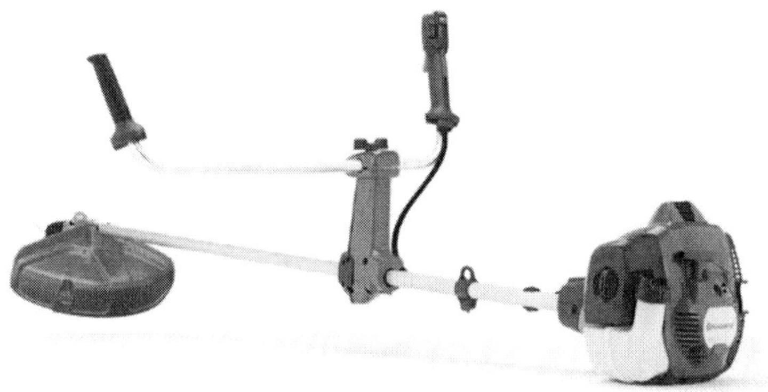

Figure 1. Petrole two-stroke brushcutter Husqvarna 525RX: Output: 1000 W. Engine maximum brake power: 1.3 hp. Number of rotations: 6500 rpm.

Figure 2. Process of Husqvarna 525RX octave spectrum vibration speed measuring.

3. Findings

Figure 3 demonstrates the results of measuring the octave spectrum of vibrations on the handle in the absence of PPE (curve 1). Curve 2 shows the change in vibration characteristics affecting the hands of the trimmer operator using the NTOT series of anti-vibration experimental gloves (the historical name of the series of products designed to protect hands from vibrations). The results show that the permissible vibration velocity standards [5] (curve 3, figure 3) are exceeded only in the octave band with an average geometric frequency of 125 Hz by 7 dB. The corrected L_{vcor} vibration level was 5 dB higher than the standard (112 dB – is standard). The use of protective equipment reduced the impact of the brushcutter on the hands in the octave band with exceeding the permissible values of vibration speed to 113 dB. At the same time, the L_{vcor} using gloves was equal to 114 dB. The curves shown in Figure 3 show that the oscillation characteristics of the L_{125} handle in the octave band with an average geometric frequency of 125 Hz are decisive in exceeding the standard indicators.

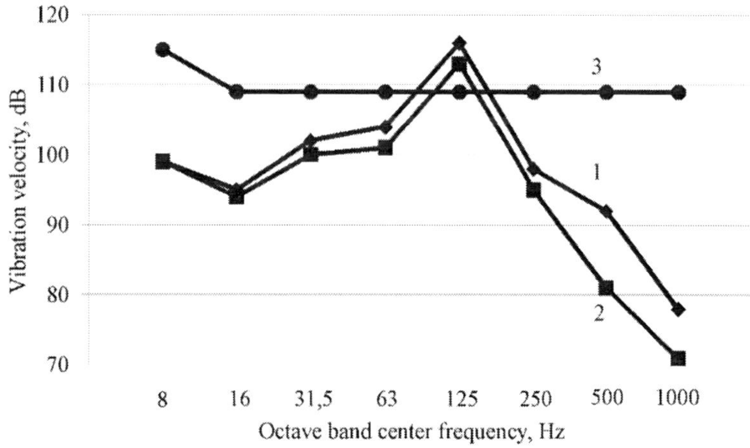

Figure 3. Vibration characteristics of the Husqvarna 525RX brushcutter (1) and vibration levels on the operator's hands with the use of PPE in octave bands (2), vibration rate. Table 1. SanPiN (Sanitary Regulations and Norms) 2.2.2.540-96 (3).

Table 1 and table 2 provide data on the effectiveness of the protective properties of anti-vibration gloves and gloves of various designs. The experimental method is described in the article [6]. This information allows you to make calculations and compare the applicability of PPE when working with a brushcutter. Table 3 demonstrates the values of the vibration velocity in the defining octave band and the corrected vibration level when using gloves of various designs. These data show that the use of 1, 4 and 6 designs bring the L_{125} to the specified level.

Table 1. Effectiveness of protective properties of various designs anti-vibration mittens under the pressure 100 N.

Object	Protective element	Protective properties efficiency, dB in octave frequency bands, Hz						
		8; 16	31.5	63	125	250	500	1000
GOST (State Standard) 12.4.002-97 Type 2a	Up to 8 mm	1.0	2.0	2.0	3.0	4.0	5.0	6.0
		Mittens						
Rep. of Belarus	Foam rubber	1.0	2.0	2.0	3.0	5.0	6.0	8.0
Gevorgyan Vibroton (sole proprietorship)	Rubber insert / porous material	1.0	2.0	2.0	4.0	8.0	9.0	9.0

Vostok-Servis-Spetckomplekt (corporation)	Tubular	1.0	2.0	2.0	4.0	5.0	8.0	10.0
Turbo Onega LLC "OPTIMIKST LTD"	Comb.-UV	1.0	2.0	3.0	5.0	8.0	10.0	11.0
Turbo Siberia LLC "OPTIMIKST LTD"	Comb.-UV	1.0	2.0	3.0	4.0	7.0	10.0	12.0
Turbo Classic LLC "OPTIMIKST LTD"	Comb.-UV	1.1	2.3	2.8	4.6	6.2	9.3	10.2
Turbo Donbass LLC "OPTIMIKST LTD"	Comb.-UV	1.0	3.0	4.0	6.0	8.0	11.0	12.0

Table 2. Effectiveness of protective properties of various designs anti-vibration gloves under the pressure 100 N.

Object	Protective element	Protective properties efficiency, dB in octave frequency bands, Hz						
		8; 16	31.5	63	125	250	500	1000
GOST (State Standard)12.4.002-97 Type 2a	Up to 8 mm	1.0	2.0	2.0	3.0	4.0	5.0	6.0
		Gloves						
X-MARINA S-Gloves	Polyethylene foam	1.0	1.8	2.0	4.0	4.5	5.8	13.6
TEGERA EJENDALS	Polyethylene foam	1.0	2.0	2.0	3.0	5.0	8.0	9.0
VibraGuard Ansell	Gelform	1.2	2.0	2.1	4.0	4.5	7.5	9.1
NTOT AMID LLC "OPTIMIKST LTD"	Comb.-UV	1.2	2.1	2.3	4.1	5.1	6.8	7.8
HTOT 3 LLC "OPTIMIKST LTD"	Comb.-UV	1.0	2.0	3.0	5.0	6.0	9.0	10.0
NTOT LORD LLC "OPTIMIKST LTD"	Comb.-UV	1.0	2.0	3.0	6.0	7.0	8.0	11.0
Vibrostat-03 Amparo "Green Shore XXI Century", Holding Company, LLC	Airgel	1.0	2.0	3.0	5.0	7.0	10.0	12.0

Table 3. Effectiveness of protective properties of various designs anti-vibration gloves under the pressure 100 N.

№	Glove design	L_{125}, dB	L_{vcor}, dB
0	Handle (gloves free)	116	117
1	Vibrostat 03	111	112
2	NTOT AMID	112	113
3	TEGERA	113	114
4	NTOT 3	111	112
5	X-MARINA	112	113
6	NTOT LORD	110	111
7	VibraGuard	112	113
	Normal level	109	112

However, L_{125} cannot be lowered to the limit level. The minimum impact on the hands in this octave band, according to the calculated data, is provided by the 6th design of gloves.

4. Conclusion

The results show a significant protective effect due to the use of anti-vibration gloves when working with the Husqvarna 525RX brushcutter. It can be assumed that a similar effect of reducing the level of acoustic radiation and vibration is possible when using lawnmowers of this class, including those manufactured by other companies.

References

[1] 2018 SAPE-2018: Luchshie razrabotki SIZ na strazhe zdorov'ya i bezopasnosti truda [SAPE-2018: Best PPE designs on the guard of health and safety at work] *Gornaya promyshlennost'*, **2(138)** 102 [In Russ.]

[2] Lazarenkov A M 2019 A study of working conditions in foundries when performing welding operations *Foundry production and metallurgy* **3** 163–165 DOI: 10.21122/1683-6065-2019-3-163-165

[3] Fed'ko V T , Tomas K I and Sapozhkov S B 1998 Protecting the surfaces of welded components against molten metal splashes in CO_2 welding *Welding International* **12(1)** 58–62

[4] Hewitt S *at al.* 2014 Anti-Vibration Gloves? *Ann. Occup. Hyg.* **59(2)** 1–15 DOI: 10.1093/annhyg/meu089

[5] SanPiN 2.2.2.540-96 1997 *Hygienic requirements for hand tools and organization of work* (Moscow: Minzdrav Rossii) p 18 [In Russ.]

[6] Smirnov V V, Buldakov P Y, Buharov S N, Viunenko Yu N and Hlopkov E A 2017 Comparative research of efficiency of anti-vibration mittens gloves *Memoirs of the Faculty of Physics* **5** 1–4

IOP Conf. Series: Materials Science and Engineering **939** (2020) 012035 doi:10.1088/1757-899X/939/1/012035

Features of microrelief formation during laser treatment of Cu-Cr-Zr alloy surface for diffusion welding

Yu Khomich[1], T Malinskiy[1], V Rogalin[1,3], I Kaplunov[2] and A Ivanova[2]

[1] Institute of Electrophysics and Electric Power, Russian Academy of Sciences, Dvorzovaya Naberezhnaya 18 191186, St. Petersburg, Russia
[2] Tver State University, Zhelyabova Str. 33 170100, Tver, Russia

[3] E-mail: v-rogalin@mail.ru

Abstract. The effect of UV laser pulses of nanosecond duration on surface of Cu-Cr-Zr alloy samples was investigated. Noticeable changes in the morphology of the surface layer after irradiation were found at pre-threshold values of the energy density of ~ 0.2 – 0.8 J/cm^2. Changes are traces of thermoplastic deformation resulting from influence of a concentrated energy flow and appears in irregular elevation of the irradiated surface area of the sample by up to 1 μm and in some cases – more. A developed metal surface, deformation of the crystal structure of the surface layer can support the diffusion process during diffusion welding. For energy densities of the order of 1 J/cm^2 or more, optical breakdown with the formation of a crater on the metal surface occurred. The mechanical impulse of a laser plasma when exposed to a metal surface prevents the thermomechanical expansion of the material, and the effect of raising the surface of the samples was not observed.

1. Introduction

Diffusion welding is a method of joining of different materials, which usually cannot be joined in another way. This process is extremely important for high-tech industries using the most advanced materials on the limit of their physical capabilities. The surface condition has an important influence on the processes of mutual adhesion of the materials being welded [1–2]. Processing with intense laser radiation actively effects on the adhesive properties of materials [3–7]. Adhesion, as is known [2], is determined primarily by diffusion phenomena, and pulsed laser irradiation contributes to a sharp change in the diffusion parameters, up to an increase in diffusion coefficient by up to 6 orders of magnitude [8–9]. These processes occur in the surface layer, the modification of which can have a significant effect on the formation of micro- and nanostructures, largely determining the adhesion properties of the surface. With this treatment, the effective surface area markedly increases.

The direct laser nanostructuring method, due to its comparative simplicity and high efficiency, is one of the most promising [4–6]. This method has only one stage, which consists in processing the surface of the material with a focused laser beam and does not require the use of any masks or other supporting means.

The main amount of work on direct laser nanostructuring was performed earlier using pico- and femtosecond lasers [10–13]. However, for practical applications, it is of interest to use cheaper and easier to use sources, such as nanosecond lasers. The papers [14–16] give examples of the formation of nanostructures on the surface of various materials: polymers, semiconductors, metals and ceramics

Content from this work may be used under the terms of the Creative Commons Attribution 3.0 licence. Any further distribution of this work must maintain attribution to the author(s) and the title of the work, journal citation and DOI.
Published under licence by IOP Publishing Ltd

obtained using these lasers; moreover, the main attention was paid to nanostructures formed as a result of optical breakdown, accompanied by ablation effects. In this paper, we studied microrelief formation features during laser surface treatment of the Cu–Cr–Zr copper alloy, which is often used for diffusion welding [17], and the main attention is paid to the pre-ablation mode of material processing, which is characterized by processes occurring in the condensed state of the metal.

2. Experimental technique

The laser source was a solid-state pulsed periodic NdYaG laser (third harmonic with a wavelength of $\lambda = 355$ nm). The choice of the UV radiation range is explained by the fact that in this range most construction materials have significant absorption. The magnitude of the quantum of radiation provides a slightly larger value of the depth of the processed material, compared with sources of visible and infrared radiation; at the same time, the atmosphere is transparent to UV radiation.

The pulse duration was $\tau = 10$ ns, and the energy in the pulse reached 8 mJ, and the pulse repetition rate was 10 Hz. In the experiment, the energy and the number of acting radiation pulses were changed. After exposure, the surface morphology of the sample was studied on an optical three-dimensional profilometer – Zygo NewView 7300 and scanning electron microscope (SEM) JEOL JSM 6610LV.

The initial content of alloying additives in the alloy (Cr ~ 1%, Zr ~ 0.1 - 0.2%) was measured by the X-ray diffraction method using an attachment to JEOL JSM 6610LV. The reflection coefficients of polished samples of copper and its low-alloy alloys were measured using a Cary-5000 UV-VIS-NIR spectrophotometer with UMA prefix. At the working wavelength, the studied samples reflected ~ 10%. The accuracy of reflection coefficient measurements was 1%.

The sample was mounted normally to the optical axis of the laser on a three-coordinate subject table. It consisted of two linear translators 8MT180 and one 8MT173, operated by the controller 8SMC1-USBhF (Standa Ltd, Lithuania). The radiation energy density on the sample was changed both by changing the energy of the exposing laser pulse and by changing the size of the laser beam cross section by shifting the surface of the sample from the focal plane of the lens. The number of laser pulses exposing on the sample was controlled by an electromechanical shutter.

3. Experimental result and discussion

In the work, two pronounced regimes of exposure of radiation on the surface of the material were observed. This is a previously relatively well-studied ablation regime [18–21] and a little-studied pre-ablation regime in which the threshold of optical breakdown is not yet reached and the main processes of exposure occur mainly without changing the phase state of the material.

In the ablative mode, the interaction of radiation with the surface of the sample occurred in the well-known mode, many times described in the literature [22]. The exposure was accompanied by optical breakdown, in which a plasma torch and crater formation on the surface of the sample occur. Figure 1a shows a microphotograph of the affected area. The profilogram (figure 1b) shows that a cavity is observed in the centre, and along the edges, there is a roller formed by the material evaporated from the crater zone and deposited onto the surface by the flow of material from the centre to the edges of the molten zone.

At lower, subthreshold values of the radiation power density ($E < 0.8 – 0.9$ J/cm^2), an irregular lifting in the sample surface was observed in the exposure zone. After a single exposure at a power density ($E = 0.2 – 0.3$ J/cm^2), it was not possible to detect surface lifting. However, because the effect is cumulative in nature, after exposure to 30 laser pulses, this was recorded. Figure 2a shows a microphotograph of the surface area of the sample after exposure. It can be seen that in addition to cleaning the surface, at the centre of the spot irregular rise of the surface level with a maximum of 30 nm was observed (the profile is shown in figure 2b). It is clearly visible that the scratch, remaining after polishing, and passing through the centre of the spot is almost invisible in the exposure zone. With increasing power density, this effect becomes more noticeable.

With an increase in power density to a level ($E \sim 0.8$ J/cm^2), the effect of exposure becomes more noticeable even after a single exposure (figure 3). In this case, the lifting height of some surface areas already reached the level of 1 μm, sliding along grain boundaries, which is characteristic of plastic deformation at elevated temperatures was observed [23]. At the same time, traces of boiling of a thin layer of instantly solidifying liquid metal appeared on the surface, manifesting themselves in the form of characteristic microcraters ~ 1–3 μm in size. Noticeable changes have also occurred in the elemental composition of the surface layer — the presence of oxygen was revealed. In the centre of the spot, its share was ~ 1% by weight, and up to ~ 4–5% along the edges. Accordingly, the relative concentration of the main element of the alloy, – copper, slightly decreased. However, a significant dispersion was observed in different spots.

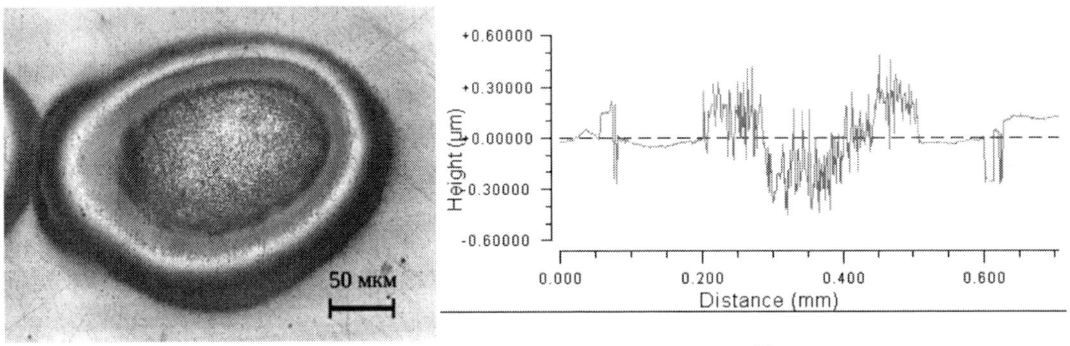

a) b)

Figure 1. Cu-Cr-Zr alloy surface after exposure to a UV pulse (λ = 355 μm, 10 ns, E = 2.15 J/cm^2, 30 pulses, f = 10 Hz): a) microphotograph, optical microscope, Zigo; b) surface profile, Zigo.

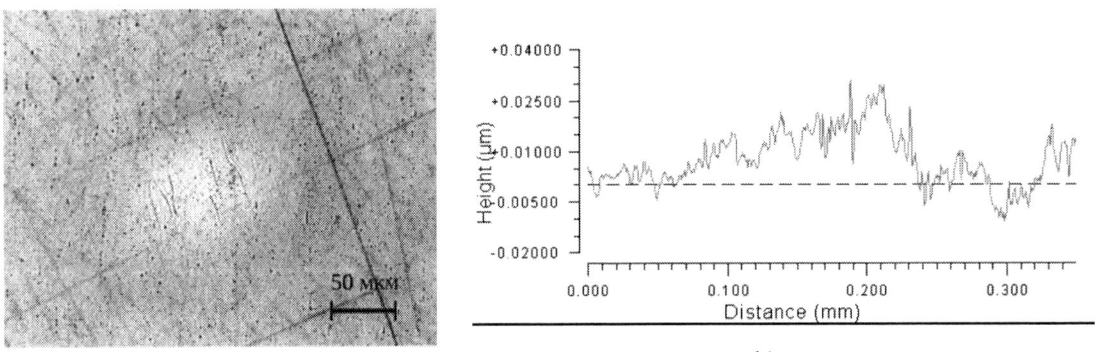

a) b)

Figure 2. Cu-Cr-Zr alloy surface after exposure to a UV pulse (λ = 355 μm, 10 ns, E = 0.2 J/cm^2, 30 pulses, f = 10 Hz): a) microphotograph, optical microscope, Zigo; b) surface profile, Zigo.

An increase in the number of exposing pulses led to a noticeable intensification of the observed effect (figure 4). Faults distinctly appeared in the exposing zone along the boundaries of grains unevenly extruded from the mass of the sample. Inside some of them, traces of crystallographic sliding, noticeably changing the slope when passing through the grain boundary, were clearly observed. The height of some sections of uplift also reached ~ 1 μm. In the region of slide bands and grain boundaries, the material was "laser etched" by subsequent pulses, that is, the initially sharp edges of slide bands and grain boundaries seemed to melt slightly, obtaining round shape, but the character of the picture was not distorted. However, the proportion of oxygen in the composition of the surface layer of the alloy did not noticeably increase.

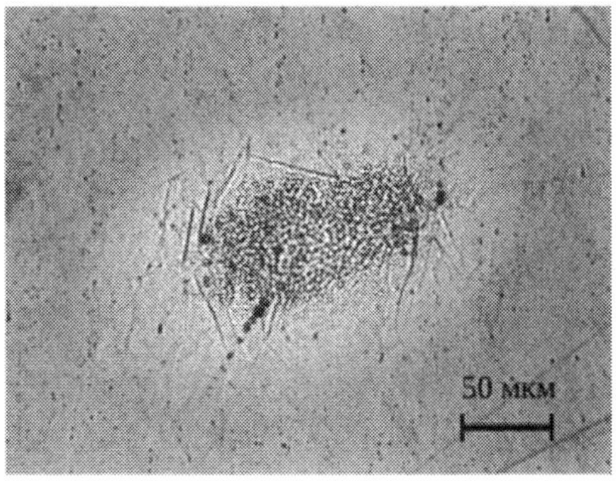

a)

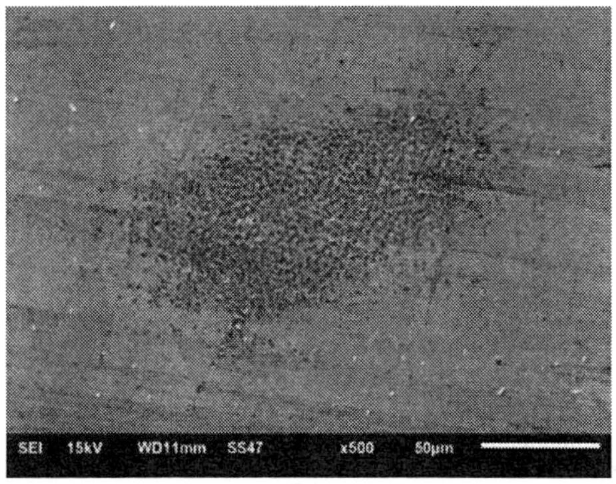

b)

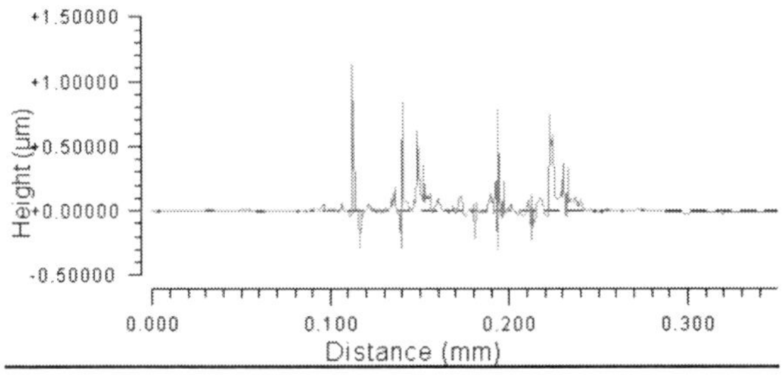

c)

Figure 3. Cu-Cr-Zr alloy surface after exposure to UV pulse (λ = 355 μm, 10 ns, E = 0.8 J/cm^2, 1 pulse):
a) microphotograph, optical microscope, Zigo; b) SEM, c) surface profile, Zigo.

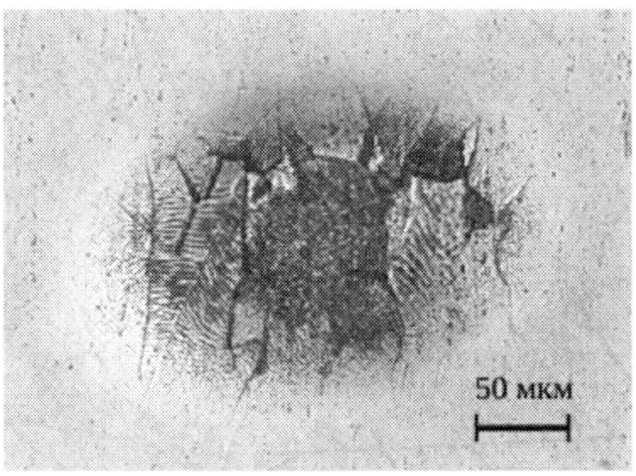

a)

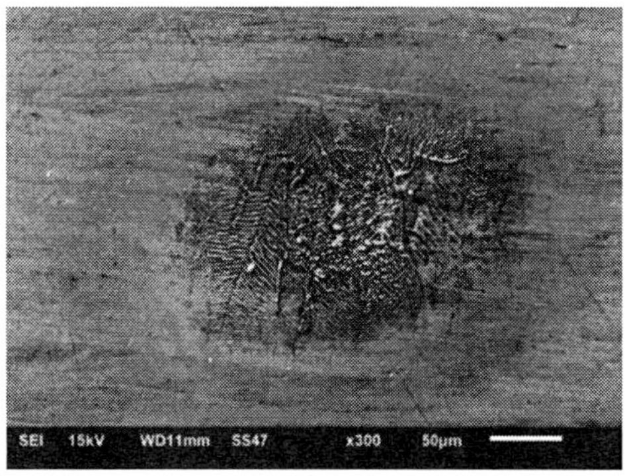

b)

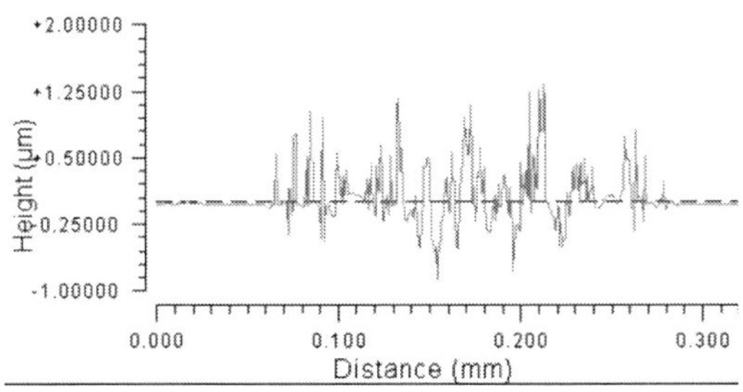

c)

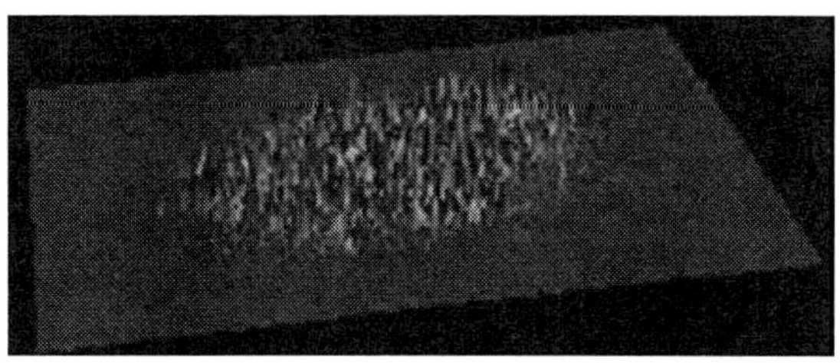

d)

Figure 4. These Cu-Cr-Zr alloy surface after exposure to a UV pulse (λ = 355 μm, 10 ns, E = 0.8 J/cm^2, 25 pulses, f = 10 Hz): a) microphotograph, optical microscope, Zigo; b) SEM, c) surface profile, Zigo; d) 3D image.

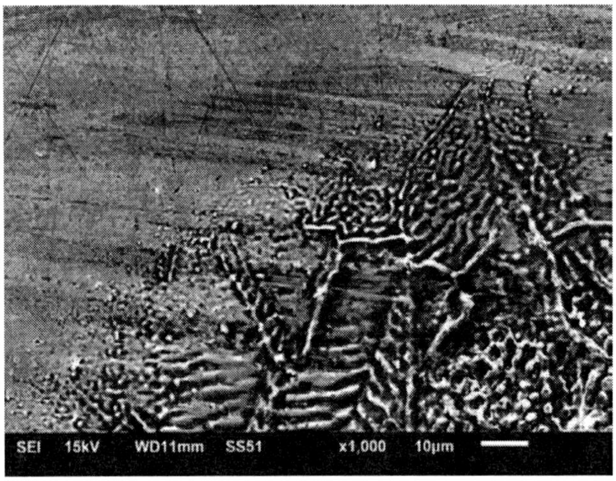

Figure 5. SEM microphotograph of a fragment of a portion of the exposure zone, shown in figure 4.

This picture is clearly visible on the periphery of the exposure zone, where the radiation power density was naturally slightly lower than in the centre of the spot, which was densely dotted with continuous microcrater formations, masking the traces of plastic deformation. This is clearly seen in figure 5, which shows a SEM microphotograph of a fragment of the exposure zone shown in figure 4.

4. Conclusions

On the samples of Cu-Cr-Zr alloy, after exposure with a UV laser pulse of nanosecond duration at pre-threshold values of the energy density of ~ 0.2 – 0.8 J/cm^2, noticeable changes in the morphology of the surface layer were found. They are traces of thermoplastic deformation that has occurred as a result of exposure to a concentrated energy flow, and appear in the irregular rise of the exposed region of the surface of the sample by up to 1 μm, and in some cases – more. In the exposure zone, classical manifestations of high-temperature plastic deformation are observed – grain-boundary sliding and traces of crystallographic slip. The effect is cumulative in nature, that is, the height of the uplifts increases with increasing number of exposing pulses. It is assumed that the obtained metal-developed surface should be very attractive for subsequent diffusion welding. Significant distortions of the crystal structure of the surface layer of the metal, which have arisen, should support diffusion process.

Traditionally, in experiments on laser exposure on the surface of solids the focus was on the phenomenon of optical breakdown of the material and its consequences. For energy densities of the order of 1 J/cm^2 or more, optical breakdown with the formation of a crater on the metal surface occurred. The mechanical impulse of a laser plasma, when exposed to a metal surface, prevents the thermomechanical expansion of the material, and therefore, similar effects were not previously observed. And the appearance of a laser crater accompanying an optical breakdown, the depth of which is noticeably higher than the lifting height, excludes the detection of these effects.

5. Acknowledgments

The study was performed using certified spectrophotometric measurement methods and validated equipment at the accredited interdepartmental academic and test laboratory of semiconductor material and dielectrics "Single Crystals and Stock on their Base" of the National University of Science and Technology MISiS and the Center for collective use of Tver State University and supported by the Ministry of Science and Higher Education of the Russian Foundation within a government contract for scientific work. The authors are grateful to S.I. Mikolutsky for active participation in the work, Corr. RAS Yamschikov V.A. for participating in the discussion of the results.

References

[1] Lyushinskij A V 2016 *Svarochnoe proizvodstvo* **7** 17 [In Russ.]

IOP Conf. Series: Materials Science and Engineering **939** (2020) 012035 doi:10.1088/1757-899X/939/1/012035

[2] Lee L H *Fundametals of adhesion* 1991 (New York: Springer Science+Business Media)
[3] Makarov G N 2013 *Physics-Uspekhi* **56**(7) 643
[4] Khomich V Yu and Shmakov V A 2015 *Physics-Uspekhi* **58**(5) 455
[5] Tokarev V N, Shmakov V A, Yamschikov V A, Khasaya R R, Mikolutskiy S I, Nebogatkin S V and Khomich V Yu 2010 *Congress Proceedings 29th International Congress on Applications of Lasers and Electro-Optics* (ICALEO 2010) (Anaheim, CA) 1257
[6] Khomich V Yu and Shmakov V A 2011 *Doklady Physics* **56**(6) 309
[7] Homich Yu V, Malinskij T V, Mikoluckij S I, Rogalin V E, Yamshchikov V A, Kaplunov I A and Ivanova A I 2020 *Fotonika i informacionnaya optika* (Moscow: MIFI) p 191 (In Russ.)
[8] Gurevich M E, Larikov L N, Mazanko V F, Pogorelov A E and Fal'chenko V M 1978 Metallofizika **73** 80 (In Russ.)
[9] Rovinskij R E, Rogalin V E, Rozenberg V M, Teplickij M D 1980 Fiz. i him. obrab. mater. **3** 7 (In Russ.)
[10] Vestentoft K, Olesen J A, Christensen B H and Balling P 2005 *Appl. Phys. A* **80** 493
[11] Vorobyev A Y, Guo C 2005 *Phys. Rev. B* **72** 195422
[12] Ostendorf A, Koch J, Korte F and Chichkov B N 2004 *Proc.SPIE* **5448** 1
[13] Lazare S, Tokarev V N, Sionkowska A, Wisniewski M 2005 *Appl. Phys. A* **81** 465
[14] Mikolutskiy S I, Khasaya R R, Khomich Yu V and Yamshchikov V A 2018 *J. Phys.: Conf. Ser.* **987** 012007
[15] Guillemot F, Prima F, Tokarev V N, Belin C, Porte-Durrieu M C, Gloriant T, Baquey C and Lazare S 2004 *Appl. Phys.* A **79** 811
[16] Lapshin K E, Obidin A Z, Tokarev V N, Homich V Yu, Shmakov V A and Yamshchikov V A 2007 *Rossijskie nanotekhnologii* **2**(11–12) 50 [In Russ.]
[17] Nikolaev A K and Kostin S A 2012 *Med' i zharoprochnye mednye splavy* (Moscow: DPK Press) [In Russ.]
[18] Lunney J G and Jordan R 1998 *Applied Surface Science* **127–129** 941
[19] Mannion P T, Magee J, Coyne E, O'Connor G M and Glynn T J 2004 *App. Surface Science* **233** 275
[20] Nolte S, Momma C, Jacobs H, Tünnermann A, Chichkov B N, Wellegehausen B and Welling H 1997 *J. Opt. Soc. Am. B* **14** 2716
[21] Chichkov B N, Momma C, Nolte S, von Alvensleben F, Tünnermann A 1996 *Appl. Phys. A* **63** 109
[22] Anisimov S I, Imas Ya A, Romanov G S and Hodyko Yu V 1970 *Dejstvie izlucheniya bol'shoj moshchnosti na metally* (Moscow: Nauka) [In Russ.]
[23] Straumal B B 2003 *Fazovye perekhody na granicah zyoren* (Moscow: Nauka) [In Russ.]

Using waste heat for heat supply and greating a single steam network with power generation due to pressure differences

E P Kondurov[1], P A Kruglikov[1], I V Shvetsov[2,4] and D V Zhelnov[3]

[1] I. I. Polzunov Scientific and Development Association on the Research and Design of Power Equipment, 3/6 Atamanskaya str., 191167 St. Petersburg, Russia
[2] Yaroslav-the-Wise Novgorod State University, ul. B. St. Petersburgskaya, 41, 173003 Veliky Novgorod, Russia
[3] Department of Expertise and Certification of the Union "St. Petersburg Chamber of Commerce and Industry", pr. Lesnoy 20, St. Petersburg, Russia

[4] E-mail: Igor.Shvetsov@novsu.ru

Abstract. In modern economical conditions external heat suppliers often become quite expensive for a consumer enterprise due to their monopoly, so the task of reducing the amount of heat received from them to the minimum, and of using one's own capabilities to the maximumarises. In this article, using a particular example, the authors considered certain technical characteristics of using waste process heat at the enterprise while transferring the enterprise's heat supply system to its own heat sources, as well as creation of a single steam network with electricity production due to pressure differences.

1. Introduction

Refusal from an external heat source, or significant change of the heat amount received from it, due to the use of one's own heat, results in the change of heat supply system topology. It is obvious that the network of pipelines should be aligned with the new mutual arrangement of sources and consumers. Otherwise, throughput with existing network pipes at certain areas may be insufficient to supply consumers with heat completely. According to the terms of reference we studied the resources to transfer the enterprise's heat system to its own heat sources, refusing from heating the network water in a boiler thermal power plant. Herewith, own heat sources mean the utilization of process heat by the enterprise, and an available boiler room [1–5].

2. Results and discussion

2.1. Analysis of capabilities to use waste heat for heat supply of the enterprise

One of particular tasks was to study the capability to use the waste process heat of workshops for heating these workshops, namely:

– in the amount of 10 Gcal / hr at each input of subscribers 911 and 918 (hereinafter the numbering of subscribers corresponds to the accepted numbering at the enterprise) by heating the return network water at existing costs in the housings, or by installing a special mini-boiler (see figure 1);

– in the amount of 40 Gcal/hr by the mini-boiler scheme with the generation of this heat in the pipeline of the return network water of the united consumer 1015 (see figure 2).

Content from this work may be used under the terms of the Creative Commons Attribution 3.0 licence. Any further distribution of this work must maintain attribution to the author(s) and the title of the work, journal citation and DOI.
Published under licence by IOP Publishing Ltd

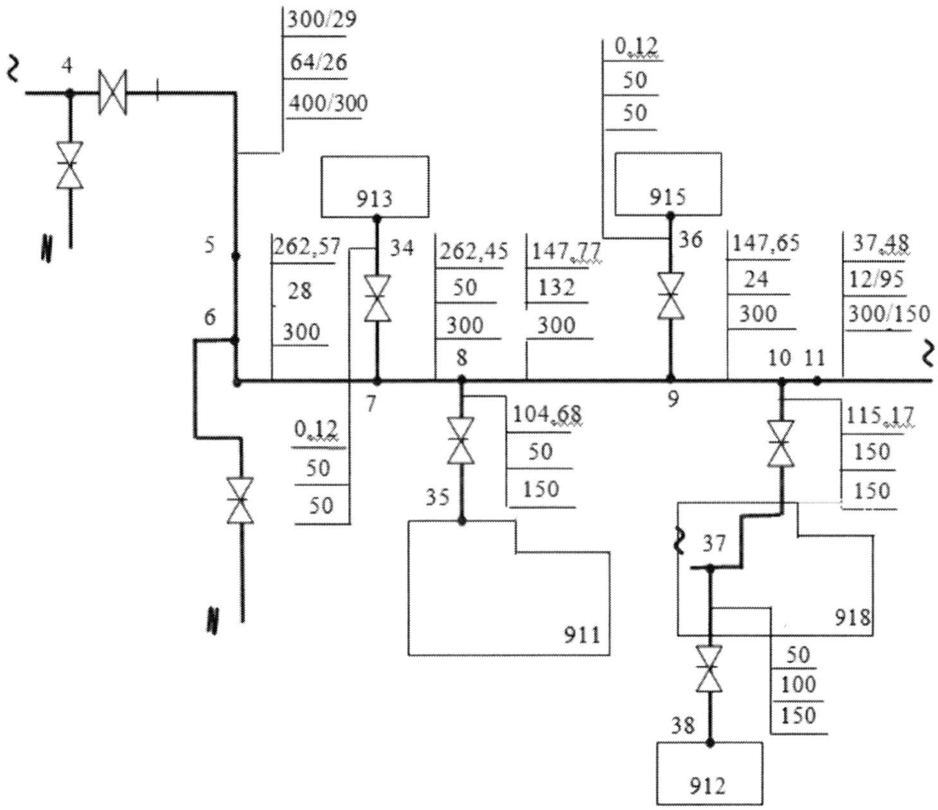

Figure 1. Section of the heat supply scheme in the area of subscribers 911 and 918: (on the picture: in order to simplify, only the pipelines of the return network water are shown; the numbers column given in the scheme indicate, top down, the values of network water consumption in t/hr, the length of the pipeline section in m, the pipe diameter in this section in mm).

As the study has shown, the heat use of subscribers 911 and 918 is quite possible, taking into account the following heat engineering characteristics of these subscribers.

At a temperature schedule of 150/70 °C the required consumption of the network water of subscribers 911 and 918 amounts to, correspondingly, 114,68 and 115,17 t/hr, and heat supply in the amount of 10 Gcal/hr for each flow results in the return water heating of about 87 °C.

With a calculated pressure at the outlet of these subscribers about 4 at, and the corresponding saturation temperature of 143 °C, heating at 87 °C results in impermissible water boiling during most of the heating season.

Installation of a mini boiler for subscribers 911 and 918, when passing through it 250 t/hr of water, is technically feasible, but hardly appropriate, as the amount of heat consumed by the network depends on the outdoor temperature, and the amount of waste process heat is an independent value. So, during most of the heating season waste-heat recovery is not possible in the network.

The most rational solution of the problem of recovery of this amount of heat is its input to the return network water in the network locations with high consumption. The closest locations to subscribers 911 and 918 are pipelines sections between nodes 4, 5, 6 and 6, 7 (figure 1), in which nominal (at a temperature schedule 150/70 °C) consumption is, correspondingly, 360 t/hr and 262,57 t/hr.

In case of installing a heat exchanger between nodes 4, 5, 6 heating of network water will be just 55,5 °C, and temperature at the output from the heater will not exceed 125,5 °C, which gives the minimum reserve value of about 15 °C before boiling. Herewith, all available heat can be used throughout all heating season.

In a temperature schedule 150/70 °C installation of a heat exchanger at the section between nodes 6 and 7 is practically excluded due to relatively low consumption of the network water and impossibility, in this regard, to prevent its boiling.

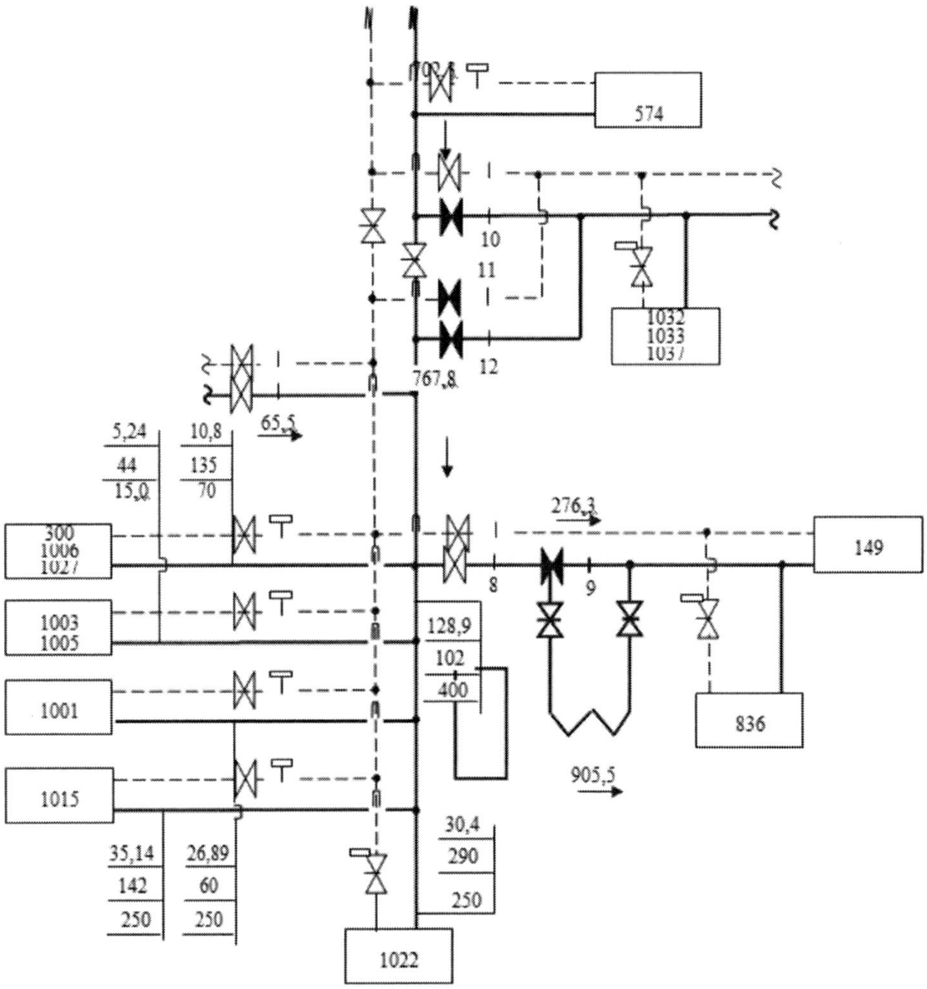

Figure 2. Section of the heat supply scheme in the area of united consumer 1015.

However, situation can change at a different temperature schedule, and consumption of water will also change. For example, at a temperature schedule 115/70 °C, the consumption of network water through the section between nodes 6 and 7 will become equal to 466,3 t/hr and the value of water heating will reduce to 43 °C.

It is more difficult to organize utilization of 40 Gcal/h in united consumer 1015 (see figure 2).

In addition to inexpedience of using a mini boiler for the above reasons, the solution of this problem is complicated by the necessity of output of 40 Gcal/h by high consumption of water (550 t/h at a temperature schedule 150/70 °C). In the nearest section of a main trunk pipeline the existing consumption 126, 9 t/h is significantly less.

The only possible solution is an artificial increase of water consumption through sections of trunk pipelines of return water, adjacent to the united consumer, with installation of network water heaters on these sections.

The most appropriate solution seems to install heaters on a bypass of the section of the trunk pipeline of return network water between shutoff valves 8 and water discharge to united consumer (1032, 1033, 1037) with overlapping this pipeline with valve 9 and closing valves 10, 11, 12, as figure 2 shows. Thanks to the mentioned overlaps, it is possible to increase the consumption of return network water through the heat exchanger from 140,2 t/h to 905,5 t/h. To increase reliability, it is recommended to install several heaters connected in parallel.

The indicated increase of network water flow rate through separate trunk pipelines will be most likely to result in the necessity of increasing pipes diameters in individual sections.

2.2. On the problem of creating a unified steam network 40 AT at the enterprise, with electricity generation due to pressure differences 40/27/1,2 AT

One of the tasks of creating a unified steam network *P*-40 was to determine the incut point *ROU* 40/27 at a specific section with determination of maximum flow rate capability *ROU*, with steam supply from a thermal power plant, disconnected consumers *P*-40, and backpressure above *ROU* 29 at.

When analyzing the distribution of steam consumption set by the terms of reference for steam consumers *P*-40, we paid attention to a high value of steam consumption through *ROU* 40/27 (to 135 t/h), which is a tangible potential loss.

The use of a steam turbine instead of *ROU* could allow taking off power, the maximum value of which is:

$$N = G \times \Delta h \times \eta = 2925 \text{ kW}; \tag{1}$$

where: G = 375 kg/s (135t/h) is a maximum steam consumption through the turbine;

Δh = 120 kJ/kg is a heat drop at the turbine in the modes with the indicated steam consumption for *ROU*;

η = 0,65 is an overall efficiency of the turbine-generator unit (adopted by analogues).

A finished steam turbine of this or similar power, produced by domestic enterprises for steam parameters ROU 40/27, is not known to the authors. Therefore, with the implementation of the idea of installing a turbine instead of ROU, it seems that difficulties will arise, because it will be necessary to develop a new turbine project, which, of course, will significantly increase the cost of its creation and, despite the obvious economic effect, and taking into account the unstable economic situation in the country, lead to an unacceptable payback period.

Regarding the question of defining a maximum flow rate capacity *ROU* at pipelines geometry set by the terms of reference, and backpressure above *ROU* 29 at, in the first place this value depends on the flow rate capacity of throttle control valves used in *ROU*. In standard *ROU*, closest in its characteristics to the working conditions in the structure of the set network P-40, a maximum flow rate capacity of throttle control valve 995-150 is

$$K_{V\max} = 282 \text{ m}^3\text{/h}, \tag{2}$$

which provides a maximum steam consumption under the modes considered:

$$G = 3.6\sqrt{\frac{\Delta P \cdot F^2 \cdot 2}{\sum \zeta \cdot v}} = 96 \text{ t/h}, \tag{3}$$

where: ΔP = 12 kg/cm^2 is a given minimum pressure difference between a source (a thermal power plant) and a consumer (network П-27);

IOP Conf. Series: Materials Science and Engineering **939** (2020) 012036 doi:10.1088/1757-899X/939/1/012036

F = 0,07 m^2 is a pipe cross-sectional area of the given diameter in the considered section;
v – m^3/kg, a specific volume of steam at the inlet to the throttle control valve *ROU*;
$\sum\zeta = \sum\zeta_r + \sum\zeta_{val}$ is a total coefficient of hydraulic resistance of the route, reduced to F = 0,07 m^2;
$\sum\zeta_r$ is a coefficient of hydraulic resistance of the pipe in section 11;
$\sum\zeta_{val}$ is a coefficient of hydraulic resistance of the valve;
K_{Vmax} is a maximum valve flow rate capacity.

The steam consumption defined above is significantly less than 135 t/h required by the task. However, you should notice that the pressure drop 40/27 given by *ROU* is not characteristic of *ROU*, and it may be provided not by a throttle control valve, but by a usual control one which widens greatly the flow rate capacity of such *ROU*.

For the considered task we can recommend a two-seat control valve 14*c*-76-64 Dy400 as a possible way of throttling steam, for it can provide a maximum pass of 180 t/h of steam.

The optimal *ROU* incut point for decreasing hydraulic losses in the section before it must be located directly in point 2.

Regarding the installation of a turbogenerator instead of *ROU*, we should notice the following. Considering a significant difference in steam parameters from different producers, it is necessary to install a mixing device in point 2 in front of the turbogenerator. After mixing, steam parameters in point 2 will be P = 32 at, *ROU* = 411÷417 °C. For comparison, nominal steam parameters at inlet of serial anti-pressure turbines to average parameters of a similar enterprise are P = 3,5 *MPa*, T = 435 °C, what follows that with the existing parameters such a turbine will be working under off-design modes at low efficiency. If it is impossible to operate the turbine with such steam parameters, it becomes necessary to redevelop the turbines or develop a new design.

Total steam consumption for a turbine from three producers is 210÷240 t/h. Nominal steam consumption for an anti-pressure turbine with a drop 35/11 *MPa* and power 6 MW is within 90÷100 t/h, i.e. it is possible to install two turbines. During a scheduled repair of one of steam producers, it allows operating one turbine for 100% of power, which is more economical than the operation of one more powerful turbine under partial loads.

In the single steam network scheme *P*-40 with two turbines, it seems interesting to install one of the turbines in direct proximity to steam producers of workshop *AK*-72 and boiler room k. 1101, which allows avoiding unnecessary heat hydraulic losses, when transporting steam to point 2.

In the single steam network scheme *P*-40 with the turbogenerators, it is necessary to provide for the presence of ROU installed parallel to the turbines, which will provide consumers with reduced pressure steam during planned or emergency shutdowns of the turbines. For this is it possible to use available *ROU*, and, when it is necessary, to provide for installation of new ones. It seems most convenient to locate *ROU* in direct proximity to the turbines. In this case switching the steam supply to consumers from the turbine to *ROU* does not influence the working mode of the rest of the reduced pressure network.

The presence of a reduced temperature steam source (ammonia production units) in the network significantly decreases the potential of fresh steam, compared with the turbine. In this regard, it is necessary to consider the option of using higher temperature steam for turbines with working drop 40/12 at, and lower temperature steam should be mixed with turbine exhaust steam to obtain steam of 27 at.

Turbine differential pressure, power and quantity selection should be made, considering reduced pressure steam production needs. For this it is necessary to analyze annual schedules for steam production and consumption *P*-40, *P*-27 and *P*-12.

3. Conclusion

1. Based on the calculation study, the options for the most rational organization of the use of waste process heat of workshops for the needs of the enterprise's heat supply with the placement of network water heaters on main pipelines with artificial increase in some cases of network water flow through these heat exchangers are substantiated.

IOP Conf. Series: Materials Science and Engineering **939** (2020) 012036 doi:10.1088/1757-899X/939/1/012036

2. When creating a single steam network *P*-40 at the enterprise, it seems appropriate:

– in case of installation of ROU 40/27 instead of the throttle-control valve ROU, to use the two-seat control valve 14s-76-64 Dy400, providing the required consumption of steam;

– when using a turbogenerator instead of a ROU, to install two turbogenerators in the immediate vicinity of the steam producers, namely: at the AK-72 workshop and boiler room k.1101.

References

[1] *The role of instrumentation and control systems in power uprating projects for nuclear power plants* 2008 (Vienna: IAEA) NP-T1.3 [In Austria]

[2] Davydenko N N et al 2010 Modernizaciya protochnoj chasti CND parovoj turbiny K-500-6,4/3000 dlya povysheniya ee moshchnosti [Modernization of the flow section of the low-pressure cylinder of the K-500-6.4 / 3000 steam turbine to increase its capacity] *Thermal Engineering* **9** 21–25 [In Russ.]

[3] *Reactor Core Thermal Power Uncertainty Calculation Unit 1* 2009 LLC Calculation LE-0113 (Exelon Generation Company) Rev. 0

[4] Cotton K and Schofield P 1971 Analiz izmeneniya rabochih harakteristik parovyh turbin [Analysis of changes in the performance of steam turbines] *Power Machines and Plants* **2** 62–72 [In Russ.]

[5] Belyakov V N et al 2019 *IOP Conf. Ser.: Mater. Sci. Eng.* **656** 012008

IOP Conf. Series: Materials Science and Engineering **939** (2020) 012037 doi:10.1088/1757-899X/939/1/012037

The universal algorithm for calculating SPICE models' parameters of semiconductor diodes

A F Korchagin[1]

[1] Yaroslav-the-Wise Novgorod State University, ul. B. St. Petersburgskaya, 41 173003 Veliky Novgorod, Russia

E-mail: Alexandr.Korchagin@novsu.ru

Abstract. The methods and procedures for calculating SPICE models' parameters of semiconductor diodes for the forward and reverse branches of the VAC are described. Recommendations for planning the experiment and selecting initial approximations of parameters for forward and reverse switching of the diode are given. Existing methods for calculating SPICE models' parameters do not allow obtaining their statistical characteristics, and for some parameters, methods for estimating them from experimental data have not been developed at all. This article offers a universal method for evaluating all the necessary parameters of SPICE models of semiconductor diodes and their statistical characteristics.

1. Introduction

The identification task is to determine the parameters of the diode model based on the results of measurements of the forward and reverse branches of the VAC diode, respectively. Most publications devoted to methods of parameter identification [1-5] focus on nonlinear optimization algorithms based on a large number of measurements.

The complexity of the problem of estimating parameters from experimental data, as a problem of nonlinear optimization, is due to the following factors:

– large dimension of the optimization space;

– the complexity of setting the initial approximation vector for the general case (some parameters, such as saturation currents and recombination current parameters, may differ by several orders of magnitude for different devices);

– the need for a large number of measurements;

– difficulty in choosing the most stable and fast-acting optimization method.

The main disadvantages of the known methods for estimating parameters of SPICE models are that they are not universal, do not have sufficient stability, do not offer methods that reduce the number of measurements and, at the same time, increase the accuracy of parameter estimation, and do not provide statistical characteristics of the obtained parameter estimates. In addition, some methods for estimating the parameters of SPICE models use various assumptions to simplify calculations, for example, dividing the VAC into sections within which some parameters can be ignored. Such assumptions are often insufficiently justified, and in some situations this approach may lead to large errors in the estimation of parameters.

The absence of statistical characteristics of the obtained estimates actually makes it impossible to compare the values of the parameters of individual devices and does not allow comparing the average

Content from this work may be used under the terms of the Creative Commons Attribution 3.0 licence. Any further distribution of this work must maintain attribution to the author(s) and the title of the work, journal citation and DOI.
Published under licence by IOP Publishing Ltd

parameters for batches of devices. Therefore, the identification and investigation of statistical properties of the obtained parameter estimates is extremely important.

2. Calculating SPICE models' parameters based on experimental data

2.1. Measurement errors

In this paper, the parameters are calculated based on the full SPICE model of the diode, taking into account the additive measurement errors in the form of the following expression for the VAC.

$$I = F\,(V_{p\text{-}n}, B) + e, \tag{1}$$

where F(·) – the nonlinear function of the voltage on p-n junction Vp-n coefficient vector *B*, *B={IS, N, ISR, NR, IKF, VJ, M, RS}T* is a vector of model parameters representing a set of parameters for a SPICE model of the diode for the direct branch and *B={IBV, NBV, BV, NBVL, IBVL, RS}T* for the backward branch, *e* is the absolute measurement error of the reverse current.

The absolute measurement error of current is expressed in terms of the relative measurement error *E* as

$$e = E \cdot I$$

Therefore, in model (1), the measurement error is not consistent, since it is proportional to the measured current. To reduce the measurement error to a consistency, we logarithm the value of the diode current and as a result for small relative errors we get

$$\ln(I + EI) = \ln(I) + \ln(1 + E) \cong E. \tag{2}$$

Therefore, for logarithmic current values, the absolute measurement error is equal to the relative measurement error and does not depend on the current value of the diode. Then equation (1), taking into account the expression (2), takes the form

$$\ln(I) = \ln[F(V_{p\text{-}n}, B)] + E, \tag{3}$$

where E – observation error in logarithmic current values with σ_E^2 variance, which does not depend on the current value.

2.2. Parameter calculation

The regression analysis procedure [6] can now be used to estimate the parameters of model B (3). The most plausible estimates of parameters B can be obtained using an iterative procedure as in:

$$\hat{B}^s = \hat{B}^{s-1} + \rho^{s-1}\left(\left\{\sum_{j=1}^{n} P^T\left(V_{p-n,_j}\right)P\left(V_{p-n,_j}\right)\right\}^{-1}\right)^{s-1} \times$$
$$\times \sum_{j=1}^{n} P\left(V_{p-n,_j}\right)\left\{ln(I_j) - ln\left[F\left(\hat{B}^{s-1}, V_{p-n,_j}\right)\right]\right\}, \tag{4}$$

where s – iteration index, ρ – coefficient chosen for the best convergence of the procedure, $P(V_{p-n})$ – a vector function defined by derivatives of the $F(\cdot)$ function as follows:

$$P(V_{p-n}) = \left\{\frac{1}{F(B,V_{p-n})}\left[\frac{\partial F(B,V_{p-n})}{\partial B}\right]\right\}|B = \hat{B}.$$

After receiving parameter estimates as an estimate of the variance of the observation error $\sigma_E{}^2$, the $s_E{}^2$residual variance is calculated as in:

$$s_E{}^2 = \sum_{j=1}^{n}\{ln(I_j) - ln[F(\hat{B}, V_{p-n,j})]\}^2.$$

The covariance matrix of parameter estimates is calculated using the formula

$$V_B \approx s_E{}^2\{\sum_{j=1}^{n} P^T(V_{p-n,j}, \hat{B}) P(V_{p-n,j}, \hat{B})\}^{-1}. \quad (5)$$

The presence of a covariance matrix (5) of parameter estimates allows us to obtain confidence intervals for each parameter and compare the values of the corresponding parameters for different diodes in the batch. If some parameters for different diodes are statistically indistinguishable, this means that they are not informative in terms of, for example, their use in solving diagnostic problems.

2.3. Initial approximations

Procedure (4) is very sensitive to initial approximations of parameters and is divergent if the initial approximation is not selected correctly. Therefore, to obtain estimates of the initial approximation, a statistical modeling method was used, including the following set of steps.

1) for each parameter, the interval for its change is defined.

2) a random vector of B parameters, whose elements are in the allowed ranges, is generated.

3) for the obtained values of B parameters, the sum of deviation squares from the model is calculated as in

$$\Delta(B) = \sum_{j=1}^{n}\{ln(I_j) - ln[F(B, V_{p-n,j})]\}^2. \quad (6)$$

4) Steps 2–3 are repeated a specified number of times, and the B^0 initial approximation of the parameters is determined from the condition

$$\Delta(B^0) = \min_B \Delta(B). \quad (7)$$

To obtain a sufficiently reliable initial approximation, the number of statistical tests must amount to at least 106–107.

When selecting the initial parameter values for the reverse branch of the VAC, it is necessary to determine the breakdown voltage with an accuracy of at least ± 0.001% using experimental methods, since when the value of the parameter BV differs from the true value by more than 0.001%, the phenomenon of overflow appears when calculating the reverse current of the diode.

2.4. Diode current calculation

The calculation of parameter estimates by formula (4) assumes at each iteration s the calculation of the diode current in accordance with equation (1) $I = F(Vp\text{-}n, B, I)$, the right part of which depends on the current value I. The most stable solution of this equation is provided by an iterative procedure as in

$$I^r = I^{r-1} + k[I^{r-1} - F(\hat{B}, V_{p-n}, I^{r-1})], \quad (8)$$

where r – iteration index.

The value of the k coefficient affects the stability of the iterative procedure (8), which is especially important in this case, since this procedure is included as an internal loop in the procedure (4). Studies of the procedure stability (8) for various parameter values have shown that this iterative procedure may diverge or loop in some situations. According to the results of the conducted research, the k coefficient was chosen to be equal to π/12 in order to ensure absolute convergence and eliminate looping of the procedure (8). As an initial approximation, the current value was taken as

$$I^0 = F(\hat{B}, V_{p-n})\big|_{RS=0}$$

3. Experiment plan for estimating parameters

Usually, the minimum number of measurements was assumed to be 30–40. The location of the measurement points (the experiment plan) *VD, j (j=1, ..., n)* in the measurement range is very critical. Matrix F

$$\Phi = \sum_{j=1}^{n} P^T\left(V_{p-n,_j}, \hat{B}\right) P\left(V_{p-n,_j}, \hat{B}\right) \tag{9}$$

It is called the information matrix of the experiment and actually determines the value of the covariance matrix of parameter estimates (5). If the measurement points are not located correctly, the information matrix may be poorly conditioned, which is often the reason for failures when trying to calculate parameter estimates using the formula (4). Obviously, the wider the measurement range (planning area), the more accurately the parameters are estimated, so the maximum value of the VD voltage was determined in accordance with the permissible power dissipation of the diode. The experiment plan in this case is a set of control points $O = \{V_{D,1}, V_{D,2}, \ldots, V_{D,n}\}$, where n – the number of measurements.

The D-optimality criterion assumes such an arrangement of measurement points in the Σ=($V_{D,min}$÷$V_{D,max}$) planning area that the determinant of the V_B matrix is minimal, that is, the D-optimal plan minimizes the volume of the ellipsoid dispersing of coefficient estimates as in $D = det\{V_B\}$.

Thus, the D-optimal plan is expressed as follows:

$$det\{V_B(O^o)\} = \min_{B\in\Sigma} det\{V_B(B)\} = \min_{B\in\Sigma} det\{(\Phi)^{-1}\} \tag{10}$$

where the F value is determined by the expression (9), and $O^o = \{V_{D,1}{}^o, V_{D,2}{}^o, \ldots, V_{D,n}{}^o\}$ – optimal plan in terms of the D-optimality criterion.

In this paper, the optimal experimental plans were obtained by numerical methods, by minimizing the average parameters determinant of the covariance matrix parameters.

Taking into account the large dimension of the optimization problem (30–40 variables), the following algorithm for finding the optimal plan of the experiment is proposed.

1) Set the number of control points n and define an arbitrary initial plan of the experiment $O^1 = \{V_{D,1}^1, V_{D,2}^1, \ldots, V_{D,n}^1\}$.

2) Select the first control point to search for a particular extremum $V_{D,i}$, (i = 1, …, n).

3) Solve a particular extreme problem for one variable point of the $V_{D,i}$ control and for other fixed points, design a plan O^2as in

$$O^2 = \left(V_{D,1}^2, V_{D,2}^2, \ldots V_{D,i}^2 \ldots, V_{D,n}^2\right) = \min_{V_{D,i}\in\Sigma} \overline{det\{(\Phi)^{-1}\}} \tag{11}$$

4) Repeat the procedure in item 4 for all points of the plan.

5) Repeat items 3–4 until the specified calculation accuracy is reached, the specified number of cycles is obtained, or the specified counting time expires.

The calculations performed for the construction of optimal experimental plans showed that their use increases the accuracy of estimating the parameters of the direct branch of the VAC by an average of 2–3 times compared to the uniform location of control points and with the same number of measurements.

Optimal planning is particularly effective when estimating the parameters of the reverse branch. In this case, as a rule, about 10% of the control points are located on the edges of the planning area, and the vast majority of the remaining plan points are grouped in a very short area with a length of approximately (0.2–0.5)% of the entire planning area adjacent to the $V_{D,max}$. point. The $V_{D,max}$. point itself must exceed the V_B parameter by (10–15)% , which is determined by the permissible dispersing power of the diode, otherwise this parameter will not be able to be determined accurately.

Thus, using a uniform plan with a number of measurement points for the reverse branch, even 200–300, often makes it impossible to calculate all the parameters of the SPICE model, since very few or even none of the measurement points fall into the critical planning area. At the same time, using the optimal plan in this situation allows you to calculate all the parameters with accuracy (1–10) %, even for 30– 40 measurement points.

4. Measurement and calculation methods for the SPICE-model's parameters of the diode

Based on the results obtained, the methodology for measuring and calculating the parameters of the SPICE model of the diode can be presented in a fairly general form by the following sequence of operations.

1) Determination of possible changes' range in the parameters of the SPICE model for the forward and reverse branches of the VAC for a given type of diode based on a priori information.

2) Determination of the optimal experimental plans for the forward and reverse branches of the VAC based on the expression (10) and the described procedure (11). The designed experiment plan will be used for all diodes of this type.

3) Performance of measurements of the current of a specific diode at specified voltages in accordance with the designed experimental plan.

4) Calculation of the initial approximation of parameters by statistical modeling in accordance with expressions (6) and (7). At this stage, you may need to adjust the parameter ranges.

5) Calculation of parameter estimates by formula (4) and covariance matrix of parameters by formula (5). At this stage, you may need to adjust the initial approximation of the parameters [7].

5. Conclusion

The measurement and calculation methods of the SPICE-model's parameters of the diode described in this paper were used to calculate the parameters of the SPICE-model of three types of diodes: KD221, KD411 and KD213. The experimental plans were designed for each type of diodes (thus, three such plans were obtained in total). All parameters of the SPICE model for the forward and reverse branches of the VAC and their covariance matrices were calculated for each diode. The accuracy of parameter estimation ranged from 0.05% to 1%, depending on the type of a parameter. Thus, the proposed method provides a fairly accurate estimation of all parameters of the SPICE model and their covariance matrix for different types of diodes, which confirms its practical significance.

References

[1] Wiele van de F, Engl W L and Jespers P 2012 *Process and Device Modeling for Integrated Circuit Design* (Netherlands: Springer) p 876

[2] Doganis K and Scharfetter D L 1983 *IEEE Trans. on Electron Devices* vol ED-30 **9** 1219

[3] 2005 *The Design Center. Circuit Analysis Reference Manual* (MicroSim Corp.)

[4] Paasschens J C J 2001 *Parameter extraction for the bipolar transistor model Mextran, level 504* Unclassified Report NLUR 2001/801 (Philips Nat. Lab.)

[5] McAndrew C C 1998 Practical modeling for circuit simulation *IEEE Journal of SolidState Circuits* vol 33 **3** 439–448

[6] Bard J 1979 *Nelineinoe otsenivanie parametrov* [Nonlinear parameter estimation] ed V P Gorsky (Moscow: Statistika Publ) p 349 [In Russ.]

[7] Popov S A, Petrov M N, Korchagin A F, Kapralov A S and Kozhevnikova V A 2005 Otsenivanie parametrov ekvivalentnoi skhemy mnogopoliusnikov s pomoshch'iu izmeritel'nogo kompleksa [Estimation of an equivalent multipolar circuit parameters using a measuring complex] *Information Technologies and Mathematical Modeling (ITMM-2005): Proceedings of the IV All-Russian Scientific and Practical Conference (18–19 November 2005 г.)* (Tomsk: Tomsk State University Publ.) part 1 p 143 [In Russ].

IOP Conf. Series: Materials Science and Engineering **939** (2020) 012038 doi:10.1088/1757-899X/939/1/012038

A method of processing semi-finished products from an alloy of titanium nickelide TN-1

S V Kryuchkov[1,4], V G Malinin[2], N P Bogdanov[1], M N Konovalov[1] and O A Malukhina[3]

[1] Ukhta State Technical Uiversity, 13, Pervomayskaya ul., Ukhta, 169300, Republic of Komi, Russia
[2] Orel State Agrarian University, 69, General Rodin ul., Orel, 302019, Russia
[3] Yaroslav-the-Wise Novgorod State University, 41, B. St. Peterburgskaya ul., Veliky Novgorod, 173003, Russia

[4] E-mail: kruk06@yandex.ru

Abstract. An experimental method was obtained for treating semi-finished products from titanium nickelide alloys TN-1 by twisting a cylindrical sample in the austenitic state at a temperature of T = 493 K to a predetermined angular strain, after which the sample was rigidly clamped, fixing the angular deformation of the sample, and cooled to a temperature of T = 293 K. At a given temperature, the rigid jamming of the sample is eliminated, while the isothermal return of the part of the deformation obtained by twisting in the austenitic state occurs. Then the sample is heated to a temperature of T = 493 K, and there is a shape memory effect (SME), manifested in the form of restoration of angular deformation. A systematic experimental study was carried out of the dependence of the influence of the complex regime of thermal-mechanical action containing the stages of generation and relaxation of stresses on the shape memory effect. It has been experimentally established that SME during heating after cooling in the direct MT range under pinched conditions can be observed with zero deformation effect in the final heating interval.

1. Research Methods

Non-traditional functional and mechanical properties of shape memory alloys (SMAs) are used to solve numerous practical problems: in power equipment, in robotics elements, in medicine, etc. [1]. The variety of various options for using materials with shape memory in industry makes it possible to talk about the relevance of the problem. In the vast majority of cases, materials when implementing reversible shape memory exhibit qualitatively similar behavior. The latter circumstance allows us to study the general laws of the effects of martensitic inelasticity on specific model materials, such as TiNi. Products from this material are difficult to manufacture. Therefore, the process of setting the desired shape for the semi-finished product is an important technological stage in the production of devices from alloys with the shape memory effect.

This article discusses a method for producing an SME by heating after cooling in the range of direct martensitic transformation (MT) in pinched conditions with zero deformation effect on the final heating interval.

Content from this work may be used under the terms of the Creative Commons Attribution 3.0 licence. Any further distribution of this work must maintain attribution to the author(s) and the title of the work, journal citation and DOI.
Published under licence by IOP Publishing Ltd

Several methods are known for processing semi-finished products for materials with thermo-elastic martensitic transformations (TMT), which allow the formation of the shape memory effect (SME).

The first method is plastic deformation of the material by thermal cycling through the intervals of martensitic transitions of the TN-1 alloy under a constant voltage of various sizes [2].

The second method consists in plastic deformation of the material by thermal cycling through the intervals of martensitic transitions of the TN-1 alloy under constant voltage at the stage of cooling the material, and in the unloaded state at the stage of heating [3].

The closest analogue to the processing of semi-finished products, which is considered in this post, is the third method of processing semi-finished products from alloys with thermoelastic martensitic transformations – includes preliminary thermocyclic training of an alloy with thermoelastic martensitic transformations in torsion mode.

Thermal cycle mode: heating in the unloaded state in the temperature range from T = 298K to T = 500K, and cooling to T = 298K is performed under load [4].

Analyzing the data of three methods of processing materials with TMT, we can conclude that in all methods after training, there is a noticeable reversible deformation, manifested in the form of SME at the heating stage and its accumulation at the cooling stage, in the third method there were deformation effects associated with SME at heating in the unloaded state, with plasticity of direct transformation (PDT) during cooling under load through the interval of direct martensitic transformation (MT).

2. Research Results

The technical result of the experiment is to obtain an SME upon heating after cooling in the interval of direct MT in pinched conditions with zero deformation effect on the final heating interval.

This result is achieved by the method of processing semi-finished products from a titanium nickelide alloy TN-1 by pre-cooling and heating through the intervals of direct and reverse martensitic transformations in the torsion mode. The difference from the considered methods of processing materials is that in the austenitic state at a temperature of T = 493 K, the sample is isothermally twisted to a predetermined angular deformation value, after which the sample is rigidly clamped, fixing the angular deformation of the sample, allowing it to move freely in the axial direction. Then the sample is cooled to a temperature of T = 293 K, where the rigid jamming of the sample is eliminated. In this case, the isothermal return of the part of the deformation obtained by twisting in the austenitic state occurs. Subsequent heating of the sample to a temperature of T = 493 K is accompanied by a shape memory effect, manifested in the form of restoration of angular deformation.

For testing, material was selected in the form of continuous cylindrical samples from an alloy close to equiatomic TH-1 with a length and diameter of the working part of 33 and 4 mm, respectively. The characteristics of temperature martensitic transitions were determined under torsion conditions at a shear stress of $\tau = 50$ MPa and amounted to $M_H = 350$ K, $M_K = 300$ K, A_H=360 K, A_K=420 K.

Before the test, the samples were annealed (in a muffle furnace in a special container) at a temperature of 723 K for one hour, followed by cooling to room temperature in air. The experimental part of the work was carried out on a special installation designed and manufactured at Ukhta State Technical University [5–7].

Before each experiment, 3–4 thermal cycles in the free state were carried out with each sample. The readiness of the sample for testing was evidenced by the absence of residual effects of reversible deformation on the resulting diagram $\gamma = \gamma$ (T).

The sample with the right end is fixed with two screws in the grip, rigidly connected to the right support. The left end of the specimen is fixed in the same way at the end of the shaft, which is able to freely rotate and move axially in the bearings of the bearings. The torque is transmitted through a pulley, rigidly fixed to the shaft by a keyed connection, on which a strong thread is wound with a load suspended at the end, while it is possible to change the direction of twisting of the load. The twist angle of the sample is determined by the scale of the measuring device. A computer program controls the measurement of the angle of rotation of the cross section of the sample φ and determines the relative twist angle according to the formula $\theta = \varphi/l$. In this case, the accuracy of measuring the angle φ was

0.009 rad, and the torque was 0,01 N · m. The installation allows the sample to report a torque of up to 5 N · m. Heating was carried out by an electric furnace. The heating and cooling rate of the sample is set by the computer and is regulated by the sensor control program based on a calibrated chromel-kopel thermocouple, the junction of which was connected to the working part of the sample. To control the uniformity of heating the sample, two thermocouples were installed at different points of its working part. The sensor at a given heating rate records the temperature with an accuracy of 2 K.

The calculation results were considered as the average of the measurements for three samples made from the same bar with a relative error not exceeding 0.1%. When assessing errors, we took into account both random errors of multiple (obeying the Gaussian distribution) and errors of single (obeying the uniform distribution) measurements. To determine the total error of the result, the law of addition of independent quantities (errors) was used, which is also true for the addition of confidence intervals. Therefore, the confidence interval of the quantity x measured in a series of experiments is written as follows, $\Delta\tilde{x} = \sqrt{\Delta\tilde{x}_{re}^2 + \Delta\tilde{x}_{em}^2}$, $\Delta\tilde{x}_{re}$ – the confidence interval corresponding to the random error of multiple measurements, $\Delta\tilde{x}_{em}$ is the confidence interval corresponding to the error of single measurements. For a confidence probability of $\alpha = 0.95$ and three times the measurement, the Student coefficient was chosen equal to a value of 4.3.

In addition to the integral characteristics of M and φ, we used calculated estimates of the values of shear strains and shear stresses determined in the approximation of an ideal plastic body [8, 9] and assigned to the external fiber of the working part of the sample.

At the same time, the following working hypotheses were used as the basis for the estimated estimate:

1. The hypothesis of flat sections is fulfilled: the cross section of the beam – flat and perpendicular to the longitudinal axis, remains flat and perpendicular to the longitudinal axis even after deformation.
2. The radius drawn from the center of the cross-section of the beam after deformation remains a straight line (does not bend).
3. The distance between the cross sections, the diameters of the cross sections after deformation do not change, and the axis of the beam is not curved.

To obtain the result of an experimental study, a sample in the austenitic state at a temperature of 493 K was loaded with a given torque, setting the corresponding values of the tangential stresses and angular deformation.

The shear strain in the outer fiber was estimated by the formula (1):

$$\gamma = (d \cdot \theta/2) \cdot 100\ \%, \tag{1}$$

where γ is the angular deformation, d is the diameter of the working part of the sample, θ is the twist angle. Moreover, the error in measuring shear strain did not exceed 0.05%.

According to [9], the estimation of the tangential stresses in the approximation of an ideally plastic body during torsion of continuous cylindrical samples was determined by the formula (2)

$$\tau = 12 \cdot \mathrm{M}/\pi \cdot d^3, \tag{2}$$

where M is the torque, d is the diameter of the working part of the sample; τ – tangential stresses.

Next, the sample was rigidly clamped, with a fixed angular deformation of the sample using a device that works on the principle of a vise and consists of two flat jaws with an adjustment knob that moves the clamping screw. With it, the sponges diverge and contract, fixing the side surfaces of the pulley as tightly as possible (creating the effect of hard pinching of the sample). Elastic pads on the jaws evenly distribute the pressure force along the side surface of the pulley. The base of the device is attached using the adjusting screws to the bed of the installation, which allows you to configure the device for samples of various lengths. After this, the sample was cooled to a temperature of T = 293 K. At this temperature, the rigid jamming of the sample is eliminated, and a part of the deformation obtained by twisting in the austenitic state is isothermally returned. Then the sample in a free state is heated to a temperature of T = 493 K, and at the same time there is a shape memory effect, manifested in the form of restoration of angular deformation. In the coordinates of the angular deformation - temperature, the dependences of the deformation responses for this method of processing the semi-finished product are presented (Figure 1). Here: γ_A is the angular deformation due to the load of the sample at T = 493 K. γ_M is the deformation due to unloading at T = 293 K. γ_{SME} is the deformation due to the shape memory effect when heated in the temperature range 293–493 K.

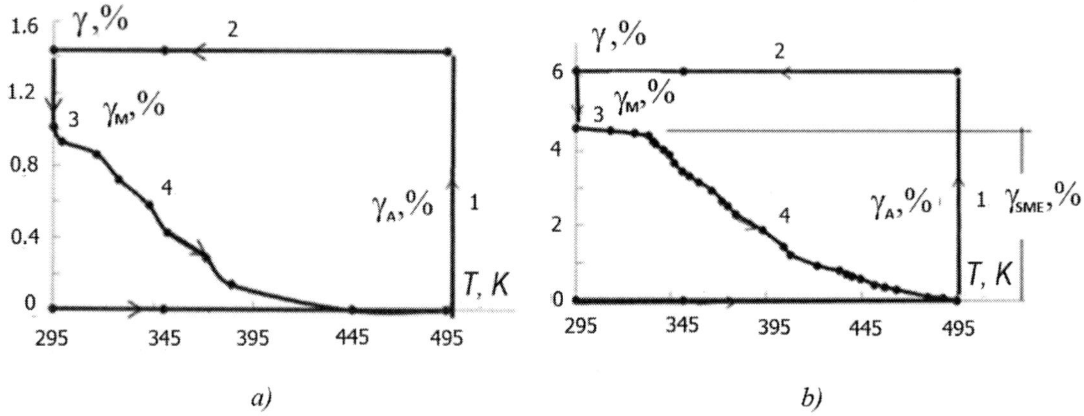

Figure 1. Temperature dependence of angular deformation for loading stages: (1) loading in austenite to τ = 100 MPa (a) and τ = 300 MPa (b) and pinching; (2) – cooling in the temperature range 493 ÷ 293 K in a pinched state; (3) – removal of pinching condition; (4) – heating in the temperature range 293 ÷ 493 K in the free state.

It should be noted that the repeated cycle leads to a similar "picture", which cannot be said for other methods of processing semi-finished products, where during the thermal cycling through the intervals of martensitic transitions in the loaded state noticeable irreversible deformations accumulate. As a rule, irreversible deformations accompanying martensitic transitions are undesirable and, as a result, worsen the dimensional stability of SME materials. It has been experimentally established here that the production of SME upon heating after cooling in the direct MT interval under pinched conditions is observed with zero deformation effect in the final heating interval.

Table 1 below shows the deformation responses as the average of the measurements for three samples at different levels of loading intensity.

Table 1. Deformation responses at various levels of loading intensity.

Deformation response	Loading Intensity τ, МПa					
	100	200	250	300	350	400
γ_A, %	1.44±0.06	3.52±0.08	4.60±0.09	6.11±0.13	10.20±0.15	11.65±0.15
γ_M, %	0.43±0.03	0.65±0.05	0.58±0.06	1.51±0.08	1.60±0.08	3.31±0.10
γ_{SME}, %	1.01±0.03	2.87±0.05	4.02±0.07	4.60±0.07	7.60±0.10	8.34±0.11

3. Conclusion

As mentioned above, the process of setting the desired shape for the semi-finished product is an important technological stage in the production of devices from alloys with the shape memory effect. The application of the proposed method is possible in metallurgy, in particular to products from alloys with a shape memory effect, and can be used in power engineering, construction, instrument making, and medicine.

References

[1] Tihonov A C, Gerasimov A P and Prohorova I P 1981 *Primemnenie Effecta Pamjati Formi v Sovremennom Mashinostroenii* [Appplication of the Shape Memory Effect in Modern Engineering] (Moscow: Mashinostroenie) p 80

[2] Andronov I N, Bogdanov N P, Vlasov V P and Severova N A 1998 *Bulletin of Tambov State University* Vol 3 **3** 236–238

[3] Andronov I N, Verbakhovskaya R A, Ovchinnikov S K and Severova N A 2007 Effects of reversible shape memory and thermocyclic return of deformation in TN-1 alloy *Zavodskaya Laboratoriya* [Industrial Laboratory] vol 73 **2** 64–67

[4] Patent 2310696 2007 MPKC22F 1/18, publ. 20.11.2007. *Bul.* **32**

[5] Andronov I N, Churilina I V, Kryuchkov S V and Bogdanov N P 2019 The effect of preliminary thermal cycling under load on the magnitude of the deformations driven by the shape memory effect *Zavodskaya Laboratoriya. Diagnostics of materials* vol 85(7) 55–63 [In Russ.]

[6] Kryuchkov S V Bogdanov N P, Malinin V G, Savich V L and Konovalov M N 2019 *IOP Conf. Ser.: Mater. Sci. Eng.* **656** 012029

[7] Vlasov V P, Andronov I N, Kakulia Yu B 1996 Utility Model Certificate No. 1538 of the Russian Federation, G01N 3/08. Installation for testing samples of materials with a complex stress state Application 94007969/28, 05. 03. 1994, publ. 01/16/1996

[8] Likhachev V A, Kuz'min S A and Kamentseva Z P 1987 *Effect Pamyati Formy* [Shape Memory Effect] (Leningrad: LGU) p 216

[9] Likhachev V A and Malinin V G 1993 *Strukturno-analiticheskaya Teoriya Prochnosty* [Structural Analytic Theory of Strength] (Saint-Petersburg: Nauka) p 471

IOP Conf. Series: Materials Science and Engineering **939** (2020) 012039 doi:10.1088/1757-899X/939/1/012039

Assessment of the anisotropic wood strength on local crushing

B V Labudin[1], T A Nikitina[1,3], E V Popov[1], K A Varenik[2] and V I Novoselova[1]

[1] Northern (Arctic) Federal University named after M.V. Lomonosov, Severnaya Dvina Emb., 17 163002, Arkhangelsk, Russia

[2] Yaroslav-the-Wise Novgorod State University, B. St.-Peterburgskaya St., 41 173003, Veliky Novgorod, Russia

[3]E-mail: t.a.nikitina@inbox.ru

Abstract. Modern structural analysis standards for wooden and glued laminated building structures of civil and industrial buildings, bridges, electric power facilities, communications, etc., on the one hand, have differences, and on the other hand, general design principles (SP). However, the design resistances values are given only for the force vector acting along and across the fibers. The designer has to get the intermediate design resistances values according to the formulas of the 30s, very approximately, and sometimes incorrectly. For example, for cross laminated timber (CLT). Based on previously performed experimental researches and the results of calculations, general patterns of the resistance to local crushing changes were obtained for samples taken from glued laminated timber structures elements (glulam). General approaches have been developed to assess the strength of glulam ($\sigma_{cr,\alpha}$) and its design resistances to local crushing ($R_{cr,\alpha}$) with varying static geometrical parameters of the hard stamp ($\psi = d/l$) depending on the external forces vector direction with respect to the wood fibers – lamellas (α). All this makes the research present-day and focused on improving reference documents.

1. Introduction

It is known that the physicomechanical properties of wood were first studied in the 19th century by Savart (1830) and Saint-Venant (1856), taking it as orthotropic material in experiments on torsion, bending, and compression. In Russia, D.I. Zhuravsky for the first time standardized the wood strength characteristics for the analysis of the Nikolaev railway wooden bridge trusses.

This question arises again with the advent of new wood-composite materials, since modern standards [1] do not reflect the entire completeness of parameters, impacts and operating conditions. The need to solve this issue is also due to the need to analyze the crushing area in the wooden structures nodes [2]. Most of the researches devoted to contact tasks relate to dowel pin and multi-pin connections, connectors in timber and glulam structures.

The following hypotheses and assumptions are made in the article for a layered anisotropic material that is heterogeneous in structure: glued laminated timber is considered as a homogeneous material due to the infinitely small thickness of the adhesive layers of the glulam array; the difference in elastic characteristics in lamellas (layers), randomly or organizedly oriented in the cross section of the glulam package, are within the statistical error, regardless of the axes t, r or t_r, i.e. we average the physical and mechanical characteristics by the influence of local defects in glued laminated timber (knots, slanting, non-glueing, etc.) that are within the normal range, we neglect it; stress and strain components are

Content from this work may be used under the terms of the Creative Commons Attribution 3.0 licence. Any further distribution of this work must maintain attribution to the author(s) and the title of the work, journal citation and DOI.
Published under licence by IOP Publishing Ltd

assumed to be connected linearly, i.e. Hooke's law is fair; initial stresses arising during the manufacturing and operation (temperature, humidity, etc.) are taken into account at the stage of constructive analyze according to the standarts. Thus, the physical model will correspond to a particular case of anisotropy – the transtropic object [3, 4, 5].

Research goal: to obtain standard and design strength characteristics for glued laminated timber under local stamped load impacts.

2. Research methodology

Testing of timber from factory elements of glulam structures was carried out on prismatic samples with dimensions b x l = 60 x 100 mm at the base and height h = 90 mm (figure 1). Lamella thickness was 20 ± 0.5 mm, use FRF-50 glue. The stamp dimensions to the sample width ψ ratio was 0.25; 0.50; 0.75; 1.00; the inclination angle of the effective load in relation to the layers α ranged from 0° to 90° (0°; 15°; 22,5°; 30°; 45°; 60°; 67, 5°; 90°).

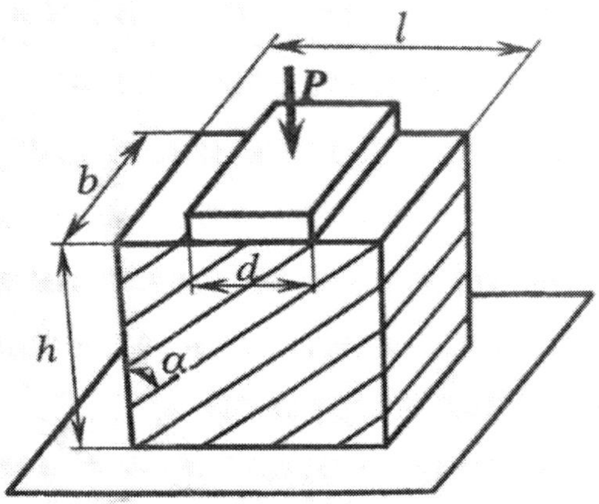

Figure 1. The shape and dimensions of the glulam timber samples.

For the test we used a press MUP-50 with a 25-ton scale. The loading rate was within 15 ± 2,5 kN/min. Each point had at least 5 samples. During testing, for each sample, the deformation of the stamp relative to the stationary part of the installation was measured. The deformation of the connection was measured using a dial indicator with graduations of 0.01 mm. The indicator was attached to the fixed bed of the machine, and the pin abutted against the movable element, which fit snugly against the stamp and moved with it. The range of the indicator in 1 cm allowed to leave it on the sample until the latter was destructed.

The start of the samples destruction was recorded on a testing machine with a distinctive stop of the load growth. At $\alpha < 45°$, the start of the destruction moment matched to a further load decrease. At $\alpha > 45°$, a further load increase began after the destruction moment.

3. Results view

According to the tests, the destruction is mainly in the areas adjacent to the stamp. Forms of destructions are very diverse and depend on the angle between the fibers directions and the force action, the complex and very heterogeneous material under the stamp, the heterogeneity of the wood structure, the initial micro destructions and defects that occur during the making of samples.

The destruction of the samples occurred as a result of the longitudinal bending of separate fibers, the indentation of late wood into earlier wood, tearing across the fibers and splitting. With an increase in α to 45°, the destruction occurred as a result of tearing across the fibers and splitting, and for α more than 45° as a result of deformation of weaker layers, i.e. by analogy with compression across the fibers. It is specific to "clean" wood. Therefore, the moment of samples destruction from α to 45 can be considered as the shear strength, and the transition to the area of the non-linear relationships between deformations and load can be considered for samples with α more than 45°. The test results are given in the work [6].

In assessing the results of statistical processing of the glulam timber strength to local crushing it can be observed that there are high accuracy rates (p = 1.5 ... 6.7%) obtained over the entire range of angles α and variation coefficients that do not go beyond 20%.

It was possible to identify regularities of change in the design resistance to crushing at an angle to the fibers (lamellas) $R_{cr,\alpha}$ after statistical processing of the results and determination of the average values of tensile strength $\sigma_{cr,\alpha}$ depending on the ratio ψ and angle α.

For this purpose, we marked the mechanical characteristics $\sigma_{cr,\alpha}$ for Y and the stamp dimensions to the sample dimensions ψ ratio for X. We got the relationship equations between the values of Y and X by using the method of least squares (OLS).

To address this issue we used a set of standard programs to get the unknown coefficients values A and B. The approximating expressions, some of which are linear equations, and some are non-linear equations (exponential and logarithmic):

$$Y = AX+B; \quad Y=AX^B; \quad Y=AX^X; \quad Y = A + B/X;$$
$$Y = 1/(AX + B); \quad Y = X/(AX + B); \quad Y = AlnX + B. \tag{1}$$

The experimental pairs X and Y were used to solve each of the equations. 16 relationship equations were considered. The values of X, Y, α, coefficients A, B for seven relationship equations are given in this research.

According to the table the following equations $Y = AX+B$ (linear) and $Y = AlnX + B$ (non-linear) are most accurately show the quantitative changes in $\sigma_{cr,\alpha}$ depending on ψ for constant environmental parameters (t = 20°C, W = 60%). Based on the smallest error for the equation $Y = AlnX + B$ was obtained at angles of 45° ... 90°, we take this relationship equation as the baseline one for subsequent calculations.

We obtained a function for defining of tensile strength (crushing) of glulam timber for various ψ and α by using the numerical coefficients values of the relationship equations and the double approximations method.

$$\sigma_{cr,\alpha} = B + (-AlnX), \tag{2}$$

here $B = Ne^{-n\alpha} = 47.352e^{-0,0324\alpha}$; $A = M\alpha^{m} = 34.173\alpha^{-0,462}$.

In formula (2), the second term ($AlnX$) takes into account an increase in the tensile strength of glulam timber with a decrease in the relative crushing area size depending on the angle α. For $X = d / l$ = 1, the second term is equal to zero, and expression (2) determines the ultimate compressive strength (crushing) over the entire surface. For d/l < 1, it has negative values. Of course, all the changing parameters (ψ, α) indicate a sharp difference between the obtained the strength and anisotropy values of glulam elements.

Table 1. The approximating relationship equations of the tensile strength (crushing) with changing the relative stamp dimensions and fiber (lamellas) angle.

α, degree	X	Y	Approximating equations	Equation error
			$Y = -16.000X + 58.000$	0.000
			$Y = 58.919 \cdot 0,715^X$	0.1683
	0.25	54.0	$Y = 1/(0.007X + 0.0166)$	0.3445
0.0	0.50	50.0	$Y = -8.420 \ln X + 43.017$	0.8858
	0.75	46.0	$Y = 43.090X^{-0,175}$	1.0494
	1.00	42.0	$Y = 40.564 + 3.569/X$	1.6579
			$Y = X/(0.0255X + 0.0023)$	3.7320

			$Y = -15.829 + 38.786$	1.7899
			$Y = 40.691 \cdot 0.562^X$	2.0735
	0.25	34.0	$Y = 1/(0.0214X + 0.0226)$	2.4268
22.5	0.50	–	$Y = -7.781\ \ln X + 23.891$	2.4406
	0.75	29.4	$Y = 23.701X^{-0.281}$	2.26102
	1.00	21.3	$Y = 21.327 + 3{,}271/X$	2.8765
			$Y = X/(0.0504X + 0.0070)$	6.9281
			$Y = -15.520X + 31.350$	1.2508
			$Y = 34{,}216 \cdot 0.462^X$	1.6707
	0.25	26.2	$Y = 1/(0.0395X + 0.0240)$	2.3224
30.0	0.50	24.9	$Y = -7.697\ \ln X + 17.095$	2.0800
	0.75	20.9	$Y = 16.891^{-0.378}$	2.4301
	1.00	14	$Y = 15.246 + 3.074/X$	2.7546
			$Y = X/(0{,}0771X + 0{,}0147)$	14.3296
			$Y = -9.880X + 22.200$	0.9060
			$Y = 22.971 \cdot 0.548^X$	0.7423
	0.25	20.5	$Y = 1/(0.0371X + 0.0411)$	0.6165
45.0	0.50	16.7	$Y = -5{,}510\ \ln X + 12.764$	0.4668
	0.75	13.6	$Y = 12.960^{-0.378}$	0.4184
	1.00	13.3	$Y = 10.123 + 2.454/X$	0.5412
			$Y = X/(0.0857X + 0.0104)$	1.3024
			$Y = -9.160X + 12{,}600$	0.6417
			$Y = 14.758 \cdot 0.263^X$	0.2958
	0.25	11.0	$Y = 1/(0.2108X + 0.0353)$	0.1981
67.5	0.50	7.2	$Y = -5.064\ \ln X + 3.878$	0.1243
	0.75	5.3	$Y = 4.187^{-0.717}$	0.2548
	1.00	4.0	$Y = 2.192 + 2.248/X$	0.3542
			$Y = X/(0.3015X + 0.0675)$	10.4701
			$Y = -7.680 + 10.300$	0.4550
			$Y = 12.344 \cdot 0.2415^X$	0.1585
	0.25	8.8	$Y = 1/(0.2850X + 0.0343)$	0.4030
90.0	0.50	6.1	$Y = -4.205\ \ln X + 3.012$	0.1489
	0.75	4.0	$Y = 3.247^{-0.756}$	0.3928
	1.00	3.1	$Y = 1.654 + 1.846/X$	0.4468
			$Y = X/(0.3950X + 0.0919)$	13.7157

Based on the results expression (2) can be written in the general equation form (3) that allows to calculate the changing values of $\sigma_{cr,\alpha}$ in the presence of data at three characteristic points for angles 0°, 45°, 90°, plot the graphs on these points:

$$\sigma_{cr,\alpha} = \frac{\sigma_{cr,0}}{1+\beta\alpha^m} + (-D ln\psi) \tag{3}$$

here $\sigma_{cr,\alpha}$ is the compressive strength (crushing) of glulam timber along the fibers, MPa; α is the inclination angle of the external force vector with respect to the direction of the fibers (layers); ψ – relative crushing area sizes, $\psi = d/l$; $\beta = (c - в)\sigma_0/45^m$; $c = 1/\sigma_{45}$; $в = 1/\sigma_0$; $m = \log(a - в)/(c - в)$; $a = 1/\sigma_{90}$; D is a coefficient that takes account of the strength change depending on the angle α, (obtained after processing the tests results); $D = 100 / (p\alpha + f) = 100 / (0.164\alpha + 9.15)$.

We used expression (3) to plot the dependence which show the relationship between $\sigma_{cr,\alpha}$ and the relative stamp dimensions ψ and α (figure 2). The graphs data clearly show the boundary character of the angle $\alpha = 45°$, because the concavity of the curves changes to convexity when passing through this value. Also this is an additional confirmation of the transition from brittle fracture to plastic fracture.

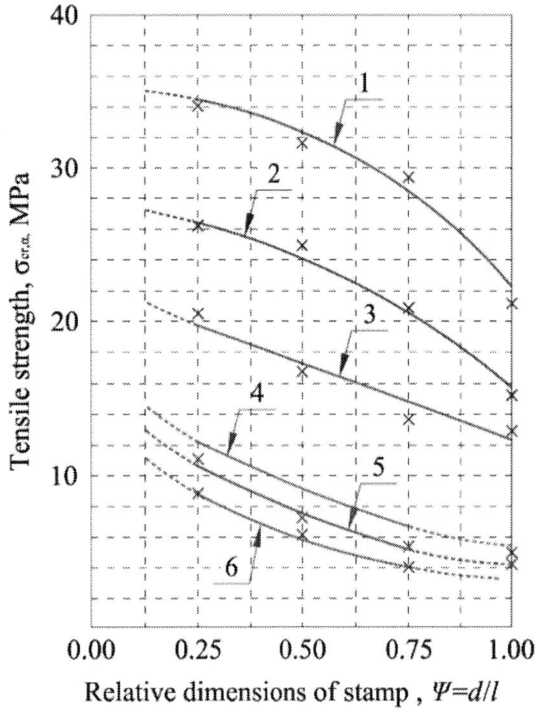

Figure 2. The dependence of the glulam timber tensile strength and ψ for various angles α: 1 – 22.5°; 2 – 30.0°; 3 – 45.0°; 4 – 60.0°; 5 – 67.5°; 6 – 90.0°.

Due to the fact that we conducted limited tests along the fibers (lamellas), we will use our data and take conditionally $\sigma_0 = 42$ MPa to draw the anisotropy curves using the formulas [1] and the formulas proposed in the present researches. The calculation results in comparison with the tests for $\psi = 1.0$ are given in table 2. Data analysis shows we obtained the closest values in almost the entire range of angles with our formula (3). Results according to the formulas of E.A. Ashkenazi [3] have quite acceptable values, which are widely used in modern scientific and technical literature to assess the strength of wood and wood materials.

Table 2. Comparison of test and design data of the wood ultimate tensile strength.

№	$\sigma_{cr,\alpha}$, MPa	α, degree							Note
		0	15	30	45	60	75	90	
1	Test data	42.0	27.9	14.6	12.3	4.6	4.0	3.1	Table 1 [6]
2	$\sigma_{cr,\alpha} = \frac{\sigma_{cr,0}}{1+\beta\alpha^m} + (-Dln\psi)$	42.0	28.1	15.3	9.0	5.8	4.1	3.0	Formula 3
3	$\sigma_\alpha = \frac{\sigma_0}{cos^4\alpha + Bsin^2 2\alpha + Csin^4\alpha}$	42.0	36.1	19.8	9.0	4.9	3.4	3.0	[3]
4	$\sigma_\alpha = \frac{\sigma_0}{1+(\frac{\sigma_0}{\sigma_{90}}-1)\,sin^3\alpha}$	42.0	34.3	16.0	7.5	4.4	3.3	3.0	SP 64.13330.2017

Based on the test data [6], and formula (3) we plotted the dependence of the anisotropy $\sigma_{cr,\alpha}$ on ψ and α (figure 3). The upper and lower boundaries of the confidence interval ($\sigma^{up}_{cr,\alpha}$ and $\sigma^{l}_{cr,\alpha}$) were defined with a confidence probability of 0.95.

$$\sigma^{up}_{cr,\alpha} = \overline{\sigma_{cr,\alpha}} + tS; \qquad \sigma^{l}_{cr,\alpha} = \overline{\sigma_{cr,\alpha}} - tS.$$

Here $\overline{\sigma_{cr,\alpha}}$ – is the arithmetic mean; S is the standard deviation; t is the student`s t-test.

Based on the graphs (figure 3) all the curves fall within the confidence interval boundaries and the actual scattering area of the test results with a probability $p_p = 0.95$. Only with relative stamp dimensions equal to 0.75 and 0.5, the curves approach the lower confidence border, which is quite acceptable, because quantitative results of σ_α belong to the material margin of safety.

Short-term values of tensile strengths can serve as the basis for obtaining the design resistance to crushing of glulam timber on the considered static effects.

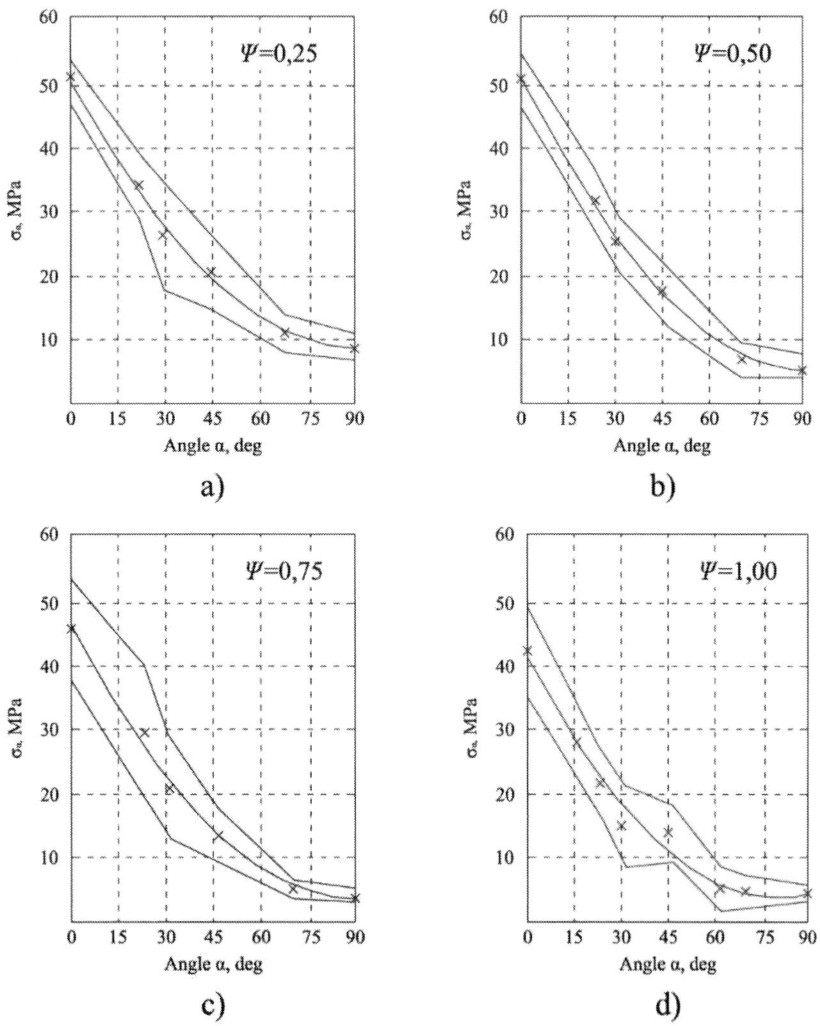

Figure 3. Dependence of the tensile strength $\sigma_{cr,\alpha}$ anisotropy of glulam timber upon crushing (compression) on the inclination angle of the layers (fibers) α and stamp dimensions ψ: *a* – 1.00; *b* – 0.75; *c* – 0.50; *d* – 0.25; 1 – average values of $\sigma_{cr,\alpha}$; 2 – curves drown by the formula (3).

In light of the above, and also assuming that the design resistances change will not qualitatively differ from the short-term resistances change of glulam timber, expression (3) can be transformed by introducing a reliability coefficient in the right-hand side, but use the standard design resistances values in the left-hand side.

Then the formula for determining the design resistance, taking into account the magnitude of the load applied through the stamp, depending on its relative dimensions and the inclination angle of the force vector with respect to the fibers (lamellas), will take the form:

$$R_{cr,\alpha} = R_0/\,(1 + \beta\alpha^m) + (-1/\,K_\alpha)\,D\,\ln\psi, \tag{4}$$

here R_0 is the design crushing resistance along the fibers, MPa; $\beta = (C - B)\ R_0/45^m$; $m = \lg\,(A - B)/\,(C - B)$; $A = 1/\,R_{90}$; $B = 1/\,R_0$; $C = 1/\,R_{45}$; $D = 100/\,(0.164\alpha + 9.15)$; $\psi = d/l$ при $(0,1 \leq \psi \leq 1)$.

We omit the methodology for calculating the design resistances [6] and give below, figure 4 shows the dependences $R_{cr,\alpha} = f(\alpha, \psi)$, plotted for five relative dimensions of the crushing area variants with standard design resistances of glulam timber R_0 = 14.0 MPa, R_{90} = 1.8 MPa, R_{45} = 4.2 MPa, taking into account the results smoothing in the range of angles α from 30° to 60°.

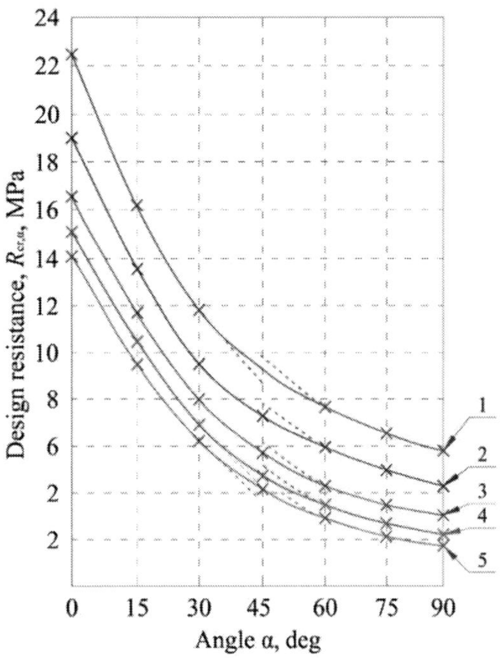

Figure 4. Change in the design crushing resistance of glulam timber at an angle α to the layers (fibers) depending on the relative dimensions of the crushing area ψ: 1 – 0.10; 2 – 0.25; 3 – 0.50; 4 – 0.75; 5 – 1.00.

4. Conclusions

1) Based on the above methodology and formulas, it is possible to obtain design characteristics for glued laminated timber and other wood-composite anisotropic materials, taking into account the size of the local impact on the crushing (compression) areas at various angles to the anisotropy axes or layers.

2) It was established experimentally that the nature of fracture during local crushing in the angle range 0 ... 45° is brittle, 45 ... 90° is plastic, which depends on the properties of the wood work under load as an anisotropic (transtropic) material.

References

[1] SP 64.13330.2017 *Derevyannye konstrukcii* [Wooden structure] updated version SNiP II-25-80 with changes [In Russ.]

[2] Barkov M S 2013 *Sovershenstvovanie uzlovyh sopryazhenij derevyannyh elementov cherez*

upory peremennoj zhestkosti dlya bol'sheproletnyh ferm: dissertaciya ... kandidata tekhnicheskih nauk: 05.23.01 [Improvement of nodal connections of wooden elements through variable stiffness stops for large-span trusse] (Krasnoyarsk) RGB OD 61 13-5/2033 p 212 [In Russ.]

[3] Ashkenazi E K 1978 *Anisotropy of wood and wood materials* (Lesn.prom-st') p 224

[4] Labudin B V 2007 *Sovershenstvovanie kleenyh derevyannyh konstrukcij s prostranstvenno-regulyarnoj strukturoj*: monography (Arhangel'sk: Arhang.gos.tekhn.un-t) p 267

[5] Labudin B V 1978 *Eksperimental'no-teoreticheskoe issledovanie perekrestnyh balok iz kleenyh derevyannyh elementov*: dis. ...kand. tekhn. nauk [Experimental and theoretical study of cross beams made of glued wooden elements] (Leningrad: LISI) p 144 [In Russ.]

[6] Labudin B V, Orlovich R B and Bazenkov T N 1991 *Soprotivlenie elementov kleenoj drevesiny mestnomu smyatiyu pod uglom k voloknam* [The resistance of glued laminated timber elements to the local crushing at an angle to the fibers] *Lesnoj zhurnal* **3** 59–63 [In Russ.]

[7] Znamenskij E M, Pyatikrestovskij K P and Gorbatova N I 1982 Sovremennoe sostoyanie zarubezhnyh i otechestvennyh norm proektirovaniya derevyannyh konstrukcij *Stroit. Konstrukcii* ser. 8 **4** 73

[8] Karlsen G G, Bol'shakov V V and Kagan M E 1975 *Konstrukcii iz dereva i plastmass* (M.: Strojizdat) p 686

[9] Lekhnickij S G 1977 *Teoriya uprugosti anizotropnogo tela* (M.: Nauka) p 416

[10] Sen-Venan V 1961 Memuar o kruchenii prizm *Memuar o izgibe* ed G Yu Dzhanelidze (M.: Fizmatgiz)

[11] Labudin B V, Veshnyakov A V, Popov V D and Yakovlev V V 1992 Novoe uzlovoe soedinenie sterzhnej dlya derevyannyh nesushchih konstrukcij *Lesnoj zhurnal* **4** 108–112

[12] Panferov K V 1960 Smyatie i szhatie derevyannyh elementov poperek volokon pri dlitel'noj nagruzke *Voprosy primeneniya dereva i plastmass v stroitel'stve* (M.: Strojizdat) pp 62–75

[13] 1981 Rekomendacii po ispytaniyu soedinenij derevyannyh konstrukcij *CNIISK* (M.: Strojizdat) p 41

[14] Slavik Yu Yu 1985 Vliyanie osnovnyh ekspluatacionnyh faktorov na bezopasnost' derevyannyh konstrukcij *Issledovaniya v oblasti derevyannyh konstrukcij* (M.: CNIISK) 107–117

[15] Sobolev Yu S 1979 Drevesina kak konstrukcionnyj material (M.: Lesn. Prom-st') p 246

[16] SP 35.13330.2011 Mosty i truby [Bridges and pipes] updated version SNiP 2.05.03-84* with changes, date of int. 2011/05/20 [In Russ.]

Field dependence of magnetoelectric effect in the electromechanical resonance region on the Permendur – Quartz – Permendur structures

V M Laletin[1], N N Paddubnaya[1] and D A Filippov[2,3]

[1] Institute of Technical Acoustics National Academy of Sciences of Belarus, Generala Lyudnikova Ave., 13, 210009, Vitebsk, Belarus
[2] Yaroslav-the-Wise Novgorod State University, B. St. Peterburgskaya, 41 173003, Velikiy Novgorod, Russia

[3] E-mail: Dmitry.Filippov@novsu.ru

Abstract. The results of an experimental study of the field dependences of the magnetoelectric effect on three-layer Permendur-Quartz-Permendur structures in the region of electromechanical resonance are presented. It has been established that at the electromechanical resonance there exists a bias magnetic field region in that an anomalous behavior of the field dependence of the magnetoelectric coefficient are observed. This effect is explained due to the presence of the *ΔE* effect in the magnetostrictive material.

1. Introduction

Compositional multiferroics have recently attracted increasing attention of researchers, as evidenced by the growth of publications on this subject [1]. The potential applications of such materials are possible due to the magnetoelectric (ME) effect, which consists in changing the polarization of a substance under the influence of a magnetic field (direct ME effect) and, conversely, in changing the magnetization under the influence of an electric field (converse or inverse ME effect). Due to the presence of the ME effect based on them, it is possible to create devices in which both electric and magnetic fields are controlled simultaneously [2–5]. Recently, a large number of new composite ME materials based on the magnetostrictive and piezoelectric phases have been synthesized. The ME effect is a new property of such materials, or as they say, *product properties* [6], and arises due to the mechanical interaction of the magnetostrictive and piezoelectric phases of the composite. The main characteristic of the ME effect is the magnetoelectric voltage coefficient (MEVC) α_E, which is determined by the ratio:

$$\alpha_E = \frac{<E>}{H}, \qquad (1)$$

where $< E >= U/\,^{p}t$ is the average electric field strength in the sample, $\underline{U}$ is the electric voltage arising on the sample under the influence of an alternating magnetic field *H*, ^{p}t is the thickness of the piezoelectric. According to the theory of the ME effect presented in [6, 7], the MEVC value at the low-frequency region is determined by the expression:

Content from this work may be used under the terms of the Creative Commons Attribution 3.0 licence. Any further distribution of this work must maintain attribution to the author(s) and the title of the work, journal citation and DOI.
Published under licence by IOP Publishing Ltd

$$\alpha_E = \frac{{}^pY\,{}^pd_{31}({}^mq_{11}+{}^mq_{12})}{{}^p\varepsilon_3\varepsilon_0\left(1-K_p{}^2\left(\frac{{}^mY\,{}^mt}{{}^mY\,{}^mt+{}^pY\,{}^pt}\right)\right)}\cdot\frac{{}^mY\,{}^mt}{{}^mY\,{}^mt+{}^pY\,{}^pt}, \qquad (2)$$

where ${}^pY$, ${}^mY$ are the Young's modules of the piezoelectric and magnetostrictive phases; ${}^pd_{31}$, ${}^mq_{11}$, ${}^mq_{12}$ are the piezoelectric and piezomagnetic coefficients; ${}^pt$, ${}^mt$ are the thickness of the piezoelectric and the magnetostrictive layers; ${}^p\varepsilon_3$ is permittivity, ε_0 is dielectric constant and $K_p{}^2 = \frac{{}^pY({}^pd_{31})^2}{{}^p\varepsilon_3}$ is the squared coefficient of electromechanical coupling.

Using the fact that $K_p{}^2 << 1$, we can rewrite equation (2) in the first approximation in the form

$$\alpha_E = \frac{{}^pY\,{}^pd_{31}\,{}^mq_{11}}{{}^p\varepsilon_3\varepsilon_0}\cdot\frac{{}^mY\,{}^mt}{{}^mY\,{}^mt+{}^pY\,{}^pt}. \qquad (3)$$

In equation (3), the piezomagnetic coefficients ${}^mq_{11}$ and ${}^mq_{12}$ are defined as follows:

$${}^mq_{11}(H_{bias}) = \left.\frac{\partial\lambda_\parallel}{\partial H_1}\right|_{H=H_{bias}}, \quad {}^mq_{12}(H_{bias}) = \left.\frac{\partial\lambda_\perp}{\partial H_1}\right|_{H=H_{bias}} \qquad (4)$$

where $\lambda_\parallel$ and $\lambda_\perp$ are longitudinal and transverse magnetostriction, H_{bias} is bias magnetic field. In equation (3) terms ${}^mq_{11}$ and ${}^mq_{12}$ have a strong dependence on magnetic field and a term ${}^mY$ has a weakly magnetic field dependence. The remaining terms in expression (3) for the MEVC are independent of the bias field. Thus, the field dependence of the MEVC is determined by the dependence of the piezomagnetic modules on the bias field. The field dependence of MEVC is important when creating devices based on the ME effect. Since the ME effect in composites is associated with mechanical vibrations, a peak increase in MEVC occurs in the region of electromechanical resonance [9]. Despite the fact that the main contribution to the dependence of the magnitude of the effect on the magnetic field is made by piezomagnetic coefficients, the dependence of the Young's modulus on the magnetic field or the so-called Delta E effect makes the significant contribution to magnitude of the effect only in the field of electromechanical resonance [10]. Currently, based on the *Delta E* - effect, highly sensitive magnetic field sensors have been created [11,12]. Studies have shown that the main frequency of electromechanical resonance for a sample in the form of a rectangular plate with planar vibrations is determined by the expression [13]:

$$f_{res} = \frac{1}{2L}\sqrt{\frac{{}^mY\cdot 2\,{}^mt+{}^pY\cdot{}^pt}{{}^m\rho\cdot 2\,{}^mt+{}^p\rho\cdot{}^pt}}, \qquad (5)$$

where L is the sample length. When the bias field H_{bias} changes, the Young's modulus of the magnet ${}^mY$, changes too, which leads to a change in the frequency of the electromechanical resonance. This makes it possible to determine the change in the Young's modulus of a magnet from a change in the frequency of electromechanical resonance [14]. In the present work, the influence of the magnetic field on the MEVC value and the Q factor of the structure in the region of electromechanical resonance are analyzed.

2. Samples and Methodology

The objects of research were the layered structures of permendur – quartz – permendur in the form of rectangular plates 20.0 × 4.5 mm in size. The schematic drawing of the structure is presented on figure 1.

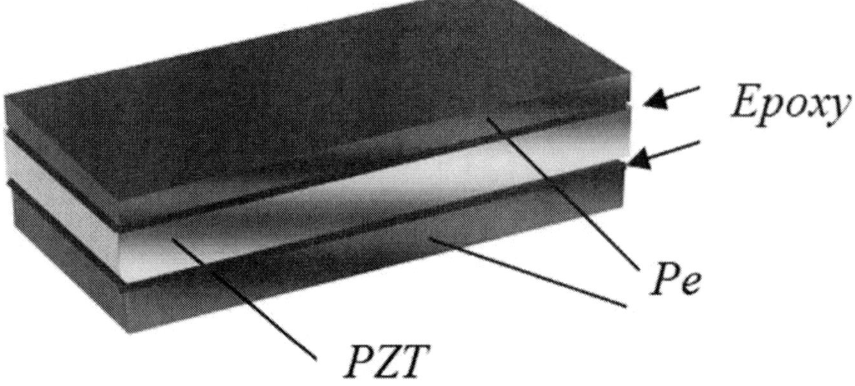

Figure 1. The schematic drawing of the asymmetric three layered structure.

A quartz plate (X-cut) with a thickness of ^{p}t=0.5 mm was used in a piezoelectric, and two rectangular permendur plates with a thickness of ^{m}t=0.16 mm each were used as a magnetostrictive material. Epoxy adhesive was used as a mechanical connection between layers. The thickness of the polymer layer in the obtained samples did not exceed 5 μm. The ME effect was studied by detecting an alternating voltage on a sample when it was placed in a bias field and an alternating magnetic field directed along the long side of the sample.

3. Results and Discussion

The results of studies of the field dependence of the ME coefficient at a frequency of 1 kHz are presented in figure 2. The field dependence of the ME coefficient has a typical form for magnetostrictive-piezoelectric composite materials. The maximum magnitude of the ME coefficient was 1.9 V/A with a bias field $H_{bias} \simeq 10$ kA/m. A study of the frequency dependences of the ME effect showed that they have a pronounced resonance, whose amplitude and frequency change with a change of the bias field. Figure 3 shows the frequency dependence of the ME coefficient in the magnetization field corresponding to the maximum of the effect. The resonance was observed at a frequency of 153.37 kHz, the resonance value of MEVC was equal to 920 V/A at a quality factor Q = 980.

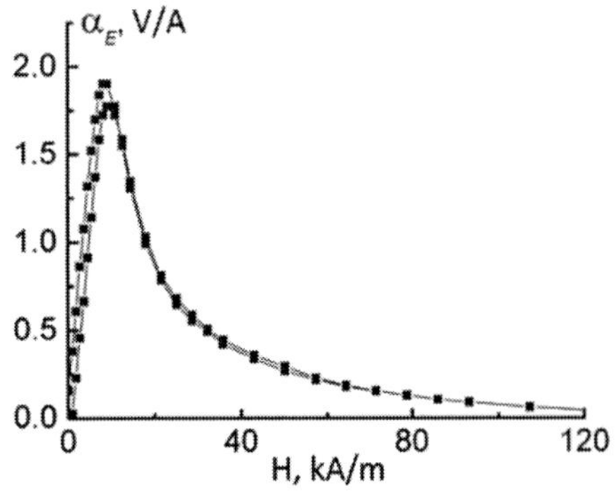

Figure 2. The field dependence of low frequency MEVC.

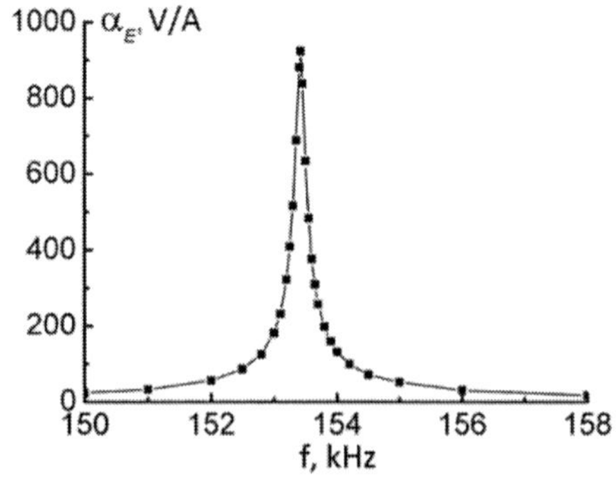

Figure 3. The frequency dependence of MEVC at bias magnetic field H_{bias}=22 kA/m.

IOP Conf. Series: Materials Science and Engineering **939** (2020) 012040 doi:10.1088/1757-899X/939/1/012040

Figure 4 shows the dependence of the resonance frequency on the bias field. As can be seen from figure 4, in fields up to 16 kA/m, the value of the resonance frequency is less than the resonance frequency of the sample at zero magnetic field. This is the region of negative *Delta* E – effect. In the bias fields above 16 kA/m, at increasing the magnetic field, the resonant frequency increases and reaches saturation at a magnetic field near 100 kA/m. This is the region of the positive *Delta E* – effect. This is the region of positive *Delta E* – effect.

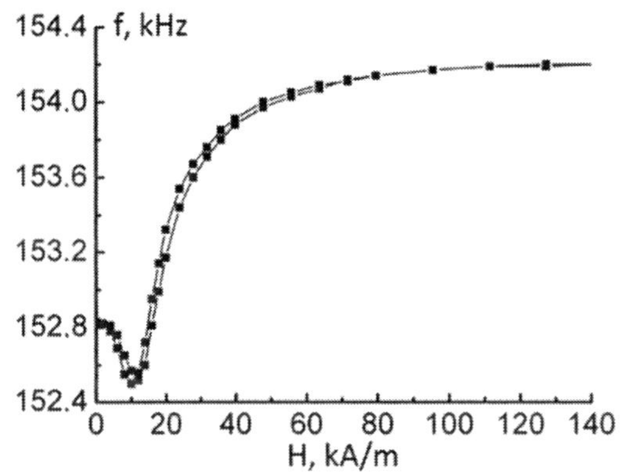

Figure 4 The field dependence of resonance frequency

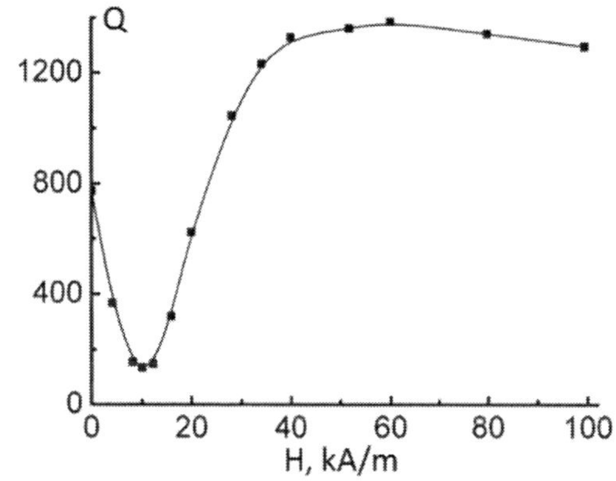

Figure 5 The field dependence of Q-factor

Figure 5 shows the dependence of the Q factor on the bias field. In this case, the quality factor was defined as the ratio of the resonant frequency to the width of the resonance line at the level of 0.707. As follows from a comparison of figure 4 and figure 5, the field dependence of the Q factor qualitatively reflects the field dependence of the resonant frequency.

In the region of negative *Delta E* – effect, the Q factor decreases by more than four times. As is known, the quality factor is determined by losses in the structure. Thus, it can be argued that the processes associated with the negative *Delta E*-effect are accompanied by an increase in losses in the structure. The dependence of the resonance frequency and the Q factor on the magnetization field leads to the fact that the field dependence of MEVC in the region of electromechanical resonance has significant differences from the field dependence of MEVC outside the resonance region. Figure 6 and figure 7 show the field dependences of MEVC at frequencies f_1 =151 kHz and f_2 =155 kHz, located below ($f_1 < f_{res}$) and above ($f_2 > f_{res}$) the region of resonant frequencies shown in figure 4. In this case, the field dependences have the form characteristic of the low-frequency ME effect, but with a large MEVC value. The maximum signal is observed at the field $H_{bias} \simeq$10 kA/m, while the MEVC value at the maximum is more than 50 times its low-frequency value, reaching α_1 = 158 V/A at a frequency f_1 =151 kHz and α_2 = 98 V/A at a frequency f_2 =155 kHz against its low-frequency value α_{low}=1.9 V/A at a frequency f = 1 kHz. This increase in the coefficient is due to the approach of the frequency to its resonance value and the associated resonant increase in MEVC. A change in MEVC with a change in the magnetization field in this case is due to a change in the piezomagnetic coefficients ${}^{m}q_{11}$ and ${}^{m}q_{12}$. The mechanisms associated with a change in the quality factor due to an increase in losses with a negative *Delta E* – effect have not yet been manifested, apparently, due to the still small amplitude of the oscillations.

As the frequency approaches its resonance value, the amplitude of the oscillations increases, which leads, on the one hand, to an increase in the magnitude of the effect, and simultaneously to an increase in losses, and, as a result, to a decrease in the Q factor in the magnetic field region of the negative *Delta E*-effect, which leads to a decrease in MEVC. Figure 8 and Figure 9 show the field dependences of MEVC at a frequency f_0 =152.8 kHz, corresponding to the frequency of electromechanical resonance in the zero bias field and in the mode of continuous tuning to the resonant frequency.

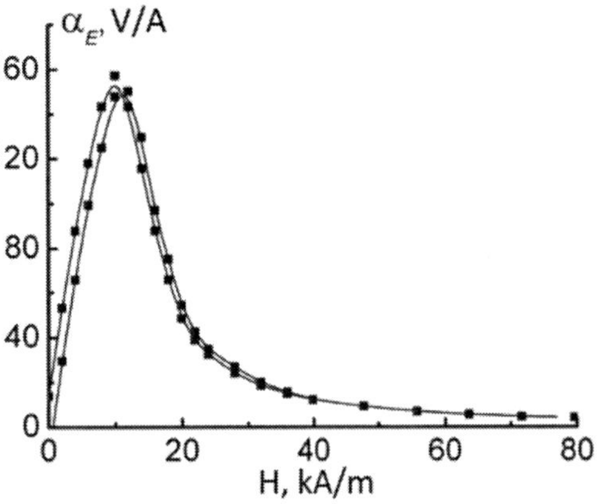

Figure 6. The field dependence of MEVC at frequency f_1=151 kHz.

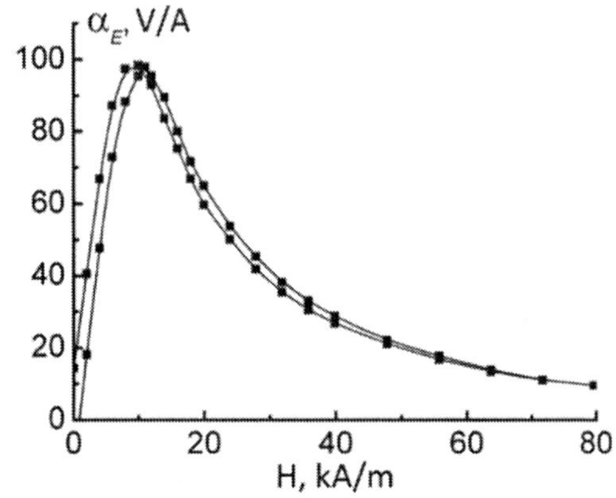

Figure 7. The field dependence of MEVC at frequency f_2=155 kHz.

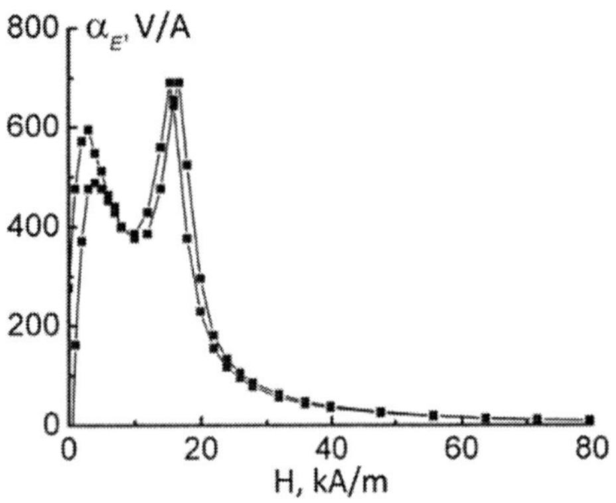

Figure 8. The field dependence of MEVC at frequency f_0=152.8 kHz, corresponding to the frequency of electromechanical resonance in the zero bias field.

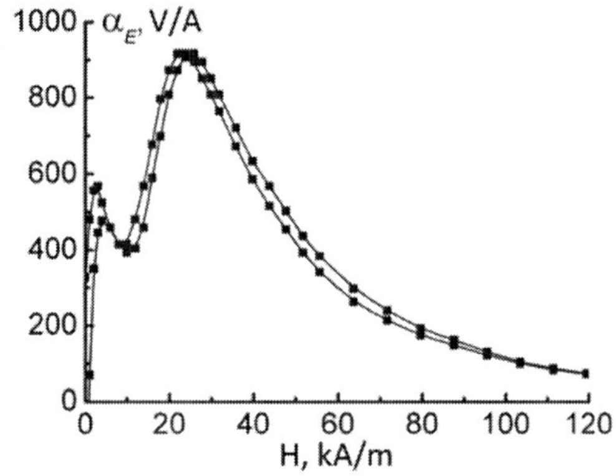

Figure 9. The field dependence of MEVC in the regime of continuous tuning to the resonant frequency.

A characteristic feature of these dependencies is the presence of two maxima. This anomalous behavior of the MEVC coefficient is associated with the presence of two mechanisms - a change in the piezomagnetic coefficients due to a change in magnetostriction with a change in the magnetic field, and a change in the quality factor of the system due to the negative *Delta E* – effect. Therefore, it can be argued that the presence of two maxima in the field dependence of the MEVC coefficient is a consequence of the field dependence of the piezomagnetic coefficients and quality factor.

4. Conclusion

A change in the Young's modulus of the permendur in a magnetic field leads to a change in the resonance frequency and quality factor of the structure, which is accompanied by an anomalous behavior of the field dependence of the ME effect in the region of electromechanical resonance in the range of magnetic fields corresponding to the presence of a negative *Delta E* – effect. Outside the region of electromechanical resonance, the field dependences of the ME coefficient have a typical form characteristic of the low-frequency ME effect.

Acknowledgment
The work was jointly supported by the Russian Foundation for Basic Research and Belarussian Foundation for Basic Research (Russian project no. 18-52-00021 and Belarussian project no. F18R-300).

References

[1] Pyatakov A P and Zvezdin A K 2012 *Physics Uspekhi* **55** *557–581*
[2] Bukharaev A A, Zvezdin A K, Pyatakov A P and Fetisov Y K 2018 *Phys. Uspekhi* **61** 1175 – 1212
[3] Spaldin N A and Ramesh R 2019 *Nature Materials* **18** 203
[4] Sadovnikov A V, Grachev A A, Beginin E N, Sheshukova S E, Sharaevskii Yu P and Nikitov S A 2017 *Phys. Rev. Applied* **7** 014013
[5] Song C, Cui B, Li F, Zhou X and Pan F 2017 *Prog. Mat. Sci.* **87** 33
[6] Van Suchtelen J 1972 *Philips Res. Rep.* **27** 28–37
[7] Filippov D A 2004 *Technical Physics Letters* **30** 983–986
[8] Filippov D A, Galichyan T A and Laletin V M 2013 *Physics of the Solid State* **55** 1840–1845
[9] Filippov D A, Bichurin M I, Petrov V M, Laletin V M and Srinivasan G 2004 *Physics of the Solid State* **46** 1674–1680
[10] Laletin V M and Filippov D A 2018 *Technical Physics* **63** 186–189
[11] Spetzler Benjamin, Kirchhof Christine, Quandt Eckhard, McCord Jeffrey and Faupel Franz 2019 *Phys. Rev. Applied* **12** 064036
[12] Spetzler Benjamin, Golubeva Elizaveta V, Müller Cai, McCord Jeffrey and Faupel Franz 2019 *Sensors* **19** 4769
[13] Filippov D A, Laletin V M and Firsova T O 2014 *Physics of the Solid State* **56** 980–984
[14] Laletin V M, Filippov D A, Mozzharov S E and Manicheva I N 2018 *Technical Physics Letters* **44** 281–283

Physical and chemical processes in the contact zone when processing metals with a blade tool

V N Latyshev[1], **A G Naumov**[1] **and I V Shvetsov**[2,3]

[1] Ivanovo state University, ul. Ermaka 39, 153025 Ivanovo, Russia
[2] Yaroslav-the-Wise Novgorod State University, ul. B. St. Petersburgskaya, 41, 173003 Veliky Novgorod, Russia

[3] E-mail: Igor.Shvetsov@novsu.ru

Abstract. The article considers the theory of radical-chain mechanism of formation of lubricant films on tribos-connected metal surfaces when cutting metals. It is shown that chemically active particles formed by the destruction of molecules of lubricant and cooling technological means (LCTM) in the contact zone play a decisive role in the synthesis of lubricants. Mechanisms of formation of chemically active atoms, radicals and ions and factors influencing these processes are presented. It has been established that the main of them can include the temperatures that occur in the contact zone, exoelectrons issued by freshly opened metal surfaces, as well as the catalytic effect of the juvenile surfaces themselves. On this basis, various methods of pre-activation of LCTM are proposed in order to obtain the required quantity and type of chemically active particles.

1. Introduction

This scientific work is dedicated to the memory of Latyshev Vladimir Nikolaevich (28.09.1929 – 03.05.2018), who was a major Soviet and Russian scientist, a public figure, an honorary figure of science and technology of the Russian Federation, a Doctor of Technical Sciences, a Professor, a Rector of Ivanovo State University (1974–2000), a creator of the scientific school for the study of lubricant and cooling technological means in the processing of materials cutting, an author of more than 400 scientific articles, monographs and educational manuals, more than 100 author's certificates and patents for inventions. The positive solution to the problem of synthesis and use of lubricant and cooling technological means (LCTM) is largely determined by the level of theoretical researches in this area. The cutting process imposes specific features on the mechanism of action of the external environment, which is determined, first of all, by high unit pressures and temperatures on the working surfaces of the cutting tool, the presence of juvenile non-oxidized metal surfaces, the flow of chemical reactions and electrochemical processes on the edge of the section of the tool-processed material, the high deformation rate of the cut metal and the short contact time between the chips and the machined surface and the cutter.

2. Results and discussion

The results of experimental and theoretical studies allow us to describe the mechanism of the lubricating action of the environment when cutting metals based on modern representations of physics, chemistry and mechanics of the blade cutting process and to outline ways of synthesis of effective LCTM. This is

Content from this work may be used under the terms of the Creative Commons Attribution 3.0 licence. Any further distribution of this work must maintain attribution to the author(s) and the title of the work, journal citation and DOI.
Published under licence by IOP Publishing Ltd

primarily due to the study of a complex and, at the same time, unique mechanism of interaction of the external environment with metal surfaces. The main reason for the specific mechanism of action of the environment in metal cutting is the presence in the contact area of the cutter with the processed material of physically pure neo-oxidized surfaces, extremely active chemically, high temperatures and pressures, as well as unusual catalytic properties of juvenile surfaces, formed when metal bonds break.

This surface comes into contact while cutting either with natural environment (air with different physical and chemical properties) or with artificially brought external environments. At the same time, the interaction of the juvenile surface with components of the external environment can be characterized by various sorbation processes, first of all, physical and chemical adsorption, as well as absorption. The characteristics of the contact zone are such that the molecules of the external environment, undergoing destruction, break down into atoms, ions and radicals, that are also chemically active particles and are able to enter into chemical interaction with freshly opened metal surfaces.

Taking into account the above-mentioned features of the contact zone, it is likely that at the contact boundary the tool-processed material will be dominated by the formation of new chemical compounds as a result of chemical reactions between juvenile metal surfaces and components of the external environment, i.e. iono- and chemosorption processes. At the same time, the physical and the chemical activity of these surfaces is such that with their participation it is possible to initiate the course of chemical reactions with components of the external environment, the thermodynamic possibility of which is unlikely under normal conditions. It is known that the number of product reactions in the cutting zone is more than 2.5 times higher than the same figure obtained under the same conditions but outside the contact zone (figure 1).

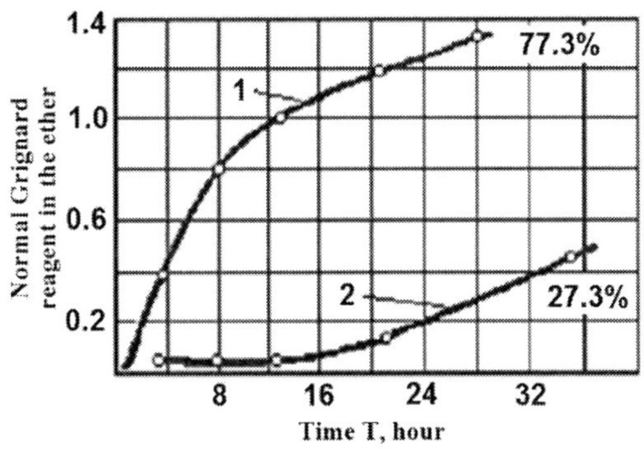

Figure 1. Activating effect of the cutting process on the reaction of chlorbenzene formation with magnesium in the ether: 1 – in the contact area; 2 – in a flask.

Thus, when considering the mechanism of lubricating chemical action LCTM should take into account the process of formation of surface radicals, the speed of chemical reactions, catalytic phenomena on micro-sites of incisor and shavings, the consequences of chemical reactions (film formation, the warmth of exothermic effects of reactions).

In accordance with modern views on the mechanism of heterogeneous chemical reactions, the interaction of metal surfaces with the external environment depends on a number of conditions, including the electronic structure of reacting molecules, the presence of unpaired electrons, the type and strength of bonds in molecules of various LCTM compounds. It is known that the chemical reaction and the process of the formation of lubricants on the contact sites of the incisor and shavings includes several successive stages: the impact of molecules and atoms with electrons, the formation of surface radicals, the period of their existence in a free active state, the elementary act of chemical reaction. Consider the factors that influence the kinetics of chemical reactions in metal cutting.

The first and necessary condition for chemical reactions is the collision of molecules or atoms of reacting components with juvenile surfaces. When cutting metals at high speeds, atoms or LCTM molecules are in a gas-like state, and metal surfaces are in an elastoplastic, chemically activated state. Chemical reactions between them will be possible if the colliding molecules or atoms have an excess of free energy.

It is known from physical chemistry that the smaller the difference in the level of direct and reverse reactions is, that is, the less energy activation will be and the greater the proportion of molecules and atoms will collide with the surface and between each other, the higher the rate of chemical reaction. Thus, one of the means of intensification of chemical reactions in friction and cutting is the selection of reacting substances that have low activation energy.

The speed of the chemical reaction can be increased by catalysts. Catalysts reduce the energy of activation of chemical reactions and thus contribute to an increase in the number of active molecules and atoms. Metals and their oxides, as well as carbides, borids, nitrids and other compounds can be used as catalysts. In addition to positive catalysts that accelerate the speed of chemical reactions, there are also negative catalysts, inhibitors, that slow down the speed of chemical reactions.

The catalytic effect of various metals on the decomposition process of LCTM and oils has been shown in a number of our earlier researches. Thus, the method of differential-thermal analysis (DTA) has established that when LCTM are heated, these substances during decomposition form intermediate chemical compounds, which differ in their properties from the original components. Metals present during thermal decomposition of LCTM in the cutting process can have a significant impact on the physical and chemical properties of the formed intermediate substances, on the speed of their formation (figure 2).

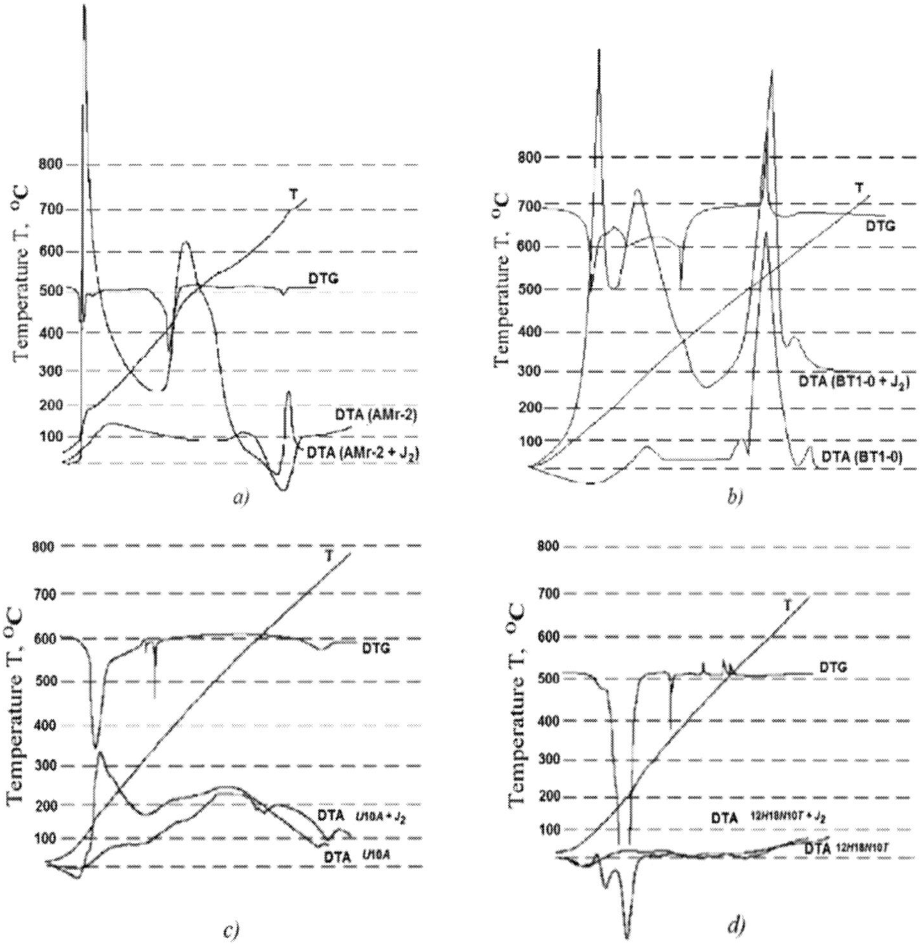

Figure 2. DTA results on the interaction of crystalline iodine with various metals: a) *VT*1-0; b) Amg-2; c) 12H18N10T; d) U10A.

Calculating the order of chemical reactions in the interaction of different metals with microcapsulated LCTM showed that for all peaks on DTA-curves studied samples of microcapsules (MC, MMC) and their constituent parts (table 1) this indicator turned out to be fractional. Here: microcapsule – particle LCTM, which has a size of 0.1–50 microns, surrounded by an impenetrable shell, MC – microcapsule, MMC – microcapsule, which has a magnetic and sensitive component. Therefore, all processes with thermal destruction of LCTM occur not sequentially but in complex parallel paths. Moreover, the increase in the heating temperature often complicated the nature of destructive processes, which indicates the reactions that occur both inside the polymer shells of microcapsules, and between the polymer and the nucleus, polymer and metal, the nucleus and metal, as well as all components with the atmosphere of the surrounding air.

Table 1. Estimated order of chemical reactions of MC and their individual components in the presence of different metals.

Metal sample	MC shells	MC magnetic shells	MCI-40A	MMC-50I-40A	MC CCl_4	MMC-50 CCl_4
1	2	3	4	5	6	7
–	0.806	$\frac{1.096}{0.714}$	$\frac{0.545}{1.833}$	$\frac{1.430}{0.489}$	0.548	$\frac{0.870}{0.596}$
MVP	0.708	$\frac{1.333}{0.625}$	–	–	0.595	0.676
12H18N10T	1.275	$\frac{1.031}{2.211}$	1.235	$\frac{1.529}{0.932}$	–	–
Steel 45	1.080	1.714	$\frac{1.367}{0,486}$	$\frac{1.140}{0.172}$	–	–
VT-6	1.048	1.167	$\frac{1.385}{1.478}$	$\frac{3.640}{1.265}$	–	–
R-9	0.923	$\frac{0.973}{0.560}$	$\frac{0.720}{0.500}$	$\frac{1.125}{2.333}$	0.755	$\frac{0.537}{1.682}$

* the fraction numerator indicates the order of the low-temperature peak reaction on the DTA curve, in the denominator – of the high-temperature.

Thus, the task of fusion of effective LCTM for mechanical processing of metals depends to a large extent on the intermediate substances generated during the thermal decomposition of LCTM, free radicals and applicable catalysts. In most cases, the cutting tool or the material being processed represents a complex set of catalysts, including oxides, carbides, nitrides and other compounds.

Aggregate state of the applied environment and temperature is the other factor on which the activity of chemical reactions depends. The gas and gas-shaped atmosphere contributes to the intensification of chemical processes, as they contain a greater number of potential active molecules and radicals involved in interaction with contact surfaces of metals. In the study of the effectiveness of the sprayed LCTM, it has been established that the specific surface of the air-liquid high-dispersal mixture with the size of the LCTM particles of 12–25 microns is approximately 700 times greater than the specific surface of the unsprayed liquid (1 g of sprayed liquid has the same surface as 27 kg of unsprayed). Thus, the gas and the locomotive atmospheres are the most favorable in terms of acceleration and intensity of chemical reactions and the formation of protective films.

The temperature effect on the speed of chemical reactions is also very large. In cutting conditions, due to the uneven distribution of temperature by width and length of shaving contact with the incisor,

IOP Conf. Series: Materials Science and Engineering **939** (2020) 012041 doi:10.1088/1757-899X/939/1/012041

as well as the presence of zones of local flares, chemical reactions can occur on separate areas of the incisor. This was recorded in our studies when cutting steel 45 using different LCTM. In most cases, the temperature in the cutting zone is almost always sufficient for thermal destruction of technological means and chemical reactions. As a result of uneven distribution of temperature and specific pressures along the length of contact on different parts of the front surface, different compounds are formed: oxides, sulfides, chlorides, metal diiodides. X-ray analysis and electromicroscopic studies have shown that in the areas directly adjacent to the cutting edge, films of sulfides, chlorides, diiodides are formed, and metal oxides – at the points of separation of shavings from the cutter (figure 3).

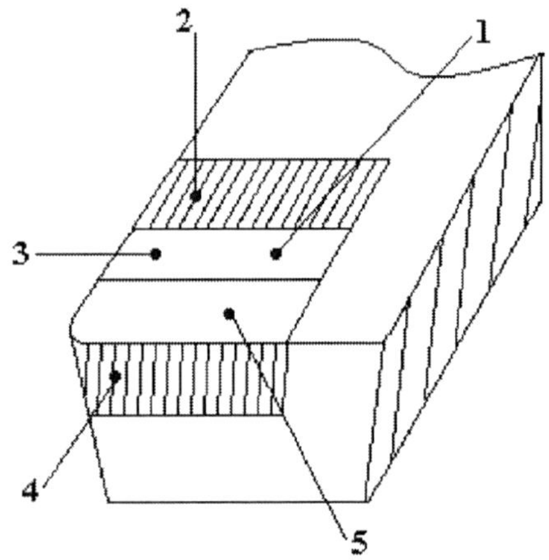

Figure 3. The layout of the films on the instrument's contact areas: 1 – $MeCl_2$, 2 – Me_xO_y, 2 – MeS, 4 – Me_xO_y, 5 – MeJ_2.

The next factor to consider when compiling LCTM formulations for cutting metals is the relationship between cutting speed and chemical reaction rate. The deformation speed while cutting metals is 10^5–10^6 sec^{-1} in average, and the rate of impact of reacting particles or the time of the elementary act of chemical reaction is 1018 sec^{-1}; the speed of oxidative processes is even higher. From the comparison it follows that the rate of chemical reaction is much higher than the rate of deformation when cutting metals, that is, the act of elementary chemical reaction has time to occur.

An extremely important factor determining the chemical activity of LCTM is the ability of its molecules to break down into reaction particles – atoms, ions and radicals, as only with continuous generation of chemically active particles is possibly branching, continuous, chemical reaction, and, therefore, maintaining on the contact sites stable chemical films, sufficient for a tangible impact on the tribological environment. The process of surface radicals in cutting is carried out in a variety of ways, including by breaking the bonds in the LCTM molecule with free valences, as a result of interaction with electrons emitted by juvenile surfaces, with environment molecules, thermal pyrolysis, etc.

In the first case, the free valences formed on the surface of the shavings are similar to free radicals, as they can interact with LCTM molecules. At the same time, the LCTM molecule, which has a small activation energy, is torn apart and the resulting radical saturates valence, i.e. chemosorbs. A distinctive feature of the cutting process is that it continuously forms free valences, that is, generates "superficial" radicals. As a result of the chemical interaction between radicals and LCTM atoms with the juvenile surface, its unbalanced force field (free valence) is extinguished, a lubricated chemical film is formed and additional heat from the exothermal reaction is released.

The described case of radical formation seems to be typical in cutting with low speeds of structural steels, as well as in the processing of light and colored alloys. The process of generating radicals depends in this case on the force field of the juvenile surface, the ratio of geometric parameters of crystal lattice of deformable metal and the size of LCTM molecules.

The second case of destruction of neutral molecules of the external environment on atoms and radicals is carried out as a result of their interaction with electrons or other energy particles, including quantums of light, issued by juvenile surfaces as a result of thermoelectronic or photoelectronic emission. For example, as a result of the interaction of exoelectrons with neutral LCTM molecules, energy is transferred from electrons to LCTM molecules, activation of the latter and their disintegration with the formation of radicals and radical groups, the activity of which is caused by the presence of an electron with uncompensated spin on the outer shell. In general, such processes in the formation of peroxide radicals can be described as follows

$$H_2O + hv(e) \rightarrow H_2O^* + hv(e)$$
$$\downarrow$$
$$H. + OH.$$

In the interaction of hydroxyl radicals, hydrogen peroxide is formed, the effect of which while cutting is caused by the ability to release active oxygen

$$OH. + OH. \rightarrow H_2O_2$$
$$2H_2O_2 \rightarrow 2H_2O^* + O_2$$
$$O_2 + hv(e) \rightarrow O_2^{*.} + hv(e)$$
$$\downarrow$$
$$O. + O.$$

where $hv(e)$ is the energy (electron) emitted by the juvenile surface; H_2O^*, O^* - - excited water and oxygen molecules; H., OH., O. – chemical radicals.

Temperature, light irradiation (electromagnetic waves), the absolute magnitude of the electrons' output and its energy are the main characteristics determining the intensity of the emission of electrons, and therefore the process of generating radicals. The electron output is several electron-volts and is minimal for metals such as potassium, sodium, lithium, and for stainless and heat-resistant materials containing, for example, nickel output of the electron 2–2.5 times higher.

The energy of the electron is one of the main criteria for the formation of radicals. The results of theoretical and experimental studies have shown that electrons simulated from the juvenile surface of metal accelerate in the electrostatic field generated by friction. At the same time, their energy increases and becomes enough to destroy neutral lubricant molecules into radicals. B.I. Kostetsky and his collaborators, using the electronic paramagnetic resonance method (EPR), found that complex organic radicals are formed when metals are frictioned in organic lubricants. Studies have established (figure 4) that an increase in the spectrum of the reference was observed during the joint heating of nitroxil and magnetic microcapsules (MMK) This indicated a decrease in the free bonds of nitroxyl as a result of chemical reactions. In the study of the mixture of nitroxyl-ozone-containing magnetic microcapsules (OMMK) spectrum was a triplet characteristic of a low concentration of nitroxyl radicals, due to the increased number of radicals, released during thermodestruction of OMMK.

Thus, the less electron output is and the higher its energy is, the higher the probability of LCTM molecule decay with the formation of radical and chemical film synthesis.

The third way to get radicals is to heat the components of LCTM to high temperatures, i.e. in thermal pyrolysis. This method of obtaining radicals is especially convenient for the decomposition of organic liquids: oils, hydrocarbon mixtures, surface-active substances. In this case, when heated to high temperatures, the vibrational movement of atoms in molecules increases and at high enough amplitudes of vibrations the bonds between atoms are broken. This case of radical generation is typical, perhaps when cutting heat-resistant and stainless steels, as well as cutting at higher speeds.

Once the radicals have formed, two things must be taken into account: their life expectancy and the type of radical. Researches carried out by Science Schools of N.N. Semenova and V.II. Kondratieva show that the life expectancy of radicals is different and depends on the temperature, type of radical, surrounding atmosphere and compiles on average 0.01–0.4 seconds for hydroxyl, which is several

IOP Conf. Series: Materials Science and Engineering **939** (2020) 012041 doi:10.1088/1757-899X/939/1/012041

orders of magnitude longer than the time needed for the formation of a chemical compound.
Depending on what type of compounds break down into atoms and radicals, the latter can be of two types: simple and complex. Inorganic aquatic CF (cutting fluid is a special case of LCTM) form simple radicals when cutting (monoradicals, biradicals and trnradicals) and organic lubricants form complex radicals. Simple radicals studied in this work include chlorine, sulfur, phosphorus, iodine, and complex - radicals formed in pyrolysis surfactant, oils, microcapsules.

Structural formula nitroxyl	Spectra EPR		
	Nitroxyl	Nitroxyl and MMK	Nitroxyl and OMMK

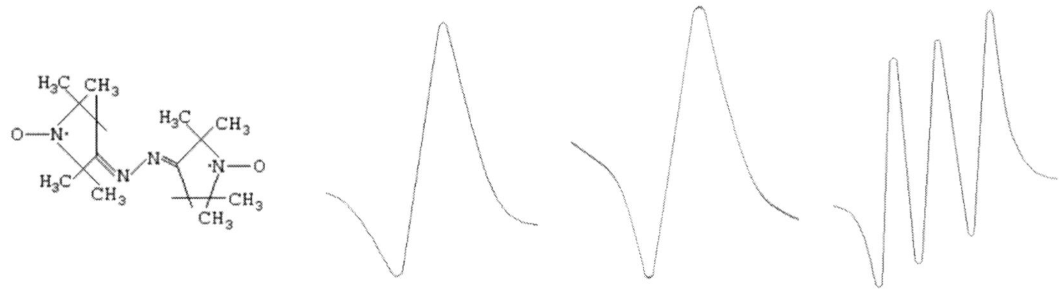

a) *b)* *c)* *d)*

Figure 4. The ability of ozone-containing magnetic microcapsules to initiate radical chain reactions: a – a structural formula of nitroxyl; b-d – spectrums of electronic paramagnetic resonance: b – nitroxyl; c – nitroxyl with magnetic microcapsules: d – nitroxyl and ozone-containing magnetic microcapsules.

Chemical reactions that occur when cutting using CF and gas are, in most cases, spontaneous and radically chained. This has been shown in our works by the example of spontaneous formation and decomposition of hydrogen peroxide when cutting alloys on an iron basis, autocatalytic reactions of nitrogen oxides with titanium and solid alloys, chemical interaction of iodine and sulfur vapors respectively with titanium and nickel alloys.

The study of the mechanism of chemical interaction of radicals of the environment with metal surfaces, first of all, oxygen and water with juvenile shavings and cutters, allows to outline ways of purposeful regulation of chemical reactions in cutting, the possibility of creating defensive films at the expense of these reactions on the cutting tool and shavings. Here are some things to consider:

First, it is necessary to keep in mind the mechanism by which the lubricant molecule spontaneously collapses in the cutting zone on atoms and radicals of free "superficial" valence, electron or electromagnetic wave, thermal pyrolysis. The first mechanism of action (destruction of free valencies) is mainly typical in cutting with low speeds (no thermostimulated emission of electrons), and the destruction of molecules of lubricant by electron (quantum of light) or pyrolysis occurs at medium to high cutting speeds. The destruction of LCTM molecules by electrons is most intense when cutting steel with a ferritic structure (maximum emission) and minimal when cutting heat-resistant alloys that have a solid solution structure (minimum emission). In the latter case, the breakdown of LCTM molecules into radicals is mainly due to thermal pyrolysis.

Secondly, it is necessary to rationally select components of LCTM, keeping in mind, first of all, two parameters of the molecule: the type and strength of the connection between atoms. These statistical characteristics can be judged in advance on the possibility of the formation of free radicals and atoms and their reactionary ability. Compounds of type - 0 - 0 - , *I-I*, *NO*, *CN* are the most favorable from the point of view of chemical connections.

Thirdly, consider the possibility of activating the necessary (positive) chemical reactions by various methods: ionization of the air-liquid mixture by electric field, activation of the gas mixture by light of optimal wavelength, activation of reactions by ultrasonic vibrations, electric current, etc. (figure 5). The correct selection of environmental components and the appropriate activation method can trigger the necessary chemical reaction and obtain chemical compounds with predetermined properties on tribo-conjugated metal surfaces.

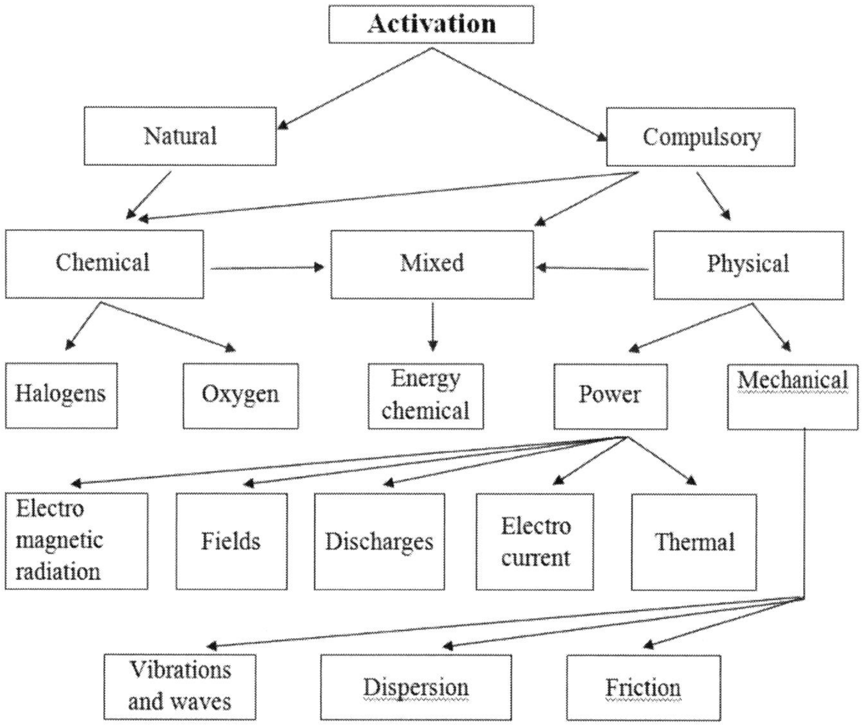

Figure 5. A physical model that describes activation.

In our previously published works, the influence of various films and individual chemical compounds (oxids, sulfides, chlorides, iodides, etc.) on friction and wearout of the cutting instrument, established a correlation between the type of chemical compound and the friction factor, found out the effect of protective films on fluctuations in the forces and temperature of the cutting, identified the optimal components of LCTM for various contact pairs of metals.

The theory of radical-chain mechanism of lubricating films formation is tested in the cutting of metals with cooling – lubricant with various substances, including atomic oxygen, peroxide compounds, ozone, surface-active substances with additives of haloid compounds, other compounds [1–3]. The mechanism of spontaneous formation of hydrogen peroxide in metal cutting has been studied in detail, the conditions of its formation and decomposition have been revealed, catalytic processes of hydrogen peroxide decay on juvenile surfaces of various metals have been studied and stabilizers of H_2O_2 decomposition have been proposed in order to maintain some excess of CF. From the point of view of the mechanism of radical-chain reactions, a theoretical explanation of the effective action of sprayed and ionized CF and oils when cutting some steels and alloys is given.

Based on experimental data, the first approximation established the connection of the internal structure of the environmental molecules with the parameters of the cutting process: the durability of the cutters and the cutting temperature (figure 6).

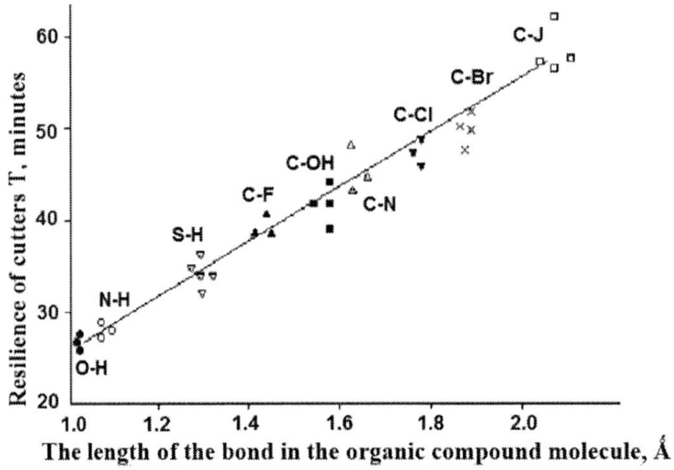

Figure 6. Effect of the bond length in the LCTM molecule on the instrument durability.

For various chemical compounds formed on heavily loaded contact sites of the cutter and shavings, a crystal and chemical parameter is calculated, which determines the ease of sliding of interatomic planes and is actually a characteristic of frictional properties of films. The resulting experimental data allowed in general form to draw up a scheme of LCTM action during the cutting process, to assess the impact of their individual species (oxidizers, emulsions, oils, chemically active liquids, etc.) on the durability of fast-cutting and solid-alloy instruments (figure 7).

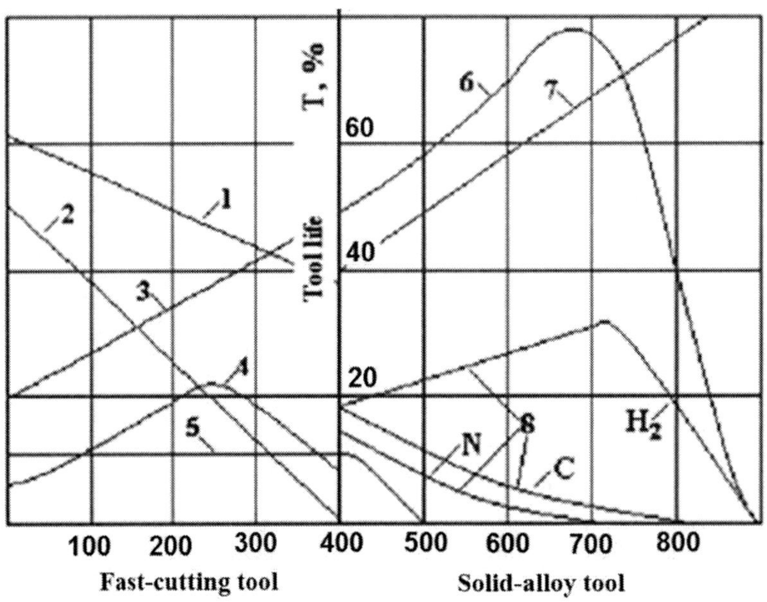

Figure 7. A generalized chart of the impact of LCTM on the durability of cutting tools: 1 – oxidized oils with high pressure additives, 2 – oxygen-containing emulsions, 3 – mineral oils, 4 – oil emulsions, 5 – solid lubricants (MoS_2, S, C), 6 – oxygen-containing LC (O, O_2, H_2O_2, 7 – LC with chemically active radicals (S, P, J, Cl), 8 – diffuse environments.

IOP Conf. Series: Materials Science and Engineering **939** (2020) 012041 doi:10.1088/1757-899X/939/1/012041

In general, the theory presented in this paper about the lubricating chemical action of the environment in metal cutting is in following: the successive stages of the mechanism of destruction of cutting fluid molecules into atoms and radicals have been revealed, a radical-chain mechanism of chemical reactions in cutting (on the example of hydrogen peroxide), a general scheme of selection of components of cutting fluid (based on the internal structure of molecules and atoms) is established and ways of activating and regulating the necessary chemical reactions are outlined, followed by the formation of protective films on the contact surfaces of the cutters and shavings.

3. Conclusion

The obtained results provide a number of considerations about the scientific and technical basis for the synthesis of effective lubricants and cooling environments.

1) In the process of cutting lubricant functions are performed not by the original components of LCTM, but by intermediate reaction particles (atoms, ions and radicals), formed by the interaction of neutral molecules with the juvenile surface. At the same time, the development of a radical-chain reaction requires a small concentration of reactionary particles. Therefore, when formulating LCTM should assume that its basis should be a neutral liquid (for performing transport functions, cooling, detergent action) and a small percentage of reaction additive. As it was established in this work, peroxide and inorganic oxidizers can be such additives for iron alloys, in the processing of titanium alloys – iodine additives, for nickel alloys sulphur additives are optimal.

Assessment of the effectiveness of cutting additives is useful to carry out on various modeling installations: pyrometers (identification of the fundamental possibility of reaction) and devices providing the formation of juvenile surfaces, for example, on friction machines in the conditions of scuffing.

2) The introduction of reaction additives in LCTM should not always be carried out during the process of its manufacture. For example, the half-life of such a reactionary additive as benzoyl peroxide at 95 °C is 1 hour, so it can break down into radicals even before cutting. Similarly, ozone and hydrogen peroxide have high rate of decomposition in water and water cutting fluids, hence zoning of cutting fluid and dissolution of peroxide is advisable to produce immediately before the use of cutting fluid, i.e. in the workplace.

Broming, chlorinating, iodation and sulphation of technological means can be done in advance, as the lubricating functions of these additives are manifested at elevated temperatures and in contact with juvenile metal surfaces. When introducing such additives into the LCTM, it is necessary that they are organically linked to the molecules of the main substance and they aren't a mechanical impurity.

3) Synthesized LCTM can be activated to quickly decompose their molecules in the cutting area into chemically active particles. These methods include different methods of physical, chemical and physicochemical effects (see figure 4).

4) LCTM can move from one aggregate state to another when cutting. Studies show that the effectiveness of lubricating chemical action of the environment depends on the intensity of the molecule's decay into radicals and the speed at which they move to the juvenile surface. As a rule, in viscous, dense liquids, the disintegration of molecules into radicals is difficult, as the latter can not move away from each other, that is, their recombination occurs. In this regard, the gas or locomotive atmosphere is more favorable for chemical radical reactions, because the rate of movement of free radicals in gases is 160 times higher than in liquids.

Thus, in view of the foregoing, the following indicative scheme for developing effective LCTM can be offered. A contact pair of metals is set – a contact cutting tool and a processed part (chemical composition, physical-mechanical and chemical properties, changes in properties under the influence of temperature). Next, they determine the work of the electron output, its possible energy, the estimated temperature range of the technology, the ratio of geometric parameters of the crystal lattice of the processed metal and the parameters of the internal structure of LCTM. The original components of the LCTM are installed – the base and additive responsible for the lubricating action. The basic principles of choosing such an additive are following: the distance between atoms in the molecule, the strength of the connection between them, the possibility of destruction on radicals due to electronic impact, pyrolysis, the force field of the juvenile surface, chemical activity in relation to the material being processed. In conclusion the LCTM test takes place: the fundamental possibility of chemical reactions, critical temperatures, heat reaction effect, i.e. the behavior of instrumental and processed materials in relation to the components of LCTM is assessed by its chemical activity, as well as the change in the tribological situation of the contact zone is assessed.

References

[1] Naumov A G, Latyshev V N, Radnuyk V S and Naumova O A 2015 Tribological Properties of Iodine as a CuttinggFluid Component During Metal Cutting *J of Friction and Wear* vol 36 **2** 184–188 DOI: 10.3103/S1068366615020129

[2] Nosenko V A, Naumov A G and Mitrofanov A P 2014 Chemical Reaction of Impregnators in an Abrasive Tool with Metal Blanks *Russian Engineering Research* vol 34 **10** 649–652 DOI: 10.3103/S1068798X14100116

[3] Naumov A G, Vereschaka A A, Batako A D and Vereschaka A S 2016 System of high-performance cutting with enhanced combined effect of cooling and lubrication medium based on Ranque-Hilsch Effect *Pocedia CIRP* **57** 457–460 DOI: 10.1016/j.procir.2016. 11.079

Structure formation and functional properties of thin-gauge tapes made of copper-based alloys obtained by melt spinning

O V Letenkov[1,2]

[1] Yaroslav-the-Wise Novgorod State University, ul. B. St. Petersburgskaya, 41 173003 Veliky Novgorod, Russia

[2] E-mail: Oleg.Letenkov@novsu.ru

Abstract. The article presents evidence proving that the alloys of the CuZnAl system, unlike the CuAl (Ni, Mn) system, have a smaller contact area with the quenching disk. It is caused by the condensation of Zinc vapors from the melt on its quenching surface. In addition, it is shown that Bronze alloys, in comparison with Brasses, are preferable for obtaining tapes by spinning. The main patterns of changes in critical points and the value of the Shape Memory Effect (εSME), depending on the types of heat treatment are also analyzed in this paper. Exposure of the alloy to the mold setting temperature of T = 800÷850 °C is accompanied by grain growth, increase in critical points, and decrease in the value of Shape Memory Effect. In this regard, the holding time at the mold setting temperature should not exceed t = 3÷5 min.

1. Introduction

Due to their unique properties, materials with the Shape Memory Effect are widely used in many areas of technology. Among these materials, Copper-based alloys occupy a special place. They are characterized by the ability to restore significant (up to 8–15%) inelastic deformations, corrosion resistance, high technological properties, relative cheapness, high thermal conductivity and, as a result, rapid response to temperature changes. Along with massive materials, a worthy place in this group is occupied by thin-gauge tapes obtained by melt spinning. The microcrystalline structure of tapes provides a variety and severity of such phenomena as Shape Memory Effect, Multiple Shape Memory, Reversible Shape Memory, etc.

The application field of such materials in engineering is rapidly expanding due to the development of new alloys and significant progress in their production technology. The ability to change form in a certain temperature range and a small cross-section area distinguish the tapes as promising materials for the production of heat-sensitive elements. The widespread use of thin-gauge tapes of Copper alloys is hampered only by the lack of knowledge of their functional properties. This restricts information about the kinetics of the formation of the tape structure in both isothermal and thermocyclic conditions. It is impossible to design and develop devices based on the work of tapes possessing Shape Memory Effect without obtaining this data.

In order to increase the efficiency of practical use of such materials in technology, it is necessary to find new ways to manage their special properties. For this reason, there has recently been an increased interest in studying the structural state and functional properties of thin-gauge tapes made of materials with Shape Memory Effect. The thermal stability and mechanical resistance of various structural states

Content from this work may be used under the terms of the Creative Commons Attribution 3.0 licence. Any further distribution of this work must maintain attribution to the author(s) and the title of the work, journal citation and DOI.
Published under licence by IOP Publishing Ltd

and their role in the implementation of martensitic transformations (MT) responsible for the complex properties of shape memory are still the most essential factors in this process.

2. Results and discussion

The above-mentioned points indicate the relevance of developing original methods and creating experimental conditions for studying the functional properties of thin-gauge tapes, as well as identifying patterns of their mechanical behavior under various types of external influence that are of scientific and practical interest [1, 2, 3].

The objective of the research is to study the structure formation and forming functional properties of thin-gauge tapes of Copper alloys for their use as heat-sensitive elements. In accordance with the objective, the following tasks are formulated:

- to investigate the influence of technological factors on the formation of the tape structure in the conditions of high-speed solidification of the melt;
- to study the influence of various types of thermal and thermomechanical processing on the functional properties of thin-gauge tapes;
- to set optimal modes for creating the Shape Memory Effect in the unloaded state, which ensure high performance of tapes in bending conditions with multiple implementation of shape memory properties;
- to develop the basics of engineering calculations for selecting the modes of creating the Shape Memory Effect for the operation of a thermosensitive element in a given deformation space.

Experimental studies were conducted on two specially created installations that allow obtaining tapes of the required geometry and measuring their functional properties under bending conditions during thermal cycling, both in the unloaded state and under the influence of constantly applied stress. Diagrams of these installations are given.

Traditional methods of measuring physical (electrical resistance) and mechanical (tensile testing) properties were implemented in the course of the experiment. Research methods utilizing optical metallography, X-ray diffraction analysis and electron microscopy are described in this paper. Theoretical research was carried out using the methods of mathematical planning of the experiment.

Alloy systems possessing the Shape Memory Effect (SME) were used as objects for research:

CuZnAl (Cu-14Zn-7,6Al; Cu-21Zn-5,8Al; Cu-10,2Zn-8,4Al) и CuAl(Ni,Mn) (Cu-13,5Al-4,8Ni; Cu-12,5Al-4,5Mn; Cu-14,7Al-4,3Ni-1Ti).

Here and subsequently, throughout the paper, the chemical composition of the alloys is given in % by weight. The tapes were divided into the following quality categories: I – absence of transverse pores, smooth edges; II – random pores along the length of the tape, minor edge defects (burrs); III – through porosity (waste materials).

The results of experiments have shown that a steel disk, in comparison with the copper one, under the same casting conditions, always forms a tape of a smaller thickness (figure 1). At the same time, the interval of variation in casting modes for obtaining tapes of categories I and II on a steel disk, in comparison with the copper one, is shorter, especially from the side of low pressure of the ejecting gas.

It was found that at pressures $(0.13–0.15)•105 \leq P \leq (0.32–0.35)•105$ PA, the thickness of the tape when casting on a copper disc can be approximated with sufficient accuracy by a linear dependence on the pressure. When the pressure increases $P > 0.35•105$ PA, the dependence of the tape thickness deviates from the linear one. In this case, the casting process is always accompanied by the appearance of a "back" release of the melt and a decrease in the quality of the tape surface.

The quality of the contact (internal) surface of the tapes is mainly determined by the contact interaction (adhesion) of the disk surface with the melt. Copper disk, unlike steel, when casting both types of alloys, provides a tighter contact with the melt. At the same time, the Brasses, unlike Bronze, in all cases contacts the quenching disk on a smaller area.

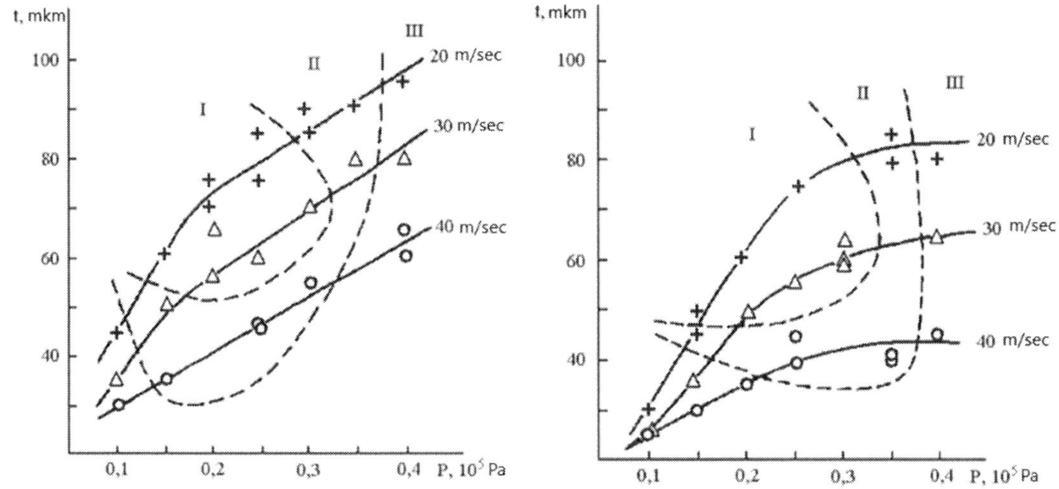

Figure 1. The influence of the pressure of the ejecting gas and the casting speed on the thickness and quality of the tapes when spinning the melt on: copper disk (a); steel disk (b). I, II, III – areas of tape thickness of the corresponding categories. Alloys of the CuAl system (Ni, Mn) (bronze), nozzle b = 0.6 mm, gap h = 0.25 mm.

It is also found that the main cause of low adhesion of the Brass tape is the condensation of Zinc vapors from the melt on the quenching surface of the disk, which leads to a violation of their mutual thermal contact.

The results obtained made it possible to reasonably select the intervals of technological factors variation and ultimately obtain tapes of categories I and II in order to further optimize the spinning process.

The regression equations describing the influence of the main technological factors on the thickness of the tape (Y) are derived using the experiment planning method. When casting alloys of the CuAl(Ni,Mn) system, this dependence is represented by a polynomial of the first degree:

$$Y = 57.5 - 16.3x1 + 15x2 + 5x3 + 8.8x4 \tag{1}$$

For alloys of the CuZnAl system, this dependence has the form:

$$Y = 64.9 - 13.5x1 + 10.2x2 + 7.6x4\,, \tag{2}$$

where: xi is the encoded (–1÷ +1) value of the Xi factor. The following technological factors were selected: X1 – circumferential speed of the disk (20÷40) m / s; X2 – pressure of the ejecting gas (0.15÷0.35)•105 PA; X3 – the gap between the nozzle and the quenching surface of the disk (0.2÷0.3) mm; X4 – the width of the nozzle gap (0.3÷0.5) mm.

Based on equations 1, 2, it is concluded that the thickness of the Brass tape is independent of the gap (factor X3) and the values of all coefficients for variable factors are relatively low compared to the Bronze tape. The results obtained are an additional confirmation of the weak adhesion of the Brass melt to the quenching surface of the disk.

The presence of texture is noted as one of the features of the structural state of the tapes. In cross-section, the grains have a columnar shape and are located perpendicular to the contact surface of the tape. In the longitudinal section, there is a tendency to tilt the crystal axis at an angle of 15÷20 in the direction of the tape movement during casting.

X-ray diffraction analysis of the tapes showed that in all cases, the axis of the preferred orientation of the grains perpendicular to the outer surfaces of the tape is the direction [110] β1.

Studies conducted using metals with a different crystal lattice from the Body-Centered Cubic alloys – BCC (Cu, Al, Zn) revealed a correlation between the type of crystal lattice at the time of crystallization, the preferred orientation of the grains and the main direction of heat removal.

It is shown that the grains with the orientation [111]HCC and [001]GPU are located perpendicular to the outer surfaces of the tapes (the main direction of the heat removal). On the basis of the Le Chatelier-Brown principle, possible causes of this effect are established. The size of the Shape Memory Effect was determined by bending. The deformation was calculated from the outer layer of the sample under the assumption of an ideal plastic body.

In the process of spinning the melt, it was noticed that the tapes are on the surface of the disk for a very short time ($\tau \approx 0.5 \div 2$ micron) and leave its surface in a "hot" state. As a result, there is a strong distortion of the tape from interaction with air, requiring additional heat treatment to correct the shape.

Heat treatment was carried out by holding at the temperature of stable β-phase existence (T≈850 °C) with subsequent quenching in water. Experiments have shown that with increasing exposure time, all characteristic temperatures increase, with the greatest change in the M_l temperature, and the value of the Shape Memory Effect decreases.

X-ray analysis revealed that the phase composition of all alloys after this type of heat treatment does not change, so the change in the properties of the tapes is associated with a change in their microstructure: grain growth, reduction of the number of quenching defects, etc. Grain in CuZnAl alloys has the maximum tendency to grow.

It is shown that intensive grain growth is observed in all alloys at the initial stages of exposure ($\tau \approx 3 \div 5$ min). With a further increase in the holding time, the grain growth occurs at a very low rate. Optical analysis indicates that a noticeable decrease in the grain growth rate, as well as a change in the functional properties of the tapes, are observed when the average grain size becomes commensurate with the thickness of the tape itself. In this case, the outer surface of the tape is a natural obstacle to further increase in grain size.

It is established that the thermal stability of Copper alloys significantly depends on their chemical composition. For example, in conditions of artificial ageing (T≤300 °C), the CuZnAl system has minimal thermal stability, since changes in the functional properties in the alloys of this system are observed after heating to T≈120÷150 °C, and exposure for t = 3 hours at T = 300 °C leads to the complete decay of the matrix phase. Alloys of the CuAl system (Ni,Mn) retain their functional properties up to temperatures T≈200÷250 °C.

CuAlMn system is the most thermally stable of all the studied alloys. The main reason for this is the presence of Manganese, which lowers the temperature of eutectoid transformation ($\beta \leftrightarrow \alpha + \gamma 2$) and the release of the Cu2MnAl during the ageing phase (Geissler phase), which tests materials in areas depleted by Aluminum, which appear in the material due to the diffusion decay of the matrix phase.

X-ray structural analysis has shown that in metastable Copper-based alloys, diffusion processes occurring at the initial stages of artificial ageing do not cause the formation of equilibrium phases (α, γ2, AlNi, etc.), but contribute to the formation of a different type of martensite in relation to the existing one. So, if the structure before heat treatment included martensite with an increased electronic concentration, after the artificial aging, the formation of martensite is characterized by a reduced electronic concentration, and vice versa. Both processes cause the Martensitic Transformation (MT) temperature range to expand and the Shape Memory Effect to decrease. Equilibrium phases appear at later stages of ageing.

It is shown that depending on the chemical composition of the alloy, natural ageing affects the characteristic temperatures in different ways. In the CuZnAl alloy system, all temperatures during ageing remain almost at the same level. In CuAlNi alloys with relatively low temperatures of Martensitic Transformation – MT ($A_t \leq 100 \div 115$ °C) there is an expansion of the temperature range of MT.

However, as a result of thermal cycling through the MT interval, all points return to their original values. At relatively high MT temperatures ($A_t \geq 120$ °C), there is no change in MT temperatures during natural ageing.

The X-ray diffraction analysis shows that changes in the properties of tapes under natural aging conditions occur as a result of diffusion processes that lead to the formation of Guigner-Preston zones in the structure of alloys. The possible causes of this phenomenon are analyzed and recommendations on the choice of material for thermosensitive elements working in reusable and single-use devices are offered.

The methods of creating the Shape Memory Effect in the free state in relation to the chemical composition and loading conditions are studied. The Reversible Shape Memory was set by active deformation of the tapes in the martensitic state, as well as by heating in the deformed and pinched state (the mode of generating reactive stresses). Heating was carried out to temperatures no higher than T=300 °C, since at these temperatures the matrix phase decay processes are intensified. Before testing, the samples were deformed in the martensitic state by the value of Extension Limit (εEL) around the mandrel at an angle of 90°, fixed in this position and subjected to heat treatment under various temperature and time factors.

The advantage of creating the Shape Memory Effect by heating in the deformed and pinched state in comparison with isothermal active deformation is shown. For all alloys, the Shape Memory Effect is formed with a pronounced maximum when heated under conditions of reactive stress generation (figure 2). Moreover, the formation time ε_{RSM}^{max} [RSM = Reversible Shape Memory] decreases both with an increase in the heating temperature and with an increase in the Aluminum content in the alloy.

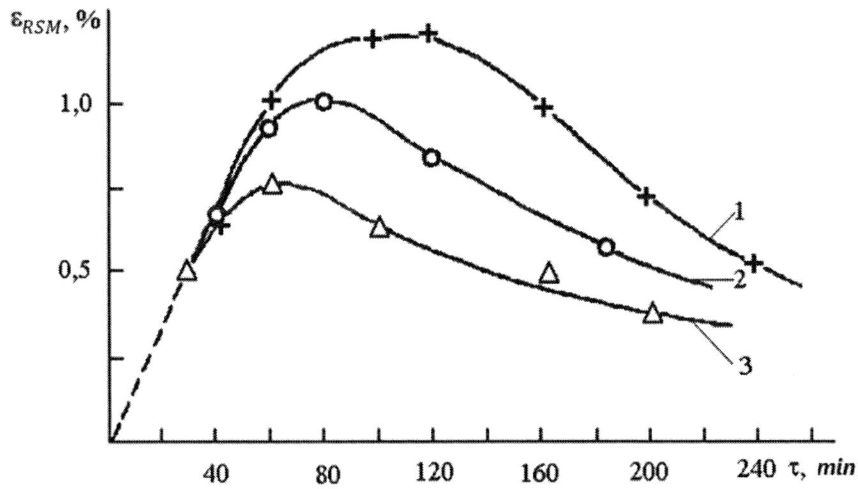

Figure 2. The influence of heating time on the Shape Memory Effect value in alloys: Cu-21Zn-5,8Al (1); Cu-14Zn-7,6Al (2); Cu-10,2Zn-8,4Al (3). T = 130 °C.

Thus, the presence of Aluminum in the alloy reduces the thermodynamic stability of the β1-phase. It was found that with an increase in the processing temperature and a decrease in the A_t temperature, the stability of the Shape Memory Effect formed in this way increases.

With an increase in the duration of exposure at the temperature of the Shape Memory Effect setting, the interval of reversible shaping gradually shifts towards pre-deformation (figure 3).

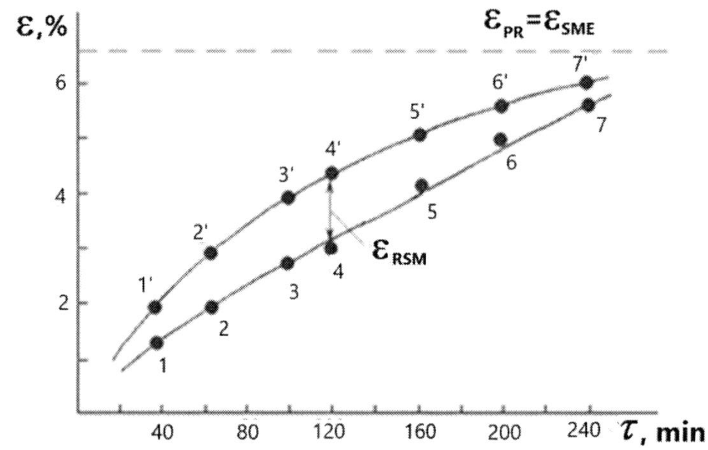

Figure 3. Changes of the Shape Memory Effect value (*a*); location of the area of reversible shaping (*b*) in the alloy Cu-10,2Zn-8,4Al. Ageing at T = 130 °C.

The identified feature makes it possible to create a Shape Memory Effect in any deformation space of 0÷εEL, which is important for the application of this effect in engineering.

It was found that the stability of the Shape Memory Effect created in this way depends on the processing temperature and the A_t temperature.

The reduction in the Shape Memory Effect when thermocyclocyclating in the free state has been determined as the common feature of all alloys. The greatest changes are observed at the initial stages of cycling (8–12 cycles). If the temperature of the Reversible Shape Memory (RSM) formation is close to the terminal temperature of the reverse martensitic transformation MT (A_t), consequently, thermal cycling of 4÷10 cycles through the MT interval results in the complete degradation of the RSM effect.

3. Conclusion

The speed of the quenching surface of the disk (casting speed) has the greatest influence on the thickness of the tape. It was found that the thickness of tapes of satisfactory quality is in a certain range t ≈ (40÷120 microns).

Natural ageing causes an expansion of the MT temperature range and a decrease in the Shape Memory Effect in alloys of the CuAl (Ni, Mn) system. Increasing the ductility of the alloy (CuZnAl system) reduces the tendency to change properties. It is established that the main mechanism for the creation of the Shape Memory Effect is the diffusion of alloy components in the areas of reactive stress initiated by preliminary deformation of martensite. The value of the reversible deformation of the Shape Memory Effect depending on the duration of heating is described by a maximum curve, and the value ε_{RSM}^{max} [RSM = Reversible Shape Memory] is determined only by the degree of pre-deformation of the Extension Limit (εEL) and is observed in the middle part of the deformation space 0÷εEL. The revealed regularities make it possible to create the Reversible Shape Memory in any area of the 0÷εEL deformation space, which is important for applying this effect in engineering.

The stability of the Shape Memory Effect depends on the A_t temperature and decreases when the temperature rises. This fact limits the upper range of performance temperatures of Copper alloys at the level of T≈100÷120 °C.

References

[1] Kruglov A I, Zubarev Y M, Zhelnov D V and Shvetsov I V 2019 *IOP Conf. Ser.: Mater. Sci. Eng.* **656** 012028

[2] Shvetsov I V, Zhelnov D V and Zubarev Y M 2019 *IOP Conf. Ser.: Mater. Sci. Eng.* **656** 012048

[3] Shvetsov I V, Vasilkov D V and Tarikov I Ya 2019 *IOP Conf. Ser.: Mater. Sci. Eng.* **656** 012049

IOP Conf. Series: Materials Science and Engineering **939** (2020) 012043 doi:10.1088/1757-899X/939/1/012043

Research of the dependence of the moisture regime in the walls of gas concrete on the characteristics of the external finishing

V I Loganina[1,2] **and M V Frolov**[1]

[1] Penza State University of Architecture and Construction, Street Titov, 28, 440028 Penza, Russia

[2] E-mail: loganin@mail.ru

Abstract. The laws of moisture condensation in the external enclosing structures of aerated concrete blocks of grades D350-D600, finished with various types of dry building mixtures, are investigated. The mathematical dependence of the temperature of the onset of condensation in the aerated concrete enclosing structure on the thermal conductivity and vapor permeability of the outer finish coating is obtained. It was revealed that when using the developed calcareous composition with the use of ash aluminosilicate microspheres for the finishing of aerated concrete blocks of grades D350–D600, moisture condensation begins at a lower outdoor temperature.

1. Intorduction

In order to reduce energy consumption for heating newly constructed buildings, the requirements for thermal protection of their external building envelopes are being tightened. In this regard, in recent years there has been an increase in the output of aerated concrete blocks, allowing the construction of single-layer walls with sufficiently high heat-shielding properties. Walls of aerated concrete blocks are most often finished with dry building mixes (DBM), which improve the decorative qualities of external fencing and provide them with protection from various climatic influences [1].

If the DBM is incorrectly selected, detachment of fragments of the external finishing coating and the appearance of fine cracks on the coating are observed [2]. Moisture condensation on the plaster – aerated concrete boundary is one of the main causes of the destruction of the finish coating.

In the outer wall for minimize the amount of condensing moisture and reduce the likelihood of its occurrence, each subsequent layer in the direction from the inner to the outer surface should have greater vapor permeability and lower thermal conductivity compared compared with the previous layer.

This condition is difficult to apply to the exterior finish of aerated concrete, because a decrease in thermal conductivity and an increase in the vapor permeability of plaster composites can cause a deterioration in the operational properties of the resulting coatings. But the approximation value indicators of the vapor permeability and thermal conductivity of the finishing coatings to values indicators of the vapor permeability and thermal conductivity aerated concrete will significantly reduce the probability of moisture condensation [3].

Content from this work may be used under the terms of the Creative Commons Attribution 3.0 licence. Any further distribution of this work must maintain attribution to the author(s) and the title of the work, journal citation and DOI.
Published under licence by IOP Publishing Ltd

2. Materials and methods

For decoration of exterior wall from aerated concrete a DBM was used. DBM made based on hydrated lime, modifying additive based on a mixture of hydrosilicates and calcium aluminosilicates, ash silica-alumina microspheres, white cement, milled aerated concrete production waste, Melflux 2651 F, VINNAPAS 8031 H, sodium oleate [4, 5].

Coatings based on the developed DBM are characterized by high vapor permeability $\mu = 0.150$ mg / (m · h · Pa), low thermal conductivity $\lambda = 0.137$ W / (m · K), a density close to the density of aerated concrete $\rho = 650$ kg / m^3 and good operational properties [6, 7]. It is assumed that the use of this DBM will reduce the likelihood of condensation and minimize its amount.

The probability of moisture condensation was determined using the Fokin-Vlasov graphoanalytic method [8, 9]. In this method, the distribution of E_i and e_i over the thickness of the enclosure at a certain temperature t_{ext} is built. The disadvantage of this method is the dependence of the conclusion about the possibility of condensation on the choice of the calculated outdoor temperature. In various works, the average temperature of the coldest five-day period, the average temperature of the coldest month, and the average temperature of the entire period of the year with negative average daily temperatures are used for this. Depending on the temperature adopted, conclusions about the presence or absence of moisture condensation in this enclosure can vary greatly.

In the works of Kupriyanov V.N. it was proposed to assess the probability of moisture condensation in the external enclosure the temperature of condensation onset $t_{o.c.}$ [10, 11]. This temperature is equal to the maximum temperature at which moisture condensation begins in the external fencing . This technique has been used in research. Design scheme of the wall under study is shown in figure 1.

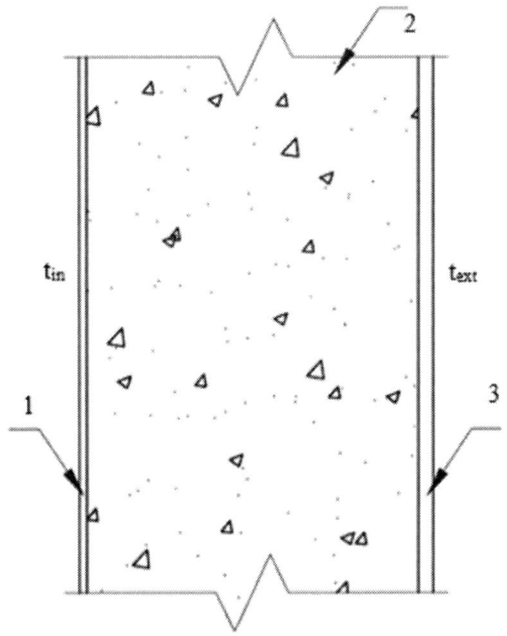

Figure 1. The design scheme of the wall: 1 – layer 1, interior decoration; 2 – layer 2, aerated concrete; 3 – layer 3, exterior finish.

As an interior finish (figure 1, layer 1), cement-slag plaster was adopted. Aerated concrete blocks of grades D350, D400, D500, D600 are accepted as the main material of the wall (figure 1, layer 2). As the exterior finish (figure 1, layer 3), 3 types of DBM were used:

– cement-sand plaster: density $\rho = 1800$ kg / m^3, thermal conductivity coefficient $\lambda_A = 0.760$ W / (m · K), vapor permeability coefficient $\mu = 0.090$ mg / (m · h · Pa);

– Knauf GRUNBAND: $\rho = 1100$ kg / m^3, $\lambda_A = 0.350$ W / (m · K), $\mu = 0.100$ mg / (m · h · Pa);

– developed dry mix DBM: $\rho = 650$ kg / m^3, $\lambda_A = 0.155$ W / (m · K), $\mu = 0.150$ mg / (m · h · Pa).

The calculated parameters of the indoor air are taken equal to the parameters of the indoor air for residential buildings: temperature t_{in} = 20.0 °C; relative humidity φ_{in} = 55% [12]. The calculated parameters of outdoor air: average temperature of the heating period $t_{h.p.}$ = –2.5 ° C; the duration of the heating period $z_{h.p.}$ = 200 days; the humidity zone is dry [13].

3. Research results

In the work, the influence of the external finishing coating on the temperature of the beginning of condensation $t_{o.c.}$ for walls of buildings located in the city of Voronezh is assessed.

For walls made of aerated concrete blocks of grades D350, D400, D500, D600 for the conditions of the city of Voronezh, the minimum allowable thickness of the aerated concrete layer in building envelopes was preliminarily determined (table 1).

Table 1. Aerated concrete layer thickness, m.

Aerated concrete brand			
D350	D400	D500	D600
0.40	0.40	0.50	0.60

To simplify further, the following conventions are used in the work for various of enclosing structures:

$$\frac{x}{y} \tag{1}$$

where x is the density of aerated concrete;
y – the density of the outer finishing layer.

As an example, the humidity mode in the aerated concrete structure 400/1800 are investigated. Each wall layer is characterized by its vapor permeability resistance $R_{p.r.}$ and heat transfer resistance $R_{t.r.}$, determined by the formulas:

$$R_{p.i.} = \frac{\delta_i}{\mu_i} \tag{2}$$

Where δ_i is the thickness of the layer, m;
μ_i – coefficient of vapor permeability of the layer, mg / (m · h · Pa).

$$R_{t.i.} = \frac{\delta_i}{\lambda_i} \tag{3}$$

where λ_i is the thermal conductivity of the layer, W / (m · K).

Figure 2 shows a graph in which the abscissa shows the vapor permeation resistance of the entire enclosure $\sum R_{p.i} / R_{p.en.}$ in relative units, and the ordinate axis shows the heat transfer resistance $\sum R_{h.i} / R_{h.en.}$ of the entire fence in relative units.

In figure 2, the inner surface of the enclosure is indicated by point A, the outer surface by point O. For the embodiment of the enclosure 400/1800, the relationship between the heat transfer resistance and the resistance is a broken line with a kink at point V, the boundary between the interior and aerated concrete, and a kink at point B, the border between aerated concrete and exterior finish (figure 2, curve 1). Points V and B are characteristic for schedules of all considered fencing options.

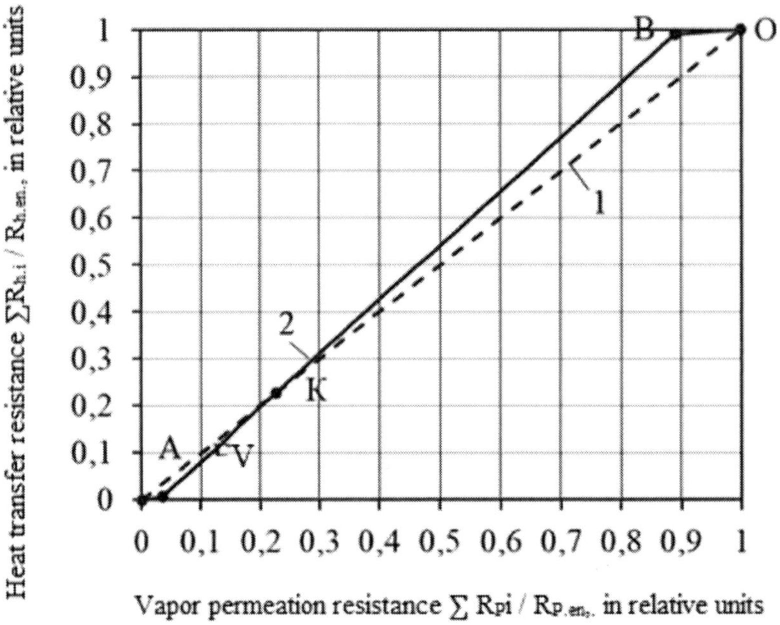

Figure 2. The graph of the relationship between the resistance to heat transfer and resistance to vapor permeation: 1 – 400/1800; 2 – diagonal of proportionality.

We draw an auxiliary diagonal of proportionality connecting the points A and O (figure 2, curve 2). This line shows the relationship between heat transfer resistance and vapor permeation resistance, typical for single-layer structures. At the intersection of curve 1 and 2, we mark point K. The segment AVK is located below the proportional diagonal, therefore, the inequality will be fulfilled throughout the length of this zone in the wall:

$$\frac{\sum R_{p.i.}}{R_{p.en.}} > \frac{\sum R_{h.i.}}{R_{h.en.}} \quad (4)$$

The segment of the KBO is located above the proportional diagonal, therefore, the following inequality will be fulfilled for the entire length of this zone in the wall:

$$\frac{\sum R_{p.i.}}{R_{p.en.}} < \frac{\sum R_{h.i.}}{R_{h.en.}} \quad (5)$$

The most likely condensation plane in the enclosure will be point B, which is the boundary between the aerated concrete and the exterior, as this point is farthest up from the proportional diagonal. In point B, for this enclosure, the following equality holds:

$$\frac{\sum R_{h.i.}}{R_{h.en.}} - \frac{\sum R_{p.i.}}{R_{p.en.}} = max \quad (6)$$

Let's look at fragments of the graphs of the relationships between heat transfer resistance and vapor permeability for enclosure 400/1800, 400/1100, 400/650 within 0.8 $<\sum R_{pi} / R_{p.en.} <1.0$, 0.8 $<\sum R_{h.i.} / R_{h.en.} <1.0$ (figure 3).

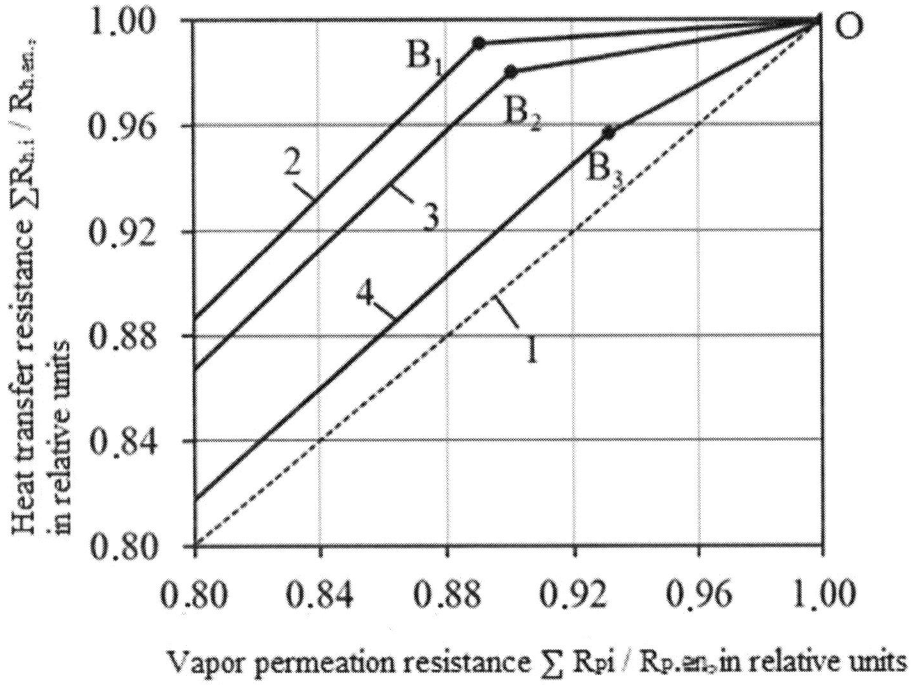

Figure 3. A fragment of the graph of the relationship between the heat transfer resistance and the vapor permeation resistance in the range of 0.8 <$\sum R_{p.i} / R_{p.en}$ <1.0, 0.8 <$\sum R_{h.i.} / R_{h.en.}$ <1.0: 1 – single-layer fencing; 2 – 400/1800; 3 – 400/1100; 4 – 400/650.

When using cement-sand plaster as the exterior finish, point B_1 is located as far as possible from the diagonal of proportionality (figure 3, curve 2).Based on this, it can be assumed that moisture condensation in the enclosure 400/1800 will begin at the highest temperature. When using DBM Knauf GRUNBAND, point B_2 is located closer to the proportional diagonal in comparison with point B_1 (figure 3, curve 3).

Based on this, it can be assumed that moisture condensation in the enclosure 400/1100 will begin at a lower outdoor temperature, while points B_1 and B_2 are close, therefore, the difference between $t_{o.c.}$ for these enclosure will not be significant. When using the developed DBM as an external finish, point B_3 is located as close as possible to the proportional diagonal (figure 3. curve 4). Based on this, it can be assumed that moisture condensation in the 400/650 enclosure will begin at the lowest temperature.

To determine the temperature of the onset of condensation $t_{o.c.}$. for each of the studied structures, the profiles of saturated water vapor pressure E_i and partial pressure of water vapor e_i were constructed. Then, $t_{o.c.}$ was determined by successive approximation. The results of the calculations are presented in figure 4.

Moisture condensation in the enclosing structure 600/1800 starts at – 8.0 °C (figure 4, curve 1). When using DBM Knauf GRUNBAND, the temperature of the onset of condensation $t_{o.c.}$. decreases by only 1 °C to –9.0 °C (figure 4, curve 2), when using the developed DBM, the temperature of onset of condensation $t_{o.c.}$. decreases by 3.6 °C to –11.6 °C (figure 4, curve 3). With a decrease in the density of aerated concrete is increasing the onset temperature of condensation $t_{o.c.}$. Condensation onset temperature $t_{o.c.}$ in the enclosure 350/650 lower by 6 °C than the temperature of the beginning of condensation $t_{o.c.}$ in the enclosing structure 350/1800.

It has been established that in building enclosing structure plastered with a cement-sand composition, condensation will form at average monthly temperatures in December, January and February in walls from aerated concrete blocks of grades D350, D400.

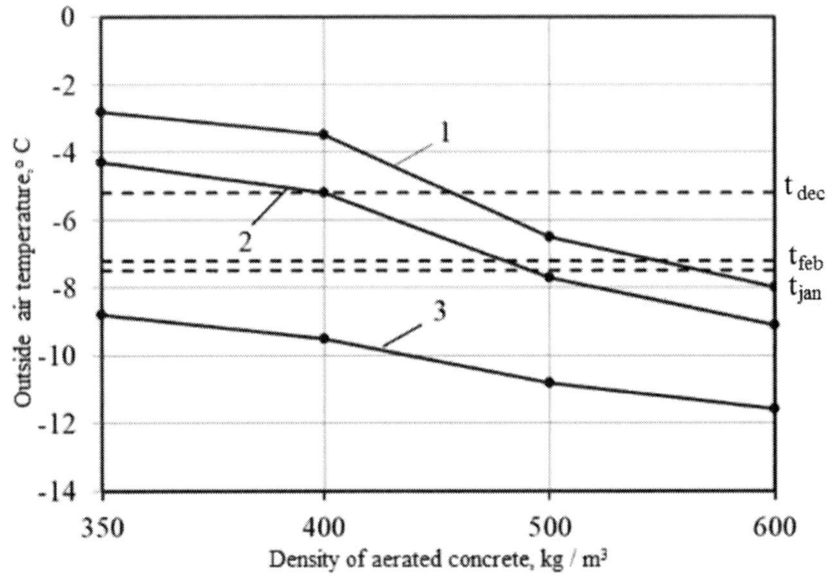

Figure 4. The dependence of the temperature of the onset of condensation $t_{o.c.}$ from the density of aerated concrete: 1 – cement-sand plaster; 2 – Knauf GRUNBAND; 3 – developing DBM.

Condensation also forms at average monthly temperatures in January and February in walls from aerated concrete blocks of grades D500. In the enclosing structures finished by the developed DBM, conditions for the formation of condensate will not be created.

4. Conclusion

The conducted studies prove that due to the use of the developed DBM for decorating walls from aerated concrete blocks of grades D350–D600, the temperature of the onset of condensation $t_{o.c.}$ is reduced. This will significantly limit or completely eliminate the formation of condensate in such building envelopes.

References

[1] Greenfeld G I 2011 *Engineering solutions to ensure energy efficiency of buildings. Finishing masonry from autoclaved aerated concrete: a Training manual* (St. Petersburg: Publishing House of the Polytechnic University) p 130

[2] Vatin V I, Gorshkov A S and Glumov A V 2011 Effect of physical, technical and geometrical characteristics of plasters on the walls of homogeneous humidity conditions of concrete blocks *Civil Engineering Journal* **1** 28–33

[3] Kornienko S V, Vatin N I, Petrychenko M R and Gorshkov A S 2015 Evaluation humidity conditions multilayer wall structure in the annual cycle *Construction of unique buildings and structures* **6** 19–33

[4] Loganina V I and Frolov M V 2019 Research of Cracking Resistance of Thermal Insulation Coatings for Aerated Concrete *Materials Science Forum* **974** 458–463

[5] Loganina V I and Frolov M V 2019 *IOP Conf. Ser.: Mater. Sci. Eng.* ***471*** *032010*

[6] Loganina V I, Frolov M V and Skachkov Yu P 2018 Substantion of selection of componets at creation of thermal insulating dry building mixtures *International Journal of*

Engineering &Technology **7(4)** 4341–4344
[7] Loganina V I, Frolov M V and Skachkov Yu P 2016 Lime composition for the walls of buildings made of aerated concrete *Advances in Engineering Research* **93** 165–169
[8] Guo Xing-guo and Chen You-ming 2007 Computational analysis of heat and moisture performance of multilayer wall in south China. *Journal of central south university of technology* **14(3)** 106–110
[9] Su X and Ang H 2003 A method to solve the multilayer non-continuity problem for heat and moisture transfer in building structures *Communications in numerical methods in engineering* **19(12)** 933–944
[10] Kupriyanov V N 2014 The influence of the outer facing layers on the processes of condensation of vaporous moisture in building envelopes *Bulletin VRO RAASN* **17** 275–282
[11] Kupriyanov V N, Safin I Sh and Shamsutdinov M R 2012 The influence of the design of the fence on the condensation of vaporous moisture *Housing* **6** 29–31
[12] SP 50.13330.2012. Thermal protection of buildings. Updated edition of SNiP 23-02-2003
[13] SP 131.13330.2018 "SNiP 23-01-99 Construction climatology"

IOP Conf. Series: Materials Science and Engineering **939** (2020) 012044 doi:10.1088/1757-899X/939/1/012044

Convergent use of neural network information technologies for monitoring and evaluating the performance of digital control operators of situational centers at critical infrastructure facilities

E L Loginov[1,6], V V Grigoriev[2], P A Bojko[3], V S Balandin[4] and G S Ivanovskij[5]

[1] Expert and Analytical Service, Situation and Analytical Center of the Ministry of Energy of Russia, 42 Schepkina, Moscow, 129090, Russia
[2] Faculty of International Economic Relations, MGIMO (University) of the MFA of Russia, 76 Vernadsky Prospect, Moscow, 119454, Russia
[3] Department of Finance and Credit, Russian State Humanities University, 6 Miusskaya sq., Moscow, 125993, Russia
[4] Corporate Finance and Corporate Management Department, Financial University under the Government of the RF, 49 Leningradsky Pr., Moscow, 125167, Russia
[5] Faculty of International Economic Relations, MGIMO (University) of the MFA of Russia, 76 Vernadsky Prospect, Moscow, 119454, Russia

[6] E-mail: loginovel@mail.ru

Abstract. In the technology under consideration is proposed the convergent adaptation of neuro-monitoring testing services and packages of intelligent information coming to an individual by neuro-monitoring readout of the interests, emotions and mood of the observed standardized conditional operator with the measuring of workload exposed on the main sensor perceptional channels of a concrete digital management operator within smart community, meta-program and others. There are formed the conditions for the provision of a human with information taking into account the characteristics of the individual intelligence structure: its unity, integrity, and systematization within the behavioral model aimed at the duty performance. At the same time the information packages are formed in an automated way correlating with his personal socio-political position and attitude to the policy of the agents of control of smart-community segments with the future forecast of the possible deviant expressions in the network.

1. Introduction

Nowadays digital management operator's network access to the virtual reality is more and more becoming not only the way of getting information and doing management but also, which is by the way highly important, an interactive immersion into artificially constructed virtual worlds. For instance, a situational center operator of a nuclear power plant, a space system, a military compound is a digital management engineer with increased intellectual workload and spends a lot of time in a state in which his inner world is affected by the virtual reality within smart community and with feedback effects [1]. That poses high risks associated with the performance of duty which are attributable to

Content from this work may be used under the terms of the Creative Commons Attribution 3.0 licence. Any further distribution of this work must maintain attribution to the author(s) and the title of the work, journal citation and DOI.
Published under licence by IOP Publishing Ltd

changes in the psyche as well as in the way of ambient reality interpretation and in the patterns of situational behavior of a standardized conditional operator of digital management [2]. Neuro-monitoring in the forms of its various tools allows the construction of individual strategies of immersion into virtual environment taking into account the cognitive and psycho-semantic characteristics of a specific engineer staff representative [3].

2. Electronic semantization of states of consciousness and psyche as the ground of the imprinting of a duty-centered behavior pattern in the mind of a digital operator

The development of such technology within the application of neuro-monitoring technologies for the effective solution of the difficult cognitive problems, which are important in the swart-community by the immersion of digital management operators into the virtual reality, that more and more often satisfies their basic psychological needs, that can be hardly fulfilled in the real life, was stimulated by the necessity of the development of a constructive digital toolset for the risk reduction [4]. This toolset uses the new opportunities as a result of electronic semantization and semantic interpretation of states of consciousness and psyche including uncertain meanings, emotional responses and cognitive activity [5].

Technological abilities of the electronic semantization of states of consciousness and psyche allow the identification, i.e. mathematical description and analysis of the specific digital management operator's behavior by his work in the cyber-physical system, which is recorded with the use of electronic monitoring tools [6].

The characteristics of conscious and psyche of a standardized conditional digital management operator determine his approaches to the handling of intellectual information which is primary in the virtual environment. In perspective, in the feedback mode they also take part in the development of consciousness of digital management operators within smart-community as a complex quasi-reality image. In this case the human is understood as a person comprehending the intellectual information and at the same time as the user of electronic communication services in a broad sense.

Electronic semantization is a semantics-oriented technology of data organization which allows the "package" within mathematical categories of the characteristics of operators of digital management with all its psycho-semantic and other peculiarities. It also includes prejudices and phobias, unconscious predilections and operational characteristics of thinking, for example, the virtual avatar characteristics, alter-ego, personal web-character in web-forums, messengers, chats, portals, blogs and social networks.

3. Agent based models allowing the regulation of acts of a standardized conditional operator

Agent based methods of the creation of a bank of neuro-descriptions, social descriptions and the descriptions of behavior of the digital management operators within smart-community in the amounts of information of communication networks consist in the construction of a computational model. This model itself represents a combination of agents each with a certain set of features including social characteristics and political preferences, as well as the interests of digital management operators with an increased intelligent workload and, in the first place, the interests of getting a reinforcing self-assessment of the chosen intelligent information package.

Agent based model is the ground of tuning of information systems that implement the elements of virtual reality including complex-structured optical, audio, audio-visual signals, virtual presence effect and others, subconscious and unconscious stereotypes and schemes of functioning of the formalized and non-formalized network collectives. This computational model allows conducting simulations of various real events in regards to the standardized conditional operator and his access to the intelligent network business-services, for instance, to the three-dimensional virtual platforms, which together form an online business office.

Here it's especially important that the information coming to the human forms a system of relations between a person and the ambient reality including the socio-political one and his participation in the processes of the formation of intelligent information electronic content. This is particularly relevant in regards to the generation of a network dynamic electronic content, which is created and constantly modified by the digital management operators.

The target activation of a strong emotional response and concentration can substantially change the forms of the expressions of irritation and aggression and will help to break the psycho-emotional tension of a standardized conditional digital management operator bot by the participation in the street riots but by playing virtual games, "shooting games" or online strategies of virtual space conquest.

4. The formation of cognitive communication functional elements in a system "human – artificial intelligence – machine environment"

For the increase of social effectiveness of control of digital management operators in the virtual reality it's necessary to identify the linkage between the package of information which is interesting to operators including their behavior in the cyber-physical system which is recorded by the electronic monitoring tools with a concrete module of his intelligent predilections in the digital virtual environment. It's necessary, for instance, to find out the indicators of aggressiveness and rigidity of a standardized conditional digital management operator which makes it more efficient to use electronic neuro-monitoring and other testing services including those with his participation in multi-user competitive games.

In other words, it's necessary to form functional elements of the complex cognitive communications in an integrated sociotechnical model "human – artificial intelligence – machine environment", which are rapidly adapting to the intellectual level of the identified, i.e. already studied concrete digital management operator with his personal characteristics, socio-political preferences and interests within smart-community.

The connection of human minds and artificial agents observed here allows the "package" of the studied digital model of a standardized conditional operator and his "digital-twin" in computer networks [7].

An important role here is played by neuro-interfaces – detectors, chips, programmed controllers and others, which allow to receive information about processes that mediate the thinking activity in a remote way [8–10]. Meanwhile, some kind of problem is the non-invasive, i.e. remote, detection of commands given down by the brain of a concrete operator within smart-community [11].

It is proposed to use automated detection and interpretation of mental commands according to indirect signs on the base of monitoring services of typical computers: the fixation of the character and dynamism of keyboard clicks and work with computer mouse, the identification of audio- and video-signals, the frequency of visits to the sites and of sending information messages. That will allow to find out the interrelationships and interferences of one logic constructions (views, predilections) and others. I.e. without the use of medical-nature sensors that are difficult to install and operate.

In this respect it is proposed to study the problem of formation of the authorization effect in the system of social communication and getting of information, i.e. the process of knowledge appropriation and filling them with a personal meaning within a duty-centered behavior model in the condition of mixing of information coming to a human from various perceptional flows.

It is proposed to compare the engineer staff personal data with the amounts of intelligent information needed for the solution of complex cognitive problems of smart community with the developed profiles of reflective matrices according to different extents of the identified operator's life satisfaction as the basic element from which the process of living in an artificially-initiated reality starts. Thus the possibility for forecast is created, including that of non-standard behavior of the receivers of information coming from any possible source within artificial segments of psyche maintained by machines and synchronized with the natural psyche.

5. The imprinting in the human mind of a behavior model of intelligent activity for the solution of complicated cognitive problems of smart-community

The goal is the imprinting in the human mind on the base of his "digital twin" of a behavior model of intelligent activeness for the solution of complicated cognitive problems of smart-communities with the preset characteristics of the impact made on the operators including his behavior in the cyber-physical system. This human more and more often realizes his interests immersing into the virtual reality which not only creates a comfortable business environment but also more and more often becomes a stimulator and a substitute of real life.

It is necessary for the digital management operators within smart-community to create individual imprinting constructions taking into account cognitive, nervous and emotional or other physiological responses of the organism and the management of the quasi-individual trajectories of behavior of the members of socio-economic processes with an increased intelligent workload. It's required, for example, for the maintenance of transformation of a socio-politically active standardized digital management operator from an opponent to the government making deviant acts to the governments client and in the future to partnership with the government.

I.e. it's needed to choose a toolset of management of quasi-individual behavior trajectories of the members of socio-economic processes with an increased intelligent workload. For this goal there can be used the peculiarities of dominant typological regularities in the operation of the cognitive mechanisms by the concurrent stimulation with the use of several perceptional channels of the process of formation of the stable artificial connections for the neuro-cognitive programming of the patterns of situational behavior. To reduce the risks of deviant acts there can be used the cognitive impacts of the virtual environments including the forms of expression of the effect of the person's identification with his digital image present in the global networks with regards to the peculiarities of his representational system and arising from this individual patterns of the attitude to the command signals from governmental structures taking into account the main perceptive channels of a concrete person.

The detection of cognitive and psycho-semantic peculiarities of the identified, in other words, the described and analyzed concrete digital management operator can be most effectively realized on the base of the study of the features of clusters of the expression of several significant factors of informational and cognitive character. In particular, instead of the operator we can put the voter as the participant of the mechanism of electronic democracy in the political market.

The informational base for this can be formed by analyzing the presence of digital management operators within smart-community in the electronic social networks, in the Internet and in other communication systems [12].

Thus, a stable convergence is reached of the factors that are difficult to compare. These are the factors of socio-political preferences and interests of the digital management operators including their behavior when working in the cyber-physical system, recorded with the use of electronic monitoring tools. Information packages brought to a human and adapted to his personality ("digital twin") form a certain mindset with the expansion of the positive intelligent component, of the stimulation of the cognitive activity and of positive emotions.

These effects are reached taking into account the indicators determined in the global information networks and databases which characterize the module of the psycho-metric condition of the digital management operators within smart-community including those of suggestibility and susceptibility in the concrete moment of time in the future.

6. The formation in mind of a standardized conditional operator of an image of quasi-reality helping the solution of the complex cognitive goals

On this base is realized the formation of the neuro-descriptions, social descriptions and semantic descriptions of a standardized conditional digital management operator's behavior within digital behavior with the access to the possibility of finding an algorithmic consensus as the matrix of the interests of agents. This matrix of interests allows to find the solutions that meet the criterion of convergence in regards of the adaptation of the intelligent information packages taken as the elements of a digital quasi-reality with the forecast of emotional unconscious responses. The search for a solution is carried out on the base of forming a digitalized portrait of a standardized conditional operator including the characteristics of his thinking activity, memory, attention which affects his choice of the electronic content including various forms of the avatar-mediated network interactions.

When possible this process is realized with the use of neuro-interfaces: detectors, chips, programmed controllers and others which allow receiving data on the processes mediating the thinking activity in a remote way.

I.e. it's necessary to receive data in the electronic information spaces for the filling of databases containing the needed information with numeral, categorical and temporal attributes which are most significant for the interpretation of reality and the creation of behavioral models for common and emergency situations. It's also worthwhile to use the information from systems allowing facial recognition, human speech understanding and interpretation of the physical activity and emotional condition of the digital management operators within smart-community and to find out his attitude to the ambient reality.

The use of neuro-monitoring testing services as elements of the complex cognitive communications in an integrated sociotechnical system "human – artificial intelligence – machine environment" with a semantic resonance is realized taking into consideration the digitalization of the standardized conditional digital management operator with the network identification of his orthodox intelligent position or interests in the unclear environments of the "over-noised" electronic content. Here the intelligent position and interests can be described with the values – numeric parameters of the matrix of results of the neuro-monitoring testing services and the analysis of the materials from the space of electronic content of neuro-descriptions, social descriptions and the descriptions of the semantics of behavior of the operators.

In the present circumstances it seems to be appropriate to activate the processes of the formation of the profile of psycho-semantic qualities and reflexive matrices depending on the various levels of the life satisfaction of the identified conditional digital management operator as the basic point from which living in an artificially tuned reality can be started. In this way the possibility is provided of the operation of "gliding" of the network formalized and non-formalized groups towards the mutual encouragement of acts beyond the expressions of the behavior model impeding the concentration on the duty performance. The living of an artificially tuned reality is realized as a package of the information adapted to the human together with his described cognitive, nervous, emotional and other physiological responses of the organism. At the same time the nervous-linguistic programming and other similar technologies in the deep field of imprinted images with a respective sense transcription are maintained by the computer technologies and synchronized with the natural psyche with overcoming the disagreement of experiences – the effect of tuned reality with personal experience kept in mind.

7. Conclusion

The virtual reality more and more often satisfies the basic psychological needs at a high level that can be hardly reached in the real life. This determines the positions of the digital management operators within smart-community and the consequential process of formation of the electronic content of intelligent information which is the ground for their solution of complicated and important for smart-community cognitive goals.

I.e. the conditions are formed for the highly effective process of operation of the elements of typical form of thinking in regards to the positioning of a standardized conditional digital management operator in a certain segment of the intelligent information consumption. This information includes not only the professional skills but also socio-political views, stimulating the participation in the groups of some kind. The formation of the information electronic content resulting from here is the ground for the solution of the complex cognitive problems important for the smart-community with the new form of thinking – the artificial psyche maintained by the computer technologies and synchronized with the natural psyche.

Acknowledgments

The article was written with the financial support of the Russian Fundamental Research Fund (project № 19-07-01066 A «The strategy of implementation of elements of the Russian digital economy for optimization of interaction between the aggregated groups of economic agents on the base of development of digital asset logistics and intelligent mobility»).

References

[1] Loginov E and Shkuta A 2017 Artificial intelligence in the government bodies *Gos. Sl.* **5** 24–29
[2] Chasovskij N and Chasovskij P 2019 On the interference between the virtual and the real world in the modern communication space *Vest. Chel. Gos. Un.* **1** 151–58
[3] Kovsh E, Ermakov P and Vorob'eva E 2016 The reflection of the level of aggressiveness and hostility in the caused brain activity of men by the assessment of the emotionally colored stimuli *Vest. Len. Gos. Un.* vol 4 **1** 55–66
[4] Loginov E, Rajkov A and Shkuta A 2018 The use of neuro-technologies by the programming of the cognitive-behavioral stereotypes of acts of a person for the stable functioning of the systems of civil governance *Nejr.: Raz., Prim.* **9** 34–45
[5] Ageev A, Loginov E and Shkuta A 2020 Neuro-operating of behavior of cognitive agents on the base of electronic semantic interpretation of the states of consciousness and psyche with the effects of immersion, presence and integration with the virtual reality *Mikr.* **1** 5–12
[6] Ageev A, Loginov E, Shkuta A and Derkach A 2019 The network based cognitive management of the complex structures with a political component in the unclear information environments *Mikr.* **5** 5–13
[7] Rajkov A 2014 The cognitive programming *Ekon. Str.* **4** 108–13
[8] Loginov E, Grigoriev V, Shkuta A, Bortalevich V and Sorokin D 2019 Intelligent monitoring, modelling and regulation information traffic to specify the trajectories of the behavior of organizational agents in the context of receipt of difficult-interpreted information *IOP Conf. Ser.: Mater. Sci. Eng.* **516** 012015
[9] Andreev I and Nazarova L 2014 The neuro-psychic aspects of Internet *Psih. and Psihot.* **7** 701–15
[10] Loginov E, Grigoriev V, Shkuta A, Sorokin D and Bortalevich V 2019 The use of artificial intelligence's elements to block the manifestations of individuals' behavioral activity going beyond the quasi-stable states *IOP Conf. Ser.: Mat. Sci. and Eng.* **516** 012028
[11] D'yakov S 2016 Projective semantics of a subject's psychical self-organization *Prob. Sovr. Ped. Obr.* **53(1)** 279–90
[12] Lojko V, Totuhov K, Kushnir N and Kushnir A 2016 The analysis of factors and indicators of the Internet impact on the intellect with the use of methods of DATA MINING *Polit. Set. El. Nauc. Zhur. Kub. Gos. Agr. Un.* **12**

IOP Conf. Series: Materials Science and Engineering **939** (2020) 012045 doi:10.1088/1757-899X/939/1/012045

The results of the application of thermo-time processing and resonant-pulsating refining in the Siberian region of the Russian Federation

D A Lubyanoy[1,4], E Yu Pudov[1], A V Markidonov[2] and A B Efremenkov[3]

[1] Prokopyevsk Branch of T.F. Gorbachev Kuzbass State Technical University, Nogradskaya str., 32 653039, Prokopyevsk, Kemerovo Oblast, Russia
[2] Россия, Novokuznetsk Institute (Branch) of Kemerovo State University, 23 Tsyolkovskogo str., Novokuznetsk, 654041, Russia
[3] Yaroslav-the-Wise Novgorod State University, B. St. Peterburgskaya str., 41 Veliky Novgorod, 173003, Russia

[4] E-mail: lubjanoy@yandex.ru
[5] E-mail: abe@novsu.ru

Abstract. Currently, in the Siberian region of the Russian Federation, thermo-time processing of the melt and the technology of resonant-pulsating refining of metal are widely used. The article highlights the results of the development and implementation of thermo-time processing and resonant-pulsed refining to improve the properties of the metal. Thermo-time treatment, as a rule, is carried out in induction and electric arc furnaces. It was found that thermo-time treatment allows to significantly improve the quality of the metal, improve the mechanical properties of the products. Resonant-pulsating refining combines melt processing with argon purge and vibrational frequency spectrum processing. The impact of this type of metal blowing on the gas content, the microstructure of cast iron, its mechanical properties, and also the operational properties of the products obtained is evaluated. It was established that the strength properties of cast iron increased by 40–65 MPa, hardness increased by 20 to 30 HB, density – by 100–300 Kg/m3. In addition, it should be noted that the use of the technologies presented in the work can significantly reduce the harmful effects of phosphorus. The operational durability of cast iron products has reached the best domestic and foreign indicators. The developed technologies allowed the implementation to achieve the best performance in the industry with minimal cost. These technologies can be widely used in foundries, electric steelmaking and oxygen-converter shops.

1. Introduction

To increase the service life of machine parts of mining and metallurgical equipment, which is the most important problem of mechanical engineering, technologies for the thermo-time treatment of metal [1, 2] and resonant-pulsed refining (RPR) [3, 4] have been developed and introduced into production. The low life of the products significantly reduces the economic efficiency of technologies and leads to their high metal consumption.

Content from this work may be used under the terms of the Creative Commons Attribution 3.0 licence. Any further distribution of this work must maintain attribution to the author(s) and the title of the work, journal citation and DOI.
Published under licence by IOP Publishing Ltd

2. Results and Discussion

For the optimal technology for producing cast irons with the necessary properties, it is necessary to determine the optimal modes of iron smelting in electric furnaces [1, 2], the content of alloying elements [5], and the rational technology for out-of-furnace metal processing. Under the conditions of AO EVRAZ ZSMK, gray cast irons for heat-resistant castings and white cast irons [6] for wear-resistant mining equipment (brands IChH17, IChH17N2, IChH22N2MD2F) are produced in a foundry equipped with IChT-10M induction furnaces with innovative sintering technology melt.

It should be noted that vanadium contained in cast iron can significantly reduce the manganese content in heat-resistant cast irons. The task is to determine the concentration of manganese in cast iron with an optimal content of both titanium and vanadium. and optimally alloyed gray cast irons, which have maximum operational stability in the manufacture of replaceable metallurgical equipment, such as slag mills (figure1) and molds [7].

Figure1. Slag for mining and smelting enterprises.

It is known that the properties of used white cast irons are determined by the chemical composition of cast iron, manufacturing technology, the amount and size of carbides [1]. The paper [6] emphasizes the study of the effect on the mechanical properties and workability of white cast irons of carbon, chromium, molybdenum and phosphorus, while the effect of titanium and vanadium has not been studied enough, both in studies of the properties of white cast irons and optimally doped gray heat-resistant cast irons [7–9].

The melting of white alloyed cast iron of the IChH17, IChH17N2 and IChH22N2MD2F grades is performed in a hot crucible after the complete discharge of the previous cast iron melting. Nickel is an undesirable impurity in cast iron IChH17, its permissible content is not more than 0.1%. When melting IChH17, cast iron waste IChH17N2 and IChH22N2MD2F cannot be used. When smelting cast iron of the IChH17N2 and IChH22N2MD2F grades, it is possible to use both domestic production waste and IChH17 cast iron in the charge. The components of the charge during the smelting of these grades are loaded into the furnace in the following order: 1) a combined return of their own production of the same grades is loaded onto the bottom of the hot crucible; 2) load all the estimated amount of pig iron; 3) load carbon ferrochrome FH800; 4) load steel scrap and then layer by layer ferrochrome and steel scrap in a ratio of 300/1000. When smelting cast iron IChH22N2MD2F, granular nickel, ferrochrome and ferromolybdenum are loaded into the filling. In the first half of the melting (before melting in a crucible 3–4 tons of liquid metal), the next portion of the charge is loaded into the furnace after the previous one is lowered into the liquid metal. In the second half of the melting, the charge is loaded into the furnace for a well-heated, settled, but not yet melted, previous portion of the filling. Slag during the smelting process is downloaded after the first half of the charge is melted and at the end of the process before adjusting the chemical composition of the metal. When smelting these cast irons for wear-resistant mining equipment, thermo-time processing of the melt (TTP) is mandatory [1, 2]. The essence of the operation lies in the fact that during heating to certain temperatures and holding the melt, homogenization occurs in it and a "short-range order" of the melt is achieved, which improves the mechanical properties and performance of products from wear-resistant cast iron for mining equipment. The temperature of TTP is in the range of 1480–1540°C, the exposure time is 10–15 minutes. When smelting cast iron, IChH17 provide heating up to 1520–1540°C, cast iron IChH17N2 1500–1520°C and cast iron IChH22N2MD2F 1480–1500°C. When smelting cast iron IChH22N2MD2F, overheating of metal above 1500°C is not allowed to avoid subsequent transcrystallization in castings. Deoxidation of the metal in the furnace is carried out by the addition of silicocalcium 10–15 minutes before release from the furnace. Metal is poured into molds from a 2-ton casting bucket. The metal is poured into molds in the presence of a slag cover of molten iron in the casting ladle. The temperature of the metal before leaving the furnace is within the range for IChH17 1520–1540°C, for IChH17N2 1500–1520°C and for IChH22N2MD2F 1400–1420°C. In the ladle before casting, the temperature data must be in the range for IChH17 1460–1480°C, for IChH17N2 1440–1460°C, and for IChH22N2MD2F 1360–1380°C. During the production of wear-resistant cast irons for mining equipment, the chemical composition of cast iron and temperature during smelting are controlled. In the cast iron IChH17 and IChH17N2, the content of carbon, manganese, silicon, chromium and nickel is controlled.

Under the conditions of AO EVRAZ ZSMK, studies were conducted to assess the effect of alloying with titanium and vanadium of gray cast iron [5, 8]. Studies have shown an increase in tensile strength of cast iron.

Figure 2. Large titanium and vanadium carbonitrides in a phosphide eutectic P = 0.30% before purging x800.

As industrial tests have shown, an increase in the strength properties of cast iron leads to a decrease in erosion of products by a falling stream of molten metal. The use of alloying in the conditions of AO EVRAZ ZSMK increased the heat resistance of products [5, 8]. It was noted that the purge of gray cast iron alloyed with titanium and vanadium, nitrogen by the RPR method allows modification of the phosphide eutectic (figure 2, 3) with smaller carbonitrides than before the purge of the metal (figure 4). At the same time, the harmful effect of phosphorus is neutralized. It was established that the strength properties of cast iron increased by 40–65 MPa, hardness increased by 20 to 30 HB, density – by 100–300 Kg/m^3.

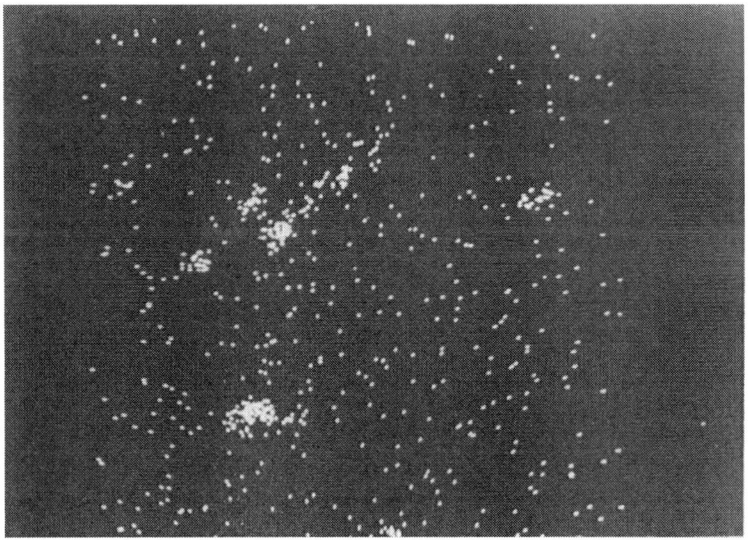

Figure 3. Small carbonitrides in a phosphide eutectic P = 0.30% after purging in resonance mode x1200.

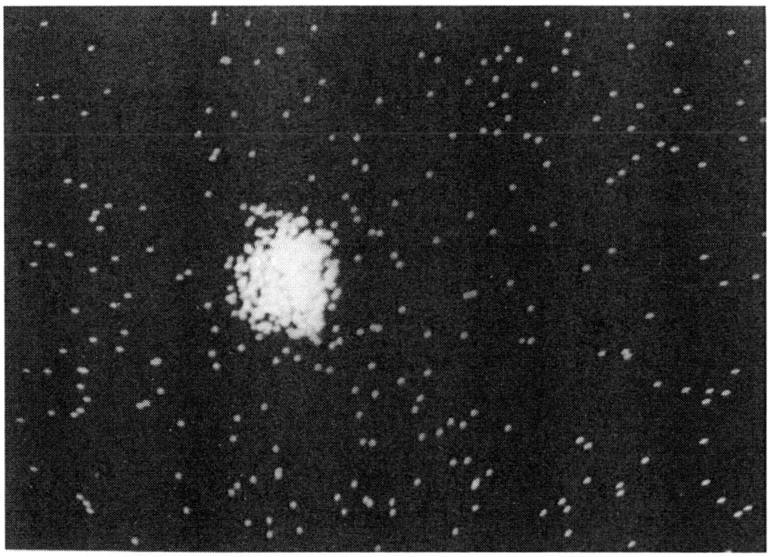

Figure 4. Large carbonitrides in cast iron before purging x1200.

The test results showed that a high content of vanadium and titanium in cast iron leads to a decrease in the fluidity of the metal, and leads to the formation of porosity in the castings. To increase the fluidity, ferrophosphorus (P = 17%) was used in the ratio of 800 kg of alloy to a 90-ton bucket. Industrial tests have shown that shrinkage effects and micropores are practically absent in castings doped with phosphorus due to better filling of the molds. The indicated effect was observed when the phosphorus content in cast iron exceeds 0.21% [3].

The analysis of the main phases of cast iron: titanium carbide, phosphide eutectic, and pearlite matrix was carried out by microstructural analysis [3]. When working out the optimal chemical composition, it is necessary to strive for each specific case to obtain a favorable microstructure of cast iron [8–15]. It was found that naturally alloyed cast iron also has all the characteristics necessary to increase the service life of wear-resistant and heat-resistant products: high strength and wear resistance, and in the presence of phosphide eutectic, also good fluidity.

Products obtained from cast irons subjected to thermo-time treatment and resonantly pulsating refining are characterized by low production costs. These products are widely used at enterprises in Siberia and the Urals. In papers [16–20], other ways to increase the operational reliability, durability, and wear resistance of products were also considered. Based on the results of this work, it should be noted that the issue of improving the wear and tear and durability of products must be approached comprehensively, taking into account, on the one hand, the technology of their production, properties of the materials of the products, and on the other hand, taking into account the optimization of operating conditions.

3. Conclusion

It has been revealed that the developed technologies for thermo-time processing of the melt and resonant-pulsating refining allow producing high-quality equipment from gray and white alloyed cast irons. Products made using this technology satisfy customers' requirements. The technologies were introduced on large induction furnaces IChT-10м and in the mold shop in the production and manufacture of small, medium and large castings. Heat-resistant castings made using these technologies have the best performance in the industry. Thus, microalloying of cast iron with titanium and vanadium significantly increases the strength properties of cast iron, thereby significantly increasing the wear resistance of products operating under conditions of abrasive wear and high temperatures. These cast irons are widely used at enterprises in the Urals and Siberia.

References

[1] Tsepelev V S, Selyanin I F , Lubyanoj D A, Baum B A and V'yukhin V V 1995 *Steel in Translation* **5** 42
[2] Lubyanoi D A 2005 Alternative cast irons production technologies for Kuzbass metallurgy *8th Korea-Russia International Symposium on Science and Technology (Tomsk, 2004)*: *collected papers* (Tomsk Polytechnic University Publ) 131–33
[3] Lubyanoi D A, Perehodov V G, Foigt D B and Buimov D V 2019 Opyt primeneniya rezonansno-pul'siruyushchego rafinirovaniya v AO «YEVRAZ ZSMK» [Experience in the use of resonant-pulsating refining at AO EVRAZ ZSMK] *Ferrous Metals* **6** 31–33
[4] Andreev V V, Lubyanoi D A, Samsonov Y N, Kaminskaya I A and Lubyanaya S V 2014 *Metallurgist* **58** 492
[5] Lubyanoi D A, Gorkavenko V V, Makarov E S, Kaminskaya I A, Frolov A G and Yakovenko N A 2002 *Metal Science and Heat Treatment* **44** 452
[6] Trukhin V and Trukhin A 2014 *Bulletin of Kuzbass State Technical University* **2** 58
[7] Samsonov Yu 2007 *Foundry Production* **4** 9
[8] Valeev D, Zinoveev D, Kondratiev A, Lubyanoi D and Pankratov D 2020 *Metals* **10** 322
[9] Seidu S O and Ogunniyi I O 2013 *Materials Research* **16** 145
[10] Fras E, Gorny M, Kapturkiewicz W and Lopez H F 2008 *Tsinghua Science & Technology* **13** 177

[11] Ahamed M S, Kumar Y V, Rahman J F, Rakesh S G and Bharat V 2014 *International Journal of Engineering Research &Technology* **3** 424
[12] Fras E, Lopez H, Kawalec M and Gorny M 2015 *Metals* **5** 256
[13] Klancnik U, Habjan J, Klancnik G and Medved J 2017 *Journal of Thermal Analysis and Calorimetry* **127** 71
[14] Riposan I, Chisamera M, Stan S and Bartsow M 2011 *China Foundry* **8** 228
[15] Kumruoglu L C 2009 *Materials and Design* **30** 927
[16] Gabelli A, Morales-Espejel G E and Ioannides E 2010 *Evolution* **2** 25
[17] Mohamed A, Sassi S, Paurobally M 2018 *Shock and Vibration* 1913289
[18] Kuzin E, Gerike B, Mamaeva M and Singh K 2019 *E3S Web of Conferences* **105** 03011
[19] Mamaeva M and Kuzin E 2019 *Matec Web of Conferences* **297** 03006
[20] Saidi L, Ben Ali J and Fnaiech F 2014 *ISA Transactions* **53** 1650

Tensor characteristics of intrinsic scattering magnetic field as an effective tool for strength diagnostics of steel products

V G Malinin[1,4], N A Malinina[1], V V Malinin[2], A A Dimov[1] and O A Malukhina[3]

[1] Oryol State Agrarian University, 69, General Rodina St., Orel, 302019, Russia
[2] Tupolev PJSC, 17, Academician Tupolev Embankment, Moscow, 105005, Russia
[3] Yaroslav-the-Wise Novgorod State University, 41, ul. B. Sankt-Peterburgskaya, Veliky Novgorod, 173003, Russia

[4] E-mail: malinin.mvg@yandex.ru

Abstract. On the basis of the methods of structural-analytical mesomechanics, tensor characteristics of the intrinsic magnetic field of scattering are formulated, their experimental substantiation is given, and studies allowing performing the operational quality control of steel products of responsible use are given.

1. Introduction

The method of magnetic memory of metals allows studying the structural and technological heredity of engineering products quickly and with good reliability. Based on this method, the authors developed a methodology for controlling the quality of new products [2, 3], with the aim of determining their predisposition to be damaged during operation.

This article presents a development of the magnetic memory method, introduces tensor parameters of the intrinsic scattering magnetic field (ISMF) based on which experimental studies of the zones of structural concentrators over the entire working part of steel bolts are performed. The data on the zones of structural concentrators in new bolted products are presented, which made it possible to develop an operational methodology for monitoring their quality.

2. Structural characteristics of the magneto-mechanical effect

In [1–5], the feasibility of using the magnetic memory method to solve the problem of determining the stress-strain state (SSS) of materials and products was convincingly substantiated, which allows us to analyze the development of zones of structural concentrators (ZSC) in the process of material loading based on the measurement of the intrinsic scattering magnetic field (ISMF) parameters. In view of the foregoing, an important task is the creation of experimental-theoretical models that reflect the basic laws of evolution of structural stresses based on recording parameters of the ISMF. It is necessary to introduce appropriate conceptual representations and mathematical tools.

Let us consider a possible solution to the problem. Functional relationships between parameters of different physical contents can exist only for mathematical objects of the same tensor valency according to the Curie Inhibition principle. We will consider the most common version of the problem for definiteness, when the structural and stress-strain state can be characterized by tensors of zero, first and second valency.

Content from this work may be used under the terms of the Creative Commons Attribution 3.0 licence. Any further distribution of this work must maintain attribution to the author(s) and the title of the work, journal citation and DOI.
Published under licence by IOP Publishing Ltd

2.1. *Scattering intrinsic magnetic field vector*

By measuring the magnetic field strength H_i, at various points of the deformed body, it is possible to determine the distribution of H_i over the coordinates of the test article. Experiments were performed on thin steel plates to verify the existence of ISMF in the sample, after deformation of the material as a vector object. We studied plates without concentrators and with various macroconcentrators in the form of central round and elliptical holes, side cuts of various configurations made of steel 3, steel 45, steel 09G2S. We used a commercially available specialized monitoring device ISMF developed by the enterprise OOO "Energodiagnostika" and a three-component flux-probe. In particular, a magnetometric stress concentration meter IKN-3M-12 was used, with a microprocessor-based recording device. H_i measurements were performed at various points in the sample, and the components of the H_i vector were scanned at each point, turning the sensor by 30°, 45°, 60°, and 90° degrees. Each experiment was repeated three times. The measurement results confirmed the invariance of H_i, i.e. the law of conservation of the tensor of the first rank was satisfied when the coordinate system was rotated

$$H_i^{'} = a_{ik} H_k, \tag{1}$$

where α_{ik} - are guide cosines. The measurement error according to the estimate of the invariant (H_i H_i) 1/2 did not exceed (1-3)%, and the scatter for individual H_i components did not exceed 5%. Moreover, the scatter was the smallest in those places where there was significant plastic deformation. The data obtained convince that the methods of measuring ISMF completely satisfy the tasks of studying the structural-mechanical state of the material.

2.2. *Distortion tensor of the intrinsic scattering magnetic field*

The next stage of experimental research was to measure the components of the ISMF over the entire working part of the deformable plate during loading. The dimensions of the working part of the plate were 60x120 2 mm. A square grid was applied with a cell size of 5 x 5 mm. The components of the vector H_i were measured at the grid nodes; the total number was 11 x 11 = 121 knots. Three measurements of the H_i components were performed in each node, and then the usual statistical processing was performed. The whole diagram of plate tension was obtained under conditions of step loading; the number of steps was 10-12, up to failure. The total number of measurements in each experiment was in the range of 1.5–2 thousand. The obtained results on the distribution of the vector field H_i made it possible to introduce into consideration the ISMF distortion tensor as a gradient from the intensity vector Hi by the formula

$$H_{ik} = \nabla_i H_k, \tag{2}$$

where ∇_i – Nabla operator [2].

Processing of the experimental data using the formula (2) made it possible to obtain an experimentally substantiated mathematical object characterizing the heterogeneity of the ISMF. The next step was to verify the parameters of Hik (2) experimentally to preserve H_{ik} as a tensor of the second rank when the coordinate system is rotated by angles: 30°, 45°, 60°, and 90° degrees. The results of processing the obtained data showed satisfactory compliance with the conservation law

$$H_i' = a_{ip} a_{kq} H_{pq}. \tag{3}$$

The scatter in the values of the first, second, and third H_{ik} invariants was within 15%, which confirmed the tensor character of the H_{ik} parameters. It should be noted that these experimental studies required the creation of original snap-ins, each experiment lasted 4–5 hours. A special experimental study was carried out, confirming the tensor character of the introduced parameter H_{ik}, called the distortion of the intrinsic scattering magnetic field.

Based on the H_{ik} distortion tensor, symmetric H_{ik}^s and antisymmetric H_{ik}^a tensors were introduced:

$$H_{ik}^s = \frac{1}{2}(\nabla_i H_k - \nabla_k H_i); \; H_{ik}^a = \frac{1}{2}(\nabla_i H_k - \nabla_k H_i). \tag{4}$$

Introduction H^s_{ik} allows the development of symmetric structural mechanics in accordance with the traditional symmetric mechanics of a deformable solid. The introduction of the parameter H^a_{ik} opens up the possibility of developing structural models that take into account moment structural stresses and the corresponding bending-torsion strains.

2.3. Tensor intensity of intrinsic scattering magnetic field

From the total distortion tensor H_{ik} (3), we can distinguish the physical significant tensor Z_{ik}, called the tensor intensity of the ISMF in the form:

$$Z_{ik} = Z_{13}\delta_{i1}\delta_{k3} + Z_{23}\delta_{i2}\delta_{k3}. \quad (5)$$

where $Z_{13} = H_{13}; Z_{23} = H_{23}; \delta_{ik}$ – is Kronecker symbol. Considering that the symmetric stress tensor is used in this work, it is advisable to construct a symmetric tensor $Z^s_{ik} = S_{ik}$:

$$S_{ik} = \frac{1}{2}H_{13}(\delta_{i1}\delta_{k3} + \delta_{i3}\delta_{k1}) + \frac{1}{2}H_{23}(\delta_{i2}\delta_{k3} + \delta_{i3}\delta_{k2}). \quad (6)$$

Cubic equation for determining the main values $S_{ik}(k = 1,2,3)$ is written as

$$S_k^3 - J_1^S S_k^2 - J_2^S S_k - J_3^S = 0;$$

where $J_1^S = S_{ii} = 0;\ J_3^S = |S_{ij}| = 0;\ J_2^S = \frac{1}{2}S_{ik}S_{ik} = \frac{1}{4}(S_{13}^2 + S_{23}^2);\ S_{13} = H_{13}; S_{23} = H_{23}.$

Solving the cubic equation, we obtain the following principal values of the tensor S_{ik}

$$S_{1,3} = \pm\frac{1}{2}\sqrt{S_{13}^2 + S_{23}^2};\ S_2 = 0;\quad S_1 \geq S_2 \geq S_3.$$

Intensity S_i of the tensor S_{ik}, and also the parameter S^{τ}_{max} is written as:

$$S_i = \sqrt{J_2^S} = \frac{1}{2}\sqrt{S_{13}^2 + S_{23}^2};\qquad S^{\tau}_{max} = \frac{1}{2}(S_1 - S_3) = \frac{1}{2}\sqrt{S_{13}^2 + S_{23}^2}. \quad (7)$$

Analyzing relations (7), it can be noted that the symmetric part of the tensor intensity of ISMF (6) has only the second invariant different from zero $\frac{1}{2}S_{ik}S_{ik}$, through which the main values S_1, S_3 are expressed and also the parameters S^{τ}_{max} and S_i.

2.4. Structural criterion of the limiting state of a material in stress concentration zones (SCZ)

Let us give comments confirming the appropriateness of the introduced tensor intensity ISMF Z_{ik} in the form (7) for the formulation of the criterion for destruction reflecting the influence of zones of structural concentrators [6]. The invariant $S_i = \frac{1}{2}(S_{ik}S_{ik})^{1/2}$ in the vicinity of the zone of the structural concentrator, with the accuracy up to a constant coefficient $\left(1/\sqrt{2}\right)$, coincides with the K_{in} parameter called the gradient of the normal component of the intrinsic magnetic field K_{in} [6–9]. The K_{in} parameter in the magnetic memory method (MMM) plays a key role as a diagnostic parameter for assessing the state of a metal in stress concentration zones (SCZ) [6, 7]. Numerous experimental studies have made it possible to establish a relationship between the dislocation density ρ, strain ε, and the ISMF parameter K_{in}.

Figure 1 and figure 2 show the dependence of the dislocation density ρ on the degree of plastic deformation ε, and figure 3 the dependence of the relative deformation ε (1) and the gradient of the normal component of the K_{in} scattering magnetic field (2) on the tensile stress σ in the macroplastic region [6].

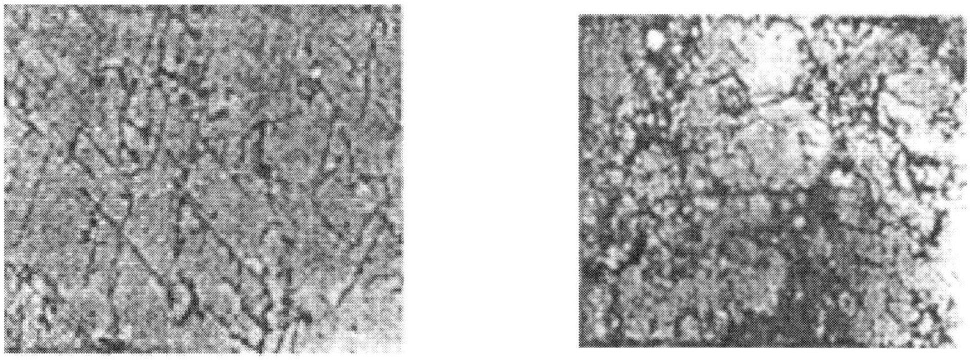

a) *b)*

Figure 1. The dislocation structure of steel 10 after plastic deformation ε under static loading: $a - \varepsilon = 2\%$, ×20000; $b - \varepsilon = 6\%$, ×20000 [6].

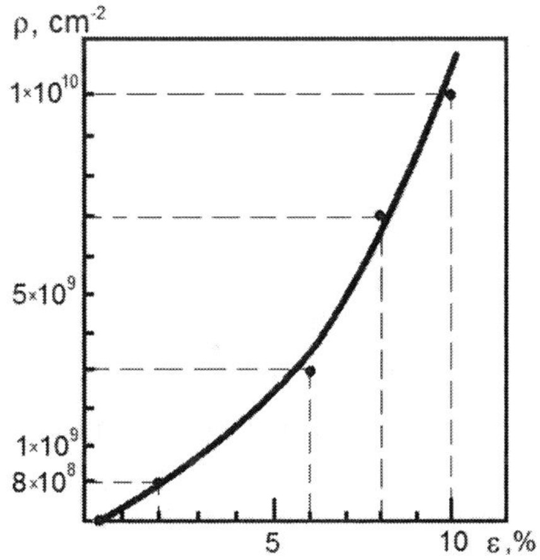

Figure 2. Dependence of the dislocation density ρ along the line of stress concentration ($H_p = 0$) on the degree of plastic deformation ε [6].

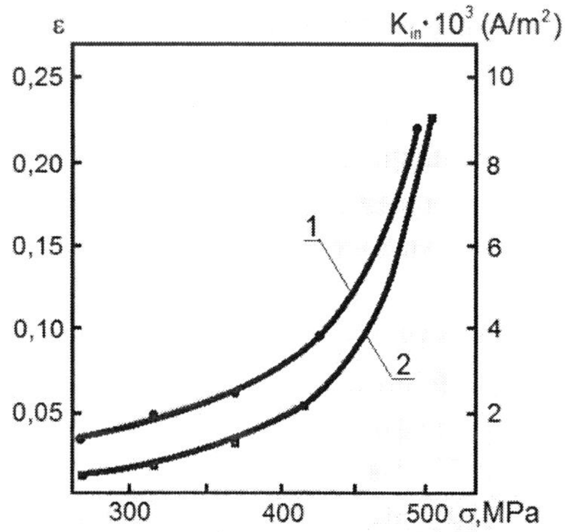

Figure 3. Dependences of the relative elongation ε (1) and the gradient of the normal component of the K_{in} scattering magnetic field (2) on tensile stresses σ in the region of plastic deformation [6].

Comparing the experimental data presented in figure 2 and figure 3, it can be noted that the K_{in} parameter, strain ε and the corresponding dislocation density ρ, as a result of the action of actual stresses, are integral characteristics of the state of the metal of the product in SCZ [6]. It is known that the main sources of damage occurrence and development are SCZ, in which the processes of fatigue, corrosion and creep develop most intensively. As it is noted in [6], many years of experience in industrial research have revealed the existence of stable lines of sign reversal of the normal component of the ISMF stress H_p (line $H_p = 0$) in local zones of metal damage. The geometrical place of the points where the parameter H_p changes sign ($H_p = 0$) is called the line of stress concentration (LCN) in [6]. It is experimentally substantiated that LCN indicates the place of the most probable formation of microcracks and in the subsequent development of macrocracks.

A strength analysis of the quality of manufactured steel bolts for critical purposes is performed in the next section, based on the introduced tensor parameters.

IOP Conf. Series: Materials Science and Engineering **939** (2020) 012046 doi:10.1088/1757-899X/939/1/012046

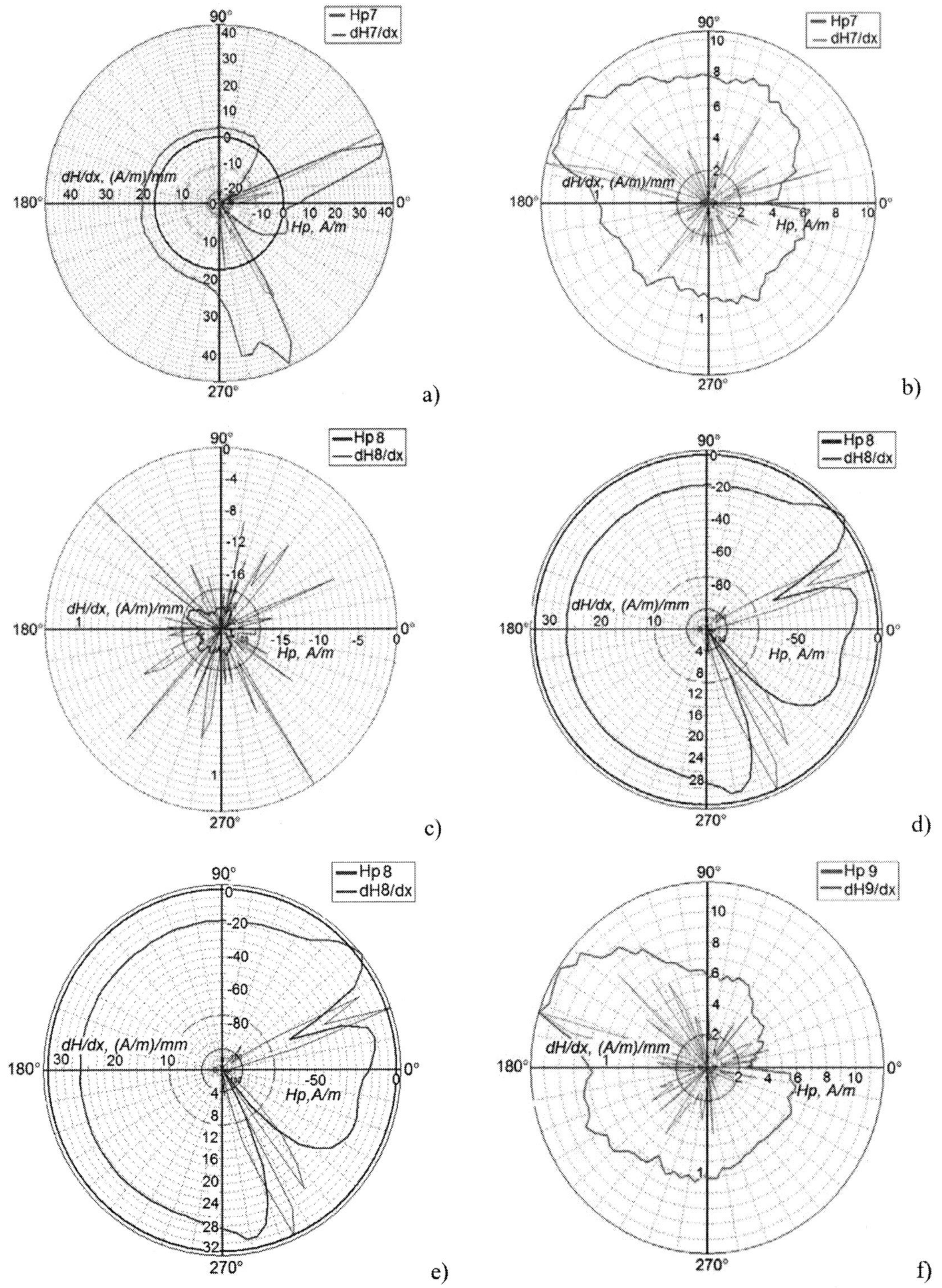

Figure 4. Distribution of ISMF components (thick lines) and their gradients (thin lines) directed: a, b – along the axis of the bolt, c, d – along the normal to the surface, e, f – along the tangent to the diameter of the bolt, positions a, c, e – relate to a bolt with an increased density of defects, positions b, d, f – correspond to the permissible degree of heterogeneity of the structure.

3. The study of the quality of steel bolts

Two batches of new steel bolts from steel 40X GOST 4543 in the amount of 10 pieces were taken for the study, which passed the test control required by the adopted technology and were found suitable for use in critical structures. The necessary equipment was made for diagnostics using the magnetic memory method, a technique was developed and research was carried out. In each batch of 5 samples, one bolt was discovered having a pronounced defective structure more than an order of magnitude higher than the corresponding magnetic parameters characterizing the structural state of the remaining products.

A voltage concentration meter IKN-3M-12 was used for research, complete with a 12-channel flux-gate transducer and the "MMP-Sistema" software product. The following parameters were chosen as diagnostic parameters: the module of the vector of the measured intrinsic magnetic field (H), which made it possible to determine the coordinates of the point of the dangerous section of the bolt and the components of the magnetic distortion tensor, which characterize the gradients of the corresponding components of the ISMF vector, reflecting the most probable direction of crack occurrence. The measurement results are shown in the figures 4 a, b, c, d, e, f.

Analyzing the results presented in figure 4, it can be noted that in a dangerous section of a bolt with an increased defect density (figure 4, positions a, c, e) compared to a similar section of a bolt where there are no anomalies in the measured magnetic parameters (figure 4, positions b, d, f) differ qualitatively, both in the nature of the distribution of parameters of the ISMF, and in quantitative characteristics. Two practically mutually perpendicular directions with a pronounced localization of the module of the vector ISMF are clearly revealed. The greatest difference is revealed by the parameter Si. The modulus of vector intensity S_i in a defective product is more than by 27 times than in a product of magnetless anomalies. The modulus of the intrinsic magnetic scattering field in these sections has a strong localization and differs by almost 10 times.

4. Conclusions

The tensor models of the ISMF are formulated and experimentally substantiated. The proposed tensor characteristics of the ISMF allow entering diagnostic parameters, which make it possible to perform rejection and select a poor-quality product in the express control mode for the distribution of the residual magnetization, which has developed naturally in the technological process of manufacturing bolts.

It is necessary to measure all components of the intrinsic scattering magnetic field strength and the components of the ISMF distortion tensor in order to further develop the method of non-destructive testing of critical steel objects, which will make it possible to use the invariant magnetic characteristics of the structural state as criteria.

References

[1] Dubov A A and Dubova A A 2006 *Method of magnetic memory and control devices* (Moscow: "Tisso" Publ) p 332

[2] Malinin V G, Malinina N A and Dimov A A 2015 Metodika eksperimental'nikh issledovanii strukturno-makhanicheskikh kharakterisrik stal'nikh plastin s konzentratorami [Method of experimental research of structural and mechanical characteristics of steel plates with concentrators] *Actualnie problemi stroitel'stva, stroitel'noi industrii i promischlennosti* (Moscow) 65–66

[3] Dubov A A, Malinin V G, Malinin V V and Malinina N A 2009 Magnitomechanicheskii effect i ego tenzornie kharakteristiki. Diagnostica oborudovanija i konstrukzii c ispolzovaniem magnitnoi pamjati metalla [Magnetomechanical effect and its tensor characteristics. Diagnostics of equipment and structures using metal magnetic memory] *Sbornik mater. 5th mezhdunarodnaja nauchno-tekhnicheskaja conf.* (M: Energodiagnostika) pp 43–46

[4] Ivanov V E and Koveshnikov A V 2018 Longitudinal magneto-optical images of stray fields on the surfaces of hard magnetic elements *Journal of Magnetism and Magnetic Materials* **460** 153–159

[5] Ivanov V E, Koveshnikov A V and Andreev S V 2017 Osobennosti magnitoopticheskikh

izobrazhenii polei rasseyaniya magnitov razlichnikh geometricheskikh form *Fizika metallov I metallovedenie* **8(118)** 772–778
[6] Vlasov V T asnd Dubov A A 2007 Teorija prozessa "deformazija-razruschenie *Fizicheskie kriterii predelnogo sostojanija metalla* (Moscow: "Tisso" Publ) p 517
[7] Vlasov V T and Dubov A A 2004 *Fizicheskie osnovi metoda magnitnoi pamjati metallov* (Moscow: "Tisso" Publ) p 424
[8] Novopaschin M D 1987 Gradientnii kriterii tekuchesti elementov konstrukzii s konzentratorami naprajagenii [Gradient criterions for local flow of metal structure elements with stress concentrators] *Modelirovanie v mekhnike: sbornik nauchnikh trudov* (Novosibirsk) vol 1 **3** 131–140
[9] Leonov M J 1981 *Mekhanika deformazii i razruschenii* (Frunze: Ilim) p 236

Analysis of methods for calculating intra-channel boiling heat transfer in refrigerants

A A Malyshev[1], **O S Malinina**[1], **K F Kouadio**[1] **and I V Shvetsov**[2,3]

[1] ITMO University, Kronverksky Pr. 49, bldg. A, 197101, St. Petersburg, Russia
[2] Yaroslav-the-Wise Novgorod State University, ul. B. St. Petersburgskaya, 41 173003 Veliky Novgorod, Russia

[3]E-mail: Igor.Shvetsov@novsu.ru

Abstract. The paper analyzes and compares methods for calculating heat transfer during boiling of refrigerants in horizontal pipes, as well as works devoted to the study of two-phase flows in minichannels.
The main provisions of the comprehensive analysis method based on true phase parameters were formulated.
It was found that for all the values of the mass velocity, there is a significant discrepancy in the calculated data, while the method based on the comprehensive analysis provides the best agreement with the experiment.
The comprehensive method was tested in relation to the analysis of heat-hydrodynamic processes at boiling in minichannels.
Comparison of the calculation results using the method based on true phase parameters with the experimental data showed a good agreement.

1. Introduction

Boiling of liquids in pipes and channels is used in many designs of heat exchange equipment in power engineering, refrigeration, food and chemical technologies. Due to the complexity of the boiling process, heat-hydrodynamic problems in these cases are usually not solved analytically, but empirical and semi-empirical approaches are often contradictory and not reliable enough.

The problem becomes even more urgent in connection with the development of "breakthrough" technologies for heat transfer intensification in heat exchange apparatus construction and the appearance of new designs with boiling in slits and minichannels. In all works devoted to the problem of intensification, an increase in heat transfer coefficients by 40% or more is stated, but the issues of heat-hydrodynamic calculation remain open.

2. Purpose of the study

The purpose of the study is to test a comprehensive method for analyzing heat transfer during in-pipe boiling, based on true phase parameters, for boiling in pipes and minichannels.

The research objectives are:

– comparison and analysis of known methods for calculating heat transfer during in-pipe boiling of refrigerants;

Content from this work may be used under the terms of the Creative Commons Attribution 3.0 licence. Any further distribution of this work must maintain attribution to the author(s) and the title of the work, journal citation and DOI.
Published under licence by IOP Publishing Ltd

– comparison of experimental data on heat transfer during boiling in pipes and minichannels with calculation using a comprehensive method.

3. Research Methods

To calculate the heat transfer during the boiling of liquids in pipes, in most cases, dependencies are used to calculate the average heat transfer coefficients along the length of the pipe ("average heat transfer formulas"). In this case, as a rule, the given phase parameters are used, calculated from the equations of the material and heat balances.

The main consumption parameters are as follows:

Mass vapor content:

$$X = \frac{M''}{(M'+M'')}, \tag{1}$$

where, M – total mass flow rate of the mixture; $M'' = M \cdot X$ – mass flow rate of vapor; $M' = M \cdot (1 - X)$ – mass flow rate of the liquid.

Volume consumption vapor content:

$$\beta = \frac{1}{\left\{1+\left[\frac{(1-X)}{X}\right]\frac{\rho''}{\rho'}\right\}}. \tag{2}$$

Mass rate:

$$w\rho = \frac{M}{f_{sec}}. \tag{3}$$

Circulation rate:

$$w_0 = \frac{M}{f_{sec} \cdot \rho'}. \tag{4}$$

Superficial vapor velocity:

$$w_0'' = \frac{w\rho x}{\rho''}. \tag{5}$$

Superficial liquid velocity:

$$w_0' = \frac{w\rho(1-x)}{\rho'} = w_0 \cdot (1-x). \tag{6}$$

Mixture velocity:

$$w_{MIX} = w_0'' + w_0'. \tag{7}$$

A widely used formula for calculating average heat transfer is an equation of the form [1]:

$$\alpha = A \cdot q^{0,6} \cdot (w \cdot \rho)^{0,2} \cdot d_{out}^{-0,2}. \tag{8}$$

As in most methods of this type, the effect of $q^{0,6}$ bubble boiling and $w\rho^{0,2}$ convection are expressed in constant values. This formalization of the process does not take into account the change in the heat transfer mechanism associated with a change in the ratio of velocity and heat flow.

From this point of view, the approach of S. S. Kutateladze is interesting, which provides for a variable influence of α_0 bubble boiling and α_{con}.convection.

$$\alpha = \alpha_{con} \cdot \sqrt{1 + \left(\frac{\alpha_{b.o}}{\alpha_{con}}\right)^2}. \tag{9}$$

One of the promising directions for the development of heat-hydrodynamic analysis of two-phase flows in a confined space is an integrated approach based on the interrelated analysis of the following characteristics [2]:

– true phase parameters (phase slip);
– two-phase flow modes;
– local heat transfer;

– pressure losses during the two-phase mixture flow.

The main true parameters are:

True volume vapor content:

$$\varphi = \frac{f''}{f_{\text{sec}}}, \tag{10}$$

where, f'' – a part of the pipe's section filled with vapor, $f_{сеч}$ – the total cross section of the pipe.

True liquid velocity:

$$w' = \frac{w_0'}{(1-\varphi)}. \tag{11}$$

True vapor velocity:

$$w'' = \frac{w_0''}{\varphi}. \tag{12}$$

Slip coefficient:

$$S = \frac{w''}{w'}. \tag{13}$$

Finding the true parameters is a separate rather complex thermophysical problem. Methods for calculating the true parameters for boiling in pipes with a diameter of more than 6 mm are presented in [2].

In particular, for refrigerants that are boiling in pipes, it is recommended that:

$$\beta - \varphi = 0{,}06 \cdot \beta \cdot (1-\beta)^{0{,}5} \cdot \left(\frac{Fr_0}{Re_0}\right)^{-0{,}23} \cdot \left(\frac{P_0}{P_{\text{cr}}}\right)^{-0{,}15}, \tag{14}$$

where, β – volume consumption vapor content; the Fr_0 and Re_0 criteria are calculated by circulation rate (4).

Using a comprehensive approach, an equation was obtained for calculating the heat transfer coefficients for wave and stratified modes when boiling R12 and R22 in pipes [2]

$$\alpha = 0{,}695 \cdot (w \cdot \rho)^{0{,}85} \cdot q^{0{,}047} \cdot \left(\frac{P_0}{P_{\text{cr}}}\right)^{-1{,}1}, \tag{15}$$

The flow modes map [3] and the dependence (14) were used for calculating the true volume vapor content.

To calculate the local heat transfer in the ring mode, the following equation was obtained:

$$\alpha_{\text{ring}} = \alpha' \sqrt{\left[1 + 36{,}5 \cdot 10^{-9} \left(\frac{w'' \cdot r \cdot q'}{q}\right)^{1{,}5} \cdot \left(\frac{\alpha_{\text{b.o}}}{\alpha'}\right)^2\right]}, \tag{16}$$

where, α' – the convective component of single-phase liquid convection, $\alpha_{b.o.}$ – the bubble boiling component [2].

The comparison of the data obtained by means of equations for calculating the average heat transfer and the calculations made with the help of the comprehensive method using true phase parameters is of considerable interest.

4. Results and discussion

Figure 1 shows the result of comparing calculations using equations (8), (9), (15) with experimental data from H. Yusida and S. Yamoguchi [4].

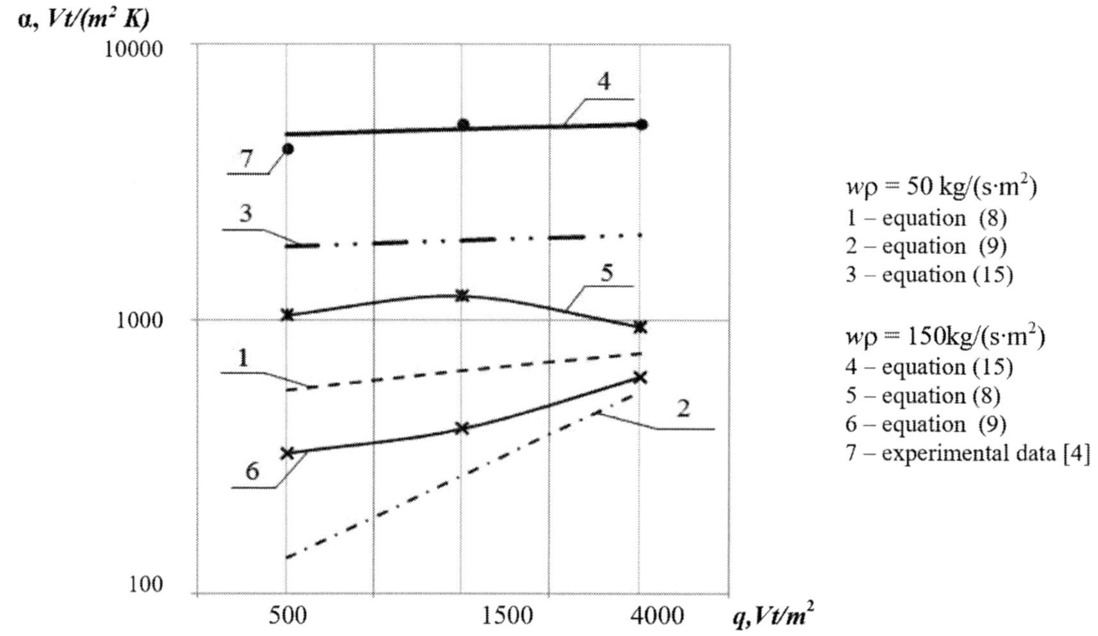

Figure 1. Comparison of calculation results using equations (8), (9), (15) with experimental data by H. Yusida and S. Yamoguchi at $t_0 = +10°C$, $d_0 = 6$ mm [4].

The presented data make it possible to conclude that at the mass velocity $w\rho = 50$ kg/(s · m²) at $q <$ 3000 W/m², the differences in the calculated values of α obtained from the formulas for the average heat transfer coefficients are very significant. The explanation for this is that each of the equations was obtained for the boiling conditions in pipes $d_0 > 15 \div 20$ mm and do not take into account the specifics of the flow in the pipe $d_0 = 6$mm. At the same time, as the diameter decreases, the difference between true and consumption parameters increases, so the use of the latter leads to significant errors.

At $w\rho$=50 kg/(s · m²) and the values of the mass consumption vapor content X=0,1 ÷ 0,8, a wave flow mode was detected using the mode diagram [3]. Using equation (15), local values of heat transfer coefficients averaged in the range X (line 3) were calculated.

The maximum deviations of the values obtained by the formula (15) from the calculation by the equations for the average heat transfer coefficients at q > 2 kW/m² are within 30%. At the same time, the types of dependencies differ qualitatively, which is explained by the difference between true and consumption parameters.

According to the mode map, at $w\rho$ = 150 kg/(s · m²) under experimental conditions, the ring flow mode is observed. The calculation using equation (15) (line 4) gives a good agreement with the experimental data [4]. The calculation results for equations (7) and (8) are 45–55% lower than the experimental data. This situation is explained by the following. Firstly, the formulas for average heat transfer coefficients do not take into account flow modes and were obtained mainly at mass velocities $w\rho$ = 50 ÷ 100 kg/(s×m²) in pipes with a diameter d_0 = 10 ÷ 20 mm. For these conditions, the ring mode, which is characterized by the highest values of heat transfer coefficients, is usually not achieved. In addition, as in the previous case, the difference between the true parameters (equation (15) and the expenditure parameters (equations (8) and (9)) is significant, the latter parameters, under certain conditions, have a limited physical meaning.

Approbation of the comprehensive method based on true parameters and developed for pipes in relation to boiling in minichannels is of particular interest.

Minichannel technologies with phase transitions of working substances are currently a very promising direction of heat exchange apparatus construction [5–8]. Minichannels began to be widely used in power engineering, chemical technology, computer systems, and compact air conditioning systems. Minichannel technologies in refrigeration are of highly relevant today.

According to [7], the heat transfer coefficient in minichannels is 2.5 times higher than in plate devices, and the amount of working substance that provides similar performance is up to 4 times lower. In comparison with plate heat exchangers with minichannels, they have improved weight and size characteristics, have increased strength and manufacturability.

However, the calculation of the heat-hydrodynamic characteristics of evaporators with boiling in minichannels is an open task of thermophysics.

The minichannel flow mode map was used for analysis [8].

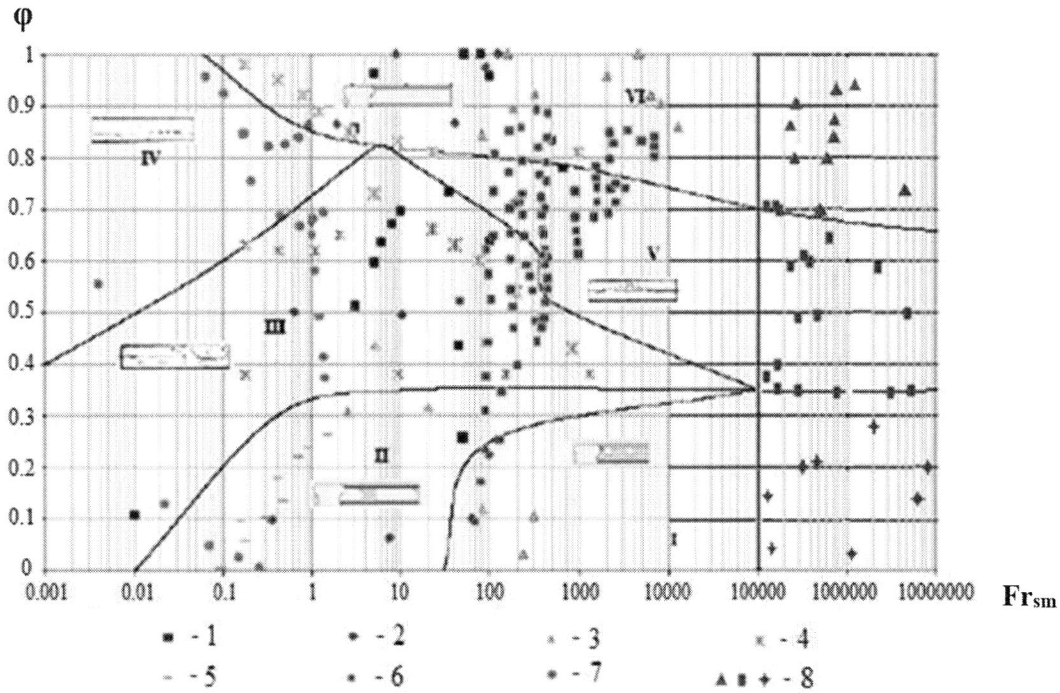

1 – Zan's data	4 – Malyshev's data	7 – Gardienko's data
2 – Kosterin's data	5 – Avdeev's data	8 – the authors' data
3 – Hugerdon's data	6 – Hovalyg's data	

Figure 2. The unified diagram of two-phase flows in a confined space.

The unified diagram includes two ranges of Froude criteria:

– $Fr_{sm} < 10^5$ – pipe operating range $d_0 = 6 \div 20$ mm;

– $10^5 < Fr_{sm} < 10^7$ – minichannel operating range $d_h = 0.5 \div 1.6$ mm.

The diagram unifies:

– experimental data on the pipes with the R12, R22, R134a, R717 refrigerants, water at $w\rho = 50 \div 250$ kg / (s·m^2) t_0 = from +20 to –20°C

– experimental data on the minichannels with the R134a, R410a refrigerants, at $w\rho = 110 \div 650$ kg/(s·m^2), t_0 = from –10°C to + 20°C.

For comparison, figure 3 shows experimental data on the true volume vapor content when boiling R12 in a pipe with a diameter of 6 mm at a boiling temperature of –10°C and for refrigerant R134a boiling in a minichannel with an equivalent diameter of $d_h = 0.5$ mm.

At the same speeds, the values of φ during boiling in the pipe are 6 ÷ 12% higher than in the minichannel, although they are qualitatively consistent with each other.

From the above, it follows that the difference between the true volume vapor content and the flow rate is greater in the minichannel than in the pipe. This indicates a greater difference between the true and superficial phase velocities in minichannels compared to pipes. Therefore, the use of the comprehensive method of analysis is most relevant for minichannels.

In minichannels, the difference between true and consumption parameters decreases with increasing speed. Figure 3 shows that the values of φ approach β as the speed increases. At mass velocities of $w\rho$ > 500 kg/(s*m^2) and at values of mass consumption vapor content X > 0.6, it is legitimate to use a homogeneous model (φ = β) using the consumption parameters.

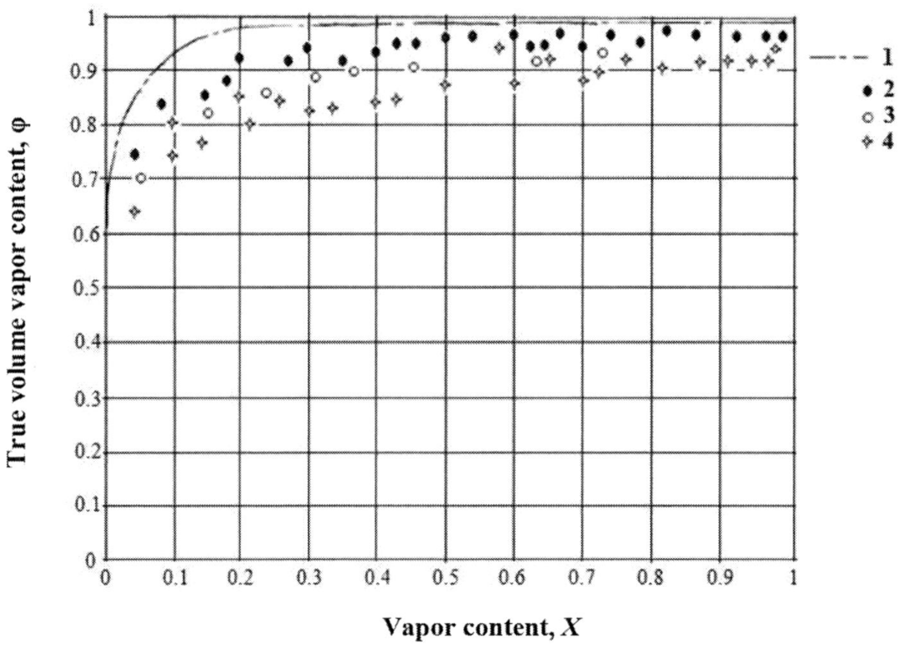

1 – The homogeneous flow φ=β
2 – The uthors' experimental data at $w\rho$ = 450 kg/(s·m^2)
3 – The data of [2] at $w\rho$ = 120 kg/(s·m^2) of the pipe d_0 = 6 m
4 – The uthors' experimental data at $w\rho$ = 120 kg/(s·m^2)

Figure 3. The comparison of experimental data for a d_o = 6 mm pipe and a minichannel with d_h = 0.5 mm at a temperature of t^0 = –8°C.

In relation to the flow in minichannels, the equation given below provides a sufficient degree of accuracy (17).

$$\varphi = \beta - 0{,}6 \cdot \beta \cdot (1-\beta)^{0.5} \cdot \left(\frac{\left(\frac{\sigma \cdot We}{g \cdot d_h^2 \cdot \rho'}\right)}{Re_{CM}}\right) \cdot \left(\frac{P_0}{P_{cr}}\right). \quad (17)$$

where, Weber's criterion: $We = \frac{\rho' \cdot D_h \cdot w_{sm}^2}{\sigma}$.

IOP Conf. Series: Materials Science and Engineering **939** (2020) 012047 doi:10.1088/1757-899X/939/1/012047

Based on the comprehensive method using the mode map (figure 2), the equation for calculating the true vapor content (17) and the equation (16) for local heat transfer in the ring mode, the heat transfer coefficients for boiling of R134a in the minichannel were calculated for the conditions [7].

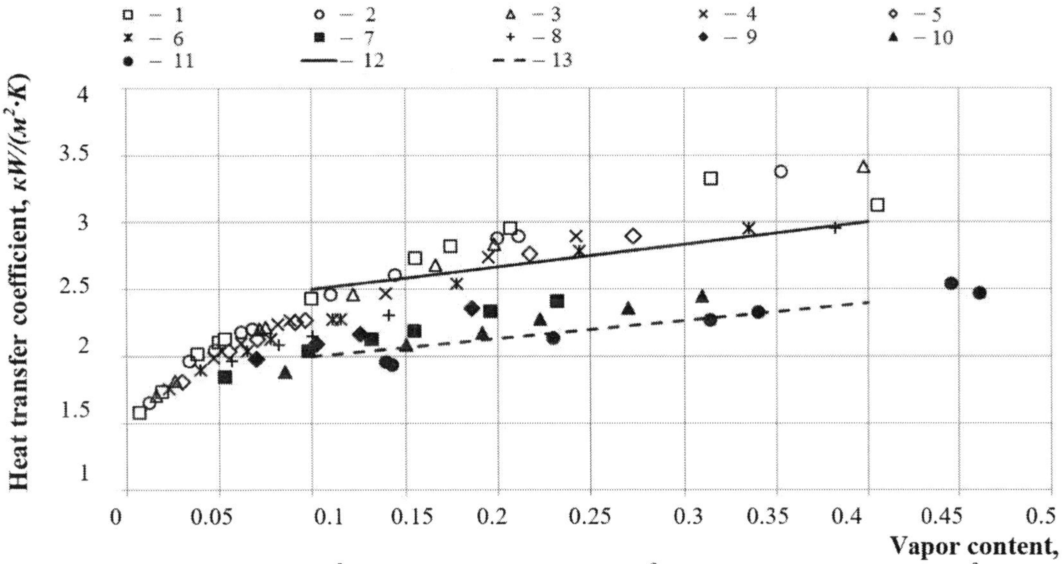

1 – G = 632.5 kg/(m^2·s) 2 – G = 567.5 kg/(m^2·s) 3 – G = 525.7 kg/(m^2·s)
4 – G = 464.3 kg/(m^2·s) 5 – G = 416.0 kg/(m^2·s) 6 – G = 365.5 kg/(m^2·s)
7 – G = 214.2 kg/(m^2·s) 8 – G = 303.8 kg/(m^2·s) 9 – G = 253.4 kg/(m^2·s)
10 – G = 153.3 kg/(m^2·s) 11 – G = 105.4 kg/(m^2·s) 12 – G = 630.0 kg/(m^2·s)
13 – G = 155.0 kg/(m^2·s)

Figure 4. The comparison of experimental data [7] with the calculation according to equations (16) and (17) for: R134a, t^0 = +30°C, d_h = 0.5 mm.

The comparison results show that in the range of x = 0.04 ÷ 0.4, the calculation using equations for local heat transfer with true velocity values describes experimental data on boiling in the minichannel with an average accuracy of 10%, both at mass velocities G = 630 kg/(m^2×s) and at G = 155 kg/(s × m^2).

5. Conclusion

1. When making the calculation of heat transfer during the boiling of refrigerants in small-diameter pipes, the best result is provided by the dependence of local heat transfer for the ring mode using the true phase parameters.
2. The developed method for calculating local heat transfer during in-pipe boiling, based on the true flow parameters, describes experimental data on boiling in minichannels with satisfactory accuracy.
3. The development of a comprehensive method for studying heat transfer and pressure losses using true parameters in the flow area in minicannels at negative temperatures, as well as on surfaces with intensified coatings and turbulators is a promising direction for further research.
4. The optimization of the heat-hydrodynamic parameters is an important research direction.

References

[1] Shurshev V F 2005 Postroenie modeli protsessa teplootdachi pri kipenii smesi kholodil'nykh agentov vnutri gorizontal'noi truby [Designing a model of the heat transfer process during boiling of a mixture of refrigerating agents inside a horizontal pipe] *Mathematical Methods in Engineering and Technology-MMTT-18. Proc. XVIII Int. Sci. Conf.* ed of V S Balakirev Kazan: Publishing House of Kazan State Technological University) vol 9 Section 11 pp

121– 122 [In Russ.]

[2] Malyshev A A, Mamchenko V O and Kisser K V 2016 Teploobmen i gidrodinamika dvukhfaznykh potokov khladagentov [Heat transfer and hydrodynamics of two-phase flows of refrigerants] (Saint-Petersburg: ITMO University) p 116 [In Russ.]

[3] Shurshev V F and Umerov A N 2005 Identifikatsiia rezhimov techeniia dvukhfaznykh potokov kholodil'nykh agentov i ikh smesei s ispol'zovaniem fazovykh diagramm [Identification of flow modes of two-phase flows of refrigerating agents and their mixtures using phase diagrams] *Bulletin of ASTU* **2(25)** p 224–231 [In Russ.]

[4] Iusida X and Iamoguchi S 1970 Teploobmen pri dvukhfaznom techenii freona-12 v gorizontal'noi trube [Heat transfer in two-phase flow of freon-12 in a horizontal pipe] *Achievements in the field of heat transfer* ed V M Borishansky (Moscow: "Mir" Publ.) pp 253–271

[5] Niño V G, Hrnjak P S and Newell T A 2020 Characterization of Two-Phase Flow in Microchannels *ACRC TR-202* October, p 5

[6] Yan C *et al.* 2013 Research on the Flow Boiling Characteristics of Water in a Multi-Furcated Tree-Shaped Mini-Channel *Advanced Materials Research* vol 629 691–698

[7] Khovalyg D M and Baranenko A V 2015 Two-phase flow dynamics during boiling of R134a refrigerant in minichannels *Technical Physics* vol 85 **3** 34–41

[8] Malyshev A A, Zaitsev A V, Kuadio K and Kisser K V 2019 Eksperimental'noe issledovanie gidrodinamicheskikh kharakteristik pri kipenii khladagentov v minikanalakh s ispol'zovaniem istinnykh parametrov faz [Experimental study of hydrodynamic characteristics during boiling of refrigerants in minichannels using true phase parameters] *Omsk Scientific Bulletin, series: Aviation-Rocket and Power Engineering* vol 3 **2** 44–50 [In Russ.]

IOP Conf. Series: Materials Science and Engineering **939** (2020) 012048 doi:10.1088/1757-899X/939/1/012048

Heat transfer through a double-glazed window by radiation

V A Maiorov[1]

[1] Yaroslav-the-Wise Novgorod State University, ul. B. St. Peterburgskaya, 41 173003 Veliky Novgorod, Russia

E-mail: Vitaly.Mayorov@novsu.ru

Abstract. A method of separately combined analysis of the process of radiant-convective heat transfer through the central part of the double-glazed windows is developed. The problem of heat transfer by radiation through a double-glazed window is formulated and a solution is obtained in an analytical form. The solution is applicable for all types of double-glazed windows with a given number, characteristics and location of low-emission coatings. The possibilities of reducing radiant heat losses through double-glazed windows due to an increase in the number of glasses and the use of glasses with low-emission heat-protective coatings are analyzed in detail.

1. Introduction

Windows are the weakest link among building envelopes in terms of thermal performance. The existing methodology for calculating the thermotechnical characteristics of translucent structures does not allow us to get a complete picture of the features of the processes of radiant and convective heat transfer through windows, their mutual influence on each other and the influence of numerous individual parameters [1]. As a result, such results do not make it possible to consciously predict the optimal combination of the parameters of double-glazed windows to achieve their best thermal properties. Such actions can only be carried out if there are analytical results.

The main element of translucent structures are double-glazed windows - bulk products consisting of two or three sheets of glass, interconnected by distance frames and sealants, forming hermetically sealed chambers filled with dried air or other gas. The main types and designs of double-glazed windows approved are shown in figure 1.

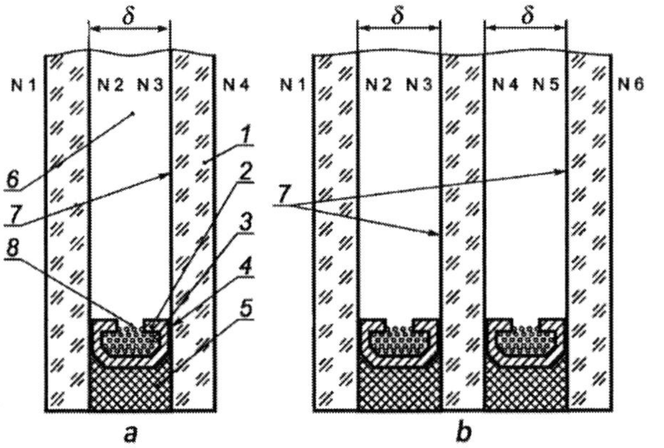

Figure 1. Designs of single-chamber (a) and two-chamber (b) double-glazed windows: 1 – glass; 2 – distance frame; 3 – desiccant; 4 – non-hardening sealant; 5 – curable sealant; 6 – inter-glass space (gas layer – chamber); 7 – recommended options for the location of the low-emission coating in case of its use; 8 – dehydration holes; **No. 1**–**No. 6** – numbering of glass surfaces in the direction from the outer surface to the inner; δ is the distance between the glasses.

Content from this work may be used under the terms of the Creative Commons Attribution 3.0 licence. Any further distribution of this work must maintain attribution to the author(s) and the title of the work, journal citation and DOI.
Published under licence by IOP Publishing Ltd

IOP Conf. Series: Materials Science and Engineering **939** (2020) 012048 doi:10.1088/1757-899X/939/1/012048

Double-glazed window chambers can be filled with dried air, as well as inert gases or their mixture (argon Ar, krypton Kr). At production of double-glazed windows various types of glasses are applied: colorless sheet; tempered, heat-strengthened, low-emission with a soft coating, low-emission with a hard coating, as well as sun-protection, laminated and other types of glasses. The application of a low-emission coating to colorless glass leads to an increase in its reflectivity in the region of near and far infrared radiation [2, 3].

2. Results and discussion

When calculating heat losses through a double-glazed window, the heat flux in individual sections of radiation-convective heat transfer (on the outer and inner surfaces of the double-glazed window, in the inter-glass gap) is usually expressed as the sum of the convective and radiant components. In this case, the value of each of these components at a known temperature difference Δt in a separate section of heat transfer is calculated using the coefficients of convective α_c and radiant heat transfer α_r [1]:

$$q_c = \alpha_c \Delta t \,; \; q_r = \alpha_r \Delta t \,. \tag{1}$$

However, due to changes in both the components of the heat flux and the relationship between them at different sites of heat transfer through the window, it is rather difficult to draw a conclusion about the magnitude of each of these components and the influence of the main process parameters on them.

In order to get a complete picture of all the properties of complex radiation-convective heat transfer through a window, the method of separately combined analysis of this process is used here. In this case, only radiant heat transfer through the window is investigated at first. Then only the convective component of the process is considered. Finally, the measure and nature of the interaction of the radiant and convective components are clarified when they are combined into a joint process of radiation-convective heat transfer through the window.

The problem is formulated as follows. A room with an internal surface of area F_i is separated by a building envelope from the external environment with surface F_o (figure 2). A single-chamber double-glazed window is installed in the opening of the building envelope. The areas of all four glass surfaces are the same, and the emission coefficients of the four glass surfaces are different and equal to ε_1, ε_2, ε_3, ε_4, respectively. The temperature of the interior surfaces of the room is the same and equal to t_i. The surface temperature F_o of the environment is also the same and is equal t_o, with $t_o < t_i$. From a room into the environment, heat is transmitted through a double-glazed window by radiation. It is necessary to determine the glass temperatures t_I, t_{II} and the density of the radiant heat flux q_r through the window. Further, the subscripts indicated in Arabic numerals refer to the parameters of the glass surfaces according to that adopted in figure 1 the numbering of the surfaces, and the lower indices indicated by Roman numerals refer to the parameters of the whole glass according to the numbering in figure 2. In the analysis of radiant heat transfer, it is assumed that the temperatures of both surfaces of each glass are equal to the temperature of the whole glass: $t_1 = t_2 = t_I$, $t_3 = t_4 = t_{II}$.

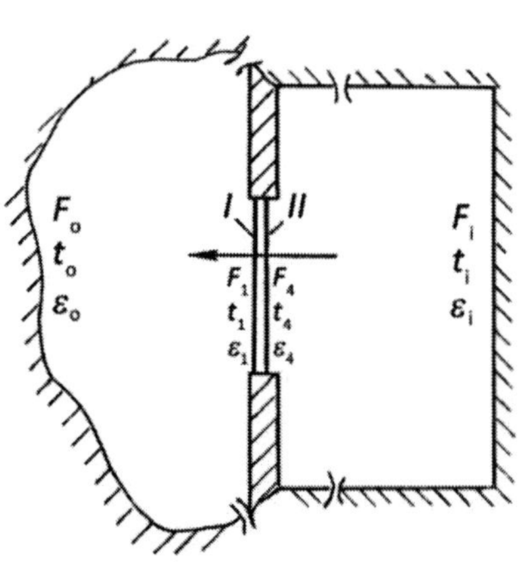

Figure 2. A physical model of heat transfer by radiation from a room into the environment through a single-chamber double-glazed window.

It should be noted that the temperatures t_o and t_i are the radiation temperatures of the surfaces. In general, they differ from the ambient temperature to the environment and the indoor air temperature t_i. But the temperatures of the internal surfaces of the room and the air temperature are usually the same. For the external environment, the equality of radiation temperature and air temperature is carried out with dense clouds in the sky. Here we accept the equality of these temperatures.

Two radiating surfaces – the surface of the room F_i and the surface of the inner glass F_4 – form a closed system, and the surface of the glass F_4 is flat, has no concavities and does not radiate on itself. For such a system, the density of the heat flux transmitted by radiation from the inner surface of the room to the inner (second) glass is determined by the expression [4]:

$$q_{r.4i} = \varepsilon_{4i} C_o \left[\left(\frac{T_i}{100} \right)^4 - \left(\frac{T_{II}}{100} \right)^4 \right]. \tag{2}$$

Here T_i and T_{II} are the indoor surface temperature and glass temperature are expressed in a thermodynamic scale (hereinafter, the temperatures t and T with the corresponding indices denote the temperatures of the corresponding surfaces, expressed in practical (t,°C) and thermodynamic (T, K) scales); $C_o = 5.67$ W/(m^2·K^4) is the emissivity of a completely black body.

The value ε_{4i} is the reduced emission coefficient of the emitting system, including F_i and the fourth surface F_4 of the double-glazed unit:

$$\varepsilon_{4i} = \left[\frac{1}{\varepsilon_4} + \frac{F_4}{F_i} \left(\frac{1}{\varepsilon_i} - 1 \right) \right]^{-1}. \tag{3}$$

This characteristic, in addition to the values of F_4 и F_i, depends on the emission factors of the glass surface ε_4 and the room surface ε_i. Considering the fact that in the vast majority of practical problems, the inner surface of the room F_i is much larger than the glass surface F_4, we get $F_4/F_i \rightarrow 0$. As a result, for all these cases, expression (3) is significantly simplified

$$\varepsilon_{4i} = \varepsilon_4, \tag{4}$$

and the reduced emission coefficient of the radiating system F_i and F_4 is determined only by the value of the emission coefficient ε_4 of the fourth surface of the glass packet. It should be noted that the result (4) is obtained from (3) also in the case $\varepsilon_i = 1.0$.

The radiating surface F_0 of the environment and the outer surface F_1 of the glass packet also form a closed radiating system. Applying similar reasoning and assumptions, we obtain the following expression for calculating the density of the radiant heat flux scattered into the environment

$$q_{r.o1} = \varepsilon_1 C_o \left[\left(\frac{T_I}{100} \right)^4 - \left(\frac{T_o}{100} \right)^4 \right], \tag{5}$$

in which ε_1 is the emission coefficient of the first glass surface in the glass packet.

Two glasses form a radiating system in the form of two infinite parallel plates. The temperatures of the plates are T_I and T_{II}, and the emission coefficients of the surfaces are ε_2 and ε_3, respectively. The radiant heat flux density in this system is calculated using the expression:

$$q_{r.23} = \varepsilon_{23} C_o \left[\left(\frac{T_{II}}{100} \right)^4 - \left(\frac{T_I}{100} \right)^4 \right]. \tag{6}$$

The reduced emission coefficient ε_{23} of such a radiating system is determined by the formula

$$\varepsilon_{23} = \left(\frac{1}{\varepsilon_2} + \frac{1}{\varepsilon_3} - 1 \right)^{-1}. \tag{7}$$

In the steady state process of radiant heat transfer through a double-glazed window, the values of all heat fluxes are the same:

$$q_{r.o1} = q_{r.23} = q_{r.4i} = q_r . \tag{8}$$

Expressions (2), (5) and (6), taking into account relation (8), make up a system of three equations with respect to three unknowns q_r, T_I, T_{II}. Here we apply the solution method used in the calculation of heat transfer by radiation through a system of screens, since glasses in a double-glazed window for thermal radiation play the role of screens. We also note that this method is applicable for double-glazed windows with any number of glasses – one, two, three or more.

We begin the solution by determining the density q_r of the radiant heat flux. For this, from equations (2), (5) and (6), taking into account relation (8), we express the temperature differences in each of the three heat transfer sections:

$$\left(\frac{T_i}{100} \right)^4 - \left(\frac{T_{II}}{100} \right)^4 = \frac{q_r}{C_o} \frac{1}{\varepsilon_4}; \quad \left(\frac{T_{II}}{100} \right)^4 - \left(\frac{T_I}{100} \right)^4 = \frac{q_r}{C_o} \frac{1}{\varepsilon_{23}}; \quad \left(\frac{T_I}{100} \right)^4 - \left(\frac{T_o}{100} \right)^4 = \frac{q_r}{C_o} \frac{1}{\varepsilon_1}. \tag{9}$$

A summation of these relations excludes unknown temperatures T_I, T_{II} and leads to an expression that includes a single unknown quantity - the density of the radiant heat flux q_r:

$$\left(\frac{T_i}{100} \right)^4 - \left(\frac{T_o}{100} \right)^4 = \frac{q_r}{C_o} \left(\frac{1}{\varepsilon_1} + \frac{1}{\varepsilon_{23}} + \frac{1}{\varepsilon_4} \right) = \frac{q_r}{C_o} \left(\frac{1}{\varepsilon_1} + \frac{1}{\varepsilon_2} + \frac{1}{\varepsilon_3} + \frac{1}{\varepsilon_4} - 1 \right) \tag{10}$$

Hence we find the q_r value:

$$q_r = \frac{C_o}{\left(\frac{1}{\varepsilon_1} + \frac{1}{\varepsilon_2} + \frac{1}{\varepsilon_3} + \frac{1}{\varepsilon_4} - 1 \right)} \left[\left(\frac{T_i}{100} \right)^4 - \left(\frac{T_o}{100} \right)^4 \right] \tag{11}$$

After that, from the equations (9) it is possible to determine the temperatures of the glasses T_I, T_{II}, since q_r is a known quantity:

$$\left(\frac{T_I}{100} \right)^4 = \left(\frac{T_o}{100} \right)^4 + \frac{q_r}{C_o} \frac{1}{\varepsilon_1}; \quad \left(\frac{T_{II}}{100} \right)^4 = \left(\frac{T_i}{100} \right)^4 - \frac{q_r}{C_o} \frac{1}{\varepsilon_4} \tag{12}$$

Having determined q_r, T_I and T_{II}, it is possible to calculate the values of the coefficients of radiant heat transfer for each of the heat transfer sites through the window: $\alpha_{r.o1}$ – for the site, external glass – environment; $\alpha_{r.23}$ – for inter-glass space; $\alpha_{r.4i}$ – for the plot, the inner surface of the room is the inner glass. As initial we apply the correlations:

$$q_r = \alpha_{r.o1}(t_I - t_o) = \alpha_{r.23}(t_{II} - t_I) = \alpha_{r.4i}(t_i - t_{II}) \tag{13}$$

Using for each of these sections the corresponding expressions (2), (5) and (6) for calculating q_r, we finally obtain:

$$\alpha_{r.o1} = \frac{1}{R_{r.o1}} = \frac{\varepsilon_1 C_o \left[\left(\frac{T_I}{100} \right)^4 - \left(\frac{T_o}{100} \right)^4 \right]}{(t_I - t_o)} \qquad \alpha_{r.23} = \frac{1}{R_{r.23}} = \frac{\varepsilon_{23} C_o \left[\left(\frac{T_{II}}{100} \right)^4 - \left(\frac{T_I}{100} \right)^4 \right]}{(t_{II} - t_I)}. \tag{14}$$

Here $R_{r.o1}$, $R_{r.23}$ are thermal resistance to radiant heat transfer of the respective sections.

The value of the total thermal resistance to radiant heat transfer is also of interest:

$$R_r = \frac{t_i - t_o}{q_r} = R_{r.o1} + R_{r.23} + R_{r.4i} = \frac{1}{\alpha_{r.o1}} + \frac{1}{\alpha_{r.23}} + \frac{1}{\alpha_{r.4i}} \tag{15}$$

The expression for calculating the density of radiant heat flux through a two-chamber double-glazed window is similarly derived:

$$q_r = \frac{C_o}{\left(\frac{1}{\varepsilon_1} + \frac{1}{\varepsilon_2} + \frac{1}{\varepsilon_3} + \frac{1}{\varepsilon_4} + \frac{1}{\varepsilon_5} + \frac{1}{\varepsilon_6} - 2 \right)} \left[\left(\frac{T_i}{100} \right)^4 - \left(\frac{T_o}{100} \right)^4 \right]. \tag{16}$$

Correlations (11) – (15) for a single-chamber double-glazed window and a similar correlation (16) for a two-chamber double-glazed window allow one to determine all the required characteristics of radiant heat transfer through such double-glazed windows and to reveal the influence of independent parameters on them. For this process, such parameters are temperatures t_i, t_o and the number of emission factors corresponding to each variant. For a single-chamber double-glazed window there are four of them ε_1, ε_2, ε_3, ε_4.

It should be noted that numerical calculations by analytical formulas (11) – (16) are currently extremely easy to perform, for example, using spreadsheets. In order to get an idea of the effect of outdoor temperature on the numerical values of the characteristics of radiant heat transfer, in table 1 and in figure 3 – figure 4 shows some results. They are obtained for a single-chamber double-glazed window with colorless glasses. The emission coefficients of all four surfaces are the same $\varepsilon_1 = \varepsilon_2 = \varepsilon_3 = \varepsilon_4 = 0.837$. The temperature inside the room is taken constant $t_i = 20$°C. Ambient temperature t_o varies from 20 to –40°C.

Table 1. Variation in the characteristics of heat transfer by radiation through a single-chamber double-glazed window with uncoated glasses ($\varepsilon_1 = \varepsilon_2 = \varepsilon_3 = \varepsilon_4 = 0.837$) with a decrease in the outdoor temperature t_o.

t_i [°C]	t_o [°C]	q_r [W/m²]	t_I [°C]	t_{II} [°C]	$R_{r.o1}$ [m²·K/W]	$R_{r.23}$ [m²·K/W]	$R_{r.4i}$ [m²·K/W]	R_r [m²·K/W]	$\alpha_{r.o1}$ [W/m²·K]	$\alpha_{r.23}$ [W/m²·K]	$\alpha_{r.4i}$ [W/m²·K]
20	19.999	0.002	20.00	20.00	0.209	0.243	0.209	0.661	4.78	4.11	4.78
20	10	14.36	13.28	16.95	0.228	0.256	0.212	0.696	4.38	3.91	4.71
20	0	27.28	6.79	14.12	0.249	0.269	0.216	0.733	4.02	3.72	4.64
20	–10	38.86	0.58	11.51	0.272	0.281	0.218	0.772	3.67	3.55	4.58
20	–20	49.19	–5.35	9.12	0.298	0.294	0.221	0.813	3.36	3.40	4.52
20	–30	58.36	–10.97	6.95	0.326	0.307	0.224	0.857	3.07	3.26	4.47
20	–40	66.47	–16.25	4.99	0.357	0.320	0.226	0.903	2.80	3.13	4.43

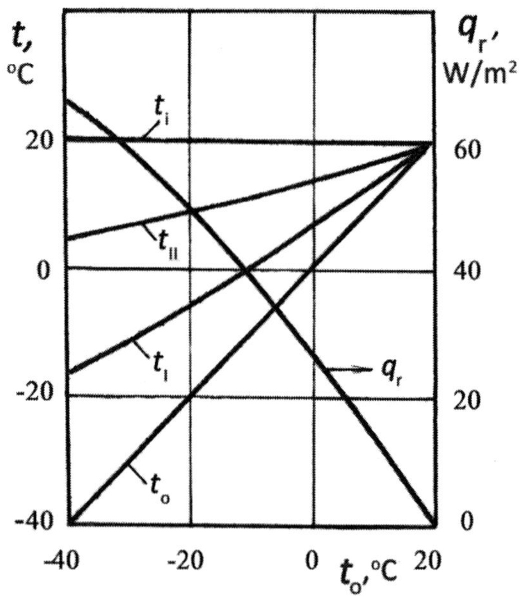

Figure 3. Variation in the temperatures t_I of the outer and t_{II} inner glasses, as well as the density q_r of the radiant heat flux in a single-chamber double-glazed window with decreasing temperature t_o the outside air.

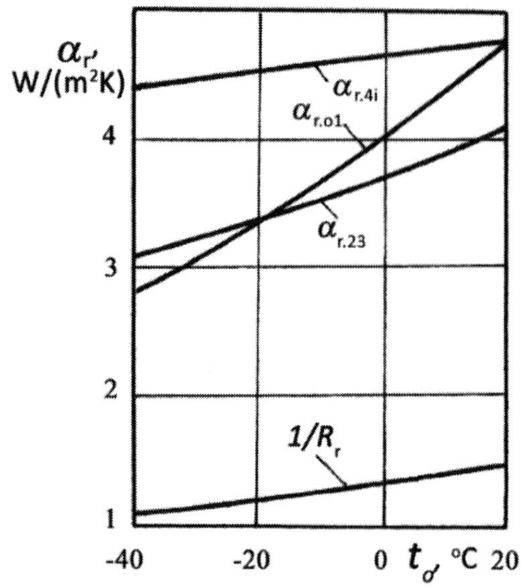

Figure 4. Dependences of the coefficients of radiant heat transfer at each of the heat transfer sites on the ambient temperature t_o (explanation in the text).

As the ambient temperature t_o decreases, the radiant heat flux through a single-chamber double-glazed window rapidly increases and reaches q_r = 58.36 W/m2 at t_o = –30°C.

A decrease in the temperatures t_I of the inner and t_{II} of the outer glasses with decreasing ambient temperature occurs according to close-linear dependences (figure 3). The temperature differences (t_i – t_{II}), (t_{II} – t_I), (t_I – t_o) in individual sections correspond to the vertical distance between the lines t_i, t_{II}, t_I, t_o. The difference in Δ t_i values in individual sections is determined by the difference in the coefficients $\alpha_{r.i}$ of the radiant heat transfer to them – these values are interconnected by the inverse correlation $\alpha_{r.i} \cdot \Delta t_i$ = q_r = const.

The change in the coefficients of radiant heat transfer in individual areas is shown in figure 4. A common feature of these three coefficients is that they all decrease with decreasing outdoor temperature. Of particular interest are the quantities $\alpha_{r.4i}$ and $\alpha_{r.o1}$. They are calculated by correlations (14), in which for the given case $\varepsilon_1 = \varepsilon_4$ = 0.837. It should be noted the numerical values of these quantities. For the coefficients of internal and external radiant heat transfer from colorless (ε = 0.837) glasses, the standard [1] values are recommended: $\alpha_{r.4i}$ = 4.4 W/(m^2·K), $\alpha_{r.o1}$ = 3.0 W/(m^2·K). From the data in table 1 it follows that the standard values of the coefficients $\alpha_{r.4i}$ and $\alpha_{r.o1}$ for these conditions are achieved by lowering the temperatures of the inner glass to 4 °C and the outer glass to –12 °C, respectively.

The curves $\alpha_{r.4i}$ и $\alpha_{r.o1}$ exit from a common point at t_o = 20°C. The subsequent difference in these values is determined only by the difference in surface temperatures between which radiant heat transfer is carried out. From a comparison of the further course of these curves, we can conclude: the coefficient of radiant heat transfer decreases both with a decrease in the maximum temperature of two surfaces and with an increase in the temperature difference of these surfaces.

Coefficient $\alpha_{r.23}$ of radiant heat transfer between surfaces decreases with decreasing arithmetic mean temperature of surfaces, which occurs both with a decrease in their maximum temperature and an increase in the difference in surface temperatures. For thermal resistance to radiant heat transfer between surfaces under such conditions, the opposite dependence is observed – its value increases. As a result, with an increase in the temperature difference on the double-glazed window due to lowering the ambient temperature t_o, the R_r value of the total thermal resistance to radiant heat transfer also increases – the data in table 1. The reciprocal of $1/R_r$, equal to the coefficient of radiant heat transfer through the double-glazed window, decreases with decreasing outdoor temperature. This dependence is shown as Curve 4 in figure 4.

There are two ways to reduce heat loss by radiation through a double-glazed window.

The first way to reduce radiant heat loss through a double-glazed window is to increase the number of glasses. Glasses act as screens for thermal radiation and an increase in their number reduces the radiant heat flux. This effect is clearly seen when comparing the values of the radiant heat flux through one-layer q_{1r}, two-layer q_{2r} and three-layer q_{3r} glazings at the same temperatures t_i and t_o. Using expressions (11) and (16), we obtain for these conditions the following relation between the indicated quantities:

$$q_{1r}/q_{2r}/q_{3r} = \left(\frac{1}{\varepsilon_1}+\frac{1}{\varepsilon_2}\right)^{-1} \Bigg/ \left(\frac{1}{\varepsilon_1}+\frac{1}{\varepsilon_2}+\frac{1}{\varepsilon_3}+\frac{1}{\varepsilon_4}-1\right)^{-1} \Bigg/ \left(\frac{1}{\varepsilon_1}+\frac{1}{\varepsilon_2}+\frac{1}{\varepsilon_3}+\frac{1}{\varepsilon_4}+\frac{1}{\varepsilon_5}+\frac{1}{\varepsilon_6}-2\right)^{-1}. \quad (17)$$

We accept the simplest option when all the glasses are the same and colorless - for all surfaces ε = 0.837. In this version, the quantitative ratio between the values is as follows: $q_{1r}/q_{2r}/q_{3r}$ = 0.418/0.265/0.193 = 1.58/1.0/0.728 = 2.166/1.373/1.0. It follows that the addition of a second ordinary glass to a double-glazed window reduces the radiant heat flux by 1.58 times, the addition of a third of the same glass additionally reduces it by another 1.373 times.

The second way to reduce radiant heat loss through a double-glazed window is to use glasses with low-emission heat-shielding coatings. For ordinary window glass, the value of the emission coefficient for the low-temperature thermal radiation corresponding to room temperatures is currently accepted as standard ε = 0.837 [1]. The low-emission coating is applied to the surface of colorless glass and has an emission coefficient substantially less. The emission coefficient of the surface of the glass with a soft low-emission coating should be no more than 0.07, for example, ε = 0.05.

The use of low-emission coatings opens up practically unlimited possibilities for reducing the radiant heat flux through the glass packet. We will evaluate these possibilities on the example of a single-chamber double-glazed window. Expression (11) for calculating the density of the radiant heat flux through a single-chamber double-glazed window includes four emission factors. Table 2 shows the results of calculating the characteristics of radiant heat transfer through a single-chamber double-glazed window depending on the number and location of low-emission coatings. The calculations are performed according to the relations (11) – (15). Initial option 1 is a double-glazed window with uncoated glasses: all four emission factors are the same ε = 0.837, heat flux q_r = 27.28 W/m^2. Options 2 – 5 for comparison: three uncoated surfaces with coefficient ε = 0.837, on the fourth is a low emission coating ε^* = 0.05, heat flux q_r^* = 4.57 W/m^2. Options 2 to 5 differ in the location of the low emission coating. The use of one low-emission coating ε^* = 0.05 increases the resistance to radiant heat transfer of each of the sections by about 16 times. In all these cases, the heat flux remains the same. In accordance with expression (11), the location of the low-emission coating (on the 1st, 2nd, 3rd or 4th surface) does not affect the value of the radiant heat flux. A change in the location of the low-emission coating causes a change in the temperature of the glasses in all variants except options 3 and 4. In these variants, a change in the location of the low-emission coating inside the gas-filled chamber from the 2nd surface to the 3rd does not entail any changes.

Table 2. The influence of the number and location of low-emission coatings on the characteristics of radiant heat transfer through a single-chamber double-glazed window at constant t_i =20°C, t_o =0°C.

Option	ε_1	ε_2	ε_3	ε_4	q_r [W/m^2]	t_I [°C]	t_{II} [°C]	$R_{r.o1}$ [m^2·K/W]	$R_{r.23}$ [m^2·K/W]	$R_{r.4i}$ [m^2·K/W]	R_r [m^2·K/W]
Both glasses without low-emission coating											
1	0.837	0.837	0.837	0.837	27.28	6.79	14.12	0.249	0.269	0.216	0.733
One low-emission coating ε* = 0.05											
2	**0.05**	0.837	0.837	0.837	4.57	17.91	19.04	3.924	0.247	0.210	4.381
3	0.837	**0.05**	0.837	0.837	4.57	1.17	19.04	0.257	3.914	0.210	4.381
4	0.837	0.837	**0.05**	0.837	4.57	1.17	19.04	0.257	3.914	0.210	4.381
5	0.837	0.837	0.837	**0.05**	4.57	1.17	2.52	0.257	0.295	3.830	4.381
Two low-emission coatings ε* = 0.05											
6	**0.05**	**0.05**	0.837	0.837	2.49	10.19	19.48	4.092	3.727	0.210	8.029
7	0.837	**0.05**	**0.05**	0.837	2.49	0.64	19.48	0.258	7.562	0.210	8.029
8	0.837	0.837	**0.05**	**0.05**	2.49	0.64	10.86	0.258	4.103	3.668	8.029
9	**0.05**	0.837	0.837	**0.05**	2.49	10.19	10.86	4.092	0.268	3,668	8.029
One low-emission coating with a different ε* value on the 3-d surface											
10	0.837	0.837	**0.10**	0.837	8.19	2.09	18.27	0.256	1.975	0.211	2.441
11	0.837	0.837	**0.15**	0.837	11.15	2.84	17.64	0.254	1.328	0.212	1.795
12	0.837	0.837	**0.20**	0.837	13.59	3.45	17.12	0.254	1.005	0.212	1.471

According to expression (11), the use of one such coating in a single-chamber double-glazed window reduces the radiant heat flux by a factor of 6:

$$q_r/q_r^* = \left(\frac{4}{\varepsilon}-1\right)^{-1} \Big/ \left(\frac{3}{\varepsilon}+\frac{1}{\varepsilon^*}-1\right)^{-1} = 5.97. \tag{18}$$

In options 6 – 9, a low-emission coating ε * = 0.05 is applied on two surfaces. Options differ in the location of these coatings. The use of two coatings ε * = 0.05 reduces the heat flux compared to the initial version by 11 (10.95) times:

$$q_r/q_r^{**} = \left(\frac{4}{\varepsilon}-1\right)^{-1} \Big/ \left(\frac{2}{\varepsilon}+\frac{2}{\varepsilon^*}-1\right)^{-1} = 10.95. \tag{19}$$

In options 10–12, one low emission coating with a different ε* value is used. Comparison of options 1, 4, and 10–12 makes it possible to evaluate the effect of the emission coefficient of a single coating located on the 3rd surface on the value of the radiant heat flux through a double-glazed window.

For a two-chamber double-glazed window, in accordance with the expression (16), the use of one low-emission coating ε * = 0.05 leads to a decrease in the radiant heat flux by 4.64 times:

$$q_r/q_r^* = \left(\frac{6}{\varepsilon}-2\right)^{-1} \Big/ \left(\frac{5}{\varepsilon}+\frac{1}{\varepsilon^*}-2\right)^{-1} = 4.64. \tag{20}$$

In a two-chamber double-glazed window, two coatings ε* = 0.05 reduce the radiant heat flux by 8.27 times, three such coatings reduce the radiant heat flux by 11.91 times.

IOP Conf. Series: Materials Science and Engineering **939** (2020) 012048 doi:10.1088/1757-899X/939/1/012048

It should be noted that in versions with any number of low-emission coatings in a two-chamber double-glazed window, a change in the location of the coating inside the gas-filled cavities from the 2nd surface to the 3rd or from the 4th surface to the 5th does not cause any changes in the process characteristics.

3. Conclusion

The presented method for solving the problem of heat transfer by radiation through a double-glazed window allows us to obtain analytical expressions for calculating the density of radiant heat flux, glass temperatures, and radiant heat transfer coefficients in each of the individual sections of heat transfer through a double-glazed window. An analysis is made of the influence of the number, quality and location of low-emission coatings on the characteristics of the process of radiant heat transfer through double-glazed windows. It is shown that a change in the location of the low-emission coating from one surface to another inside gas-filled chambers does not cause changes in the characteristics of the radiant heat transfer process.

References

[1] ISO Standard No. 15099 2000 *Thermal performance of windows, doors and shading devices – Detailed calculation*

[2] Gläser H J 2008 History of the development and industrial production of low thermal emissivity coatings for high heat insulating glass units *Applied Optics* **47** 193–199

[3] Mayorov V A 2018 Window Glasses: State and Prospects *Optics and Spectroscopy* **124** 559–573

[4] Kutateladze S S 1990 *Teploperedacha i gidrodinamicheskoye soprotivleniye: Spravochnoye posobiye* [Heat Transfer and Hydrodynamic Resistance: A Reference Guide] (Moscow: Energoatomizdat) [In Russ.]

Heat transfer through a double-glazed window by convection

V A Maiorov[1]

[1] Yaroslav-the-Wise Novgorod State University 41, B. St. Peterburgskaya str., 173003 Veliky Novgorod, Russia

E-mail: Vitaly.Mayorov@novsu.ru

Abstract. The formulation and method of solving the problem of convective heat transfer through a double-glazed window are presented. A description is given of the influence of the main parameters of the glass on the mechanism and quantitative characteristics of the process. The presence of the optimal thickness of the gas-filled chamber, at which its thermal resistance has a maximum value, is revealed. The possibilities of reducing convective heat losses through double-glazed windows by increasing the number of glasses and the use of low-heat-conducting gas filler are analyzed in detail. The main characteristics of the process of convective heat transfer for single and double chamber double-glazed windows with various gas fillers are given.

1. Formulation of the problem

The physical model of the process of convective heat transfer through a vertical single-chamber glass unit is shown in figure 1. Chamber width δ. The indoor air temperature is constant t_i =20°C, the outdoor temperature is t_o. The temperature of the inner surfaces of the glasses t_2 and t_3. Due to the presence of a temperature difference between the glasses in a gas-filled chamber, free-convection circulation occurs – the gas rises up near the warm glass and drops down near the cold glass. Coefficient $\alpha_{c.23}$ characterizes the intensity of convective heat transfer between glasses. External $\alpha_{c.o1}$ and internal $\alpha_{c.4i}$ heat transfer coefficients take into account only the convective component of heat transfer. Here, the normative values [1] of the last two coefficients are accepted: $\alpha_{c.o1}$ = 20 W/(m^2·K), $\alpha_{c.4i}$ = 3.6 W/(m^2·K). Glasses have the same thickness δ_g = 4 mm and the thermal conductivity coefficient λ_g = 1 W/(m·K). For calculation of convective heat transfer transmitted through a single-chamber double-glazed window of heat flow

$$q_c = \frac{t_i - t_o}{R_c} \quad (1)$$

it is necessary to determine the components of the total thermal resistance of convective heat transfer:

$$R_c = \frac{1}{\alpha_{c.o1}} + \left(\frac{\delta}{\lambda}\right)_g + \frac{1}{\alpha_{c.23}} + \left(\frac{\delta}{\lambda}\right)_g + \frac{1}{\alpha_{c.4i}} = R_{c.o1} + R_{c.23} + R_{c.4i} + 2\left(\frac{\delta}{\lambda}\right)_g . \quad (2)$$

Under the accepted conditions, all components, except $R_{c.23}$, in this expression are constant values: $R_{c.o1}$ = 0.05 m^2·K/W; $R_{c.4i}$ = 0.278 m^2·K/W, $(\delta/\lambda)_g$ = 0.004 m^2·K/W.

The remaining term of the thermal resistance of convective heat transfer through a gas-filled chamber

Content from this work may be used under the terms of the Creative Commons Attribution 3.0 licence. Any further distribution of this work must maintain attribution to the author(s) and the title of the work, journal citation and DOI.
Published under licence by IOP Publishing Ltd

$$R_{c.23} = 1/\alpha_{c.23} \tag{3}$$

contains an unknown value of the coefficient $\alpha_{c.23}$ of free convective heat transfer through the gas layer.

The intensity of the convective heat transfer through the glass packet under the accepted conditions is influenced by the thickness of the gas layer δ, the outside temperature to and the type of gas filler. It is required to find out the influence of the indicated parameters, to determine the thermal resistance of convective heat transfer through the glass packet and to choose the optimal combination of parameters to reduce the intensity of this process.

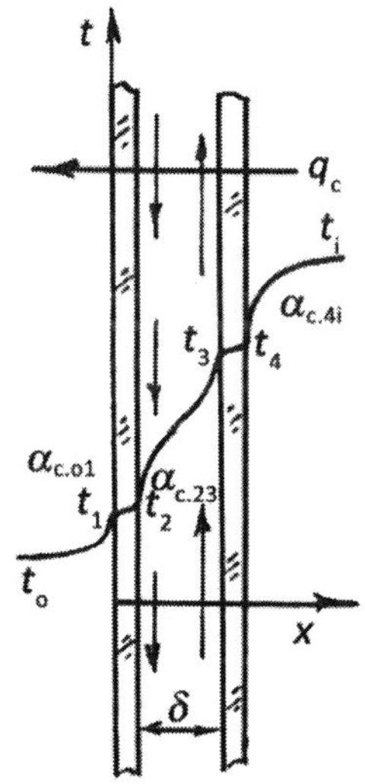

Figure 1. Physical model of the process of heat transfer by convection through a single-chamber double-glazed window.

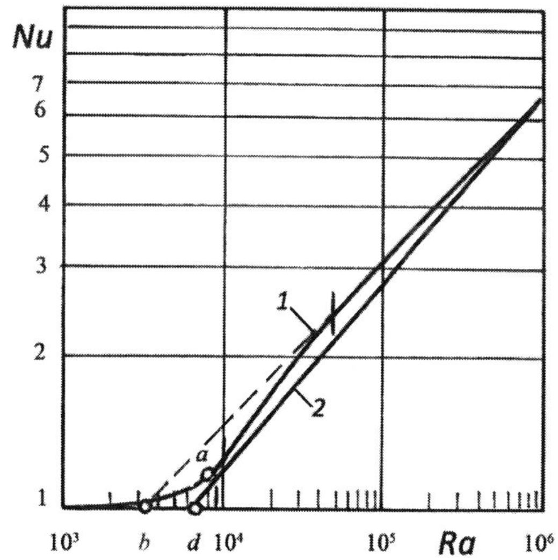

Figure 2. Dependences of *Nu* on *Ra* for free convective heat transfer in a vertical cavity: *1* – set of equations (4); *2* – equation (7).

2. Features of free convective heat transfer through a gas-filled chamber

The intensity of the free convective heat transfer in the vertical gas layer $\alpha_{c.23}$ is recommended to be described by a set of criteria equations [1]:

$$Nu = 1 + 1.7596678 \cdot 10^{-10} Ra^{2.2984755}, Ra \le 10^4; \tag{4a}$$

$$Nu = 0.028154 Ra^{0.4134}, 10^4 \le Ra \le 5 \cdot 10^4; \tag{4b}$$

$$Nu = 0.0673838 Ra^{1/3}, 5 \cdot 10^4 \le Ra. \tag{4c}$$

Here the Nusselt criterion is

$$Nu = \frac{\alpha_{c.23}\delta}{\lambda}, \tag{5}$$

The Rayleigh criterion

$$Ra = Gr\,Pr = \frac{g\beta\rho^2\delta^3 c(t_3 - t_2)}{\mu\lambda} = \frac{g\beta\delta^3(t_3 - t_2)}{\nu^2} Pr. \tag{6}$$

These criteria include: $g = 9.81$ m/s^2 – acceleration of gravity; β is the temperature coefficient of volume expansion (for gases $\beta = 1/T = 1/[273.15 + (t_2 + t_3)/2]$); ρ is the gas density; c is the heat capacity of the gas; λ is the thermal conductivity of the gas; μ, ν are the dynamic and kinematic viscosity coefficients of the gas. The Rayleigh criterion is sometimes written in the form of a product of the Grashof criteria Gr and Prandtl $Pr = \mu c/\lambda$. The physical properties of the gas are calculated at the average gas temperature in the cavity.

The set of equations (4) was obtained in [2] as a result of generalization and analysis of all available relevant experimental data for vertical gas interlayers, the ratio of whose height to thickness δ is greater than 40. Solid line *1* in figure 2 correlates this data. Equations (4a)–(4c) are the equations of the individual parts of curve *1* in the corresponding sections. The boundaries of these areas are marked. A large number of significant digits in equations (4a)–(4c) are used to precisely match the individual parts of curve *1* at the boundaries of the sections.

For comparison, figure 2 shows a straight line *2*, which corresponds to the approximate criteria equation:

$$Nu = 0.035(Gr \cdot Pr()^{0.38}) \tag{7}$$

It follows from expression (5) that the Nusselt criterion has the physical meaning of the ratio of the magnitude of the heat flux transferred by convection to the magnitude of the heat flux transmitted by the thermal conductivity through the stationary gas filling the interlayer. From the data in figure 2, it follows that $Nu = 1$ for values of Ra ‹1000. In this range of Ra numbers in the cavity, there is a slow circulating laminar gas flow in the form of a single cell. Such circulation occurs at any non-zero temperature difference ($t_3 - t_2$) of the glasses. The temperature distribution in the gas gap is linear, and the heat flux through the gap is equal to that which is transmitted by the thermal conductivity through the stationary gas filling the gap. This mode is called conductive. As the Ra number increases, the intensity of gas circulation increases and convective heat transfer begins to appear. The temperature distribution in the gap gradually deviates from linear. The number Nu deviates from unity and slowly increases. In this case, a stable laminar single-cell flow regime is maintained until, at approximately Ra = 6000, a secondary cell begins to form in the flow. With a further increase in the Ra number, the flow structure gradually becomes more complicated and the intensity of convective heat transfer increases. A detailed description of the change in the flow structure in the vertical gap is given in [3]. It should be noted that for double-glazed windows with a maximum temperature difference between indoor and outdoor air, the upper limit of the numbers Ra is in the range of 6000–10000.

The value of $R_{c.23}$ of thermal resistance is the main characteristic of convective heat transfer through the gas layer. In view of (3) and (5), we can write:

$$R_{c.23} = (1/\mathrm{Nu})\cdot(\delta/\lambda). \tag{8}$$

It follows that the reference for comparison is the thermal conductivity resistance δ/λ of the gas layer. With an increase in the Nusselt number, the value of $R_{\kappa.23}$ decreases in comparison with this standard.

It follows from expressions (4)–(6) that the value of the Nusselt criterion is influenced by the thickness of the gas layer δ, the temperature difference between the glass surfaces ($t_3 - t_2$), and the physical properties (type) of the gas filler. Let us find out the influence of each of them.

2.1. The effect of the thickness of the gas layer on its thermal resistance

In figure 3, curve *1* shows the dependence of the thermal resistance of free convective heat transfer through a vertical air gap on its thickness. Constant parameters are selected in accordance with the recommendations [1]: the physical properties of air are determined at an average temperature of 10°C; the temperature difference of the inner surfaces of the glasses ($t_3 - t_2$) = 15°C. With increasing δ, the number Ra increases in proportion to δ^3. For small values of δ, the number $Nu = 1$ and, therefore, the value $R_{c.23} = \delta/\lambda$ increases linearly with increasing δ in the conductive heat transfer mode. Then, as the Nusselt number increases caused by the growth of δ, a smooth deviation from the straight line occurs with a transition through a pronounced maximum and subsequent decrease and reaching a constant value in the horizontal section. For values of $\delta > 30$ mm, the value of $R_{c.23}$ remains constant and does not depend on the thickness of the gas layer. This horizontal section corresponds to the criterion equation (4c), according to which the number Nu increases in proportion to δ.

On curve 1, two points are of particular interest. The first point a is the maximum value of the thermal resistance $R_{c.23.\text{опт}}$ at the optimum thickness $\delta_{opt} = 16.48$ mm of the gas layer. The slope of the curve at the maximum point is zero and it is equal to the slope of the curve in the horizontal section at $\delta > 30$ mm. Therefore, the value of δ_{opt} corresponds to a value of the number Ra_{opt} at which the slope of curve *1* in figure 2 is 1/3, which is equal to the slope of its straight section at $Ra > 5 \cdot 10^4$. For the data in figure 2 we get a fixed value $Ra_{opt} = 8104$. $Nu_{opt} = 1.17$ corresponds to it – point *a* in figure 2.

The second point *b* corresponds to the minimum value $\delta_{min} = 12.17$ mm of the thickness of the gas layer. For $\delta > \delta_{min}$, a change in the thickness of the gas layer does not cause a significant change in $R_{c.23}$. A decrease in the thickness of the interlayer less than δ_{min} leads to a rapid linear decrease in its thermal resistance. Therefore, for the accepted conditions, a double-glazed unit with a gas interlayer with a thickness of at least δ_{min} should be used. Point *b* is located near curve *1* at the intersection of its linear ascending section with a dashed horizontal line. The corresponding δ_{min} value of $Ra_{min} = 3267$ is obtained from the condition $Nu_{min} = 0.0674 Ra_{min}^{1/3} = 1$. This is the Ra_{min} value in figure 2 is depicted by point b on the abscissa. The ratio $\delta_{opt}/\delta_{min} = (Ra_{opt}/Ra_{min})^{1/3} = 1.354$ remains constant.

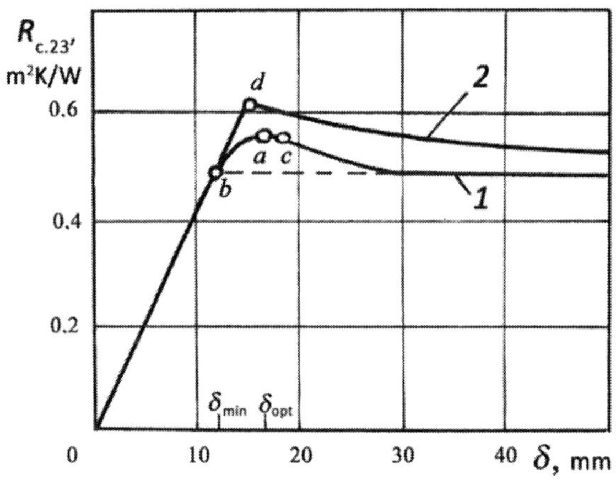

Figure 3. The dependence of the thermal resistance of free convective heat transfer through a vertical air gap on its thickness: *1* – when using equations (4) in the calculations; *2* – when using equation (7).

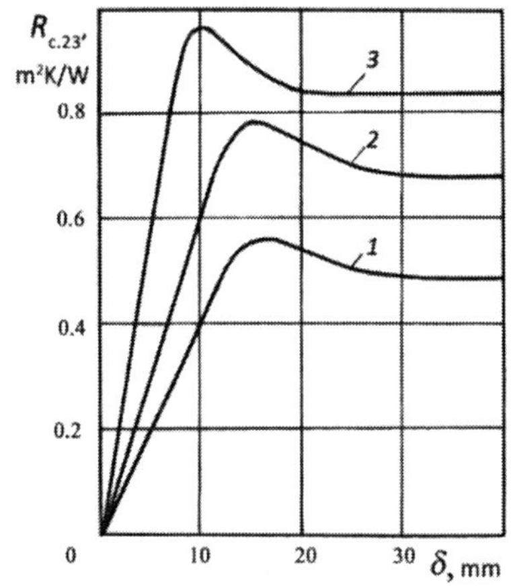

Figure 4. The effect of the physical properties of gas on the dependence of the thermal resistance of free-convective heat transfer between glasses on the distance between them: *1* – air: *2* – argon; *3* – krypton.

An increase in the thickness of the gas-filled chamber above the optimal value δ_{opt} leads to a decrease in the resistance of convective heat transfer and is therefore undesirable. Judging by curve *1* in figure 3, it is difficult to determine the exact boundary of the maximum allowable thickness. For example, if we take the value $\delta_{max} = 18$ mm (point *c*), then we get $Ra_{max} = 10554$, $Nu_{max} = 1.297$.

Using the data in figure 3, the following can be used to determine the range of variation of the permissible parameters of the double-glazed unit:

$$3267 < Ra < 10554. \tag{9}$$

For comparison, curve *2* in figure 3 shows the dependence obtained using the approximate criteria equation (7). It consists of two sections: the initial increasing linear section at point *d* is abruptly replaced by a falling section. The value of the number Ra_d = 6782 corresponding to the abscissa δ_d = 15.53 mm of the break point *d* is obtained from the condition $Nu_d = 0.035(Ra_d)^{0.38} = 1.0$ – point *d* on the abscissa axis in figure 2. Approximate curve *2* basically reproduces the form of the exact dependence *1*. In the range of δ> 15 mm, approximate curve *2* gives overestimated $R_{c.23}$ values with an error of about 10%.

2.2. The effect of the physical properties of gas

To increase the thermal resistance of the gas layer, gases with low thermal conductivity should be used. Table 1 shows the physical properties of such gases at a temperature of 10°C. Figure 4 reflects the influence of the physical properties of gases on the value of thermal resistance of freely convective heat transfer through a vertical gas layer. The temperature difference of the glass surfaces in all cases is assumed to be constant $(t_3 - t_2)$ = 15°C. With a decrease in the thermal conductivity of gases, the slope of the initial linear portion of the curves increases, the maximum acquires a more pronounced character, but the value corresponding to the maximum of the optimum thickness δ_{opt} of the gas layer decreases. Ultimately, with a decrease in the thermal conductivity of gases, higher values of thermal resistance are achieved for any thickness of the gas interlayers. The values of the characteristic quantities δ_{opt} and $R_{c.23.opt}$ for these conditions are indicated in table 1. It should be noted once again that with a decrease in the thermal conductivity of gases, the value δ_{opt} of the optimal thickness of the gas layer also decreases. Reducing the optimal thickness of the gas layer allows you to both reduce the amount of expensive inert gas (krypton), and create multi-chamber double-glazed windows of acceptable thickness.

Table 1. The physical properties of gases [1].

Gas	t	$v \cdot 10^5$	Pr	λ	ρ	$\mu \cdot 10^5$	c	δ_{opt}	$R_{c.23.opt}$
	[°C]	[m²/s]		[W/m·K]	[kg/m³]	[N·s/m²]	[J/kg·K]	[mm]	[m²·K/W]
Air	10	1.429	0.711	0.02496	1.232	1.761	1008	16.48	0.565
Argon	10	1.274	0.667	0.01684	1.699	2.164	519	15.6	0.792
Krypton	10	0.674	0.653	0.00900	3.560	2.400	245	10.3	0.976

2.3. The effect of the temperature difference of the glass surfaces $(t_3 - t_2)$

The data in figure 4 were obtained at a constant temperature difference between the glass surfaces $(t_3 - t_2)$ = 15°C. An increase in the temperature difference of the glass surfaces causes a linear increase in the number *Ra* and a corresponding increase in the intensity of free convective heat transfer. The shape of the curves in figure 4 and the characteristic values of these curves δ_{opt} and $R_{c.23.opt}$.

Figure 5 shows the dependence of the optimal thickness of the gas interlayers on the temperature difference on them. For each curve, the relation is true

$$(\delta_{opt})^3 \cdot (t_3 - t_2) = Ra_{opt} \cdot (v^2/g\beta Pr) = C, \quad (10)$$

in which the constant *C* depends on the type of gas. The dashed line marks the values of δ_{opt} at a normalized value$(t_3 - t_2)$ = 15°C, for which the results are shown in the previous figure 4. An increase in the temperature difference between the glasses leads to a decrease in the optimal distance between them. In this case, the corresponding value of $R_{c.23.opt}$. also decreases.

More clearly the effect of the temperature difference between the glasses on the thermal resistance of convective heat transfer between them is shown in figure 6. Here, in the form of curves *1*–*4*, using four air-filled double-glazed units with a distance between the glasses of 10, 15, 20, and 30 mm as an example, it is shown how an increase in the temperature difference causes a decrease in $R_{c.23}$. Moreover,

IOP Conf. Series: Materials Science and Engineering **939** (2020) 012049 doi:10.1088/1757-899X/939/1/012049

the greater the distance between the glasses, the earlier it occurs and the more sharply this decrease is caused by an increase in the intensity of gas circulation between the glasses. As a result, at$(t_3 - t_2) > 25^{\circ}$C, the distance between the glasses practically ceases to affect the value of thermal resistance.

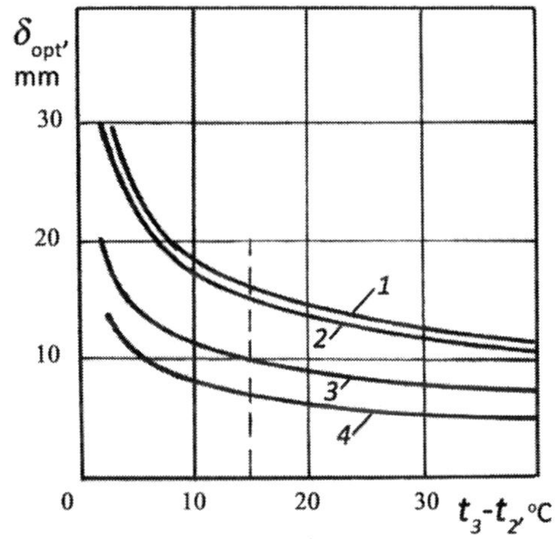

Figure 5. The dependence of the optimal distance between glasses on the temperature difference between them: *1* – air; *2* – argon; *3* – krypton; *4* – xenon.

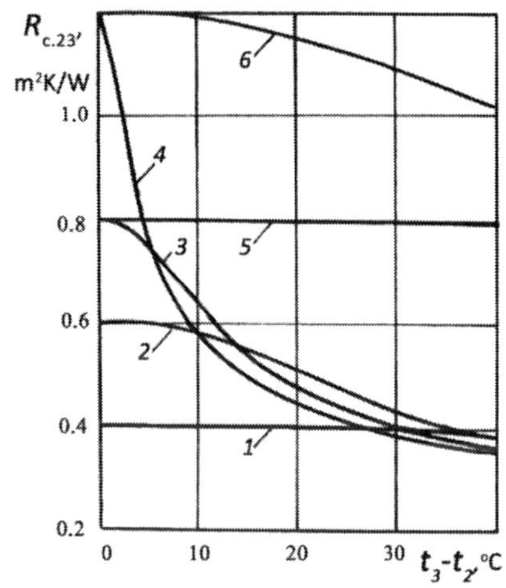

Figure 6. The dependence of the thermal resistance of the air chamber on the temperature difference on it: *1* – δ = 10 mm; *2* – δ = 15; *3* – δ = 20; *4* – δ = 30; *5* – 2δ = 20; *6* – 2δ = 30 mm.

Such an undesirable effect of the temperature difference on the value of thermal resistance can be reduced if the gas layer with glass is divided into two parts of equal thickness. Then, on each of the halves, half of the total temperature difference is realized. This is the second way to increase the thermal resistance of convective heat transfer through the glass. In figure 6 curves *5* and *6* depict the total thermal resistance of two air chambers divided in half with a total thickness of 20 and 30 mm, respectively, depending on the value of the total temperature difference on them. These curves should be compared with curves *3* and *4* for undivided air chambers with a thickness of 20 and 30 mm, respectively. The difference between the pairs of these curves is very significant. The thermal resistance of a bisected chamber with a total thickness of 20 mm remains constant (curve *5*), when, as a similar value of an undivided air chamber of the same thickness (curve *3*), it halves with an increase in the temperature difference from zero to 40°C. An even more significant difference with increasing temperature difference is observed between curves *6* and *4*.

3. Results and discussion

To solve the formulated problem (1) – (6) by determining the characteristics of convective heat transfer transmitted through a single-chamber glass packet of heat flow, it is necessary to calculate the thermal resistance of convective heat transfer through the gas interlayer $R_{c.23}$. To do this, under the conditions adopted in the statement of the problem for calculating $R_{c.23}$, you must first determine the temperatures of the inner surfaces of the glasses t_2 and t_3.

Here, to solve, we use the method of successive approximations. Calculations are performed using spreadsheets.

We enter in the table the values of all constant parameters: δ, t_i, t_o, $\alpha_{c.o1}$, $\alpha_{c.4i}$, δ_g, λ_g, type of gas filler. In a first approximation, we choose arbitrary values of the temperature of the inner surfaces of the glasses

t_2 and t_3 from the interval between t_o and t_i. For this, for example, you can use the condition for the equality of temperature drops in all sections: $t_3 - t_2 = t_2 - t_o = t_i - t_3$. We calculate the average temperature of the gas layer $(t_2 + t_3)/2$. For this temperature, we select from the tables or calculate the physical properties of the gas filler λ, ν, Pr. Then we calculate the numbers Ra (6), Nu (4) and the value $R_{c.23}$ (8). Next, we calculate the individual components and the total value R_c (2) of the total thermal resistance of convective heat transfer. By the formula (1) we find the value of the convective heat flux q_c. After that, we determine the temperature of all glass surfaces:

$$t_1 = t_o + q_c \cdot R_{c.o1};\quad t_2 = t_1 + q_c \cdot (\delta/\lambda)_g;\quad t_3 = t_2 + q_c \cdot R_{c.23};\quad t_4 = t_i - q_c \cdot R_{c.4i}\,. \tag{11}$$

The calculation of the first approximation is completed. The values of the temperature of the inner surfaces of the glasses t_2 and t_3 obtained in the first approximation are used for repeated calculations in the second approximation. The calculations are repeated until in two successive approximations the values of the total thermal resistance of convective heat transfer R_c do not coincide with an accuracy of the third significant digit. The solution quickly converges. Usually no more than three to four approximations are required.

Similarly, calculations are performed for a two-chamber double-glazed window. In this case, in the case of an equal distance between the glasses and the same type of gas filler, the temperature differences on both gas layers are assumed to be the same. The physical properties of the gas are calculated at an average temperature of two gas interlayers.

Tables 2, 3 shows examples of calculating by method of successive approximations the main characteristics of convective heat transfer through single- and two-chamber double-glazed windows, respectively, at constant t_i =20°C, t_o =0°C for two types of gas filling: air, argon. The final results are shown on the last line of each option. In a first approximation, the temperature selection of the inner surfaces of the glasses was carried out from the condition of equality of the temperature difference in all areas: $t_3 - t_2 = t_2 - t_o = t_i - t_3$.

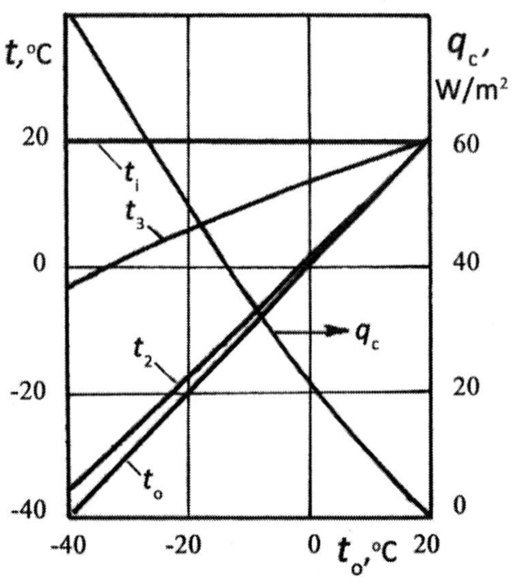

Figure 7. Changes in the temperatures t_2 and t_3 of the inner surfaces of the glasses and the density of the convective heat flux q_c in a single-chamber air-filled double-glazed window with decreasing temperature t_o of the outside air.

Figure 7 shows how the temperature of the inner surfaces of the glasses and the density of the convective heat flux in a single-chamber double-glazed window filled with air change with a decrease in the temperature of the outside air. The distance between the glasses is δ = 15 mm. The remaining parameters are indicated in the statement of the problem.

Table 2. The main characteristics of convective heat transfer through a single-chamber double-glazed window with a distance between glasses of δ = 15 mm at constant t_i =20°C, t_o =0°C.

Approximation	t_o [°C]	t_1 [°C]	t_2 [°C]	t_3 [°C]	t_4 [°C]	t_i [°C]	Ra	Nu	R_c [m²·K/W]	q_c [W/m²]
Air										
1st	0.00	6.67	6.67	13.33	13.33	20.00	2712	1.014	0.929	21.54
2nd	0.00	1.08	1.16	13.93	14.02	20.00	5244	1.062	0.901	22.19
3rd	0.00	1.11	1.20	13.75	13.84	20.00	5156	1.060	0.903	22.16
4th	0.00	1.11	1.20	13.76	13.85	20.00	5161	1.060	0.903	22.16
Argon										
1st	0.00	6.67	6.67	13.33	13.33	20.00	3200	1.020	1.209	16.54
2nd	0.00	0.83	0.89	15.34	15.40	20.00	6987	1.121	1.131	17.69
3rd	0.00	0.88	0.96	15.02	15.09	20.00	6804	1.113	1.136	17.61
4th	0.00	0.88	0.95	15.04	15.11	20.00	6816	1.114	1.135	17.62

Table 3. The main characteristics of convective heat transfer through a two-chamber double-glazed window with the same distance between the glasses δ = 15 mm at constant t_i =20°C, t_o =0°C.

Approximation	t_o [°C]	t_1 [°C]	t_2 [°C]	t_3 [°C]	t_4 [°C]	t_5 [°C]	t_6 [°C]	t_i [°C]	Ra	Nu	R_c [m²·K/W]	q_c [W/m²]
Air												
1st	0.00	5.00	5.00	10.00	10.00	15.00	15.00	20.00	2036	1.007	1.533	13.04
2nd	0.00	0.65	0.70	8.49	8.54	16.32	16.38	20.00	3186	1.020	1.518	13.17
3rd	0.00	0.66	0.71	8.47	8.53	16.29	16.34	20.00	3178	1.020	1.518	13.17
Argon												
1st	0.00	5.00	5.00	10.00	10.00	15.00	15.00	20.00	2402	1.010	2.103	9.51
2nd	0.00	0.48	0.51	8.90	8.94	17.32	17.36	20.00	4044	1.034	2.062	9.70
3rd	0.00	0.48	0.52	8.88	8.91	17.27	17.31	20.00	4029	1.034	2.063	9.70

For both types of double-glazed windows filled with krypton, very large values of the Nusselt number lead to a significant decrease in resistance to convective heat transfer. This is due to the fact that the *Ra* numbers in these cases significantly exceed the permissible values determined by condition (9) due to the fact that the accepted value of the thickness of gas-filled chambers δ = 15 mm is one and a half times higher than the optimal value δ_{opt} = 10.3 mm for a single-chamber double-glazed window. For double-glazed windows filled with air and argon, the thickness of gas-filled chambers δ = 15 mm used in the calculations is quite close to the optimal values.

4. Conclusion

A method for solving the problem of convective heat transfer through a double-glazed window by the method of successive approximations is developed. The solution is applicable for double-glazed windows with any number of glasses. The presented solution method allows you to perform a detailed analysis of the influence of the thickness of the gas layer, the physical properties of the gas filler and the temperature difference between the glasses on the thermal resistance of convective heat transfer through the glass packet. The main characteristics of the process for single and double chamber double-glazed windows with various gas fillers are given.

IOP Conf. Series: Materials Science and Engineering **939** (2020) 012049 doi:10.1088/1757-899X/939/1/012049

References

[1] *Thermal performance of windows, doors and shading devices – Detailed calculation* 2000 ISO Standard 15099

[2] Wright J L 1996 A Correlation to Quantify Convective Heat Transfer Between Vertical Window Glazings *ASHRAE Transactions* **102** 940–946

[3] Wright J L, Jin H, Hollands K G T and Naylor D 2006 Flow visualization of natural convection in a tall, air-filled vertical cavity *Int. J. Heat Mass Transfer* **49** 889–904

Assessment results of the dependence of the processed surface quality on the applied tool material

B Ya Mokritsky[1], E C Sitamov[1], I G Arendatelev[2,4] and I M Zawadzki[3]

[1] Komsomolsk-on-Amur State Technical University, prosp. Lenina, 27 681013 Komsomolsk-on-Amur, Russia
[2] Yaroslav-the-Wise Novgorod State University, ul. B. St. Petersburgskaya, 41, 173003 Veliky Novgorod, Russia.
[3] Novgorod chemical-industrial College, ul. Kochetova, 28, 173025 Veliky Novgorod, Russia

[4] E-mail: Igor.Arendatelev@novsu.ru

Abstract. Prior studies of surface roughness generated by machining the cutting edge of the tool does not remove the need for additional studies under specific processing conditions, for example, turning a specialized corrosion resistant stainless steel grade 09H17N7U. There are no regulatory recommendations to ensure the necessary quality of detail in the roughness parameters for this steel (pre-existing are outdated). The scope of application of this steel is rapidly expanding due to its unique performance properties, physical and mechanical characteristics. The authors performed such a study in relation to turning with various tool materials.

1. Introduction

In the manufacture of parts made of hard-to-handle materials, the manufacturer simultaneously has to solve several problems. For example, reducing the share of tool costs in the prime cost of a part manufacturing and ensuring the required quality of the part [1, 2, 3]. We considered a decrease in the share of tool costs by choosing the most rational metal-cutting tool and tool material. Of all the parameters characterizing the surface quality of the part, we examined the roughness of the treated surface. This indicator is significant from the perspective of many researchers. As a metal-cutting tool, standard interchangeable carbide inserts of domestic production are considered. As a substrate of instrumental material we used instrumental carbide grade VK8 (as a basic option). Several coating options applied to this substrate are considered.

Corrosion-resistant specialized stainless steel of the 09H17N7U grade, which is used for a number of critical parts of underwater vehicles and has been widely used in a number of industries due to its unique properties, was selected as a hard-to-handle material. The properties of this steel are partially reflected in table 1. These properties are mainly determined by the features of its chemical composition, for example, the large presence of chromium and nickel.

Table 1. Physico-mechanical characteristics of steel grade 09H17N7U.

Tensile strength, MPa	Yield strength, MPa	Elongation at break, %	Impact strength KCU, kJ / m^2	Heat treatment
830	735	12	490	Hardening and high tempering

Content from this work may be used under the terms of the Creative Commons Attribution 3.0 licence. Any further distribution of this work must maintain attribution to the author(s) and the title of the work, journal citation and DOI.
Published under licence by IOP Publishing Ltd

IOP Conf. Series: Materials Science and Engineering **939** (2020) 012050 doi:10.1088/1757-899X/939/1/012050

Previous recommendations on the processing of this specialized steel are not applicable today due to the fact that they are outdated, if only because there is no longer the metal-cutting equipment that they were oriented to. Therefore, developing new recommendations is necessary. We have undertaken development of new recommendations only for longitudinal turning with standard interchangeable inserts when machining rigid (diameter about 200 mm, length 800 mm) workpieces on universal turning equipment (machine model 16K25) when the workpiece is installed in a three-jaw turning chuck and taper center of the tailstock of the machine. The use of a steady rest at half the length of the workpiece did not affect the results. The results were obtained with the following parameters of the cutting mode: cutting speed of 50 m/min, longitudinal feed rate of 0.21 mm/rev, cutting depth of 1 mm. The roughness of the treated workpiece surface was measured with a TR200 instrument using a TA-620 model rack. This device is widely used in experiments.

2. Results and discussion

1) The following tool materials were used:

a) VK8 + TiCN (5 μm) + (TiAl)N (3 μm) + Al_2O_3 (5 μm) + TiC (5 μm) (*the number in parentheses indicates the thickness of the coating layer*);

b) VK8 + Al_2O_3 (2 μm) + TiCN (5 μm) + (TiAl)N (3 μm) + TiN (3 μm);

c) VK8 + (TiAl)N (3 μm) + Al_2O_3 (3 μm) + (TiAl)N (3 μm) + Al_2O_3 (3 μm);

d) VK8 + TiN (3 μm) + TiC (3 μm) + TiN (3 μm) + TiC (3 μm);

e) VK8 +TiC (3 μm) + TiN (3 μm) + (TiAl)N (2 μm);

f) VK8 + TiCN (2 μm) + TiC (3 μm) + TiN (1,5 μm);

g) VK8 + TiC (1,5 μm) + TiN (3 μm);

h) VK8 + TiN (2 μm) + TiC (5 μm);

i) VK8 + TiCN (0,5 μm) + TiN (1 μm);

j) VK8 + TiN (0,5 μm) + TiC (1 μm);

k) VK8 (basic option).

We selected these coatings of the VK8 substrate due to the fact that there were opportunities and equipment for their manufacture.

2) According to the capabilities of the TR200 instrument, the following six roughness parameters were used for roughness control: R_a – average arithmetic deviation of the profile, R_q – standard deviation of the profile, R_z – height of the roughness of the profile by ten points, R_t – total height of the roughness, R_p – height of the largest protrusion of the profile, R_v – the highest profile height.

3) The studies were performed on the example of turning with corrosion-resistant specialized stainless steel 09H17N7U at a cutting speed of 50 m/min, a longitudinal feed of 0.21 mm/rev, a cutting depth of 1 mm.

4) Interchangeable standard carbide inserts of square shape, rhombic shape, pentahedral shape, and trihedral shape were used.

5) The surface roughness was measured without removing the workpiece from the machine immediately on the surface obtained after 3 minutes turning. The surface area on which the roughness was measured was noted. Then the roughness was measured in the same areas with the workpiece removed from the machine. It was established that the roughness values coincided or differed in the third decimal place.

Similar measurements were performed on the surface formed by the cutting insert, which had some (about 0.3 mm) wear on the back surface. It was established that the trends in roughness values with a worn (0.3 mm) insert and an insert that worked for 3 minutes remained.

This research methodology largely coincides with the well-known [4–7] but differs in the specificity of the processed material and the applied instrumental materials.

A general idea of the test bench intended for these studies is given in figure 1.

The roughness parameters determined by the TR200 device for processing with different cutting inserts are shown in table 2. The model TR200 roughness control device with a model TA-620 stand was used (TR200 operating mode – cyclic display of 2.5 mm values, (*n * cutoff*) n * cutting off step, 5, international standard ISO 4287) during turning (cutting conditions parameters cutting speed 50 m/min, workpiece speed 160 rpm, longitudinal feed rate 0.21 mm/rev, cutting depth 1 mm) specialized stainless steel 09H17N7U.

Figure 1. Photo of the test bench: on the left is a workpiece; on the right is the TA-620 rack with the TR200 device.

Table 2. The results of assessing the impact of tool material (and the shape of an interchangeable carbide insert) on the quality of the processed surface by its roughness.

No of insert	Information about the material and parameters of the interchangeable insert	Roughness parameters and its values, μm	
		Parameters	Values
21	MC2210	R_a	5.030
		R_q	5.806
		R_z	21.31
		R_t	28.87
		R_p	11.14
		R_v	10.17
19	VK8 heat-sharpened and polished with a hole (circumference diameter 17.5 mm)	R_a	8.605
		R_q	9.433
		R_z	29.90
		R_t	31.42
		R_p	16.20
		R_v	13.69
18	VP1255 (fine-grained base of the TT7K12 carbide (Co – up to 12%, WC = 81%; TiC = 4%; TaC = 3%) + three-layer coating (TiC-TiCN-TiN))	R_a	6.204
		R_q	7.070
		R_z	24.06
		R_t	28.05
		R_p	12.03
		R_v	12.02
20	Square insert with a hole (diameter of the circumference 17.5 mm)	R_a	9.283
		R_q	10.67
		R_z	36.30
		R_t	39.74
		R_p	16.73
		R_v	19.56

5.5-1	5-faceted insert, coated with VK8	R_a	7.10
		R_q	8.25
		R_z	28.95
		R_t	30.97
		R_p	16.28
		R_v	12.66
5.5-2	5-faceted insert coated with VK8 + (TiAl)N (3 μm) [deposition of coating layers by ion bombardment condensation]	R_a	8.127
		R_q	8.791
		R_z	24.266
		R_t	27.00
		R_p	10.90
		R_v	13.35
23	Square insert without hole (diameter of the circumference 16.5 mm, mark 1c)	R_a	9.95
		R_q	9.135
		R_z	32.96
		R_t	35.02
		R_p	20.89
		R_v	14.06

The information in table 2 is difficult to analyze. To facilitate its comprehension, the authors suggest using a technique for comparing parameters, namely, a comparison by the relative value of the comparison as the ratio of the differences between the maximum and minimum values to the maximum value:

$\mathbf{B_1}$ = the maximum and minimum values/the maximum value.

Let us explain this by comparing the values of R_a for the surface roughness obtained by processing with an insert No. 19 (see table 2) or an insert No 5.5-1. For the insert No19, R_a = 8.605 μm. For the insert No 5.5-1, R_a = 7.1 μm. Then

$\mathbf{B_1}$ = (8.605 – 7.1)/8.605 = 17.5.

The number 17.5 to some extent shows how the values of R_a are relatively different from each other. For convenience, this figure can be converted into conditional percentages by multiplying **B1** by a conditional 100%, i.e.:

$\mathbf{B_1}$ × 100% = $\mathbf{B_2}$ = 1750%.

The same calculation for R_p show the following:

$\mathbf{B_1}$ = (16,28 – 16,2)/16,28 = 0,5. $\mathbf{B_2}$ = 50%.

Accordingly, one can conclude that for these inserts the difference in the parameter R_p is much smaller than in the parameter R_a.

Examples of similar calculations for other parameters are given in table 3.

Table 3. Examples of a comparative assessment of the difference in parameters, μm, surface roughness of the part after processing with different inserts.

	Insert No 19, VK8 heat-sharpened and polished with a hole (circumference diameter 17.5 mm)	Insert No 5.5-1 5-faceted, coated with VK8	The relative value of **B1** differences in roughness parameters	The value of B2 differences in roughness parameters, %
R_a 17,5 μm	8.605	7.10	17.5	1750
R_q, μm	9.433	8.25	12.5	1250
R_z, μm	29.9	28.95	3.2	320
R_t, μm	31.42	30.97	1.4	140
R_p, μm	16.2	16.28	0.5	50
R_p, μm	13.69	12.66	7.5	750

The results obtained allow for the following conclusions:

1) The surface roughness parameters depend on many factors associated with the properties of the material of the cutting inserts and its geometry.

2) The values of the roughness parameters are not equivalent when using inserts of different shapes and geometry, the presence of coating on the plate also matters. This is most noticeable when comparing the roughness parameters of inserts having different shapes and with different instrumental materials.

3) Of above-mentioned inserts, the best roughness parameters of the processed surface of the workpiece are achieved using the insert MC2210. Moreover, this same insert not only provides better roughness, but also a quite acceptable period of working capacity (180 minutes before reaching a wear of 0.5 mm). For comparison, a 4-faceted insert, material VK8, heat-sharpened and polished with a hole (the circumference diameter is 17.5 mm) has a working period of 160 minutes.

3. Conclusion

This research proves by the example of turning steel of 09H17N7U grade steel that there is a interrelation between the quality (six roughness parameters) of the processed surface and the operational properties of the tool material. From the considered nomenclature of materials and geometry of interchangeable cutting inserts, the most efficient were identified. It allows using the results as recommendations and recommending the proposed methodology for choosing an efficient instrumental material.

The scientific novelty of research is establishing the interrelation between the quality of the processed surface (in terms of roughness parameters) and the tool material used (as well as the shape of the interchangeable cutting insert and the topography of its front surface) during turning of specialized stainless steel 09H17N7U. The practical significance of the research consists in the selection of the most informative roughness parameters, in the development of a methodology for selecting tool material for the required quality of the processed surface.

References

[1] Kruglov A I, Zubarev Y M, Zhelnov D V and Shvetsov I V 2019 *IOP Conf. Ser.: Mater. Sci. Eng.* **656** 012049

[2] Shvetsov I V, Zhelnov D V and Zubarev Y M 2019 *IOP Conf. Ser.: Mater. Sci. Eng.* **656** 012048

[3] Shvetsov I V, Vasilkov D V and Tarikov I 2019 *IOP Conf. Ser.: Mater. Sci. Eng.* **656** 012049

[4] Basova E V, Alekseeva R E 2015 Description of the roughness of a surface processed by the method of high speed milling *Proceedings of Nizhny Novgorod State Technical University* **1** 95–100

[5] Nguyen Van Quang, E Yu Kuznetsov and A S 2013 Pit Culture People Influence of cutter wear on the roughness of the treated surface *Proceedings of Tula State University* **1** 206–209

[6] Ovsyannikov V E and Tereshchenko V Yu 2013 Correlation analysis of the surface roughness profile *Izvestiya TulSU: Technical Sciences* **10** 16–21

[7] Sergeev A S, Plotnikov A L, Makarov A M and Uvarova T V 2014 Features of constructing a mathematical model for calculating the roughness parameter for multi-stage processing of carbon steels with end mills *Reference Book Engineering J.* **12** 22–8

IOP Conf. Series: Materials Science and Engineering **939** (2020) 012051 doi:10.1088/1757-899X/939/1/012051

Results of testing the rationality of tool materials using the experimental method and the method of simulation modeling

B Ya Mokritsky[1], E C Sitamov[1], V N Belyakov[2,4] and I M Zawadzki[3]

[1] Komsomolsk-on-Amur State Technical University, pr. Lenina, 27 681013 Komsomolsk-on-Amur, Russia
[2] Yaroslav-the-Wise Novgorod State University, ul. B. St. Petersburgskaya, 41 173003 Veliky Novgorod, Russia
[3] Novgorod chemical-industrial College, ul. Kochetova, 28 173025 Veliky Novgorod, Russia

[4] E-mail: Valeriy.Belyakov@novsu.ru

Abstract. Growing requirements for physical and mechanical characteristics and operational properties lead to the development of ever new structural materials, the range of which is constantly expanding. This complicates the selection of rational tool materials required for machine processing, which requires considerable time. Currently, there are no regulatory recommendations for such work, so there is a need for a quick and less costly choice. Experimental studies, trial and error methods are not suitable for this. Therefore, it is necessary in a universal way, in the use of a model of the cutting process, which has not yet been developed due to the complexity of the simultaneously occurring processes. The absence of such a model of the cutting process can be temporarily replaced by computer simulation modeling. This work presents such modeling in the Deform software environment as applied to process turning of corrosion-resistant specialized stainless steel.

1. Introduction

The lack of a mathematical model of the cutting process with a blade tool impedes solution of many problems in improving the design of metal cutting tools and their tool material. The available examples are of private nature. This prompts tool designers and researchers to solve these problems without using a mathematical model. At least using computer simulation modeling. There are certain opportunities for this. First of all, this is the ANSYS software environment. There are simplified versions of it, for example, the Deform software environment. Its Deform-3D software package was developed by an American company Forming Technologies Corporation (SFTC) and is based on the use of finite element analysis and is aimed specifically at the simulation of metalworking technological processes. With regard to solving problems of improving a metal cutting tool, there are various kinds of publications, including those with the participation of the authors [1, 2].

2. Results and discussion

The objective of this work is to determine, by means of simulation modeling, the most rational coating for tooling solid carbide of the VK8 grade in relation to turning process of the corrosion-resistant

Content from this work may be used under the terms of the Creative Commons Attribution 3.0 licence. Any further distribution of this work must maintain attribution to the author(s) and the title of the work, journal citation and DOI.
Published under licence by IOP Publishing Ltd

specialized stainless steel 09H17N7U. This steel is in demand in many industries. All previously available recommendations for it on the use of the processing tools and on the cutting conditions are out of date. There is a need to develop new recommendations for modern operating conditions.

To simplify presentation of the results, the following information is provided:

a) the same operating conditions of the tool during full-scale experiments (cutting speed 50 m/min, feed 0.21 mm/rev. workpiece, cutting depth 1 mm) using a screw-cutting lathe;

b) for a replaceable square solid carbide plate with a central hole (the diameter of the circumscribed circle is 17.5 mm) with various options for the tool material, namely:

- VK8 + TiCN (5 μm) + (TiAl) N (3 μm) + Al2O3 (5 μm) + TiC (5 μm) (*the number in brackets indicates the thickness of the given coating layer*);
- VK8 + Al_2O_3 (2 μm) + TiCN (5 μm) + (TiAl)N (3 μm) + TiN (3 μm);
- VK8 + (TiAl)N (3 μm) + Al_2O_3 (3 μm) + (TiAl)N (3 μm) + Al_2O_3 (3 μm);
- VK8 + TiN (3 μm) + TiC (3 μm) + TiN (3 μm) + TiC (3 μm);
- VK8 + TiC (3 μm) + TiN (3 μm) + (TiAl) N (2 μm);
- VK8 + TiCN (2 μm) + TiC (3 μm) + TiN (1.5 μm);
- VK8 + TiC (1.5 μm) + TiN (3 μm);
- VK8 + TiN (2 μm) + TiC (5 μm);
- VK8 + TiCN (0.5 μm) + TiN (1 μm);
- VK8 + TiN (0.5 μm) + TiC (1 μm);
- VK8 (basic version).

It was these coatings on the VK8 substrate that we selected due to the fact that there were opportunities and equipment for their manufacture.

To measure the achieved wear rate, the processing was interrupted every 15 minutes. Wear was measured on a Micro Vu Sol 161 multisensor measuring center (video measuring machine). An example of plotting the dependence of the wear rate $\mathbf{h_w}$ on the time **t** of the operation of a solid carbide plate of grade VK8 is shown in figure 1.

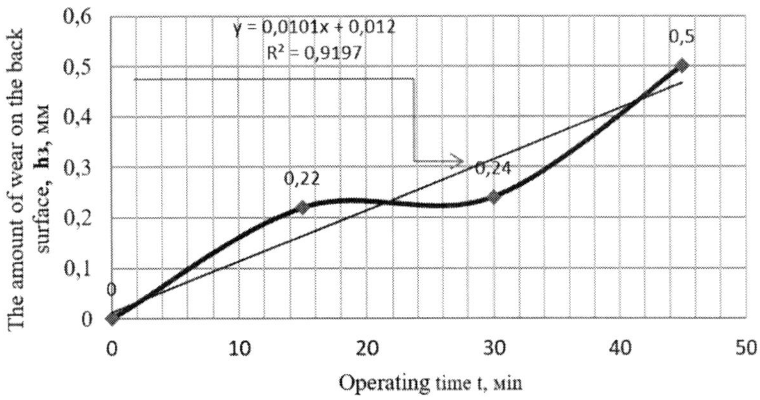

Figure 1. An example of plotting the dependence of the wear rate $\mathbf{h_w}$ on the time **t** of the operation of a solid carbide (VK8) tool (cutting mode v = 50 m/min, s = 0.21 mm/rev, t = 1 mm).

On a bold, straight line, the corresponding wear values are indicated, this is 0.22 mm after 15 minutes of operation of the cutting plate, 0.24 mm after 30 minutes and 0.5 mm after operating for 45 minutes. A thin line draws a straight line, which, with a dispersion of $R^2 = 0.9197$ (see the upper left corner of figure 1) represents a linear approximation of the experimental curved line. In the same place, in the upper left part of figure 2, an equation describing this approximating line is given.

The results obtained experimentally were systematized by arranging them according to the ranking according to the operating time until the wear of the plates reaches 0.5 mm along the rear face. The results that were obtained by simulation modeling, were also systematized in a similar way.

The adequacy of the results of simulation modeling and full-scale experiments were compared according to the position in the ranking of these tool materials.

Graphs similar to the graph shown in figure 2 were obtained from experimental studies of several different interchangeable cutting plates. Part of the equations describing the dependence of the wear rate on the tool operating time obtained during field experiments are shown in table 1.

Table 1. Examples of equations describing approximating equations for a number of experimentally tested plates (with dispersion R^2, $\mathbf{h_w}$ is the wear rate, **x** is the current value of the plate operating time).

Number of the tested plate	Equation and its dispersion R^2
1	$h_3=0.00000007x^3 - 0.00002x^2+0.0044x$ $R^2=0.9806$
2	$h_3 = 0.000001x^3 - 0.0001x^2 + 0.005x$ $R^2 = 0.9973$
2.1	$h_3 = -0.0003x^2 + 0.0224x$ $R^2 = 1$
2.2	$h_3 = 0.0011x^2 - 0.0293x$ $R^2 = 1$
4	$h_3 = 0.00002x^3 - 0.0014x^2 + 0.0311x$ $R^2=1$

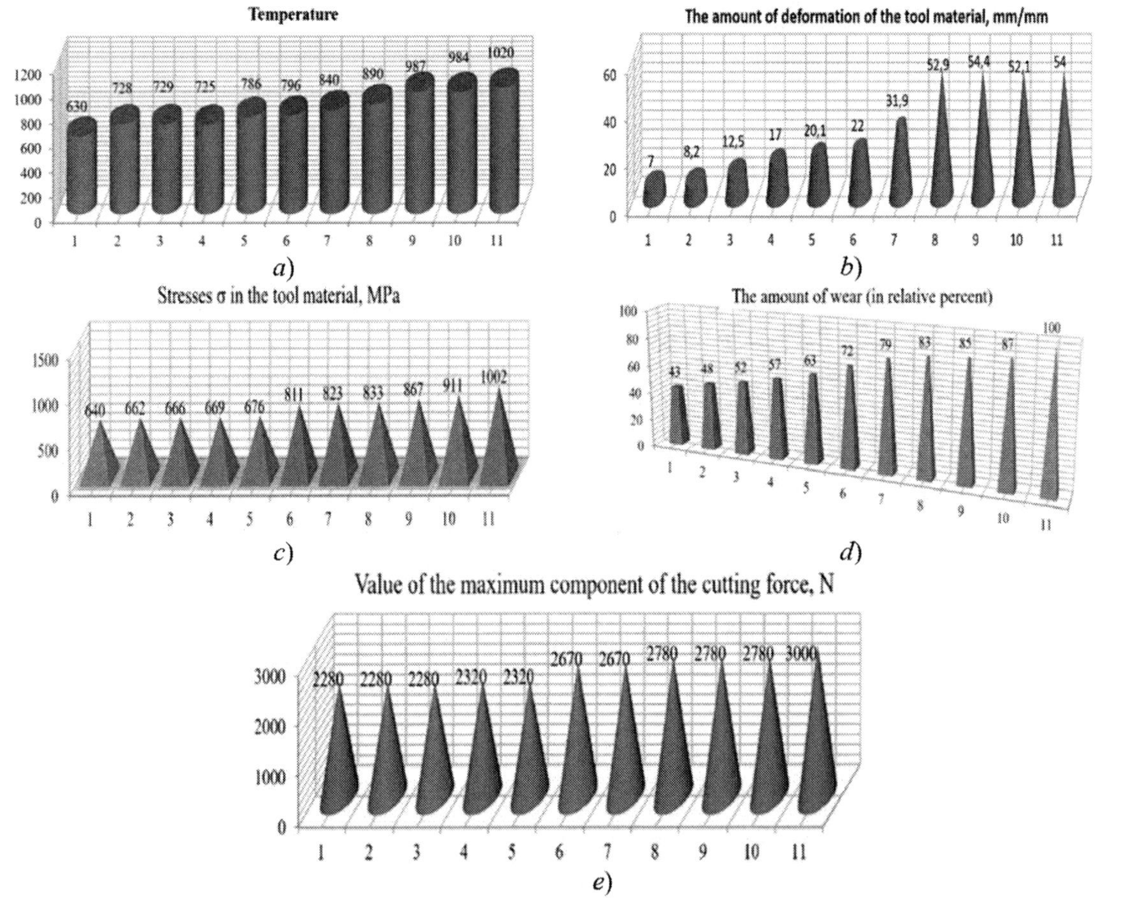

Figure 2. Graphic illustration of simulation modeling results (vertically – parameter values, horizontally – conditional numbers of tool materials.

The above stated information was obtained during experimental studies. This made it possible to build different tool materials (indicated above VK8 with different ten coatings) by ranking. For ranking, we used the indicator "operation time to reach wear of 0.5 mm." These data are shown below and used to assess the adequacy of similar ranking in simulation modeling by indicators (some of them are mentioned above) "temperature", "stress", "strain", "tool wear". In figure 2, this ranking is shown by numbers affixed to the corresponding tool material.

Figure 2 shows a direct correlation of the wear rate with the magnitude of the stress in the tool material and with the cutting temperature. A more complex correlation is noted between the wear rate and strain of the tool material. To understand the reasons for such a more complex correlation, an analysis of additional parameters was performed, namely, "total cutting force", "strain rate of the tool material" and "maximum stress in the tool material". It is established that this is due to the feature of the properties of the coatings used.

Since one of the objectives of this article is to assess the adequacy of the results obtained by simulation modeling, the results of such control are given below. The results are presented in the form of a comparison of the location of tool materials in the ranks (in the order of magnitude) for each of the considered parameters (temperature, strain, stress, wear). This is summarized in table 2.

Table 2. The result of comparing the data obtained experimentally and by simulation modeling of tool materials for the considered parameters.

Tool material	Results of the simulation modeling The serial number of the location of the tool material by parameter				Experimental result
	temperature	stress	strain	wear	
Row number / column number	2	3	4	5	6
A	1	1	1	1	1
B	2	2	2	2	2
C	3	3	3	3	3
D	4	4	4	4	4
E	5	5	5	5	5
F	6	6	6	6	6
G	7	7	7	7	7
H	8	8	8	8	8
I	9	9	9	9	9
J	10	10	10	10	10
K	11	11	11	11	11

Annotations:

1) The number (serial number) in columns 2, 3, 4, 5 indicates the location of the tool material in the randometric row for this parameter based on the condition, the smaller the number, the more rational the parameter value with the respect to the steel turning process 09H17N7U according to the results of simulation modeling.

2) The number (serial number) in column 6 indicates the location of the tool material in the randometric row according to the wear rate of the tool material, based on the condition, the smaller the number, the more rational this tool material (in one of its operation modes).

3) Letters A – J indicate tool materials. This is a VK8 grade solid carbide with different coating options, namely (the thickness of the coating layer is indicated in brackets with a dimension of microns):

A – VK8 + TiCN (5 μm) + (TiAl)N (3 μm) + Al_2O_3 (5 μm) + TiC (5 μm);

B – VK8 + Al_2O_3 (2 μm) + TiCN (5 μm) + (TiAl)N (3 μm) + TiN (3 μm);

C – VK8 + (TiAl)N (3 μm) + Al_2O_3 (3 μm) + (TiAl)N (3 μm) + Al_2O_3 (3 μm);

D – VK8 + TiN (3 μm) + TiC (3 μm) + TiN (3 μm) + TiC (3 μm);

E – VK8 + TiC (3 μm) + TiN (3 μm) + (TiAl)N (2 μm);

F – VK8 + TiCN (2 μm) + TiC (3 μm) + TiN (1,5 μm);

G – VK8 + TiC (1,5 μm) + TiN (3 μm);

H – VK8 + TiN (2 μm) + TiC (5 μm);

I – VK8 + TiCN (0,5 μm) + TiN (1 μm);

J – VK8 + TiN (0,5 μm) + TiC (1 μm);

K – VK8.

The data shown in table 2 indicate the coincidence of the experimental data and the data obtained by simulation modeling. Based on this, it was concluded that simulation in the selected software environment can be used to achieve the goal, the results of simulation modeling are adequately confirmed by the experimental data.

At the same time, an additional study of the results of simulation modeling by the parameters of "wear", "temperature", "stress", "strain" was carried out to establish the trend of their change over time. In figure 4, this is illustrated for one of the tool materials for a limited period of time.

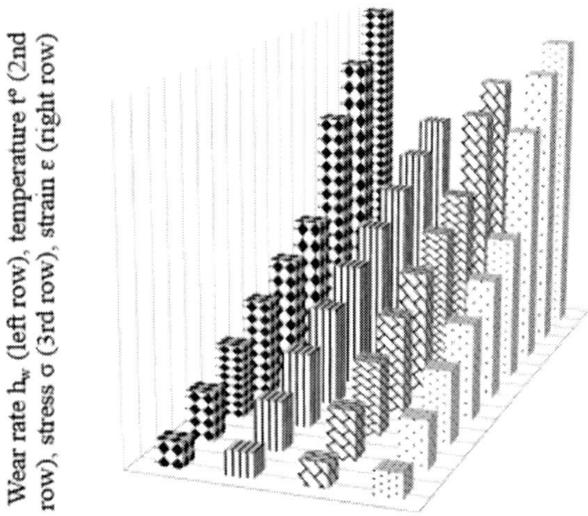

Figure 4. Illustration of the relationship between the trend of wear and with other parameters studied (temperature, stress, strain) using an example of one of the tool materials in the initial cutting period (up to 600 seconds).

From the data from figure 4 it follows that all four parameters tend to increase in time. This does not contradict the prevailing notions. But the growth rate of these parameters is slightly different. The most intensive growth rate is that of the "wear" and "strain" parameters. According to the "wear" parameter this is due to the well-known different nature of the increase in wear rate in the tool running-in the steady-state cutting area. In terms of the "strain" parameter, this is explained by the fact that, despite a balanced temperature increase (second row on the left), strain develops more rapidly in this tool material than temperature and stress do, which is associated with low heat resistance of the TiC and TiN layers.

It should be noted that a long time ago Betaneli A.I. showed that the relationship between stresses and strains is described by the equation (generally accepted designations) of the general form:

$$\sigma_x = \frac{E}{(1-\mu^2)}\left(\varepsilon_x + \mu\varepsilon_y\right) ;$$
$$\sigma_y = \frac{E}{(1-\mu^2)}\left(\varepsilon_x + \mu\varepsilon_y\right) ;$$
$$\tau_{xy} = \frac{\gamma_{xy}}{2(1+\mu)}E.$$

In our studies, this is exactly what is illustrated in figures 3 and 4.

Based on this, we can conclude that simulation modeling is a reliable tool for obtaining information on comparative assessment of the performance of various tool materials. It can significantly reduce the time and cost of choosing rational tool materials. Partially materials were published in [3, 4, 5].

3. Conclusion

It has been established that with regard to turning process of corrosion-resistant specialized stainless steel 09H17N7U, the most rational are tool materials VK8 + TiCN (5 μm) + (TiAl) N (3 μm) + Al_2O_3 (5 μm) + TiC (5 μm); VK8 + Al_2O_3 (2 μm) + TiCN (5 μm) + (TiAl) N (3 μm) + TiN (3 μm); VK8 + (TiAl) N (3 μm) + Al_2O_3 (3 μm) + (TiAl) N (3 μm) + Al_2O_3 (3 μm). This is mainly explained by the fact that the coating is multilayered (microcracks growth is inhibited at each interface between the coating layers). But the architecture of these coatings is also important.

Instrumental coatings, in comparison with the VK8 substrate, significantly reduce the levels of temperature, stress, and strain. The maximum cutting temperature and the maximum rate of its growth was noted for tool material VK8 without coating.

The nature of the course of strain processes when using a tool with a coating is higher than when using a tool without a coating. This allows one to predict a decrease in cutting force and, accordingly, an increase in the wear resistance of a coated tool.

References

[1] Vereshchaka A, Mokritsky B, Mokritskaya E, Sharipov O and Ohanyan M 2017 Two-component end mills with multiplayer composite nanostructured coatings as a viable alternative to monolithic carbide end mills *Mechanics & Industry* **18** 705 DOI: 10.1051/meca/2017052
[2] Grigoriev S N, Tabakov V P and Volosova M A 2011 *Technological methods for increasing the wear resistance of contact pads of cutting tools: monograph* (Stary Oskol: TNT) p 380
[3] Kruglov A I, Zubarev Y M, Zhelnov D V and Shvetsov I V 2019 *IOP Conf. Ser.: Mater. Sci. Eng.* **656** 012028
[4] Shvetsov I V, Zhelnov D V and Zubarev Y M 2019 *IOP Conf. Ser.: Mater. Sci. Eng.* **656** 012048
[5] Shvetsov I V, Vasilkov D V and Tarikov I Ya 2019 *IOP Conf. Ser.: Mater. Sci. Eng.* **656** 012049

IOP Conf. Series: Materials Science and Engineering **939** (2020) 012052 doi:10.1088/1757-899X/939/1/012052

The influence of micro- and nanodots lubricating and cooling technological means (LCTM) on the process of blade cutting of materials

A G Naumov[1], V V Novikov[1], O A Naumova[1] and I V Shvetsov[2,3]

[1] Ivanovo state University, ul. Ermaka, 39, Central Federal district, 153025 Ivanovo, Russia
[2] Yaroslav-the-Wise Novgorod State University, ul. B. St. Petersburgskaya, 41, 173003 Veliky Novgorod, Russia

[3] E-mail: Igor.Shvetsov@novsu.ru

Abstract. Improving the efficiency of cutting tools and the quality of the processed surfaces in the course of metal blade processing is largely determined by the intensity of adhesive interactions between the tribo-conjugated surfaces of the tool and processed materials. This is due to the high chemical activity of the juvenile surfaces formed in the process of chip separation in tribo-conjugated contact. As it is illustrated by theory and practice, one of the main mechanisms effecting on the amount of adhesion is the formation of separation lubricant films at the interface between the tool and the processed material by introducing external technological means into the contact zone (lubricating and cooling technological means – LCTM).

1. Introduction

According to the theory of radical chain mechanism, juvenile surfaces formed during friction and cutting of metals not only show great activity to physical adsorption and chemical reactions, but also emit a stream of energy particles (electrons, radiation quanta), the intensity of which is at its maximum at the moment of opening the layers. These particles have a significant impact on the flow of chemical processes occurring on the contact surfaces of the cutting tool and the processed material. As a result of their interaction with the molecules of the external environment, the latter are activated and decay with the formation of reaction particles – active atoms, radicals and radical groups, whose activity is due to the presence of an electron with uncompensated spin on the valence shell. Having a high chemical activity, these particles interact with juvenile surfaces, resulting in intensive formation of compositions (secondary structures), the main function of which is to separate clean metal surfaces, i.e. to reduce adhesion interactions.

The formation of chemical radicals can occur both naturally and forcibly. The general patterns of the processes of forced radical formation do not differ from the natural ones. As a result of external effects on LCTM, its components receive additional energy, which transfers them into a metastable (intermediate) state, which is characterized by a weakening or partial violation of intramolecular bonds of LCTM elements. One of the most important conclusions of this theory is that the lubricant oil films are formed only in the contact zone, where there are necessary conditions for its implementation.

Content from this work may be used under the terms of the Creative Commons Attribution 3.0 licence. Any further distribution of this work must maintain attribution to the author(s) and the title of the work, journal citation and DOI.
Published under licence by IOP Publishing Ltd

Therefore, only a small amount of the external environment is sufficient to initiate the lubricating effect.

This condition has been repeatedly confirmed both directly by the authors of this work, as well as by other domestic and foreign researchers and manufacturers. The conducted research on the effect on the cutting process of sprayed lubricants and coolants has found that, to ensure the lubricating effect, it is sufficient to supply 0.5–3.0 g/hour of sprayed oil or 50–150 g/hour of sprayed water-based lubricants and coolants (1 g of sprayed liquid has the same surface as 27 kg of unsprayed). When studying the effect of water on the steel cutting process, it was found that a drop of water has the same effect on the cutting process as a stream with a flow rate of several liters.

In this work, the quality of tool materials is understood as the compliance of the controlled parameters of the physical and mechanical characteristics of tool materials with those values that determine the performance properties of a metal-cutting tool or, comparing them with each other, we can give preference to those that have advantages in performance characteristics [1]. The paper also uses approach and software environments that allow, in comparison with other specified approaches and software environments, to more accurately model the stress state of the tool material and, thereby, increase the accuracy of designing tool parameters for specific operating conditions [2].

2. Results and discussion

Our research [3], aimed at studying the lubricity of nanoscale iodine-containing structures pre-formed in the surface of tools, revealed a significant improvement in the characteristics of the cutting process and increasing the performance of high-speed tools in various machining operations (figure 1).

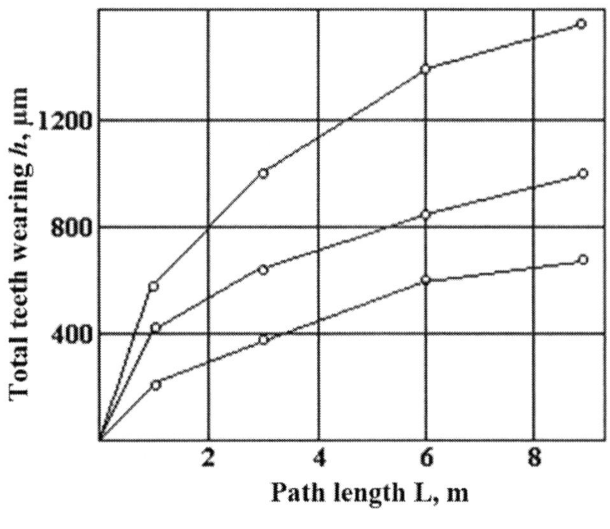

Figure 1. Dynamics of changes in the total teeth wearing of P9 steel cutters from the length of the cutting path when milling grooves in U8 carbon steel without the use of LCTM: 1 – without additional hardening; 2 – hardened by chemical-heat treatment; 3 – hardened by chemical-heat treatment with additional introduction of iodine vapors in the amount of 0.05 %, V = 1.6 m/s, S = 315 mm/min.

The second conclusion is based on the position that the volume of lubricating oil films formed as a result of radical chain reactions must be sufficient for the implementation of the lubricating action, i.e. it is directly dependent on the number of active atoms and radicals in the contact zone. Under otherwise identical conditions, the concentration of active particles is determined by the binding energy in the molecules of the external environment and the work of the electron output of LCTM components.

In figure 2 there are presented the results of studies on the use of LCTM microdoses as a technological tool, each of them had a shell made of an impervious material – in the form of microcapsules, the sizes of which varied from 1 to 45–50 microns [4]. To increase the efficiency of such LCTM microcapsules were given the ability to direct movement to the wedge of the tool. In the course of research, it was found that the optimal consumption of microcapsules was in the range of 5.5–8.5 g/hour. Thus, the reduction in the amount of LCTM supplied to the contact zone, compared to traditionally used tools, was more than 10,000 times, while an increase of 1.5–2.0 times was recorded in the durability of tools.

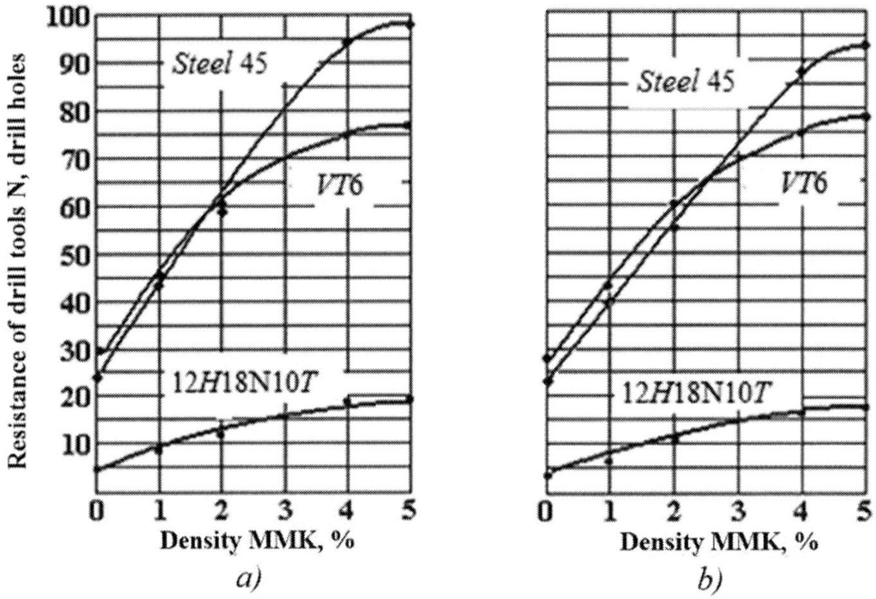

Figure 2. Correlations of drill resistance on drill tools of R6M5 steel on the concentration of microcapsules in the carrier: a) at V = 4.71 m/min; b) at V = 6.28 m/min.

The presented correlations clearly show that there is a certain range of microcapsule concentrations, in which the durability of tools increases with an increase in the amount of LCTM. Overcoming a certain critical value of the concentration of microcapsules leads to stabilization of resistance indicators and with a further increase in their number, the stability of the cutting tool practically does not change.

As noted above, for the active formation of separation lubricant films, it is necessary that there is a sufficient number of active particles involved in this process in the contact zone. The problem of intensifying the formation of reaction particles in the contact zone was solved by pre-transmitting energy to individual elements of the process medium by affecting LCTM by various physical and chemical methods – physical and chemical activation of LCTM [5]. As a result, LCTM molecules pass into an excited state before the contact zone and, further natural energy impact on them from the contact zone, becomes quite sufficient for their destruction with the subsequent formation of chemically active atoms and radicals. LCTM activation methods are most relevant for substances with high values of binding energy between atoms in molecules.

Recently, much attention has been paid to the use of atmospheric air activated by corona discharge as LCTM [6]. The observed effect is due to facilitation units between the tool and the material being treated lubricating effect of the oxide structures, the initiator of education which is active oxygen and hydroxyl groups, the resulting impact on air corona discharge. However, the authors of the research do not take into account that the air composition is quite complex.

The main components of air can be divided into three groups: constant, variable and random. The first are oxygen (about 21%), nitrogen (about 78%) and inert gases (about 1%). The content of these components is almost independent of the place where the sample was taken. The second group includes carbon dioxide (0.02–0.04%) and water vapor (up to 3%). The content of random components (the third group) depends on local conditions. Often there are mixed products of industry and transport, products of life of biological objects (microorganisms, plants, animals, etc.).

Our research has determined that each of these air components in one way or another affects the characteristics of the blade cutting process. Thus, it was found that the presence of moisture in the air contributes to a significant increase in the resistance of cutting tools, and the maximum increase is observed at a low concentration of water vapor (figure 3). Pre-activation of humidified air by corona discharge leads to an even greater increase in the durability of tools.

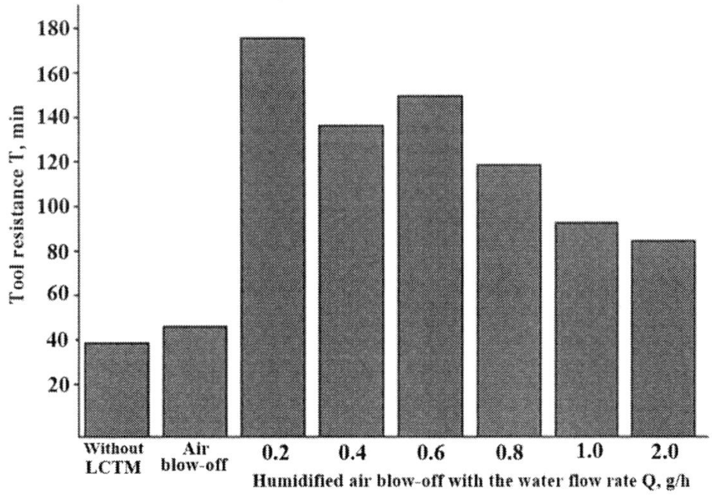

Figure 3. High-speed tools resistance when cutting steel 45 with humidified air blow-off at the contact zone $V = 1.2$ m/s, t = 0.5 mm, $S = 0.1$ mm/rev.

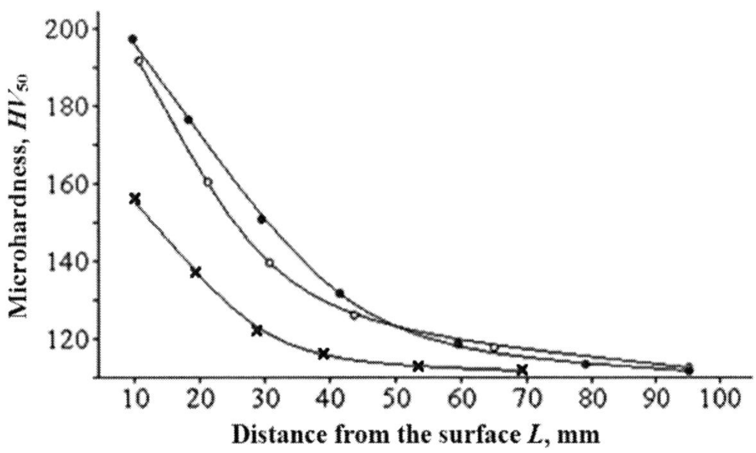

Figure 4. The size and microhardness changes of the plastic deformation zone when turning titanium *VT*1-0, $V = 0.5$ m/s, $t = 0.5$ mm, $S = 0.1$ mm/rev: • – without LCTM, × – blow-off with compressed air, × – blow-off with compressed air enriched with water vapor.

The presence and size of plastic deformation zones is one of the main parameters that characterize the intensity of adhesive interactions between the working surfaces of the tool and the processed material. Under certain conditions, this indicator can also be used to evaluate the effectiveness of separating lubricant films. Thus, the analysis of the experimental curves presented in figure 4 shows that the use of humidified air flow as LCTM reduces the depth of distortion of the matrix structure of the processed material. This can be interpreted as a decrease in the strength of adhesion. At the same time, the experiments recorded a decrease in the microhardness of the distorted standing, which indicates a quantitative decrease in the points of contact surfaces setting.

A similar improvement in the characteristics of the cutting zone and tool stability is observed in the case of introduction of organic matter into the air stream – industrial oil I-20A in concentrations of 0.1–1.0 g / hour. There was an increase in the resistance of the incisors, a decrease in chip thickening and roughness of the treated surface. Under other identical conditions of the experiments, it can be stated that the introduction of microdose of organic substances with a high lubricating capacity into the air improves the tribological state of the contact zone. For figure 5 presents a tribogram of the friction process according to the disk-disk scheme of 45 steel by 45 steel, hardened to a hardness of 57 units by *HRC*. As follows from the figure, the introduction of 0.5 g/hour of I-20A oil into the air flow effectively reduces the friction moment (zone "*b*") compared to dry friction (zone "*a*") to a value comparable to that of 100% I-20A oil. Pre-ionization of the air-oil flow (zone "*c*") leads to a further reduction of the friction moment and at the same time improves the dynamics of the contact interaction processes by stabilizing it.

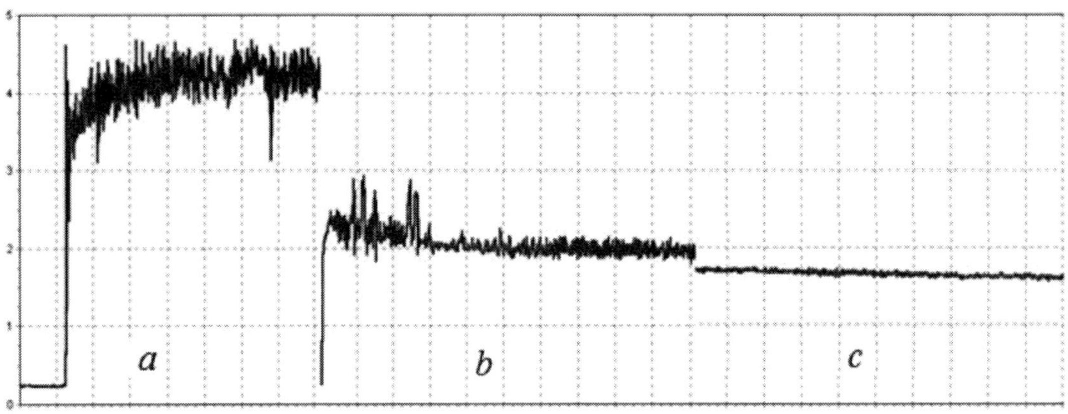

Figure 5. Tribogram of the friction moment of 45 steel on 45 steel hardened under loads: *a*) 2.6 *MPa*, *b*), *c*) 3.6 *MPa*, $V = 0.5$ m/s.

The study of the effect of the main components of air (oxygen and nitrogen), as well as their quantity, on the parameters of the cutting process during shaping was carried out on the author's installation, located under the vacuum cap of the vacuum post *VUP-4* (figure 6). Rotation from the electric motor 1 through the reduction gear 2 is transmitted to the drive shaft 11, which is a screw with a nut. Upon rotation of the shaft results in translational movement of a nut fixed thereto a horizontal guide 8. On this guide, a plate with a vertical guide 5 is fixed, on which a cutter 7 is installed using clips 6, the vertical movement of which is regulated by a micrometer screw 4.the Test sample 10 is fixed motionless on the base plate 3 with clips 9.

In the course of research, the influence of air, as well as nitrogen and oxygen separately on the formation of separation films at different pressures (from normal to 10^{-5} mm Hg) was studied by the amount of cutting forces, chip roots, development and change of microhardness of plastic deformation zones.

The materials used were 45 steel, *12H18N10T* austenitic stainless steel, and *VT1-0* titanium alloy. The samples were 0.8 mm thick plates. Cutting was carried out with cutters made of high-speed steel *R6M5* at a cutting speed of $V = 2$ mm/s and a cutting depth of $t = 0,05$ mm.

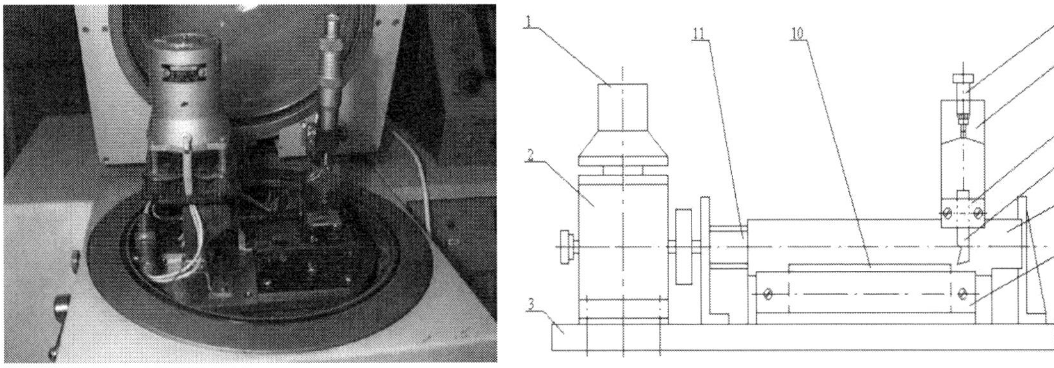

a) *b*)

Figure 6. Physiacal appearance (*a*) and installation diagram (*b*) for controlled atmosphere cutting.

In the course of research, it was found that changes in the type of gaseous LCTM and its amount have a noticeable effect on facing separation processes and cutting characteristics.

Figure 7 shows images of the facing roots obtained by the falling cutter method when turning the VT1-0 titanium alloy. It follows from the photos that a decrease in the pressure in the vacuum chamber, and, consequently, the amount of potential lubricant (in these cases, air), significantly leads to an increase in the facing thickening coefficient ξ and a decrease in the conditional shear angle β.

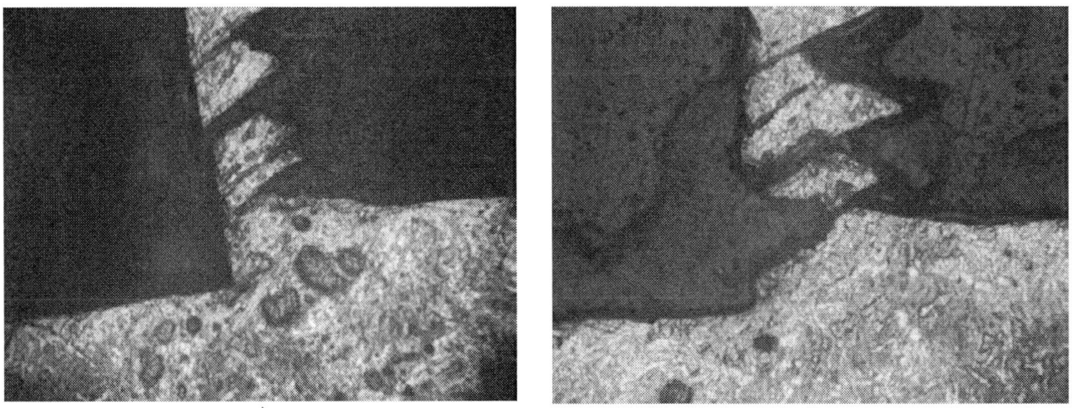

a) *b*)

Figure 7. Pictures of facing roots obtained during free turning in a vacuum chamber of *VT1-0* alloy with a *R6M5* steel cutter when using air as LCTM: *a)* at normal pressure; *b)* at a pressure of 10^{-5} mm Hg. $V = 2$ mm/s, $t = 0.1$ mm x 300.

The theory of radical chain reactions describing the formation of separation lubricant films on tribo-conjugated surfaces when using air as LCTM is based on physical and chemical phenomena involving oxygen and its oligomers. More recent studies conducted by the authors of this work have found that in this case, when describing the physical and chemical processes in the contact zone, it is necessary to take into account the fact that the amount of nitrogen is more than 3 times higher than the concentration of oxygen [7]. When studying the behavior of nitrogen in the composition of air used as

IOP Conf. Series: Materials Science and Engineering **939** (2020) 012052 doi:10.1088/1757-899X/939/1/012052

an external technological medium, it was determined that the amount of energy released in the contact zone during the destruction of metal during cutting is quite sufficient to transfer the nitrogen molecule to an excited state and its subsequent decay. Therefore, the combined effect of air on the kinetics of contact interaction is a more complex process, in which not only oxygen, but also nitrogen takes part.

These studies have shown that the use (separately) of both oxygen and nitrogen as LCTM has a noticeable effect on chip separation processes (figure 8).

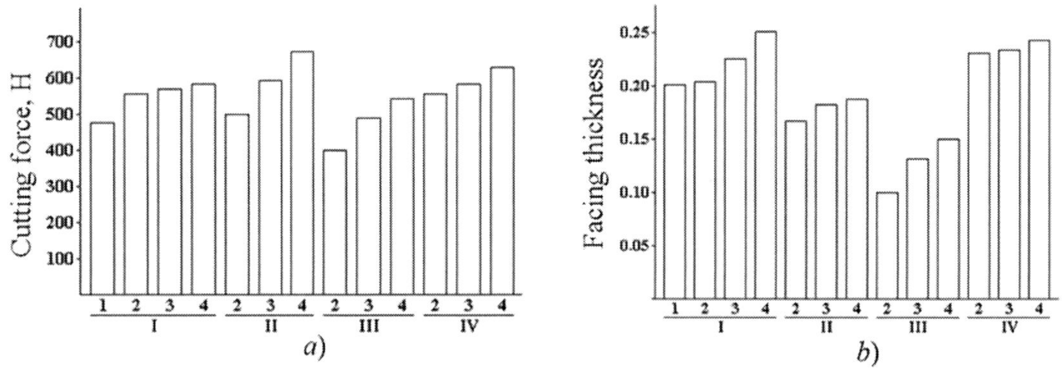

Figure 8. The results of cutting forces studies (*a*) and facing thickening (*b*) during free turning in a vacuum chamber of steel 45 when using air (I), oxygen (II), nitrogen (III), helium (IV) as LCTM at a pressure of: 1 – normal, 2 – 10^{-1} mm Hg, 3–10–2 mm Hg, 4–10–4 mm Hg V = 2 mm/s, t = 0.1 mm.

A detailed analysis of the presented results showed that in the case under consideration, pure oxygen had worse cutting forces than nitrogen and air at normal pressure. In our opinion, this is due to the lack of oxide films formed, as a result of which the adhesive interactions between the working surfaces of the tool and the processed material remain high. This confirms the previously obtained data on the study of the lubricity of ozone-containing microcapsules. Research has found that to effectively improve the characteristics of the cutting process (conditional shear angle, relative shear value, chip shrinkage, friction coefficient) and the stability of cutting tools, additional introduction of an oxygen – ozone oligomer in the composition of ozone-containing microcapsules is necessary. At the same time, it is determined that the number of additional microcapsules introduced depends on the brand of the processed material. Thus, when turning 40X and *12H18N10T* steels, the effective concentration of ozone-containing microcapsules in the LCTM (distilled water) used was 4%, for *VT5-1* – 2%, *VT6* – 1%.

When using nitrogen as LCTM (figure 8), the minimum values of cutting forces and chip thickening coefficient are marked. This can be explained by changing the matrix structure of the surface of the processed material as a result of the introduction of nitrogen atoms into it with the formation of new compounds. During chip separation, freshly uncovered metal surfaces emit electrons, the concentration of which is quite high in the absence of passivating oxide films. The energy of these electrons is sufficient to transfer nitrogen molecules to an excited state and convert some of them into atoms and radicals. An important role in this process is played by the chemical activity of the resulting juvenile metal surfaces. The result of such physical and chemical transformations is the formation of nitride compounds of the Fe_nN_m type (ξ – and ε-phases) with orthogonal and hexagonal lattices in the contact zone. This was established by electron microscopy using replicas extracted from the incised side of the facing. Experiments have not revealed the presence of a cubic γ ′ – *phase* (Fe_4N). Apparently, this is due to a rather narrow area of homogeneity of this phase in the *Fe-N* system, or its number is very small. At the same time, electronograms recorded new phases on the front surface of the incisor, the decoding of which showed the presence of individual inclusions of nitride compounds of the Fe_nN_m type, most of which also have a hexagonal structural lattice.

For a pair of metals of the same name with a simple cubic lattice, the friction resistance increases and decreases when metals with different crystal lattice structures or metals with a hexagonal crystal friction. Therefore, the newly formed nitride phases should contribute to improving the tribological situation of the contact zone, which is recorded in the conducted studies. On the other hand, the formation of nitride phases will inevitably lead to an increase in surface hardness, and this, in turn, will change the chip separation process, shifting it towards cutting more brittle materials, which are characterized by a decrease in the amount of longitudinal chip shrinkage, its broadening and thickening, as shown in figure 8, *b*.

The use of ionized air increases the effect of its components, both by contributing to the intensification of the formation of lubricant films between tribo-conjugated surfaces, and by modifying the phase composition of the metal surfaces of the contact zone.

In our opinion, the relatively low cutting forces (approximately at the air level) recorded in the case of using helium are due to the following. The small size of helium (1.37 Å), for example, compared to the nitrogen molecule (4.45 Å), allows it to penetrate the surface of the processed material quite freely. Distortions of the matrix lattice and the appearance of stress concentrators caused by the introduction of helium lead to changes in the physical and mechanical characteristics of the surface layer of the processed material and, first of all, its hardness. This facilitates the chip separation process similar to the use of nitrogen as LCTM.

3. Conclusion

Thus, the research has established that for effective impact on the tribological environment of the contact zone when cutting metals, the required amount of external technological tools can be significantly reduced. When using air as LCTM, including air activated by the action of electrical discharges, the cumulative effect is an integral indicator of the effectiveness of its components, which are included in both the first and second and third groups of elements and substances that make up the air. At the same time, the effectiveness of such LCTM when cutting various materials can be regulated by varying the number of these components.

References

[1] Kruglov A I, Zubarev Y M, Zhelnov D V and Shvetsov I V 2019 *IOP Conf. Ser.: Mater. Sci. Eng.* **656** 012028

[2] Shvetsov I V, Mokritsky B Y and Zhelnov D V 2019 *IOP Conf. Ser.: Mater. Sci. Eng.* **656** 012047

[3] Latyshev V N, Naumov A G and Radnyuk V S 2007 Tribologicheskie svojstva soedinenij joda, predvaritel'no sformirovannyh v poverhnosti bystrorezhushchih instrumentov [Tribological properties of iodine compounds preformed in the surface of high-speed tools] *Trenie i iznos* vol 28 **5** 441–448

[4] Naumov A G 2002 Uluchshenie ekologii processov lezvijnoj obrabotki metallov [Improvement of ecology of processes of blade processing of metals] *Stanki i instrument* **7** 9–13

[5] Latyshev V N and Naumov A G 2006 *Smazochno-ohlazhdayushchie tekhnologicheskie sredstva: Spravochnik* [Activation of STS In the book Lubricating and cooling technological means: Guide] ed L V Khudobin (Moscow: Mashinostroenie) p 543

[6] Bedunkevich V V 2003 Improving the efficiency of cutting tools made of superhard materials and cermets based on the method of dry electrostatic cooling *WPI Proc. of the Universities Mechanical Engineering* **7** 41–47

[7] Latyshev V N, Naumov A G, Mineev L I and Demyanovsky N A 2007 Osobennosti formirovaniya vtorichnyh struktur na tribosopryazhennyh metallicheskih poverhnostyah s uchastiem ionizirovannogo vozduha [Features of secondary structures formation on tribo-conjugated metal surfaces with the participation of ionized air] *Metalloobrabotka* **1** 9–12

Modifications of lightweight concrete for use in housing construction

O S Naumova[1]

[1] Yaroslav the Wise Novgorod State University, ul. B. St.Peterburgskaya, 41 173003, Veliky Novgorod, Russia

E-mail: osn5@yandex.ru

Abstract: The article analyzes the types, methods and characteristics of lightweight concrete modification in order to improve their strength and other characteristics. To substantiate the relevance of the study, the specificity and importance of the use of lightweight concrete in housing construction was determined, the evolution of the use of lightweight concrete in the construction of buildings and structures for various purposes was studied, the main types and characteristic features of lightweight concrete types depending on their composition, specifics of the technology for their production, usage and subsequent operation of structures and elements were examined. The classification of lightweight concrete structure modifiers is developed; their influence on the properties of concrete mixtures is studied.

1. Introduction

Currently, the use of lightweight concrete in the construction of both low-rise and high-rise buildings is becoming increasingly significant. Lightweight concrete is used as an independent structural material for the construction of load-bearing structures of houses and as heat-insulating materials. It is almost impossible to overestimate their importance for modern construction; therefore, for the development of house building, economical lightweight concrete with improved characteristics based on various modifications is needed.

The use of lightweight concrete on porous aggregates as a structural material has been known since ancient times. In particular, in the territory of modern Armenia, ancient structures have survived to the present day, during the construction of their arches and walls lightweight lime-based concrete was used.

It is known that in Italy in the V century hydraulic structures were built on the basis of lightweight concrete by adding crushed and lumpy light aggregate of volcanic origin (tuff or pumice) or artificially prepared (brick) to lime. For example, Pliny proposed mixing 1 part of lime, 2 parts of pozzolan and 1 part of crushed tuff. And such a recipe for several centuries was the only one in the construction of hydraulic structures, since one of the important properties of concrete is the ability to harden not only in air but also under water.

The use of low thermal conductivity of lightweight concrete began much later; approximately in the XIX century, not only monolithic structures, but also individual artificial stones began to be made from them. Initially, lime was used as a binder, but with the development of the chemical industry, it began to be replaced with Portland cement, which made it possible to increase the strength characteristics and hardening speed.

Content from this work may be used under the terms of the Creative Commons Attribution 3.0 licence. Any further distribution of this work must maintain attribution to the author(s) and the title of the work, journal citation and DOI.
Published under licence by IOP Publishing Ltd

IOP Conf. Series: Materials Science and Engineering **939** (2020) 012053 doi:10.1088/1757-899X/939/1/012053

The development of the fuel and metallurgical industries gave impetus to the use of various types of slag as porous aggregates for lightweight concrete. In addition, a serious impetus to the development of construction from lightweight concrete structures was given by the production of concrete mixtures using machines with the subsequent molding of artificial concrete stone.

At the end of the XIX century, Professor H. A. Zhitkevich recommended the use of brick gravel, coke, and slag from boiler furnaces and blast furnaces as porous aggregates for concrete. He also recommended the composition of lightweight concrete and defined the values of their thermal conductivity coefficients. In 1912, Professor H. A. Belelubetskiy proposed walls 45 cm thick made of brick and slag concrete stones. Engineers S. L. Prokhorov and M. N. Smirnov proposed the production of multi-hollow slag concrete stones, which, after they were improved in 1926, were called “peasant” stones. They also began to produce slag concrete stones using a wooden hand machine, which was improved by S. A. Torletsky in 1924.

The beginning of the trends of lightweight cellular concrete modifying based on the introduction in their composition of the porous filler can be attributed to the 1930s. The founder of this direction of concreting development is considered to be B. G. Skramtaev, whose ideas were supported and developed by Professor N. A. Popov, who experimentally proved the effectiveness of this direction. In addition, V.G. Dovzhik, I.N. Akhverdov, A.I. Vaganov, N.P. Bleshchik and others made an invaluable contribution to the development of lightweight concrete technology [1–6].

2. The main types of lightweight concrete used in modern housing construction

Lightweight concretes belong to the group of lightweight materials and are called porous concretes. They are classified quite extensively on the basis of various parameters, while it is necessary to take into account the impossibility of compiling the final version of the classification due to the appearance of new types of porous concrete, for which there is still no clear manufacturing technology, which allows them to be produced only in limited quantities.

Particularly, in accordance with the established ideas and existing practice, porous concrete can be of the following types:

– traditionally, it is recognized that lightweight concrete is concrete based on binding and porous aggregates; the porosity of the material is mainly due to the porous structure of the aggregate grains;

– large-porous (sandless) concrete on porous or dense aggregates is concrete based on a binder and porous or dense (ordinary) aggregate, while the porosity of the material is largely due to pores in the spaces between grains as a result of the low content of small fractions of aggregate;

– porous lightweight concrete is concrete based on binder and porous aggregates, the porosity of which is ensured by both the porous structure of the aggregate grains and the pores in the cementing binding aggregate, formed by introducing foam into the mixture or expanding the mixture with gas-forming additives, or introducing additives involving air into the mixture;

– cellular concrete is concrete made of a ductile fine-grained mixture in which pores are formed as a result of introducing foam into the mixture or swelling of the mixture with the addition of blowing agents [2].

The main types of concrete macrostructure are presented in figure 1.

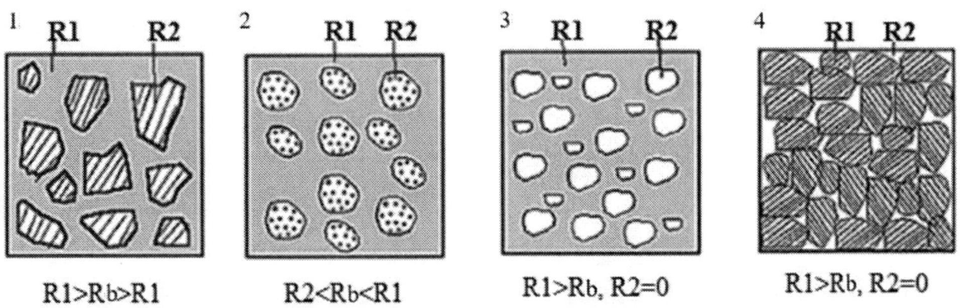

Figure 1. The main types of concrete macrostructure: 1– dense, 2 – dense with porous aggregate, 3 – cellular, 4 – granular, R1 and R2 – strength of concrete components, Rb – average structure strength.

IOP Conf. Series: Materials Science and Engineering **939** (2020) 012053 doi:10.1088/1757-899X/939/1/012053

For clarity, the classification of lightweight concrete by type of binder and by type of large porous aggregate is shown in figure 2.

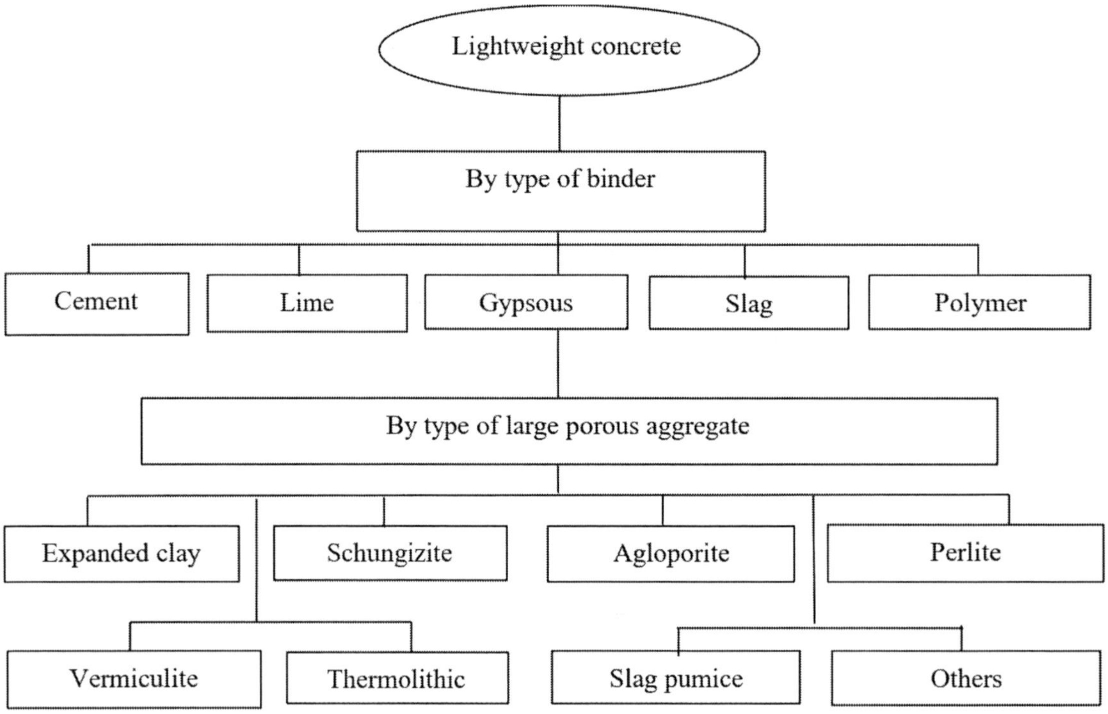

Figure 2. Classification of lightweight concrete by type of binder and by type of large porous aggregate.

For porous lightweight concrete, instead of the structure it is allowed to indicate the type of blowing agent in the name of concrete. Depending on which large porous aggregate is used, the following types of lightweight concrete are defined:

– expanded clay concrete (concrete on expanded clay gravel);
– schungizite concrete (concrete on schungizite gravel);
– agloporite concrete (concrete on agloporite crushed stone);
– slag pumice concrete (concrete on slag pumice crushed stone and gravel);
– perlite concrete (concrete on expanded perlite crushed stone, etc.;

From the point of view of application, lightweight concrete based on porous aggregates should be classified as:

– structural, used for the construction of load-bearing walls, so they are subject to increased requirements for strength and resistance to low temperatures;
– thermal insulating, used as an insulator, as well as for the construction of structures, which are subject to increased requirements for thermal insulation;
– structural and thermal insulating intended for the construction of load-bearing walls, ceilings, interior partitions, fences, therefore, they are subject to serious requirements for both strength and thermal insulation.

The pores in lightweight concrete can be filled with either gas or air, so they are respectively called aerated concrete or foam concrete. The classification of blowing agents can be represented in the form of a scheme shown in figure 3.

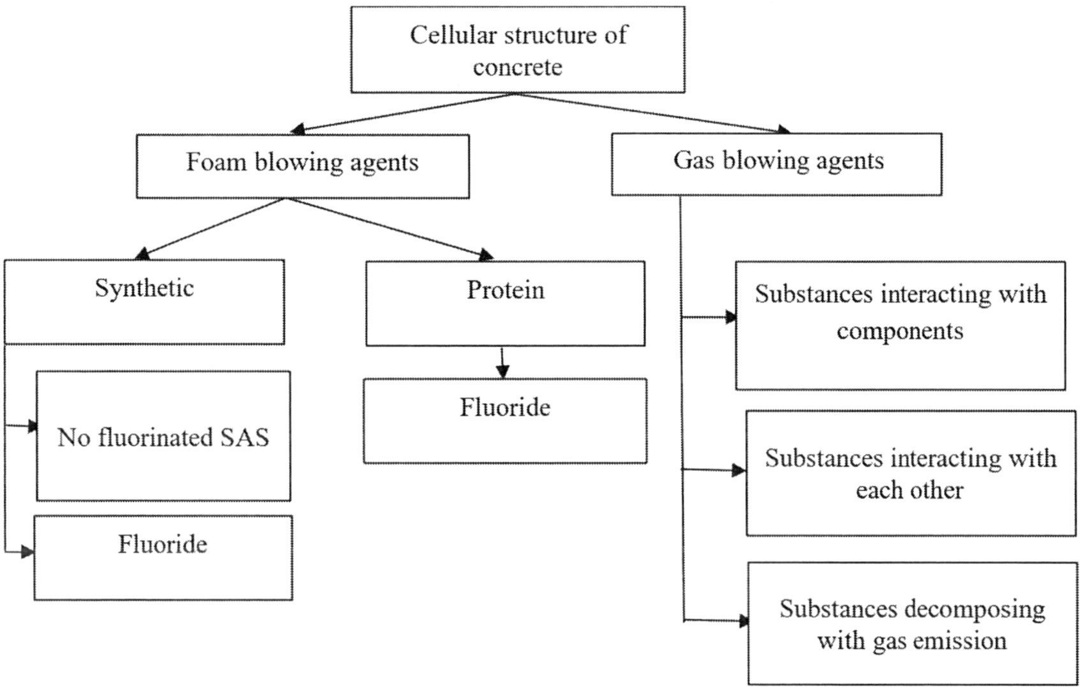

Figure 3. Classification of blowing agents of cellular concrete.

The process of concrete porization is carried out by introducing into the composition of the previously prepared foam (sandless mixtures) or gas-forming (sand mixtures) or air-entraining additives (mixtures with sand and without sand).

It should be noted that as a result of the analysis of the initial components in the production of light concrete, binders with various hardening methods (air, hydraulic, autoclave) are used, the choice of which is determined by the required concrete grade.

3. Study of possible modifications of lightweight concrete

Modification of the structure of lightweight concrete has now acquired significant proportions due to the need to increase strength and other technical and operational parameters, including reducing costs for production and sale.

Particularly, in comparison with ordinary equal-strength heavy concretes, modified light concretes on average have the following characteristics

– the density is reduced by 20–50%, it can significantly reduce not only the mass of the structures themselves, but the total load of the building or structure on the foundation, as well as reduce the consumption of concrete and reinforcement;

– reduction of levels (upper and lower) of the area of micro crack formation boundary to 25%, which increases the efficiency of prestressing of reinforcement in structures and the ultimate tensile strength of concrete;

– increase the coefficient of dynamic hardening and ultimate deformability in shock loading up to 25%, i.e. increased resilience, which is especially important to ensure the safety of pile heads and the durability of foundation grouts;

– increasing frost resistance by 2–3 grades and water resistance by 1–3 grades, which entails the possibility of increasing the durability and reliability of structures in particular, and buildings in general, improving operational properties;

– a decrease in the thermal conductivity coefficient by more than 3 times, due to which an increase in the heat efficiency of building envelopes and buildings as a whole is achieved;

– increased resistance to fire and ultra-high temperatures by an average of 0.4 hours, which seriously increases the fire resistance of structures in bearing capacity, positively affecting the fire safety of the building [3].

The process of modifying construction materials is carried out to improve their operation properties due to the introduction into their structure of special function additives. In concretes, modifying additives are used, which operating principle is to change the structure through chemical, physical or physical-chemical interaction [4, 8].

For clarity, the classification of modifiers is shown in figure 4. Structural additives contribute to increasing the strength properties of composite materials in concrete, both light and heavy, by reducing strains that can develop under the influence of stresses that arise during operation.

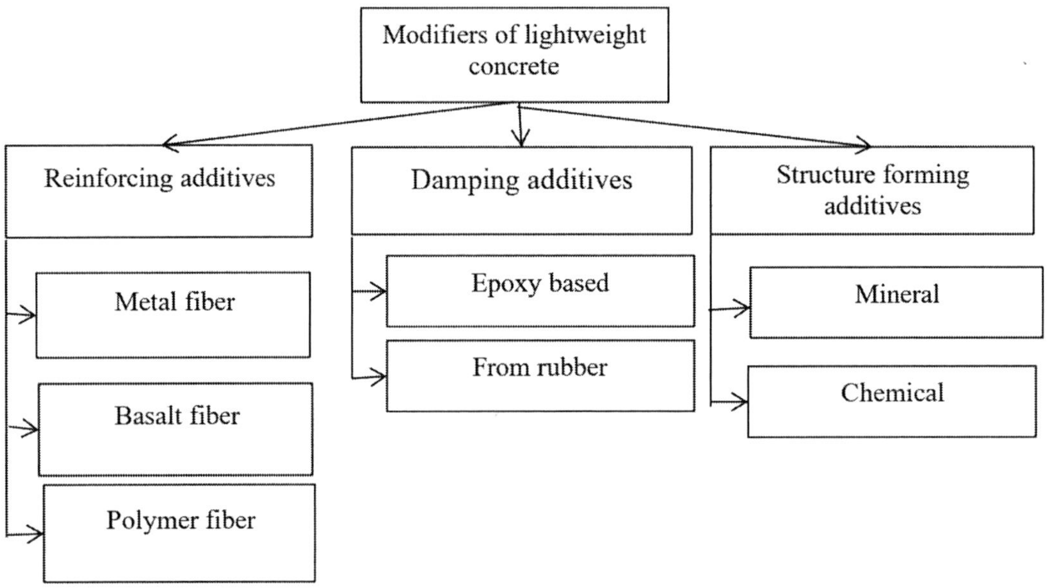

Figure 4. Modifiers of lightweight concrete classification.

The most used in the production of lightweight concrete are hydraulic binders based on Portland cement clinker, in particular, Portland cement, quick-hardening cement, slag Portland cement, quick-hardening slag Portland cement [4].

The choice of cement brand is necessary based on the conditions of hardening of concrete. In particular, for the production of structural thermal-insulating concrete with heat-moisture steaming, it is necessary to use cements M400 and higher, the cement consumption in the end is up to 380 kg, which entails a significant increase in the cost of the structures under construction. The problem arises of the need to save raw materials, the solution of which is possible by reducing the binder in combination with a decrease in the density of lightweight concrete.

It is possible to solve the above problem when using Portland cement for the production of lightweight concrete, to which slags and ashes are added as having lower thermal conductivity due to the amorphous glassy structure. As a result, an artificial stone made from slag Portland cement will have improved thermotechnical properties; in addition, resistance will increase in aggressive environments.

As analysis has shown, it will reduce cement consumption by about 20–30% due to the binder vibratory grinding. Additional cement vibratory grinding with a wet (with water) or dry method, as well as with additives that contribute to a decrease in the water-holding capacity of cement and water demand, respectively, will significantly improve the uniformity and workability, and hence the formability of the concrete mixture.

The use of plasticizers in the composition of concrete mixture allows for temporary liquefaction of the mixture, but without the need to increase the consumption of binder and water. It has been experimentally determined that when adding to the mixture up to 3% of the cement mass of superplasticizers in equal plastic mixtures, water consumption decreases to 30%, and cement consumption – up to 20%.

As a result of adding various hydrophobizing silicone additives in liquid form and a number of other chemical additives to the concrete mixture, its plasticizing properties increase and, as a result, not only water demand is reduced by 12%, but also cement consumption is reduced by 5-7%, there is increase in homogeneity, average density, etc. Additives of this property make it possible to modify the structure of a lightweight concrete mixture, which, together with hydrophobization, entails a decrease in capillary suction and water absorption, leading to an increase in frost resistance and durability. The disadvantage of using the above additives is an increase in the hardening time of the concrete mixture [6].

A modification of lightweight concrete is polystyrene concrete, which includes porous polystyrene granules, water, as well as an air-entraining additive as a porous aggregate. The use of this aggregate makes it possible to obtain concrete with an extremely low density in combination with resistance to decay, hydrophobicity, the highest indices of load-bearing characteristics, thermal insulation, fire protection, sound insulation, frost resistance and periods of freezing and thawing, which positively affects the operation life. This type of concrete is characterized by good strength, but at the same time ease of processing.

Lightweight self-compacting concrete is another modification of concrete mixes, and especially of heavy self-compacting concrete. They are made using polyacrylate esters and polycarboxylates, do not require forced compaction during installation without compromising the quality of concrete structures. A feature of lightweight self-compacting concrete is the use of volcanic tuff as a filler (sand with a fraction of 0–5 mm, crushed stone with a fraction of 5-10 mm) with the addition of dense sand, because otherwise, the concrete mixture cannot provide the rheological stability of the concrete mixture, i.e. delamination effect due to the ascent of individual expanded clay grains [7].

To increase the frost resistance of lightweight concrete, it is possible to introduce silicon-organic surface-active additives into the lightweight concrete mixture that affect the reduction of capillary porosity of concrete due to the different water requirements of the applied porous aggregates.

4. Conclusions

As practice shows, the use of lightweight concrete in housing construction allows saving not only time, but a significant amount of money. In addition, being a heat-efficient material, they provide an opportunity to save on heat supply.

The indisputable advantage of such concretes is their fire safety, since, for example, foam concrete with a thickness of 150 mm protects against fire for four hours. Even with a smaller thickness, the material will not burn, split, or emit toxic gases.

Despite the indisputable advantages, lightweight concretes also have drawbacks: abrasion, water absorption, etc., the reduction of their negative impact is possible due to the use of improved concrete based on the use of modifying additives, provided that the requirements for the manufacturing technology, transportation, laying and aging are adhered to.

References

[1] Kastornykh L I and Sinitsina N A 2014 Issledovaniye svoystv legkikh samouplotnyayushchikhsya betonov [Study of the properties of lightweight self-compacting concrete] *Bulletin of SUSU. Series «Construction Engineering and Architecture* **14(4)** 47–51 [In Russ.]

[2] Yarmakovskiy V N 2010 Modifitsirovannyye legkiye betony razlichnykh vidov dlya ograzhdayushchikh i nesushchikh konstruktsiy zdaniy [Modified lightweight concrete of various types for building envelopes and supporting structures] *Scientific works of the 2nd Inter. Conf. on Concrete and Reinforced Concrete: "Concrete and Reinforced Concrete – Development Paths"* **4** 176–186 [In Russ.]

[3] Gagarin V G 2009 Teplofizicheskiye problemy sovremennykh stenovykh ograzhdayushchikh konstruktsiy mnogoetazhnykh zdaniy [Thermophysical problems of modern wall enclosing structures of multi-storey buildings] *ACADEMIA. Architecture and construction* **5** 297–305 [In Russ.]
[4] Golik V I and Komashchenko V I 2017 Ferruginous quartzite processing waste as a source of additional metal recovery and backfilling *Gornyi Zhurnal* **3** 43–47 [In Russ.]
[5] Inozemtsev A S 2015 Poisk effektivnykh metodov povysheniya konstruktsionnykh svoystv vysokoprochnykh legkikh betonov [Search for effective methods to increase the structural properties of high-strength lightweight concrete] *Molodoy uchenyy* vol 17 **97** 133–137 [In Russ.]
[6] Pshin'ko A N, Krasnyuk A V and Bychkov S A 2010 Analiz tekhnologiy proizvodstva i sposobov modifikatsii legkikh betonov na neorganicheskikh vyazhushchikh i zapolnitelyakh [Analysis of production technologies and methods for modifying lightweight concrete with inorganic binders and aggregates] *Science and Transport Progress. Bulletin of the Dnipropetrovsk National University of Railway Transport* **35** 150–155 [In Russ.]
[7] Burgos D, Guzmán A, Delvasto S and Hossain K M A 2017 The use of a volcanic material as filler in self-compacting concrete production for lower strength applications *Materiales De Construccion* vol 67 **325** e111 DOI: 10.3989/mc.2017.09315
[8] Nikolenko S D, Sushko, E A, Sazonova S A, Odnolko A A and Manokhin V Y Behaviour of concrete with a disperse reinforcement under dynamic loads 2017 *Magazine of Civil Engineering* **75(7)** 3–14

Improving the efficiency of planing operations when processing large parts based on gas analysis

O V Nikulenkov[1]

[1] Yaroslav-the-Wise Novgorod State University, ul. B. St. Petersburgskaya, 41 173003 Veliky Novgorod, Russia

E-mail: Oleg.Nikulenkov@novsu.ru

Abstract. The paper presents a comprehensive study of the process of diffusion of impurities, the formation of gases and their mass transfer in the gas-air space of the treatment zone. In addition, the author discusses changes in the state of the cutting process, including the state of the cutting tool and the surfaces of the processed products. Accordingly, the main scientific result is the development of methods and means of controlling the processes of mechanical processing based on mathematical models. The use of such systems in automated production allows you to increase productivity and reduce processing costs by increasing the reliability of processing at higher cutting conditions, timely change of substandard tools, reducing product scrap and tool consumption. In addition, to increase the reliability of the processing systems by timely replacement of extremely worn or broken tools on a backup tool.

1. Introduction

The growth in the automation of production processes in mechanical engineering places high demands on the means of diagnosis, which are used for the operational control of the condition of equipment, the detection and localization of faults. For metalworking equipment, one of the urgent issues of increasing its reliability and improving the quality of products is the diagnosis of the condition of the cutting tool and the rapid detection of the initial stage of critical wear, chips or breakage.

Known methods for evaluating machining processes, based on measuring cutting forces, vibroacoustic emission, thermoEMF and others, do not provide complete information about ongoing processes or do not fully reveal their potential capabilities, so their use in automated control and monitoring systems is limited. An analysis of the works [1, 2, 3, 4] devoted to methods of control of machining processes, as well as information-measuring systems, shows that much attention is paid to the issues of ensuring a given or optimal reliability of monitoring the state of the cutting tool, as well as the accuracy of measurements. The smallest group consists of works in which the quality of the surfaces of the processed products is controlled, although this parameter is one of the main ones. Thus, the scientific problem discussed in this paper is very topical. The paper is devoted to the establishment of interconnections between physicochemical and thermal deformation processes, as well as the condition of the cutting tool and the quality of the processed material with gas formation parameters.

Content from this work may be used under the terms of the Creative Commons Attribution 3.0 licence. Any further distribution of this work must maintain attribution to the author(s) and the title of the work, journal citation and DOI.
Published under licence by IOP Publishing Ltd

IOP Conf. Series: Materials Science and Engineering **939** (2020) 012054 doi:10.1088/1757-899X/939/1/012054

2. Results and discussion

The occurrence of high temperatures in the cutting zone contributes to the diffusion of gas impurities from the removed chips and the formation of volatile compounds upon contact of the processed material with atmospheric air. The temperature field arising in a solid under the action of a moving or motionless heat source of any shape acting temporarily or continuously can be obtained as a result of one or another combination of temperature fields arising under the action of a point system of instant heat sources. The main foci of heat during cutting are, as you know, the plane of shear in the sheared deformable layer of material and the friction zone of the front and rear surfaces of the tool with descending chips and the workpiece, respectively.

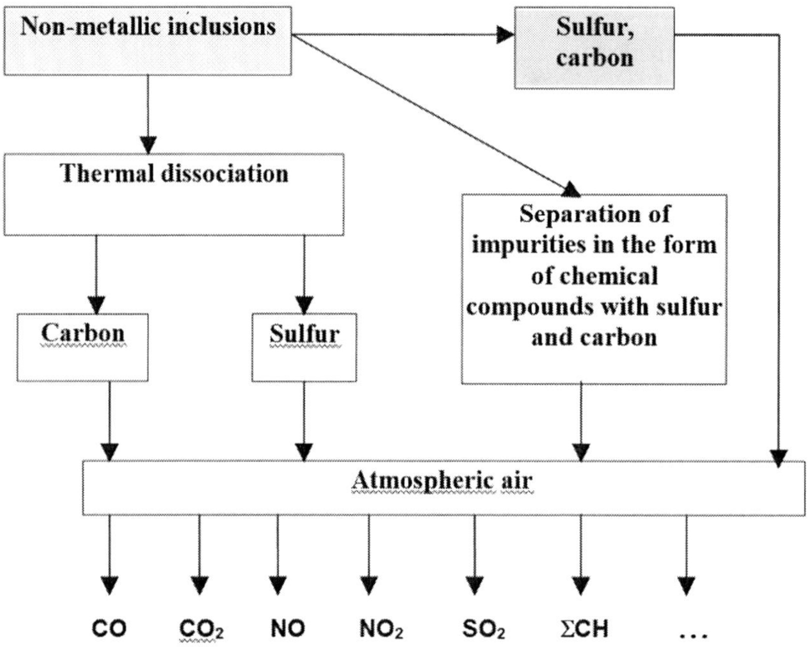

Figure 1. The main processes of burnout of impurities with the formation of volatile compounds.

Figure 1 shows the main processes that lead to the mass transfer of diffusing elements in the material with the formation of volatile compounds in a gaseous medium. The theory of thermal processes in technological systems during cutting is only part of the general theory of heat conduction and physical processes, which presents expressions for determining various quantities, including cutting forces:

P_x and P_y – components of the cutting force, determined in the conditions of the experiment, according to the reference or literary sources:

$$P_X = Cp_X * t^{1,2} * S^{0,55} \quad (1)$$

$$P_Y = Cp_Y * t^{0,9} * S^{0,75} \quad (2)$$

$$P_Z = Cp_Z * t^{1} * S^{0,75} \quad (3)$$

The value of the normal component of the cutting force P_N is determined by the formula

$$P_N = \sqrt{P_X^2 + P_Y^2}, \tag{4}$$

where P_x and P_y are the constituent cutting forces, determined by reference data, by the conditions of the experiment, or from literature, for example, [1]:

The values of the coefficients Cp_X, Cp_Y and Cp_Z are also determined by reference data or by formulas. For example, for cast iron

$$Cp_Z = const\ HB^{0.55}, \tag{4}$$

where Cp_Z is a constant component;

HB – Brinell hardness.

Based on this, the mathematical expressions are presented below, which reflects the diffusion processes of impurity atoms without taking into account their interaction with the cooling medium under various metal processing conditions.

$$N(\mathrm{s},\tau)_D = No^{*}\mathrm{erfc}\left(\frac{s\cdot\sin\phi}{2\sqrt{\frac{t\cdot\eta}{\upsilon\cdot\sin\phi}D_0\cdot\exp\left(\frac{E_D}{RT_D}\right)}}\right) \tag{5},$$

where τ is the diffusion time, sec;

t is the cutting depth, mm;

v is the cutting speed, m / min;

η is the coefficient of longitudinal shrinkage of chips;

No is the initial concentration of impurities in the metal;

erfc is an additional function of Gaussian errors;

Do is the diffusion constant;

E_D is the thermal energy of deformation;

R = 8.314 J / mol · K is the gas constant.

Figure 2 shows the heat distribution during chip formation, described by expression (5).

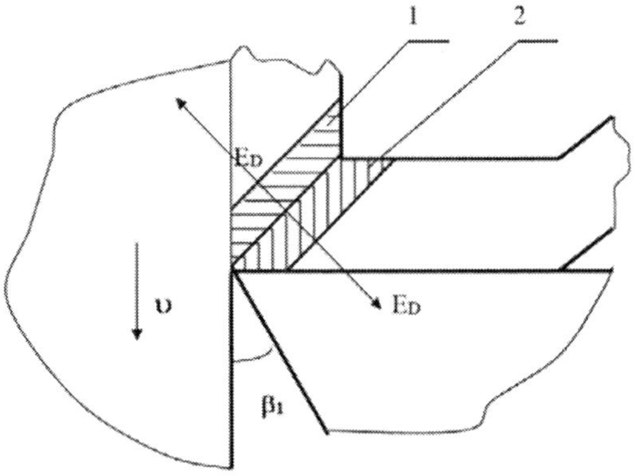

Figure 2. Heat distribution during chip formation: 1 – elementary volume before deformation, 2 – elementary volume after deformation.

Thus, the obtained mathematical model allows us to determine the concentration of diffusing impurities under any processing conditions of various materials. The developed technique allows to determine the areas of the processed surface with deviations from the specified hardness. This technique is based on the analysis of the physicochemical properties of the specific gases generated in the cutting zone during processing of materials. Three main groups are identified as controlled gases: those formed by combining the air components with impurities burnt out from the processed material (carbon oxides, sulfur oxide, hydrocarbons), as well as compounds formed by combining atmospheric oxygen and atmospheric oxygen. The results obtained during the experiments revealed a change in the concentration of a number of controlled gases in the treatment zone, reflecting a change in hardness and confirming the presence of abnormal areas in the surface layer of the processed material. When planing with a cutting speed v = 0.07 ... 0.4 m / s, feed S = 10 ... 20 mm / dv.x, cutting depth t = 0.5 ... 2 mm, a different concentration level is noted formed nitrogen oxides in the processing zone of various materials.

Table 1 presents data on the concentration of nitric oxide in the processing of materials with various physicochemical properties.

Table 1. Change in the concentration of nitric oxide in the processing of materials with various physicochemical properties.

№	Material	Concentration of nitrogen oxides
1	Stainless steel	15
2	Steel 40H	3
3	Steel 18 HGT	13
4	Steel 38H2MUA	15
5	Grey cast iron	26
6	Aluminum alloy	3

The hardness of the material can be estimated when there is a dependence of the concentration level of a gas on a diagnostic parameter. During the research, a large-sized part made of gray cast iron was processed with a surface area of approximately 5 m^2, the hardness of which should not be lower than HB 170 ... 180. Objective control of the surface hardness of a workpiece with a large processing area is difficult or impossible. A change in the concentration of gases formed during the cutting process is a consequence of the scatter in the physicochemical parameters of the workpiece, including the presence of hardness anomalies. The concentrations of released gases measured during the planing indicate the presence of anomalies in the hardness of the surface of the workpiece, which is confirmed by the results of hardness measurements by standard methods. The calculation of the linear correlation coefficient between the concentration of carbon monoxide CO in the treatment zone and the hardness of cast iron according to the results of planing a large cast iron table of a machine on a planing machine is presented below.

Table 2. Calculation of linear correlation coefficients between the concentration of carbon monoxide CO and the hardness of cast iron.

HB	ppm	X^2	Y^2	$X_i\ Y_j$
163	5.6	20.8	31.4	25.5
165.2	5.2	45.3	27.04	35.0
162.7	6.2	17.9	38.4	26.2
160.5	2.8	4.2	7.8	5.8
162.7	3.8	17.9	14.4	16.1
165.8	3.4	54.6	11.6	25.1
165.8	5.2	40.8	27.04	33.2
167.2	6.6	76.2	43.6	57.6
∑	47.8	354.0	282.3	303.1

where *X* is the average value of the measured Brinell hardness, *HB*;

Y – averaged over two histograms, the value of the level of CO emission during planing finish processing of the table, ppm;

X_i, Y_j are the arithmetic mean of hardness and concentration, respectively.

Based on the data given in table 2, we determine the value of the variances: $S_X = 2.84$, $S_Y = 2.45$. The correlation coefficient according to the results of measurements of hardness and concentration level in the processing zone of carbon monoxide is: $R_{XY} = 0.63$.

The value of the sample function $T = 3.69$, i.e. T > t. From this we can conclude that the measured values of the hardness of cast iron and the concentration of gases formed in the treatment zone (i.e., *X* and *Y*) are connected by a linear dependence. The dependence of the concentration of carbon monoxide on the hardness of the processed material is presented in table 3.

Table 3. The dependence of the concentration of carbon monoxide on the hardness of the processed material.

Material	Hardness, HB	N / N_o
Steel 10	97	0.74
Steel 20	126	0.67
Steel 20H	134	0.66
Steel 40	179	0.62
Steel 45	180	0.61
Steel 50	191	0.60
U8	210	0.56
08H18N10T	165	0.56

Thus, the experiment to determine the concentration of carbon oxides in the cutting zone when machining a large part and measuring the hardness of the surface of the material confirmed the fundamental possibility of diagnosing the processed material (in this case, gray cast iron), based on the analysis of the gas-air medium during the machining of metals (in this case in the process of cutting metals by planing).

In this case, restrictions on the cutting force and power were taken depending on the parameters *t* (cutting depth), *v* (cutting speed) and *h* (tool wear), determined from experimental studies based on gas analysis. The most optimal when using this method, as experiments show, are speeds from 0.33 to 0.42 m/sec. The use of recommended cutting speeds depending on the type of processing prevents the acceleration of wear of the cutting tool and reduces the auxiliary time when processing the workpiece itself.

When processing large table blanks, the gas-air medium was continuously monitored in the processing zone with registration of anomalous zones on the surface at each pass of the planing cutter. After processing and recording the anomalous zones, the table surface hardness was monitored and a conclusion was drawn on the suitability of this product for use, further or repeated processing, which made it possible to increase the monitoring efficiency in the manufacture of products. Theoretical studies and practical results obtained show that the greatest and most reliable information about the state of the cutting process of materials is obtained when using equipment for a comprehensive analysis of the gas-air environment. A study of the influence of the state of the material on the gas formation process with the implementation of this method was carried out when machining large-sized tables of heavy machines, which are a large casting with an area of about 5 m^2, on a planing machine. Based on the overall dimensions of the workpiece, when monitoring the entire machined surface, it is necessary to carry out about 2000 hardness measurements. This is due to the fact that with one measurement, an area of 2500 mm^2 is required. The proposed method allowed to reduce the number of measurements from 100 to 200 (in the anomalous zones detected by gas analysis). The proposed method was also used in the control of large-sized bushings of compressor cylinders during processing on boring machines in mechanical repair work.

3. Conclusion

The interrelations of cutting conditions and the quality of the processed material with the parameters of gas generation during planing are determined. The main physicochemical processes during mechanical processing are investigated and identified, which allow to effectively monitor the condition of the workpiece material, based on gas analysis in the chip formation zone. The mathematical dependence of the diffusion of impurities on the processing parameters is established, which allows controlling the planing process and ensuring the quality of the surface layer of the processed products. In the process of research it was found that the concentration of formed gases generated during machining increases with increasing surface hardness of the processed products. Recommendations are developed for controlling the surfaces of products in planing operations, based on the analysis of gas formation in the cutting zone, which made it possible to increase labor productivity and processing quality by evaluating the working capacity of the cutting tool and the characteristics of the processed material.

References

[1] Shvetsov I V, Belyakov V N, Zhelnov D V, Letenkov O V and Nikulenkov O V 2018 Diagnostics of wheel life in grinding *Russian Engineering Research* vol 38 **8** 610–13
[2] Kruglov A I, Zubarev Y M, Zhelnov D V and Shvetsov I V 2019 *IOP Conf. Ser.: Mater. Sci. Eng.* **656** 012028
[3] Shvetsov I V, Zhelnov D V and Zubarev Y M 2019 *IOP Conf. Ser.: Mater. Sci. Eng.* **656** 012048
[4] Shvetsov I V, Vasilkov D V and Tarikov I Ya 2019 *IOP Conf. Ser.: Mater. Sci. Eng.* **656** 012049

IOP Conf. Series: Materials Science and Engineering **939** (2020) 012055 doi:10.1088/1757-899X/939/1/012055

Features of the operation of a refrigeration screw compressor for capacity control

A N Noskov[1], M Shaposhnikova[1], I G Arendatelev[2,4] and I M Zawadzki[3]

[1] Saint-Petersburg National Research University of Information Technologies, Mechanics and Optics, 49, 197101 Kronverksky pr., St. Petersburg, Russia
[2] Yaroslav-the-Wise Novgorod State University, ul. B. St. Petersburgskaya, 41, 173003 Veliky Novgorod, Russia
[3] Novgorod chemical-industrial College, ul. Kochetova, 28, 173025 Veliky Novgorod, Russia

[4] E-mail: Igor.Arendatelev@novsu.ru

Abstract. To ensure the required temperature of the cooled object by maintaining the heat balance between the heat gain and the refrigerating capacity of the refrigerating machine, a compressor performance control system is used. The article presents the dependencies of the change in efficiency coefficient at full and partial performance with joint regulation of productivity and geometric compression ratio. The laws of the optimal change in the geometric degree of compression with a change in performance are given. The experimental characteristics of a screw compressor running on R22 freon were used.

1. Introduction

Oil-flooded screw compressors (OSC) are widely used in various fields which use artificial cold [1]. In order to maintain the temperature of the cooled object in a predetermined temperature range by ensuring the heat balance between the heat influx and the refrigerating capacity of the refrigerating machine when the temperature of its operation changes, a system for controlling the performance of an oil-flooded screw compressor is used.

There are the following methods for regulating the performance of refrigeration screw compressors: using external devices; changing rotor speed; using built-in devices that change the volume of the cavity of the screws when they are cut off from the suction window.

- using external devices;
- changing rotor speed;
- using built-in devices that change the volume of the cavity of the screws when they are cut off from the suction window.

Regulation with the use of external devices is carried out by bypassing the compressed refrigerant or by suction throttling.

A method for controlling OSC performance by changing the rotor speed in the range of optimal peripheral speed is quite economical, since the geometric compression ratio does not change. Currently, regulation of rotor speed using frequency converters has become widespread. In [2, 3], performance control was considered by changing the rotational speed of the screws, but with

Content from this work may be used under the terms of the Creative Commons Attribution 3.0 licence. Any further distribution of this work must maintain attribution to the author(s) and the title of the work, journal citation and DOI.
Published under licence by IOP Publishing Ltd

a decrease in the rotational speed, with a small volumetric capacity of the screw compressor, the operational efficiency is significantly reduced due to large internal refrigerant leaks.

One of the advantages of screw compressors over other types of compressors is the ability to control the cooling capacity of the built-in devices.

The problem of choosing a performance regulator and geometric compression ratio of a screw compressor, which ensures efficient operation of the chiller while reducing productivity and changing the boiling and condensation temperatures, is relevant.

2. Results and discussion

Widespread regulation by one spool leads to significant loss of work due to a decrease in the geometric degree of compression ε_G with a decrease in productivity [2]. This paper presents the dependences of the compressor efficiency at 100% capacity with a constant geometric compression ratio $\varepsilon_G = 5.0$ and when adjusting the geometric degree by changing the cylindrical part of the discharge window. When the compressor is operating at full capacity with an SRM D standard profile with external screw diameters of $D_1 = D_2 = 204$ mm on *R*22 freon at a boiling point of 10 °C and a condensation temperature of 350 °C, the indicator efficiency of the compressor without adjusting the geometric compression ratio by 4.75% higher than εG-controlled compressor.

A decrease in compressor efficiency was due to a decrease in the discharge window area.

The performance control by the slide valve and the thrust bearing has a limited area of variation in the geometric degree of compression, which does not allow to obtain the optimal law of its change with a decrease in productivity [2, 4].

To solve the problem of regulating the capacity and geometrical compression ratio of a screw compressor, which ensures the efficient operation of the refrigerating machine, it is necessary to choose the design of the capacity regulator and geometrical compression ratio to minimize losses from mismatch between the internal pressure of the refrigerant vapor and the condensing pressure. It is also necessary to evaluate the energy efficiency of the proposed regulator when regulating compressor performance.

3. Objects of study and research methods

The authors propose the use of a performance and geometrical compression ratio regulator for a screw compressor, consisting of a slide valve and two rotary valves. Such a regulator allows one not only to regulate the geometric degree of compression, but also to obtain the necessary law of variation of ε_G with a decrease in productivity. Figure 1 shows the design of such a regulator.

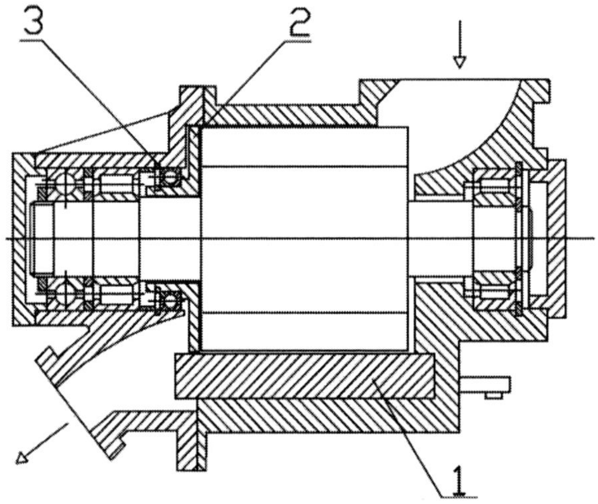

Figure 1. The design of the regulator of performance and the geometric degree of compression by the valve and rotary dampers: 1 – spool; 2 – rotary shutters; 3 – the bearing of rotary shutters.

To evaluate the performance of a refrigerating machine with an oil-flooded screw compressor and a regulator consisting of a slide valve and rotary dampers, it is necessary to analyze the characteristics of a screw compressor with full and partial performance in various modes of operation.

At the department of Low-Temperature Energy of Saint-Petersburg National Research University of Information Technologies, Mechanics and Optics an experimental study of refrigeration OSC on R22 freon was carried out with the possibility of changing the performance and geometric compression ratio and its characteristics were obtained.

Studies of the refrigerating OSC were carried out at the experimental bench. The experimental bench is a cascade-refrigerating machine. The upper branch of the cascade consists of an oil-flooded screw compressor *S* 3-900 from Kühlautomat (Germany), a condenser, a control valve and a condenser-evaporator. The condenser-evaporator is a condenser for the lower branch of the cascade. An experimental OSC and a water heat exchanger enter the lower branch of the cascade-refrigerating machine.

Tests of the experimental compressor were carried out on a gas ring. After compression in the experimental compressor, the refrigerant vapor is throttled by the valve and cooled in the water heat exchanger and condenser-evaporator, not reaching the saturation state, after which it is sucked into the compressor. The suction temperature is set by cooling the refrigerant vapor with an upper stage-refrigerating machine. The suction and discharge valves set suction and discharge pressures.

An asynchronous electric motor was used as the drive of the experimental OSC. The bench was equipped with the necessary measuring equipment. Standard pressure gauges with a maximum permissible measurement error of 0.4%, and temperature determined the refrigerant vapor pressure at the nodal points by laboratory thermometers with a division value of 0.10 °C.

To determine the mass flow rate of the refrigerant vapor, a narrowing device, a diaphragm, was used. The rotational speed of the electric motor shaft of the experimental OSC was determined by a frequency meter using an optoelectronic pair.

The power consumed by the experimental OSC was determined by the electric power of the electric motor, by means of a measuring complex for measuring current, voltage and power with a maximum permissible measurement error of 0.5%. When determining the power on the compressor shaft, catalog data on the coefficient of efficiency of the electric motor were used. The root mean square relative errors in determining the effective power *Ne* (power on the compressor shaft) do not exceed 3%. The refrigerant used was *R22* freon.

The experimental screw compressor has the following characteristics: the number of teeth of the lead and driven screws 4/6; outer diameters of screws 160 mm; screw length 144 mm; rotational speed of the drive screw is 49.2 s^{-1}; theoretical volumetric capacity is 0.0841 m^3/s; asymmetrical screw tooth profiles RDC Turbocompressor (Kazan, Russia). A slide valve and interchangeable inserts simulating rotary dampers carried out performance control. The position of the spool and the interchangeable inserts provided a different geometric degree of compression. The compressor efficiency was estimated by the efficiency factor $\eta_e = N_s/N_e$, where N_s is isentropic power.

4. Research results and discussion

As a result of an experimental study of a screw oil-flooded compressor, the values of the efficiency factor were obtained from the degree of pressure increase at full capacity and relative productivity equal to 75% and 50% of the maximum.

At the first stage of the experiment, the injection window remained unchanged, with the initial value of the geometric compression ratio $\varepsilon_G = 2.6$. The obtained dependences η_e, at a boiling point of $t_0 = -20$ °C, on the external compression ratio $\pi_н = p_к/p_в$, where $p_н$ and $p_в$ are the discharge and suction pressures, are shown in figure 2. The current value of the geometric degree of compression changes along the line a – b (figure 3).

At full capacity, the maximum η_e is achieved at $\pi_н = 3$, which corresponds to the coincidence of the discharge pressure and the internal compression pressure. The maxima η_e with a decrease in productivity are shifted towards lower values of $\pi_н$ due to a decrease in ε_{Gi}.

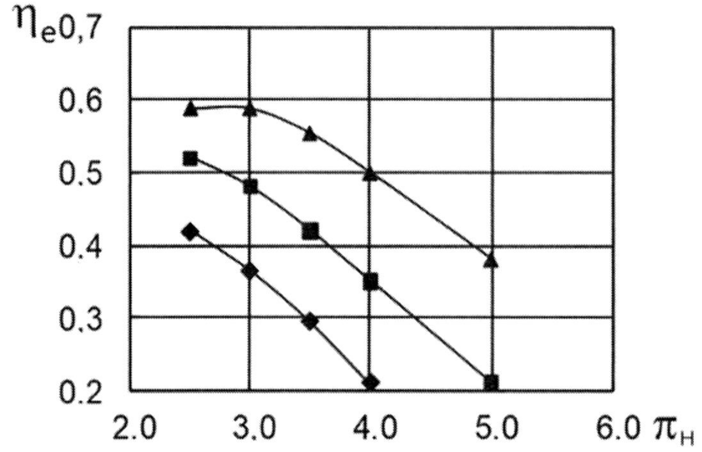

Figure 2. The dependence of the efficiency factor η_e of the screw compressor from the external compression ratio $\pi_н$ when controlled by one valve with a constant geometric compression ratio ε_G ▲ – $V_{Ti}/V_T = 1.0$; ■ – $V_{Ti}/V_T = 0.75$; ◆ – $V_{Ti}/V_T = 0.5$.

At the second stage of the experiment, the injection window was changed so that the ratio of the volumes of the paired cavity at the beginning and at the end of the compression process is the current value of the geometric degree of compression ε_{Gi}, with a relative volumetric productivity of 75% and 50%, it remains equal to 2.6 (line a – c in figure 3).

As a result of the experiment, an insufficient increase in η_e was obtained, which is explained by significant losses during expulsion of steam into the suction chamber during the contraction of the paired cavity until it is cut off from the suction window [5, 6]. The increase in pressure in the cavity at the beginning of the compression process and the decrease in the internal compression ratio is $\pi_a = p_a/p_в$, where p_a is the pressure in the cavity at the moment of its connection with the discharge window.

At the third stage of the experiment, in order to determine the optimal law for changing the current value of the geometric compression ratio ε_{Gi} from the relative productivity V_{Ti}/V_T, a study was carried out with various discharge windows. A graph of ε_{Gi} versus V_{Ti}/V_T at which the compressor's efficiency at a boiling point $t_0 = -20$ °C is maximal, is presented in figure 3 (line a – d).

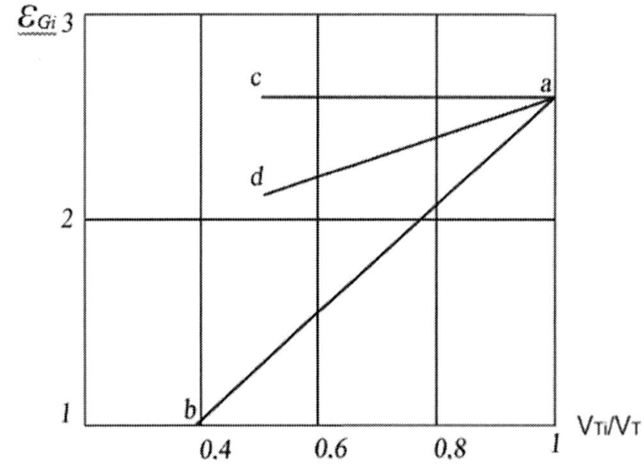

Figure 3. The dependence of the change in the geometric degree of compression ε_G on the relative productivity V_{Ti}/V_T *a – b –* ε_G *= const*; *a – c –* ε_{Gi} *= const*; *a – d –* ε_{Gi} *= var*.

Figure 4 shows the experimental dependences of the efficiency factor of the compressor η_e on the external compression ratio π_H at the boiling point of the refrigerant t_0 = – 20 °C when working with a relative capacity of V_{Ti}/V_T = 1.0; 0.75 and 0.5, where V_T is the total theoretical volumetric capacity, V_{Ti} is the partial theoretical volumetric capacity. The current value of the geometric compression ratio varied along the line *a – d*. This ensured the equality of the external and internal compression ratio.

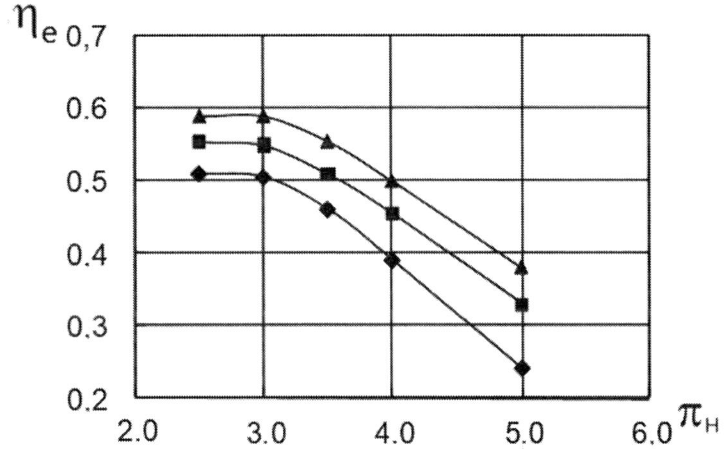

Figure 4. The dependence of the efficiency factor η_e of a screw compressor on the external compression ratio π_K when changing the geometric compression ratio ε_{Gi} along the line *a – d*, the designations are the same as in figure 2.

Figure 5 shows the dependences of η_e on the relative productivity V_{Ti}/V_T at different boiling points and performance control with one valve. Figure 6 shows the same dependences when controlling the capacity with the slide valve and rotary dampers when ε_{Gi} changes along the line *a – d*.

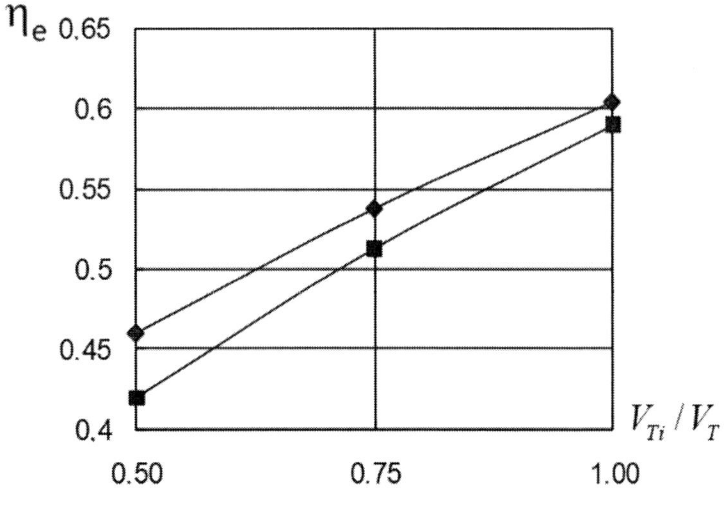

Figure 5. The dependence of the efficiency factor η_e of a screw compressor on the relative volumetric productivity V_{Ti}/V_T when controlling with one spool. ◆ – t_0= –10 °C; ■ – t_0= –20 °C.

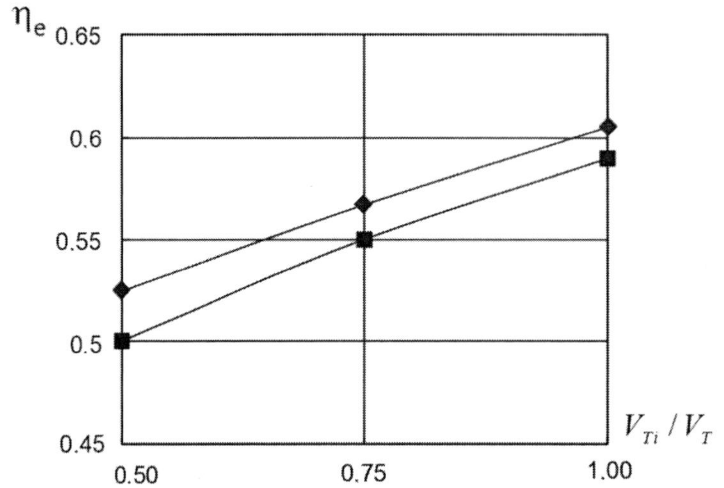

Figure 6. The dependence of the efficiency factor η_e of a screw compressor on the relative theoretical volumetric capacity V_{Ti} / V_T when changing the geometric compression ratio ε_{Gi} along the line $a - d$, the designations are the same as in figure 5.

From these dependencies it can be seen that the most effective way to regulate the performance with a decrease in the condensation temperature is the slide valve with rotary valves. Moreover, the change in the current geometric degree of compression occurs along the line $a - d$.

Compared to regulating the productivity by one valve with a relative productivity of 75%, the efficiency factor η_e increases by 5.4% (from 0.538 to 0.567) at a boiling point of $t_0 = -10$ °C and by 7.2% (from 0.513 to 0.55) at a temperature of $t_0 = - 20$ °C. At 50% productivity, the efficiency factor η_e increases by 14% (from 0.46 to 0.525) at a boiling point $t_0 = -10$ °C and by 19% (from 0.42 to 0.5) at a of temperature $t_0 = -20$ °C.

The obtained dependencies can be used in solving the problem of evaluating the operation of a refrigerating machine in a spread of external and internal parameters. In this case, a promising method is the method of the theory of entropy potentials [7].

5. Conclusion

When ensuring the efficient operation of the refrigerating machine when regulating performance, an effective regulator is a regulator consisting of a slide valve and rotary dampers. With a decrease in productivity by 75%, the value of efficiency factor increases to 7.2%, and with a decrease in productivity by 50% – up to 19% compared with a regulator consisting of one valve with a constant geometric compression ratio.

References

[1] Stosic N 2004 Screw Compressors in Refrigeration and Air Conditioning *Int. J. of HVACR Research* No **10(3)** 233–63

[2] Fleming J S, Tang Y and Anderson H 1994 Optimisation Techniques Applied to the Design of a Refrigeration TiN Screw Compressor *Int. Compressor Engineering Conf.* 1049

[3] Disha K S, Patel H K and Darshna M J 2015 Comparative Study of Slide Valve and Variable Frequency Drive for Screw Compressor Control System *Indian J. of Applied Research* vol 5 **1** 4–6

[4] Stosic N, Kovacevic A and Smith I K 1998 Modelling of Screw Compressor Capacity Control *Int. Engineering Conf.* p 1307 URL: http://docs.lib.purdue.edu/icec/1307

[5] Noskov A N and Steshina Y G 2016 Features of the suction process of a refrigeration screw compressor *Collection of Reports of the VI Int. Scientific and Technical Conf. "Kazakhstan-Cold 2016"* (Almaty) pp 108–11
[6] Noskov A N, Zimkov A A and Tarasenkov D S 2015 Calculation of the loss of operation of a refrigeration screw compressor during the injection process when controlling compressor performance *Proceedings of the 7th ISTC "Low-temperature and food technologies in the 21st century"* (St. Petersburg) pp 169–72
[7] Lazarev V L 2015 Representative Information Models for Monitoring and Control in the Conditions of Uncertainty *Proc. of the 18th Int. Conf. on Soft Computing and Measurements* (Russia, Saint Petersburg: Publ. IEEE) pp 54–7

IOP Conf. Series: Materials Science and Engineering **939** (2020) 012056 doi:10.1088/1757-899X/939/1/012056

Development of a biomechanical robotic orthopedic apparatus with a corset based on composite materials

V I Novikov[1], A G Muravyev[2,3,4,6], S A Muravyev[3], D S Stepura[1] and N S Shmakova[5]

[1] LLC «Orthopedic Small Enterprise «ORTEZ», Solnechnaya str., 3 115516, Moscow, Russia

[2] Yaroslav-the-Wise Novgorod State University, B. St. Peterburgskaya str., 41, 173003 Veliky Novgorod, Russia

[3]LLC «TECHBIONIK» 42, Bol'shoy Blvd., bld.1, 4/1448/7, the territory of the Innovation Center "Skolkovo", 121205, Moscow, Russia

[4]AO SPA «Kvant», 73, B. St. Peterburgskaya str., 173001 Veliky Novgorod, Russia

[5]AO «Technopark Slava» 20, Nauchnyi Proyezd, bld.2, 117246, Moscow, Russia

[6] E-mail: alexandr.muravev@yandex.ru

Abstract. The article presents the results of a comprehensive study on the creation of a biomechanically based robotic orthopedic apparatus (ROA) with a lack of functionality for limb removal. A number of biomechanical schemes of the biotechnological system "man-ROA" have been developed, as well as a control system for the motor functions of ROA. Electromyographic signals were chosen as controllers. A microprocessor control system has been developed. The typical actions of the ROA user are defined. The software of a neuro-controlled node has been developed. Design, manufacture and debugging of a neuro-controlled unit and ROA in general was completed. The use of composite materials made it possible to improve the biomechanical characteristics of walking and reduce the energy consumption of the entire biotechnological system.

1. Introduction

The problems of efficient reproduction of human locomotion and, above all, walking with the help of mechanical devices such as orthopedic devices in patients with disorders of the musculoskeletal system have a deep history. Beginning in the middle of the 20th century, when interest in the problems of movement efficiency increased, questions arose and definitions of the energy efficiency of walking of two-legged walking devices. In particular, in [1, 2], with parametric optimization of movement models by the criterion of energy consumption, it was determined that walking anthropomorphism is of great importance. Attention is also paid to reducing energy consumption during rational oscillatory movements of the body and limb. In this case, the moments of force in the joint, providing the movement of the links of the model, are significantly reduced. It was noted that optimization of walking speed has a big impact.

Mathematical modeling of two-legged walking conducted by V.V. Beletsky [1,2] showed that the so-called "comfortable" movement, when the center of mass moves linearly, is significantly less energetically beneficial than uncomfortable movement, when there is a movement of the center of mass similar to anthropomorphic walking, characterized by rocking movements of both the trunk and legs. In

Content from this work may be used under the terms of the Creative Commons Attribution 3.0 licence. Any further distribution of this work must maintain attribution to the author(s) and the title of the work, journal citation and DOI.
Published under licence by IOP Publishing Ltd

this case, the required energy costs are almost halved. The simulation showed that the use of passive springs as energy storage is rational and energetically beneficial. This confirms the need for energy recovery when walking with the help of elastic elements.

Based on these works, it can be concluded that anthropomorphic movements during locomotion are carried out by more than 60% due to the rationality of biomechanical characteristics and only less than 40% due to the direct action of the moments of force in the joints. Walking is more economical, the less connections are imposed on it. For each connection made, you have to pay extra energy to maintain this connection.

Thus, it is advisable to achieve a reduction in energy consumption for movement when walking anthropomorphic mechanisms due to rational vibrational movements of links having a certain mass, optimal impulse application of torque in joints and integration of elastic elements for certain parameters of legs and speed.

Such conclusions served in the future as the basis for conducting mathematical modeling and studying the influence of the parameters of prosthetic and orthopedic products on the biomechanical characteristics of walking. The results of these studies are presented, including in [3,4,5].

A comprehensive analysis of the results of these studies made it possible to take a fresh look at the influence of biomechanical characteristics on walking support, which found its application in this work.

Of particular importance, these biomechanical characteristics become necessary when it is necessary to restore the functions of the lower limb partially lost as a result of various diseases and deformations of an orthopedic nature. In this case, a combination of "living" and "non-living" appears, segments, links, and joints.

Consider the case when one of the lower limbs is affected in combination with lesions of the spine in the lumbar region, which is widespread in a number of diseases and deformities. In addition, similar orthopedic pathologies of the lower limbs can be caused by a number of neurological diseases.

In this case, flaccid paresis, paralysis of various degrees of manifestation, accompanied by pathological settings, contractures in the joints of the lower extremities, secondary deformations appear. The greatest loss of walking opportunities is manifested in case of muscle damage in the hip joint, when it is difficult to push the lower limb into the phase of the back push and carry it forward to the phase of transfer. In some cases, flaccid paralysis is combined with spasticity of individual muscles of the lower extremities. As a result, the patient has difficulty and limited movements necessary for walking, delay, lack of skills that provide locomotion.

Consider which muscles carry out movements in large joints of the lower limb, as well as the movement of the spine, which was one of the objectives of this work. This allowed us to further determine the most rational places for the installation of sensors and was used in the design of the corset.

Of greatest interest are movements in the hip joint. Hip joint flexion mainly involves the iliopsoas, sartorius, muscle tensioner of the wide fascia, scallop, and thigh muscle. The extension involves the gluteus maximus, biceps femoris, semitendinosus, semi-membranous, large conductive.

We also indicate the muscles that perform certain movements of the spine. During extension v trapezius, posterior dentate muscles, upper and lower plaster muscles of the neck and head, cruciate, transverse, short muscles of the back (transverse, interspinous, lifts of the ribs and the group of the occipital-iliac muscles).

When bending - sternocleidomastoid, scalene, long muscle of the head and neck, rectus abdominis, oblique abdominal muscle, iliac lumbar.

In severe cases, as a rule, with the consequences of spinal injuries accompanied by spinal cord injury – a combination of a back injury and spinal cord injury to the spinal cord, the motor functions of both lower extremities are affected. In these cases, medical exoskeletons are currently used. But in cases of moderate severity or relatively uncomplicated cases, in the above series of diseases and deformities, when one limb is affected, their use is not advisable from a medical, biomechanical or economic point of view. In [6], the view was expressed that the widespread introduction of exoskeleton requires discussion and the work of an interdisciplinary team.

Well-known medical exoskeletons do not give to the patient himself to participate properly in the process of locomotion, to use the biomechanical characteristics inherent in man to ensure movement in conjunction with the device. Such patients partially retained motor functions, that is, they have mild, moderate degrees of damage, with lesions localized mainly at the level of the hip joint in combination with lesions of the spine at the level of the pelvic and lower thoracic spine.

For most diseases and deformities, the consequences of which are paresis and paralysis, these exoskeletons cannot be used, since in addition to activating motor functions, the main therapeutic factor is the provision of orthopedic correction of the limb, which is absent in them. Most exoskeletons cover both legs and several joints, and not one of the most affected or on which the walking process depends more. On one joint, the use of exoskeleton is very rare. For example, its use on the ankle joint is known [7].

Also, as a rule, elastic elements are not built into medical exoskeletons, they do not have a rational use of mass-inertial characteristics, and they do not take into account the possibility of the influence of the construction scheme on the normalization of biomechanical characteristics. The operation of many types of medical exoskeleton requires an operator accompanying the patient.

Instead of exoskeleton, it is more rational to limit oneself to the use of biomechanically sound orthopedic devices. However, their designs must combine:

– rational kinematic construction scheme;
– rational mass inertial characteristics;
– built-in elastic elements;
– devices that reproduce motor functions due to built-in drives and an external energy source.

Existing traditional orthopedic devices do not fully meet biomechanical requirements. In addition, indications for their appointment are limited due to a number of inherent disadvantages. Including they have a large mass and are not convenient enough to use. As for the ability to reproduce motor functions that complement the residual muscle capabilities, they are absent in known orthopedic devices.

2. Biomechanical scheme of a robotic orthopedic apparatus

We conducted a comprehensive study on the creation of a biomechanically grounded robotic orthopedic apparatus (ROA) for people with disabilities and patients with lesions of the lower limb and lower torso who have a lack of functionality to carry the limb into the transfer phase. Such a removal is normally carried out due to the action of muscle forces on flexion in the hip joint at the beginning and middle of the transfer phase and the action of muscle forces on extension at the end of the transfer phase until the start of the front push. In this period, the mass inertial characteristics of the limb, which provide the pendulum motion, and the location of the rotation axes of the joints, and the potential energy accumulated due to the elastic properties of the ligaments, muscles, and tendons, passing into the kinetic at the moment the limb is torn off, simultaneously affect the limb.

A number of biomechanical schemes of the biotechnological system "man-ROA" have been developed. So, to determine the effect of the mismatch of the hinge axes with the axes in the joints, a developed biomechanical scheme is proposed that combines the links of the orthopedic apparatus with segments of the lower limb and its analysis is carried out. In orthopedic devices on the lower limb, the projections of the hinge axes on the sagittal plane, as a rule, do not coincide with the projections of the axes of the corresponding joints of the orthopedic device. This is due to both the absence of knee nodes providing rotation of the hinge along the hypocycloid, which is characteristic of the movements of the knee joint, and the need to create under-stability in the knee joint when walking, as well as a number of clinical indications. Errors arising during the assembly of the orthopedic apparatus also lead to mismatch of the axes.

The difference in the location of the axes of the joints of the orthopedic apparatus and joints of the limb causes when walking the displacement of the links of the apparatus relative to segments of the limb. This leads to piston-like movements of the limb in the sleeves, causes a change in the load on various areas of the limb, affects the functional length of the limb in the biomechanical "man-orthosis" system and the provision of movements in the joints. Therefore, it was considered advisable to consider in more detail the effect of the mismatch between the axes of the hinges of the orthopedic apparatus and the joints of the limb on the biomechanical system, and based on the analysis of this problem, develop some recommendations that can be used in the assembly and creation of designs of ROA.

As a result, the optimal location of the hinges relative to the joints of the limb was revealed, both from the standpoint of under-stability, and from the point of view of changing longitude dimensions in the biokinematic chain.

The masses of links combining the masses of segments of the affected limb and the mass of sections of ROA located between the joints are selected. Moreover, the use of the original composite material sleeves as materials made it possible not only to lighten the ROA mass, but also to regulate the mass-inertial characteristics by providing oscillatory movements of the tibia links with the foot, bringing them closer to normal.

In addition, the use of composite material allowed the construction of the sole of the ROA foot sleeve in the form of a spring. This, in turn, made it possible to enhance the repulsion of the distal ROA with the limb installed in it in the phase of the back push, thereby improving the biomechanical characteristics of walking and reducing the energy consumption of the entire biotechnological system. In addition, ROA uses elastic polymeric four-leafed ankle joints, and an elastic element is installed in the hip joint, which is compressed during re-extension in the sagittal plane in the hip joint from the vertical posteriorly at the end of the foot support phase. At the very beginning of the transfer phase, the elastic element is unclenched, and the accumulated potential energy passes into the kinetic energy, developing a moment of forces that contributes to the removal of the apparatus with the limb in the sagittal plane. The appearance of the ROA frame is shown in figure 1.

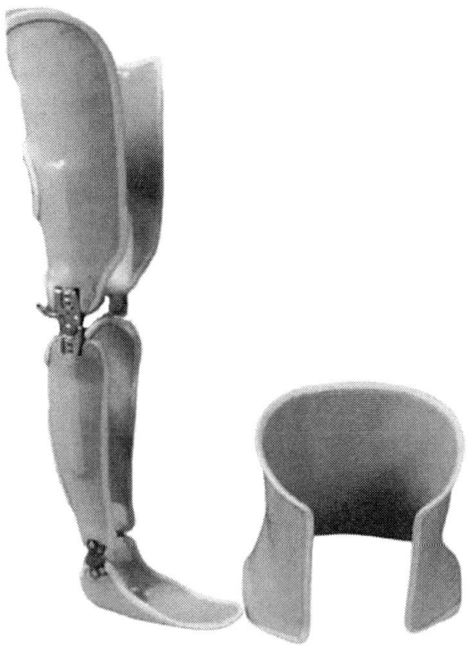

Figure 1. ROA frame made of composite material.

3. The control system for the motor functions of ROA

Particular attention in this work was paid to the study of the ROA motor function control system and the selection of its most rational option. Known control systems used in lower limb prostheses, medical exoskeletons, usually assume control programmed depending on the signals of pressure sensors located under their plantar surface or on myotonic signals of limb sections, as well as movement programs similar to the norm taking into account pathology. As for the control from electroencephalographic signals, as noted [8], the development of brain-computer interfaces for controlling movements of the lower limbs using imaginary leg movements is complicated in that the data analysis procedure must take into account a large number of artifacts in the recorded EEG signals caused by electronics executive mechanisms of movement, intense body movements and tonic muscle activity.

Our analysis showed that it is advisable to use electromyographic signals as control signals, which confirms the conclusions made in [7]. However, when developing motor functions from an external energy source relative to the hip joint, it is advisable to combine electromyographic signals with the signals of pressure sensors under the sole of the ROA foot sleeve, which was not previously used.

The developed microprocessor control system for ROA with an external energy source (EES) provides the implementation of a sensor feedback system, setting up settings and work patterns using a local user myointerface.

In this case, for control, bioelectric signals obtained from the skin electromyographic electrodes from the muscles of the trunk and pelvic region, depending on their safety and activity, were used.

In order to develop a preliminary design of a neuro-controlled node, its structural scheme was developed and the components included in its composition were selected.

The main components of the neuro-managed node are:

– control board with a microcontroller;
– drive system of the push mechanism;
– control sensors;
– sensors and feedback sensors;
– elements of a local control system;
– an external source of energy.

It is assumed that in a neural-controlled node fed by an external energy source, using the elements of the local control system, the user will select the operating mode of this neuro-controlled node.

At the same time, depending on the selected operating mode, when a signal is received from the control sensors, the microcontroller of the control board will transmit control commands to the drive system of the jog mechanism.

The position of the jog mechanism, in turn, will be determined and corrected by the microcontroller based on the signal of the sensors and feedback sensors.

The main component of the neuro-controlled node, which mechanically provides bending and extension of the hinge joint of the biomechanical robotic orthopedic apparatus, is the drive system of the push mechanism.

Initially, a worm gear motor was considered as a drive for the jog mechanism.

However, despite its advantages of a constant torque, the drive of the jog mechanism based on a worm gear motor would require its installation directly in the hip joint, which would significantly increase the external dimensions of this joint and exclude the possibility of using standard joints offered by the industry .

In this regard, a linear actuator with a DC collector motor was chosen as the drive of the jog mechanism.

To provide feedback on the position of the linear actuator rod, options for using a potentiometer and incremental encoder were considered as feedback sensors.

Although the potentiometer provides an absolute value of the position of the actuator rod, proportional to the change in the electrical resistance of the potentiometer, its use significantly increases the longitudinal overall dimension of the actuator.

Therefore, an incremental quadrature encoder mounted on the shaft of the actuator motor was selected as a feedback sensor.

Due to the stable discharge characteristics and high specific energy intensity, a lithium-iron-phosphate storage battery is selected as an external energy source.

To select a control board with a microcontroller, we analyzed the characteristics of several debug boards that are most suitable for solving the required control tasks.

According to the results of the analysis, a board based on the Atmel ATmega 328 microcontroller is selected as a control board with a microcontroller.

The capabilities of the control board did not allow direct connection of the jog mechanism drive, therefore, for direct drive, a high-power motor controller was introduced into the structural diagram of the neuro-controlled unit.

For additional control of the operation of the drive system of the jog mechanism, a current sensor is used installed on the power line of the motor controller.

Two laid on active electromyographic electrodes, which should be fixed on top of the user's control muscles to capture their bioelectric signal, are selected as sensors for controlling a neuro-controlled node.

Of the electromyographic electrodes produced by the industry, the electrodes produced by SPE "Galateya" corresponded most fully to the solution of control problems.

To improve the quality of control of a neuro-controlled node, two corrective pressure sensors, which are supposed to be placed in the forefoot and heel of the artificial foot of a biomechanical robotic orthopedic apparatus, are additionally added to the structural diagram.

Two versions of these control pressure sensors are proposed.

In one design they are push limit switches, in the other they are film load resistors.

To implement a local control system for a neuro-controlled node, the typical actions of a user of a biomechanical robotic orthopedic apparatus are determined.

These included turning on and off a neuro-controlled node, switching operating modes (walking mode, "sit down" and "stand up" modes) and settings for the operating speed of the jog mechanism. A schematic diagram of a neuro-controlled ROA node is presented in figure 2.

To develop the software of a neuro-controlled node, the algorithms of its operation are determined when performing basic actions in the three modes described above.

When you turn on a neuro-managed node, its initialization is performed. At the same time, the values of the pre-set constants of the settings of the sensors and the time parameters of the node's operation are loaded into the microcontroller's memory. The drive of the push mechanism with control of the current strength and the position of the rod is displayed in the starting position of the walking mode.

Then begins the cyclic execution of the main block of the control program. In this case, the signals are read from the operation mode switching buttons, the potentiometer for setting the speed of the jog mechanism drive, as well as from the first control electromyographic electrode and pressure sensor located in the toe of the biomechanical robotic orthopedic apparatus.

If the values of the signals from the first control electromyographic electrode and the pressure sensor located in the toe of the biomechanical robotic orthopedic apparatus exceed the values of the respective settings, the first phase of the step begins.

In this case, the drive of the push mechanism starts bending the hip joint with control of the current strength, position of the rod, time and signals from the second control electromyographic electrode and pressure sensor located in the heel of the biomechanical robotic orthopedic apparatus.

If the values of one of the above signals of the corresponding setting are exceeded, the second phase of the step begins.

In this case, the drive of the push mechanism starts the extension of the hip joint with control of the current strength, position of the rod, time and signals from the first control electromyographic electrode and pressure sensor located in the toe of the biomechanical robotic orthopedic apparatus.

If the signal values are exceeded from the first control electromyographic electrode and the pressure sensor located in the toe of the biomechanical robotic orthopedic apparatus of the corresponding setting, the first phase of the step begins. Otherwise, at the end of the second phase of the step, the neuromodule awaits the initiating signal to execute the first phase.

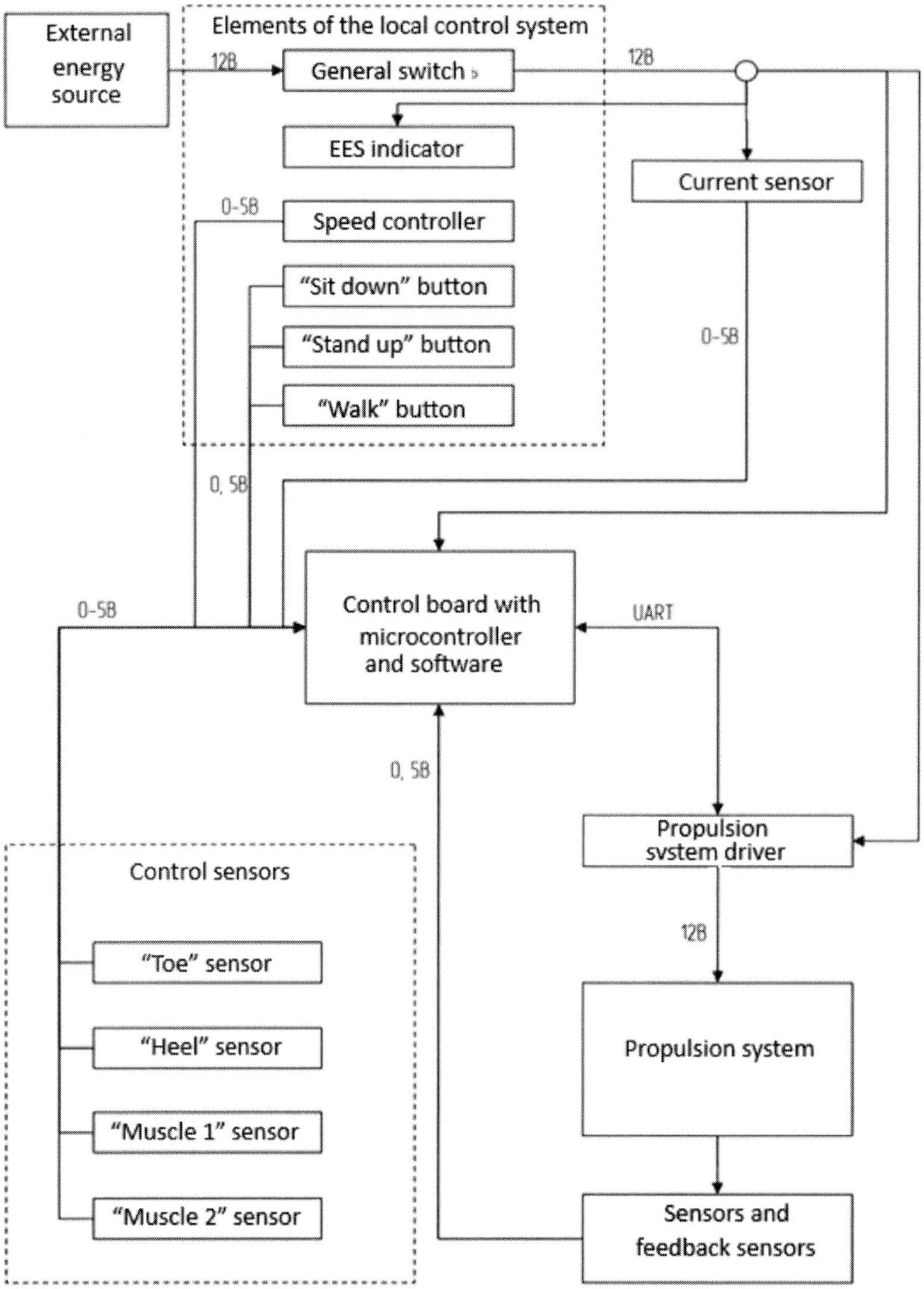

Figure 2. Schematic diagram of a neuro-controlled ROA node.

When the operating mode is switched to the "sit down" mode, the drive of the push mechanism starts bending the hip joint to an angle of 90 degrees with control of the current strength, position of the rod, and time.

When the operating mode is switched to the "stand up" mode, the drive of the jogging mechanism begins to extend the hip joint to the starting position of the walking mode with control of the current strength, position of the rod, time.

To return to walking mode, the user must press the appropriate button.

In order to debug a manufactured prototype of a neuro-controlled assembly, the drive of its push mechanism was fixed on the design model of a biomechanical robotic orthopedic apparatus made on the basis of the scientific and production laboratory of orthoses of OOO «Orthopedic Small Enterprise «ORTEZ».

During debugging, the values of the constants of the sensors operation settings and the time parameters of the node operation were adjusted. As a result of debugging, the operation of the neuro-controlled node is ensured in the planned mode.

Based on the conducted research, the design and manufacture of ROA was performed.

4. Conclusion

The developed ROA provides:

– controlled translational-translational movements of the thigh segment into the phase of transfer of the lower limb, initiated by bioelectric signals and carried out by a neuro-controlled node;

– the implementation of orthopedic correction, fixation, unloading of the affected limb;

– improvement of elastic, kinematic, mass-inertial biomechanical characteristics of ROA;

– minimization of cosmetic damage;

– the presence of an elastic element in the hip node;

– both walking and patient sitting on a chair, stool, etc.

At the same time, in ROA with EES it will be necessary to carry out:

– energy recovery due to elastic elements;

– the implementation of the front plantar region of the sleeve of the foot in the form of a spring, which will allow the patient to reduce energy consumption for movement, reduce energy consumption of renewable energy sources and the weight of the wire;

– rational distribution of the load on the affected limb due to the manufacture of individual sleeves of the foot, lower leg, thigh;

– creating orthopedic correction and reducing the likelihood of secondary deformations, increased comfort;

– reducing the effects of piston movements and thereby reducing trauma to the skin, circulatory disorders;

– a rational kinematic construction scheme, providing under-stability in the knee joint when walking;

– a decrease in the mass of ROA in comparison with traditional ones;

– the possibility of thermoforming of thermowells as medical indications and anthropometric parameters change, carried out at a temperature of less than 90°C.

Thus, the studies and design developments made it possible for the first time to develop a system of bioelectric control of the motor functions of the hip joint of a robotic orthopedic apparatus. An integrated approach to the design of ROA from the standpoint of biomechanics has improved walking characteristics and reduced energy consumption; the mass of the battery, drive ROA.

The results of the work performed allow us to talk about the prospects of the selected direction of rehabilitation of disabled people and patients with injuries of the musculoskeletal system and the need for further research and practical application of ROA.

The cluster approach, which brought together specialists from organizations of various fields of activity, made it possible to carry out a comprehensive review of multidisciplinary problems "at the intersection of sciences" and find promising solutions in research and the creation of innovative prosthetic and orthopedic products.

References

[1] Beletsky V 1984 *Dvunogaya khod'ba. Model' i zadachi dinamiki i upravleniya* [Bipedal walking.

Model and problems of dynamics and control] (Moscow: Science) p 286 (In Russ.)
[2] Formalsky A M 1982 *Peremeshcheniye antropomorfnykh mekhanizmov* [Relocation of anthropomorphic mechanisms] (Moscow: Science) p 368 (In Russ.)
[3] Klein A B and Novikov V I 1981 Vliyaniye raspolozheniya tazobedrennogo sharnira v protezakh posle vychleneniya bedra na energotraty pri khod'be [Influence of the location of the hip joint in prostheses after articulation of the hip on energy consumption when walking] *Protezirovaniye i protezostroyeniye* **56** 26–32 (In Russ.)
[4] Farber B S, Plastinin M V, Novikov V I and Sukhoverkhova A I 1988 Issledovaniye vliyaniya skhemy postroyeniya ortopedicheskogo apparata na energotraty pri khod'be [Study of the influence of the construction scheme of an orthopedic apparatus on energy consumption when walking] *Protezirovaniye i protezostroyeniye* **84** 70–78 (In Russ.)
[5] Novikov VI 1987 Biomekhanicheskiye predposylki postroyeniya ortopedicheskikh apparatov nizhnikh konechnostey [Biomechanical prerequisites for the construction of orthopedic devices of the lower limbs] *Protezirovaniye i protezostroyeniye* **80** 24–30 (In Russ.)
[6] Ashraf S G 2018 Robotic exoskeletons: The current pros and cons *World J Orthop* **9** 112–119
[7] Koller J R, Remy C D and Ferris D P 2018 Biomechanics and energetics of walking in powered ankle exoskeletons using myoelectric control versus mechanically intrinsic control *Journal of NeuroEngineering and Rehabilitation* **15** 42
[8] *BCI-controlled exoskeleton: neurorehabilitation for patients with impaired lower limb function* [Electronic resource] https://bioelectric.hse.ru/en/news/228980748.html (Accessed 03/13/2020)

Design hydrodynamic analysis of cavitation in narrow channels of the open-pit dump truck's hydraulic system

D A Panasenkov[1,4], **A P Zaycev**[1,5], **A B Kartashov**[1,6], **N A Pikalov**[1,7], **D M Dubinkin**[2,8] **and A B Efremenkov**[3,9]

[1] Wheeled Vehicles Department, Bauman Moscow State Technical University, 2nd Baumanskaya St., 5 105005, Moscow, Russian Federation
[2] Department of Metal Cutting Machines and Tools, Kuzbass State Technical University, Vesennyaya St., 28 650000, Kemerovo, Russian Federation
[3] Yaroslav-the-Wise Novgorod State University, B. St. Peterburgskaya str., 41 173003, Veliky Novgorod, Russia

[4] E-mail: forestbol@bk.ru
[5] E-mail: zaycev.a.p@bmstu.ru
[6] E-mail: kartashov@bmstu.ru
[7] E-mail: nikitapikalov@bmstu.ru
[8] E-mail: ddm.tm@kuzstu.ru
[9] E-mail: abe@novsu.ru

Abstract. The paper deals with the processes that occur when the hydraulic fluid flows in the channels of the hydraulic block of the brake drive of the open-pit dump truck. The main attention is paid to the phenomenon of cavitation caverns emergence and hydro-mechanical losses depending on the hydraulic fluid parameters and the geometric parameters of hydraulic channels. The SST-HL and SSG turbulence models are presented in this paper. The Rayleigh Plesset cavitation model built into the ANSYS CFX software package is also used.

1. Introduction

When designing the hydraulic drive of the open-pit dump truck, it is often necessary to ensure the most compact and efficient placement of cartridge valve equipment that facilitates the operation of brake mechanisms, hydraulic cylinders for folding the frame and lifting the cargo platform. In order to minimize the range of hydraulic components, they are combined into hydraulic blocks. The main task in the development of such blocks is to ensure the speed of the braking system, steering and lifting of the cargo platform. This task is particularly acute when developing the hydraulic system of open-pit dump trucks with articulated frame, because the efficiency and safety of using such vehicles in mining works depends largely on the speed and reliability of the components of the overall hydraulic system.

Electromagnetic spool distributors are widely used in the hydraulic system of open-pit dump trucks, since they allow implementing various control schemes [1, 2]. In addition to the friction process, which leads to the gradual destruction of the moving parts of the spool distributor [3], there is an inevitable erosion wear that occurs due to the fact that metal particles enter the spool. One of the reason why metal particles get into the hydraulic fluid is cavitation erosion [4]. During cavitation,

Content from this work may be used under the terms of the Creative Commons Attribution 3.0 licence. Any further distribution of this work must maintain attribution to the author(s) and the title of the work, journal citation and DOI.
Published under licence by IOP Publishing Ltd

IOP Conf. Series: Materials Science and Engineering **939** (2020) 012057 doi:10.1088/1757-899X/939/1/012057

there is a violation of the continuity of the fluid flow in the low-pressure zone, accompanied by local hydraulic shocks that occur when cavitation caverns disappear. Prolonged exposure to such erosion leads to the gradual destruction of the channel walls and the separation of small metal particles that fall into the spool distributors of the hydraulic drive, which has an erosive effect on the spool components. In [5] it is shown that the wear effect in hydraulic valves leads to deterioration of the spool distributor, reduces its service life, and results in leaks in the system.

Since the erosive effect on the spool is a consequence of cavitation in hydraulic channels, special attention should be paid to reducing the probability of cavitation phenomena when developing hydraulic blocks. Thus, the paper [6] demonstrates a computational model for evaluating cavitation-erosion damage in the solenoid valves of a diesel injector. In [7], the authors managed to reduce the calculation error when modeling the cavitation process in the throttle channel by approximating the design model to the actual operating conditions of the hydraulic drive.

It should be noted that the process of selecting rational parameters of hydroblock channels is complicated by a large number of them, since the phenomenon of cavitation depends on: the geometric parameters of the channels, the material of the hydraulic block housing, the pressure of the hydraulic fluid, the type of hydraulic fluid and the quality of the internal surfaces of the channels (roughness).

The tests carried out in [8, 9, 10] showed that the choice of product material has a significant impact on the process of cavitation, and in [11] it is shown that the stage of computer simulation of the cavitation process is necessary to exclude the method of trial and error during design and reduces the time of the development process in general.

One of the criteria for effective operation of the brake hydraulic drive is its speed that is closely related to such characteristics of the working fluid flow as the flow rate and fluid consumption. In [12, 13, 14, 15], studies showing the relationship between the parameters of the fluid flow: fluid consumption, pressure drop, speed and the parameters of the hydraulic fluid itself: temperature and viscosity are presented. Analysis and calculation of the geometry of hydraulic channels, taking into account the parameters of the hydraulic fluid, makes it possible to select the necessary value of the diameters of the working channels in order to achieve the maximum speed of the brake mechanism.

2. Evaluation of the hydraulic fluid and channel cavitation parameters of the hydraulic block

As previously noted, when designing the hydraulic block of the brake drive of the open-pit dump truck, it is necessary to take into account a significant number of factors that affect the operation of the executive mechanisms. Figure 1a shows a geometric model of the developed hydraulic block for the open-pit dump truck with a total weight of 75 tons. The channel parameters of this block were determined to predict the nature of the fluid flow at different viscosity values, which, in turn, depend on the temperature. Pressure values on the block channel walls were also obtained, which can be used in strength calculations for rational selection of the hydraulic block housing material and, as a result, reduce the value of deformation to reduce the volume of leaks. In addition, the influence of roughness at different viscosity values on output parameters (hydro-mechanical losses) is considered. The latter will help you choose the best way to process the block channels to find a balance between cost, efficiency, and the impact on the probability of cavitation.

Figure 1b shows a diagram of the section of the hydraulic block for which the calculated study was performed. The red color shows the channels that will be used in the model, provided that the valves 2, 3, 4, 5 are normally open, and the liquid flows through the valve 1 with a constant flow rate and pressure of 70 bar. In addition to calculating the resistances and studying the influence of viscosity on the flow of the working fluid in the channels, the dangerous section of the hydroblock channel is studied in order to predict the occurrence of cavitation.

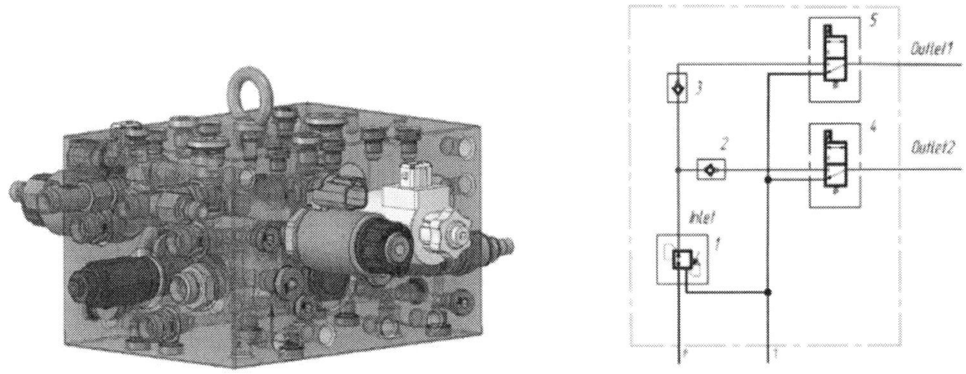

a) *b)*

Figure 1. The hydraulic block of the open-pit dump truck: *a* – the geometric model; *b* – the hydraulic diagram of the studied section of the hydraulic block.

2.1. Description of theoretical prerequisites

In order to determine the nature of the fluid flow in the channels of the hydraulic unit at the design stage and to prevent undesirable phenomena such as cavitation, as well as to account for hydro-mechanical losses, the design hydrodynamic analysis (CFD analysis) in the Ansys CFX 19.2. software package was performed.

In the first approximation, the SST-HL (Shear Stress Transport) model was chosen as the design turbulence model, which simultaneously uses two turbulence models of k-ε and k-ω types. The k-ω turbulence model behaves most reliably in the boundary layer, and the k-ε turbulence model shows itself best in the situation of transition of the liquid flow to a free flow. The HL modification, which is an addition to the SST turbulence model, allows the most accurate modeling of the separation of the liquid flow from the channel wall.

In the second approximation, the Reynolds stress model, namely the SSG (Reynolds-stress) was used as the design model. This model is used for calculations of more complex engineering problems, since the k-ε and k-ω models have significant disadvantages in real turbulent flows. One of the most important functions in the Reynolds stress model is the pressure-strain correlation, Φ_{ij}. It is used to stimulate turbulence to an isotropic state by redistributing Reynolds stresses.

The pressure strain member can be divided into two parts:

$$\Phi_{ij} = \Phi_{ij,1} + \Phi_{ij,2} \tag{1}$$

where, $\Phi_{ij,1}$ is the "slow" component, also known as the "return to isotropy" term; $\Phi_{ij,2}$– the "fast" component of the equation.

The SSG model, unlike other Reynolds stress models where the pressure-strain correlation is linear, uses quadratic pressure-strain correlation relationships:

$$\Phi_{ij,1} = -\rho\varepsilon[C_{s1}a_{ij} + C_{s2}\left(a_{ik}a_{kj} - \tfrac{1}{3}a_{mn}a_{mn}\delta_{ij}\right)] \tag{2}$$

$$\Phi_{ij,2} = -C_{r1}Pa_{ij} + C_{r2}\rho kS_{ij} - C_{r3}\rho kS_{ij}\sqrt{a_{mn}a_{mn}} + \\ +C_{r4}\rho k\left(a_{ik}S_{jk} + a_{jk}S_{ik} - \tfrac{2}{3}a_{ki}S_{ki}\delta_{ij}\right) + C_{r5}\rho k\left(a_{ik}\Omega_{jk} + a_{jk}\Omega_{ik}\right) \tag{3}$$

where ρ – density, Ω_{ij} – vorticity tensor, a_{ij} – anisotropy tensor, Φ_{ij} – pressure-strain correlation, $\overline{u_i u_j}$ – Reynolds stress tensor, ε – dissipation rate of turbulent kinetic energy, ρ – fluid density, $C_{r1}, C_{r2}, C_{r3}, C_{r4}, C_{r5}, C_{s1}, C_{s2}$ – constants, S_{ij} – average strain rate, δ_{ij} – the Kronecker coefficient, k –

average turbulent kinetic energy, U_i, U_j – speeds.

In addition to the turbulence model, the cavitation model was also used in the design of the hydraulic block, in the channels of which cavitation may occur. When estimating cavitation, a preliminary calculation for a channel with a convergent solution without a cavitation model was carried out. In the AnsysCFX, the Rayleigh-Plesset model was used as a cavitation model, implemented in a multiphase structure as an interphase mass transfer model.

2.2. Description of the design model of the hydraulic block channels

Figure 2 shows a model of the studied section of the hydraulic block channels (see figure 1b). To build a better-quality design finite element grid, channels that do not affect the calculation results were removed from the geometric model, but these channels were used in the calculations to determine the pressure on the hydroblock's walls.

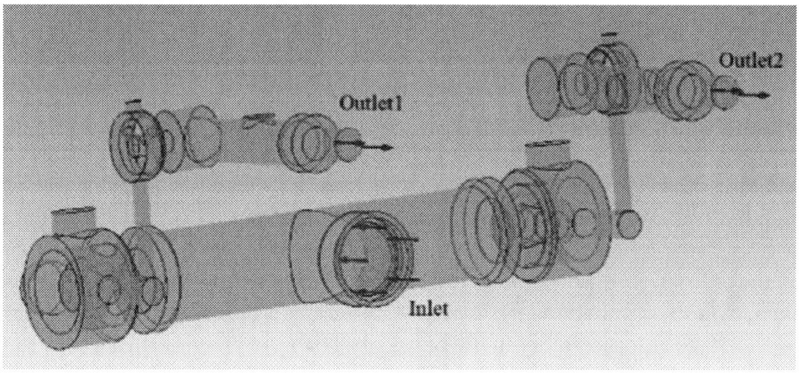

Figure 2. The hydraulic block channels.

When constructing the finite element grid (see figure 3), the “Inflation” function was used in the Ansys CFX software package along the channel walls to more accurately determine the fluid flow regime in the boundary layer and to obtain more accurate solutions using the selected SST-HL and SSG turbulence models.

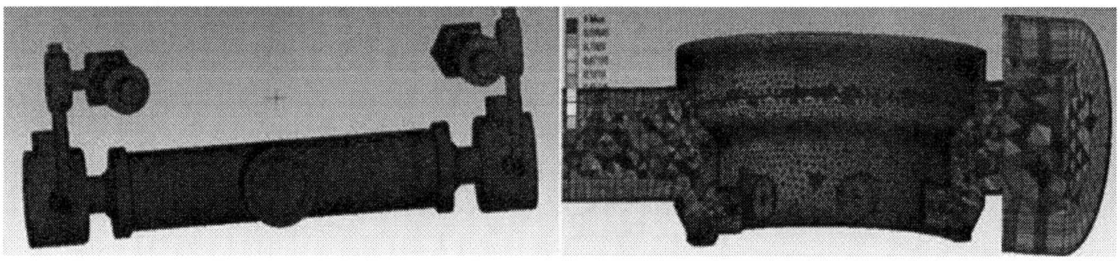

Figure 3. The finite element grid.

Setting the fluid pressure parameters at the inlet and its flow rate at the outlet is one of the reliable ways to set the boundary conditions with the highest probability of convergence. The inlet pressure “TotalPressure” equal to 70 bar, and the “MassFlowRate” Outlet1 and Outlet2 (see figure 2) of mass flow rate were set as boundary conditions. The flow rate at the outlets was set to be the same, since the valves that are considered fully open in our model must open synchronously with each other, while having the same resistance at the outlet of the hydraulic block. To determine the hydro-mechanical losses in the channels of the hydraulic block, the value of channel wall roughness was considered in the settings of the design model. The calculation was performed in a stationary setting, the process was considered to be adiabatic.

As a result of the calculations, the pressure distribution along the channel wall, the flow line, the pressure and speed in the cross section of the channels at the inlet and outlet are obtained for estimating hydro-mechanical losses.

Figure 4 shows a section of distribution block channels. This section describes an area outlined in green that has undergone several changes during development, such as a change in the channel diameter (2 and 3 mm) and a change in roughness. Calculations were also made for mass expenditures of 0.25 and 1 kg/s. Such modes of operation are dangerous, because there is a high probability of cavitation or even supercavitation, that is a mode in which cavitation caverns completely fill the channel, and the pressure characteristic is greatly deteriorated, the process of destroying the walls of channels and clogging the hydraulic system with small metal particles are also possible.

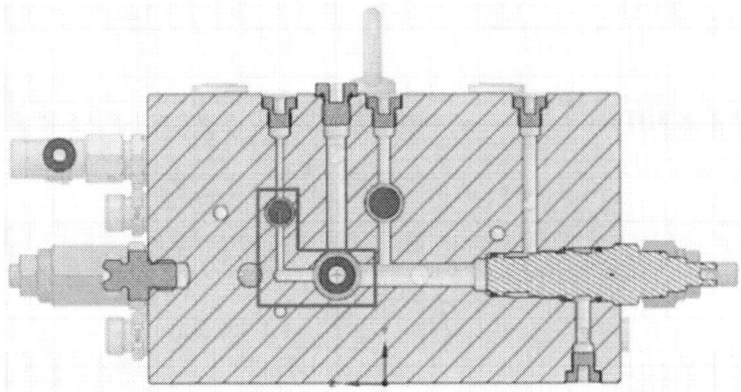

Figure 4. The section of the hydraulic block channel.

The channel with a high probability of cavitation was calculated separately using a finer finite element grid. Other sections of the hydraulic block do not have a significant impact on the processes taking place in the channel, but their accounting would significantly complicate calculations and require large computing resources.

2.3. Analysis of cavitation process occurring in hydraulic channels

During the initial analysis of the cavitation process, taking into account the temperature equal to 100°C, the mass flow rate of 1 kg/s, the wall roughness value of 6.3 microns, and the saturated steam pressure value equal to 1250 PA, the values of the volume fraction of gas in the hydraulic channel were obtained, which are shown in figure 5. Areas with a large value of the volume fraction of gas indicate the presence of a cavitation process.

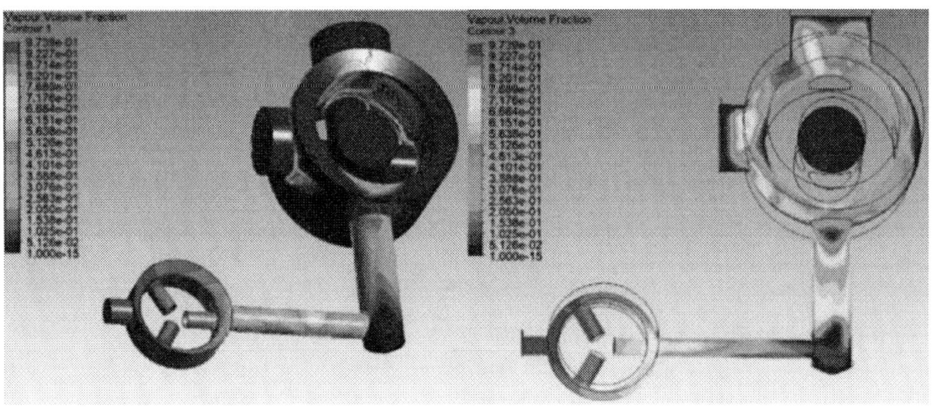

Figure 5. The calculation results for estimating the cavitation process.

The influence of geometric parameters of the channel on the occurrence of cavitation is also calculated. The temperature of the hydraulic fluid was assumed to be 100 °C, the pressure was 70 bar and the roughness was 6.3 microns with a reduced mass flow rate of 0.25 kg/s.

Initially, when designing, the diameter of the channel leading to the valve was chosen equal to 2 mm. During the calculation, the value of the volume fraction of gas in the supply channel indicated the presence of cavitation in it (figure 6).

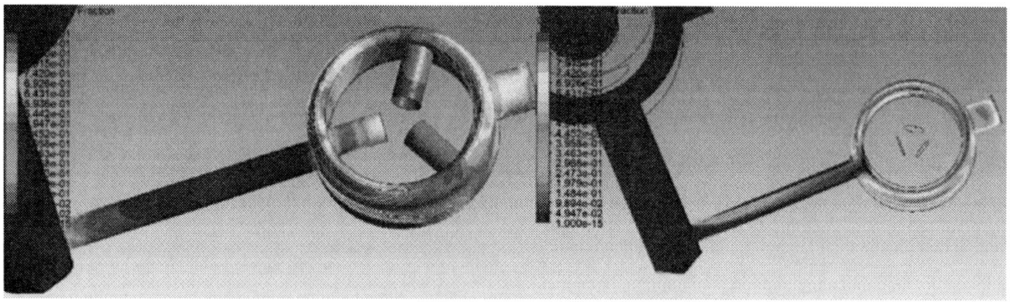

Figure 6. Cavitation in a channel with a diameter of 2 mm.

In order to reduce the formation of cavitation caverns, the diameter of the supply channel was increased to 3 mm based on the results of calculations. The nominal flow rate was limited to 30 *l/min*, based on the operating parameters of the valve used, which will be described in more detail in paragraph 2.4.

2.4. Findings

The calculation was made for working channels (see figure 2) with channel wall roughness of 3.2, 6.3 and 12.5 microns, at temperatures of minus 10, plus 30, 60 and 100°C, with a set inlet pressure of 70 bar and outlet flow of 30 l/min. The maximum temperature equal to 100 °C is taken into account in the calculation, despite the fact that this temperature regime is not working for the hydraulic system of the open-pit dump truck. The values of viscosity and density of the hydraulic fluid at different temperatures were established empirically and are presented in table 1.

Table 1. The empirical dependence of the dynamic viscosity and density of the hydraulic fluid on the temperature at a pressure of 70 bar.

The average value of the fluid temperature, °C	Minus 10	30	60	100
Dynamic viscosity, Pa * s	0.62	0.06	0.02	0.007
Density, kg / m^3	894	864	840	810

Based on the above-given conditions, the speed and pressure values were obtained, averaged over the area of the input and output sections depending on the temperature, that is, the fluid viscosity and the wall roughness, which are presented in table 2.

Table 2. The values of the hydraulic fluid flow speed and pressure.

The hydraulic fluid temperature, °C	Roughness, mcm	Speed, m/s			Pressure, bar.		
		Inlet	Outlet1	Outlet2	Inlet	Outlet1	Outlet2
−10	3.2	4.92	10.28	10.29	69.87	43.20	43.29
	6.3	4.92	10.28	10.29	69.87	43.19	43.29
	12.5	4.92	10.28	10.29	69.87	43.17	43.27

IOP Conf. Series: Materials Science and Engineering **939** (2020) 012057 doi:10.1088/1757-899X/939/1/012057

30	3.2	4.84	10.31	10.31	69.89	48.92	48.46
	6.3	4.84	10.31	10.31	69.89	48.90	48.44
	12.5	4.84	10.31	10.31	69.89	48.86	48.40
60	3.2	4.82	10.49	10.42	69.89	48.66	48.55
	6.3	4.82	10.49	10.42	69.89	48.61	48.50
	12.5	4.82	10.50	10.43	69.89	48.51	48.38
100	3.2	4.81	10.70	10.57	69.89	48.68	48.67
	6.3	4.81	10.60	10.55	69.89	48.39	48.63
	12.5	4.81	10.63	10.69	69.89	48.35	48.19

Based on the results obtained, it is possible to estimate how much influence a certain roughness value has on the output parameters at the appropriate temperature and dynamic viscosity.

Figure 7 shows the flow lines with the distribution of the fluid speed along the channels and the pressure along the wall for the selected roughness at different temperature conditions and a roughness value of 6.3 microns.

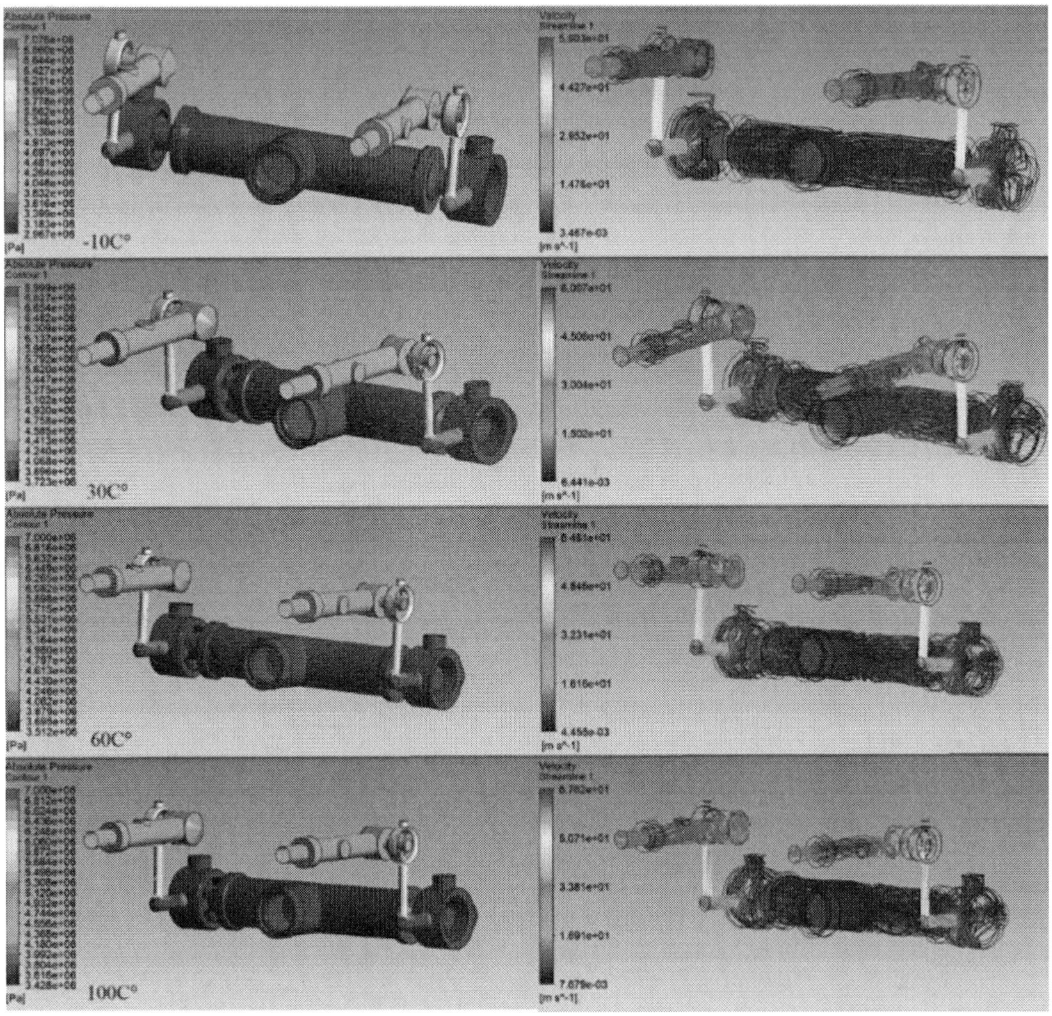

Figure 7. The pressure distributions on the wall and the flow lines.

IOP Conf. Series: Materials Science and Engineering **939** (2020) 012057 doi:10.1088/1757-899X/939/1/012057

The speed value indicated on a scale related to the flow lines, it can be concluded that the value of the maximum fluid flow speed can be reached at the maximum temperature in the working channel, i.e. at the minimum viscosity value.

Below are the results of calculating the cavitation of the selected channel at a saturated steam pressure of 1250 Pa. Figure 8 shows that the intensity of cavitation formation on the channel walls is significantly higher at a roughness of 12.5 microns than at 6.3 and 3.2 microns, and the differences between 6.3 and 3.2 microns are insignificant. For these reasons, the choice of roughness in the design was made in favor of 6.3 microns.

Figure 8. Cavitation on the channel walls of the distribution block.

When modeling fluid flow in the hydraulic block, cavitation in the operating temperature range from minus 10 to 60°C was not observed. Besides, the formation of cavitation near the channel walls at the temperature of 100°C was minimized by selecting the optimal roughness values and geometric dimensions of the channel.

3. Conclusion

As a result of the design hydrodynamic analysis of the developed distribution hydraulic block of the open-pit dump truck, the intensity of cavitation caverns occurrence was revealed at various parameters, such as surface roughness, temperature regime, fluid flow rate and the supply channel diameter. The dependence of losses on the values of roughness and temperature (viscosity) is also established. The obtained data being taken into consideration, it was possible to significantly reduce cavitation at 100 °C, thereby providing a greater cavitation margin in the open-pit dump truck hydraulic system, to avoid the phenomenon of supercavitation and to exclude it at other temperature conditions, as well as to select the necessary geometric parameters of the hydraulic block channels and the hydraulic fluid of the open-pit dump truck hydraulic system.

Acknowledgements

This study was conducted at the Bauman Moscow State Technical University financially supported by the Russian Ministry of Education and Science (Agreement about federally funded grants provided in the form of subsidies No. 14.577.21.0287 (Unique identifier of the work: RFMEFI57718X0287).

References

[1] Tamburrano P, Plummer A R, Distaso E and Amirante R 2018 A Review of Direct Drive Proportional Electrohydraulic Spool Valves: Industrial State-of-the-Art and Research Advancements *Journal of Dynamic Systems Measurement and Control* **141(2)** 020801

[2] Jerry Boza 2016 Design and Validation of an Electro-Hydraulic Pressure-Control Valve and Closed-Loop Controller *Western Michigan University* **12** p 128

[3] Jin-gang Liu, Gao-sheng Wang, Tian-heng Peng and Sheng-qiang Jiang 2019 Numerical

Simulation of Solid Particle Erosion in Aluminum Alloy Spool Valve *Multiscale and Multiphase Computational Particle Technology* vol 2019, 9465406 DOI: 10.1155/2019/9465406
[4] Jasionowski R, Polkowski W and Zasada D 2016 The Destruction Mechanism of Titanium Subjected to Cavitation Erosion *Key Engineering Materials* **687** 117–122
[5] Yuan-Jian Yang, Peng Weiwen, Debiao Meng and Shun-Peng Zhu 2014 Reliability analysis of direct drive electrohydraulic servo valves based on a wear degradation process and individual differences *Proceedings of the Institution of Mechanical Engineers Part O Journal of Risk and Reliability* **228(6)** 621–630
[6] Kayakol N 2016 Cavitation modelling in micro channels *Conference Paper* May 2016 p 7
[7] Lomakin V O, Kuleshova M S and Kraeva E A 2015 Fluid flow in the throttle channel in the presence of cavitation *Dynamics and Vibroacoustics of Machines, Procedia Engineering* **106** 27–35
[8] Rudolf P, Juliš M, Klakurková L, Gejdoš P and Hudec M 2019 *IOP Conf. Series: Earth and Environmental Science* **240** 062057
[9] Rooze J *at al.* 2012 Hydrodynamic cavitation in micro channels with channel sizes of 100 and 750 micrometers *Microfluid Nanofluid* **12** 499–508
[10] Araz Sheibani Aghdam, Morteza Ghorbani, Gokberk Deprem, Fevzi Çakmak Cebeci and Ali Koşar 2019 A New Method for Intense Cavitation Bubble Generation on Layer-by-Layer Assembled SLIPS *Scientific Reports* **9** 11600
[11] Ji Pei, Majeed Koranteng Osman, Wenjie Wang, Desmond Appiah, Tingyun Yin and Qifan Deng 2019 A Practical Method for Speeding up the Cavitation Prediction in an Industrial Double-Suction Centrifugal Pump *Energies* **12** 2088
[12] Saeed Sh, Aboul-Fotouh T M and Ashour I 2016 A Current Viscosity of Different Egyptian Crude Oils: Measurements and Modeling Over a Certain Range of Temperature and Pressure *J. Pet Environ Biotechnol* vol 7 **6** 1000305
[13] Knežević D M, Lovrec D, Mitar J and Karanovic V 2009 Determination of Pressure Losses in Hydraulic Pipeline Systems by Considering Temperature and Pressure *Strojniski Vestnik* **55(4)** 237–243
[14] Knežević D and Savić V 2006 Mathematical modeling of changing of dynamic viscosity, as a function of temperature and pressure, of mineral oils for hydraulic systems *Mechanical Engineering* vol 4 **1** 27–34
[15] Bair S and Michael P 2010 Modelling the pressure and temperature dependence of viscosity and volume for hydraulic fluids *International Journal of Fluid Power* vol 11 **2** 37–42

Strength of multiferroic layered structures in position sensor structures

R V Petrov[1,5], O V Sokolov[1], M I Bichurin[1], A R Petrova[2], S Bozhkov[3], I Milenov[3] and P Bozhkov[4]

[1] Yaroslav-the-Wise Novgorod State University, 41, ul. B. St. Peterburgskaya, Veliky Novgorod, 173003, Russia
[2] Moscow Aviation Institute (National Research University), 4 Volokolamskoye shosse, 125993, Moscow, Russia
[3] University of Transport, 158 Geo Milev str., 1574 SPZ Slatina, Sofia, Bulgaria
[4]Olimex Ltd, 2 Pravda str., 4000, Plovdiv, Bulgaria

[5] E-mail: roman.petrov@novsu.ru

Abstract. The article is devoted to modeling and calculating the strength of multiferroic layered structures in the designs of position sensors used in various practical applications, including CPS sensors. Progress in the study of new multiferroic layered structures in which a giant magnetoelectric effect has been discovered has opened up completely new prospects for the design of position sensors. The studies carried out open up obvious prospects for creating economical and cheap sensors for use in CPS structures, as well as electric motors for determining angular positions and other automated systems. The strength test of magnetostrictive-piezoelectric two-layer structures metglas/PZT, metglas/($LiNbO_3$ cut Y+140°), metglas/GaAs was carried out in the bending mode of the magnetoelectric effect. It was found that in all cases the strength condition is satisfied with a very good margin. Thus, the reliability of position sensors with a magnetoelectric structure is fully ensured. The studies carried out open up obvious prospects for creating economical and cheap sensors for use in CPS structures, as well as electric motors for determining angular positions and other automated systems.

1. Introduction

Position sensors are one of the main components used in automation devices and for the automation of various technological processes. Position sensors are widely used in the automotive industry as crankshaft position sensors (CPS). For the designs of CPS sensors, inductors on a metal magnetic core and Hall sensors in a miniature design are most often used [1]. Progress in the study of new multiferroic layered structures in which the giant magnetoelectric (ME) effect was discovered [2] has opened up completely new prospects for the design of position sensors [3]. The studies carried out open up obvious prospects for creating economical and cheap sensors for use in CPS structures, as well as electric motors for determining angular positions [4] and other automated systems. New materials such as self-biased magnetoelectric structure have opened up unique possibilities for improving automotive sensors [5].

For use in technical devices, strong and reliable materials are required. In addition to direct strength tests, methods of modeling and predicting the conditions of destruction of the objects under study are widespread. A clear advantage of the modeling method is its low cost and speed of obtaining results.

Content from this work may be used under the terms of the Creative Commons Attribution 3.0 licence. Any further distribution of this work must maintain attribution to the author(s) and the title of the work, journal citation and DOI.
Published under licence by IOP Publishing Ltd

The requirements for sensors that are used in cars are of a very high level, as it relates to human safety. Sensors are subjected to various types of mechanical stress, they operate in a wide temperature range, in addition, there are effects of electric and magnetic fields. The reliability of such a sensor should provide a continuous vehicle range of hundreds of thousands of kilometers. Thus, preliminary verification of the ME structure for strength is an urgent problem.

Various materials are used to manufacture multiferroic layered structures with magnetoelectric effect [6]. For example, composite structures based on PZT piezoelectric ceramics and magnetostrictive materials such as nickel or metglass alloy are used. Layered multiferroic structures based on monocrystalline substances and magnetostrictive metals or alloys, including such as Terfenol-D. Multiferroic means that in these materials several types of "ferro" ordering coexist simultaneously: ferromagnetic and ferroelectric. For research we have chosen the following materials PZT, $LiNbO_3$, GaAs, metglas. By connecting the magnetostrictive component and the piezoelectric component, the required multiferroic medium is obtained.

The purpose of the article is to calculate the strength of multiferroic layered structures in the designs of position sensors in order to understand the limits of the operation of these sensors.

2. Modeling and strength analysis

For calculation and modeling, the theoretical approach for calculating the ME of composites is used, given in [7]. Information from articles [8-10] was used as the initial data. The parameters of the piezoelectric composite were taken from the website of OOO Avrora-ELMA.

The most severe operating conditions occur in the resonant mode. The strength test of magnetostrictive-piezoelectric two-layer structures metglas/PZT, metglas/($LiNbO_3$ cut Y + 140°), metglas/GaAs in the bending mode of resonance of the magnetoelectric effect is carried out below.

Total composite thickness

$$t = {}^{p}t + {}^{m}t. \tag{1}$$

Volume fractions of piezoelectric and ferromagnet

$${}^{p}v = \frac{{}^{p}t}{t}$$

$${}^{m}v = \frac{{}^{m}t}{t}. \tag{2}$$

Effective density of the composite

$$\rho = {}^{p}v\,{}^{p}\rho + {}^{m}v\,{}^{m}\rho. \tag{3}$$

We draw the axis X along the neutral line of the composite beam:

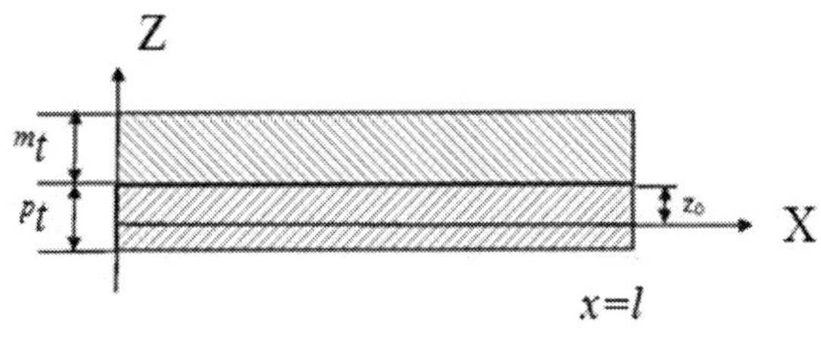

Figure 1. The position of the interface between the piezoelectric and magnetostrictive phases relative to the neutral line in a two-layer composite.

Longitudinal component of the strain tensor

$$S_1 = -z\frac{\partial^2 w}{\partial x^2}, \tag{4}$$

where w is the lateral displacement.

Longitudinal component of the stress tensor in a piezoelectric

$$^{p}T_1 = \bar{c}_{11}^{D} S_1 - \bar{h}_{31} D_3, \tag{5}$$

the third component of the electric field strength vector

$$E_3 = -\bar{h}_{31} S_1 + \bar{\beta}_{33}^{S} D_3, \tag{6}$$

where

$$\begin{aligned} \bar{c}_{11}^{D} &= \left({}^{p}s_{11}^{E} - \frac{d_{31}^2}{\varepsilon_{33}^{T}\varepsilon_0} \right)^{-1} \\ \bar{h}_{31} &= \bar{c}_{11}^{D} d_{31} \\ \bar{\beta}_{33}^{S} &= \frac{1+\bar{h}_{31} d_{31}}{\varepsilon_{33}^{T}\varepsilon_0} \end{aligned} \tag{7}$$

Longitudinal component of the stress tensor of metglass

$$^{m}T_1 = {}^{m}Y^{B}\left(S_1 - q_{11} h_1\right), \tag{8}$$

where ${}^{m}Y^{B} = \frac{{}^{m}Y}{1-{}^{m}K_{11}^2}$, ${}^{m}K_{11}^2 = \frac{{}^{m}Y q_{11}^2}{\mu\mu_0}$ – the square of the magneto-mechanical coupling coefficient.

We substitute (4) into (8) and get

$$^{m}T_1 = -z\,{}^{m}Y^{B} \frac{\partial^2 w}{\partial x^2} - \bar{q}_{11} h_1, \tag{9}$$

where $\bar{q}_{11} = {}^{m}Y^{B} q_{11}$.

Torque

$$M = \int_{z_0-{}^{p}t}^{z_0} bz\,{}^{p}T_1 dz + \int_{z_0}^{z_0+{}^{m}t} bz\,{}^{m}T_1 dz = -b\frac{\partial^2 w}{\partial x^2} D - b\,{}^{p}t^2 \langle h_{31}\rangle D_3 - b\,{}^{m}t^2 \langle q_{11}\rangle h_1, \tag{10}$$

where

$$\begin{aligned} \langle h_{31}\rangle &= \frac{1}{{}^{p}t^2} \int_{z_0-{}^{p}t}^{z_0} z\bar{h}_{31} dz = \frac{2z_0 - {}^{p}t}{2\,{}^{p}t}\bar{h}_{31} \\ \langle q_{11}\rangle &= \frac{1}{{}^{m}t^2} \int_{z_0}^{z_0+{}^{m}t} z\bar{q}_{11} dz = \frac{\bar{q}_{11}\left(2z_0 + {}^{m}t\right)}{2\,{}^{m}t} \end{aligned}, \tag{11}$$

total cylindrical stiffness of the composite beam

$$D = {}^{p}D + {}^{m}D, \tag{12}$$

where

$$\begin{aligned} {}^{p}D &= \frac{1}{3}\bar{c}_{11}^{D}\,{}^{p}t\left({}^{p}t^2 - 3\,{}^{p}t z_0 + 3z_0^2\right) \\ {}^{m}D &= \frac{1}{3}\,{}^{m}Y^{B}\,{}^{m}t\left({}^{m}t^2 + 3\,{}^{m}t z_0 + 3z_0^2\right) \end{aligned}, \tag{13}$$

b – the width of the magnetoelectric composite.

Let us find the voltage across the piezoelectric

$$U = \int_{z_0 - {}^p t}^{z_0} E_3 dz = {}^p t^2 \langle h_{31} \rangle \frac{\partial^2 w}{\partial x^2} + {}^p t \langle \beta_{33}^S \rangle D_3, \tag{14}$$

where

$$\langle \beta_{33}^S \rangle = \frac{1}{{}^p t} \int_{z_0 - {}^p t}^{z_0} \bar{\beta}_{33}^S dz = \bar{\beta}_{33}^S. \tag{15}$$

Hence, we express the electrical displacement in the piezoelectric

$$D_3 = \frac{U}{{}^p t \langle \beta_{33}^S \rangle} - \frac{{}^p t \langle h_{31} \rangle}{\langle \beta_{33}^S \rangle} \frac{\partial^2 w}{\partial x^2} \tag{16}$$

and substitute in (10)

$$M = -bt^3 \langle c_{11} \rangle \frac{\partial^2 w}{\partial x^2} - \frac{b\, {}^p t \langle h_{31} \rangle}{\langle \beta_{33}^S \rangle} U - b\, {}^m t^2 \langle q_{11} \rangle h_1, \tag{17}$$

where

$$\langle c_{11} \rangle = \frac{1}{t^3} \left(D - \frac{{}^p t^3 \langle h_{31} \rangle^2}{\langle \beta_{33}^S \rangle} \right). \tag{18}$$

The position of the interface between the piezoelectric and magnetostrictive phases relative to the neutral line $z_0$is determined from the minimum condition $\langle c_{11} \rangle$

$$z_0 = \frac{\left(\bar{c}_{11}^D \, {}^p t^2 - {}^m Y_B \, {}^m t^2\right)\langle \beta_{33}^S \rangle - \bar{h}_{31}^2 \, {}^p t^2}{2\left({}^m Y_B \, {}^m t + \bar{c}_{11}^D \, {}^p t\right)\langle \beta_{33}^S \rangle - 2\bar{h}_{31}^2 \, {}^p t}. \tag{19}$$

Transverse force

$$V = \frac{\partial M}{\partial x} = -bt^3 \langle c_{11} \rangle \frac{\partial^3 w}{\partial x^3}. \tag{20}$$

Bending vibration equation

$$\rho bt \frac{\partial w^2}{\partial \tau^2} = \frac{\partial V}{\partial x}. \tag{21}$$

Substituting (20) into (21), we obtain

$$t^2 \langle c_{11} \rangle \frac{\partial^4 w}{\partial x^4} + \rho \frac{\partial w^2}{\partial \tau^2} = 0. \tag{22}$$

Time shift is harmonic $w \sim e^{i\omega\tau}$, therefore

$$\frac{\partial^4 w}{\partial x^4} - k^4 w = 0, \tag{23}$$

where the wavenumber

$$k = \left(\frac{\rho}{t^2 \langle c_{11} \rangle} \omega^2 \right)^{\frac{1}{4}}. \tag{24}$$

General solution of the equation of motion (23)

$$w = C_1 \cosh(kx) + C_2 \sinh(kx) + C_3 \cos(kx) + C_4 \sin(kx). \tag{25}$$

Open circuit condition

$$\int_0^l D_3 dx = 0 \tag{26}$$

We integrate (14) over x

$$Ul = {}^p t^2 \langle h_{31} \rangle \frac{\partial w}{\partial x}\bigg|_0^l = {}^p t^2 \langle h_{31} \rangle k \left[C_1 r_2 + C_2 (r_1 - 1) - C_3 r_4 + C_4 (r_3 - 1) \right], \tag{27}$$

where

$$\begin{aligned} r_1 &= \cosh(kl) \\ r_2 &= \sinh(kl) \\ r_3 &= \cos(kl) \\ r_4 &= \sin(kl) \end{aligned} . \tag{28}$$

Boundary conditions for rigidly restrained one end of the beam and free fixing of the other end of the beam

$$\begin{aligned} & w(0) = 0 \\ & \frac{\partial w}{\partial x}(0) = 0 \\ & V(l) = 0 \\ & M(l) = 0 \end{aligned} . \tag{29}$$

Combining boundary conditions (29) with (27), we obtain a linear system of five inhomogeneous algebraic equations with respect to five unknowns C_1, C_2, C_3, C_4, U

$$\begin{aligned} & C_1 + C_3 = 0 \\ & C_2 + C_4 = 0 \\ & C_1 r_2 + C_2 r_1 + C_3 r_4 - C_4 r_3 = 0 \\ & -t^3 \langle c_{11} \rangle k^2 (C_1 r_1 + C_2 r_2 - C_3 r_3 - C_4 r_4) - \frac{{}^p t \langle h_{31} \rangle}{\langle \beta_{33}^S \rangle} U - {}^m t^2 \langle q_{11} \rangle h_1 = 0 \\ & Ul = {}^p t^2 \langle h_{31} \rangle k \left[C_1 r_2 + C_2 (r_1 - 1) - C_3 r_4 + C_4 (r_3 - 1) \right] \end{aligned} . \tag{30}$$

Taking into account that $r_1^2 - r_2^2 = 1, r_3^2 + r_4^2 = 1$, we will find a solution for this system

$$C_1 = -C_3 = -\frac{l\,{}^{m}t^2 \langle q_{11}\rangle \langle \beta_{33}^S\rangle (r_1 + r_3)}{2k\left(\langle c_{11}\rangle klt^3 \langle \beta_{33}^S\rangle (1 + r_1 r_3) + {}^{p}t^3 \langle h_{31}\rangle^2 (r_1 r_4 + r_2 r_3)\right)} h_1$$

$$C_2 = -C_4 = \frac{l\,{}^{m}t^2 \langle q_{11}\rangle \langle \beta_{33}^S\rangle (r_2 - r_4)}{2k\left(\langle c_{11}\rangle klt^3 \langle \beta_{33}^S\rangle (1 + r_1 r_3) + {}^{p}t^3 \langle h_{31}\rangle^2 (r_1 r_4 + r_2 r_3)\right)} h_1 \quad . \tag{31}$$

$$U = -\frac{{}^{m}t^2\, {}^{p}t^2 \langle q_{11}\rangle \langle h_{31}\rangle \langle \beta_{33}^S\rangle (r_1 r_4 + r_2 r_3)}{\langle c_{11}\rangle klt^3 \langle \beta_{33}^S\rangle (1 + r_1 r_3) + {}^{p}t^3 \langle h_{31}\rangle^2 (r_1 r_4 + r_2 r_3)} h_1$$

Then the ME-coefficient for voltage

$$\alpha_E = \frac{\overline{E}_3}{h_1} = -\frac{{}^{m}t^2\, {}^{p}t^2 \langle q_{11}\rangle \langle h_{31}\rangle \langle \beta_{33}^S\rangle (r_1 r_4 + r_2 r_3)}{t\left[\langle c_{11}\rangle klt^3 \langle \beta_{33}^S\rangle (1 + r_1 r_3) + {}^{p}t^3 \langle h_{31}\rangle^2 (r_1 r_4 + r_2 r_3)\right]} . \tag{32}$$

To account for losses in the calculation, you need to put $\omega = 2\pi\left(1 + \frac{1}{2Q} i\right) f$, where the quality factor of the resonance is $Q = 51$.

Figure 2 shows the dependence of the magnetoelectric voltage coefficient on the frequency of the alternating magnetic field.

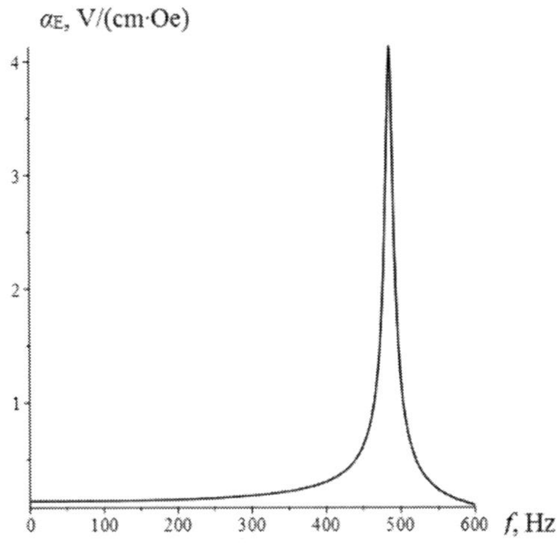

Figure 2. Dependence of the magnetoelectric voltage coefficient on the frequency of the alternating magnetic field.

The calculation used the following material parameters of metglass 2605S3A: ${}^{m}\rho = 7290\ \frac{kg}{m^3}$, ${}^{m}Y = 1.0 \cdot 10^{11}\ Pa$, $\mu = 10^4$, $q_{11} = 3.5 \cdot 10^{-9}\ \frac{m}{A}$, thickness ${}^{m}t = 6.9 \cdot 10^{-5}\ m$; material parameters PZT: ${}^{p}\rho = 7510\ \frac{kg}{m^3}$, ${}^{p}s_{11}^{E} = 1.53 \cdot 10^{-11}\ \frac{m^2}{N}$, $\varepsilon_{33}^{T} = 1750$, $d_{31} = -1.75 \cdot 10^{-10}\ \frac{m}{V}$, thickness ${}^{p}t = 5 \cdot 10^{-4}\ m$; length of magnetoelectric sample $l = 2.5 \cdot 10^{-2}\ m$.

Resonance frequency for the flexural mode of the magnetoelectric effect in such a sample $f_r = 485.6\ Hz$.

Limiting voltage for PZT $\sigma_m = 80\ MPa$.

It is obvious that the maximum longitudinal stresses in PZT will occur at the resonant frequency.

Let us find an expression for the longitudinal voltage in a piezoelectric. For this, in (5) we substitute (4) and (16)

$$^{p}T_1 = \left(\frac{^{p}t\bar{h}_{31}\langle h_{31}\rangle}{\langle\beta_{33}^{S}\rangle} - z\bar{c}_{11}^{D}\right)\frac{\partial^2 w}{\partial x^2} - \frac{\bar{h}_{31}U}{^{p}t\langle\beta_{33}^{S}\rangle}. \quad (33)$$

We have already found the voltage across the piezoelectric U. Let us find the second partial derivative of the vertical displacement along the longitudinal coordinate

$$\frac{\partial^2 w}{\partial x^2} = k^2[C_1\,cosh(kx) + C_2\,sinh(kx) - C_3\,cos(kx) - C_4\,sin(kx)]. \quad (34)$$

Coefficients C_1, C_2, C_3, C_4 we also already know. Thus, we now also have an expression for the longitudinal stress in PZT.

Obviously, depending on the vertical coordinate z, the longitudinal stress reaches its maximum modulus value either on the lower face of PZT at $z = z_0 - {}^{p}t$, or on the upper face of PZT connected to the metal glass, at $z = z_0$.

Figure 3 shows the dependences of the longitudinal stress in PZT on the lower and upper faces on the longitudinal coordinate x. In the calculation, the value of the alternating magnetic field strength was taken $h_1 = 1\,Oe$.

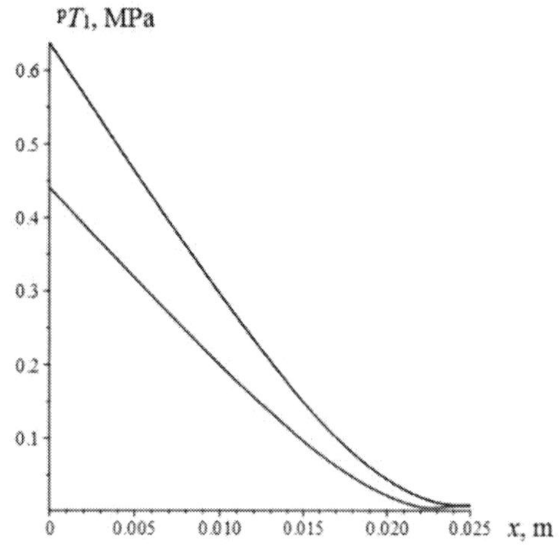

Figure 3. Dependence of the longitudinal stress in PZT on the longitudinal coordinate x (black color – on the lower edge, blue – on the upper edge).

The figure shows that the longitudinal stress in the PZT reaches its maximum value on the lower edge at $x = 0$, that is, at the clamped end. For a given PZT ${}^{p}t = 5 \cdot 10^{-4}\ m$ thickness, this maximum value ${}^{p}T_1^{max}$.

Let's do a similar calculation for several PZT thicknesses. The results of this calculation are presented in table 1.

Table 1. Calculation for PZT thicknesses.

${}^{p}t, m$	f_r, Hz	${}^{p}T_1^{max}$, MPa
10^{-3}	897.5	0.34
$5 \cdot 10^{-4}$	485.6	0.64
$3 \cdot 10^{-4}$	318.7	0.87
10^{-4}	147.5	1.24
$5 \cdot 10^{-5}$	104.0	1.14

From table 1 we can see that the dependence ${}^{p}T_1^{max}$ on ${}^{p}t$ is not too sharp and on the segment $[5 \cdot 10^{-5},\ 3 \cdot 10^{-4}]$ has a maximum that cannot greatly exceed 1.24 MPa. Therefore, it is clear that the strength condition is satisfied with a large margin ${}^{p}T_1^{max_m}$. In the course of the study, mathematical modeling was carried out in an applied engineering package of computer modeling. Figure 4 shows the result of modeling the mechanical stresses in MPa for the ME material metglas/PZT in the flexural mode of resonance for the PZT layer.

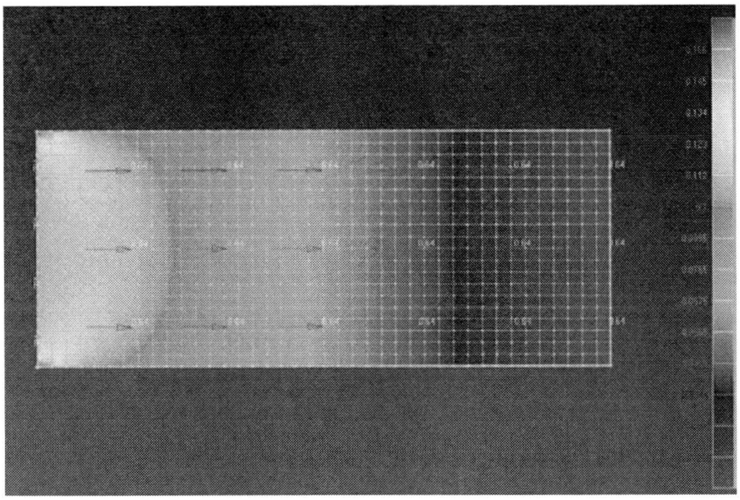

Figure 4. Result of simulation of mechanical stresses (VonMises Stress) in MPa for ME material metglas/PZT at flexural resonance mode for PZT layer 0.5 mm thick. The scale shows the value of mechanical stress in the metglas/PZT material in MPa, arrows indicate the value and direction of the force.

Now let's make a similar calculation for a magnetoelectric composite, in which a Y+140 ° cut of lithium niobate is used instead of PZT. Material constants for Y+140 ° cut of lithium niobate ${}^{p}\rho = 4647\ \frac{kg}{m^3}$, ${}^{p}s_{11}^{E} = 6.91 \cdot 10^{-12}\ \frac{m^2}{N}$, $\varepsilon_{33}^{T} = 49.8$, $d_{31} = -2.724 \cdot 10^{-11}\ \frac{m}{V}$. According to some data, the limiting stress for Y+140 ° cut of lithium niobate $\sigma_m = 550\ MPa$, according to other data, it does not exceed the maximum yield stress for lithium niobate 110 MPa.

The results of this calculation are presented in table 2.

Table 2. The results of the calculation for a magnetoelectric composite.

${}^{p}t, m$	f_r, Hz	${}^{p}T_1^{max}, MPa$
10^{-3}	1666.5	0.56
$5 \cdot 10^{-4}$	844.4	1.05
$3 \cdot 10^{-4}$	520.5	1.59
10^{-4}	211.7	2.86
$5 \cdot 10^{-5}$	141.6	3.01
$3 \cdot 10^{-5}$	114.2	2.72

From table 2 we can see that the dependence ${}^{p}T_1^{max}$ on ${}^{p}t$ is not too sharp and has a maximum on the segment $[3 \cdot 10^{-5},\ 10^{-4}]$ that cannot greatly exceed 3.01 MPa. Therefore, it is clear that the strength condition ${}^{p}T_1^{max_m}$ is satisfied with a very good margin even if the ultimate stress is 110 MPa.

Figure 5 shows the result of modeling the mechanical stresses in MPa for the ME material metglas/($LiNbO_3$ cut Y+140 °) at the flexural resonance mode for the lithium niobate layer.

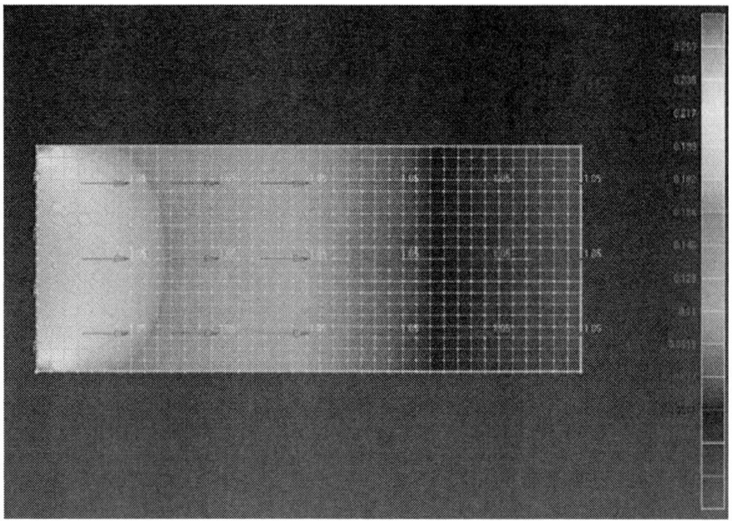

Figure 5. Result of simulation of mechanical stresses (VonMises Stress) in MPa for ME material metglas/($LiNbO_3$ cut Y+140 °) at flexural resonance mode for a 0.5 mm thick lithium niobate layer. The scale shows the value of mechanical stress in the material metglas/($LiNbO_3$ cut Y+140 °) in MPa, arrows indicate the value and direction of the force.

Finally, let us make a similar calculation for a magnetoelectric composite, in which a thin narrow strip of gallium arsenide cut from the (001) layer in the <110> direction is used as a piezoelectric. Material constants for it ${}^{P}\rho = 5320\,\frac{kg}{m^3}$, ${}^{P}s_{11}^{E} = 8.25\cdot 10^{-12}\,\frac{m^2}{N}$, $\varepsilon_{33}^{T} = 12.9$, $d_{31} = -1.347\cdot 10^{-11}\,\frac{m}{V}$. Gallium arsenide yield strength $10\ GPa$.

The results of this calculation are presented in table 3.

Table 3. The results of the calculation for a magnetoelectric composite.

${}^{P}t, m$	f_r, Hz	${}^{P}T_1^{\max}, MPa$
10^{-3}	1285.0	1.02
$5\cdot 10^{-4}$	669.8	1.84
$3\cdot 10^{-4}$	425.3	2.69
10^{-4}	185.2	4.48
$5\cdot 10^{-5}$	126.9	4.65
$3\cdot 10^{-5}$	103.5	4.18

From table 3 we can see that the dependence ${}^{P}T_1^{\max}$ on ${}^{P}t$ is not too sharp and has a maximum on the segment $[3\cdot 10^{-5},\ 10^{-4}]$, which cannot greatly exceed 4.65 MPa. Therefore, it is clear that the strength condition ${}^{P}T_1^{max}$is satisfied with a very good margin. Figure 6 shows the result of modeling the mechanical stresses in MPa for the ME material metglas/GaAs at the flexural resonance mode for the GaAs layer.

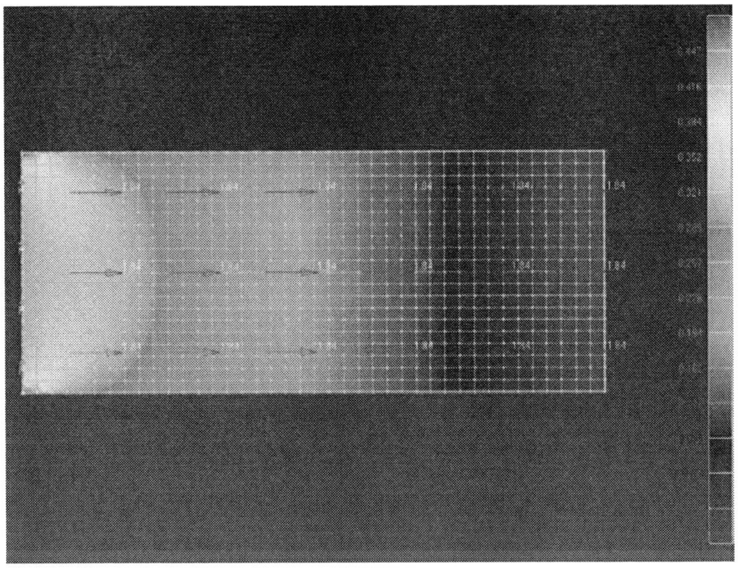

Figure 6. Result of simulation of mechanical stresses (VonMises Stress) in MPa for ME material metglas/GaAs at flexural resonance mode for GaAs layer 0.5 mm thick. The scale shows the value of mechanical stress in the metglas/GaAs material in MPa, arrows indicate the value and direction of the force.

3. Conclusion

The article analyzes the strength of multiferroic layered structures in the designs of position sensors in order to understand the limits of the operation of these sensors. The strength test of magnetostrictive-piezoelectric two-layer structures metglas/PZT, metglas/($LiNbO_3$ cut Y+140 °), metglas/GaAs was carried out in the bending mode of the magnetoelectric effect. It was found that in all cases the strength condition is satisfied with a very good margin. Thus, the reliability of position sensors with a magnetoelectric structure is fully ensured. The studies carried out open up obvious prospects for creating economical and cheap sensors for use in CPS structures, as well as electric motors for determining angular positions and other automated systems.

Aknowledgements

This research was funded by by RFBR project number № 19-58-18001. The article is related to the implementation of the project "Modeling and design of position sensors based on multiferroic layered structures" under Contract No. КП-06-Русия/20 28.09.2019, Todor Kableshkov University of Transport – Sofia, on the Bulgarian side.

References

[1] Krivoruchenko A A 2007 Contactless position sensors. Problems of choice and practice of use *Components and technologies* **1** 32–35 [In Russ.]

[2] Bichurin M I, Petrov V M, Petrov R V and Tatarenko A S 2016 *High Sensitivity Magnetometers, Smart Sensors, Measurement and Instrumentation* eds A Grosz et al. (Switzerland: Springer International Publishing) Chapter 5 pp 127–66

[3] Petrov R V, Bichurin M I, Leontiev V S, Kolesnikov N A, Bozhkov S T, Stanev L G, Pacheliev V T, Milenov I K and Bozhkov P T 2016 The crankshaft position sensor based on magnetoelectric materials *IEEE International Power Electronics and Motion Control Conference (PEMC)* 848–853 DOI: 10.1109/EPEPEMC.2016.7752105

[4] Nikitin A O, Kolesnikov N A and Petrov R V 2016 Magnetoelectric sensor for detecting the position of brushless DC motor's rotor axis *Vestnik NovSU* **7 (98)** 81–84 [In Russ.]

[5] Petrov R V, Sokolov O V, Leontiev V S, Semenov G A, Ivanov S A, Petrova A R, Bichurin M I, Bozhkov S T and Milenov I K 2019 Self-biased magnetoelectric structure for automotive sensors *Mechanics Transport Communications* vol 17 **3** 1891 [In Russ.]
[6] Bichurin M I, Petrov V M, Petrov R V and Tatarenko A S 2019 *Magnetoelectric Composites (*Pan Stanford Publishing) p 297
[7] Petrov R V, Bichurin M I and Petrov V M 2012 *Rezonansnyye effekty v magnitostriktsionno-p'yezoelektricheskikh kompozitakh dlya tverdotel'nykh elektronnykh ustroystv* [Resonance effects in magnetostrictive-piezoelectric composites for solid-state electronic devices] (Palmarium Academic Publ.) p 264 [In Russ.]
[8] Ichiro Yonenaga 2005 Hardness, yield strength, and dislocation velocity in elemental and compound semiconductors *Mater. Trans.* vol 46 **9** 1979–85 DOI: 10.2320/matertrans.46.1979
[9] Domenjoud M, Lematre M, Gratton M, Lethiecq M and Tran-Huu-Hue L 2013 Theoretical and experimental study of the electroacoustic behavior of lithium niobate under an initial mechanical stress *IEEE Transactions on Ultrasonics, Ferroelectrics, and Frequency Control* **60 (10)** 2219–2224 DOI: 10.1109/TUFFC.2013.2814
[10] Starodubtsev Yu N 2011 *Magnitomyagkiye materialy: entsiklopedicheskiy slovar'-spravochnik* [Soft magnetic materials: an encyclopedic dictionary-reference book] (Moscow: Tekhnosfera)

Utilizing factor models to simulate emergency shutdowns of power grids

M Polkovskaya[1,2], T Buzina[1] and N Fedurina[1]

[1] Irkutsk State Agricultural University named after A.A. Ezhevsky, pos. Molodezhny, Irkutsk district, Irkutsk region, Russia

[2] E-mail: polk_mn@mail.ru

Abstract. In the article, based on statistics on emergency outages in the Levoberezhniy and Pravoberezhniy districts of Irkutsk and the sums of average daily temperatures in years 2010-2017 regression models are built. As an auxiliary factor for the models built on values that were grouped by year, the seasonality index was calculated using the "Kassandra" model. In the model, built for values that were grouped by month, a trend was added in addition to the seasonal component.
A retrospective prediction of emergency outages based on the data of the Levoberezhniy district, gathered in April 2020 showed that a model with a trend and a seasonal component gives a more accurate prediction. Since the sum of the average daily temperatures can be considered a random variable, the proposed factor models can be used to predict emergency outages only with some probability. The solution of inverse problems, when the factor value is determined for a given number of emergency outages, can be used to assess the risks associated with the influence of low and high temperatures on the objects of the electric grid complex.

1. Introduction

Modern day industry, as well as many other kinds of human activities are extremely reliant on the power supply. Power outages may lead to negative consequences across all economic sectors: industry, agriculture and forestry, construction, transport, trade, healthcare, etc. That is why it is crucial to predict and prevent downtimes of various components of power grids [1– 4].

Various studies of emergency power supply shutdowns identify a multitude of factors that affect the uptime of power grid facilities [5–7]. Those factors can be divided into 4 groups: operational, climatic, external, and others [8, 9]. The operational ones include faulty equipment, high network load, operational errors, low-quality maintenance, etc. [10].

Climatic reasons are quite diverse, and the degree of their influence depends primarily on the climatic zone. Various studies examined the way power grids are affected by the following: relative humidity and air temperature, daily rainfall, wind speed, number of thunderstorm hours [8, 11, 12]; magnetic storms [13], as well as extreme natural phenomena (such as a hurricane) [5, 7].

The external factors most often involve human actions (vandalism or theft of electrical equipment, wire breakage, accidents, etc.), as well as plants, animals and birds [8, 14]. Other reasons for emergency shutdowns include those for which the classification has not been established (ignition, relay protection being triggered, high-voltage switches being disconnected, etc.

Content from this work may be used under the terms of the Creative Commons Attribution 3.0 licence. Any further distribution of this work must maintain attribution to the author(s) and the title of the work, journal citation and DOI.
Published under licence by IOP Publishing Ltd

The study [15] revealed that emergency outages are not just being influenced by climatic factors, but are, in fact, subject to seasonal patterns (in three seasons: summer, winter, autumn-spring). The seasonal component was assessed by monthly data in the study [16], and by the quarterly data in [17].

Thus, the aim of the study is to make regression models for predicting emergency outages on power grids, that would take seasonality into account. With this goal in mind, we evaluated the effect of the average monthly temperatures on the emergency shutdowns. Then, we determined the influence of the seasonal component on the series of failures of the electric network equipment. After that, we built the factoring models to simulate emergency outages taking seasonality into account. Finally, we proposed several algorithms for estimating the number of outages for a certain probability.

2. Materials and methods

In order to build models with climatic parameters for predicting emergency outages on power grids, we utilized daily data on emergency outages at substations of Southern electric networks of the Irkutsk Electric Network Company for 2010–2017. (in the Pravoberezhniy and Levoberezhniy districts of the city of Irkutsk). The climatic factor is the sum of the average daily temperatures for months (observation point at Irkutsk Airport) for the same period [18]. Correlation analysis was used in order to assess the relationship between the temperature and the failures of various elements of the electric grid. Calculations of the parameters of regression equations and selection of the components of the time series (seasonal components) were carried out using the least squares method.

3. Results and discussion

At the first stage, we estimated the effect the sum of the average daily temperatures grouped by months on the number of emergency outages. In the city of Irkutsk, the data on emergency outages is logged for two districts: Levoberezhniy and Pravoberezhniy. Since the number of annual blackouts in these areas has different trends, we performed the study not for the city of Irkutsk as a singular unit, but separately for each district. Note that in the Pravoberezhniy district in 2011 there was a decrease in the number of emergency outages (figure 1). However, the situation is the opposite in the Levoberezhniy district, as the number of emergency shutdowns for 2011 and 2012 compared to 2010 has increased more than twofold. This fact is to some extent explained by the construction of new microdistricts, which in turn led to an increase in the load on electric networks. In subsequent years, there are wave-like changes in the number of blackouts in both districts.

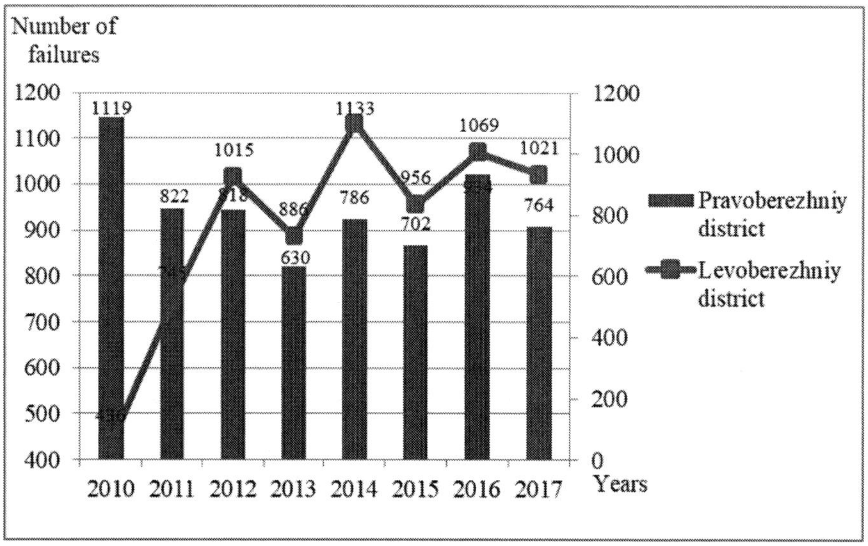

Figure 1. The number of emergency shutdowns in electric networks by districts of Irkutsk for 2010–2017.

Evaluation of the significance of the relationship between the temperature parameter and equipment failures for the entire study period showed that in the Levoberezhniy district the connection is weak and significant (R = 0.50), and in the Pravoberezhniy district it is absent (R = 0.15). Subsequently, correlation coefficients were calculated using data for a specific year and for a specific month for the entire period.

A strong significant relationship between the number of emergency shutdowns in the Levoberezhniy district (y_1) for years and the sum of average daily temperatures (x) is observed in 2010, 2014, 2015, 2017 (R = 0.86; 0.79; 0.74 and 0.83, respectively); weak significant relationship is observed in 2011 (R = 0.50) and 2013. (R = 0.55). According to the data of the Pravoberezhniy district, a strong significant relationship between temperature and blackouts was detected in 2010, 2013, 2014 and 2015 (R = – 0.64; 0.68; 0.69 and 0.77), and a weak significant – in 2016, 2017 (R = 0.53; 0.59). In this connection in 2010 is the reverse.

When calculating the correlation coefficients from monthly data on emergency outages and the sum of average daily temperatures, the following results were obtained. According to the data of the Levoberezhniy district, a strong significant relationship between the studied parameters was detected in April (R = 0.76) and June (R = 0.69), significant – in March (R = 0.44), May (R = 0.35), July (R = 0.55), September (R = – 0.40), December (R = – 0.34).

In the Pravoberezhniy district, the heat parameter has a significant effect on the accident rate in July (R = 0.87) and December (R = – 0.73). A close relationship between the number of failures and temperature is highlighted in January (R = – 0.60), February (R = – 0.63), June (R = 0.60), and November (R = 0.43). It is noteworthy that the correlation coefficients of the winter months are negative. Decreasing air temperature causes the number of accidents to rise, and, conversely the number of accidents decreases as the air gets warmer.

Based on the identified relationships, we constructed regression equations from annual and monthly data. The value chosen as a factor s the sum of the average daily temperatures for months – x, effective signs: y_1 – emergency shutdowns for years in the Left Bank district; y_2 – emergency shutdowns by years in the Right-Bank district; y_3 – emergency blackouts for months in the Levoberezhniy district; y_4 – emergency shutdowns for months in the Pravoberezhniy district. The obtained dependencies are checked for accuracy and adequacy. The significance of the regression model was checked using the Fisher F-test. Its value is found as the ratio of the variance of the initial series of observations for the studied indicator, and the unbiased estimate of the variance of the residual sequence for this model. The significance of the coefficients of the equation is checked using Student's criterion (t-statistics). The calculated values of the Fisher criterion are shown in table 1, and the tabular value F_{table} = 4.96. Since the calculated value of the F-criterion is larger than the tabulated one, the equation is considered statistically significant.

Table 1. Factor models of emergency power outages grouped by years.

Year	Equation	The calculated value of the Fisher criterion
	Levoberezhniy district	
2010	$y_1 = 36.61 + 0.022x$	28.7
2014	$y_1 = 89.70 + 0.076x$	16.7
2015	$y_1 = 73.42 + 0.066x$	12.0
2017	$y_1 = 78.56 + 0.080x$	21.5
	Pravoberezhniy district	
2010	$y_2 = 92.42 - 0.065x$	7.0
2013	$y_2 = 50.61 + 0.039x$	8.6
2014	$y_2 = 63.59 + 0.031x$	8.6
2016	$y_2 = 53.98 + 0.047x$	14.3

The adequacy of the model was tested according to four conditions: 1) fluctuations in the values of the remainder of the series have to be random (using the series criterion); 2) a series of residues has to correspond to the normal distribution law (*RS*-criterion); 3) the average value of the sequence εi needs to be equal to zero; 4) the values of the residue series have to be independent (Darbin-Watson *d*-test).

Considering significant variations in the number of emergency shutdowns between 2010 and 2012, the equations obtained for 2010 are not suitable for predicting emergency shutdowns, since, despite the high accuracy and adequacy estimates for the obtained equations, they do not reflect the actual data. It is logical that in order to obtain predictions it is better to use the regression equations for 2017 in the Levoberezhniy district and for 2016 in the Pravoberezhniy district, since the influence of unaccounted factors is more accurately reflected in these.

According to the equations given in table 2, significant regression equations were obtained only for the month of April in the Levoberezhniy district; as well as July and December in the Pravoberezhniy district. Despite the high correlation coefficient between equipment failures and the sum of temperatures, the equation for June in the Levoberezhniy district according to the Fisher criterion turned out to be insignificant, since the calculated value is lower than the table value ($F_{table} = 5.98$).

Table 2. Factor models of emergency power outages grouped by months.

Month	Equation	The calculated value of the Fisher criterion
	Levoberezhniy district	
April	$y_3 = 45.78 + 0.45x$	8.1
June	$y_3 = -230.35 + 0.63x$	5.5
	Pravoberezhniy district	
July	$y_4 = -149.32 + 0.39x$	18.2
December	$y_4 = 507.76 - 0.65x$	6.7

Next, we analyzed the presence of the seasonal component in emergency outages. We utilized the "Kassandra" model, which is based on the least squares method [19].

The equation includes three components of the initial time series x_t: trend (y_t), seasonal fluctuations (s_t), residual term (ε_t):

$$x_t = y_t + s_t + \varepsilon_t, \tag{1}$$

The trend is expressed as a polynomial in time

$$y_t = \sum_{i=0}^{n} a_i t^i; \tag{2}$$

The seasonal component s_t is the sum of strictly periodic functions with weights t_i:

$$s_t = \sum_{i=0}^{m} s_i(t) t^i. \tag{3}$$

Here, $s_i(t)$ are functions that reflect the variation in the amplitude and shape of seasonal fluctuations as a function of time. The oscillation period of the functions specified by the expansion in the Fourier series is one year

$$s_i(t) = \sum_{j=1}^{K/2} \alpha_{ij} \cos\frac{2\pi jt}{K} + \sum_{j=1}^{K/2-1} \beta_{ij} \sin\frac{2\pi jt}{K}, \tag{4}$$

where K is the number values x_t in a series for a year (for monthly data K=12).

To estimate the parameters of the model a_i, α_{ij}, β_{ij}, we used the least squares method

$$\sum_{t=1}^{T\Sigma} b_t \varepsilon_t^2 \rightarrow min, \tag{5}$$

under conditions (1) – (4). Here b_t are some positive informativeness weights used to account for the qualitative, structural changes of the series under consideration in the past and the regular effects of external factors.

In order to estimate the parameters of the trend-seasonal model we used the "Oracle" program created by I.V. Mokriy. Absolute seasonal deviations in the Levoberezhniy and Pravoberezhniy districts have similar seasonal waves (figure 2), as shown by the assessment of the seasonality index. At the same time, the amplitude of seasonality index fluctuations in the Levoberezhniy district is much higher than in the Pravoberezhniy district. The largest number of emergency shutdowns occurs from April to August, followed by a gradual decrease. This may occur due to more difficult operating conditions of the grids caused by heat, as well as increased load on the electric network, e.g. due to the use of air conditioners.

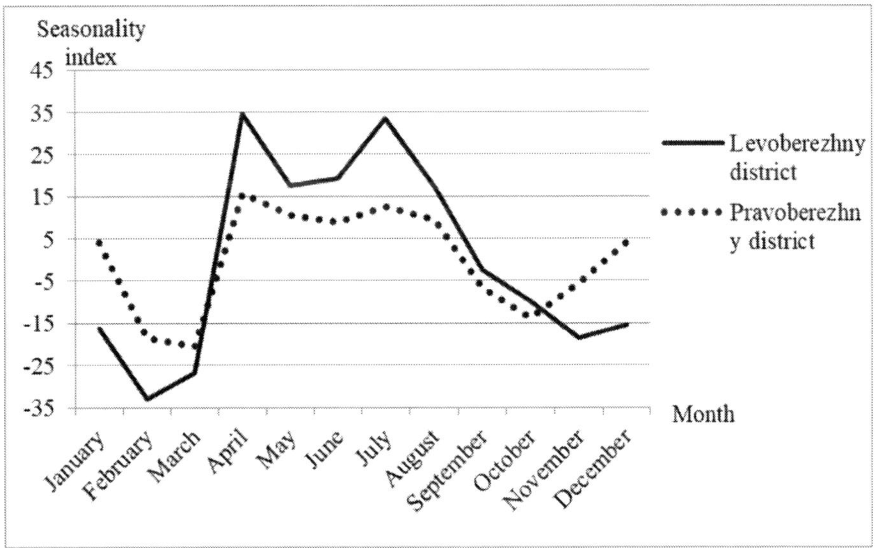

Figure 2. Seasonality index of the series of emergency outages on power grids in the Levoberezhniy and Pravoberezhniy districts of Irkutsk for 2010–2017.

The obtained seasonality indices are included as additional factors in the regression equations with the sum of the temperatures (table 3). Here s_{iL} is a series of seasonality indices obtained by months in the Levoberezhniy district, and s_{iP} is in the Pravoberezhniy district. The significance check of the equation yields the tabular value of the Fisher criterion of 4.26.

Table 3. Factor models of emergency shutdowns on electric networks with the seasonal component by years.

Year	Equation	The calculated value of the Fisher criterion
	Levoberezhniy district	
2010	$y_1 = 36.70 + 0.029x - 0.18s_{iL}$	17.3
2011	$y_1 = 62.80 - 0.015x + 1.47s_{iL}$	4.4
2012	$y_1 = 84.70 - 0.059x + 1.97s_{iL}$	9.4
2014	$y_1 = 94.03 + 0.006x + 1.35s_{iL}$	21.8
2015	$y_1 = 77.12 + 0.027x + 0.76s_{iL}$	7.5
2017	$y_1 = 82.55 + 0.031x + 0.98s_{iL}$	18.6

	Pravoberezhniy district	
2012	$y_2 = 68.20 - 0.017x + 1.31s_{iP}$	8.3
2013	$y_2 = 50.89 + 0.033x + 0.35s_{iP}$	4.3
2014	$y_2 = 64.56 + 0.015x + 0.91s_{iP}$	16.8
2015	$y_2 = 55.29 + 0.034x + 0.77s_{iP}$	10.8
2016	$y_2 = 76.89 + 0.021x + 2.23s_{iP}$	8.1

It should be noted that the seasonality factor allowed us to obtain models for a larger number of series (2011, 2012 in the Levoberezhniy district, 2015 in the Pravoberezhniy district), however, the significance of some equations according to the Fisher criterion has decreased.

Since the predictions of the models obtained for monthly values are calculated for the next year (s), we included the trend into the model as an additional factor (table 4). The number of regression dependencies with the trend and seasonal deviations for months, satisfying the criteria of accuracy and adequacy, is equal to the factor models shown in table 2. However, the standard error of the equations presented in table 4 is lower (except for the dependence of the effective sign in July for the Pravoberezhniy district, for which the standard error is 1 less than that of a similar dependence with the trend and seasonality).

Table 4. Factor models of emergency shutdowns on electric networks with the seasonal component by months.

Month	Equation	The calculated value of the Fisher criterion
	Levoberezhniy district	
April	$y_3 = 28.40 + 0.71x - 9.50t + s_{4\,L}$	6.6
June	$y_3 = -96.06 + 0.36x + 5.23\,t + s_{6\,L}$	5.7
	Pravoberezhniy district	
July	$y_4 = -175.82 + 0.42x - 0.76t + s_{7P}$	7.9
December	$y_4 = 84.97 - 0.043x - 8.20t + s_{12P}$	15.8

To compare the obtained models, we make a prediction of the number of emergency outages in April 2020 in the Levoberezhniy district and compare it with the actual value of 119 (table 5). The sum of average daily temperatures in April 2020 amounted to 218.4 °C [18].

Table 5. The results of a retrospective prediction of outages on power grids in April 2020 according to the data of the Levoberezhniy district based on factor models with and without the trend and seasonality.

Model	Forecast value	Deviation from the actual value,%
y_3=45.78+0.45x	144	21.1
y_3=28.40+0.71x-9.50t+s_{4L}	113	4.6

According to the result, a more accurate prediction is yielded by the regression model with the sum of the average daily temperatures for the month, trend and seasonal component.

It should be noted that climatic factors, including the sum of temperatures, are random variables [20]. Therefore, the obtained equations (table 1–4) allow us to simulate an effective attribute only for some probability. We propose an algorithm for modeling emergency outages on electrical networks based on factor dependencies (figure 3). According to the algorithm, the first stage determines the statistical parameters of the climatic parameter (for example, temperature). After that, using the acceptance criteria we select the distribution law and randomly model the factor values based on that. The obtained values are substituted into the regression equations, which are then used to calculate the number of failures. The results can be estimated with some probability. It should be noted that inverse problems can be solved using regression dependencies in order to obtain factor values based on the values of emergency shutdowns.

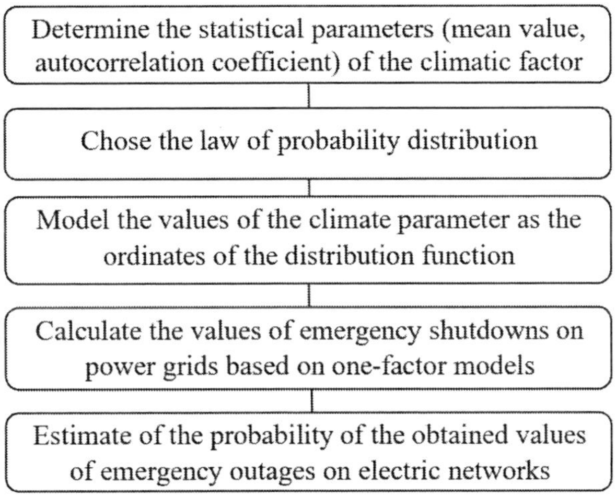

Figure 3. The proposed algorithm for modeling outages on electrical networks and the climate factor, which is a random variable, based on regression equations.

4. Conclusion

Emergency shutdowns on electric networks are influenced by various factors, the careful analysis of which allows for timely measures to be taken in order to prevent them.

When determining the regression dependences of the failures of electric grids, the sum of the average monthly temperatures was taken as a climatic factor. Factor models were built for years and months, utilizing the emergency shutdown data for 2010–2017 in the Levoberezhniy and Pravoberezhniy districts of Irkutsk.

Since the Irkutsk region is a zone with an acutely continental climate, one of the factors affecting the subject parameter is seasonality, which was simulated using the Kassandra model. In addition to the seasonal component in the model for months, a trend is added as a factor.

When comparing models with and without seasonality and trends using indicators of accuracy and adequacy, slight deviations were revealed. However, a retrospective prediction of emergency outages in the Levoberezhniy district in April 2020 showed that a model with a trend and a seasonal component gives a more accurate prediction.

The obtained regression equations can be used to simulate the number of emergency shutdowns only for a certain probability, since the sum of the average daily temperatures is a random variable. In this case, inverse problems can be solved, where the factor value will be calculated on the basis of the set value of emergency shutdowns using the regression equation. This approach can be used to assess the risks associated with the impact on the objects of the electric grid complex of low and high temperatures.

Acknowledgment

The some results study was obtained with funded by RFBR according to the research project no. 19-07-00322.

References

[1] Grakowski L, Chojnacki A, Gebczyk K and Banasik K 2019 Statistical Analysis and Modeling of the Reliability of Overhead Low Voltage Lines *Przeglad elektrotechniczny* vol 95 **12** 261–264

[2] Chatterjee S, Roy B K S 2019 Prediction of Line Contingency Catastrophic Failures Using Synchrophasor *Journal of control automation and electrical systems* vol 30 **6** 1107–1115 DOI: 10.1007/s40313-019-00501

[3] Karamov D N, Naumov I V and Perzhabinsky S M 2018 Matematicheskoe modelirovanie otkazov ehlementov ehlektricheskoj seti (10kV) Avtonomnye ehnergeticheskikh sistem s vozobnovlyaemoj raspredelennoj generatsiej [Mathematical modeling of failures of electric network elements (10 kV) Autonomous energy systems with renewable distributed generation] *Izvestiya Tomskogo politekhnicheskogo universiteta. Inzhiniring georesursov* vol 329 **7** 116–130

[4] Tabadar Z, Ansarifar G R and Pirouzmand A 2019 Probabilistic safety assessment of portable equipment applied in VVER-1000/V446 nuclear reactor during loss of ultimate heat sink accident for stress test program development *Progress in nuclear enerrgytom* **117** UNSP 103101 DOI: 10.1016/j.pnucene.2019.103101

[5] Campbell R J 2012 Weather-Related Power Outagesand Electric System Resiliency *Congressional Research Service Report* (USA: DC, Washington) pp 1–15

[6] Kazim M *et al.* 2020 Fault Detection and Localization for Overhead 11-kV Distribution Lines With Magnetic Measurements *IEEE Transactions on instrumentation and measurement* vol 69 Part 1 **5** 2028–2038 DOI: 10.1109/TIM.2019.2920184

[7] Reed D 2008 Electric utility distribution analysis for extreme winds *J. Wind Eng. Ind. Aerodyn.* **96** 123–140

[8] Vasilieva T N and Lopatin E I 2011 Analiz prichin otkazov ehklekticheskogo oborudovaniya raspredelitel'nykh setej 0,38…10 Kv [Analysis of the causes of failures of the eclectic equipment of distribution networks 0.38 ... 10 Sq] *Vestnik Ryazanskogo Gosudarstvennogo Agrotekhnologicheskogo universiteta im. P.A. Kostycheva* **3(11)** 64–66

[9] Sultan V and Hilton B A 2020 Spatial Analytics Framework to Investigate Electric Power-Failure Events and Their Causes *ISPRS International journal of geo-information* vol 9 **1** 54 DOI: 10.3390/ijgi9010054

[10] Popescu V 2011 *Journal of sustainable energy* vol 2 **4** [In Russ.] (Available at: http://www.energy-cie.ro/archives/2011/nr_4/v2-n4-5.pdf)

[11] Rybakov L M and Ivanova Z G 2015 Prognozirovaniye otkazov i planirovaniye rezerva zapasnykh elementov, apparatov i oborudovaniya raspredelitel'nykh elektricheskikh setey 10 kV [Failure prediction and reserve planning of spare elements, apparatus and equipment for 10 kV distribution electric networks] *Vestnik Chuvashskogo universiteta* **1** 104–110

[12] Shauleva N M, Zakharova A G and Starichenko D K 2010 Vliyanie pogodno-klimaticheskikh faktorov na ehkspluatatsionnuyu nadezhnost' raspredelitel'nykh setej ugol'nykh razrezov [The influence of weather and climate factors on the operational reliability of coal distribution networks] *Vestnik Kuzbasskogo gosudarstvennogo tekhnicheskogo universiteta* , **1** 114–116

[13] Gil A *et al.* 2019 Does time series analysis confirms the relationship between space weather effects and the failures of electrical grids in South Poland? *Web of Science ResearcherID u Orcid journal of mathematics in industry* vol 9 **7** DOI: 10.1186/s13362-019-0064-9

[14] Chayanam K 2005 *Analysis of Telecommunications Power Outages – Inline* director of thesis A P Snow (Available online: etd.ohiolink.edu/rws_etd/document/get/ohiou1125024491/inline)

[15] Burdukov N I, Galeev M T and Voltov I P 2017 Prognozirovaniye veroyatnosti avariynykh otklyucheniy na ob"yektakh elektrosetevogo kompleksa [Prediction of the probability of emergency shut-off on the objects of the electric network complex] *Electric power. transfer and distribution* **2(41)** 28–31

[16] Belov S I and Petrov P S 2017 Prognozirovaniye avariynykh otklyucheniy v elektricheskikh setyakh 35-220 kV [Prediction of emergency shutdowns in electric networks of 35-220 kV] *Bulletin of the Moscow State Agroengineering University named after V. P. Goryachkin* issue 4 **80** 77–82

[17] Kotov O M and Sandykov Y V 2016 Consideration of seasonality of failures of the main equipment in the reliability calculations of distribution electric networks *Trudy pervoj nauchno-tekhnicheskoj konferentsii molodykh uchenykh Ural'skogo ehnergeticheskogo instituta "Pervoj nauchno-tekhnicheskoj konferentsii molodykh uchenykh Ural'skogo ehnergeticheskogo instituta"* (Ekaterenburg) pp 165–166

[18] *Arhiv pogody v Irkutske* [Weather archive in Irkutsk] (Available at: https://rp5.ru/)

[19] Polkovskaya M N 2019 Situation management of agricultural production based on the prediction of prices for agricultural products *Proceedings of the VIth International Workshop 'Critical Infrastructures: Contingency Management, Intelligent, Agent-Based, Cloud Computing and Cyber Security* pp 83–89

[20] Barsukova M N, Ivanyo Ya M and Storublevtseva P M 2019 Modeling the income of agricultural producers taking into account the variability of meteorological conditions *Climate, ecology, agriculture of Eurasia* (Irkutsk; Irkutsk SAU) pp 182–190

Practical method of vector optimization on the example of optimization of thermal conditions

S A Popov[1], S N Kuzmenko[1] and S P Slavinsky[1]

[1] Yaroslav-the-Wise Novgorod State University, 41, B.St.Peterburgskaya st., Veliky Novgorod, 173004, Russia

[2] E-mail: Stanislav.Popov@novsu.ru

Abstract. The practical method of vector optimization is described by the example of optimization of thermal conditions in industrial premises during their design based on the construction of a multi-response regression model. A practical algorithm is proposed for determining the discrete set of Pareto optimal thermal parameters.

1. Introduction to thermal conditions

The calculation of the dissipation power of the heater and the temperature in the production premise is usually calculated according to the following formula [1]:

$$p = q(t_b - t_n), \quad (1)$$

where p *is* the heater dissipation power (kW), t_b is the indoor temperature (°C), t_n is the outdoor temperature (°C), q is the coefficient describing the thermophysical characteristics of the premise, determined experimentally (kW/°C). The adjacent premises, the temperatures in which are interdependent, and the thermal regime of which obeys formula (1), are considered as production premises. The thermal regime refers to the combination of temperatures in the studied premises.

Such a model for calculating the thermal regime is, firstly, linear, and, secondly, one-dimensional and cannot be used to calculate the thermal regime of the aggregate of production premises. The tasks of designing the thermal conditions of several production premises are multicriteria. In this case, the task of modeling and optimizing thermal conditions is to determine the values of the input independent variables, which are the dissipation power d of the heating devices, providing the required values of the output dependent variables in the form of a temperature vector inside m premises:

$$\mathbf{T} = [(t_1 - t_n),(t_2 - t_n),\ldots(t_d - t_n)]^{\mathrm{T}}, \quad (2)$$

where $(t_i - t_n)$ is the relative temperature inside the i premise, $i = \overline{1,m}$.

2. Multiple ther mal model

To describe the relationship between the values of independent variables and dependent variables (response), multi-reflection mathematical regression models are used [2]. In this case, the type of multiresponse model is determined with accuracy to constant coefficients. Based on the resulting

Content from this work may be used under the terms of the Creative Commons Attribution 3.0 licence. Any further distribution of this work must maintain attribution to the author(s) and the title of the work, journal citation and DOI.
Published under licence by IOP Publishing Ltd

model, the unknown variables of the model (indoor temperatures) are evaluated. A multiresponse model can be represented as:

$$\mathbf{T} = \mathbf{F}(\mathbf{P}, \mathbf{B}) + \mathbf{E}, \tag{3}$$

where $\mathbf{P} = \{p_1, p_2, \ldots, p_d\}^{\mathrm{T}}$ is the power vector d of heating appliances, $\mathbf{B} = \{b_1, b_2, \ldots, b_r\}^{\mathrm{T}}$ is the vector of coefficients whose values are determined by the results of experiments, $\mathbf{E} = \{e_1, e_2, \ldots, e_m\}^{\mathrm{T}}$ is the temperature error observation vector $\mathbf{T}$, $\mathbf{F}(\mathbf{P}, \mathbf{B})$ is the m-dimensional temperature function, depending on the form of the multiresponse model $\mathbf{F}$, power dissipation $\mathbf{P}$ and coefficients of model $\mathbf{B}$.

To model the dependence of temperatures on power dissipation, a linear multiresponse model was constructed with respect to the coefficients, which allows you to include in the model a large number of variables of the real process of designing the thermal regime. In addition, such linear models provide a short response time calculation.

Such a model has the following form:

$$\mathbf{T} = \mathbf{R}(\mathbf{P})^{\mathrm{T}} \mathbf{B} + \mathbf{E}, \tag{4}$$

where $\mathbf{R}(\mathbf{P}) = \left\{ \frac{\partial \mathrm{f}_1(\mathbf{P}, \mathbf{B})}{\partial \mathbf{B}}, \frac{\partial \mathrm{f}_2(\mathbf{P}, \mathbf{B})}{\partial \mathbf{B}}, \ldots, \frac{\partial \mathrm{f}_m(\mathbf{P}, \mathbf{B})}{\partial \mathbf{B}} \right\}$ is the matrix of derivatives of the model by the coefficients.

The calculation of the coefficient estimates is performed using the following formula:

$$\hat{\mathbf{B}} = \left[\sum_{i=1}^{n} \mathbf{R}(\mathbf{P}_i) \mathbf{V}_{\mathbf{E}}^{-1} \mathbf{R}(\mathbf{P}_i)^{\mathrm{T}} \right]^{-1} \sum_{i=1}^{n} \mathbf{R}(\mathbf{P}_i) \mathbf{V}_{\mathbf{E}}^{-1} \mathbf{T}_i, \tag{5}$$

where the matrix $\mathbf{V}_{\mathbf{E}}$ is the covariance matrix of the observation errors and n is the number of temperature measurements in the premises for a given power dissipation. The covariance matrix of coefficient estimates is

$$\mathbf{K} = \left[\sum_{i=1}^{n} \mathbf{R}(\mathbf{P}_i) \mathbf{V}_{\mathbf{E}}^{-1} \mathbf{R}^{\mathrm{T}}(\mathbf{P}_i) \right]^{-1}.$$

As a matrix $\mathbf{V}_{\mathbf{E}}$, its estimate is used, which is determined by the results of duplicate experiments for a given vector of independent variables.

Based on the results of observations, model (4) makes it possible to estimate temperatures in production rooms for given dissipation capacities of heating devices and perform the optimization procedure for thermal conditions.

So, the dependence of two output variables $\mathbf{T} = \{(t_1 - t_n), (t_2 - t_n)\}^{\mathrm{T}}$ on five input variables $\mathbf{P} = \{p_1, p_2, \ldots, p_d\}^{\mathrm{T}}$, d = 5, was described using a quadratic model, which takes into account the quadratic terms and interactions of the input variables, in the following form:

$$\begin{cases} t_1 = b_{10} + \sum_{i=1}^{d} b_{1i} p_i + \sum_{i=1}^{d} \sum_{j=1}^{d} b_{1ij} p_i p_j + \sum_{i=1}^{d} b_{1ii} p_i, \; i \neq j \\ t_2 = b_{20} + \sum_{i=1}^{d} b_{2i} p_i + \sum_{i=1}^{d} \sum_{j=1}^{d} b_{2ij} p_i p_j + \sum_{i=1}^{d} b_{2ii} p_i, \; i \neq j \end{cases}. \tag{6}$$

As functions $\mathbf{F}(\mathbf{P}, \mathbf{B}) = \{\mathrm{f}_1(\mathbf{P}, \mathbf{B}), \mathrm{f}_2(\mathbf{P}, \mathbf{B}), \ldots, \mathrm{f}_m(\mathbf{P}, \mathbf{B})\}^{\mathrm{T}}$ for modeling thermal regimes, quadratic relatively independent variable polynomials with all interactions can be used. In this case, function $\mathbf{R}(\mathbf{P})$ in expression (4) takes the following form:

$$\mathbf{R}^{\mathrm{T}}(\mathbf{P})=\left\{\begin{array}{c} 1, p_1, p_2, \ldots, p_k, p_1^2, \ldots, p_k^2, p_1 p_2, p_1 p_3, \ldots, p_{k-1} p_k \\ 1, p_1, p_2, \ldots, p_k, p_1^2, p_2^2, \ldots, p_k^2, p_1 p_2, p_1 p_3, \ldots, p_{k-1} p_k \\ \ldots \\ 1, p_1, p_2, \ldots, p_k,, p_2^2, \ldots, p_k^2, p_1 p_2, p_1 p_3, \ldots, p_{k-1} p_k \end{array}\right\}. \tag{7}$$

Then, a step-wise regression procedure was applied to this model with a decrease of the number of members. In the step regression procedure, coefficients are found for which the inequality holds

$$\frac{\hat{b}_i}{\sqrt{k_i}} \leq t_{1-\frac{\alpha}{2}}, \tag{8}$$

where k_i is the diagonal matrix $\mathbf{K}$ element for the i-th coefficient, $t_{1-\frac{\alpha}{2}}$ is the quantile of the student distribution at the α significance level.

Members including coefficients satisfying inequality (8) are excluded from the corresponding functions, and the coefficient estimates are recalculated, and the procedure for reducing the number of model members is repeated [2].

3. Admissible areas

There is a so-called admissible area of input variables **D** and a corresponding area of output variables that ensure that all the requirements for the thermal regime are met. This area contains a subset of *unimprovable* or *Pareto-optimal* output variables and their corresponding input variables **P**, which cannot be improved simultaneously by all optimized criteria without degrading at least one of these indicators. Among the set of Pareto-optimal variables, the final variant of the thermal regime for all rooms is then selected [3].

To solve the problem of finding an admissible area, a method is used that makes it possible to probe the set of parameters with the points of the sequence distributed in the parallelepiped (1).

The study of the parameter space consists of three stages.

1. The first stage is the compilation of test tables.
2. Select n test points $\mathbf{P}_1, \ldots, \mathbf{P}_n$ located in the general set G.
3. At each point, all local criteria $\mathrm{f}_v(\mathbf{P}_i)$, $v=\overline{1, m+d}$, are calculated. A test table (matrix) W is compiled for each test point.

The first r columns of this table represent test points $\mathbf{P}_1, \ldots, \mathbf{P}_n$. The next m columns of this table represent the values of the parameters-criteria of validity $\mathbf{T}_i=\mathbf{F}(\mathbf{P}_i)$, $i=\overline{1, n}$, calculated in accordance with expression (4). The matrix W has a dimension and is represented in the form:

The next m of these columns presents the values of the parameter criteria $\mathbf{T}_i=\mathbf{F}(\mathbf{P}_i)$, $i=\overline{1, n}$ calculated in accordance with expression (4).

This table presents the values of the parameter criteria the parameters-criteria of validity. The matrix **W** has dimensions $[n \times(r+m)]$ and is represented in the form:

$$\mathbf{W}=\left\{\begin{array}{cccccccc} p_{11} & p_{12} & \cdots & p_{1r} & \mathrm{f}_{11}(\mathbf{P}_1) & \mathrm{f}_{12}(\mathbf{P}_1) & \cdots & \mathrm{f}_{1m}(\mathbf{P}_1) \\ p_{21} & p_{22} & \cdots & p_{2r} & \mathrm{f}_{21}(\mathbf{P}_2) & \mathrm{f}_{22}(\mathbf{P}_2) & \cdots & \mathrm{f}_{2m}(\mathbf{P}_2) \\ \ldots & \ldots & \ldots & \ldots & \ldots & \ldots & \ldots & \ldots \\ p_{n1} & p_{n2} & \cdots & p_{nr} & \mathrm{f}_{n1}(\mathbf{P}_n) & \mathrm{f}_{n2}(\mathbf{P}_n) & \cdots & \mathrm{f}_{nm}(\mathbf{P}_n) \end{array}\right\}, \tag{9}$$

where n is the number of test points, m is the number of parameters-criteria of validity.

IOP Conf. Series: Materials Science and Engineering **939** (2020) 012060 doi:10.1088/1757-899X/939/1/012060

Test points represent a discrete approximation of the area of input parameters, and the values of the parameters-criteria of validity represent a discrete approximation of the area $\mathbf{F}(\mathbf{\Pi})$.

The first r columns of this matrix form a discrete admissible area **D** in the space of input parameters containing s points. The next m columns form the image of the area **D** in the space of parameters-criteria of validity $\mathbf{F}(\mathbf{D})$. At the request of the designer, k parameters can also be included in the optimization process, which are not eligibility criteria and for which technical limitations are not set, but which are desirable to optimize [4].

Usually, optimization (for example, reduction) of one particular criterion leads to an increase in other particular criteria. Points at which the vector optimality criterion **Y** is not reducible by all particular criteria at the same time are called unimprovable solutions or Pareto optimal [1]. Pareto optimality means that it is impossible to reduce the value of one of the particular criteria without increasing at least one of the others. The point **P*** is the best possible solution if there is no such **P**, that

$$\begin{cases} f_i(\mathbf{P}) \le f_{iP}(\mathbf{P}^*), & i = 1,\ldots,m; \\ f_j(\mathbf{P}) < f_j(\mathbf{P}^*) & \textit{for some } j, \end{cases} \tag{10}$$

where $f_j(\mathbf{P})$ is value of a particular optimality criterion. The Pareto-optimal area will be based on the matrix **W**, which determines the allowable set of input parameters. The search for Pareto-optimal points is performed among the rows of the matrix W in accordance with the following conditions. Point $\mathbf{P}_i$ ($\mathbf{P}_i \notin \mathbf{D}$) is Pareto optimal if there is no such point. that for all and at least one v exists. An area is called Pareto-optimal if it consists of all Pareto-optimal points.

The construction of the Pareto-optimal area begins after obtaining an admissible set **D** in the space of input parameters in the form of a matrix **W**. The first r columns of this matrix form a discrete admissible area **D** in the space of input parameters containing s points. The next m columns form the image of the area **D** in the space of parameters-criteria of validity.

By checking the conditions for a point to belong to a Pareto-optimal set for all points of the matrix **W**, only Pareto-optimal points remain in this table [5].

Consider a practical method for constructing a Pareto-optimal set of solutions (criteria must be minimized). In figure 1 the Pareto-optimal set in the criterion space is represented by a solid line in the form of the boundary of the region bounded by points **A** and **B**. Based on the analysis of the Pareto-optimal area, the decision maker (DM) determines the most preferable option $\mathbf{P}^0$.

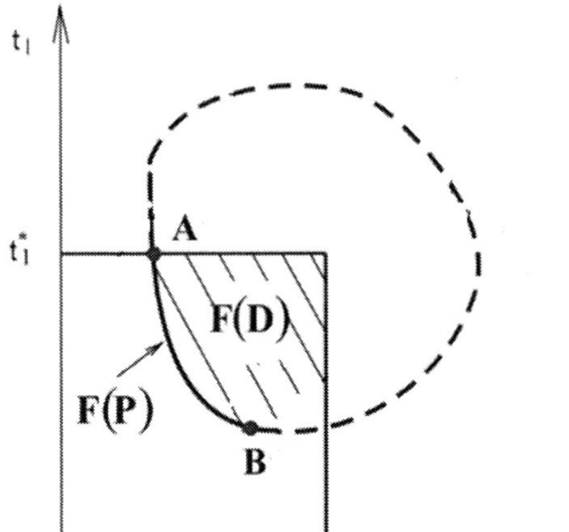

Figure 1. Formation of an admissible area of input variables $\mathbf{F}(\mathbf{D})$ by crossing the region of limiting temperatures t_1^* and t_2^* and $\mathbf{F}(\mathbf{P})$ area.

4. Construction of a discrete Pareto-optimal set on a grid

Using grids to study admissible points in a multi-criterion problem, optimization allows one to rather easily and efficiently recognize and select admissible points from the space of input parameters that belong to an admissible area. However, the admissible area D in problems with continuous parameters and criteria is continuous. Therefore, its approximation usually means the selection of such a finite subset that would approximate any value of each criterion calculated at an arbitrary point from the domain **D** with a predetermined accuracy [6].

To select test points it is advisable to use distributed in the space of a unit volume with the origin at the origin, the dimension of which is equal to the dimension of the parameter space, sequences (or grids) Q1, Q2 ... with simple algorithms for calculating the coordinates of their points

The process of selecting points $\mathbf{P}_i$ proceeds as follows. According to the Cartesian coordinates of the next grid point, $\mathbf{Q}_i = (q_{i1},\ldots,q_{id})$, $i=\overline{1,n}$, in space we find the Cartesian coordinates of point $\mathbf{P}_i = (p_{i1},\ldots,p_{id})$, where are the maximum and minimum values of the variable, respectively

$$p_{ij} = p_j^* + q_{ij}\left(p_j^{**} - p_j^*\right),\ j=\overline{1,d},\ i=\overline{1,n}, \tag{11}$$

where p^{**} and p^* are the maximum and minimum values of the variable, respectively.

For example, for one-dimensional parameter space, we generate the following grid, including 5 points in Cartesian space: Q= $\{0;0,25;0,5;0,75;1\}$. Let $p^{**}=20, p^*=5$. Then, based on the expression (8), we obtain the values of the grid points in the parameter space. For example, for one-dimensional parameter space, we generate the following grid, including 5 points in Cartesian space: Q $=\{0;0,25;0,5;0,75;1\}$. Then, based on the expression (8), we obtain the values of the grid points in the parameter space: $\{0;0,25;0,5;0,75;1\}$.

The value δ'_v must be binary rational, because otherwise, for example, when rounding, the grid points may be outside the allowable area. The coordinates of the grid points (its nodes) are defined as follows:

$$\mathbf{Q}_j = \left(p_{1i_1}, p_{2i_2},\ldots,p_{vi_v},\ldots,p_{di_d}\right).$$

The coordinates of the grid points (its nodes) are defined as follows:

$$n_v = \frac{\left(p_v^{**} - p_v^*\right)}{2\delta_v} + 1.$$

The grid node number depends on the formation order of the grid nodes and can be expressed, for example, as follows:

$$j = i_1 + n_1(i_2 - 1) + n_1 n_2 (i_3 - 1) + n_1 n_2 n_3 (i_4 - 1) + \ldots, \tag{12}$$

where $i_v = \overline{1,n_v}$ is the grid node number.

For illustration, figure 1 shows a grid constructed for three input parameters with a discreteness value. The circles show the grid node numbers. For illustration grid shows built for three parameters c with discrete values $2\delta'_1 = \frac{1}{4}$; $2\delta'_2 = 2\delta'_3 = \frac{1}{2}$. In round circles show numbers of nodes,

Figure 1 shows a grid constructed for three input parameters with a discreteness value. The circles show the grid node numbers. Tahe coordinates of the grid points (its nodes) are defined as follows:

The grid node number depends on the order of formation of the grid nodes and can be expressed, for example, as follows (12). The grid node number depends on the order of formation of the grid nodes and can be expressed, for example, as follows.

IOP Conf. Series: Materials Science and Engineering **939** (2020) 012060 doi:10.1088/1757-899X/939/1/012060

For illustration figure 2 shows a grid constructed of three input parameters with a discreteness value. The circles show the grid node positions.

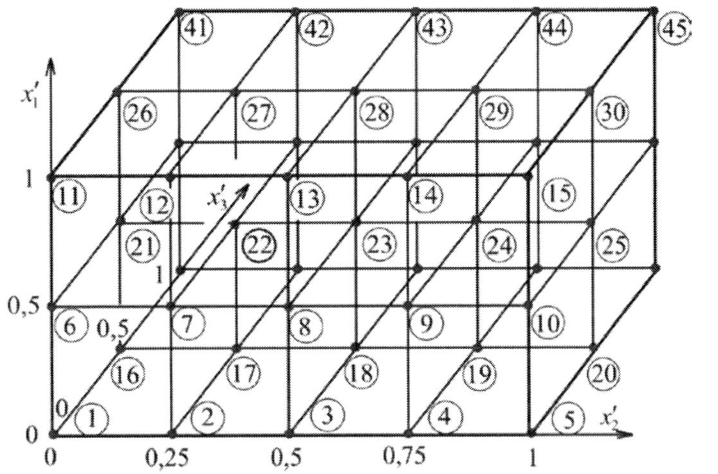

Figure 2. Grid size 5×3×3 round circles marked nodes. Point has 10 sets of 45 nodes. For example, node with coordinates (0.25; 0.5; 0.5), which coexists with the index (2,2,2), at a rate of (30), equal j = 2 + 5 * 1 + 5 * 3 * 1 = 22.

5. The procedure of vector discrete optimization

On the first stage, there is an optimization area, and in the widest space, enter the parameter to approximate the sequence, for which you can specify the total vector of the incoming parameter, where, and n – the number of usable segments.

For all permeable discs with a disconnected diaphragm, $\mathbf{F}(\mathbf{P},\mathbf{B}) = \{f_{1_i}(\mathbf{P},\mathbf{B}), f_{2_i}(\mathbf{P},\mathbf{B}), \ldots, f_{m_i}(\mathbf{P},\mathbf{B})\}^T$, where $i = \overline{1,n}$, a n change the 20-point perimeter of the socket to include 106 nods. It is impossible to explain such a large number of reflections with the help of imitation models, thereby further multiplying the multi regression line of the yielding coefficient model (8). To create a Pareto-optimized solution, suggests the following algorithm.

In terms of specificity, the suboptimal points in the area are more likely to be limited by the number of cut-outs; for partial suboptimal points, create a new cut, and then set new sections for separate partitions and semiconductor areas.

To create a Pareto-optimized solution following algorithm is necessary. On the first stage, there is an optimization area, and in the widest space, enter the parameter to approximate the sequence, for which you can specify the total vector of the incoming parameter, where, and n-the number of usable segments. For all permeable discs with a disconnected diaphragm, change the 20-point perimeter of the socket to include 106 nodes. It is impossible to explain such a large number of reflections with the help of imitation models, thereby further multiplying the multi-click regression line of the yielding coefficient model (8).

In terms of specificity, the suboptimal points of the area are more likely to be limited by the number of cut-outs; for partial suboptimal points, create a new cut, and then set new sections for separate partitions and semiconductor areas including все suboptimal точки; это значительно step up the rate economic growth множества.

The operations according to p. 1–3 continue until the specified approximation accuracy is reached, i.e. when inequalities will be satisfied,. Here is the grid cell size for the i-th input variable, is the permissible accuracy of determining the optimal point for this variable.

To construct Pareto-optimal solutions, the following algorithm is proposed.

1. At the first stage of optimization, the search area in the space of input parameters is approximated by a grid, for each node of which the vector of output parameters is calculated, where, and n is the number of grid nodes. For seven variables with a discretization of the range of variation of variables by 20 points, the grid included more than 106 nodes. It is impossible to calculate such a number of responses even with the help of a simulation model; therefore, for this, we used a multi-response regression linear model with respect to the coefficients (8).

2. For the set of vectors of output parameters over the grid obtained in step 1, by the search method of all points.

3. In the vicinity of each suboptimal point, the search area is limited to several cells of the original mesh; a new grid is constructed for a separate suboptimal point, and then these new grids for individual points are combined and a search area of complex shape of a smaller volume and with a significantly smaller sampling step is obtained, including all suboptimal points; this significantly increases the accuracy of the approximation of the optimal set.

Figure 3 shows the Pareto-optimal discrete set of output parameters. As a result, we obtain a discrete approximation of the Pareto optimal domain of solutions by a set of points, among which we can choose the preferred solution. The search for the final optimal values of the input variables was carried out among the Pareto optimal points by introducing additional restrictions. The resulting Pareto-optimal set of output parameters is shown in figure 3.

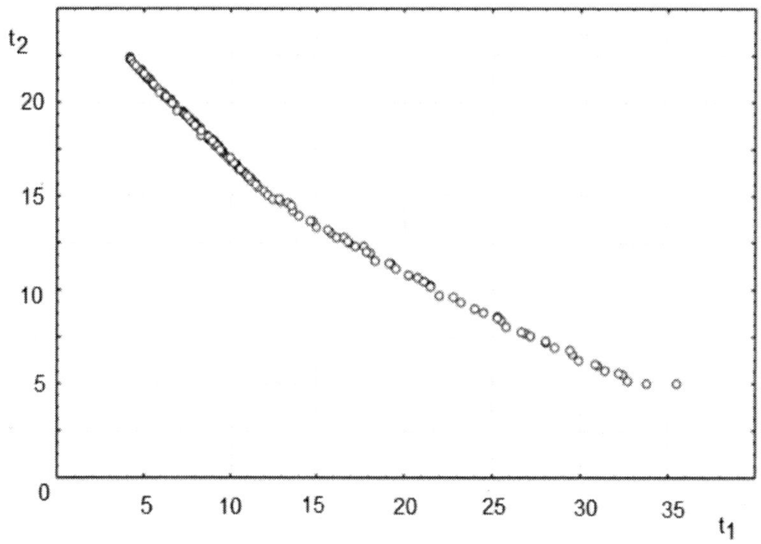

Figure 3. Discrete optimization Pareto-optimal vector procedure.

6. Conclusion

The proposed practical method for constructing an admissible set of variables is possible to obtain an admissible set in a discrete form, which makes it possible to provide an acceptable probability of an error in its construction. The optimization procedure is based on a multi-response linear with respect to the coefficients of the model, the construction of which is the first stage of multi-criteria optimization. The proposed practical vector optimization method based on multi-feedback models allows vector optimization of thermal conditions for two or more rooms.

References

[1] Rreith F and Bohn M S 1986 *Principles of Heat Transfer* (New York: Harper, Row Publishers) p 700

[2] Rencher Alvin C 2002 *Methods of multivariate Analysis A John* (Wiley & Sons, Inc. Publication, Brigham Young University) 339–345

[3] Sobol I M and Statnikov R B 2006 *Selection of optimal parameters in the problems with many criteria* (Moscow: Drofa) p 175 [In Russ.]

[4] Popov S A 2019 Determination of the admissible intervals of the input technological parameters that provide the necessary values of the parameters – acceptance criteria *IOP Conf. Ser.: Mater. Sci. Eng.* **656** 012043

[5] Popov S A and Abramov A M 2017 Multiresponse model based design parameters optimization: engineering approach *EAI Endorsed Transactions on Energy Web* vol 4 **12** e2

IOP Conf. Series: Materials Science and Engineering **939** (2020) 012061 doi:10.1088/1757-899X/939/1/012061

Dependence of the preliminary plastic deformation on the curves of chromium-nickel steel deformation

V A Popov[1]

[1] Yaroslav-the-Wise Novgorod State University, ul. B. St. Petersburgskaya, 41 173003 Veliky Novgorod, Russia

E-mail: Vladimir.Popov@novsu.ru

Abstract. The test results of pre-deformed tubular samples at various values of preliminary deformation are presented. The article describes the chamber of the apparatus, which makes it possible to test tubular samples under conditions of plane and volumetric stress states. Curves built on test results of samples with different levels of preliminary plastic deformation are presented, as well as the test results of samples deformed by the ray $\alpha = 0.5$. The experiments make it possible to determine the value of the ultimate uniform deformation after preliminary deformation or on the final segment of the multi-link loading path.

1. Introduction

The research on metal deformation after preliminary plastic deformation with a different loading path is mainly aimed at experimental verification of the fundamental principles of the theory of plasticity. On the other hand, when developing multi-operational technological processes for manufacturing products by stamping methods, it seems extremely important to know the ultimate homogeneous deformations at intermediate or final stages of deformation.

There are quite a lot of research works that describe experiments on plastic deformation of metals along broken (two- or three-link) trajectories. These works are mainly devoted to testing hypotheses and postulates of the deformation theory of plasticity. Researchers in [1, 2, 3] experimentally studied the laws of delay of scalar and vector properties of materials. In [4], according to the results of testing tubular samples along the ray paths after preliminary plastic deformation by other proportional loading paths, a significant change in the loading surface with a strong "cross effect" was noted. In [5], an experimental study of complex paths was used to evaluate various models of elastoplastic deformation. In the works by A. S. Vavakin et al [6], the elastoplastic properties of aluminum alloys were studied under proportional and broken loading trajectories.

It is noteworthy that the experiments mentioned and not mentioned here were carried out at very small values of preliminary plastic deformation (1÷3%).

Below are the test results of pre-deformed tubular samples, presented in the form of the dependence $\overline{\sigma} = \overline{\sigma}(\overline{\varepsilon})$ ("stress intensity – strain intensity") at various values of preliminary deformation.

2. Results and Discussions

Thin-walled samples (outer diameter d_H= 26.4 mm, wall thickness t = 0.3 mm) made of H18N10T-VD steel were tested in the apparatus, the test chamber of which is shown in figure 1. A detailed description of the setup is given in [7]; here we mention only some features. The named chamber is a two-layer

Content from this work may be used under the terms of the Creative Commons Attribution 3.0 licence. Any further distribution of this work must maintain attribution to the author(s) and the title of the work, journal citation and DOI.
Published under licence by IOP Publishing Ltd

vessel containing the inner case 1 and the holder 2. The pressure of the working fluid at the ends of the locking punches 12 and 13 is reversed into compressive forces that displace the holder and the inner case relative to each other, thereby providing radial support to the inner vessel.

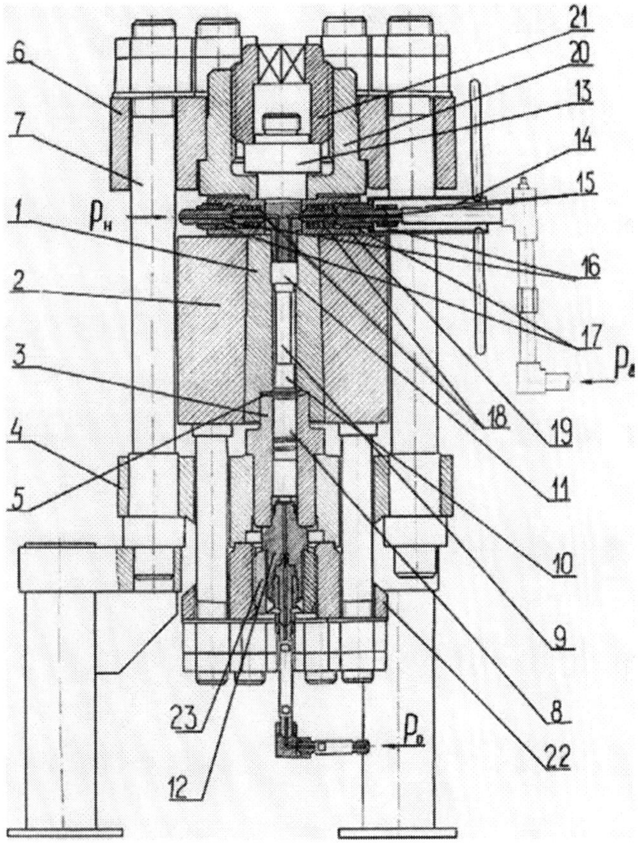

Figure 1. The test chamber.

Assembled sample 9 with upper 11, lower 10 grippers and plunger 8, transmitting axial force to the sample, is placed in the chamber cavity and loaded with internal $p_в$, external $p_н$ and pressure p_0 under the plunger. The purpose of the other parts can be seen from the figure and does not require additional explanation.

The apparatus allows testing tubular samples under conditions of plane and volumetric stress states at maximum working fluid pressures of 1000 MPa. In the test chamber, the well-known loading scheme $p_в - p_н - P$ is implemented, all three pressures taking part in creating the axial force.

From the simulacrum of the deviators of the increments of plastic deformations and stresses, it follows that when loading at $\alpha = \sigma_\theta/\sigma_z = 0.5$ (σ_θ, σ_z are the annular and axial stresses in the sample wall, respectively), the ring deformation is zero, and the average diameter of the sample does not change. This circumstance, after switching on a multiplier between the cavities of the internal and external pressures with a certain ratio of the diameters of the plungers, allowed testing the samples along the named beam under conditions of strictly proportional (stepless) loading.

Hand pumps with a plunger diameter of 4 mm were used as pressure sources.

During the tests, the samples were loaded along the aforementioned ray ($\alpha = 0.5$) to strain intensities $\bar{\varepsilon}$= 0.056; 0.140; 0.198. Then, the tubes deformed in this way were loaded along the ray at $\alpha = \sigma_\theta/\sigma_z = 2.0$ up to the formation of local deformation and fracture. These tests were carried out on air in a simple

IOP Conf. Series: Materials Science and Engineering **939** (2020) 012061 doi:10.1088/1757-899X/939/1/012061

device placed in the socket of a deadweight gauge. At the same time, in another nest of the latter there is an exemplary pressure gauge for measuring the pressure of the working fluid. During the experiment, the diameter of the sample was periodically measured with a micrometer in order to calculate the strain intensity and stress intensity. Note that under loading along this ray, the axial strain ε_z is equal to zero, therefore, from the condition of constant volume $\varepsilon_r = -\varepsilon_\theta$, thus, the strain rate is determined as follows:

$$\overline{\varepsilon} = \frac{2}{\sqrt{3}}\varepsilon_\theta \tag{1}$$

Stress intensity (when $\sigma_r = 0$) will be:

$$\overline{\sigma} = \sqrt{\sigma_z^2 - \sigma_z\sigma_\theta + \sigma_\theta^2} \tag{2}$$

The following are indicated in (1), (2): $\varepsilon_z, \varepsilon_\theta, \varepsilon_r$ are axial, annular and radial logarithmic deformations; $\sigma_z, \sigma_\theta, \sigma_r$ – axial, annular and radial stresses.

Using the calculated values of $\overline{\varepsilon}$ and σ, the curves $\overline{\sigma} = \overline{\sigma}(\overline{\varepsilon})$, were constructed which are shown in figure 2.

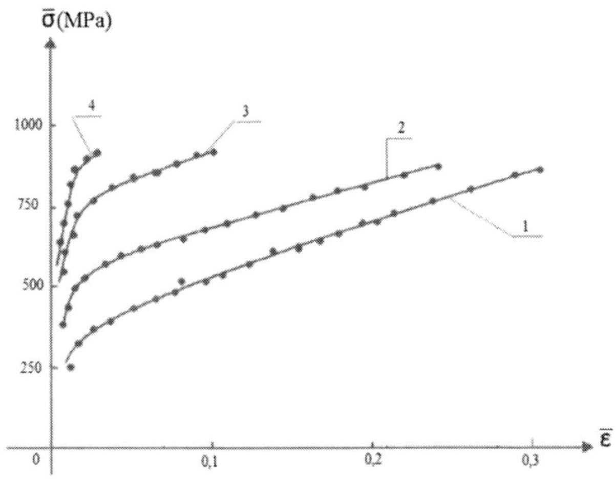

Figure 2. Curves $\overline{\sigma} = \overline{\sigma}(\overline{\varepsilon})$, constructed on the basis of test results of samples with different levels of preliminary plastic deformation: 1 – original sample; 2 – $\overline{\varepsilon} = 0.056$; $\overline{\varepsilon} = 0.140$; $\overline{\varepsilon} = 0.198$.

There is also a diagram based on the results of tests along this ray of the original, that is, previously non-deformed sample. By the type and location of the graphs, we can conclude that the higher the level of preliminary deformation, the lower is the value of the ultimate homogeneous deformation. Noteworthy is also a steeper rise in the initial portion of the diagram for pre-deformed samples. As for the stress intensity corresponding to the moment of failure (or the formation of local deformation), here the discrepancy is about one percent. This value is comparable with the experimental error.

It should be noted that preliminary and subsequent deformation of the samples was carried out under plane deformation (for $\alpha = \sigma_\theta/\sigma_z = 0.5$, $\varepsilon_\theta = 0$, for $\alpha = \sigma_\theta/\sigma_z = 2.0$, $\varepsilon_z = 0$). According to the definition by Davidenkov and Friedman, plane deformation is a "rigid" loading path, usually ending in failure without the formation of local deformation. Indeed, in the sample shown in figure 3, brought to fracture by the ray $\alpha = 0.5$, the diameter remained practically unchanged, and the fracture occurred without localization of deformations.

In the original sample deformed at $\alpha = 2.0$, one-sided swelling occurs, in other, previously deformed samples, local deformation is not observed (figure 4).

Figure 3. A sample brought to fracture during deformation along the ray $\alpha = \sigma_\theta/\sigma_z = 0.5$. On the left is a non-deformed sample.

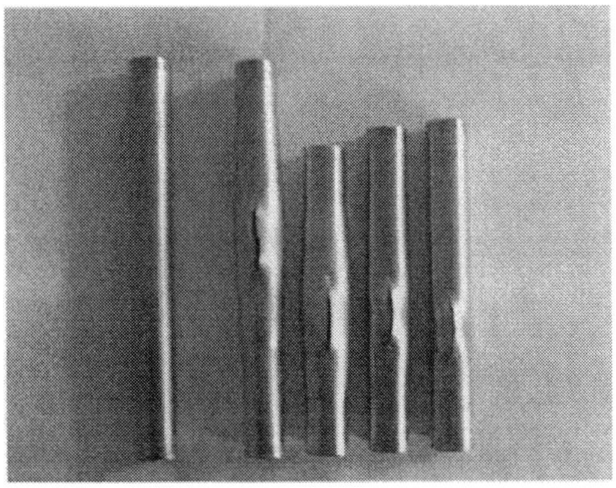

Figure 4. Samples tested under loading along the ray $\alpha = \sigma_\theta/\sigma_z = 2.0$ after preliminary deformation at $\alpha = 0.5$.

Figure 5 shows the test results of samples deformed along the ray $\alpha = 0.5$ to the level $\bar{\varepsilon} = 0.140$ and then tested at $\alpha = 2.0$ with a time delay of 10 min, 2 days, 5 months. The first two samples showed no difference. A sample that has been aged for five months has found an increase in ultimate uniform strain by 30% compared to the first. All experimental points are ideally located on one curve.

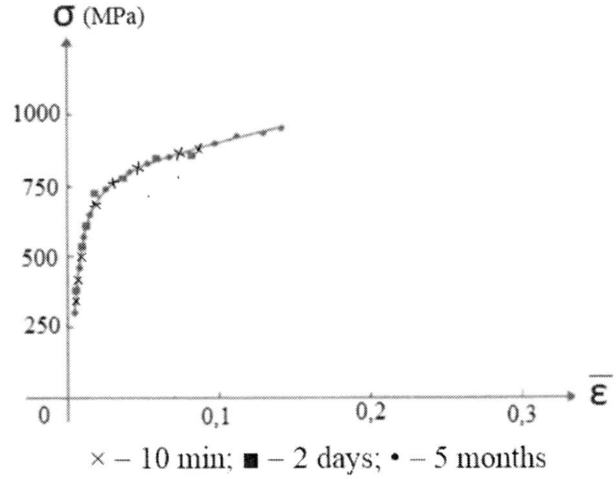

Figure 5. Samples testing results.

The samples were tested at the same values of preliminary deformation ($\bar{\varepsilon} = 0.140$) and different time delays.

3. Conclusion

The effect of preliminary plastic deformation on the nature of the exhaustion of the bearing capacity of chromium-nickel steel images when deforming them along a different loading path has been studied. The significant influence of preliminary residual strain on the value of ultimate uniform strain and the shape of the curves $\bar{\sigma} = \bar{\sigma}(\bar{\varepsilon})$("stress intensity – strain intensity") at the final stage of deformation has been experimentally established. The steeper rise in the initial part of the diagram for samples with large preliminary deformation also attracts attention.

For samples that have passed a time delay of five months after preliminary deformation $\overline{\varepsilon}$= 0.140, an increase in the ultimate homogeneous deformation by 30% is found in comparison with samples tested without a time delay.

The loading of the samples along the rays representing the state of plane deformation, the destruction is completed without any local deformation (in the form of a neck or swelling).

A new technique was used to create preliminary deformation by loading the sample along a ray at which its average diameter does not change ($\boldsymbol{\alpha = \sigma_\theta/\sigma_z = 0.5}$).

References

[1] Romanov Yu M, Sinitskiy A K 1981 Issledovaniye ustoychivosti plasticheskogo deformirovaniya tonkostennykh trub pri razlichnykh putyakh nagruzheniya [Investigation of the stability of plastic deformation of thin-walled pipes under various loading methods] *Trudy Leningradskogo politekhnicheskogo instituta* **343** 25–34 [In Russ.]

[2] Koval'chuk B I 1983 K voprosu o potere ustoychivosti plasticheskogo deformirovaniya obolochek [On the issue of the loss of stability of plastic deformation of shells] *Problemy prochnosti* **5** 11–17 [In Russ.]

[3] Ohashi, Tanaka 1981 Povedeniye malouglerodistoy stali pri plasticheskom deformirovanii vdol trekhzvennykh ortogonal'nykh trayektoriy v trekhmernom vektornom prostranstve deviatora deformatsii [The behavior of mild steel during plastic deformation along three-link orthogonal trajectories in the three-dimensional vector space of the strain deviator] *Teoreticheskiye osnovy inzhenernykh raschetov* **4** 26–33 [In Russ.]

[4] Kuznetsov N P 2000 Plasticheskoye deformirovaniye po putyam nagruzheniya, soderzhashchim promezhutochnyye razgruzki [Plastic deformation along loading paths containing intermediate unloading] *Dis.of kand. nauk* [Candidate of Science diss.] (Veliky Novgorod: NovSU) p 239 [In Russ.]

[5] Melnikov B E, Izotov I N and Kuznetsov N P 1990 Raschet i eksperimental'noye issledovaniye slozhnykh putey uprugoplasticheskogo deformirovaniya [Calculation and experimental study of complex ways of elastoplastic deformation] *Problemy prochnosti* **8** 14–17 [In Russ.]

[6] Vavakin A S, Zlatomrezhev S A, Kasatikov V N and Stepanov L P 1984 Eksperimental'noye issledovaniye uprugoplasticheskikh svoystv alyuminiyevykh splavov AMg6 i D16 pri proportsional'nom deformirovanii [Experimental study of the elastoplastic properties of aluminum alloys AMg6 and D16 under proportional deformation] (Moscow: USSR Academy of Sciences. Institute of Problems of Mechanics) p 42 [In Russ.]

[7] Popov V A 2003 Ustanovka dlya ispytaniya tonkostennykh trubchatykh obraztsov v usloviyakh ob'yemnogo napryazhennogo sostoyaniya [Apparatus for testing thin-walled tubular specimens in the conditions of volumetric stress state] *Korablestroitel'noye obrazovaniye i nauka* (St.-Perersburg: StPbSMTU Publ) 423–428 [In Russ.]

IOP Conf. Series: Materials Science and Engineering **939** (2020) 012062 doi:10.1088/1757-899X/939/1/012062

Assessment of the effectiveness of the regenerative suspension of a timber truck based on the results of simulation

V I Posmetev[1], V O Nikonov[1,2] and V V Posmetev[1]

[1] Voronezh State University of Forestry and Technologies named after G. F. Morozov, Timiryazev str., 8 394087, Voronezh, Russia

[2] E-mail: 8888nike8888@mail.ru

Abstract. The relevance of using various recuperative mechanisms in vehicle suspensions is justified in order to reduce fuel consumption and improve their environmental performance. The developed hydraulic drive scheme and a mathematical model of the movement of a forest car equipped with a suspension with a recuperative mechanism are presented. An assessment of the effectiveness of the regenerative suspension of a timber truck was carried out according to the results of simulation. Analytical dependences of the influence of time, speed and diameters of recuperative suspension hydraulic cylinders on the values of recuperative power under various conditions of movement of a forest car are obtained.

1. Introduction

Timber removal by a logging truck under logging conditions is accompanied by the occurrence of a large number of longitudinal and transverse vibrations of the sprung mass in its suspension while overcoming various obstacles that arise on the logging road. This in turn leads to the appearance of significant heat losses during the conversion of vibrational energy, causing an increase in fuel consumption. Constantly increasing requirements for fuel efficiency contribute to the development of a scientific field related to the recovery of various types of energy in the suspension of a forest car [1, 2].

Gao ZP et al. Scientists conducted a study of the management strategy for energy storage of electromagnetic active suspension in various road conditions. A complex gamma correlation coefficient for changes in energy storage productivity was revealed. Using the proposed strategy will reduce the system's energy consumption by 14.51 %, restore 2.45 % of energy, and improve the vehicle's traffic conditions [3]. Zheng P et al. In their work investigated a regenerative hydraulic-electric suspension, which allows the vehicle to accumulate vibration energy during its movement on the road. They proposed models of road profiles, optimized battery capacity and electrical load. Modeling showed that regenerative suspension can restore 100–400 watts of power with different quality of the roadbed [4]. Jia Y et al. In the article conducted an analytical and numerical study of a magnetic spring recuperative suspension, found that the amount of energy accumulated by a car when driving on an asphalt road is 100 kW [5]. Guo ZJ and others in their work investigated the structure and principle of operation of the regenerative suspension, which allows the accumulation and reuse of electromagnetic energy. Based on the simulation, the effect of vehicle speed on the power output was established. At a vehicle speed of 60 km/h, the output current of recuperated energy is 1.53 A [6]. Lv X and others in their article examined the operation of a vehicle equipped with a recuperative hydraulic suspension that allows you to convert and accumulate energy arising from colliding with obstacles into electric, identified important factors

Content from this work may be used under the terms of the Creative Commons Attribution 3.0 licence. Any further distribution of this work must maintain attribution to the author(s) and the title of the work, journal citation and DOI.
Published under licence by IOP Publishing Ltd

that affect the shock absorbing properties, as well as the amount of energy stored by the suspension [7].

Sun Z et al. In their research cited a system for recovering and reusing energy in a vehicle's suspension. By modeling in various road conditions, a comparison of the obtained recuperated power was performed [8]. In the materials of the article, Yu W et al. Led a comprehensive study of energy recovery technologies used simultaneously in a vehicle. A new power balance strategy has been developed that ensures the rational use and storage of extracted energy, as well as increasing the efficiency of conversion of this energy. The results also showed that the profit when applying the proposed strategy increases with increasing amount of recoverable energy [9]. The study of Taghavifar H and Rakheja S is aimed at studying the accurate assessment of the accumulated energy from the suspension of a vehicle in typical operating conditions. On the basis of mathematical modeling and data obtained in laboratory conditions, it was established that the amount of energy accumulated to a greater extent is affected by the number of roughnesses on the road, and to a lesser extent the speed of the vehicle [10]. Zou J et al. Conducted a study of energy recovery in an interlocking hydraulic suspension using recuperative shock absorbers. The results of mathematical modeling showed that regenerative shock absorbers can attenuate vibrations, improve driving conditions for the driver, and also accumulate energy to power the vehicle's electrical equipment [11]. The work of Zhang J et al. Is devoted to the study of the energy efficiency of vibration of a car's suspension, as well as the development of a dynamic model of energy recovery of an electromagnetic suspension system. The results obtained indicate that the proposed model makes it possible to determine the power of accumulated energy with high accuracy [12].

Zhao L et al. Conducted a study to reduce vehicle energy consumption by using a recuperative hydraulic damper in its design. Based on the developed simulation model, it was revealed that the damping characteristics of the damper meet the requirements of current standards, the accumulation of hydraulic energy is carried out during compression strokes, as well as increasing driver comfort when driving a vehicle [13]. Wen G et al. Proposed a new design of a regenerative hydroelectric suspension. Based on the developed mathematical model, an analysis of its main characteristics is performed. It was revealed that at a vehicle speed of 50 km/h the energy recovery rate was 42.3 % [14]. Wang R and other scientists in the article presented a new system for restoring vibration energy in a suspension by converting it into electric current to improve the dynamic characteristics of the entire vehicle. Based on simulation, in conditions of car movement on an asphalt road, it was revealed that this suspension has improved dynamic characteristics, as well as reduced angular and vertical accelerations of the car body [15]. Zhang H et al. Investigated the regenerative semi-active suspension of a vehicle under low-quality road conditions, as well as performed simulation and bench tests of this suspension. It was established that at the resonant frequency of the sprung mass, the amplitude of the parameters of the suspension index is ranked in decreasing order, the amount of stored energy increases with increasing number of obstacles on the road [16]. Chen L and others in their article cited the results of research and optimization based on simulation of the operation of a regenerative hydraulically interconnected suspension. The results showed that the dynamic characteristics of a car equipped with the proposed suspension are inferior to traditional interconnected suspensions, however, it can be operated in various road conditions and restore part of the vibration energy while maintaining comfortable conditions for the driver when driving [17].

Despite the significant scientific contribution of foreign researchers to the development and study of many structural designs of vehicle suspensions with the recovery of various types of energy, the results of the considered developments should not be mechanically applied to all types of vehicles due to the specific features of their use. Currently, rational circuit designs and scientific approaches have not yet been developed that are associated with the recovery of hydraulic energy in the suspensions of timber trucks. In this regard, the research carried out by the authors in this direction is relevant and timely.

The aim of the work is to evaluate the effectiveness of the regenerative suspension of a timber truck according to the results of simulation modeling, as well as to identify the dependences of the influence of time, speed and diameters of the hydraulic cylinders of the regenerative suspension on the values of the regenerative power for various conditions of movement of the timber truck.

2. Materials and methods

Based on the analysis of the results of numerous studies in studying the state of the issue, the authors proposed the design of a timber truck with a regenerative hydraulic drive, in which the suspension of the timber truck (figure 1) performs the function of the regenerative mechanism [18].

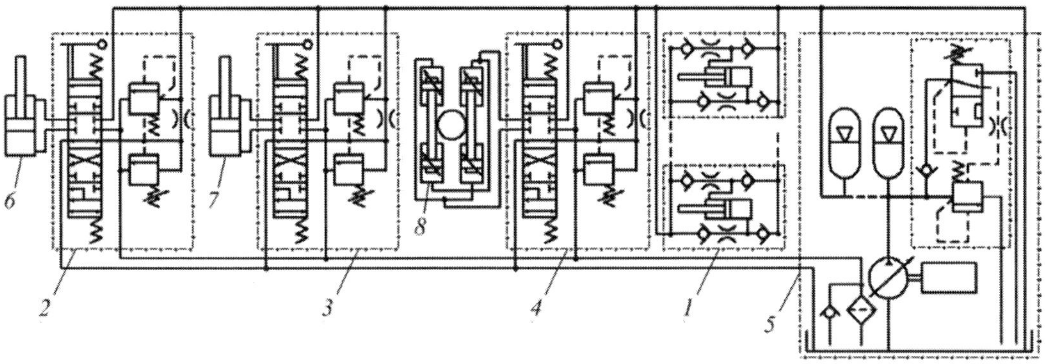

Figure 1. The hydraulic circuit with a regenerative mechanism in suspension of a forest car: *1* – recuperative suspension mechanism; *2-4* – control valves; *5* – pump-accumulator unit; *6-8* – hydraulic cylinders of the boom, stick and slewing ring.

Evaluation of the effectiveness of using a suspension with a recuperative mechanism in the design of a forest car, as well as the determination of its optimal design parameters, was carried out on the basis of a developed mathematical model for the movement of a forest car on a low-quality forest road. In the process of developing a mathematical model, the well-known methods of classical mechanics were used [19].

In the developed mathematical model, a forest car was described by a rigid body moving in three-dimensional space, perceiving force from six wheels when hitting a roughness of a forest road. The mass m of the truck and the moment of inertia J were determined at any time for the current axis of rotation. The determination of the position in space of a forest car in a mathematical model is described by the Cartesian coordinates of its gravity (x, y, z) and the deflection angles by the local coordinate system (φ_x, φ_y, φ_z). To describe the movement in the mathematical model of a forest car, a system of differential equations was used, including the basic laws of dynamics of translational and rotational motion:

$$\begin{cases} m\dfrac{\partial^2 x}{\partial t^2} = \sum\limits_{i=1}^{3} F^{x}_{\text{lef.i}} + \sum\limits_{i=1}^{3} F^{x}_{\text{rig.i}}; \\ m\dfrac{\partial^2 y}{\partial t^2} = \sum\limits_{i=1}^{3} F^{y}_{\text{lef.i}} + \sum\limits_{i=1}^{3} F^{y}_{\text{rig.i}}; \\ m\dfrac{\partial^2 z}{\partial t^2} = -m_{wh.} \cdot g + \sum\limits_{i=1}^{3} F^{z}_{\text{lef.i}} + \sum\limits_{i=1}^{3} F^{z}_{\text{rig.i}}; \\ J_{x}\dfrac{\partial^2 \phi_{x}}{\partial t^2} = \sum\limits_{i=1}^{3} M^{x}(F_{\text{lef.i}}) + \sum\limits_{i=1}^{3} M^{x}(F_{\text{rig.i}}); \\ J_{y}\dfrac{\partial^2 \phi_{y}}{\partial t^2} = \sum\limits_{i=1}^{3} M^{y}(F_{\text{lef.i}}) + \sum\limits_{i=1}^{3} M^{y}(F_{\text{rig.i}}); \\ J_{z}\dfrac{\partial^2 \phi_{z}}{\partial t^2} = \sum\limits_{i=1}^{3} M^{z}(F_{\text{lef.i}}) + \sum\limits_{i=1}^{3} M^{z}(F_{\text{rig.i}}), \end{cases} \quad (1)$$

where t – time; $F_{lef.i}$ and $F_{rig.i}$ – the forces perceived by the suspension of a timber truck from the wheels of the left "lef." and right "rig." sides; M^{i} – moments of the considered forces relative to the axis i.

To carry out the initial assessment studies, a simplified mathematical model of the wheel was used, in which the deformation of the wheel was not taken into account, and it was also taken into account that the change in height from the frame of the forest car to the supporting surface of the forest road is associated with the functioning of the suspension. The support force from the side of the wheel to the frame of the logging vehicle was determined taking into account their viscoelastic interaction according to the formula:

$$F_{\text{i}}^{z}=c_{\text{i}}\left(z_{\text{rig.i}}\left(x_{i},y_{i}\right)+R_{\text{wh.}}-z_{\text{wh.i}}\right)-d_{\text{i}}\left(\frac{\partial z_{\text{wh.i}}\left(x_{i},y_{i}\right)}{\partial t}-\frac{\partial z_{\text{wh.i}}}{\partial t}\right), \tag{2}$$

where i – the wheel number; c_i – the stiffness coefficient of viscoelastic interaction; $z_{rig.i}(x, y)$ – the vertical coordinate of the surface of the forest road under the wheel of the forest truck, equal to the coordinate of the lower point of the wheel; $z_{wh.i}$ – the vertical coordinate of the point of attachment of the wheel to the frame of the forest car; $R_{wh.}$ – the radius of the wheel of the forest car; d_i – the viscous-elastic interaction damping coefficient.

The wheels of the forest car perceive the adhesion forces to the supporting surface of the forest road, in order to prevent the forest car from tipping over, the condition for the transverse component of the adhesion forces to the forest road required for this value to be met. The traction force acting on the side of its driving wheels also affects the forest car frame. These forces, acting comprehensively on a forest car, contribute to its movement in the three-dimensional space of the mathematical model.

The system of second-order differential equations describing the system of equations of motion for a forest car does not have an analytical solution in the general case, due to the impossibility of the required substitution of nonanalytic perturbing functions acting from the supporting surface of the forest road. In this regard, to solve this system of equations with the required accuracy, a second-order numerical Runge-Kutta method is used. This method consists in discretizing the time t into equal intervals, denoted by the variable τ, with a step Δt. At each step of integration, the calculation of forces and moments acting from the wheels through the suspension onto the frame of the forest car is performed. Further, according to the known coordinates and speeds of the frame of the forest car at the current integration step, the coordinates and speeds are calculated at the next integration step. Numerical integration for the x-cartesian component is performed as follows:

$$\begin{cases} x_{\tau+1}=x_{\tau}+v_{x\tau}\cdot\Delta t+\dfrac{F_{x\tau}}{m}\cdot\dfrac{(\Delta t)^{2}}{2}; \\ v_{x\tau+1}=v_{x\tau}+\dfrac{F_{x\tau}}{m}\cdot\Delta t, \end{cases} \tag{3}$$

where x_τ, $v_{x\tau}$, $x_{\tau+1}$, $v_{x\tau+1}$ – the coordinates and speeds of a forest car of mass m along the x axis in the previous and subsequent integration steps.

In the process of calculating the previous coordinates and speeds in the following, they find the functions of the time dependence of the angles and coordinates of the location of the forest car. Next, an analysis of these functions is carried out in order to assess the characteristics of the slope, amplitude and frequency of oscillations of the frame of the forest car.

To study the rotation of a forest car on a forest road, a horizontal supporting surface was used, described by $f_i(t)$ equal to 0. To study the movement of a forest car in a straight line along a low quality forest road, the relief of the supporting surface consisting of overhangs extending from 2 to 5 m and obstacles with a length of 0.2 to 0.5 m.

Also, in the framework of the mathematical model, the function of the surface height from the coordinate of the contact point $z(x, y)$ was described as a superposition of Gaussian peaks with roughness parameters on the forest road (x_i, y_i), the height of these roughnesses H_i and the standard deviation σ_i of the roughness width:

$$z(x,y)=\sum_{i=1}^{N_{prof.}+N_{let.}} H_i \exp\left(-\frac{(x-x_i)^2+(y-y_i)^2}{\sigma_i^2}\right), \tag{4}$$

where $N_{prof.}$, N_{let} – respectively, the number of protrusions and obstacles.

Next, at each step of numerical integration, the instantaneous value of the recuperated power in the suspension of a timber truck was determined:

$$N_{rec.}(t)=\sum_{i=1}^{6}\left(v_{zwh_i}\right)^2 d_0\left(\frac{D_{hyd.}}{D_{hyd.0}}\right)^2, \tag{5}$$

where i – the wheel number of the forest car; $v_{zwh.i}$ – component of the speed of the forest truck along the z axis of the wheel mounting point; d_0 – the damping coefficient; $D_{hyd.}$ – the diameter of the hydraulic cylinder of the regenerative suspension mechanism.

After that, the calculation of the maximum $N_{rec.max.}$ and the average $N_{rec.av.}$ of the recovered power was performed according to the function $N_{rec.}(t)$:

$$N_{rec.max.}=\max_{t_{st.turn}<t<t_{en.turn}}\left(N_{rec.}(t)\right); \tag{6}$$

$$N_{rec.av.}=\frac{1}{t_{en.turn}-t_{st.tern}}\int_{t_{st.turn}}^{t_{en.turn}} N_{rec.}(t)dt, \tag{7}$$

where $t_{st.turn}$ and $t_{en.turn}$ – time points at which the turn of the forest car begins and ends.

At the end of the formation of the mathematical model, according to the indicators calculated above, acting as criteria for optimizing the parameters, the diameter of the hydraulic cylinder $D_{hyd.}$ of the regenerative suspension mechanism was determined:

$$N_{rec.av.}\left(D_{hyd.}\right)\rightarrow \max. \tag{8}$$

To implement the developed mathematical model in order to assess the effectiveness of the regenerative suspension of a forest car, a computer program was developed in Object Pascal language in the Delphi 7 programming environment, which allows to reveal the dependences of the influence of time, speed and diameters of the regenerative suspension hydraulic cylinders on the value of the regenerative power in various conditions of movement of the forest car [20].

To study the effect of the diameter of the hydraulic cylinder $D_{hyd.}$ of the regenerative suspension mechanism on the intensity of energy accumulation of the working fluid, as well as on the efficiency of smoothing the vibrations of the frame of the logging truck, we performed a series of experiments in the developed computer program in which the diameter of the hydraulic cylinder $D_{hyd.}$ was changed with a step of 10 mm from 30 to 100 mm. To study the effect of the speed v of a logging car when it turns on the values of the recovered power $N_{rec.max.}$ and $N_{rec.av.}$, a series of experiments was carried out in the developed computer program in which the speed v was varied with a step of 10 km/h in the range from 30 to 90 km/h.

3. Results

It was revealed that the maximum values of recuperated power $N_{rec.max.}$ and $N_{rec.av.}$ are observed when the diameter of the hydraulic cylinder $D_{hyd.}$ is in the range from 50 to 60 mm «figure 2, *a*». With an increase in the diameter of the hydraulic cylinder $D_{hyd.}$, the damping of the oscillations of the frame of the logging truck «figure 2, *b*, *3* – $D_{hyd.}$ = 100 mm» improves, which leads to a deterioration in the effect of energy recovery in the suspension hydraulic cylinder, due to the small amplitude of its stroke. With a decrease in the diameter of the hydraulic cylinder $D_{hyd.}$ «figure 2, *b*, *2* – $D_{hyd.}$ = 60 mm and *1* – $D_{hyd.}$ = 30 mm»", the amplitude of the piston stroke in the suspension hydraulic cylinder increases when the timber truck hits an obstacle and the recovered power increases. However, in this case, the amount of accumulated working fluid decreases with each movement of the piston of the suspension hydraulic cylinder.

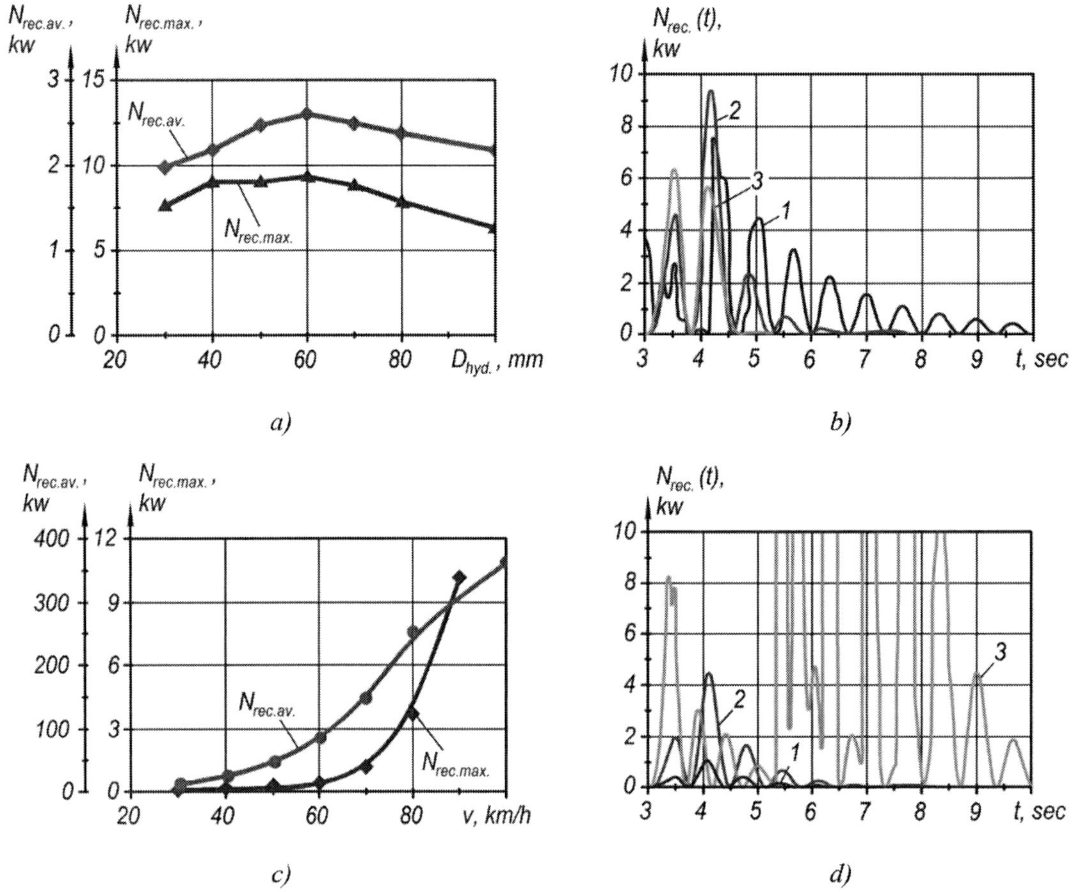

Figure 2. Dependences of the influence of time, speed, and diameters of regenerative suspension hydraulic cylinders on the values of regenerative power in various conditions of movement of a timber truck.

The results of computer experiments made it possible to determine that with an increase in the rotation speed v of the logging truck, there is a significant increase in the recuperated power $N_{rec.max.}$ and $N_{rec.av.}$ (figure 2, *c*). It was revealed that the maximum peak of recuperated power is affected by the largest peak of fluctuations in the suspension of the logging truck, obtained with the maximum mass transfer of the logging truck to one of its sides (figure 2, *d*; *1* – v = 30 km/h, *2* – v = 50 km/h, *3* – v = 90 km/h). In the process of changing the speed of the forest car when it turns from 70 to 90 km/h, an increase in regenerative power is observed from 40 to 340 kW (figure 2, *c*). For real rates of entry into the turn of a timber truck, the regenerative power will be 5–8 kW. It was also revealed that with an increase in the turn-through speed of the logging truck, there is an increase in the average recovered capacity (figure 2, *c*). At the same time, the values of the average recuperated power for the real rates of entry into the turn of the forest car are in the range from 1 to 4 kW.

4. Conclusion

Thus, the developed mathematical model, and a computer program based on it, made it possible to evaluate the effectiveness of the regenerative suspension of a timber truck in various conditions of its movement. The use of a suspension with energy recovery of the working fluid in a forest car will enable it to increase the recuperated power in the range from 1 to 7 kW when driving along a forest road with a large number of turns. In the recuperative suspension mechanism, the most rational is the use of a hydraulic cylinder with a diameter of 50 to 60 mm, which, at the highest possible speed of the forest car, will allow reaching the maximum value of the recuperated power with effective damping of the vibrations of its frame.

References

[1] Posmetev V I and Nikonov V O 2017 *World of transport and technolog. machines* **3(58)** 27–34
[2] Nikonov V O and Posmetev V I 2018 *Voronezh Scientific and Technical Bulletin* **2(24)** 20–39
[3] Gao Z P, Chen S Z, Zhao Y Z and Liu Z 2019 *Energy* **170** 521–536
[4] Zheng P and Gao J W 2019 *Mathematical biosciences and engineering* **16(6)** 6298–6318
[5] Jia Y, Li S S and Shi Y 2018 *Energies* **11(11)** 1–15
[6] Guo Z J and Chen Y C 2017 *Proceedings of the 2017 3rd international forum on energy, environment science and materials (IFEESM 2017), AER-Advances in Engineering Research* **120** 1714–1720
[7] Lv X, Ji Y, Zhao H, Zhang J, Zhang G and Zhang L *Energies* **13(441)** 1–14
[8] Sun Z, Wang R, Meng X and Jiang Q 2020 *Lecture Notes in Electrical Engineering* **586** 216–225
[9] Yu W and Wang R 2019 *Energy* **189**
[10] Toghavifar H and Rokheja S 2019 *Proceedings of the Institution of Mechanical Engineers, Part D: Journal of Automobile Engineering* **233** 2687–2700
[11] Zou J, Guo X, Abdelkareem M A A, Xu L and Zhang J 2019 *Mechanical Systems and signal processing* **127** 345–369
[12] Zhang J, Peng H, Xue J, Sun Y, Zhang J and Peng Z 2018 *Journal of Vibration, Measurement and Diagnosis* **38(4)** 772–779
[13] Zhao L, Wang Y, Song X, Liu Q and Song J 2018 *Journal of Harbin Institute of Technology* **50(7)** 66–72
[14] Wen G and Zhou C 2018 *Automotive Engineering* **40(5)** 575–583
[15] Wang R, Jiang Q, Ya Q, Chen L and Meng X 2017 *Transactions of the Chinese Society for Agricultural Machinery* **48(8)** 350–357
[16] Zhang H, Guo X, Hu S, Fang Z, Xu L and Zhang J 2017 *Transactions of the Chinese Society of Agricultural Engineering* **33(16)** 64–71
[17] Chen L, Zhang C, Wang R, Ye Q and Niu L 2017 *Transactions of the Chinese society for Agricultural Machinery* **48(1)** 303–308
[18] Nikonov V O, Posmetev V I 2018 *Energy Efficiency and Energy Saving in Modern Production and Society: proceedings of an international scientific and practical conference. Part I. Voronezh* 216–224
[19] Arnold V I 2017 *Mathematical methods of classical mechanics* 416
[20] Posmetev V I, Nikonov V O and Posmetev V V 2019 *Certificate of state registration of a computer program № 2019610199*

Estimation of impact of rock conditions on the conveyor workings geometry by means of geophysical methods

E Pudov[1], E Kuzin[1]and A Efremenkov[2]

[1] Branch of T.F. Gorbachev Kuzbass State Technical University in Prokopievsk, Nogradskaya str., 19a, 653039, Prokopievsk, Kemerovo region, Russia
[2] Yaroslav-the-Wise Novgorod State University, ul. B. S.-Peterburgskaya, 41 173003, Veliky Novgorod, Russia

[1] E-mail: pudov_evgen@mail.ru

Abstract. This paper presents the results of a comprehensive survey of a coal-bearing rock mass conducted by means of a combination of geophysical methods. The task of a comprehensive survey is to establish the relationships between the manifestations of mining and geological processes in the preparatory workings and the parameters of electromagnetic emission and high-frequency radio wave scanning. The estimation of the parameters of unsteady geophysical fields, resulting in the stress-strain state of the rock mass, was carried out using the ANGEL-M complex. The assessment of decompression and the presence of fracturing of rocks around the belt road was performed by the OKO-2 GPR system. Analysis of the results of the examination of underground mine workings using the "ANGEL-M" and "OKO-2" complexes shows a high degree of convergence of the state of the workings and instrument readings. Cave-in places of in the workings are located in high amplitude regions of radio waves. Identification of the development sites prone to geodynamic manifestations will allow to take measures for strengthening the frame support in advance, preventing deformation of the extensible belt conveyor. The implementation of a comprehensive survey will improve the reliability of transport conveyor lines and the safety of mining operations.

1. Introduction

The first experience in predicting the rock-bump hazard at mines that develop layers prone to rock burst shock has shown that conventional (direct) methods for determining the stress state of rock mass (the dust coal output method) are very time-consuming and non-operational. As a rule, their information content and reliability are limited to individual local sites, and require a large amount of drilling of trial boreholes for a comprehensive assessment of the rock-bump state over vast areas (for example, a breakage face).

In order to overcome such difficulties, the country's leading institutions (All-Russian Research Institute of Mining Geomechanics and Survey, the "VostNII" Research Center on Industrial and Environmental Safety in the Mining Industry, Kuzbass State Technical University) have developed geophysical methods for assessing the stress state of the rock mass and predicting the rock-bump hazard of coal seams and ore deposits based on the generalization and analysis of long-term observations made at coal mines in Russia. The majority of these methods have been tested for many years and have taken a worthy place in the list of the most promising geophysical methods for predicting rock-bump hazard at mining enterprises of Russia.

Content from this work may be used under the terms of the Creative Commons Attribution 3.0 licence. Any further distribution of this work must maintain attribution to the author(s) and the title of the work, journal citation and DOI.
Published under licence by IOP Publishing Ltd

One of the main advantages of geophysical methods is the ability to estimate the average stress in a sufficient volume, often without drilling trial boreholes in the rock mass, which greatly facilitates and accelerates the process of evaluating its stress state and rock-bump hazard.

2. Problem Statement

The demand for geophysical methods for monitoring the rock-bump hazard, especially at the present stage of ore and coal deposits development, is largely due to their (methods) technological capability. The qualitative changes that occurred in the element base and in computer technologies significantly contributed to the process of developing new effective technological schemes for implementing and using geophysical methods [1–7]. The list of geophysical equipment developed for the practical application of various geophysical methods for predicting the rock-bump hazard of coal seams has significantly changed and expanded.

This article contains the results of a comprehensive survey of the belt road with the use of geophysical methods on the example of the 3-2-7 "Sibirginskaya" mine.

Figure 1 presents a section of the 3-2-7 lava-mining plan, which shows a part of the belt road bearing the same name that is subject to complete cave-in in the specified places, having an intense manifestation of heaving.

The main task of a comprehensive survey is to identify the relationships between the current / past manifestations of mining and geological processes and the results of readings of such geophysical survey devices as ANGEL-M (determination of electromagnetic emission) and the OKO-2 GPR (based on high-frequency radio-wave scanning).

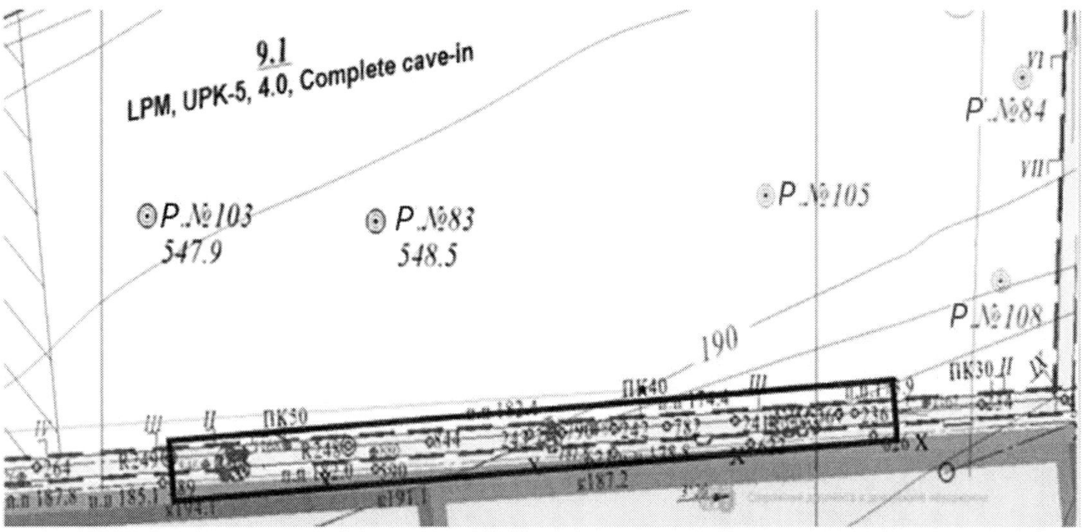

Figure 1. The section of the lava-mining plan for the 3-2-7, belt road.

The ANGEL-M device was used to measure the site of the 3-2-7 belt road at the range of 54–34 pickets.

3. Findings

A summary diagram of the ANGEL-M readings for the soil from the 3-2-7 belt road is shown in figure 2. Such a comprehensive presentation of the results makes it possible to evaluate changes in the readings of the device, both depending on the position of the device and measurements along the workings and depending on the depth. The preliminary assessment of the situation clearly indicates an increased level of readings in the contour rock mass. The influence of anthropogenic noise is highly likely.

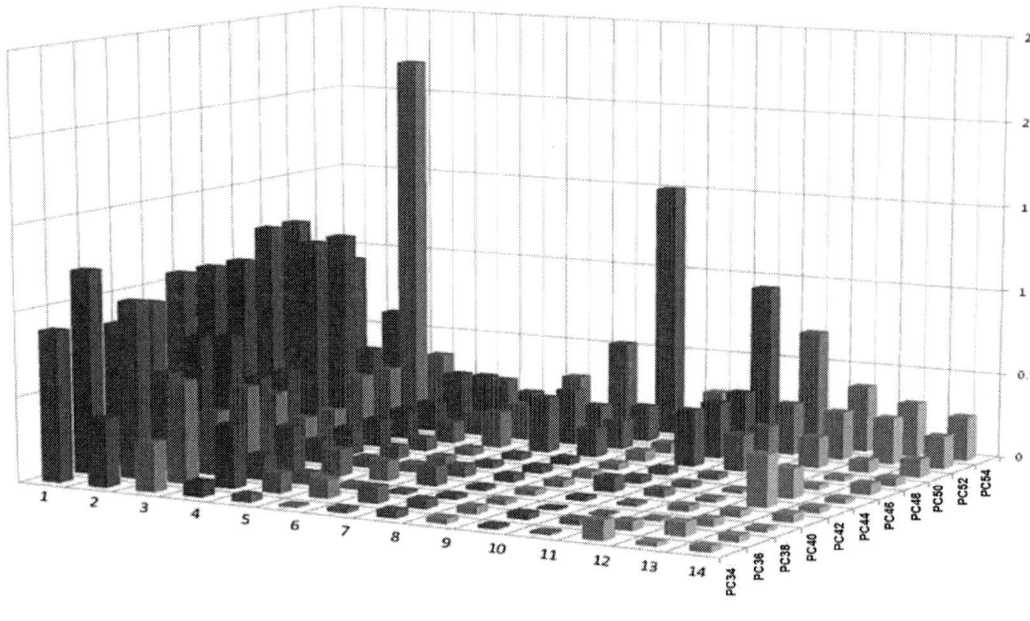

Figure 2. A summary diagram of the ANGEL-M device readings when examining the soil of the 3-2-7 belt road.

Figure 3 clearly shows the dependence of the level of the received signal not only on the depth, but also on the site in the workings being examined. When moving away from the lava breakage face, the signal level decreases.

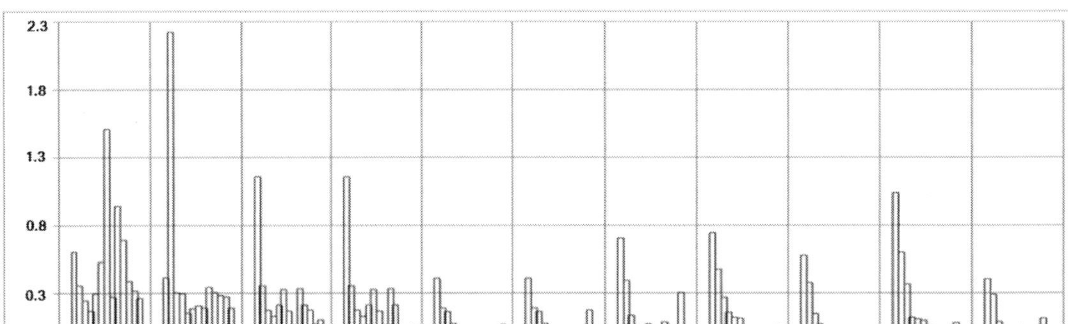

Figure 3. Picket distribution of the ANGEL-M readings with AES function depending on the depth.

Figures 4 and 5 show the total and average signal intensity received by the ANGEL-M device with the AES function.

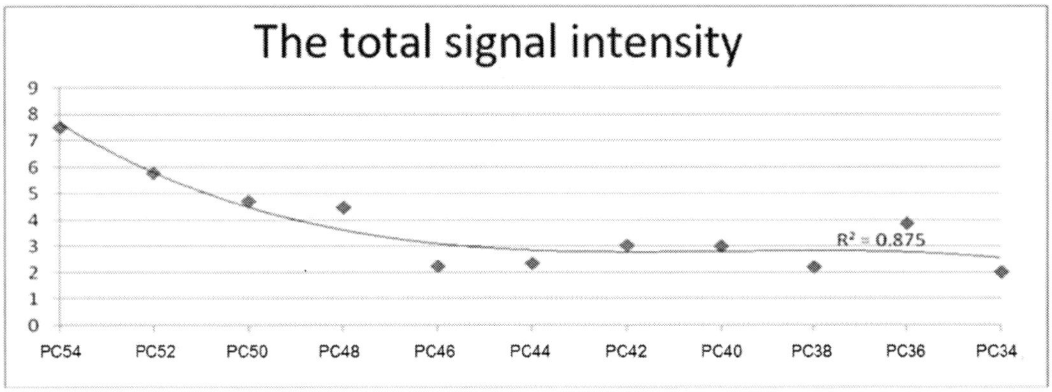

Figure 4. The total signal intensity received by the ANGEL-M device with the AES function.

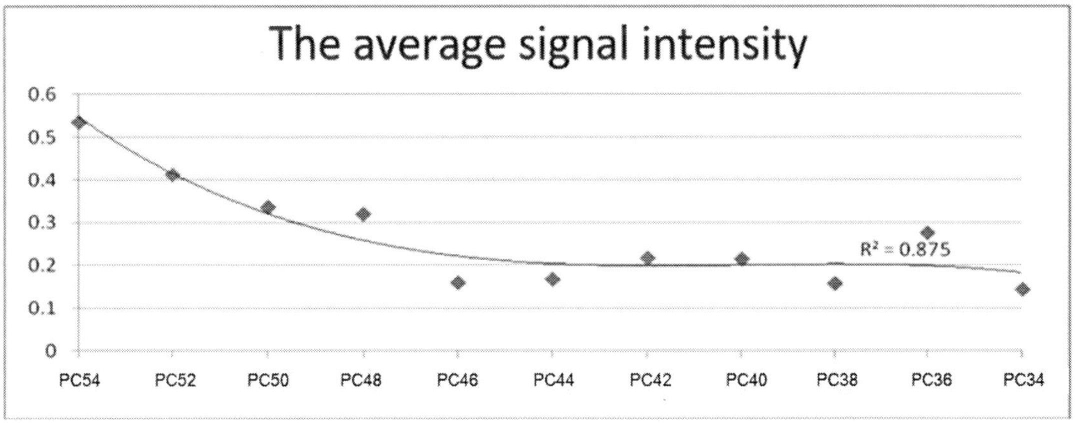

Figure 5. The average signal intensity received by the ANGEL-M device with the AES function.

The graphs constructed on the base of approximating the results also indicate an increased level of readings at a distance of up to 100 meters near the front of working face.

In this case, the maximum value of the amplitude of a high-frequency radio wave passing through a rock mass is taken as the amplitude.

The diagram on figure 6 also presents the obvious dependence of the reading values on the position relative to the developed site. At all the surveyed depths, the amplitude of the passing wave decreases with the distance from the breakage face.

Presumably, the amplitude with proper calibration for specific mining and geological conditions, it is possible to assess the stress state (tendency to cave in) of the rock mass, and consequently to timely take the necessary measures to ensure the proper level of safety.

Figure 7 and figure 8 show the values of the total and average signal amplitudes.

When comparing the results of the two devices used – ANGEL-M with the AES function and OKO-2, it is possible to conclude with a high degree of probability that the results correlate with each other.

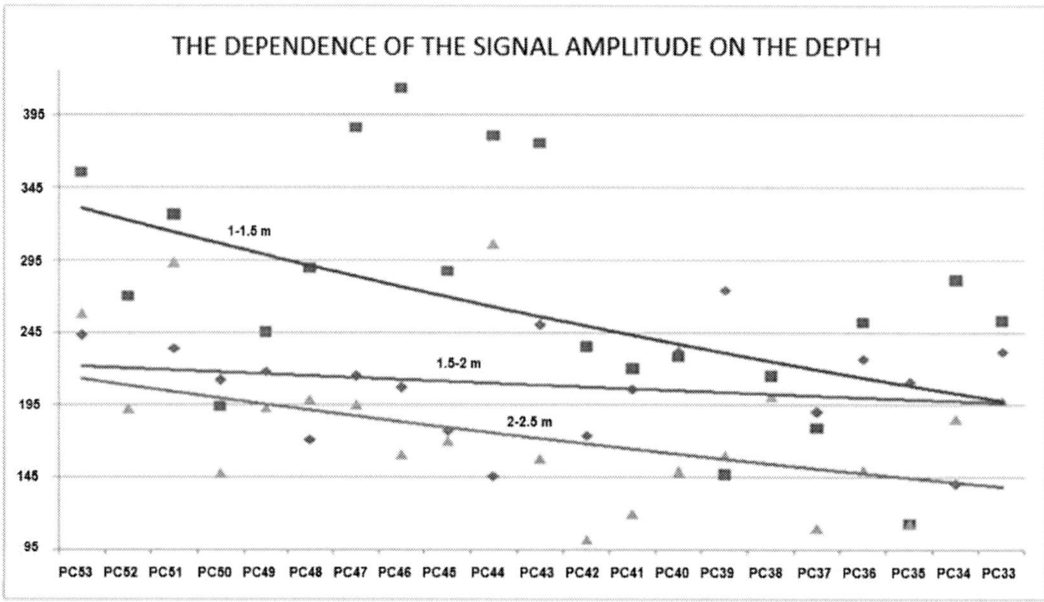

Figure 6. The diagram presenting the results of georadiolocation studies of the soil from the 3-2-7 belt road.

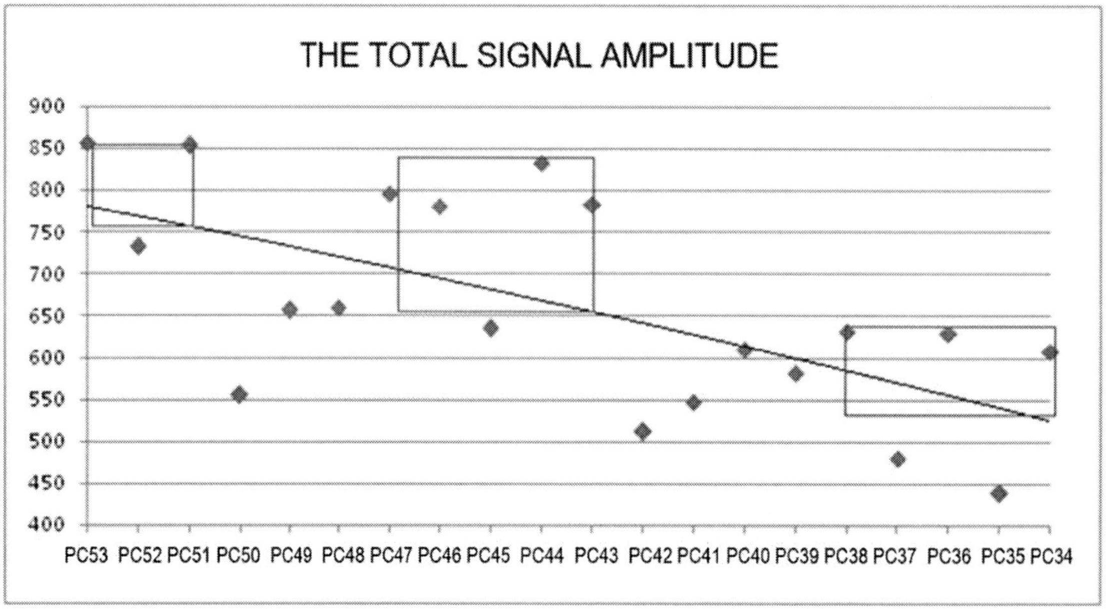

Figure 7. The values of the total signal amplitude of the georadiolocation survey of the soil from the 3-2-7 belt road.

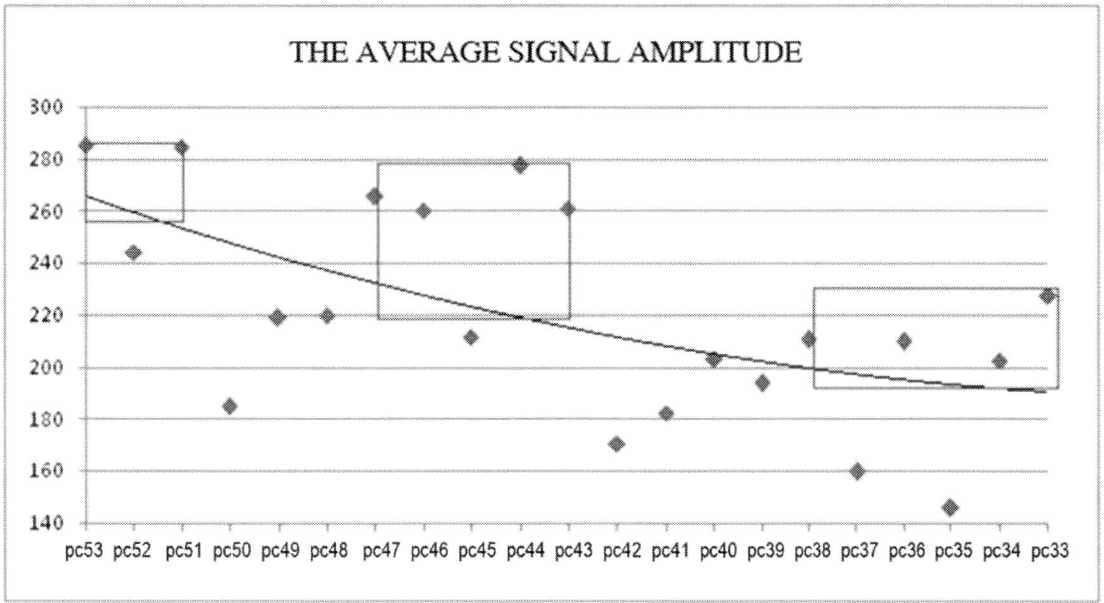

Figure 8. The values of the average signal amplitude of the georadiolocation survey of the soil from the 3-2-7 belt road.

The comparison of the state of workings presented in the plan on figure 1 and the diagrams on figure 7 and figure 8 diagrams is of particular interest. The high convergence of the cave-in sites indicated on the plan and the excess of the amplitudes of passing radio waves relative to the average (approximated) level can be clearly seen on them.

4. Conclusion

As a result of conducting a comprehensive survey using several geophysical methods, the following dependencies were identified:

– the results of surveys of workings using the ANGEL-M with the AES function and the OKO-2 devices indicate the presence of dependencies between the rock conditions and the data obtained;

– the front of the wrking face of the 3-2-7 belt road of the "Sibirginskaya" mine is detected at a distance of up to 100 m;

– the high degree of convergence of the state of workings and the readings obtained indicate the possibility of identifying high-stress and prone to cave-in zones of the rock mass;

– the excess of the amplitude of the radio wave passage in georadiolocation studies above the average value with a high degree of convergence coincides with the sites critical for the possibility of cave-in.

Based on the comparison of the geophysical survey data obtained from the wind roadway and belt road, the increased stress in the rock mass is observed at the distance of 80–120 m from the front of the working face. In places where less stress was recorded, unloading has already occurred in the form of falls and cave-in.

In this regard, it is relevant and timely to apply such a comprehensive approach for systematic monitoring of workings, identifying critical values for stresses simultaneously with readings obtained using these methods.

If there are known (identified by a series of measurements) threshold values of geophysical methods and planned measurements of indications, it is possible to determine critical areas of workings that are prone to cave in.

The analysis of soil heaving and lining deformation. As a result of the simulation analysis, it was revealed:

– internal stresses in the enclosing carbon-hydrogen mass in comparison with 35 and 55 pickets along the belt road present a similar situation. This indicates that the emerging phenomena associated with the collapse and fall of the roof, have little correlation with the progress of the working face front;

– the coefficient of residual strength is distributed in the section along the belt road in such a way that it directly indicates the possibility of deformations of workings, similar to those that have already occurred.

Figure 9, 10 shows sketches of the belt road geometry after the dynamic phenomena that occurred in March 2018.

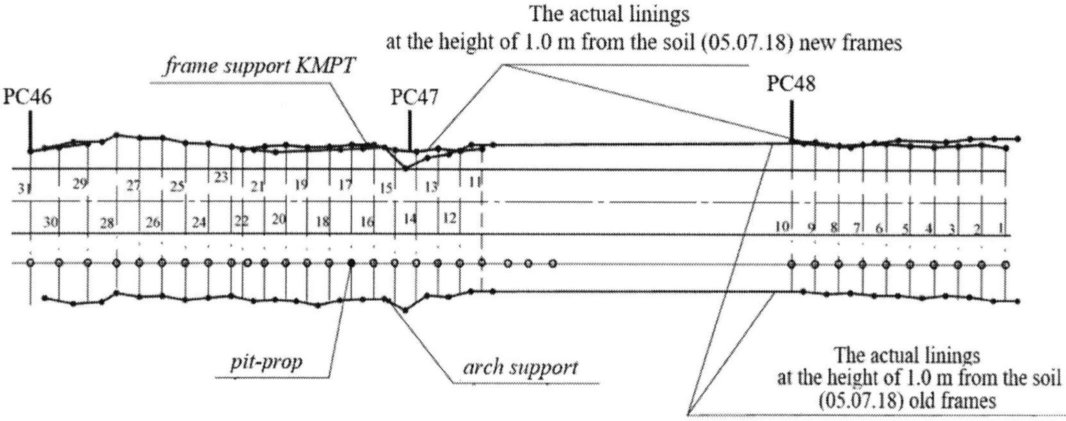

Figure 9. The geometry plan of the belt road with the designation of the axes of the reinforcing frame support (05.07.2018).

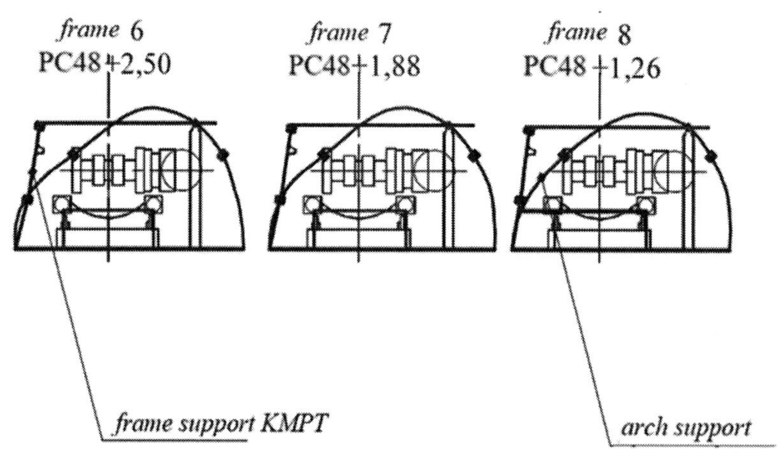

Figure 10. The geometry of the belt road in cross-section along the axes of the reinforcing frame support (05.07.2018).

The stress distribution pattern (figure 11) shows that the presence of workings and the first lava layer of 3-1-7 changes the uniformity of the resulting stresses from horizontal to vertical layers, which is due to the presence of anthropogenic voids that contribute to the discharge directions.

When comparing figure 10, 11, a complete convergence of the dynamic phenomena that occurred and the results of simulation using finite element analysis is revealed.

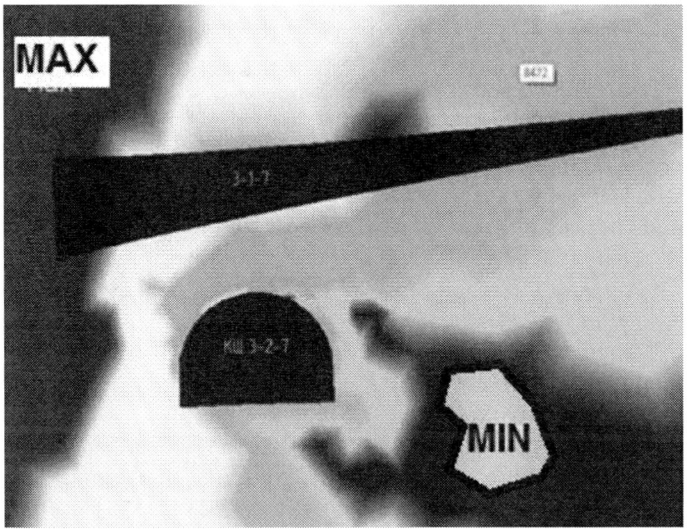

Figure 11. The pattern of relative distributions of internal stress values in the rock mass.

The results of simulation calculations indicate a critical level of internal stresses in the rock mass, which fully decompresses a significant part of the coal-bearing rock mass containing the considered workings. With such patterns of critical internal stress distributions, any dynamic phenomena can contribute to the development of destructive or critical loads for the support. Thus, it can be concluded that at a certain stage of operation of the workings under consideration, there were external factors that critically affected the increase in internal stresses in the hydrocarbon mass.

As a result of geomechanical processes that contribute to the movement of rock masses, the main vectors of free movement are the roof on the side opposite to the extraction pillar, as well as the soil – which was observed in regard to the fact of heaving.

Changes in external factors (an increase of cave-in in rock mass on the surface of the ground, the probability of the rock burst shock), which significantly increase the stresses affecting the hydrocarbon mass, could not be taken into account when making recommendations for fastening, which led to a critical discrepancy between the loads and the load-bearing capacity of the frame support.

References

[1] Sheng Li Wang, Qiu Shen Wang, Lei Li, Jing Ke Ren and Jin Ying Zhao 2012 Geological Hazard Zoing Evaluation of Typical Sections of Mountain Roads *Applied Mechanics and Materials* **170–173** 1252–1256 DOI:10.4028/www.scientific.net/AMM.170-173.1252

[2] Xu S, Xiao L, Li C and Wang M 2017 Status review on two kinds of typical geological hazard zoning techniques in China *Journal of Natural Disasters* **26** 22–31 DOI: 10.13577/j.jnd.2017.0203

[3] Leonovich A 2018 *Geophysics* **1** 32-33 DOI:10.4213/kvant397

[4] Lacazette A, Vermilye J, Sicking Ch and Geiser P 2013 *Direct Imaging of Natural Fracture Flow Networks at the Reservoir Scale* 2nd EAGE Workshop Naturally Fractured Reservoirs: Naturally Fractured Reservoirs in Real Life cp-371-00039 DOI: 10.3997/2214-4609.20132019

[5] Akhmetov E, Kasimov K and Zhumatayeva M 2020 Investigation of accidents at hydraulic structures and methods for monitoring their safety *Bulletin of Tomsk Politechnic University of Engineering Georesources* **331** 70–82 DOI: 10.18799/24131830/2020/4/2595
[6] Razorenov Yu I and Dzeranov B V 2017 Parameters of Rock Structures for Underground Ore Mining *Structural Mechanics of Engineering Constructions and Buildings* **6** 55–67 [In Russ.]
[7] Abramovich A, Pudov E and Kuzin E 2017 Prerequisites for the Establishment of the Automated Monitoring System and Accounting of the Displacement of the Roof of Underground Mines for the Improvement of Safety of Mining Work *E3S Web of Conferences* **21** 01011
[8] Mogilatov V S, Antonov E Yu and Shein A N 2019 3d tomographic inversion of tem sounding *Russian Geology and Geophysics* vol 60 **1** 97–107 DOI: 10.15372/GiG2019007
[9] Antonov E Y, Mogilatov V S and Epov M I 2019 Effect of transmitter current waveform on transient electromagnetic responses *Russian Geology and Geophysics* vol 60 **4** 492–499 DOI: 10.15372/RGG2019027
[10] Golik V I, Gabaraev O Z, Maslennikov S A, Khasheva Z M and Shulgaty L P 2016 The provision of development conversion perspectives into undeground one for russian iron ore deposits development *The Social Sciences* vol 11 **18** 4348–4351
[11] Sudakova M S, Sadurtdinov M R, Tsarev A M, Skvortsov A G and Malkova G V 2019 Ground-penetrating radar for studies of peatlands in permafrost *Russian Geology and Geophysics* vol 60 **7** 793–800 DOI: 10.15372/GiG2019059
[12] Loginov G N and Petrov A M 2019 Automatic detection of geoelectric boundaries according to lateral logging sounding data by applying a deep convolutional neural network *Russian Geology and Geophysics* vol 60 **11** 1319–1325 DOI: 10.15372/GiG2019134
[13] Umirova G K, Istekova S A and Modin I N 2018 Results of magnetotelluric surveyon reference geophysical profiles of the Karaton-Sarkamys block in Kazakhstan *Russian Geology and Geophysics* vol 59 **1** 96–104 DOI: 10.15372/GiG20180107
[14] Cheremisin A A, Vasil'ev Y V, Olenchenko V V, Epov M I, Toib R E, Shnipov I S, Shirokov S V and Boltintsev V B 2018 Potentialities of ultrawideband GPR in low-resistivity geoenvironments *Russian Geology and Geophysics* vol 59 **2** 206–215 DOI: 10.15372/GiG20180209
[15] Epov M I, Shurina E P, Shtabel E P and Shtabel N V 2016 Three-dimensional modeling of pulsed soundings using a fast Fourier transform *Russian Geology and Geophysics* vol 57 **2** 329–336 DOI: 10.15372/GiG20160209

Development of a new type of metamaterial and its application in antenna technology

A N Salakhitdinov[1] **and Kh B Mirzokulov**[1,2]

[1] Department of Telecommunication engineering, Samarkand branch of Tashkent University of Information Technology, 47 A Sh.Mirzo Street, Samarkand, Uzbekistan

[2] E-mail: liverpool_2592@mail.ru

Abstract. In this paper, we consider a method for creating left-sided metamaterials. We select a method for creating metamaterials from a system with distributed parameters that are non-resonant structures. The equation of a discrete transmission line is solved taking into account the conditions for creating a metamaterial. The resulting solution in the form of effective permittivity shows that such a medium really has a double negative value of both the dielectric and magnetic permittivity. A generalized scheme of a universal method for searching for creating metamaterials is considered.

1. Theoretical part

This part summarizes the method for creating left-sided metamaterials, and selects a method for creating metamaterials from a system with distributed parameters that are non-resonant structures. The equation of a discrete transmission line is solved taking into account the conditions for creating a metamaterial. The resulting solution in the form of effective permittivity shows that such a medium really has a double negative value of both the dielectric and magnetic permittivity. A generalized scheme of a universal method for searching for creating metamaterials is considered.

The principal possibility of creating superdirectional antennas appeared after the creation of artificial materials with unique electrophysical characteristics ($\varepsilon(\omega)$, $\mu(\omega)$) – metamaterials (MTM).

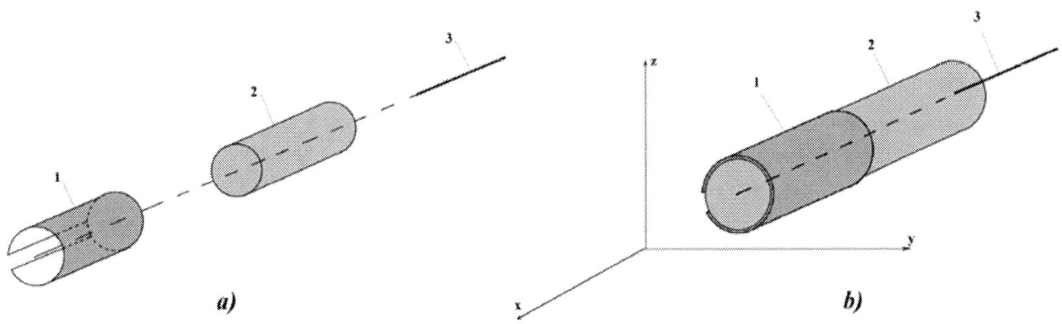

Figure 1. Elements of the metastructure cell (a), a piece of a broken single-wire communication channel (1), a metal rod (2), a dielectric filler (3); an elementary metastructure cell(b).

Content from this work may be used under the terms of the Creative Commons Attribution 3.0 licence. Any further distribution of this work must maintain attribution to the author(s) and the title of the work, journal citation and DOI.
Published under licence by IOP Publishing Ltd

Metamaterial creation can be attributed to nonlinear dynamic phenomena, since the" medium " of the metamaterial must have dispersion properties ($\varepsilon(\omega)$, $\mu(\omega)$), moreover, the higher the frequency of irradiation, the easier it is to create a metamaterial. On the other hand, it is shown below that a metamaterial appears exclusively in an open system when all elements of the metasystem are interconnected, and therefore there is a connection between the output and input of the system. It is shown below that creating a metamaterial using an open communication line is of great importance in antenna theory. Thus, the creation of a metamaterial is really a dynamic process of exposure to electromagnetic radiation with metamaterial [5]. Note that the metamaterial is located in the near zone of electromagnetic radiation for the manifestation of meta-properties.

In this paper, in contrast to [1], we draw up a diagram of elementary cells (figure 1). The scheme is voluminous, so for simplicity of calculations, an equivalent system is shown only on the z axis.

In the diagram in figure 2 the metastructure is created from pieces of a copper tube of a certain length, a single-wire cable communication line, where copper cylindrical conductors are used as metal "rods" (figure 1a), and the conductors are located strictly parallel, which again simplifies model calculations. Cylindrical segments made of polyvinyl chloride are considered as an effective medium. Since the "medium" must be continuous, all segments of the cable line and polyvinyl chloride are fixed with a silicone-based glue, which has ε_0^d very close to ε_0 polyvinyl chloride.

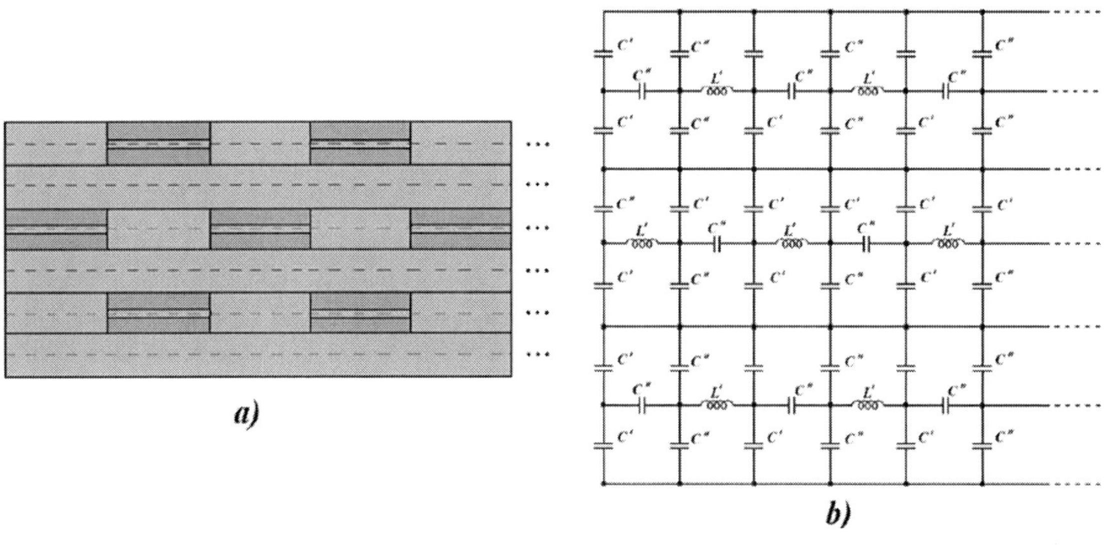

Figure 2. Principal (a) and equivalent (b) scheme of the metamaterial.

Consider a system in the form of a metamaterial consisting of elements of a long transmission line, where there are also traveling waves and reverse waves. Following the work [2], we can associate the effective parameters of the "medium" with the average values of electric and magnetic fields.

Model calculations are made using the following formulas, which are consistent with the theory of the effective medium [3, 4, 5, 6, 7]:

$$\vec{D} = \varepsilon_{eff} \vec{E}; \quad \varepsilon_{eff} = \frac{Z_0^2}{Z^2}; \tag{1}$$

$$Z(\omega) = \sqrt{\frac{\mu_{MTM}(\omega)}{\varepsilon_{MTM}(\omega)}} = Z_0\sqrt{\frac{1+\mu_{MTM}(\omega)}{\varepsilon_{MTM}+\varepsilon_{eff}}} = Z_0\sqrt{\frac{1}{\varepsilon_{eff}(\omega)}},$$

$$n_{eff}(\omega)=\sqrt{\varepsilon_{eff}}+\sqrt{\left(\varepsilon_{eff}(\omega)+\mu_{MTM}(\omega)\right)\left(1+\mu_{MTM}\right)}.$$

From these equations we find

$$\begin{cases}\mu_{MTM}(\omega)=\dfrac{1}{\sqrt{\left|\varepsilon_{eff}(\omega)\right|}}-1\\ \varepsilon_{MTM}(\omega)=\left|\varepsilon_{eff}(\omega)\right|\mu_{MTM}(\omega)\end{cases}. \quad (2)$$

The formula from the works of Nader Engheta (2) [8] is remarkable in that, as we show, it allows us to transfer the effective environment method for systems with distributed parameters (back-wave) (figures 1–2).

For a single-wire line, as we have shown [3], the following results can be obtained, summarized in table 1.

Table 1. The values of and for different values of $\left|\varepsilon_{eff}'\right|$.

Conditions	$e^{\alpha}=b+\sqrt{b^2-1}$ $\omega>2\omega_p$	$e^{\alpha}=b+\sqrt{b^2-1}$ $\omega<2\omega_p$	$e^{\alpha}=b-\sqrt{b^2-1}$, $\omega>2\omega_p$	$e^{\alpha}=b-\sqrt{b^2-1}$, $\omega<2\omega_p$
Values of ω	$\omega=2.2\omega_p$	$\omega=1.7\omega_p$	$\omega=2.2\omega_p$	$\omega=1.1\omega_p$
Values pf $\left|\varepsilon_{eff}'\right|$	0.21	0.49	1.21	0.5
$\mu(\omega)$	1.18	0.42	–0.09	0.41
$\varepsilon(\omega)$	0.25	0.21	–0.11	0.205

Table 1 shows that $\omega>2\omega_p$ a meta-environment with two negative parameters is formed $\mu(\omega)=-0.09$, $\varepsilon(\omega)=-0.11$.

In contrast to the work [6], we made an elementary cell (figure 1) and an electrical equivalent circuit (figure 2). the Distinctive feature of this scheme is that it is not a resonant structure, where there are traveling and reverse electromagnetic waves and their interaction.

Note that the properties of a metamaterial are manifested when exposed to external electromagnetic influence. Therefore, as a rule, the "metasredu" is located near the transmitting antenna (in the near field) with a frequency of at least 10 MHz and higher.

In the equivalent scheme (figure 2b), there is a phase shift (along the "y" axis), between the rows of the "x" axis and the "y" axis, located in parallel (figure 1a).

Find the total resistance Z(ω) for the equivalent circuit (figure 2b):

1) Serial connection Z (ω) on the "x" axis (figure 2):

$$Z_i(\omega)=\sum_{i=1}^{9}Z_i=9jX_L.$$

If you consider the capacity between Z_i

$$Z_{ic}=9j\left(X_L-\frac{1}{\omega C_Д}\right),$$

where $C_д$ – is the dielectric constant of polyvinyl chloride.

Since $\frac{1}{\omega C_Д} << \omega X_L$ we will not consider the member $\frac{1}{\omega C_Д}$ further.

$$Z^x_{eff}(\omega) = 9jX_L = 9Z_1.$$

2) Parallel connection on the "y" axis
Then find the value $Z_{eff}(\omega)$ by $Z_L >> Z_C$

$$\left|Z^{xy}_{eff}(\omega)\right| = \sqrt{(Z^{xy}_1(\omega))^2 + (Z^{xy}_2(\omega))^2 + \ldots + (Z^{xy}_9(\omega))^2}.$$

For our case $\left|Z^{xy}_{eff}(\omega)\right| = 5.24Z_C$.

3) Parallel connection on the "z" axis:

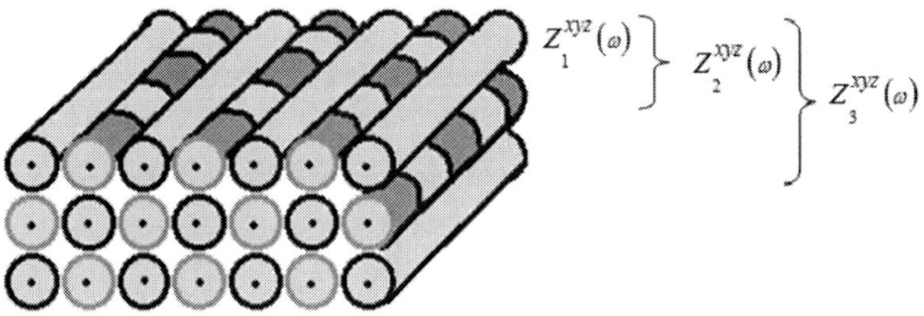

Figure 3. On the left the metastructure (figure 2), on the right, the value of the total resistance relative to the "xyz" axis.

For the first layer $Z^{xyz}_1(\omega) = \left|Z^{xy}_{eff}(\omega)\right| = 5.24Z_C$.

$$Z^{xyz}_2(\omega) = \frac{Z^{xyz}_1(\omega) \cdot Z^{xyz}_1(\omega)}{Z^{xyz}_1(\omega) + Z^{xyz}_1(\omega)} = \frac{5.24Z_C \cdot 5.24Z_C}{5.24Z_C + 5.24Z_C} = 2.62Z_C,$$

$$Z^{xyz}_3(\omega) = \frac{Z^{xyz}_2(\omega) \cdot Z^{xyz}_1(\omega)}{Z^{xyz}_2(\omega) + Z^{xyz}_1(\omega)} = \frac{2.62Z_C \cdot 5.24Z_C}{2.62Z_C + 5.24Z_C} = 1.7Z_C,$$

$$Z(\omega) = \left|Z^{xyz}_{eff}(\omega)\right| = \sqrt{\left(Z^{xyz}_1(\omega)\right)^2 + \left(Z^{xyz}_2(\omega)\right)^2 + \left(Z^{xyz}_3(\omega)\right)^2} = 6.1Z_C. \quad (3)$$

Next we find $|\varepsilon_{eff}|$ the entire metamaterial

$$|\varepsilon_{eff}| = \left|-\left(\frac{\left(1+\frac{1}{2Z(\omega)}\frac{\omega^2}{\omega_p^2}\right)}{\sqrt{\left(0{,}25\frac{\omega^2}{\omega_p^2}-1\right)}}\right)^2\right|. \quad (4)$$

For our case when $Z(\omega) = 6.1Z_C$ we find the value $|\varepsilon_{eff}|$

$$|\varepsilon_{eff}| = \left|-\left(\frac{\left(1+0.08\frac{\omega^2}{\omega_p^2}\right)}{\sqrt{\left(0{,}25\frac{\omega^2}{\omega_p^2}-1\right)}}\right)^2\right|. \quad (5)$$

When ${}^{\omega}/_{\omega_p} = 2.2$ we find the value $|\varepsilon_{eff}|$:

$$|\varepsilon_{eff}|=9.16.$$

Then using the formula (2) we find the values $\mu_{MTM}(\omega)$ и $\varepsilon_{MTM}(\omega)$

$$\begin{cases}\mu_{MTM}(\omega) = -0.67 \\ \varepsilon_{MTM}(\omega) = -6.1\end{cases}.$$

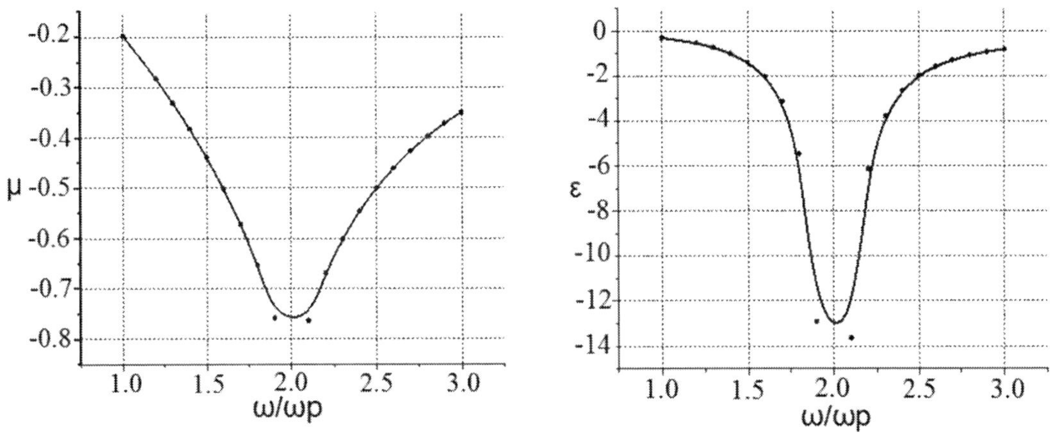

a) b)

Figure 4. Graphs of dependences of: a) magnetic permeability; b) dielectric permittivity on frequency for an equivalent circuit (figure 2).

Figure 4 shows the graphs of theoretical calculated dependences of a) magnetic permeability and b) dielectric permittivity on frequencies according to the formulas (3), taking into account (2).

It can be seen that the calculated data give double negative values (μ<0, ε<0), which indicates the possibility of creating a metamaterial in the entire assumed frequency range.

2. Practical part

During the experimental part of the work, a sample of metamaterial was created using the dynamic-evolutionary method from the elements of the communication line. The main metal element is copper. Copper tubes with a diameter of 0.6 cm are cut into parts of 0.4 cm, which are put on the communication line (figure 5b, 5c), then the elements are cut longitudinally in order to get an open communication line. The next element is a system of conductors acting as the main potential line and insulation in the form of polyvinyl chloride to create an open capacitor. The next step is to combine elements to create a metastructure with a common surface of 5.5x8.5 sm^2 (figure 5a).

The elements of the metamaterial must be arranged symmetrically and the places of breaks are located with a turnover of 180^0 between them. This is done in order to achieve phase variability. Next, a regular communication line (without a metaelement) and a line with metaelements are alternately assembled in one row. The metamaterial is made in three rows, and the order of the set of elements changes in each row. This is done in order to achieve phase variability. All elements of the metamaterial are connected with each other using a silicone-based glue.

When creating a metamaterial, you must strictly observe symmetry and order, and only in this case can you achieve maximum efficiency of the metamaterial.

IOP Conf. Series: Materials Science and Engineering **939** (2020) 012064 doi:10.1088/1757-899X/939/1/012064

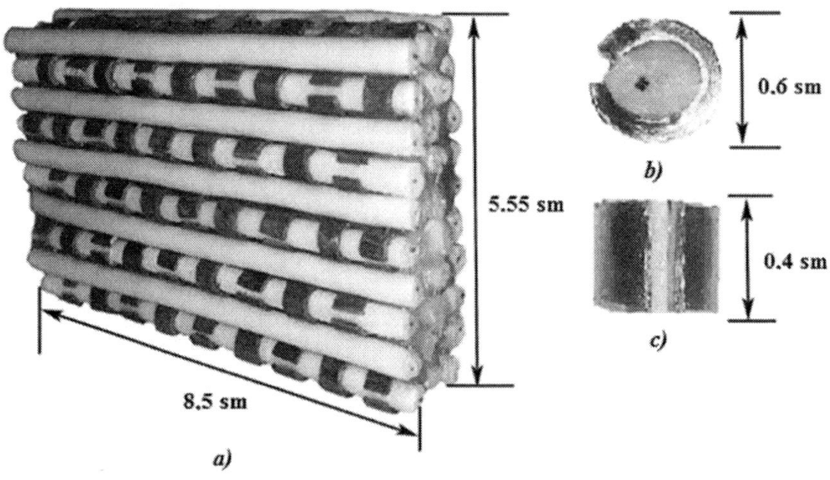

Figure 5. a) Metamaterial in the form of a piece of cable line with $Z_0 = 50\ Om$; b) metamaterial element in the form of a longitudinally cut piece of cable line top view; c) metamaterial element side view.

The installation corresponds to the schematic experiment (figure 6) in the sense that the signal generator is taken as a generator of the Rohde company & The Schwarz SM 300, which operates in the frequency range from 3 KHz to 3 GHz.

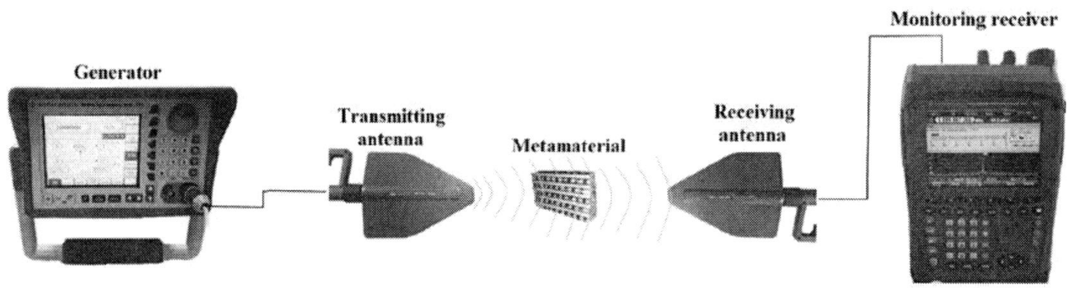

Figure 6. Installation assembled to study the effectiveness of metamaterial relative to antenna feeder devices.

When measuring, special attention should be paid to the location of the metamaterial between the transmitting and receiving antenna. When making measurements follow the following rules:

1. First of all, all elements (generator, antennas, and portable frequency analyzer) are connected using a cable ($Z_0 = 50\ Om$) with appropriate connectors.
2. The system of communication at close distances is initially adjusted without MM, to the maximum sensitivity of the receiver, in order to observe all changes when using the metamaterial.
3. Between the antennas of the transmitter and receiver is the MM without any sources connected to the MM.
4. At the same time, the metamaterial must be installed in such a way as to achieve maximum efficiency. Observations have shown that the distance from the transmitting antenna to MM varies at different frequencies.

5. We determine the signal parameters (field strength, signal strength) in a wide frequency range, first without metamaterial and then with metamaterial.

The main measured characteristic of MM is the relationship between the signal level and frequency, the intensity of the electric (or magnetic) field with a frequency in the entire measured range of waves of the transmitter from 0.6 to 2.4 GHz (figure 7), which shows the dependence of the amplified signal on the frequency of the transmitter.

Let's look at these charts in more detail:

– figure 7 shows the dependence of the signal level with frequency in the frequency range $0.6 \div 2.4\ GHz$ with a frequency range of 0.1 GHz. When comparing the amplified signal from the presence of MM, it can be seen that in the frequency range $0.6 \div 1.2\ GHz$ and $1.7 \div 2.4\ GHz$, the transmitter signal is actually amplified when present, and these amplifications reach the maximum value at frequencies $2.2 \div 2.4\ GHz$.

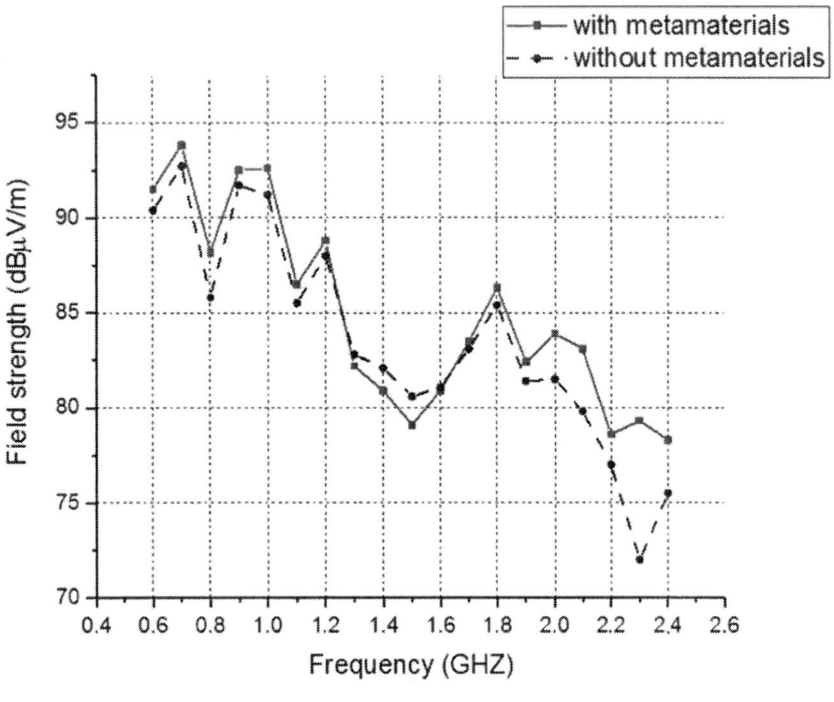

Figure 7. Results of experimental works: the solid line, the medium with the metamaterial, the dashed line environment without metamaterial.

The average gain of the transmitter is 4 dBμV/m (decibel microvolts per meter), the maximum value at 2.24 GHz is about 7 dBμV/m. Note that the gain is achieved in a wide range of frequencies $0.6 \div 1.2\ GHz$and $1.7 \div 2.4\ GHz$.

Experimental results are given for the selected model method in the frequency range of the generator $450 \div 3GHz$. The main thing: the metamaterial is not connected to any electrical circuits, and the distance between the transmitting and receiving antenna is selected, so as to get the maximum signal gain. In this research work, the signal gain of the order of more than 7 dbμv/m was obtained.

References

[1] Panchenko B A and Nefedov E I 1986 *Microstrip antennas* (Moscow: Radio i svyaz')
[2] Machulyansky A, Verbitskiy V and Yakimenko Y 2016 *Proc. of the IEEE 36th Int. Sci. Conf. Electronics and Nanotechnology (Kiev)* p 140–43

IOP Conf. Series: Materials Science and Engineering **939** (2020) 012064 doi:10.1088/1757-899X/939/1/012064

[3] Salakhitdinov A N and Mirzokulov Kh B 2019 *Dynamics of systems, mechanisms and machines* **7** 2
[4] Yashin M M, Yurasov A N, Ganshina E A, Garshin V V, Semenova D V, Mirzokulov Kh B and Danilov G E 2019 *Herald Of the Bauman Moscow state technical university* **83** 5
[5] Yurasov A, Yashin M, Ganshina E, Granovsky A, Garshin V, Semonova D and Mirzokulov Kh 2019 *J. of Phys.: Conf. Ser.* **1389** 012113
[6] Panchenko B A and Gizatulin M G 2010 *Nano antennas* (Moscow: Radiotekhnika)
[7] Marqués R, Martín F and Sorolla M 2008 *Metamaterials with Negative Parameters: Theory, Design and Microwave Applications* (USA: Wiley)
[8] Engheta N and Ziolkowski R W 2006 *Metamaterials: Physics and Engineering exploration* (USA: Wiley-IEEE Press)

IOP Conf. Series: Materials Science and Engineering **939** (2020) 012065 doi:10.1088/1757-899X/939/1/012065

The influence of coatings heat resistance on the effectiveness of protecting a solid metal surface from drops of molten metal during MAG welding

S B Sapozhkov[1,2] **and S I Arendateleva**[1]

[1] Yaroslav-the-Wise Novgorod State University, ul. B. St. Petersburgskaya, 41 173003 Veliky Novgorod, Russia

[2] E-mail: ssb@novsu.ru

Abstract. A technique is proposed for determining the heat resistance of protective coatings. The use of this technique allowed determining the most effective coatings in terms of heat resistance, for example, PZ-1 and PZ-2, which can be recommended for widespread use in welding production.

1. Introduction

One of the requirements for protective coatings [1–9] is to ensure reliable protection of welded products from drops of molten metal during MAG welding. For this purpose, heat-resistant fillers are introduced into them.

However, the methods for determining the heat resistance of protective coatings are currently insufficiently studied. This paper presents the results of a study of the influence of both the components that make up the coatings and the protective coatings themselves on the temperature of a calorimetric system with a previously known effective heat capacity, based on experimental data, coating compositions are determined that most effectively protect the surface of the welded products from drops of molten metal during MAG welding.

It is based on the method for determining the heat of solid fuels combustion. Since the specific heat of coal combustion is a known value, coal was chosen as the main component, and the coating components and coatings themselves served as additives to coal.

2. The main part

The studies were carried out on a calorimeter B – 08 – MA, it is a device consisting of the calorimeter itself and the recorder. The principle of operation of the calorimeter is to measure the temperature change of the calorimetric system with a previously known effective heat capacity when burning a strictly defined amount of the test substance. For experiments, briquettes weighing 1.0 g were made using a press, a general view of which is shown in figure 1. A loaded bomb (figure 2) was installed in the calorimetric vessel shown in figure 3, the rest of its cavity was filled with distilled water. After the explosion (burning) of the test substance in an oxygen medium, temperature changes were recorded according to a metastatic thermometer using the optical system that equipped the device.

Content from this work may be used under the terms of the Creative Commons Attribution 3.0 licence. Any further distribution of this work must maintain attribution to the author(s) and the title of the work, journal citation and DOI.
Published under licence by IOP Publishing Ltd

IOP Conf. Series: Materials Science and Engineering **939** (2020) 012065 doi:10.1088/1757-899X/939/1/012065

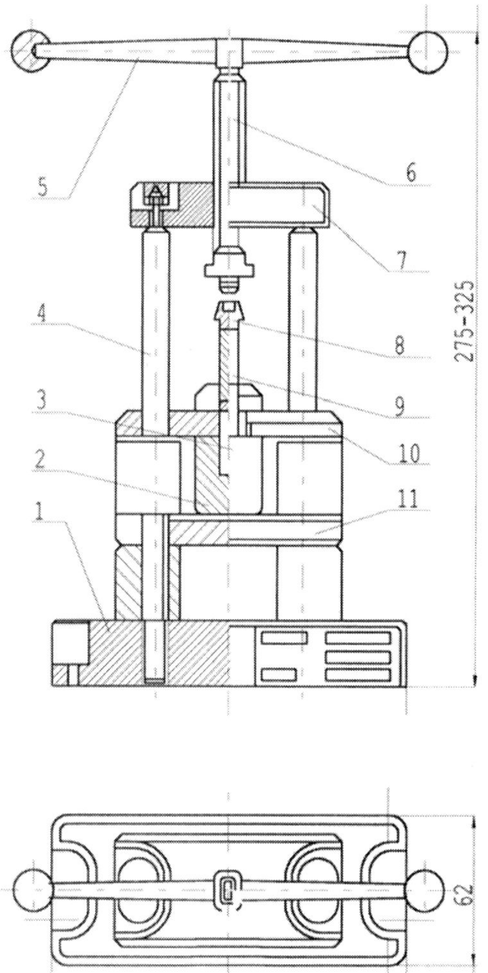

Figure 1. Briquetting press: 1 – base; 2 – matrix; 3 – needle; 4 – rack; 5 – handle; 6 – screw; 7 – a guide; 8 – pusher; 9 – a punch; 10 – a level with a groove; 11 – folding bar.

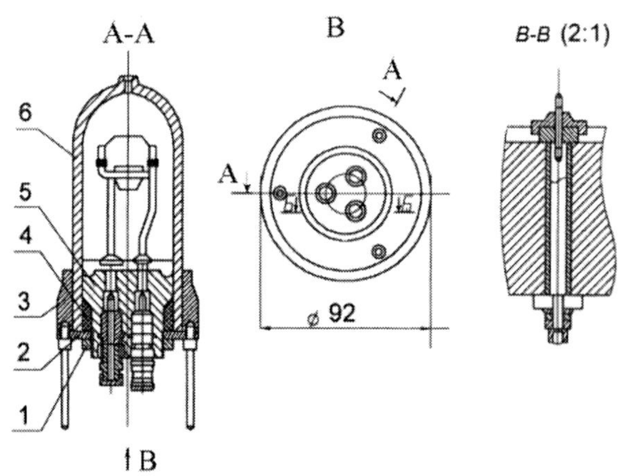

Figure 2. Calorimetric bomb: 1 – locknut; 2 – metal ring; 3 – flare nut; 4 – rubber gasket; 5 – cover; 6 – case.

IOP Conf. Series: Materials Science and Engineering **939** (2020) 012065 doi:10.1088/1757-899X/939/1/012065

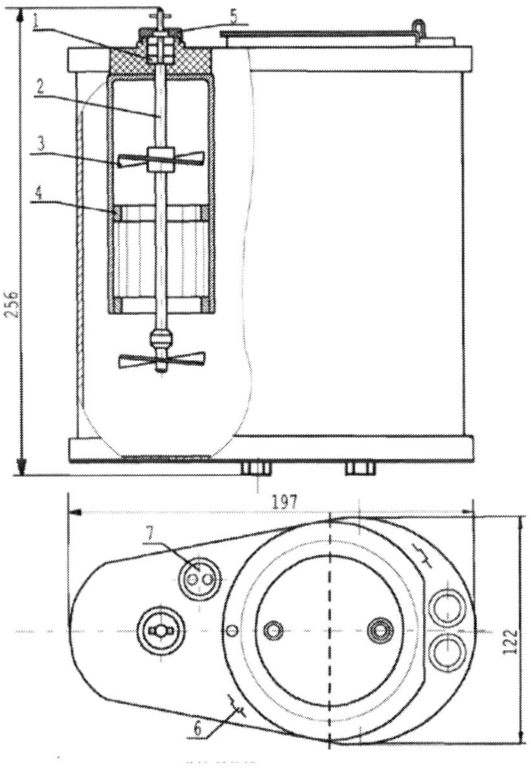

Figure 3. Calorimetric vessel: 1 – bearing; 2 – axis of the mixer; 3 – impeller; 4 – heater; 5 – a nut; 6 – ear; 7 – pin terminal.

The calorific value in the bomb of the tested analytical fuel sample (test substance) (Q_b^a) in J/kg is calculated by the formula:

$$Q_b^a = \frac{\text{с}\cdot\text{К}[(n_2+\Delta n_2)-(n_1+\Delta n_1)+\Delta n]-\text{т}_1 q}{\text{т}}, \tag{1}$$

where c – calorimetric system heat capacity, J/^{0}C;

K – thermometer scale division value, ^{0}C;

n_1– thermometer reading corresponding to the initial temperature of the main period in the divisions of the thermometer scale;

Δn_1 – correction to thermometer readings per gauge (when reading thermometer n_1) in scale divisions;

n_2 – thermometer reading corresponding to the final temperature of the main period in the divisions of the thermometer scale;

Δn_2 – correction to thermometer readings per gauge (when reading thermometer п_2) in scale divisions;

Δn – correction to thermometer readings, taking into account the heat exchange of the calorimeter with the environment in the divisions of the thermometer scale;

т_1– mass of wire taken for igniter, kg;

q – calorific value of the ignition wire burning, J/kg;

т – weight of test fuel sample, kg.

The correction to the thermometer readings, taking into account the heat exchange of the calorimeter with the environment ($\Delta\text{п}$), is calculated by the formula:

$$\Delta n = \frac{v_1 + v_2}{2} \cdot z_1 + v_2 \cdot z_2 \qquad (2)$$

where v_1 – the average rate of change of the thermometer over a half-minute interval in the initial period in the divisions of the thermometer scale;

v_2 – the average rate of change of the thermometer over a half-minute interval in the final period in the divisions of the thermometer scale;

z_1 – the number of half-minute intervals of the main period with a rapid rise in temperature; z_1 always include the first period of the main period, regardless of the magnitude of the rise for this period;

z_2 – the number of half-minute intervals of the main period not included in the number z_1 (if the total number of half-minute intervals of the main period is z, then $z_2 = z - z_1$).

The number z_1 is determined depending on the value of the criterion a, which is calculated by the formula

$$\mathrm{a} = \frac{n_a - n_1}{n_2 - n_1}, \qquad (3)$$

where n_a – thermometer reading corresponding to the temperature after 2 minutes of the main period in the divisions of the thermometer scale;

n_1 – thermometer reading corresponding to the initial temperature of the main period in the divisions of the thermometer scale;

n_2 – thermometer reading corresponding to the final temperature of the main period in the divisions of the thermometer scale.

Then z_1 is defined according to table 1.

Table 1. The dependence of the number $\mathbf{z_1}$ on the criterion *a*.

Criterion *a*	<0.50	0.50…0.64	0.64…0.73	0.73…0.8	0.82…0.91	0.91…0.95	>0.95
z_1	9	8	7	6	5	4	3

We give *an example of the calculation of combustion for a sample of pure coal*: the readings of the metastatic thermometer are given in table 2. The correction to the thermometer reading, taking into account the heat exchange of the calorimeter with the environment (Δп), was calculated by the formula (2) with previously determined criterion *a*:

$$\mathrm{a} = \frac{2.645 - 1.239}{2.720 - 1.239} = 0.95.$$

Using table 1 we find $z_1 = 4$; $z_2 = 10 - 4 = 6$.

We calculate the average rate of change of the thermometer (v_1 and v_2):

$$v_1 = \frac{1.220 - 1.239}{10} = -0.0019; \; v_2 = \frac{2.720 - 2.745}{10} = -0.0025;$$

$$\Delta n = \frac{-0.0019 - 0.0025}{2} \cdot 4 + (-0.0025) \cdot 6 = -0.058.$$

Heat capacity c = 15000 J/°C; thermometer scale division K = 1.001 °C; correction to thermometer readings per gauge $\Delta n_1 = 0$ and $\Delta n_2 = 0$; mass of wire taken for igniter, $\mathrm{т}_1$– $0.0001 \cdot 10^{-3}$ kg; calorific value of the ignition wire burning $q = 2.51 \cdot 10^6$ J/kg, weight of test fuel sample т =0.001 kg.

Next, by the formula (1) we calculate the calorific value:

$$Q = \frac{15000 \cdot 1.001(2.72-1.239-0.058)-0.0005 \cdot 10^{-3} \cdot 2.51 \cdot 10^{6}}{0.001} = 21.879 \text{ MJ/kg}.$$

Table 2. Thermometer reading.

Half-minute reading numbers			Thermometer reading in scale divisions		
Period			Period		
Initial	Main	Final	Initial	Main	Final
1	n_1 11(igniter)	n_2 21	1.221	2.430	2.720
2	12	22	1.220	1.239	2.732
3	13	23	1.223	2.542	2.732
4	14	24	1.225	2.608	2.736
5	n_a 15	25	1.227	2.645	2.738
6	16	26	1.229	2.678	2.740
7	17	27	1.231	2.690	2.742
8	18	28	1.233	2.700	2.743
9	19	29	1.236	2.710	2.744
10	20	30	1.237	2.716	2.745

The calculation data of the calorific value for pure coal and coal with the addition of protective coating components are given in table 3. Figure 4a shows histograms of the calorific value of pure coal and coal with the addition of protective coating components.

Table 3. Calculation of calorific value for pure coal and coal with the addition of coating components [1–4, 8, 9].

No	Component	Calorific value, MJ/kg
1	Coal	21.879
2	Coal + aluminum powder	21.693
3	Coal + kaolin	25.184
4	Coal + liquid glass	18.462
7	Coal + silicone cream	20.397
8	Coal + dextrin	28.925
9	Coal+ mash liquid concentrate	24.954
10	Coal + anhydrous sodium carbonate	19.375
12	Coal + silicon organic liquid	38.835
13	Coal + polyvinyl butyral	28.945
14	Coal + zircon	26.679
15	Coal + chalk	18.963

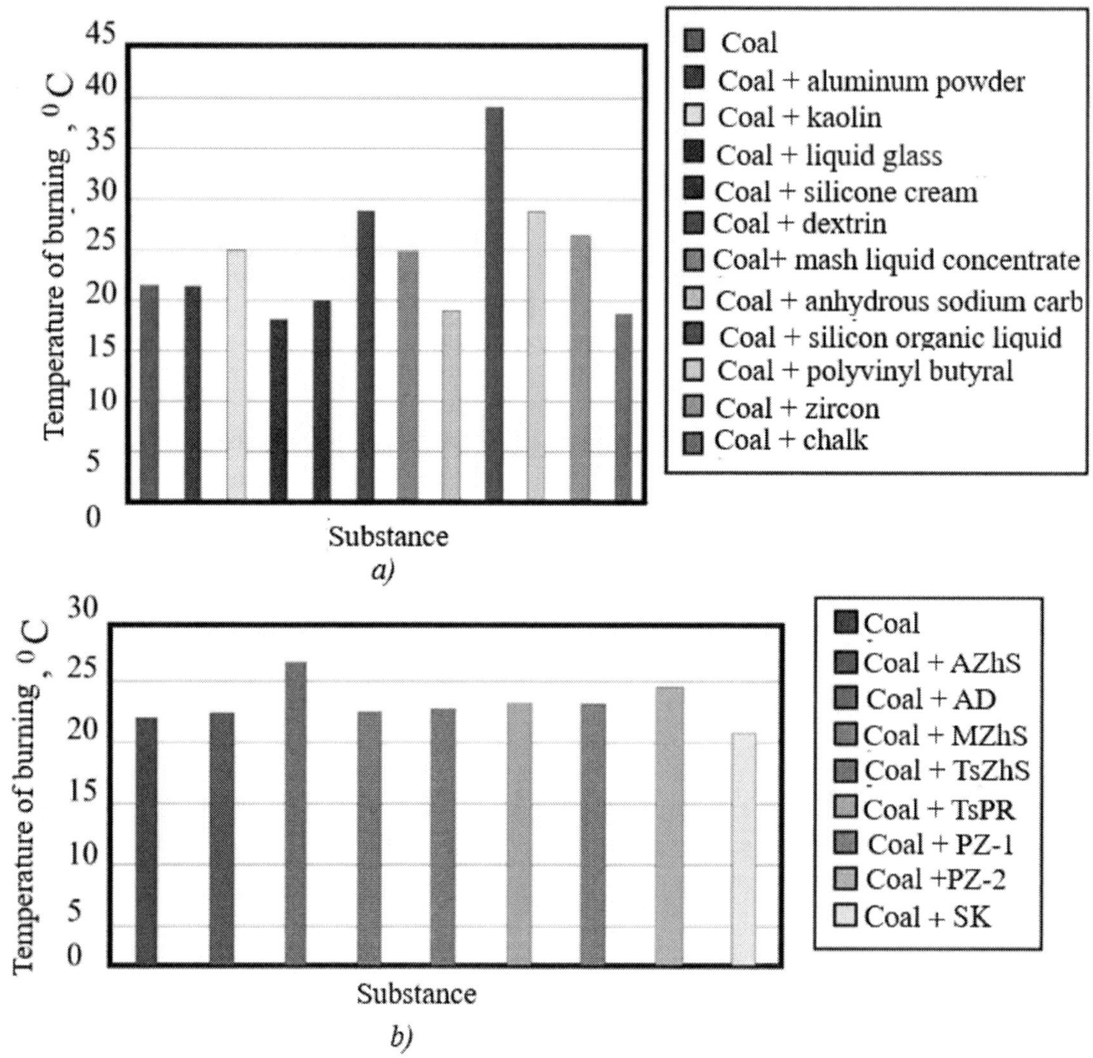

Figure 4. Histograms of the calorific value: a – coal with additives of protective coating components; b – coal with additives of protective coatings.

An analysis of the results of the studies shows the effect of various components of the protective coatings on the specific heat of combustion of coal. Such components GKZH-94 (silicon organic liquid) as polyvinyl butyral, dextrin, zircon, KBZh (mash liquid concentrate), kaolin significantly increase it, whereas silicone cream, anhydrous sodium carbonate, chalk, liquid glass, on the contrary, reduce it. Such component as aluminum powder does not have a noticeable effect. Based on these results, it can be concluded that the heat-resistant coating should contain substances having a higher calorific value. Obviously, a large role will be played by the percentage of these substances in the coating, as well as the effect on them of other substances that make up the coating. For completeness of the analysis, the influence of the compositions of known protective coatings (table 4) on the specific heat of combustion of coal was studied.

Table 4. Common protective coating compositions.

Coating	Composition
MV	Chalk – 35% (GOST 4415–75), water – 65% (GOST 2874–73)
KBZh	Sulphite-alcohol stillage concentrate KBZh – 190 g and water 1000 g
PZ-1	Sulphite-alcohol stillage concentrate KBZh – 75 g, soap – 35 g, anhydrous sodium carbonate – 25 g, water – 1000 g
PZ-2	Sulphite-alcohol stillage concentrate KBZh – 85 g, soap – 35 g, kaolin – 40 g, anhydrous sodium carbonate – 37.5 g and water – 1000 g
AD	Aluminum powder PAK-1 (GOST 5494–71) – 100 g, dextrin (GOST 6034–74) – 100 g and water – 1000 g
MZhS	Chalk – 30%, liquid glass – 65%
AZhS	Aluminum powder PAK-1 – 50 g, liquid glass – 1000 g
TsZhS	Zirkon (TsMTU 4469–54) – 30%, liquid glass – 70%
TsPR	Zirkon – 40%, polyvinyl butyral (GOST 9439–73) – 3%, solvent 646 (GOST 18188–72) – 57%
SK	Silicone cream acc.to RTU 322–57

The experimental technique is similar to that described above. The calculation data of the calorific value for a sample of pure coal, as well as coal with the addition of protective coatings are given in table 5. Figure 4b presents histograms of the calorific value of pure coal and coal with the addition of protective coatings.

Table 5. Calculation of calorific value for pure coal and coal with the addition of coatings.

No	Substance	Calorific value, MJ/kg
1	Coal	21.879
2	Coal + AZhS	22.199
3	Coal + AD	26.726
4	Coal + MZhS	22.319
5	Coal + TsZhS	22.516
6	Coal + TsPR	22.954
7	Coal + PZ-1	23.100
8	Coal +PZ-2	24.358
9	Coal + SK	20.397

As can be seen from the studies, the most heat-resistant coatings are type AD, PZ-1, PZ-2 and TsPR. The following substances give the greatest effectiveness to coatings: dextrin, zircon, kaolin, etc.

3. Conclusion

Using the technique for determining the heat resistance of protective coatings made it possible to determine the most effective coatings in terms of heat resistance: PZ-1 and PZ-2.

The following substances give the greatest effectiveness to coatings: polyvinyl butyral, dextrin, zircon, etc.

References

[1] Fedko V T, Sapozhkov S B, Sokolov P D and Yastrebov A P 2004 *Elements of theory and technology of anti-spatter surface shielding when welding in carbon dioxide:* Monograph. (Tomsk: TSU Publ.) p 138

[2] Solonenko O P, Kudinov V V, Smirnov A V, Cherepanov A N, Popov V N, Mikhalchenko A A and Kartaev E V 2005 Micro-Metallurgy of Splats: Theory, Computer Simulation and Experiment *JSME International Journal* Series B vol 48 **3** 366–380

[3] Knotek O and Elsing R 1987 Monte Carlo Simulation of Lamellar Structure of Thermally Sprayed Coatings *Surf. Coat. Technol* **32** 261–271

[4] Ghafouri-Azar R, Mostaghimi J, Chandra S and Chamchi M A 2003 Stochastic Model to Simulation of a Thermal Spray Coating *Journal of Thermal Spray Technology* vol 12 **1** 53–69

[5] Fedko V T, Tomas K I and Sapozhkov S B 1998 Protecting the surfaces of welded components against molten metal splashes in CO_2 welding *Welding International* **1** 58–62

[6] Vaz C T, Bracarense A Q, Felizardo I and Pereira Pessoa E C 2012 Impermeable low hydrogen covered electrodes: Weld metal, slag, and fumes evaluation *J Mater. Res. Technol.* **1(2)** 64–70 DOI: 10.1016/S2238-7854(12)70012-1

[7] Sapozhkov S B and Burakova E M 2018 *IOP Conf. Ser.: Mater. Sci. Eng.* **441** 012042

[8] Sapozhkov S B, Burakova E M and Tesleva E P 2016 *IOP Conf. Ser.: Mater. Sci. Eng.* **142** 012019

[9] Saraev Y N, Chinakhov D A, Ilyashchenko D I, Kiselev A S and Gordynets A S 2017 Investigation of the stability of melting and electrode metal transfer in consumable electrode arc welding using power sources with different dynamic characteristics *Welding Internat.* **31(10)** 784–790

IOP Conf. Series: Materials Science and Engineering **939** (2020) 012066 doi:10.1088/1757-899X/939/1/012066

Physicochemical patterns of the spread of a molten metal drop over a solid metal surface during MAG welding

S B Sapozhkov[1], M L Tikhonova[1] and Sh S Nozirzoda[1]

[1] Yaroslav the Wise Novgorod State University, ul. B. St.Peterburgskaya, 41 173003 Veliky Novgorod, Russia

[1] E-mail: ssb@novsu.ru

Abstract. In this paper, theoretical and experimental studies of the kinetics of the spreading of liquid drops of molten metal over a solid metal surface in the absence of extraneous (external) driving forces of hydrostatic pressure, gravity, etc. In the kinetic mode, wetting appears as the spreading of liquids on a solid surface. Wetting and spreading are spontaneous processes occurring due to a decrease in the surface energy of the Gibbs system

1. Introduction

The interaction of molten metal drops with a solid metal surface during MAG welding is a complex physicochemical problem, the scientific and applied solution of which has greatly increased in recent years due to the continuously expanding application of this type of welding in industry. The strength of the connection of a drop of molten metal with a solid metal surface depends on the area of their physicochemical interaction [2–5].

Surface phenomena and especially the phenomena of spreading (wetting) of hot metal drops on a solid metal surface during MAG welding are not well understood.

It is known that, when drops of molten metal come into contact with a solid metal surface, various physicochemical processes can occur: corrosion, adsorption decrease in strength due to a sharp decrease in free energy at the "drop-to-surface" interface, and others. The spreading of a molten metal drop on the surface plays a very important role in all these processes. Along with a purely surface distribution, molten metal atoms can penetrate a solid metal surface through regular volume diffusion, as well as through diffusion along grain boundaries and other structural defects, which in some cases leads to strong adhesion of drops to the surface. The patterns of volume diffusion have been studied in detail [5–10], while the issues of the spreading of molten metal drops over the surface during MAG welding, despite their great importance, have been paid much less attention until recently.

2. The main part

If a drop of metal in the molten state spreads over the surface of a solid metal, then the wetting pattern is affected by a number of conditions, from which it is necessary to distinguish:

1) the presence of an oxide film on the metal surface;
2) the presence of affinity of one metal to another, which determines the value of the contact angle;
3) the proximity of the melting points.

Content from this work may be used under the terms of the Creative Commons Attribution 3.0 licence. Any further distribution of this work must maintain attribution to the author(s) and the title of the work, journal citation and DOI.
Published under licence by IOP Publishing Ltd

The high affinity of metals to oxygen, up to very high temperatures, leads to the formation of oxide films on the surface in an air atmosphere. In practice, the presence of an oxide or adsorbed film in metals reduces the adhesion of drops to the surface during MAG welding.

We will consider the influence of the micro relief of a solid surface on the nature of the distribution of liquid metal. When analyzing the behavior of a liquid on a solid surface, the relationship between the quantities σ_s, σ_l and σ_{sl} is usually considered, and (σ_s and σ_l are the specific free surface energies of the solid body and the liquid at the interface with the medium in which the experiment is conducted; σ_{sl} is the free energy at the interface) (see figure 1).

It is assumed that if

$$\sigma_s > \sigma_{sl} + \sigma_l, \tag{1}$$

then, there is complete wetting, i.e. a drop of liquid will spread over a solid surface in the form of a gradually thinning phase layer. Otherwise, i.e. at

$$\sigma_s < \sigma_{sl} + \sigma_l \, , \tag{2}$$

a drop with a finite edge angle is formed. Under these conditions, the spread of liquid metal can occur only by surface diffusion.

Thus, an analysis of the wetting conditions leads to the conclusion that it is possible in principle to have two qualitatively different forms of liquid distribution over a solid surface: spreading (viscous flow in a continuous phase layer) and surface diffusion (atom migration).

However, inequalities (1) and (2) are valid only for a perfectly smooth surface of a solid body. Since in real conditions each solid surface has its specific micro relief characteristic, for the correct description of the liquid distribution it is necessary, along with the physical properties of the system, also to take into account the geometric features of the surface.

P. A. Rebinder [1] considers the additional friction force acting along the contour, the magnitude of which is related to the degree of surface roughness. This force inhibits the advancement of the front of the spreading drop and leads to the fact that the contact angle during spreading is larger than when the drop flows out (wetting hysteresis). B.V. Deryagin, while theoretically investigating the dependence of the contact angle θon the micro relief, came to the conclusion that, under the condition $K cos\theta \geq 1$, liquid can spread on a rough surface along micro cavities and grooves (K is the roughness coefficient, i.e. the ratio of the true surface to the apparent, θ is the contact angle on a perfectly smooth surface).

From the condition $\mathrm{K}\cos\theta \geq 1$ it follows that at acute wetting angles on an ideally smooth surface (i.e., at $cos\, \theta > 0$) it is in principle always possible to create such a micro relief that the liquid will spread

This position can be illustrated by the following scheme: let there be a longitudinal groove on the surface of a solid body with a cross section in the form of an isosceles triangle and with an angle ϕ at the apex (figure 1, a). Let also this liquid form an acute angle θ on a smooth surface of the same material. In the absence of extraneous forces, the spreading is thermodynamically possible when it is accompanied by a decrease in the free energy of the system.

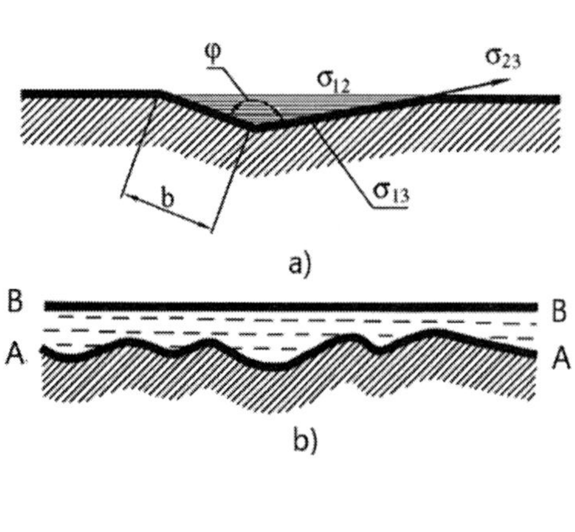

Figure 1. Schemes: *a* – to determine the conditions for the spreading of fluid along the groove; *b* – to estimate the value $\Delta\sigma$. *AA* is the surface of a solid; *BB* is a liquid surface.

In relation to the considered geometry of the groove, spreading along it will take place if

$$\sigma_s \cdot b > \sigma_{sl} \cdot b + \sigma_l \cdot b(\sin \varphi/2). \quad (3)$$

Since for a perfectly smooth surface $\sigma_s - \sigma_{sl} = \sigma_l \cdot \cos\theta$, we obtain the following equation of fluid spreading along the groove:

$$\varphi < 180^0 - 2\theta. \quad (4)$$

Relation (4) thus determines the critical value of the input angle φ at the apex of micro deeps on the surface of a solid body, at which a transition from one form of surface distribution - diffusion, to another – viscous spreading, is possible.

The loss of surface energy during spreading ($\Delta\sigma$) can be estimated as follows. Taking into account the micro relief of a solid surface (figure 1, b)

$$\Delta\sigma = \text{K}\sigma_s - (\text{K}\sigma_{sl} + \sigma_l). \quad (5)$$

In static conditions (motionless drop on a smooth surface) $\sigma_s - \sigma_{sl} = \sigma_l \cdot \cos\theta$. It follows that

$$\Delta\sigma = \sigma_l(\text{K}\cos\theta - 1). \quad (6)$$

All values included in relation (6) are available for direct experimental determination. The value of the roughness coefficient Кcan be found using micro profilograms of a solid surface, the analysis of which allows for a very accurate determination of the value К as the ratio of the actual surface area to "geometric". For example, the roughness coefficients К for various surfaces are as follows [2–3]:

in delivery condition К = 2.95;
sandblasted К = 3.21;
ground К = 2.65;
machined with an emery wheel К = 3.72.

At present, there is no unified theory of the crystal lattice; therefore, the values of the surface energy of solid bodies calculated by different authors vary greatly. The most promising is the thermodynamic method based on the analysis of surface and bulk properties of solids. On this basis, equations are derived that relate the surface tension of the crystal to the surface tension of its own melt. The equation for determining the coefficient of surface tension of solid metals at the boundary with its own melt, expressed in terms of the coefficient of surface melt, is as follows

$$\sigma_l = \sigma_s \frac{L}{\lambda_c} = \sigma_l \sqrt[3]{\left(\frac{\rho_s}{\rho_l}\right)} \frac{L}{\lambda_V}, \quad (7)$$

where σ_{sl} – the surface tension coefficient of the crystal at the boundary with its own melt;
σ_s – crystal surface tension coefficient;
σ_l – melt surface tension coefficient at the boundary with intrinsic vapor;
ρ_s and ρ_l – crystal and melt density, respectively;
λ_c – heat of sublimation;
λ_V – heat of vaporization;
L– heat of melting.

The equations of the surface tension coefficients of solid metals at the interface with intrinsic vapor have the form

$$\sigma_s = \sigma_l \sqrt[3]{\left(\frac{\rho_s}{\rho_l}\right)^2} \frac{\lambda_c}{\lambda_V}. \quad (8)$$

The values of the surface tension coefficients for iron, determined by these equations, are as follows: σ_s = 1582 MJ/m^2; σ_l = 1840 MJ/m^2; σ_{sl} = 46 MJ/m^2.

Knowing the values of the surface tension coefficients and the roughness coefficient, we can calculate the values of the wetting angle of a molten metal drop of various surfaces of the welded products by the following equation:

$$\cos\theta = \frac{\frac{\sigma_S - \sigma_{sl}}{\sigma_l} + 1}{K}. \qquad (9)$$

Table 1. Wetting angle values for various surfaces.

	Surface condition			
	In delivery condition	Ground	Sandblasted	Machined with an emery wheel
Roughness coefficient	2.95	2.65	3.21	3.72
Theoretical wetting angle, deg	51.6	46.1	55.1	60.4
The average value of the experimental contact angle, deg	53.3	47.8	57.7	63.0
Error, %	3.26	3.73	4.82	4.27

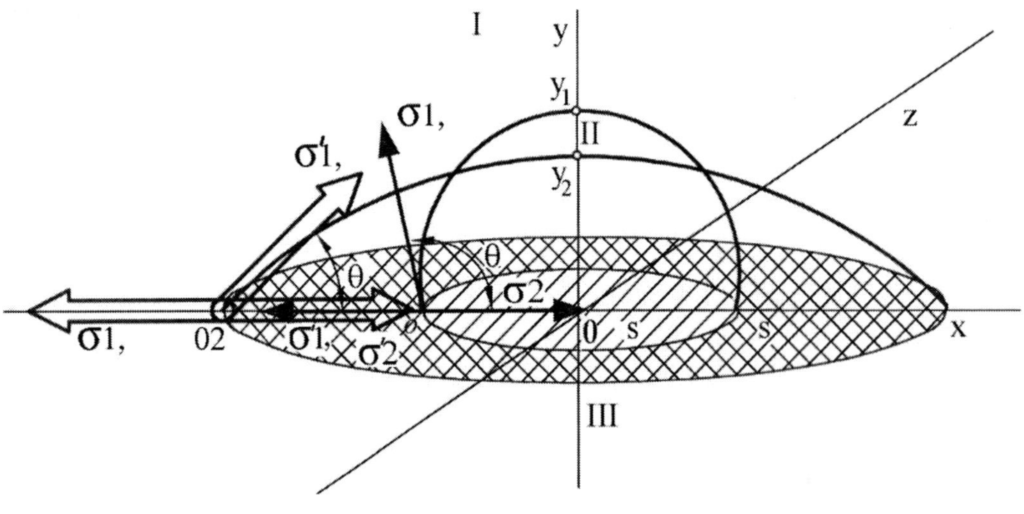

Figure 2. The spreading of the drop when changing the wetting angles and surface tension.

Table 2. The dependence of the wetting angle on the diameter of the drop print.

Diameter of a drop print, mm	The wetting angle, deg			
	The state of the surface of the welded metal			
	Ground	In delivery	Sandblasted	Machined with an emery wheel
1.8	49.0	50.0	59.5	61.0
2.0	41.0	49.0	56.0	73.0
2.2	50.7	50.0	63.5	56.0
2.5	42.0	51.3	59.5	61.25
2.7	51.0	58.0	57.3	54.0
3.0	44.25	51.0	52.5	54.5
3.2	40.0	57.0	54.5	59.0
3.5	39.0	48.0	59.0	65.35
3.8	51.0	45.0	45.0	58.5
4.0	45.6	47.5	60.5	53.22
4.3	38.0	48	58.5	57.0

To confirm the theoretical data on the spreading of a drop on the surface of the metal being welded, experimental studies were performed on the MAG welding of samples made of St3 steel in the delivery condition, machined with an emery wheel, sandblasted and ground. Welding was carried out by a PDG-508UZ semiautomatic device with a VDU-504 power source using Sv-08G2S wire with a diameter of 1.6 mm. Welding modes were $I_w = 320...330$ A, $U_w = 32...34$ V under normal conditions [3, 9].

3. Conclusion

The values of the contact angles for various surfaces of the welded products, determined by theoretical calculation and experimental studies, are given in table. 1. However, as shown by experimental data (table 2), the wetting angle θ also depends on the diameter of the drop print. With an increase in the print diameter, the wetting angle slightly decreases. On the surface treated with an emery wheel, a small angle of wettability is observed, and therefore, the adhesion force of drops of molten metal to the surface will be less. The ground surface is wetted better, which means that the complexity of manufacturing products with a ground surface increases.

References

[1] Rebinder P A 1978 *Poverkhnostnyye yavleniya v dispersnykh sistemakh* [Surface phenomena in disperse systems] (Moscow) [In Russ.]

[2] Fedko V T, Sapozhkov S B, Sokolov P D and Yastrebov A P 2004 *Elements of theory and technology of anti-spatter surface shielding when welding in carbon dioxide: Monograph.* (Tomsk: TSU Publ.) p. 138

[3] Solonenko O P, Kudinov V V, Smirnov A V, Cherepanov A N, Popov V N, Mikhalchenko A A and Kartaev E V 2006 Micro-Metallurgy of Splats: Theory, Computer Simulation and Experiment *JSME International Journal* B vol 48 **3** 366–380

[4] Knotek O and Elsing R 1987 Monte Carlo Simulation of Lamellar Structure of Thermally Sprayed Coatings *Surf. Coat. Technol* **32** 261–271

[5] Ghafouri-Azar R, Mostaghimi J, Chandra S and Chamchi M A 2003 Stochastic Model to Simulation of a Thermal Spray Coating *J. of Thermal Spray Technology* vol 12 **1** 53–69

[6] Steinke T and Bäker M 2006 Monte Carlo Simulation of Thermal Sprayed Coatings *Proc. ITSC'06* (USA, Seattle) 329–334

[7] Fedko V T, Tomas K I and Sapozhkov S B 1998 Protecting the surfaces of welded components against molten metal splashes in CO_2 welding *Welding International* **1** 58–62
[8] Vaz C T, Bracarense A Q, Felizardo I and Pereira Pessoa E C 2012 Impermeable low hydrogen covered electrodes: Weld metal, slag, and fumes evaluation *J Mater. Res. Technol.* **1(2)** 64–70 doi: 10.1016/S2238-7854(12)70012-1
[9] Sapozhkov S B and Burakova E M 2018 *IOP Conf. Ser.: Mater. Sci. Eng.* **441** 012042
[10] Saraev Y N, Chinakhov D A, Ilyashchenko D I, Kiselev A S and Gordynets A S 2017 Investigation of the stability of melting and electrode metal transfer in consumable electrode arc welding using power sources with different dynamic characteristics *Welding International* **31(10)** 784–790

Wetting the surface of the welded products with protective coatings during MAG welding

S B Sapozhkov[1,2] and O V Ushakova[1]

[1] Yaroslav-the-Wise Novgorod State University, ul. B. St. Petersburgskaya, 41 17300 Veliky Novgorod, Russia

[2] E-mail: ssb@novsu.ru

Abstract. In this paper, theoretical and experimental studies of the kinetics of spreading (distribution) of protective coatings when welding MAG on a solid metal surface in the absence of extraneous (external) driving forces of hydrostatic pressure, gravity, etc. In the kinetic mode, wetting appears as the spreading of liquids on a solid surface.

1. Introduction

The interaction of molten metal droplets during welding of MAG with a solid metal surface is a complex physicochemical problem, the scientific and applied solution of which has greatly increased in recent years due to the continuously expanding application of this type of welding in industry. Prevention of the bond strength of molten metal droplets with a solid metal surface is provided by protective coatings that are applied to a solid metal surface during MAG welding [2–5].

Wetting the surface of the metal being welded (solid metal surface) with liquid material (protective coating) during MAG welding is an indispensable condition for its formation. In the kinetic mode, wetting is represented as the spreading of liquids over a surface. Wetting and spreading are spontaneous processes occurring due to a decrease in the Gibbs surface energy [1].

The appearance, continuity, adhesive strength and protective ability of coatings depend on how well the coating wets the substrate and spreads on it.

2. Main part

Ensuring the interaction of dissimilar materials requires contact between them [6–10]. Since the action of forces – orientational, induction, dispersion – manifests itself at a distance not exceeding 0.5 nm, contact can be considered only if the gap is not more than 0.5 nm. The speed at which a specific contact is reached depends on the nature of the surface and the properties of the contacting materials. The contact of solid surfaces with gases is easily achieved, but even when using liquids there are favorable conditions for contacting. The surface topography is important in providing contact. A rough surface can be considered as a multicapillary system. The depth of inflow (rise) of the liquid H into the pores of such a substrate is determined by the forces of capillary pressure:

$$H = \frac{K\sigma\cos\varphi}{\rho r g}, \tag{1}$$

where K – is a constant;

Content from this work may be used under the terms of the Creative Commons Attribution 3.0 licence. Any further distribution of this work must maintain attribution to the author(s) and the title of the work, journal citation and DOI.
Published under licence by IOP Publishing Ltd

σ – is liquid surface tension;
cos ϕ – is capillary inclination angle (wetting angle);
ρ – is liquid density;
r – is capillary (pore) radius;
g – is gravity acceleration.

The rise time τ to establish the hydrostatic equilibrium of the liquid column can be calculated based on the Poiseuille equation:

$$\tau = \frac{2\eta l^2}{r\sigma\cos\varphi}, \quad (2)$$

where l – is capillary (pore) length.

It follows from the foregoing, that the main factors determining the completeness of contact (filling of irregularities and pores of the surface of the substrate) are the viscosity, density and surface tension of the protective coating, the size, shape and location of the surface irregularities. It is important to note that the rate of absorption of protective coatings decreases, however, the possible depth of their penetration increases with a decrease in the diameter of capillaries and cavities. The viscosity of the protective coatings deposited on the substrate varies widely – from several units Pa·s (for solutions) to many thousands Pa·s (for melts). It does not remain constant but increases rapidly when the coating is cured. This makes it difficult to achieve full contact. The penetration of coatings into the recesses of the substrate (cavity) is also prevented by the air in them (figure 1).

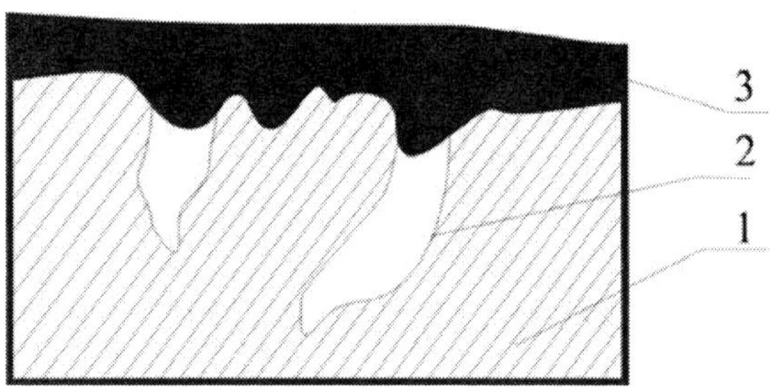

Figure 1. The penetration pattern of the protective coating in surface irregularities: 1 – substrate, 2 – air cavities, 3 – protective coating.

Since the dimensions of the contact surface significantly affect the adhesion of the coatings, there is usually an interest in increasing it. Nevertheless, even on non-porous substrates with a low degree of surface roughness, contact is never absolutely complete.

The surface tension at the solid – liquid interface is indirectly judged by wetting $\theta(\mathrm{C} = cos\ \theta)$.

It is believed that the liquids moisten the solid at $\theta < 90^0$C, and at $\theta > 90^0$C they do not wet.

At a minimum contact angle $[\theta = f(\tau)]$, favorable conditions are created for producing defect-free films due to good spreadability of the material and, therefore, high protective characteristics of the coatings are achieved.

Wettability is determined by the height h and radius r of a drop of material deposited on a substrate and examined under a microscope.

The following is calculated $tg\ \theta$:

For $\theta < 90^0$ (figure 2, *a*)
$$tg\theta = \frac{2rh}{(r^2 - h^2)}; \quad (3)$$

For $\theta > 90^0$ (figure 2, b)
$$tg\theta = \frac{\sqrt{2rh - h^2}}{r - h}. \quad (4)$$

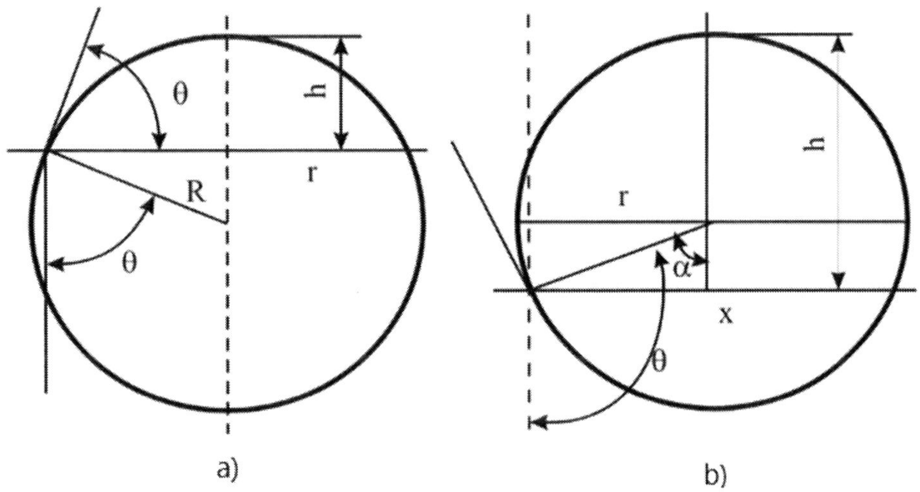

Figure 2. Scheme for determining the wetting angle θ: *a)* $\theta<90^0$, *b)* $\theta>90^0$.

Using the table of trigonometric quantities, were can find θ and C = cos θ.

The wetting angle can also be determined by graphical construction. A graph is plotted on graph paper to do this, plotting the drop radius r on the abscissa axis and height h on the ordinate axis. From point H and P, arcs are drawn with a radius of HP (figure 3) until they cross each other through the points of intersection with the vertical OA (point A). Then point A is connected to point P and at point P, the perpendicular BP is restored to the straight line AR. The wetting angle θ is between the perpendicular BP and the horizontal OP.

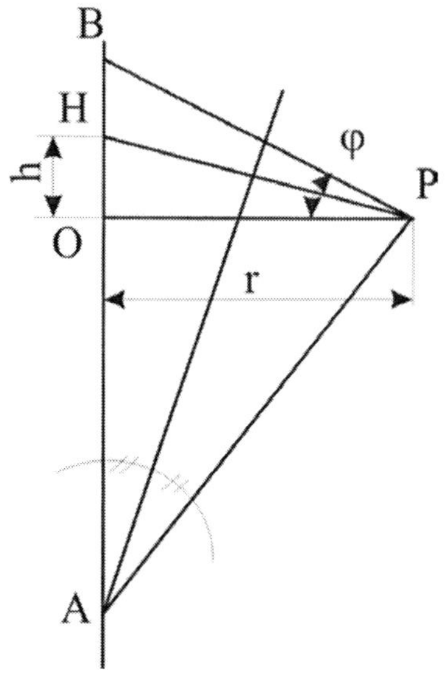

Figure 3. Determination of the wetting angle by graphical construction.

Thus, the wettability is determined by the wetting angle and is a characteristic of surface tension on the surface "coating – substrate" and can serve as an indirect characteristic of the spreadability of the protective coating.

Let us consider a drop of a protective coating with a contact angle θ placed on a solid horizontal surface in air. In the absence of external influences, the drop has the shape of a spherical segment (figure 4).

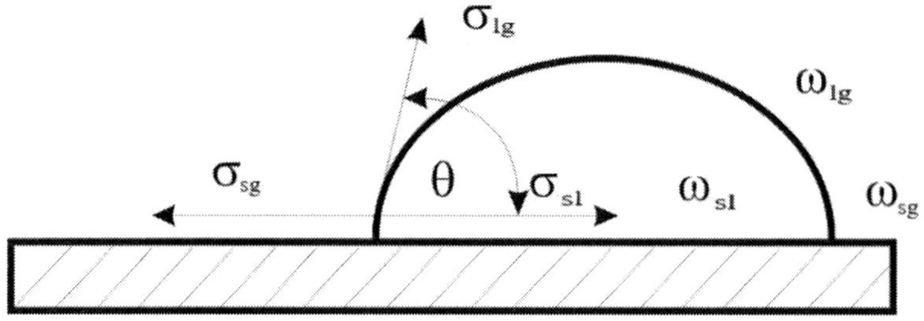

Figure 4. A drop of liquid on the surface of a solid.

If we denote the surface tension at the interface "solid body – gas", "solid body – liquid" and "liquid – gas" by σ_{sg}, σ_{sl}, σ_{lg}, and the surface area of the phases involved in wetting by ω_{sg}, ω_{sl}, ω_{lg} , respectively, we can evaluate the surface energy of the system as

$$G_{surf}= \sigma_{sg}\ \omega_{sg}+ \sigma_{sl}\ \omega_{sl}+ \sigma_{lg}+ \omega_{lg}\ . \tag{5}$$

It is always less than the sum of the surface energies of individual substances, since there is wetting of the surface, which is associated with a decrease in the surface energy of the system. To remove the droplet from the surface, work must be applied, which can be calculated as the product of the adhesion work W_a and the surface area of the liquid-solid contact ω_{sl}:

$$W=W_a\ \omega_{sl}. \tag{6}$$

In turn, the work of adhesion W_a can be determined from the well-known Young equation:

$$\Sigma_{sg} - \sigma_{sl} = \sigma_{lg} cos\ \theta; \tag{7}$$

$$W_a= \sigma_{lg}+ \sigma_{sg} - \sigma_{sl} = \sigma_{lg}(1 + cos\ \theta) \tag{8}$$

The product $\sigma_{lg}\cos\theta$ is often called the wetting energy or adhesive stress. The adhesion work is greater, the stronger the interaction of the contacting phases.

To characterize the wetting process, Schönborn introduced the spreading coefficient S, the difference between the work of adhesion and the work of cohesion $W_к$, numerically equal to 2 σ_{lg}:

$$S= W_a - W_к= W_a- 2\ \sigma_{lg}= \sigma_{lg}(\cos\theta - 1). \tag{9}$$

When completely wetted, S → 0. In this case, it is necessary that the adhesion work is 2 times greater than the surface tension of the liquid at the interface with the environment. From equation (9) it also follows that the smaller the work of fluid cohesion (less than σ_{lg}), the easier this liquid wets various solids

Another characteristic of wetting ability may be the relative work of adhesion:

$$Z_a = \frac{W_a}{W_k} = \frac{1+\cos\theta}{2}. \tag{10}$$

Depending on the interaction at the interface, the values of Z_a can vary from 1 to 0.

The work of adhesion when wetting solids is usually positive, since forces of intermolecular attraction always act between bodies of any nature. Therefore, equilibrium contact angles, as a rule, are less than 1800 (absolute non-wetting is not realized).

The surface roughness positively affects the wetting of hydrophilic bodies (θ decreases) and in most cases negatively affects the wetting of hydrophobic ones. Spreading processes are complicated in the case of highly viscous liquids. The angle θ in this case, as a rule, is high (close to 900) and often a false impression is created that the liquid does not wet the surface well. It does not take into account the fact that the equilibrium state of the contact angle for such liquids is established slowly, over time. In addition, the angle formed by spreading can easily change under the influence of mechanical influences. Spreading in a viscous regime, as well as swelling (restoring a forcefully deformed drop to its original state), is opposed by a certain mechanical force *f*, which is nothing but the strain force of the sample. It is not true thermodynamic equilibrium that intensifies, but false, inhibited:

$$\Sigma_{sg} - \sigma_{sl} = \sigma_{lg}\cos\theta + f. \tag{11}$$

If the force f is large, then the spreading, despite its economic profitability, does not occur. Therefore, it is possible to lower the viscosity of the protective coatings to improve the spreading.

"Runaway", the collection of coatings in separate islands and drops are the signs of poor wetting of the surface. This phenomenon is often observed when applying aqueous protective coatings to poorly degreased surfaces.

The wetting of 9 types of coatings was studied (see table 1) [2, 9 and 10] with various types of substrates: steel in the delivery state, sandblasted and ground surfaces and a surface with a thin layer of lubricant. Measurements of droplet sizes were carried out using an MPB-2 microscope.

Table 1. Compositions of common protective coatings.

Coating	Composition of coating
MV	Chalk – 35% (GOST 4415–75), water – 65% (GOST 2874–73)
KBZH	Concentrate of sulphite-alcohol stillage bard – 190 g and water 1000 g
PZ-1	Concentrate of sulphite-alcohol bard KBZH – 75 g, soap – 35 g, soda ash – 25 g and water – 1000 g
AD	Aluminum powder PAK-1 (GOST 5494-71) – 100 g, dextrin (GOST 6034-74) – 100 g and water – 1000 g
MZHS	Chalk – 30%, liquid glass – 65%
AZHS	PAK-1 aluminum powder – 50 g, liquid glass – 1000 g
TSZHS	Zircon (TSMTU 4469–54) – 30%, liquid glass – 70%
TSPR	Zircon – 40%, polyvinyl butyral (GOST 9439–73) – 3%, solvent 646 (GOST 18188–72) – 57%
SC	RTU Silicone Cream 322–57

The studied protective coating was applied to the sample using the pipette, the droplet size was determined according to the procedure shown in figure 1. Five drops of the test coating were applied to each sample to obtain more accurate research results. Then, according to formulas (3) and (4), the contact angle θ was calculated.

The results were processed according to the method of Ya.I. Frenkel to determine the error of the measurement results. Such processing of the results consists in assessing the value of the sample average (arithmetic mean) and determining its accuracy. The calculation was carried out according to the following formulas:

$$\bar{x} = \frac{1}{n}\sum_{i=1}^{n} x_i, \tag{12}$$

where $\bar{x}$ – is a sample mean;
n – the number of measurements;
$x_1, x_2, \dots, x_n$ – are random variables.

Since the measurement errors are random, the resulting estimate of the result $\bar{x}$ is also random. The measure of its error is the so-called selective standard of the mean:

$$S_{\bar{x}} = \sqrt{\frac{S_x^2}{n}}, \tag{13}$$

where S_x^2 – is a sample variance that can be calculated based on the same sample:

$$S_x^2 = \frac{\sum_{i=1}^{n}(x_i - \bar{x})^2}{n-1} = \frac{\sum_{i=1}^{n} x_i^2 - \frac{1}{n}\left(\sum_{i=1}^{n} x_i\right)^2}{n-1}. \tag{14}$$

The error value of the found estimate $\bar{x}$ is determined by the value of the so-called confidence interval ε_β, which "covers" the true value α with a given confidence probability, i.e.

$$\alpha = \bar{x} \pm \varepsilon_\beta. \tag{15}$$

The value of the confidence interval for sufficiently large sample sizes ($n \geq 30-50$) depends only on the accepted value β, for example:

For β=0,68 $$\varepsilon_\beta = S_{\bar{x}}. \tag{16, a}$$

For β=0,95 $$\varepsilon_\beta = 2S_{\bar{x}}. \tag{16, b}$$

For β=0,99 $$\varepsilon_\beta = 3S_{\bar{x}}. \tag{16, c}$$

The dependence on the number of experiments ε_β is also significant for values of n <30, which is most characteristic of the experiment. In this case, the tabular value of the *t*-criterion corresponding to the significance level $p = 1 - \beta$ and the number of degrees of freedom of the sample variance $f = n-1$ is used as a coefficient for calculation of ε_β:

$$\varepsilon_\beta = t(p, f)\sqrt{\frac{S_x^2}{n}}. \tag{17}$$

If the values of the boundaries of the confidence interval $\bar{x} - \varepsilon_\beta$ и $\bar{x} + \varepsilon_\beta$ have different signs, the assessment of the result is insignificant and with probability β we can assume that $\bar{x} \approx 0$. Usually β equals 0.95, less often 0.99 and 0.999.

Let us give an example of calculating the error of the wetting angle for the PZ-1 coating deposited on a sandblasted surface. After coating the samples, the following contact angles were obtained: 51.1^0; 56.9^0; 40.0^0; 38.9^0; 47.5^0. We define the measurement error according to the methodology above for β = 0.95. In this case, n=5; p=1 –β=1–0.95=0.05; f=n–1=5–1=4. Using formulas (12) ... (14) we calculate the sample mean and variance:

$$\bar{X} = \frac{51.1 + 56.9 + 40.0 + 38.9 + 47.5}{5} = 46.88°;$$

$$S_x^2 = \frac{(51.1 - 46.88)^2 + (56.9 - 46.88)^2 + (40.0 - 46.88)^2}{5} +$$
$$+ \frac{(38.9 - 46.88)^2 + (47.5 - 46.88)^2}{5} = 45.922.$$

Next, according to the Student's distribution tables, we find the value $t(pf) = t(0.05;4) = 2.78$ and by formula (17) we calculate the value of the confidence interval :

$$\varepsilon_\beta = 2.78\sqrt{\frac{45.922}{5}} = 8.42^0.$$

Estimation of the wetting angle by the formula (15) is 46.88 ± 8.42^0.

Next, $C = \cos\theta$ and the relative work of adhesion Z_a were calculated. The obtained research results show that the aqueous solution of chalk (MV) and the aqueous solution of KBZH wet the steel in the delivery state satisfactorily (the average contact angle θ_{av} for MV is 24 ... 32^0, for KBZH – 23 ... 33^0) and the ground surface (θ_{av} for MV – 22 ... 33^0, for KBZH – 35 ... 45^0) and wet sandblasted surface worse (θ_{av} at MV – 49 ... 58^0, at KBZH – 48 ... 62^0) and a surface with a thin layer of lubricant (θ_{av} for MV – 34 ... 46^0, for KBZH – 47 ... 53^0)

Coatings MZHS and AZHS wet the steel in the delivery state satisfactorily (θ_{av} fort MZHS – 68 ... 870, for AZHS – 61 ... 680) and a ground surface (θav for MZHS – 44 ... 640, for AZHS – 53 ... 64^0), wet the sandblasted surface poorly (θ_{av} for MZHS – 96 ... 124^0, for AZHZ – 88 ... 92^0), and wettability improves when these coatings are applied to a surface with a thin layer of lubricant (θ_{av} for MZHS – 29 ... 39^0, for AZHZ – 43 ... 51^0), obviously, applying of the engine oil film on the surface reduces the surface tension of the coating droplet and improves its wetting. The coatings of TSZHS and SK wet the steel in the delivery state satisfactorily (θ_{av} for TSZHS – 54 ... 88^0, for SK – 46 ... 69^0), sandblasted (θ_{av} for TSZHS – 45 ... 87^0, for SK – 61 ... 76^0) and polished (θ_{av} for TSZHhS – 45 ... 87^0, for SK - 61 ... 76^0) surfaces and better wet the surface with a thin layer of lubricant (θ_{av} for TsZhS – 17 ... 35^0, for SK – 23 ... 35^0).

AD coatings wet steel in the delivery state satisfactorily (θ_{av} = 27 ... 39^0), sandblasted (θ_{av} = 23 ... 37^0) and polished (θ_{av} = 19 ... 27^0) surfaces, and wet the surfaces with a thin layer of lubricant worse (θ_{av} = 34 ... 86^0). A coating PZ-1 has the best wetting results.

The average wetting angle of degreased surfaces does not have a large spread and is in the range of 35 ... 45^0, when applying these coatings to the surface with a thin layer of lubricant, the wetting sharply improves, and the wetting angles are in the range of 16 ... 20^0. It is obvious that the addition of surface-active substances (surfactants) to the protective coating (soap, soda ash) sharply reduces the surface tension and increases the wetting ability of the coatings. For example, the introduction of 0.1% oleic soap (sodium oleate) into water reduces its surface tension from 72 to 25 erg/cm^2.

3. Conclusion

As a result of studies for determining the wettability of the applied protective coatings on the surface of the welded products during MAG welding, it was found that satisfactory wetting is observed for the PZ-1 coating, which can be recommended for widespread use in welding production. It is possible to recommend protective coatings based on liquid glass (MZHS, AZHS, TSZHS) for introduction into production as lubricant for surfaces.

References

[1] Rebinder P A 1978 *Poverkhnostnyye yavleniya v dispersnykh sistemakh* [Surface phenomena in disperse systems] (Moscow) [In Russ.]

[2] Fedko V T, Sapozhkov S B, Sokolov P D and Yastrebov A P 2004 *Elements of theory and technology of anti-spatter surface shielding when welding in carbon dioxide: Monograph.* (Tomsk: TSU Publishing) p 138

[3] Solonenko O P, Kudinov V V, Smirnov A V, Cherepanov A N, Popov V N, Mikhalchenko A A and Kartaev E V 2005 Micro-Metallurgy of Splats: Theory, Computer Simulation and Experiment *JSME International Journal* B vol 48 **3** 366–380

[4] Knotek O and Elsing R 1987 Monte Carlo Simulation of Lamellar Structure of Thermally Sprayed Coatings *Surf. Coat. Technol.* **32** 261–271

[5] Ghafouri-Azar R, Mostaghimi J, Chandra S and Chamchi M A 2003 Stochastic Model to Simulation of a Thermal Spray Coating *J. of Thermal Spray Technology* vol 12 **1** 53–69
[6] Steinke T and Bäker M 2006 Monte Carlo Simulation of Thermal Sprayed Coatings *Proc. ITSC'06* (USA, Seattle) 329–334
[7] Fedko V T, Tomas K I and Sapozhkov S B 1998 Protecting the surfaces of welded components against molten metal splashes in CO_2 welding *Welding International* **1** 58–62
[8] Sapozhkov S B and Burakova E M 2018 *IOP Conf. Ser.: Mater. Sci. Eng.* **441** 012042
[9] Saraev Y N, Chinakhov D A, Ilyashchenko D I, Kiselev A S and Gordynets A S 2017 Investigation of the stability of melting and electrode metal transfer in consumable electrode arc welding using power sources with different dynamic characteristics *Welding International* **31(10)** 784–790
[10] Sapozhkov S B, Burakova E M and Tesleva E P 2016 *IOP Conf. Ser.: Mater. Sci. Eng.* **142** 123667

IOP Conf. Series: Materials Science and Engineering **939** (2020) 012068 doi:10.1088/1757-899X/939/1/012068

Investigation of the influence of the parameters of the laser welding technological process on the chemical composition, structural-phase state and physicochemical properties of the material of the weld, heat-affected zone, pipe body and product

V V Savin[1,2], L A Savina[1], V A Chaika[1,3], A S Martyushev[1] and S Z Lonchacov[1]

[1] Immanuel Kant Baltic Federal University, Nevsky st., 14 23616 Kaliningrad, Russia

[2] E-mail: VVSavin@kantiana.ru
[3] E-mail: v_chayka8@rambler.ru

Abstract. The aim of this work is to study the welds of titanium grade VT1-0. In this work, we study the chemical composition, microstructure of the weld, the heat-affected zone and the body of the material, mechanical tensile tests, microhardness measurements and determination of hydrogen content

1. Introduction

At present, quite a lot of welding methods have been developed: fusion, gas welding, electric arc, electric contact, plasma, high frequency current welding, electron beam, laser, each of which has its own advantages and disadvantages that determine the most effective areas of application of one or another welding method. The main requirements for modern welding methods: 1) the mechanical properties of the weld should be close to the mechanical properties of the materials being welded; 2) the structural strength of the welded joint must satisfy the requirements of the design documentation; 3) the geometric dimensions of the welded assembly must be in the tolerance field.

The laser welding method has a number of significant common technological advantages compared to alternative methods of fusion welding, and, in particular, electric arc welding [1–7]. Namely:

1. The high value of the concentration coefficient of the laser welding energy source allows welding in the deep penetration mode, reduces the volume of the weld pool and reduces the level of longitudinal and transverse deformations by several times.

2. The laser welding method allows several times to increase the speed of obtaining a welded joint and reduce the value of the zone of thermal influence.

3. Laser welding due to the deep penetration mode allows you to replace multi-pass welding in the heat conduction mode with a single-pass and thereby significantly reduce thermal investment in the welded structure.

4. The dispersion of the weld structure in the deep penetration mode is significantly higher compared to electric arc welding, which improves the mechanical properties of the weld.

5. Higher laser welding speeds reduce the level of deformation of the welded structure and increase the structural strength of the welded joint.

Content from this work may be used under the terms of the Creative Commons Attribution 3.0 licence. Any further distribution of this work must maintain attribution to the author(s) and the title of the work, journal citation and DOI.
Published under licence by IOP Publishing Ltd

IOP Conf. Series: Materials Science and Engineering **939** (2020) 012068 doi:10.1088/1757-899X/939/1/012068

Among all the materials in the production of tubes titanium is the most attractive due to its mechanical properties. Currently, there are two ways of obtaining titanium tubes: casting and welding. A method of manufacturing a casting has several disadvantages, for example, a certain percentage of defects produced during casting of titanium in forms, post-treatment of edges and obtained at the output of the sample. If you use the production of pipes using bending and subsequent welding of the seam, this method of manufacture there are no disadvantages of the previous one. But, there is another problem – it is not known whether the seam is as strong as casting?

As practice shows specific technological difficulties in welding titanium and its alloys are as follows.

1) High chemical reactivity of titanium alloys at high temperature, especially in the molten state, relative to oxygen and hydrogen. It should be noted that a significant influence on properties of welded joints of titanium alloys provides the quality of protection of the weld and the composition of the protective environment. To achieve high quality of welded joints of titanium alloys you need to protect the front side of the weld, the root pass and heat affected zone. A requirement is to protect the cooling sections of the weld metal and the HAZ (heat affected zone) to a temperature of 400...500 °C [8, 9].

2) The tendency to grain growth when heated to high temperatures, especially above 330 °C (in region β-phase).

3). Increased tendency to formation of cold cracks with increasing content in the base metal and the seam of the impurity gases. Cold cracks can occur immediately after welding, and after maturing. To eliminate the effect of these physic-chemical and metallurgical features of high-quality welded joints of titanium alloys is required as less heat input during welding. This can be achieved through the use of highly concentrated energy sources such as electron or laser beam. But electron beam welding requires a vacuum, complex tooling, so in most cases the choice is made in favor of the welding laser beam.

The use of highly concentrated thermal energy source – laser [12], which is currently available in a welding energy source that no excess of atmospheric pressure may obtain a power density of over 100 W/cm^2, allows to realize the mode of deep penetration of the material.

At the same time, laser welding has a hard heat cycle, as the heating rate of the weld can reach up to $1.4 \cdot 10^4$ deg/sec. The heating rate in the zone of thermal influence (HAZ) in this case will be $5 \cdot 10^3$ deg/s, and the cooling rate of $5 \cdot 10^2$ deg/s [13, 14].

In connection with such uneven heating between the heated and unheated sections of the metal and the weld, temporary and residual elastic stresses arise associated with the effects of thermal compression of the material of the weld and the heat-affected zone. Also, deformations and stresses arise, caused by phase or structural transformations of the solidified melt and cooled crystallized material with increased speeds. The resulting elastic stresses and plastic deformations are mainly longitudinal in relation to the weld [13]. Longitudinal residual stresses are significantly superior to other stress components, and are the main ones when welding structures up to 10 mm thick [14].

The aim of our work is to conduct experiments on the manufacture of laser-welded tubes from VT1-0 titanium alloy and to study the mechanical properties, as well as to study the microstructure of the weld and heat affected zone (HAZ).

The formation of the weld is represented by the kinetic changes in the interfacial surface of the melting front and crystallization of the pool, in theory, of interest in cross sections along the length of the pool.

Therefore, it is of particular interest to study the trend in the microstructure of the Ti ↔ Ti weld obtained by laser welding technology, HAZ, and the body of the metal being welded.

At temperatures exceeding 350 °C, titanium actively absorbs oxygen with the formation of interstitial phase having high strength and hardness (~ 2 times higher than that of titanium) and low ductility [15].

As a result, it is necessary to evaluate the change in the hardness gradient measured in the cross section of the weld: *Base metal – HAZ – weld – HAZ – base metal.*

A sharp increase in strength and a decrease in ductility caused a severe limitation of the allowable content of these gases in titanium: oxygen - up to 0.15 %, nitrogen - up to 0.05 %.

Hydrogen, in turn, even at a low content very sharply worsens the properties of titanium [12]. Although its content decreases with increasing temperature, the hydrogen in the solid supersaturated solution is released and forms a separate phase — titanium hydrides (TiH2), which greatly embrittle titanium and contribute to the formation of “cold” cracks long after welding (delayed fracture) [13]. In addition, hydrogen promotes the formation of pores. In this regard, the permissible hydrogen content in the metal is limited to 0.01 % (SST 19807).

Therefore, the last important indicator of the quality of the weld pool is an understanding of the gas saturation process, since saturation of the weld metal with oxygen, nitrogen and hydrogen sharply reduces the ductility and long-term strength of welded structures.

2. Results

Within the framework of the project, studies were carried out on the influence of the parameters of the technological process for laser welding of VT1-0 titanium samples (SST 19807-91), which relates to wrought titanium alloys with α-structure, increased ductility and low strength (σ_v not more than 700 MPa).

These alloys are characterized by high ductility, both in hot and in cold condition. This allows you to get all kinds of semi-finished products: foil, tape, sheets, plates, forgings, stampings, profiles, pipes, etc. [10, 11].

VT1-0 alloy is alloyed with carbon, nitrogen and oxygen, which are α-stabilizers (interstitial elements), as well as aluminum, as an element of substitution. Alloys with an α-structure stable at different temperatures, to which technical titanium VT1-0 belongs, are not hardened by heat treatment, and therefore have good weldability [12].

In the work we used plates of 70.0x200.0 mm with a thickness of t = 1.0; 2.0; 3.0; 5.0; 10.0 mm. Welding was performed on a fiber laser with a power of up to 6 kW in argon atmosphere at the company IRS LaserTechnology LLC (Yekaterinburg). Type of weld – butt. The technology of preparing the edges for welding – milling cutting, corresponding to a roughness class of at least 8, followed by washing in distilled water, degreasing with alcohol and drying by argon blowing. The main technological parameters are given in table 1.

Table 1. Technological parameters of laser welding.

Marking, №	The thickness of the welded plates, mm	Laser beam power P, kW	Power density in focus Wp, MW./cm^2	Welding speed Vs, mm/s
1	1.0	2.0	1.11	33.3
2	2.0	2.3	1.27	33.3
3	3.0	3.4	1.88	3.40
5	5.0	5.7	3.15	40.0
10	10.0	6.0	3.32	6.67

The microstructure was studied using a scanning station based on the SIAMS-800 solid microstructure fragment analyzer for panoramic applications in reflected and transmitted light, including an OLYMPUS BX-51 metallographic microscope and SIAMS Drive System.

Experimental studies of the mechanical properties were performed on a tensile testing machine for testing tube and flat samples at tensile temperature at room temperature RMG200-MG4 (manufacturer – Stroypribor). Tests for the statistical tension of the metal of the pipes were carried out at a temperature of 20°C according to GOST 10006 and in accordance with the requirements of GOST 6996. The test speed to yield strength < 10 mm/min, yield strength < 40 mm/min. On this tensile testing machine, samples 1, 2, 3 and 5 mm thick were tested (three samples each). Tensile testing of samples with a thickness of 10 mm was carried out on a universal testing machine UTS112-50 (manufacturer – Testsystems LLC) rated power with a force of 50 kN.

The microhardness of the material was also determined by the size of the diagonal of the imprint obtained after unloading on the PMT-3M. Load 0.5 HV, exposure time 10 sec, Vickers unit, 40x objective.

In titanium welded samples, the hydrogen content was determined by the method of 3 standards using emission spectral analysis on an ISP-51 spectrograph and a pulse discharge generator in accordance with GOST 24956.

Figure 1 shows the microstructure of the weld, HAZ, and the body of the material of the VT1-0 alloy with a thickness of 1 mm at different magnifications. The microstructure of the base metal is similar to the microstructure of an alloy of an equilibrium state annealed at temperatures of 500 ÷ 800°C, because consists of grains of α-solid solution.

The heat-affected zone is the region of the base metal heated during the welding process in the temperature range from T_p (the temperature of the onset of recrystallization) to T_m (melting point).

At the boundary of the base metal with the beginning of the heat-affected zone, a fine-grained structure appears and, as a result of polygonization processes associated with the return phenomenon and subsequent more distinct appearance of large elongated grains with a predominant direction from the weld axis to the body. As you approach the weld, i.e. with an increase in the maximum heating temperature, the grain size increases, and they become more equiaxed.

Figure 1b shows the microstructure of the HAZ and the body of the material of the VT1-0 alloy with a thickness of 2 mm, which is similar to the microstructure of a sample with a thickness of 1 mm.

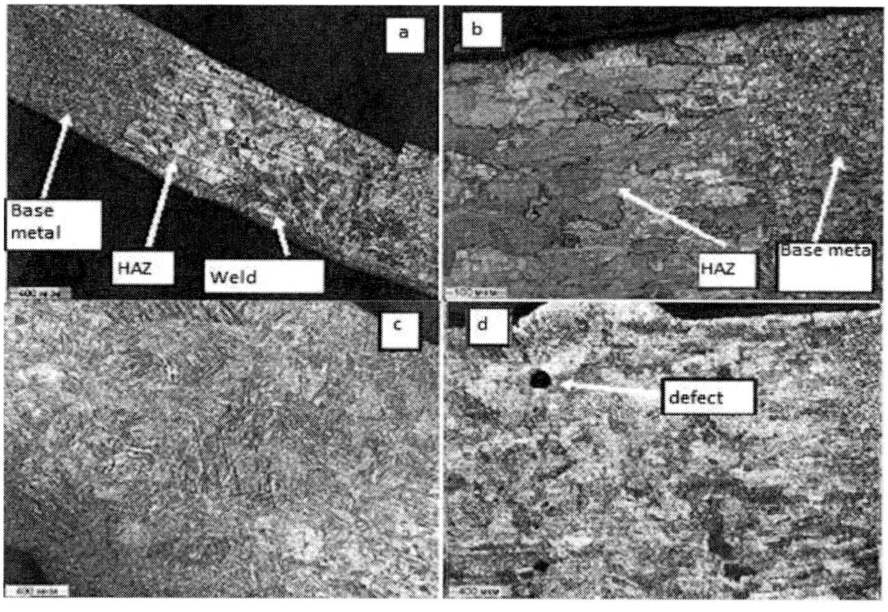

Figure 1. The microstructure of the weld, HAZ and the body of the material of the VT1-0 alloy with a thickness of 1 mm (a), 2 mm (b), 3 mm (c) and 5 mm (d) at different magnifications.

As can be seen in figure 1, the heat-affected zone of the VT1-0 alloy with a thickness of 3 mm is a needle-like structure and basket-like weaving, with the latter predominating, which may be associated with the martensitic transformation of the high-temperature α-phase during rapid cooling.

Figure 1d shows the microstructure of the weld, HAZ, and the body of the material of the VT1-0 alloy 5 mm thick at different magnifications. The appearance of defects in the form of micropores is shown. The estimated micropore size is 150 microns.

Figure 2 shows the microstructure of the weld, HAZ, and the body of the material of the VT1-0 alloy 10 mm thick at different magnifications. Micropores of 230-500 microns appear. The microstructure is characterized by large grain. However, in our case, the presence of large grains indicates that a complete reverse transition to the α-modification did not occur properly. This is the result of insufficient study of the welding technology, namely: the welding method, welding speed, linear energy costs and the nature of the thermal cycle.

Confirmation of this must be sought when determining the microhardness gradient in the cross section of the base metal, HAZ, and weld. After all, the fusion zone is crucial in assessing the weldability of titanium alloys and the serviceability of welded joints. A weld metal is a cast metal with its specific properties. The needle-like multidirectional microstructure (figure 1a) indicates a martensitic transformation, usually occurring at temperatures exceeding 882°C [9, 13, 15].

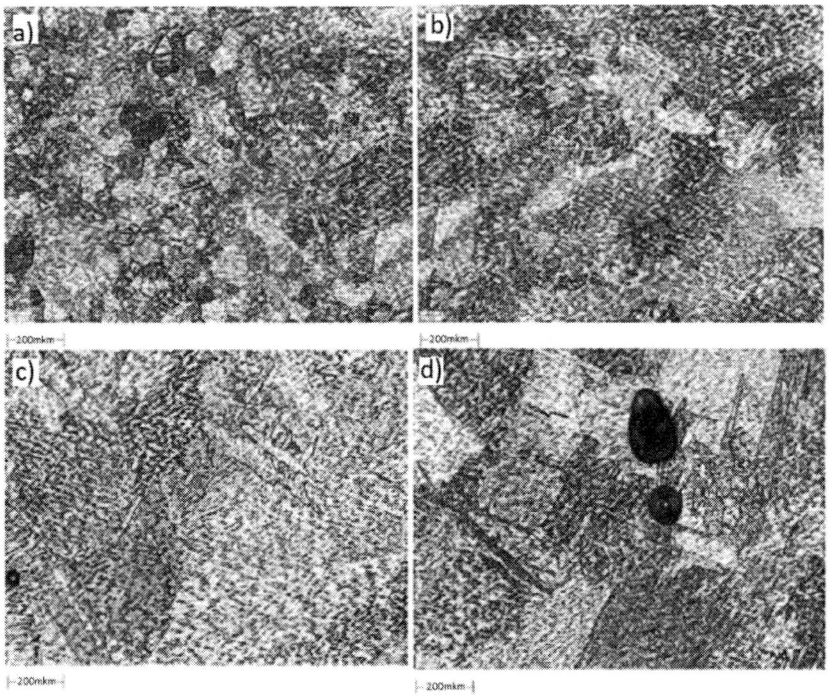

Figure 2. The microstructure of the weld, HAZ and the body of the material of the alloy VT1-0 10 mm thick with different magnifications.

The most equilibrium structures are observed in figure 1d, however, there is a lack of penetration (the presence of cavities), which indicates an insufficiently optimal technological regime when conducting the welding process.

Experimental studies of the mechanical properties were performed on an RMG200-MG4 tensile testing machine for tensile at room temperature for flat samples of 1, 2, 3, and 5 mm and on a universal testing machine UTS112-50 of 10 mm thick samples.

Diagrams of tensile tests of samples with a thickness of 1, 2, 3 and 5 mm are shown in figure 3.

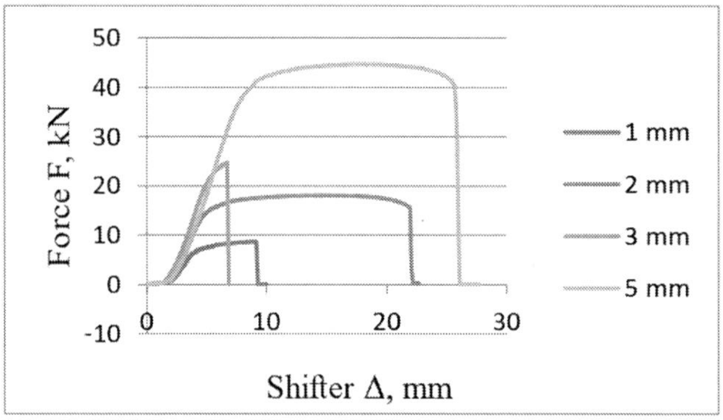

Figure 3. A typical diagram of tensile tests of laser-welded samples of alloy VT 1-0.

Table 2 presents the results of tensile tests. The results of mechanical tensile testing of welded specimens from a wrought titanium alloy with plate thicknesses from 1.0 mm to 3.0 mm show a decrease in strength properties and an increase in the ductility of the welded joint. The welded joint with a plate thickness of 5.0 mm behaves illegally – it demonstrates the highest strength and plastic properties of all the thicknesses of the tested samples.

Table 2. Results of mechanical tensile testing of welded samples of VT 1-0 alloy on a tensile testing machine RMG200-MG4.

Thickness, mm	V, MPa/s	F_o, mm^2	l_o, mm	l_k, mm	δ, %	P_{max}, H	σ_b, MPa	$P_{0.2}$, H	$\sigma_{0.2}$, MPa
1	4.00	19.45	25.0	27.11	5.32	8940	459.64	6450	331.62
2	4.00	40.0	36.0	46.02	27.78	17890	447.42	12900	322.60
3	4.00	58.9	44.0	44.51	1.15	23390	397.17	18600	316.32
5	4.00	100.5	55.0	67.71	23.15	44490	442.69	34300	341.29

As a rule, the destruction of titanium samples passed through the base metal without touching the HAZ and the weld, which indicates a complete and high-quality penetration of the weld. Judging by the resulting neck, the fracture is viscous, characteristic of an annealed state. General data obtained during the tensile test are presented in table 3.

Table 3. The results of the tensile tests of titanium samples.

Sample thickness, mm	Conditional yield strength $\sigma_{0,2}$, MPa	Tensile strength σ_b, MPa	Relative extension δ, %
1	331.62	459.64	8.88
2	322.63	447.42	27.81
3	316.32	397,17	1.15
5	341.79	442.69	23.15
10	281.40	459.64	26.6

The results of measuring the microhardness of the material by the size of the diagonal of the fingerprint after removing the load, obtained when the device was operated in manual mode, are presented in figures 4 and 5. The black line in the figures corresponds to the trend line.

As can be seen from the figures, the microhardness in the weld zone is higher than in other zones, which is explained by temperature processes and the diffusion of non-metallic impurities.

In titanium welded samples, the hydrogen content was determined (table 4).

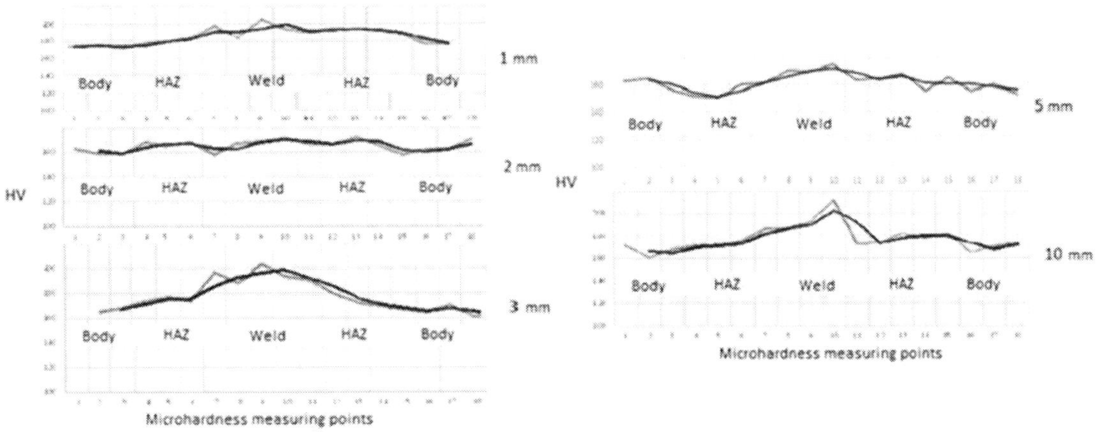

Figure 4. Histogram of the distribution of microhardness on samples with a thickness of 1 mm, 2 mm, 3 mm, 5 mm, 10 mm.

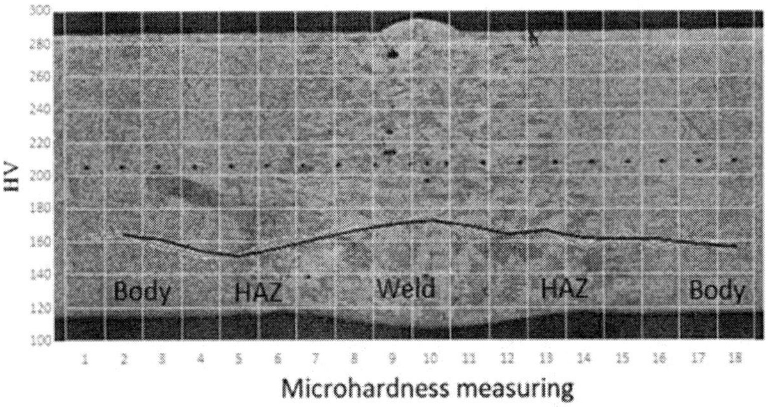

Figure 5. Overlay of the histogram of the distribution of microhardness on the image of the weld of a sample with a thickness of 5 mm.

Table 4. Percentage of hydrogen in samples of various thicknesses.

Sample thickness, (mm)	The hydrogen content, at. %	
	The seam	Body
1	0.007	0.075
2	0.005	0.006
3	> 0.1	> 0.2
5	0.003	0.004
10	0.003	0.004

As can be seen from the obtained results, the microhardness in the weld zone is higher than in the heat-affected zone and is comparable with the microhardness in the body of the product. This agrees well with the factor of thermal influence in the heat-affected zone [3, 5, 7] and quenching of the material (quenching from the liquid state) in the weld zone [5]. It should be borne in mind that during laser melting and subsequent quenching from a liquid state, a change in the total content and the probability of local redistribution of impurity interstitial atoms is inevitable [8–12]. Especially hydrogen [8, 12].

In this regard, the study conducted the study of the hydrogen content in titanium welded samples in the weld and body of the product. The research results are presented in table 4.

As can be seen from the results obtained, only the sample with a thickness of 3 mm has an overestimated hydrogen content, both in the body itself and in the weld. Exceeding the hydrogen content may result from overstating the hydrogen in the feed prior to welding. For all other samples, the hydrogen content is within the requirements of SST 19807.

An increase in the hydrogen content in titanium leads to the "loosening" of the microstructure and blurring of the grain boundaries [8, 9, 12]. This can be clearly seen on the example of figure 2c, where there is a local violation of the martensitic needle structure. However, with an increase in a certain threshold of the hydrogen content, plastic deformation may lead to a tendency to cracking and rupture.

Mechanical properties (conditional yield strength $\sigma_{0.2}$ and temporary tensile strength σ_b) of sample 3 are underestimated (see table 3) with respect to other thicknesses, which once again confirms the fact that excessive hydrogen content in the weld has a significant effect on mechanical and, as a consequence, the operational properties of titanium and its alloys.

However, the fact that the remaining samples contain hydrogen in accordance with the requirements of ND indicates that welding according to the developed technology within the project allows to obtain a high-quality weld. This fact also confirms that the deformation and rupture of the welded samples passed outside the zone of thermal influence and the weld itself.

3. Conclusions

The experiments on measuring the physicochemical and deformation characteristics, as well as the study of the HAZ microstructure and the weld seam of the VT1-0 titanium alloy, showed with full confidence that this technology is applicable for the serial production of welded titanium pipes and other structures made using a fiber laser with a power of up to 6 kW in argon. The mechanical properties of the obtained welded pipes, with such a technological process, are comparable to seamless pipes of a similar composition and thickness.

Laser welding allows you to get the narrowest welds with a more uniform equilibrium structure and a minimum width of HAZ. The influence of technological heredity in preparing edges for welding on the level of defectiveness in welding titanium structures from VT1-0 alloy should be taken into account.

The developed technology allows to fully use it for the production of welded pipes from titanium alloys of various wall thicknesses.

However, additional studies of the quality of the welds obtained in the production of pipes from titanium alloys are necessary, taking into account technological processes involving their preliminary deformation before welding.

Acknowledgments

The work was supported by the Ministry of Education and Science of the Russian Federation, unique project identification number RFMEFI57817X0252.

References

[1] Williams S, Scott G and Calder N 2007 Direct welding of aerospace alloys by diode lasers *PHOTONICS* **4** 12–17

[2] February 2008 Laser welding of products from titanium alloys by radiation of fiber lasers *RITM* 39–40

[3] Paton B E *et al.* 2009 Laser welding of titanium alloys *Automatic welding* **10** 35–39

[4] Kochergin S A and Saushkin B P 2012 The technology of laser welding in the manufacture of pressure vessels made of titanium alloys *proceedings of MGTU "MAMI"* **2(14)** vol 2 91–94

[5] Klimenov V A *et al.* 2014 Features of changing mechanical properties in the weld zone of VT1-0 submicrocrystalline titanium alloy *Reshetnev readings of the Siberian State University of Science and Technology named after Academician MF Reshetnev* (Krasnoyarsk) vol **1** pp 296–298
[6] Kapustyan A E *et al.* 2014 Obtaining welded products from sintered titanium alloys *Bulletin of the Dnipropetrovsk Scientific University* **3(51)** 84–91
[7] Grigoryanu A G, Shiganov I N and Misyurov A I 2006 *Technological processes of laser processing* (Moscow: Publ. house MTGU im. N.E. Bauman) p 664
[8] Kalachev B A 1985 *Hydrogen fragility of metals* [Moscow: Metallurgy] p 216
[9] Muravyov V I and Lonchakov S Z 2001 The influence of technological heredity on the structure and properties of weld metal during fusion welding of titanium alloys *Proc. of the 5th Meeting of Russian metal experts* 367–369
[10] Bratukhin A G *et al.* 1999 *Modern aircraft manufacturing technologies* (Moscow: Mechanical Engineering) p 832
[11] Zwicker W 1979 *Titanium and its alloys* (Moscow: Metallurgy) p 512
[12] Gurevich S M *et al.* 1986 *Metallurgy and welding technology for titanium and its alloys* ed V N Zamkova [2nd ed. add. and reslave] p 240
[13] Bratukhin A G *et al.* 1997 *Stamping, welding, soldering and heat treatment of titanium and its alloys in aircraft manufacturing* (Moscow: Mechanical Engineering) p 600
[14] Gurevich S M *et al.* 1986 *Metallurgy and welding technology for titanium and its alloys* ed. V N Zamkova [2nd ed. add. and reslave] p 240
[15] Muravyov V I 1999 Optimization of heating for stamping sheet blanks from titanium alloys *Forging and stamping* **1** 31–36

IOP Conf. Series: Materials Science and Engineering **939** (2020) 012069 doi:10.1088/1757-899X/939/1/012069

Prospects for the use of u-shaped welded pipes of steel a316l heat transfer assortment in chemical engineering

V V Savin[1], **K N Lebedeva**[2], **V V Herelovich**[3], **L A Savina**[1] **and V A Chaika**[1,4]

[1] Immanuel Kant Baltic Federal University, Nevsky st., 14 236016, Kaliningrad, Russia
[2] JSC "VNINEFTEMASH", 4th Roshchinsky passage, 19 115191, Moscow, Russia
[3] LLC "TECHNO TRADE", st. Zavodskaya, 1 238224, pos. Ozerki, Duards district, Kaliningrad region, Russia

[4] E-mail: v_chayka8@rambler.ru

Abstract. To expand the possibilities of using welded stainless steel pipes in chemical engineering, comprehensive studies of the experimental batch of straight-seam and U-shaped laser-welded tubes of A316L steel have been performed. The corrosion resistance, strength and structural-phase changes in the product material during austenization were studied: short-term annealing at 1000 °C. It is shown that longitudinal welded pipes have a high margin of ductility. This opens up prospects for their wide application for the production of bent assortment in the form of U-shaped products. It has been experimentally proved that heat treatment according to the regime of short-term austenization completely eliminates metal hardening, providing relaxation of its elastic stresses and the formation of the structural state necessary for technological redistribution of A316L steel. Heat treatment according to the mode of short-term austenization at a temperature of 1000–1050 °C for 1–5 minutes can be recommended as a mandatory technological operation that completes the metallurgical conversion of laser-welded stainless tubes of heat-exchange gauge.

1. Introduction

Seamless pipes, which are now widely used in nuclear power equipment and chemical engineering, can last at least 30 years. The introduction of new laser welding methods for stainless steels and heat-resistant alloys and automated intelligent control systems for the technological process of production of welded pipes can significantly improve the quality and technological properties of welded pipes.

In many cases, they not only are not inferior, but also surpass the similar properties of seamless pipes. At the same time, labor productivity dramatically increases, and the cost of their production is significantly reduced.

World practice confirms the widespread use of welded pipes for heat exchangers of nuclear power plants and apparatuses of the chemical industry.

The wider introduction of welded pipes in these sectors requires systematic comprehensive research, scientific generalization of the results and the introduction of appropriate amendments to existing national standards and industry standards.

The purpose of this study was to substantiate the possibility of using longitudinal welded laser welded tubes for the manufacture of U-shaped products for chemical engineering. To achieve this

Content from this work may be used under the terms of the Creative Commons Attribution 3.0 licence. Any further distribution of this work must maintain attribution to the author(s) and the title of the work, journal citation and DOI.
Published under licence by IOP Publishing Ltd

goal, comparative tests of laser-welded longitudinal welded tubes made of A316L steel (GOST 9941-81, EN10217-7) with an outer diameter of Ø25 mm and a wall thickness of 2.0 mm and U-shaped products from them were performed. Experimental samples of laser-welded tubes were made at Tekhno TUB LLC (https://technotube.ru/techno-tube/). U-shaped products from them are made by JSC "VNIINEFTEMASH".

2. Results

The objects for research were samples cut from different zones of the welded pipe and the U-shaped product in the initial and annealed states. Annealing corresponded to short-term austenization at 1000°C in air [1, 2]. The studies were carried out from the outer and inner surfaces of the pipe. The labeling of the samples is shown in table 1.

Table 1. Samples for studying the properties of laser-welded straight and bent (U-shaped) pipes made of 316 L steel.

Marking	Type of laser welded pipe	Heat treatment status	Seam Location	Research surface
6Иш		without heat treatment	with a seam	internal outdoor
62	straight pipe	heat treated	with a seam	internal outdoor
63		heat treated	seamless	internal outdoor
6НОИ		without heat treatment	seamlessly bending	internal outdoor
6н		heat treated	seam outside the bend	internal outdoor
6в	U pipe	heat treated	seam inside the bend	internal outdoor
6с		heat treated	side seam	internal outdoor

The weld structure of experimental samples of laser-welded straight-line pipes made of A316L steel is presented in figure 1. Metallographic studies were performed on a device OLIMPUS. The width of the weld varies from 0.6 mm (in the center of the section) to 1.2 and 1.3 mm (on the outer and inner surfaces of the pipe). An example of the structure shown in figure 1, indicate that the weld of the experimental pipes has an internal burr with a thickness of up to 200 μm.

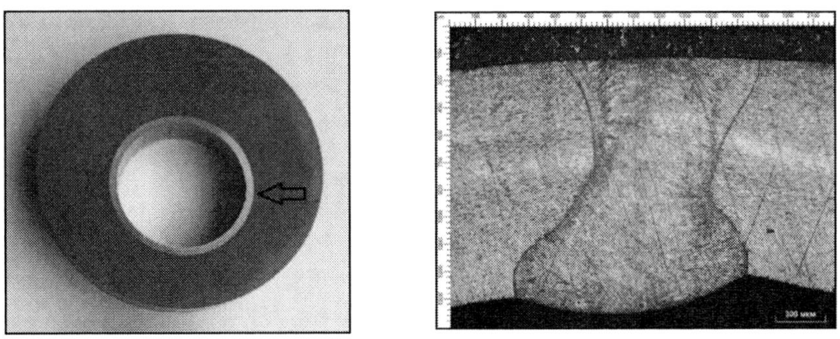

Figure 1. Sample for metallographic studies from a longitudinal welded laser welded pipe of steel grade A316L (the seam is indicated by an arrow) and the structure of the seam.

In order to compare the performance properties tested for corrosion resistance. Samples for research have dimensions (20 × 40) mm. The studied (working) surfaces were polished (to a roughness $R_a \leq 0.8$ μm) and isolated from a flat pair with a contact window (12 × 12) from the active medium in an open glass cell with a silver-silver reference electrode (see figure 2).

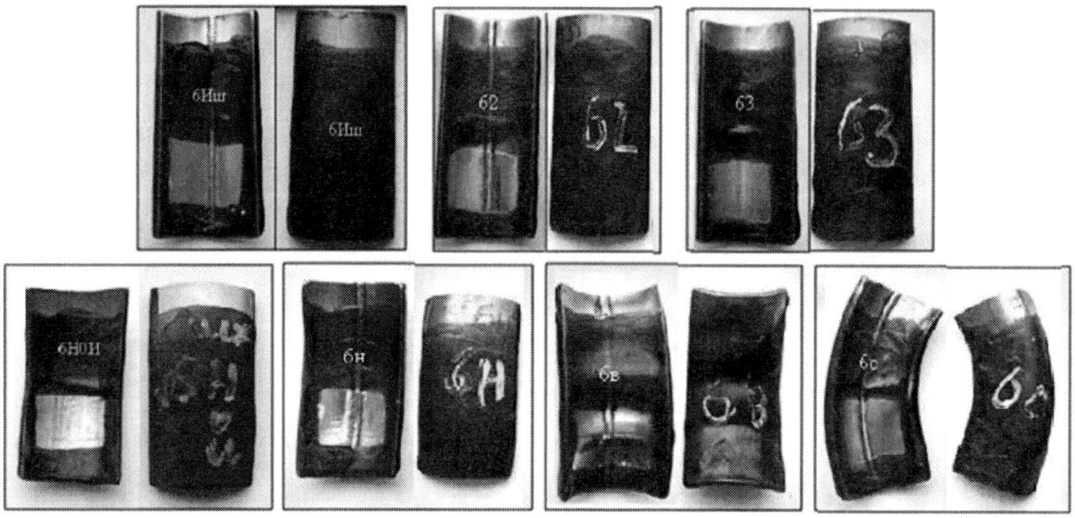

Figure 2. Samples of straight and U-shaped laser-welded tubes made of A316L steel (inner and outer surfaces) for pitting tests.

Corrosion tests for the tendency to pitting corrosion were carried out by the potentiodynamic method of measuring pitting resistance in an aqueous solution with 16.5 g/dm^3 of sodium chloride. In the potential sweep mode at a speed of 0.5 mV/s, the forward and reverse polarization curves were recorded on an IPC-Pro potentiostat with a current meter with a sensitivity of $1 \cdot 10^{-6}$ A.

On the polarization curves (an example is shown in figure3a), the corrosion characteristics (potentials) were recorded: the onset of corrosion (E_{cor}), Pitting formation ($E_{p.s.}$), repassivation ($E_{r.p.}$), According to which the levels (bases) were calculated: pitting resistance ($\Delta E_{p.s.} = E_{r.p.} - E_{cor}$) and repassivation ($\Delta E_{r.p.} = E_{r.p.} - E_{cor}$) of samples (see figure 3b). The results evaluated the resistance to local corrosion, taking into account the heat treatment mode.

The results obtained indicate that all the investigated samples of steel A316L have high corrosion resistance in a chloride-containing medium. The basis of pitting resistance varies within (400 ÷ 480) mV and can even reach limit values (up to 520 mV).

Heat treatment in the form of short-term (up to 5 minutes) austenization at 1000° C in the straight-seam case practically does not affect the tendency to pitting (local corrosion). However, samples from a longitudinal pipe from the pipe body (without a seam) show pitting resistance at a level 426 and 358 mV, and with a seam of 512 and 420 mV ($\Delta E_{p.s.}$ and $\Delta E_{r.p.}$ respectively). This indicates that the metal in the weld zone (metal after laser melting and rapid solidification in the conditions of directional heat removal – quenching from a liquid state) is more corrosion resistant. This is fully consistent with the well-known fact of increasing the corrosion resistance of a material obtained by various methods of quenching from a liquid state. Laser melting is one of the varieties of this method [3].

Annealing a U-shaped pipe reduces pitting resistance. So, without heat treatment, the basis is 521 and 386 mV, and after it – 480 and 305 mV ($\Delta E_{p.s.}$ and $\Delta E_{r.p.}$ respectively).

The difference in corrosion resistance of U-shaped pipes before and after heat treatment can be explained by metal softening as a result of austenite recrystallization and phase (α–γ) transformation.

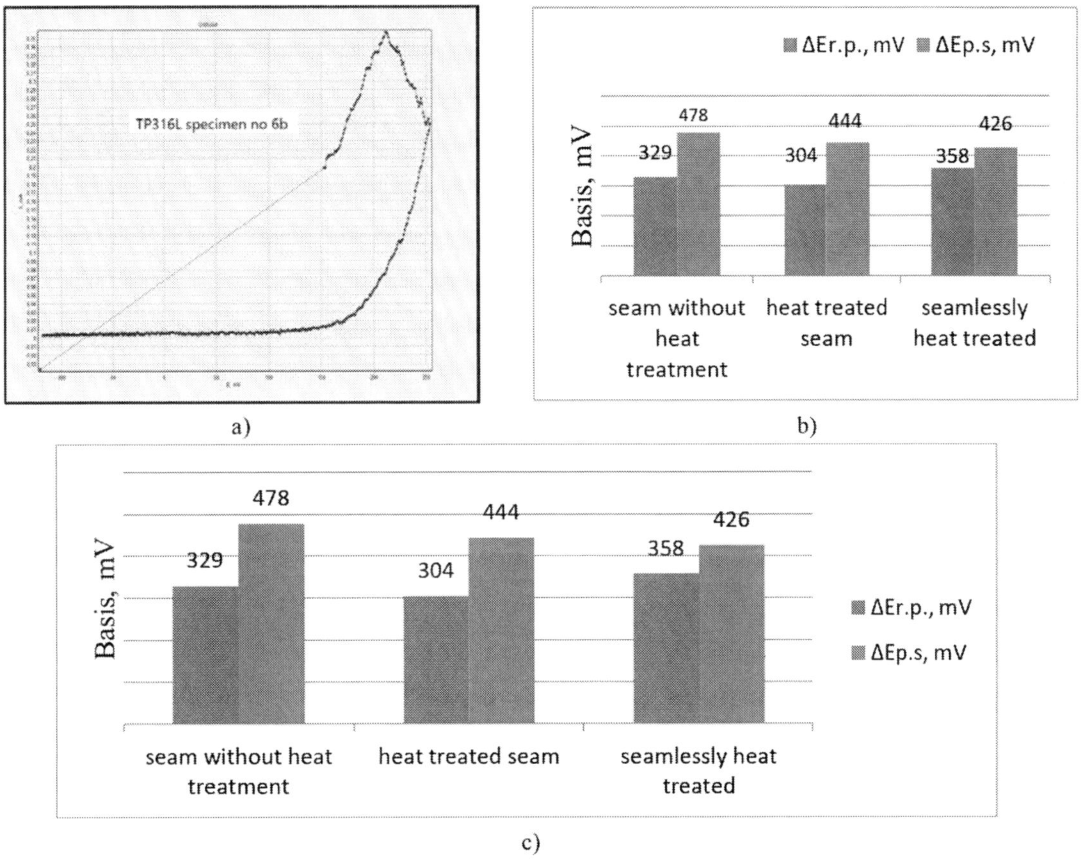

Figure 3. The results of potentiodynamic measurements of the pitting resistance of samples of longitudinal welded and U-shaped laser-welded tubes made of A316L steel: a – an example of polarization curves of potentiodynamic measurement; b – levels (bases) of repassivation and pitting resistance for longitudinal seams; c) for U-shaped samples.

To assess the effect of heat treatment on the stress state of samples (pipes), including those caused by the presence of defects of the crystal lattice of the second type (randomly distributed dislocations) [4], we analyzed changes in the position, width, and profile of X-ray lines in diffractograms. The survey was carried out on a DRON-3M diffractometer in Co-Kα radiation at points with a step of 0.2 degrees and a shutter speed of 10 seconds at a point without rotating the sample in its plane. On the samples with a size of 20 × 20 mm, the (311) line of the γ phase was recorded in the range of angles 2θ from 108.5 ° to 114.5 °. When installing the samples in the prefix GP-13, the weld was positioned vertically (dividing the beam projection in half).

The results of estimating the width of the diffraction maximum (311) of X-ray scattering by the γ-phase (austenite) are presented in table 2 (last column). The results obtained indicate a significant difference in the shape and width of the diffraction maximum of x-ray scattering by the γ phase in the samples of U-shaped welded pipes before and after heat treatment: more than 3 times. Moreover, the splitting of the $K_{\alpha 1}$ and $K_{\alpha 2}$ Co-K_{α} doublet of X-ray radiation is clearly manifested in all diffractograms of the annealed samples, which indicates a high degree of perfection of the crystal lattice of the γ-phase grain material after annealing.

The obtained graphical and numerical data illustrating the profiles and width of the X-ray lines indicate that as a result of austenization (at 1000 °C) in the samples of welded pipes of the A316L brand, internal stresses are caused by cold deformation of the strip during rolling, plastic deformation during tube forming billet and subsequent bending of welded pipes into a U-shaped product. The hardening in the metal of a welded pipe made of A316L steel is carried out by primary recrystallization (austenite), the process of which at temperatures of 1000–1050 °C occurs intensively (transiently) and takes no more than 5 minutes in time [5]. In this case, the metal strength (yield strength) is reduced by almost 2 times, the hardness level (HV) decreases from 300-350 MPa to 160-180 MPa, and ductility (relative and uniform elongation) increases by 1.5–2 times [6].

For a comparative analysis of the mechanical properties of laser-welded straight-seam pipes with a size of Ш25Ч1.5 mm and Ш25Ч2.0 mm made of a316l steel intended for producing bent products, mechanical tests were performed on a universal breaking machine by static stretching [GOST10006-80, GOST 1497-84, GOST 11701-84]. Experimental samples of pipes were divided into 2 parts: samples in the initial state; samples that have undergone heat treatment. The test results are shown in figure 4.

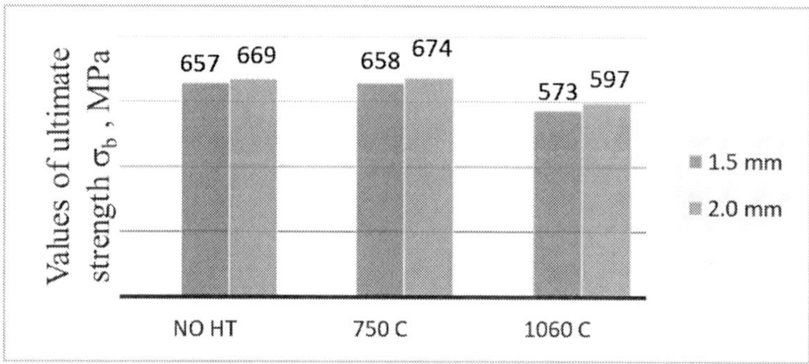

a)

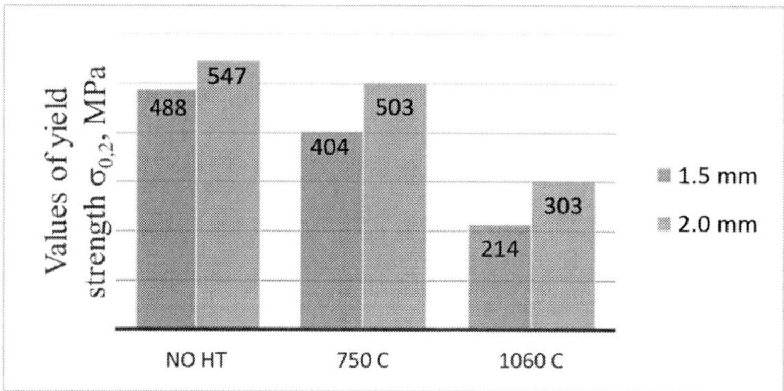

b)

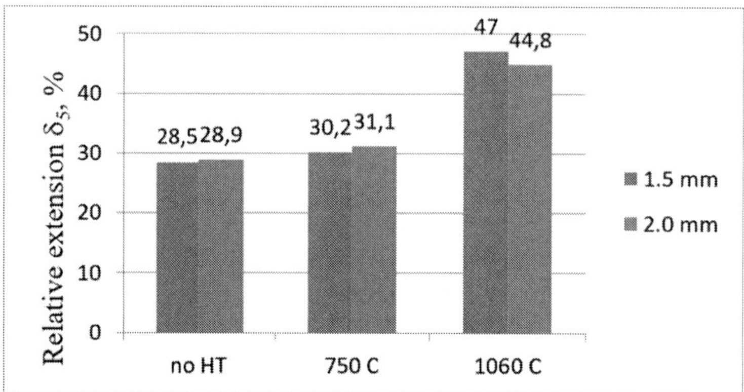

c)

Figure 4. Influence of the heat treatment mode on the mechanical properties of laser-welded straight pipes Ø25 made of a316l steel: a – strength; b – yield strength; c – ductility.

Table 2. Scheme of scientific and business experience.

N	Marking	The place where the sample is cut in relation to the bend: I-in; N-outside	External (E) and internal (I) pipe surface	The appearance of the sample	H/t, °C	Diffraction maximum shape	Width of the diffraction maximum (doublet), degrees
1	6N0I (seamless)	N	E	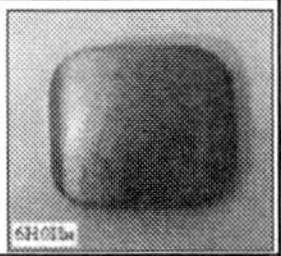	No	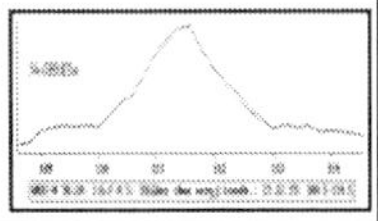	1.71±0.9
2			I			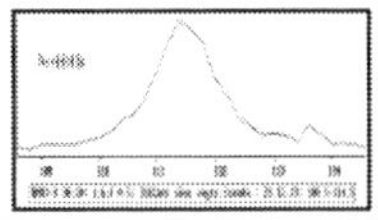	1.43±0.06
3	6nT	N	E	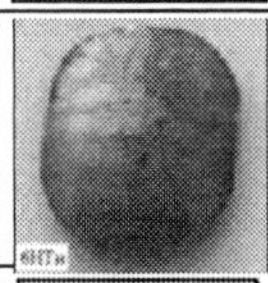	1000	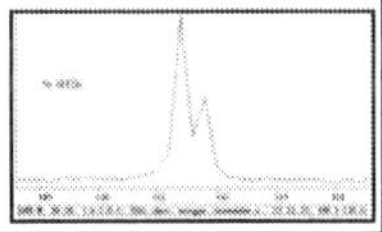	0.44±0.01
4			I	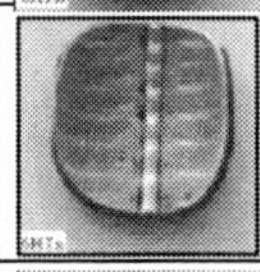		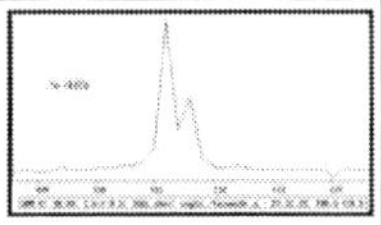	0.45±0.01
5	6вI	I	E		No	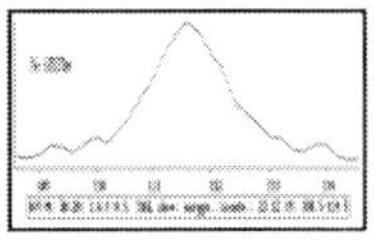	1.74±0.08
6			I	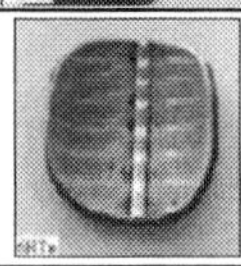		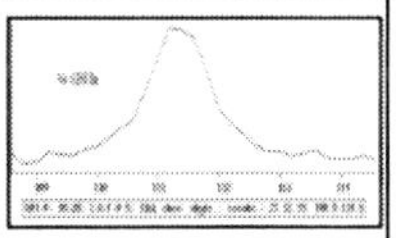	1.40±0.06
7	6вT_1	I	E		(T_1) 1000	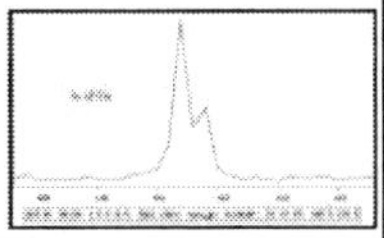	0.42±0.02
8			I			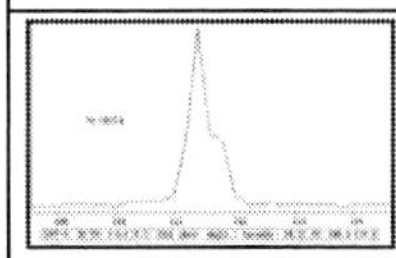	0.48±0.01
9	6вT_2	I	E	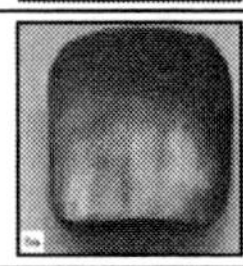	(T_2) 1060	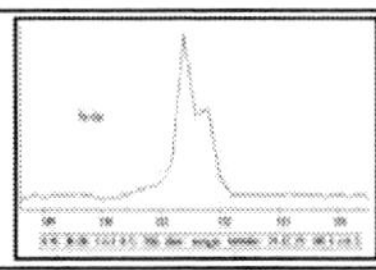	0.55±0.02
10			I	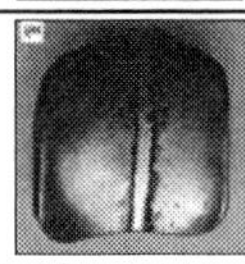		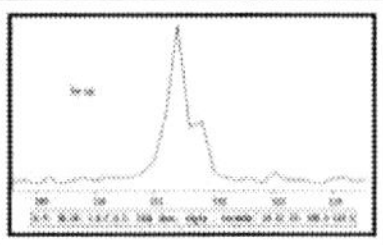	0.51±0.02

3. Conclusion

The results of mechanical and technological tests of straight-seam laser-welded pipes made of a316l steel for static stretching indicate that even if there is no final heat treatment in the technological cycle of their manufacture, welded pipes have a high margin of plasticity, which allows their applicability for the production of bent grades (U-shaped welded pipes).

Experimental data in the form of changes in the width and profile of the x-ray line of the metal of laser-welded pipes confirm the fact that the final heat treatment under the regime of short-term austenization completely eliminates the metal naklep, providing relaxation of its elastic stresses and the formation of the necessary structural state of a316l steel.

Heat treatment under the regime of short-term austenization at a temperature of 1000–1050 °C for 1–5 minutes can be recommended as a mandatory technological operation that completes the metallurgical conversion of laser-welded stainless pipes of the heat exchange range.

This heat treatment increases the corrosion resistance of the material of U-shaped welded pipes made of a316l steel to the level of their operational requirements when used in products of the heat exchange range for chemical engineering.

Acknowledgments

The work was supported by the Ministry of Education and Science of the Russian Federation, unique project identification number RFMEFI57817X0252.

References

[1] Columbier L and Gokhman I 1958 *Stainless and heat-resistant steels* (Metallurgizdat) p 241
[2] STP 26.260.484-2004 *Heat treatment of corrosion-resistant steels and iron-Nickel-based alloys in chemical engineering*
[3] Smirnova N A and Misyurov A I 2012 Features of structure formation during laser processing *Bulletin of the Bauman Moscow state technical University* **6(6)** 15–129
[4] Umansky Ya S, Skakov Yu A, Ivanov A N and Rastorguev L N 1982 *Crystallography, radiography and electron microscopy* (Moscow: Metallurgy) p 351–365
[5] Tikhonova M S and Belyakov A N 2013 Features of static recrystallization of austenitic stainless steel after intensive plastic deformation *Vesnik of Nizhny Novgorod University named after N. I. Lobachevsky* **2(2)** 92–95
[6] Gorelik S S, Dobatkin S V and Kaputkina L M 2005 *Recrystallization of metals and alloys* 3rd edition (Moscow: MISIS) p 432

IOP Conf. Series: Materials Science and Engineering **939** (2020) 012070 doi:10.1088/1757-899X/939/1/012070

Grounds and problem statement for software complex for photolithography optimization for minimization of losses in optical structures of photonic integrated circuits

A A Sharapov[1,2,3], E S Shamin[1,2], I D Skuratov[1,2] and E S Gornev[1]

[1] JSC Molecular Electronics Research Institute, 12/1 Academician Valiev Street, Zelenograd, Moscow, 124460, Russia
[2] Moscow Institute of Physics and Technology, 9 Institutskiy per., Dolgoprudny, Moscow Region, 141701, Russia

[3] E-mail: andrey.sharapov@phystech.edu

Abstract. The losses in optical components are the key obstacle in development of photonic integrated circuits (PIC). This makes the developers to include additional signal gain elements into the optical scheme. One of the major and optimizable factors of signal loss is the sidewall roughness of the optical components. After analysis of factors causing the line-edge roughness (LER), we developed the basic model of the roughness emerging process during forming of photolithography nanostructures. We proposed an approach to optimize the photolithography process in order to minimize the optical losses in fabricated optical structures. The software complex on the basis of this approach is currently under development.

1. Introduction

The continuous development in integrated circuits manufacturing is associated with technology node scaling [1]. However, the edge roughness of nanostructures cannot be decreased with the same factor [2]. Hence, the roughness-to-size ratio is growing for single elements, which makes devices on their basis unstable in terms of electrotechnical performance. Such issues are particularly relevant for structures having dimensions about the size limit of used technology.

For example, figure 1 shows the I-V curves modeled for a single FinFET having certain values of LER. Due to shortness of transverse dimensions in comparison to the roughness amplitude, the gate width is obviously subject to effects linked with the edge roughness emergence. The modeling of this structure was performed in Synopsys TCAD, and the existence of structure unevenness was simulated by applying the built-in method of edge transformation by noise function. This function is given by an inverse Fourier transform of autocorrelation function with predefined values of standard deviation σ and correlation length Λ:

$$f_{\text{autocorrelation}} = \sigma^2 \sqrt{\pi} \Lambda e^{-x^2/\Lambda^2}. \quad (1)$$

Content from this work may be used under the terms of the Creative Commons Attribution 3.0 licence. Any further distribution of this work must maintain attribution to the author(s) and the title of the work, journal citation and DOI.
Published under licence by IOP Publishing Ltd

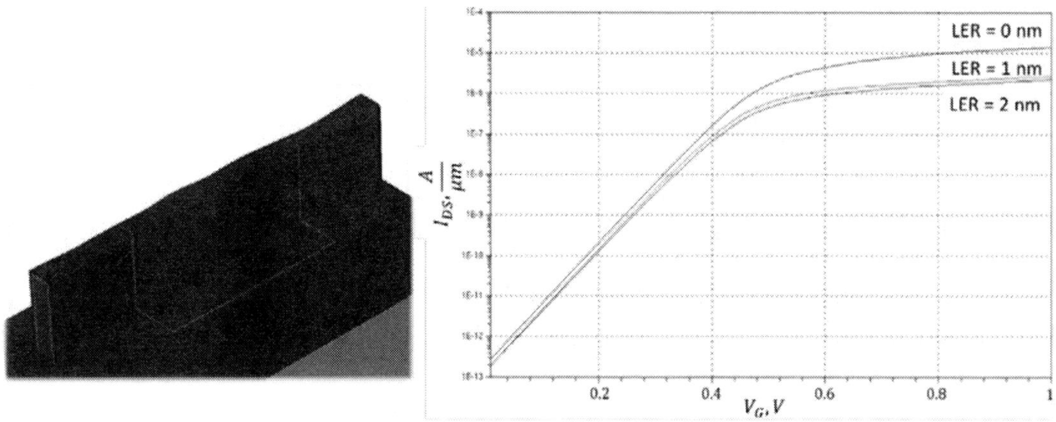

Figure 1. Scheme of FinFET and current–voltage curves modeled for fin with different line edge roughness values. Geometrical parameters of the fin structure – 8 × 40 × 100 nm; gate length – 30 nm; correlation length – 5 nm.

Not only electrotechnical performance of silicon microelectronic devices is affected by edge roughness, but this phenomenon is also an issue for silicon photonics. Despite the fact that this realm of science and technology develops rather quickly, sidewall roughness in optical waveguides which gives the main impact into the propagation losses in photonic circuit elements, continues to be a serious concern.

Due to light scattering at short-period roughness of waveguides, the loses may reach several decibels per millimeter, while total area of optical integrated circuits may span up to several square centimeters.

Losses calculation method for waveguides with account for the sidewall roughness is described in [3]. These losses for fin waveguides which are one of the most common structures in silicon photonics, are given by the following expression [4]:

$$\alpha(TE,TM) = \frac{4{,}34\,\sigma^2}{\sqrt{2}d^4\beta_{TE,TM}} g(V)\cdot f_e(x,\gamma), \tag{2}$$

where

$$g(V) = \frac{h^2V^2}{1+p^2} \text{ и } f_e(x,\gamma) = \frac{\left\{\left((1+x^2)^2+2x^2\gamma^2\right)^{1/2}+1-x^2\right\}^{1/2}}{[(1+x^2)^2+2x^2\gamma^2]^{1/2}}. \tag{3}$$

These functions in turn are defined by the set of standard parameters:

$$x = p(L_c/d)\,,\ \gamma = \frac{n_2V}{n_1p\sqrt{\Delta}},\ \Delta = \frac{n_1^2-n_2^2}{2n_1^2},$$
$$h = d\sqrt{n_1^2k_0^2-\beta^2},\ V = k_0d\sqrt{n_1^2-n_2^2} \text{ и } p = d\sqrt{\beta^2-n_2^2k_0^2}. \tag{4}$$

Here k_0 is the wavenumber in air, β is the effective propagation constant of certain mode, d is the half-width of the waveguide, L_c is the roughness correlation length, and σ is its standard deviation.

2. Theoretical description of edge roughness for waveguides formed by photolithography

All the phenomena causing the edge roughness of nanostructures during nanofabrication process, can be divided into the following groups by physical principles [5]:

1. Photonic shot noise is the lithographic phenomenon which takes place during aerial image formation.
2. Effects caused by chemical properties of resist and peculiarities of chemical reactions modifying the resist and affecting the resist imaging.
3. Effects associated with diffusion properties of photoresist.
4. Effects caused by plasma etching (also referred to as dry etching).

We suggest to implement the optimization algorithm which allows to find technology process parameters, which provide minimal losses in fabricated PIC due to decreasing of edge roughness magnitude.

2.1. Brief description of optical effects in photolithography

Photolithography process modeling is one of the key stages in microelectronic devices fabrication. It can be divided into modeling of aerial image and resist modeling.

As known from optics, description of light propagation principles through holes in the screen, are determined by three parameters, which are holes diameters, distance from the screen to observation plane, and wavelength of used light source. The three regimes of optics can be well demonstrated for the case of a screen with a single hole and monochromatic normal incident light beam. These cases are called geometrical optics, Fresnel diffraction, and Fraunhofer diffraction [6].

Practically important cases for photolithography are the diffraction optics regimes. Moreover, due to continual development in microelectronics, critical dimensions of the structures are decreasing, and the impact of diffraction effects becomes more and more significant. These effects lead to fuzzy boundaries, corner rounding, line-end shortening, and as for most critical, illumination of undesired area on the wafer [7].

Common view of intensity distribution cutline is presented on figure 2. This distribution apparently has smooth tails that exceed the bounds of initial topology elements. Further theoretical description in this work is essentially the impact analysis of these distribution tails on the structure unevenness.

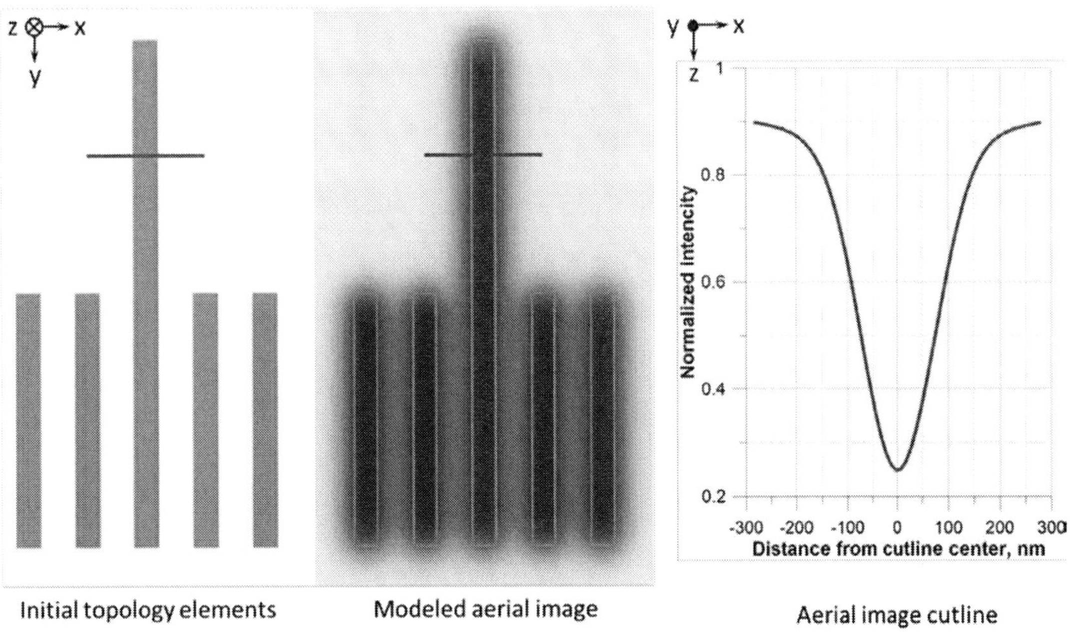

Figure 2. Common view of aerial image cutline at the position of red line. Geometrical parameters of the dark-field topology elements with 140 nm pitch: 100 nm × 1 μm for short lines, 100 nm × 2 μm for middle line. Intensity is normalized by the full illumination intensity.

2.2. *Impact of photonic shot noise into edge roughness*

First phenomenon causing the edge roughness, is photonic noise [8]. Let us consider the light source, which emits photons at the rate F per unit time and unit area. Let us also consider the binary random phenomenon of photon emission during the time segment dt, that is short enough so no more than one photon can be emitted during it, i.e. $dt \ll 1/F$. The probability of photon emission during dt is Fdt. Let us now find the probability that exactly n photons will be emitted during time interval T, which is N time periods each equal to dt:

$$P(n) = C_N^n (Fdt)^n (1 - Fdt)^{N-n}. \tag{5}$$

We can use the Poisson limit theorem for binomial distribution in (5) if $N \to \infty$ and $NFdt = TF = const$, which is a practical case for photolithography:

$$\lim_{N\to\infty} C_N^n (Fdt)^n (1 - Fdt)^{N-n} = \frac{(TF)^n}{n!} e^{-TF} \tag{6}$$

From Poisson distribution properties one can estimate expected value of photon amount, emitted during time interval T:

$$E(n) = \sum_{n=0}^{\infty} n \frac{(TF)^n}{n!} e^{-TF} = e^{-TF} \sum_{n=1}^{\infty} \frac{(TF)^n}{(n-1)!} = TF\, e^{-TF} \sum_{m=0}^{\infty} \frac{(TF)^m}{(m)!} = TF\,, \tag{7}$$

and the variance of photon amount is

$$\sigma_n^2 = E\left[\left(n - E(n)\right)^2\right] = E(n^2) - \left(E(n)\right)^2 = TF\,. \tag{8}$$

Let us switch to dimensional units to associate these estimations with physical quantities. The key quantity is light intensity, which expresses the amount of energy transferred by n photons per unit time T to unit area A:

$$I = \frac{n}{TA} \frac{hc}{\lambda} \tag{9}$$

Then at each spatial segment intensity expected value can be calculated as follows:

$$E(I) = \frac{E(n)}{TA} \frac{hc}{\lambda} = \frac{F}{A} \frac{hc}{\lambda}, \tag{10}$$

The same substitution can be used to obtain intensity standard deviation:

$$\sigma_I(I) = \frac{1}{TA} \frac{hc}{\lambda} \sigma_n = \frac{E(I)}{\sqrt{TF}} = \frac{E(I)}{\sqrt{E(n)}} \tag{11}$$

As a result, we get the relation between intensity standard deviation and its expected value for each spatial segment:

$$\sigma_I(I) = \frac{\sqrt{E(I)}}{\sqrt{TA \frac{\lambda}{hc}}} \tag{12}$$

Equation (12) allows to calculate an area in which photonic shot noise could affect the edge roughness. Expected value of intensity in (12) could be substituted for modeled values of intensity, unit area A could be substituted for a pixel area of the modeled intensity and time period T could be substituted for exposure time. Then we could build the map of intensity standard deviations over the intensity distribution. When modeling intensity distribution (also referred to as aerial image) and subsequent resist imaging using empirical models based on the data from fabrication results, one can estimate the average imaging intensity I_{imaging}, which characterizes the boundaries of the areas where resist becomes exposed (blue line at figure 3). This value is closely related to the simplest of resist models, the so-called constant threshold resist models. The average imaging intensity value applied as a threshold on the intensity distribution plot, gives an isoline, which is surely not the same as the imaging contour due to performed averaging, though it is a simple and robust estimation. Using three-sigma rule at each point of the aerial image, we can define the area of unreliable illumination as follows:

$$\begin{cases} I + 3\sigma_I(I) > I_{\text{imaging}}\,, & \text{if } I < I_{\text{imaging}} \\ I - 3\sigma_I(I) < I_{\text{imaging}} & \text{otherwise.} \end{cases} \tag{13}$$

We will have a closer look at this area in the next section.

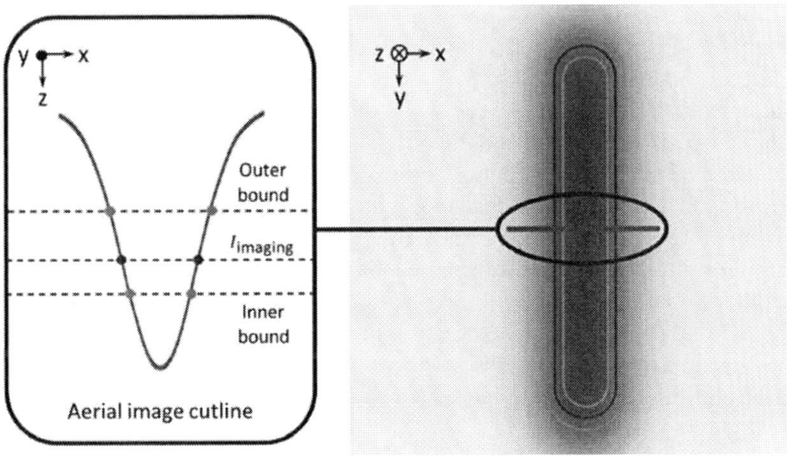

Figure 3. Schematic explanation of area of unreliable illumination definition (between green contours) for a dark-field topology element. Blue isoline corresponds to the average imaging intensity.

A common practice for microelectronic manufacturers is to calculate normalized aerial image distribution. This is appropriate for the use of empirical resist models and significantly simplifies the calculations. To get the dimensional quantities of intensity, one shall use the factor of light source power for the certain photolithography setup.

2.3. *Impact of photoresist imaging effects into edge and sidewall roughness*

Resist imaging process is enabled by illumination of resist molecules. For a positive resist, incident light breaks the molecular bonds in illuminated areas on the resist layer surface, and solubility of resist in these areas increases. For a negative resist, on the contrary, the solubility of illuminated areas decreases. Significant part of modern photoresists are so-called chemically amplified resists, which represent a mixture of a polymer and a photo-sensitive compound called photoacid generator. The latter substance decomposes upon exposure and releases acid, which dopes polymer during post-exposure baking and is being released into resist again, i.e. behaves as a catalyst. In the doped areas, polymer becomes soluble, and reacted products are removed in etching.

In all mentioned types of photoresist, it is the chemical reaction that plays the key role in imaging. This reaction is characterized by a certain activation energy.

In order that the resist in different areas of surface had similar properties in relation to incident illumination, its composition shall have as narrow distribution of molecular weights. If confined to consideration of resist as absolutely uniform substance having fixed threshold of chemical reaction activation, the ratio of this activation energy to the unit area is essentially the resist threshold exposure dose $D_{\text{threshold}}$.

To get relation with photolithography parameters, the dose can be expressed in terms of quantities introduced in the previous section:

$$D = \frac{n}{A}\frac{hc}{\lambda} = TI \qquad (14)$$

Hence, having certain exposure time T, resist illumination area matches the area, surrounded by contour, corresponding to fixed intensity value $D_{\text{threshold}}/T$. Due to the absence of lateral shifts of such contour, the line edge roughness, which is defined as standard deviation of line points from the average edge position, equals zero.

In fact, edge roughness of fabricated structures is defined by random deviations of illumination intensity in the area of unreliable illumination. For each molecule in this area, the condition to start chemical reaction $I > D_{\text{threshold}}/T$ is satisfied with certain probability. This means, that for a large number of molecules there are such ones, that are not modified, despite the fact that they are formally in the area of illumination. In case of chemically amplified resists, acid diffusion during post-exposure baking "smooths" the resist properties, and the interface between different states of photoresist becomes more even. The width of the formed structure, however, is still uneven.

Let us now consider the impact of discrete polymer-based resist structure on edge roughness. To a basic approximation, the polymer macromolecule can be represented as freely-jointed chain of monomers. This allows not to account for the peculiarities of single atom interaction while describing spatial polymer configuration. In addition to structure discreetness of the substance, one shall also take into account the fact that there is a certain variation in mass and radius of constituent molecules. Taking all this into consideration, and assuming that all the molecules within the area of unreliable illumination, form the interface and thus define the edge roughness, one can make a schematic illustration of these phenomena (see figure 4).

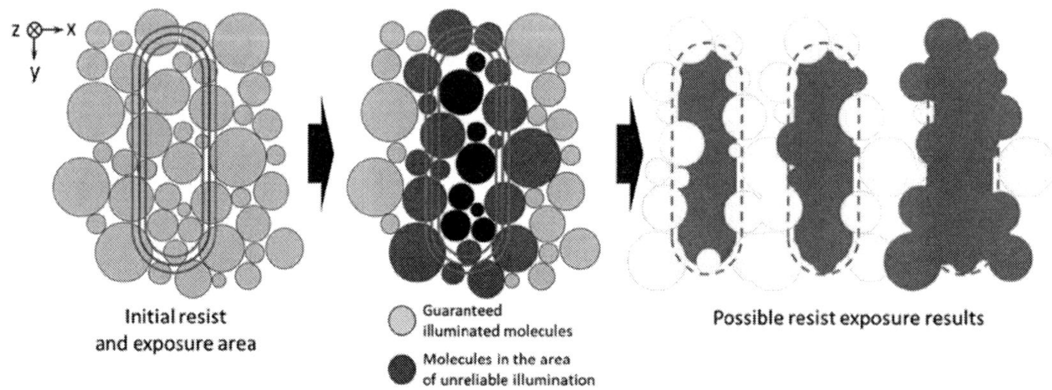

Figure 4. Schematic demonstration of impact of intensity dispersion, non-zero width of molecular size distribution, and resist discreetness, to edge roughness forming. Geometric size of molecules is increased by several decades for the sake of visibility.

Estimation of roughness characteristics for given topology and materials, including correlation length and power spectral density analysis, is performed using stochastic modeling. Input parameters for this technique are molecular mass distribution of photoresist, and resist illumination dose distribution.

Due to requirements of high etch selectivity, and in order to prevent the negative effect on the width evenness of fabricated elements, the most common is plasma etching, which forms a sidewall with roughness inherited from the top edge of the structure downwards in the material (see figure 5). Hence, it is acceptable to assume that roughness of the structure as well as its linear qualitative characteristics, stay unchanged.

Figure 5. Sidewall morphology of fin structure pictured by atomic force microscopy [9]. Edge roughness is inherited along the etching direction.

3. Optimization algorithm

Now having information on sidewall roughness of structures for certain fabrication process, one can estimate the qualitative parameters of width unevenness for waveguides, and also optical losses using equations (2) – (4). In order to decrease the resulting losses, one shall adjust certain parameters of technology process using the optimization algorithm presented on figure 6.

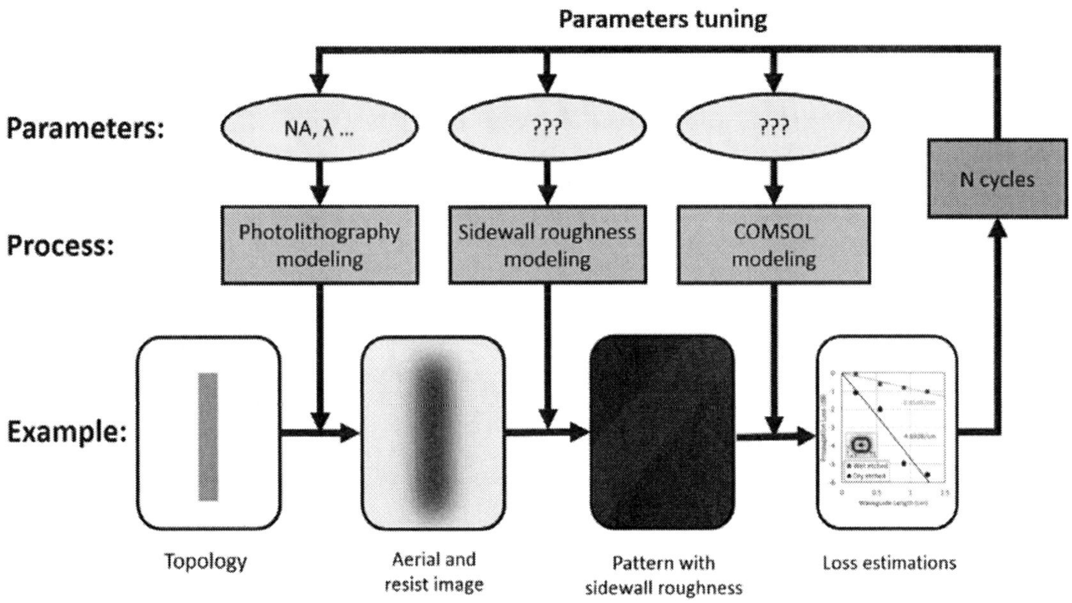

Figure 6. Scheme of suggested optimization algorithm for photolithography adjustment to minimize losses of fabricated optical elements.

Verification of suggested algorithm shall be conducted both comparing the relative optical losses in PIC before and after technology adjustment, and also by assessment of roughness for fabricated structures. The set of test structures to conduct such measurements is under development.

Sidewall roughness in silicon microelectronics is usually measured using scanning electron microscopy (SEM) [10], and the most stable on-wafer technique to estimate LER value is by calculation of the standard deviation of edge points position sampled from the SEM image [11]. Optoelectronic structures are currently also fabricated on silicon wafers, which gives a reason to use mentioned roughness analysis methods in photonic integrated circuits manufacturing.

4. Conclusion

In this work we considered how do certain processes of microelectronic technology influence the amplitude of edge roughness of fabricated structures. We suggested the algorithm of photolithography optimization to minimize the optical losses in structure of photonic integrated circuits. The implementation and verification of the algorithm suggested in this work is the subject of further research.

References

[1] International Technology Roadmap for Semiconductors 2011

[2] Krasnikov G Ya 2011 *Constructive and Technological Features of Submicron MOSFETs* (Moscow: Technosphere)

[3] Marcuse D 1969 Radiation losses of dielectric waveguides in terms of the power spectrum of the wall distortion function *Bell Syst. Technol. J.* **48** 3233–44

[4] Yap K P, Delâge A, Lapointe J, Lamontagne B, Schmid J H, Waldron P, Syrett B A and Janz S 2009 Correlation of scattering loss, sidewall roughness and waveguide width in silicon-on-insulator (SOI) ridge waveguides *IEEE J. Lightw. Technol.* **27(18)** 3999–4007

[5] Sharapov A A, Baranov G V 2018 Analysis of factors influencing the roughness of topological structures of Si microelectronics *Int. Forum "Microelectronics 2018"* (Moscow: Technosphere) 205–207

[6] Sivukhin D B 1980 General physics course. Optics vol 4 (Moscow: Nauka)

[7] Bespalov V A et al. 2013 *Introduction to photomask design for micro- and nanosystem fabrication. Cadence MaskCompose* (Moscow: BKL)

[8] Mack C 2008 *Fundamental principles of optical lithography: the science of microfabrication* (John Wiley & Sons)

[9] Baranov G V, Milenin A, Baklanov M 2016 Investigation of the impact of plasma etching steps on the roughness of the fin FET channel sidewalls in the scheme of hetero-integration *Russian Microelectronics* **45(3)** 186–190

[10] Gornev E S, Kalendin V V, Novikov Yu A, Rakov A V, Todua P A 2004 Metrology in linear measurements in nanotechnology *Materials of Electronic Technics* **2** 55–59

[11] Sharapov A A, Baranov G V 2018 Comparative analysis of estimation techniques for qualitative roughness characteristics of nanostructures *Proc. of MIPT* **2(38)** 72–79

Modification of wear-resistant coatings of the Fe-Cr-C system with spherical tungsten carbide WC-W_2C

A V Shchegolev[1,2], A V Ishkov[1] and V V Ivanayskiy[1]

[1] Altai State Agricultural University, Krasnoarmeiskiy pr., 98 656049, Barnaul, Russia

[2] E-mail: qqq681@mail.ru

Abstract. The article describes the production procedure, studies the structure, and determines the properties of new wear-resistant coatings of the Fe-Cr-C system intended for hardening the working bodies of agricultural machines. To increase the wear resistance and give new consumer qualities to such coatings, they were modified with spherical tungsten carbide WC-W2C – spherorelite.
It is shown that the introduction of the complex (WC-W2C+B4C) into the basic charge allows increasing the hardness of the deposited to 1000 ... 1200 HV100, and wear resistance by 3.1 times in relation to the base material, and by 1.3 ... 1.8 times in relation to the base (not hardened) material.

1. Introduction

To extend the service life and preserve the parameters of the details of the working bodies of agricultural machines, their surface is strengthened with coatings of high-chromium cast irons – alloys and composite materials of the Fe-Cr-C, Fe-Cr-C-B systems and others [1]. Reinforcing coatings can be applied to the part by various methods.

However, the wear resistance of these materials, in some cases (crusher hammers, chopper knives, plowshares, mixer blades, etc.), may be insufficient [3].

Existing methods to increase the wear resistance of these materials are based on: the development of new surfacing materials based on various d-metals or alloys [4]; the introduction of additional components into existing materials (nanoparticles, superhard carbides, ceramics, etc.) [5–6]; a change in the nature of contact interaction with the implementation of the Charpy effect; the formation in the coating or at its borders of new wear-resistant phases [8-11], etc.

However, the most promising, from a technical point of view, is the introduction of small amounts (up to 5 wt.%) of active additives into the finished surfacing materials, for example, the well-known dry friction modifier MoS_2 [12], which allows the known high-performance coating to be used without significant changes induction surfacing technology [1].

For the formation of hardening coatings of high chromium cast iron on the surface of the working bodies of agricultural machines by induction surfacing in industry, materials of grades ПГ-С27 and ПГ-УС25 are widely used. Previously, we have already studied similar coatings made of cast iron grade ПГ-УС25, reinforced with an additive of up to 5 wt. % B_4C [13], as well as coatings reinforced with additives up to 25 ... 30 wt. % cermet Cr_3C_2/ПГ-УС25 obtained by the SHS method [14]. The coatings were deposited with a layer of 0.2 ... 0.3 mm on the blade of the harvester knife part, while their wear resistance increased by 2.5 ... 3.1 times.

Content from this work may be used under the terms of the Creative Commons Attribution 3.0 licence. Any further distribution of this work must maintain attribution to the author(s) and the title of the work, journal citation and DOI.
Published under licence by IOP Publishing Ltd

Spherical powder of tungsten carbide eutectic $WC-W_2C$ – spherorelite is a new, promising material for hardening various parts, as well as giving new properties to known coatings based on Ni, Fe, Cr [15]. This unique product based on one of the hardest, strongest (HV = 2200 MPa, = 700 GPa) and thermally stable materials in the industry – WC, is obtained by the original technology of centrifugal spraying of its ingots. At the same time, the granules have a regular, strictly spherical shape, and due to the atomization of ingots from a $WC-W_2C$ carbide eutectic mixture additionally alloyed with d-metals, they also have a stable chemical composition and a fine-grained structure, which ensures high technological effectiveness of the powder and a guaranteed additional increase in material hardness.

However, at present, for the modification of hardening, wear-resistant coatings of the Fe-Cr-C system with tungsten carbide and, in particular, spherorelite, the industry uses more high-energy, but also much less productive than high-frequency welding deposition methods of their application: plasma, electric arc and gas-flame technology [5, 7, 8]. Therefore, the creation of wear-resistant coatings of high-chromium cast iron, hardened with spherorelite, using the method of induction surfacing is relevant.

The purpose of this work is to obtain by method of induction surfacing wear-resistant coatings of hard alloy ПГ-УС25 on steel 65Г, additionally hardened by their modification to 5 ... 10 wt. % spherorelitis КВС-5, as well as their material science research.

2. Experimental procedure

As the initial reagents used: powder of high-chromium hard alloy (cast iron), grade ПГ-УС25); grinding powder of boron carbide B_4C, grain size 10 (GOST 5744-85); spherical eutectic tungsten carbide $WC-W_2C$ – spherorelite, grade КВС-5 (experimental batch); nickel powder, thermosetting, ПТ-19Н-01; as well as borate flux for induction surfacing П-0,66 [7]. All powder metal and flux materials were dried, crushed, and sieved through sieves with a mesh, ∅ 0.350 and 0.125 mm, respectively, before preparing the charge.

Coatings were obtained by induction surfacing of a basic industrial charge, composition, wt. %: ПГ-УС25 – 85; flux П-0,66 – the rest [3]. The components were dosed by weight and mixed in a biconical mixer for 0.5 ... 1 hour. The modifier, the $WC-W_2C+B_4C$ complex, was introduced into the charge by reducing metal (up to 75 wt.%), or non-metal (up to 10 wt. %) parts.

The coatings were surfaced on an ELSIT-100/75 HDTV unit with a water-cooled one-turn slotted inductor made of a profiled copper tube, ∅ 15 mm, applying the finished mixture with a layer of 2–3 mm thickness on experimental samples made of 65Г steel, 50 × 20 × 5 mm in size, with high-frequency heating (ν = 65 ... 75 kHz) to a temperature of 1150 ... 1250 ° C, for a time of 45 ... 60 sec. The obtained samples were used to optimize the compositions and refine the surfacing technology, evaluating the state of their surface and the hardness of $HRC_э$ (TK-2 press).

Metallographic studies were carried out on a CARL ZEISS AXIO OBSERVER-Z1 M microscope and MN-6 microhardness meter according to standard ASTM methods: E 3, E 407, E 1558, E 883, E 384. X-ray phase analysis of the products was carried out on a DRON-6 diffractometer (Cu, K-radiation, λ = 1.5418 Å), phase identification – in the PDWin software package, using card files: PDF-2, ICDD.

X-ray fluorescence microanalysis was performed on a Philips SEM515 with an EDAX ECON IV microanalyzer.

Wear resistance was determined according to the original method, which combines friction tests on loosely fixed abrasive particles GOST 23.208-79 (ASTM G174-04 (2017), and according to the finger-ASTM G99 (2010), on a special laboratory tribometer, load 40 ... 45 ± 0.25 N, internal standard – steel 65Г, hardened to 55 ... 57 HRCэ.

Relative wear was determined by weighing the samples on a ВЛК-500 balance, with an accuracy of ± 0.01 g.

3. Results and discussion

The experimental batch of spherorelite КВС-5 used by us is additionally alloyed by the manufacturer up to 0.5 ... 0.8 wt. % Co, which allowed to obtain a product with a hardness of up to 3100 MPa, with the following composition, wt. %: W 95.8-96.0; C 3.8-4.0; Fe <0.11; Co 0.5-0.8; C (free) 0.02-0.05, product fraction, ∅

IOP Conf. Series: Materials Science and Engineering **939** (2020) 012071 doi:10.1088/1757-899X/939/1/012071

0.25-0.5 mm.

The difficulty of obtaining monolithic coatings from high-chromium cast iron, when modified with tungsten carbide, using high-frequency technology, is explained by the existence of an objective technical contradiction between the working temperature of this process 1100 ... 1200 °C [3], the melting points of carbides 2500 ... 2800 °C, formed in the W-C system [16], as well as melting points of 1400 ... 1500 °C of ceramic materials of the ВК type on a cobalt bond [1]. The large temperature difference and the short process time do not allow to obtain a high-quality, monolithic coating.

Trial experiments conducted by us on hardening ПГ-УС25 with КВС-5 spherorelite alone, taken in an amount up to 5 wt. %, as expected, did not give the desired effect. In the coatings obtained, the tungsten carbide granules did not form a metal bond with high-chromium cast iron and were tinted from the layer even in the process of manufacturing isolated metallographic samples. In addition, due to a sharp difference in the mechanical properties, thermal expansion coefficients, and the melting point of spherorelite and hard alloys, cracks, pores, and other macrodefects formed in the coatings upon receipt.

It is known that the modification efficiency and the quality of coatings can be improved by increasing adhesion at the boundary of spherorelite granules with high-chromium cast iron at a melting temperature of 1100 ... 1200 °C. This problem can be solved by using tungsten carbide in coatings obtained by high-frequency technology not in pure form, but as part of special complexes – mixtures with substances capable of chemically interacting simultaneously with both the surface of the modifier and the binder material, and their activity it should appear at temperatures below 1100 ... 1300 °C. Boron carbide B_4C, can become such a component of the surfacing charge, which, under the conditions of high-frequency heating, as shown earlier [1, 17], reacts with the components of the hard alloy and the deposited steel surface, forming various products in the volume and at the phase boundaries: eutectic Fe-B, borides Fe_2B, FeB, carboborides Fe, $Me(C,B)_3$, boron austenite Fe, $Me_{23}(C,B)_6$ where: Me is the alloying metal (s), which is also confirmed by the data of [9, 18] .

The high reactivity of B_4C under conditions of high-frequency heating and its introduction into the WC-W_2C+B_4C complex is also explained by the topochemical reaction between these substances and iron, steel with the formation of low-melting 960 ... 980 °C Fe-B eutectic [18]. These circumstances suggest that tungsten carbides WC, W_2C, which are part of spherorelite, are involved in physicochemical and adhesive processes at the phase boundaries.

Therefore, further modification of ПГ-УС25 cast iron coatings was already carried out by the КВС-5+B_4C complex, introducing its components into the finished charge for induction surfacing by reducing the ПГ-УС25 cast iron spherorelite content, and due to the flux for induction surfacing П-0,66 – boron carbide. The total number of components of the modifying complex in the mixture was 4 ... 12.5 wt. %, which, with the selected modification methodology, allows one to preserve the thickness, uniformity, quality, and optimal coating deposition regime in industrial conditions.

Figure 1 shows the structure of the coating obtained by modification of cast iron ПГ-УС25 complex: 3 ... 5 wt. % spherorelitis КВС-5 and 1 ... 1.5 wt. % boron carbide, with the main phases indicated on it (numbers).

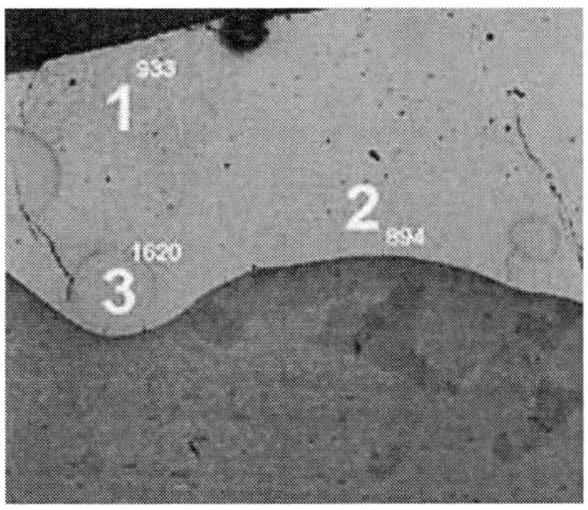

Figure 1. Structure of the coating modified by the complex: 3 ... 5 wt. % spherorelite КВС-5 + 1 ... 1.5 wt. % B_4C.

IOP Conf. Series: Materials Science and Engineering **939** (2020) 012071 doi:10.1088/1757-899X/939/1/012071

As can be seen from the micrograph shown in figure 1, when modifying high-chromium cast iron with the KBC-5 + B_4C complex, in an amount of 4 ... 6.5 wt. % it was possible to obtain a monolithic coating in which spherorelite granules formed metal bonds with a matrix metal, are firmly held in the deposited layer and do not crumble during friction.

The following main phases are formed in such a coating during induction surfacing: 1 – boron fine-dispersed eutectic from Fe and Cr carbides, with individual inclusions of crushed needle carbides Cr_7C_3, $Cr_{23}C_7$ with a hardness of up to 930 ... 950 HV_{100}; 2 – eutectic from Fe, Cr carbides, with inclusions at the fusion boundary with the base metal of primary and secondary boron dendrites, with hardness up to 850 ... 900 HV_{100}; 3 – spherical granules of carbide WC- W_2C, with a defined boundary, adhesive-bonded with a cast iron matrix, with a hardness of up to 1600 ... 1950 HV_{100}.

It is interesting to note that in the obtained coating, despite significant melting of the base metal, there are completely no dendritic structures typical for induction surfacing at the metal base / coating interface [1].

However, there are macroscopic defects in the coating: pores, cracks, the formation of which is associated with the presence of non-metallic inclusions (B_4C and products of its assimilation by surfacing material – graphite, soot).

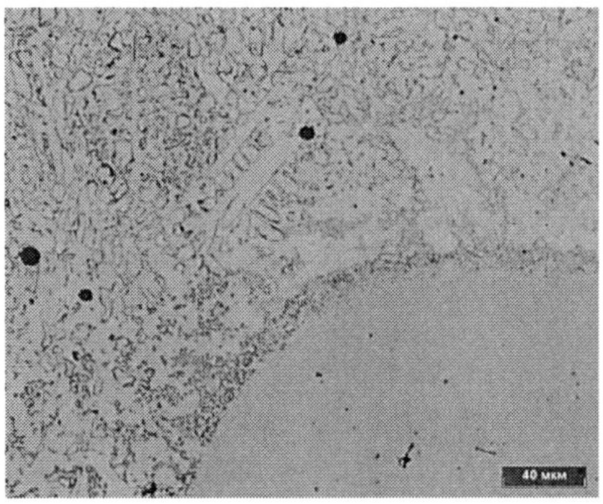

Figure 2. Fine structure of the spherorelite / high chromium cast iron interface in the complex modified coating: 5 ... 10 wt. % spherorelite KBC-5 + 1.5 ... 2.5 wt. % B_4C + 2 ... 3 wt. % ПТ-19Н-01.

To improve the quality of the coating, a softer material was additionally introduced into its composition – a thermosetting nickel powder clad with aluminum ПТ-19Н-01 in an amount of n. 2 ... 3 wt. % for the "healing" of cracks. In addition, the introduction of this component, theoretically, will increase the content of spherorelite in the material up to 8 ... 10 wt. % without deterioration of the quality of the coating [2, 4, 8].

Figure 2 shows the fine structure of the coating obtained by modification of cast iron ПГ-УС25 complex: 8 ... 10 wt. % spherorelitis KBC-5; 1.5 ... 2.5 wt. % boron carbide and 2 ... 3 wt. % nickel powder ПТ-19Н-01. As can be seen from the figure, not only crack healing occurred in the coating material, but also a chemical interaction of boron carbide with the surface of spherorelite granules is observed, leading to the formation of a new phase at the interface – a tungsten-boride eutectic, which has a characteristic skeletal morphology, which leads to increase the adhesive interaction of the WC, W_2C modifier with the coating matrix.

In table 1, the results of a chemical analysis of the composition of the phases of the coating, modified by a complex of 3 ... 5 wt. % spherorelitis + 1 ... 1.5 wt. % B_4C, at control points (see figure 3), obtained by X-ray fluorescence microanalysis.

As follows from the data in table 1, the composition of the main phases of the obtained coating – boronated (point 1) and non-borated (point 2) eutectics from Fe, Cr carbides, does not differ significantly in the main and dissolved elements (Cr, Mn, Fe, Si) . It can be seen that the boration of this phase of the material occurred mainly due to isomorphic substitution of part of the carbon in the eutectic Fe-Cr-C with boron, as evidenced by the fraction of this element reduced in 2 ... 3% in phase 1. But the introduction of B_4C into the material coating led to an increase in the eutectic content of silicon, and almost complete displacement of Nickel, which is confirmed by the data of [9].

Table 1. The results of a chemical analysis of the composition of the phases of the coating, modified by the complex composition, wt. %: 3 ... 5 - WC- W_2C + 1 ... 1,5 – B_4C.

Spectrum point (phase)	Element, Kα line	Composition, % элемента, %	
		wt.	at.
1.	C	08.91	32.81
	Si	02.24	02.32
	Cr	35.73	26.56
	Mn	02.39	01.98
	Fe	49.53	34.29
	Ni	00.09	00.07
	B	02.23	01.53
2.	C	11.13	35.81
	Si	01.24	01.71
	Cr	33.37	24.16
	Mn	01.39	00.98
	Fe	42.53	29.12
	Ni	00.99	00.65
3.	C	03.71	16.26
	Si	02.52	05.07
	Cr	24.84	26.98
	Mn	02.83	02.91
	Fe	59.38	60.05
	Co	00.55	00.39
	W	07.68	02.36
	B	01.17	00.93

The transformations occurring at the interface between spherorelite granules and high-chromium cast iron (point 3) also change the composition of the base material, and the degree of boron assimilation here is higher than in case of point 1, since almost the entire amount of carbon determined at point 3 corresponds to a stoichiometric fraction of this element in the carbide eutectic WC-W_2C [16].

However, the X-ray phase analysis of this coating did not reveal new phases of individual or mixed borides in it, which once again confirms the main form of boron in modified materials – in the composition of the eutectic Fe-C-B, Fe-Me-C-B, where: Me is the alloying metal (s) where it replaces part of the carbon. While the formation of special carbides $W_2Fe_{21}C_6$, W_2Co_4C in the coating of new phases (figure 3), indicates the occurrence of surface chemical reactions due to the participation of the low-melting Fe-B eutectic and redistribution of carbon from the carbides and 3, 4 component eutectics Fe-Cr-C- (B) [10, 17].

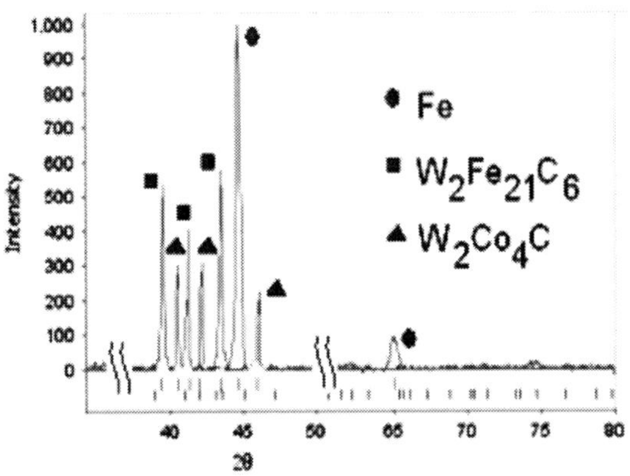

Figure 3. A fragment of an x-ray of the coating modified by the complex: 3 ... 5 wt. % spherorelitis + 1 ... 1.5 wt. % B_4C.

The absence of peaks of FeB and Fe_2B in the X-ray diffraction pattern, for example, is similar to coatings that we obtained earlier by high-frequency heating of 65Г steel in a charge containing 80 ... 85 wt. B_4C [18], as well as the absence of peaks of Cr, W borides, can be associated with a small total share of boron both in the introduced complexes (1 ... 2.5) and in the charge as a whole (0.8 ... 2 wt. .%).

The conducted studies allowed us to develop two optimized composition of the charge for industrial induction surfacing of high-chromium cast iron ПГ-УС25 hardened with spherorelite, the compositions and some properties of which are shown in table 2.

Table 2. Compositions of a charge for induction surfacing of coatings of high-chromium cast iron ПГ-УС25 hardened with spherorelite.

Charge	Ingredient	Composition, wt. %	HRCэ
1.	ПГ-УС25	80...85	60...65
	КВС-5	3...5	
	B_4C	1...1.5	
	Flux П-0,66	the rest	
2.	ПГ-УС25	75...78	50...55
	КВС-5	8...10	
	B_4C	1.5...2.5	
	ПТ-19Н-01	2...3	
	Flux П-0,66	the rest	

According to table 2, you can issue the following production recommendations. To obtain thin wear-resistant coatings on the cutting edges of parts and hardening, for example, heat-treated knives of agricultural machines, it is preferable to use composition No. 1, since the hardness of this material is 1.2 ... 1.4 times higher than the hardness of the base metal (65Г steel), which provides the blade with a self-sharpening effect and increases the service life [1]. To obtain thicker coatings (lancet paws, crusher hammers, etc.), as well as in other cases, composition No. 2 can be recommended.

4. Conclusion

It has been proposed to increase the wear resistance of Fe-Cr-C coatings obtained by induction surfacing of ПГ-УС25 high-chromium cast iron and their modification with the $WC-W_2C+B_4C$ complex.

It is shown that the introduction of the specified complex into the basic industrial charge for induction surfacing of the complex allows to increase wear resistance by 3.1 times (with respect to the base metal) and 1.3 times (with respect to the base, unmodified material), depending on the mass fraction spherorelitis in the charge.

The detected changes in the structure and properties of the coatings are explained by various processes: the formation of new phases; Charpy effect; grinding and dispersion hardening of the eutectic Fe-Cr-C during high-frequency boronation and modification.

References

[1] Ishkov A V, Ivanaysky V V and Krivochurov N T 2014 *Composite Coatings of the Fe-Cr-C-B System for Hardening Machine Parts* (Saarbrucken: LAP LAMBERT) p 286
[2] Laird G, Gundlach R and Röhrig K 2000 *Abrasion-resistant Cast Iron* Handbook Ed. American Foundry Society (Schaumburg, USA)
[3] Zum K and Eldis G 1980 Abrasive Wear of White Cast Irons *Wear* **64** 175–194
[4] Guilemany J M, Miguel J M, Vizcaino S *et al.* 2002 Role of Heat Treatments in the Improvement of the Sliding Wear Properties of Cr_3C_2–NiCr Coatings *Surface Coating Technology* **157** 207–213

[5] Kazamer N *et al.* 2016 A Comparison between Hardness, Corrosion and Wear Performance of APS-Sprayed WC-CoMo and WC-Co Coatings *Solid State Phenomena* **254** 71–76 DOI: 10.4028/www.scientific.net/SSP.254.71

[6] Hou Y, Wang Y, Pan Z and Yu L 2012 Influence of Rare Earth Nanoparticles and Inoculants on Performance and Microstructure of High Chromium Cast Iron *Journal of Rare Earths* **30** 283–288

[7] Ryabtsev I A, Panfilov A I, Babinets A A, Ryabtsev I I, Gordan G N and Babijchuk I L 2015 Structure and Abrasive Wear Resistance of Deposited Metal Hardened with Carbides of Different Types *The Paton Welding J.* **5–6** 78–82 DOI: 10.15407/tpwj2015.06.17

[8] Sahraoui T, Fenineche N-E, Montavon G and Coddet C 2003 Structure and Wear Behaviour of HVOF-sprayed Cr_3C_2–NiCr and WC–Co Coatings *Materials and Design* **24** 309–313 DOI: 10.1016/S0261-3069(03)00059-1

[9] Ivanov S G, Garmaeva I A, Guriev M A *et al.* 2015 Features of Multicomponent Saturation Alloyed by Steels *Advances in Mechanical Engineering. Lecture Notes in Mechanical Engineering* ed A Evgrafov (Switzerland: Springer International Publishing) 49–54 DOI: 10.1007/978-3-319-15684-2_7

[10] Suetin D V, Shein I R and Ivanovskii A L 2009 Structural, Electronic and Magnetic Properties of η-carbides (Fe_3W_3C, Fe_6W_6C, Co_3W_3C and Co_6W_6C) from First Principles Calculations *Physica B* **404** 3544–3549 DOI: 10.1016/j.physb.2009.05.051

[11] Weidow A and Andrén H-O 2011 Grain and Phase Boundary Segregation in WC–Co with TiC, ZrC, NbC or TaC Additions *Int. J. Refractory Met. HardMater* **29** 38–43 DOI: 10.1016/j.ijrmhm.2010.06.010

[12] Okovityi V A 2002 Plasma Wear-resistant Coatings with Inclusions of a Solid Lubricant *Welding International* vol 16 **11** 918–920

[13] Krivochurov N T, Ivanaysky V V, Ishkov A V and Schegolev AV 2018 About the Efficiensy of Various Hardening Methods the Knives of the Chopper Straw Spreader of Combine Harvester *Bulletin of Altai State Agrarian University* **9** 167

[14] Shchegolev A V *et al.* 2018 Modification of Wear-resistant Coatings of Fe-Cr-C System Based on the Cr_3C_2 Obtained with Help of SHS Method *IOP Conf. Ser.: Mat. Sc. and Eng.* **441** 012047

[15] Zhurda A P 2014 Tungsten Carbide Based Cladding Materials *The Paton Welding Journal* **6–7** 69–74

[16] Massalski T B, Okamoto H, Subramanian P R and Kacprzak L 1990 Binary Alloy Phase Diagrams *ASM International Materials Park* vol 3 3589

[17] Portnoi K I, Levinskaya M K and Romashov V M 1969 Constitution Diagram of the System Iron-Boron *Sov. Powder Metall. Met. Geram* **8** 657–659

[18] Mishustin N M, Ivanaysky V V and Ishkov A V 2012 Composition, Structure and Properties of Wear Resistant Coatings Obtained at 65Г and 50НГА Steels at Highspeed rfc-borating Process vol 320 **2** 68–72

IOP Conf. Series: Materials Science and Engineering **939** (2020) 012072 doi:10.1088/1757-899X/939/1/012072

Reducing labor input of monitoring condition of heat exchange equipment surfaces

I V Shvetsov[1,3], E V Garnyshova[2], E V Izmaylova[2], Y V Vankov[2] and A R Zagretdinov[2]

[1,3] Yaroslav-the-Wise Novgorod State University, ul. B. St. Petersburgskaya, 41 173003 Veliky Novgorod, Russia
[2] Kazan State Power Engineering University, Krasnoselskaya str., 51 420066 Kazan, Russia

[3] E-mail: Igor.Shvetsov@novsu.ru

Abstract. Deposits on surfaces of heat exchange equipment reduce the heat transfer coefficient, heat exchange efficiency, and lead to significant energy losses. The paper considers a hardware-software complex for monitoring the state of heat exchange surfaces. Studies were carried out using models of surfaces of heat exchange equipment – 400×160×2 mm steel plates, by the method of free vibrations. The studies have shown the possibility of using the method of free vibrations for the detection of deposits on the heat exchange equipment, which reduces energy losses and labor input of monitoring the condition of the surfaces of heat exchange equipment.

1. Introduction

The purpose of this work is to assess the possibility to control deposits on surfaces of heat exchange equipment by parameters of free damped vibrations. Thickness must be determined on the external side of the heat exchange surface while operating.

In the field of housing and communal services and industry the cases of operating heat exchangers and boilers without special water preparation are not rare. This causes scale on the internal side of apparatus and units (drums, pipes) which worsens heat exchange, and efficiency on the whole. Thermal conductivity of scale is more than 40 times lower than that of metals, so even a thin scale layer leads to dramatic decrease of thermal conductivity and increase of temperature of metallic surfaces being heated, which reaches dangerous values and reduces mechanical strength of the metal [1]. As a result, there occur damages of the metal. The problem of deposit control arises in case of operating heat exchangers of hot water supply system when the flow section of pipes and surfaces of heat exchange has been almost completely overgrown with deposits. So, when the thickness of a deposit layer is from 0,4 to 19,1 mm, the energy overrun is from 4 to 90% (figure 1).

2. Results and discussion

To reduce the energy losses is possible by means of well-timed diagnostics of surfaces of heat exchange equipment, in order to control the deposits thickness. For this you should have accessible techniques and equipment which allow monitoring in real time.

Content from this work may be used under the terms of the Creative Commons Attribution 3.0 licence. Any further distribution of this work must maintain attribution to the author(s) and the title of the work, journal citation and DOI.
Published under licence by IOP Publishing Ltd

IOP Conf. Series: Materials Science and Engineering **939** (2020) 012072 doi:10.1088/1757-899X/939/1/012072

Figure 1. Flow section reduction due to deposits.

Acoustic control methods are based on measurement of physical parameters associated with the phenomenon of propagation of elastic waves in a solid [2, 3]. In the base of low frequency control methods there lies correlation dependence between elastic constants of material and physical mechanical properties of an item [4–6]. A free vibration method (FV) is based on the use of elastic properties of vibrating systems performing free vibrations. To control means to excite elastic vibrations in the object being controlled, as well as to analyze characteristics of proper vibrations of the object. The method is used to detect inner defects lying at the depth of several centimeters in massive details made of materials with a high coefficient of elastic wave attenuation. Spectrum frequency range of defective and high-quality zones of items which are usually controlled by the free vibration method lies within 0,5–20 kHz. Vibrations receiving is performed with the help of a piezoelectric receiver or a microphone [7, 8].

The authors developed a hardware and software complex which includes a device for controlling deposits thickness on heat exchange surfaces (figure 2) and two systems – registration of signals and their processing.

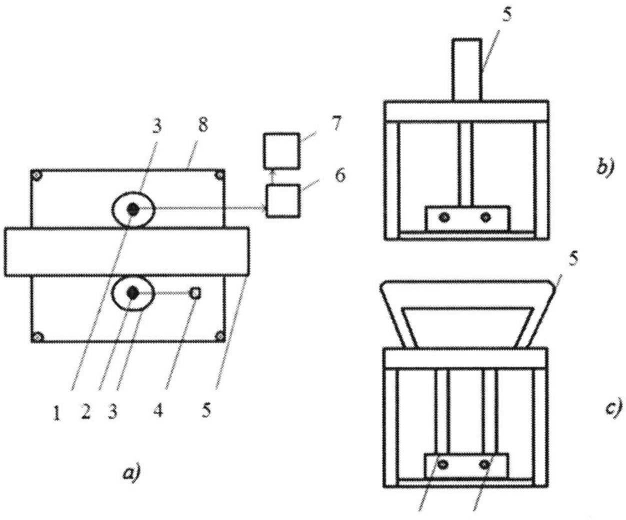

Figure 2. Device for controlling deposits thickness on heat exchange surfaces: *a*) view from above, *b*) side view, *c*) front view.

The numbers in figure 2 mark:

1 – piezoelectric sensor (microphone); 2 – hammer; 3 – openings; 4 – toggle switch; 5 – handle; 6 – analog to digital converter; 7 – computer; 8 – case.

IOP Conf. Series: Materials Science and Engineering **939** (2020) 012072 doi:10.1088/1757-899X/939/1/012072

The registration system includes a control object (a plate being investigated) which is fixed around the perimeter, a sensor, and a hammer operated by a toggle switch. The processing system includes an analog to digital converter (ADC) and a computer with a specialized software installed on it.

The device works as follows: there are two openings, namely, a toggle switch and a handle, on the case. The handle allows the device to be replaced. In the first opening (figure 2) there is a hammer operated by the toggle switch. In the second opening there is a piezoelectric sensor connected with an ADC. The device case is placed on the first diagnosed site of the heat exchange surface. Then, at pressing the toggle switch, the hammer excites vibrations on the heat exchange surface, and the piezoelectric sensor receives an analog signal. Then the toggle switch is turned off. The constant strength of impact is provided by the abduction of the hammer at the same distance. After exciting the vibrations, the analog signal from the piezoelectric sensor goes to the ADC, at its coming out the digital signal enters the computer RAM, where, due to the special software (figure 3) written in a graphic programming environment LabVIEW of the Firm National Instruments, for which the authors obtained the certificate of state registration of a computer program [9], it is recorded, analyzed, and the sound signal is recorded.

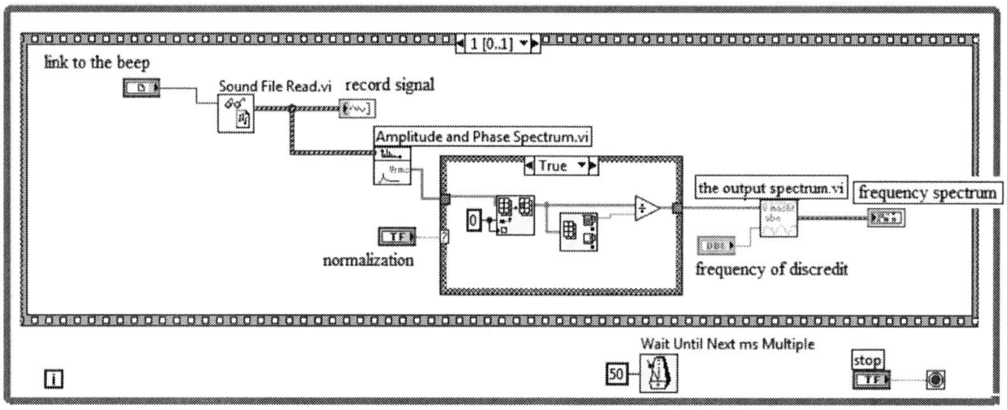

Figure 3. Part of program block-diagram.

Analysis program works as follows: signals are read, amplitude and frequency spectrum is formed for each signal, spectra are compared by correlation coefficients, the obtained coefficients related to one series of experiments are averaged, the averaged coefficients are compared with the confident interval boundary. Finally, the conclusion about the presence of deposits is made.

After diagnosing the first site, the device is replaced to the following site of the heat exchange surface.

As a heat exchange surface model, a steel plate which was 400 mm long, 160 mm wide and 2 mm thick was used. Investigations were made using a clean plate, as well as plates with various thicknesses of deposits.

While studying acoustic characteristics, the tenfold determination of frequencies of proper vibrations of the plate was being carried out for each plate type. The plate was struck, amplitude and frequencies characteristics of the response were measured by the microphone located in the control point.

After the experimental series a layer of deposits was placed on the plate, and investigations were repeated [10].

Figure 4 shows the example of the recorded signal and its amplitude spectrum, figure 5 gives the example of the current spectrum, the reference one, and the difference between them [11]. The plate with deposits is presented by the current spectrum, the clean one is presented by the reference spectrum.

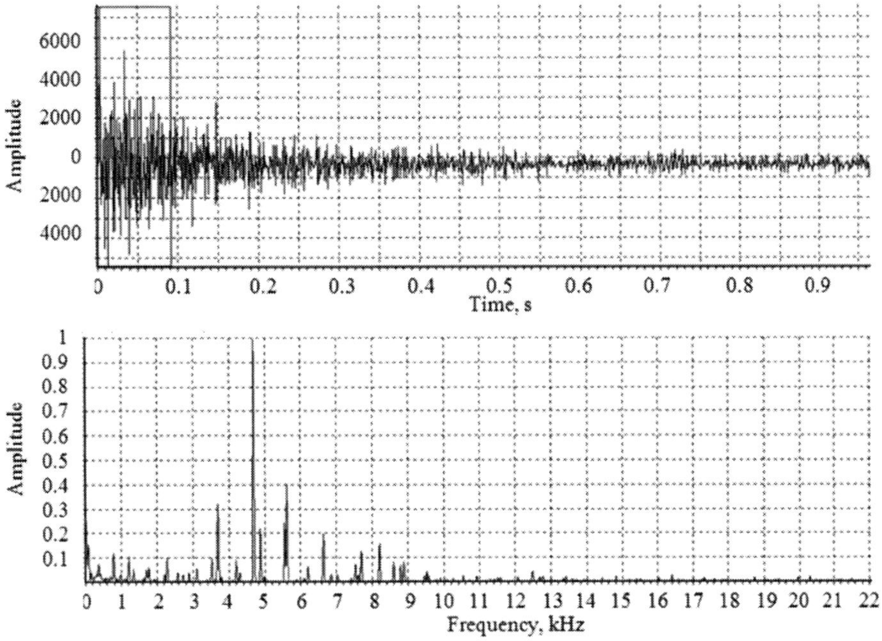

Figure 4. Signal and its amplitude spectrum.

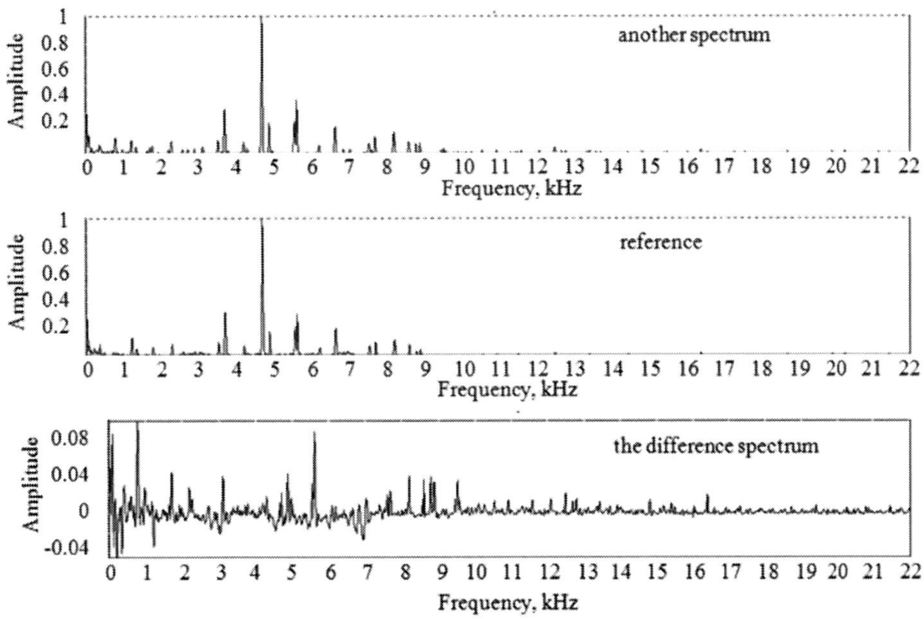

Figure 5. Example of current spectrum, reference spectrum and difference between them.

Criteria used for comparison, such as the Spearman Correlation Coefficient (figure 6) and the Fisher Sign Statistics (figure 7) showed that their values were in the confidence interval $P = 0.95$, which was constructed by each criterion.

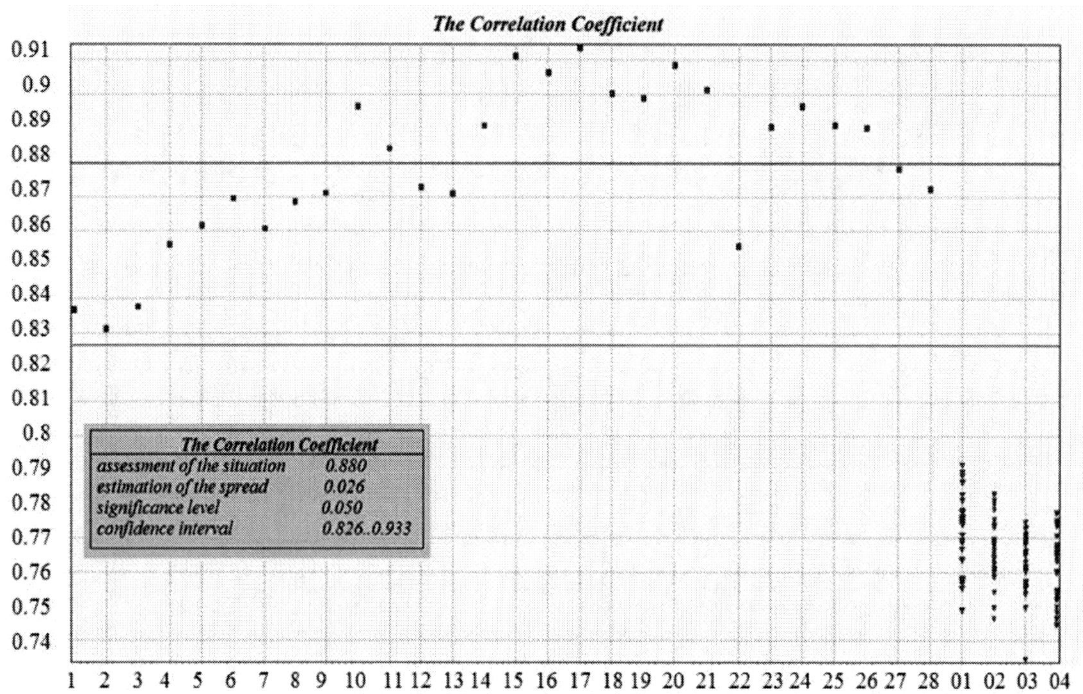

Figure 6. Spearman Correlation.

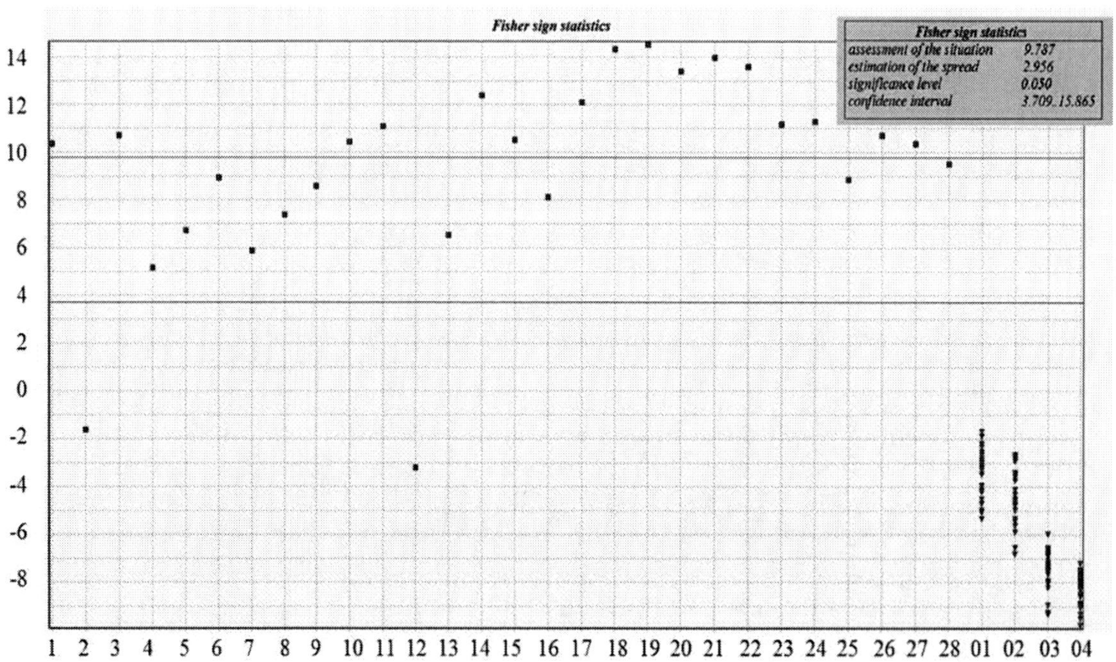

Figure 7. Fisher statistics.

Spearman nonparametric ranking is found:

$$r = 1 - \frac{6}{n\ (n^2 - 1)} \sum_{i=1}^{n} (ranka_i \ - \ ranka_{si})^2 \ ,$$

where *rank* a_i – rank of amplitude a_i in the variation row of amplitudes of the checked spectrum (i.e. the number of the place which this amplitude takes among all amplitudes of the given spectrum sorted in ascending order); *rank* a_{si} – the same for the reference spectrum.

Fisher sign statistics S is calculated as the number of frequencies of positive amplitude of the difference spectrum (half of the number of frequencies of zero amplitude is added to this sum).

The analysis of spectra of plates vibrations shows that the presence of deposits on the plate surface significantly changes its spectrum of proper vibrations in the range from 20 Hz to 10 kHz. The obtained results showed that the method of free vibrations of the plate with deposits at the experimental complex provided the preset accuracy and reliability, this makes it possible for the method to determine the thickness of deposits of heat exchange equipment.

3. Conclusion

The studies showed the possibility to use the method of free vibrations to detect deposits on heat exchange equipment thanks to the developed hardware and software complex which allows to reduce the labor input of monitoring the condition of surfaces of heat exchange equipment.

References

[1] Saifullin E R, Vankov Y V, Izmaylova E V and Bazukova E R 2018 *J. Phys.: Conf. Ser. 1058* 012058

[2] Aleshin N P 1989 *Metody akusticheskogo kontrolya metallov* [Methods of acoustic control of metals] (Moscow: Mashinostroenie) p 456

[3] Vankov Yu V 2004 *Pribory i Sistemy Upravleniya* **(2)** 53–58

[4] Vankov Yu V, Ziganshin Sh G, Izmailova E V and Zagretdinov A R 2015 *IOP Conf. Ser.: Mater. Sci. Eng.* **86** 012036

[5] Zagretdinov A R, Vankov Y V and Ivshin I V 2016 *The Analysis of Frequencies of Own Insulators Fluctuations for the Solution of a Problem of their Control Procedia Engineering* **150** 2334–9

[6] Vankov Yu V, Ziganshin Sh G, Izmailova E V and Serov V V 2015 *IOP Conf. Ser.: Mater. Sci. Eng.* **86** 012035

[7] Ziganshin Sh G, Vankov Yu V and Izmaylova E V 2015 *Kontrol tekhnicheskogo sostoyaniya truboprovodov akusticheskimi metodami* [Monitoring the technical condition of pipelines using acoustic methods] (Kazan: Publ. house of Kazan state power engineering University) p 156

[8] Vankov Yu V 2003 *Nizkochastotnye metody kontrolya. Metod svobodnykh kolebaniy* [Low-frequency control methods. The method of free oscillations] (Kazan: Publ. house of Kazan state power engineering University) p 140

[9] Izmaylova E V, Vankov Yu V, Garnyshova E V, Ziganshin Sh G and Zagretdinov A R 2019 *Certificate of state registration of the computer program no 2019615470 "2RSoundSA"* (Registered in the register of computer programs on 26.04.2019.) [In Russ.]

[10] Izmaylova E V, Politova T O, Garnyshova E V and Garifullin M Sh 2019 *IOP Conf. Ser.: Earth and Envir.l Sc.* **288** 012091

[11] Garnyshova E V and Izmaylova E V 2019 Kontrol tolshchiny otlozheniy na teploobmennom oborudovanii [Monitoring the thickness of deposits on heat exchange equipment] *Proc. Int. Sc. Tech. Conf. "Radioelectronics, electrical engineering and power engineering"* (Moscow: MEI) p 303

IOP Conf. Series: Materials Science and Engineering **939** (2020) 012073 doi:10.1088/1757-899X/939/1/012073

Vibroacoustic method of pressure control

I V Shvetsov[1,3], A R Zagretdinov[2], V E Izmestyeva[2], E V Izmaylova[2] and Sh G Ziganshin[2]

[1,3] Yaroslav-the-Wise Novgorod State University, ul. B. St. Petersburgskaya, 41 173003 Veliky Novgorod, Russia
[2] Kazan State Power Engineering University, Krasnoselskaya str., 51 420066 Kazan, Russia

[3] E-mail: Igor.Shvetsov@novsu.ru

Abstract. The method was experimentally tested on a pipe with excessive internal pressure. It is established that the main frequency of its free-damp vibrations increases with increasing internal pressure of the pipe. The method of conducting experimental studies is described, and the main results are presented.

1. Introduction

The equipment of most technological systems operates under excessive steam water, oil, gas, air, fuel oil and other media pressure. Such equipment can include pipelines, vessels, reservoirs, tanks, etc. The reliability and efficiency of such systems undoubtedly depends on proper pressure control and regulation.

Pressure measurement methods are usually based on comparing the forces of the measured pressure with the forces of:

- the forces of fluid (mercury, water) column pressure of the corresponding height;
- the forces developed during the deformation of elastic elements (springs, membranes, barometric and aneroid capsule, metal bellows and manometric pipe);
- the forces of loading weight;
- the elastic forces that occur when certain materials are deformed and cause electrical effects.

In accordance with these methods, pressure measuring instruments can be divided into fluidal, deformational, piston-type and electric [1]. The most widespread in the industry are strain measuring devices.

The reliability of these pressure measuring devices depends largely on the operating conditions. An aggressive environment, extremely high or low temperatures, ripples and contaminations can cause device failure. In addition, their installation always requires an end-to-end embedding in the control object, which reduces the operational reliability of the equipment. The presence of these disadvantages contributes to the development of new methods of pressure control.

The internal pressure of the working medium influences the natural vibrations of pipelines and vessels [2–4]. The pressure of the process product can be determined by changing the natural frequency of vibrations of the equipment. Such pressure monitoring can be arranged by installing microphones or vibroacoustic sensors on the external surface of pipelines and vessels.

Content from this work may be used under the terms of the Creative Commons Attribution 3.0 licence. Any further distribution of this work must maintain attribution to the author(s) and the title of the work, journal citation and DOI.
Published under licence by IOP Publishing Ltd

The vibroacoustic method does not require end-to-end insertion into the control object. It is acceptable for equipment working with aggressive and polluted environments.

The method can be organized basing on the analysis of the natural noise of the pumped product [5]. The flow noise always contains separate spectral components that are caused by resonant vibrations of the equipment [6].

The vibroacoustic method of pressure measurement has been tested in the course of experimental studies. An experimental installation has been developed, the block diagram and photo of which are shown in figure 1 and 2, respectively.

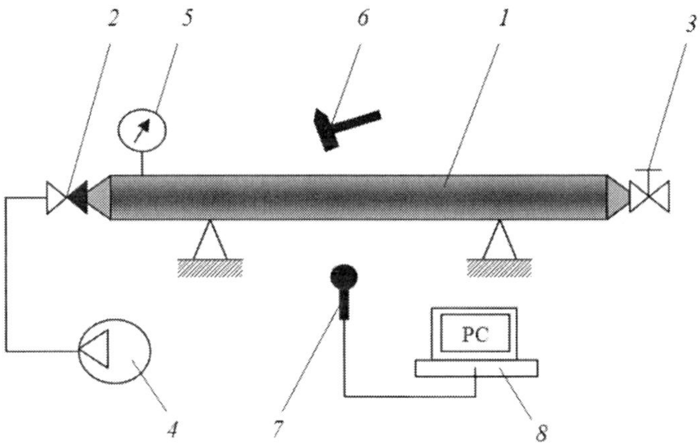

Figure 1. An experimental installation block diagram: 1 – pipe (length 2 m, outer diameter 0.159 m, wall thickness 6 mm), 2 – check valve, 3 – gate valve, 4 – compressor, 5 – pressure gauge (accuracy class 2.5, measuring limit 10 bar), 6 – hammer, 7 – microphone, 8 – personal computer.

Figure 2. The picture of the experimental installation.

The experimental installation functions as follows.

Compressor 4 creates excessive air pressure in pipe 1. The air pressure is controlled using a pressure gauge 5. Hammer 6 in the pipe 1 excites freely damping vibrations. The sound of the impact is recorded by a microphone 7 and then in a personal computer 8.

IOP Conf. Series: Materials Science and Engineering **939** (2020) 012073 doi:10.1088/1757-899X/939/1/012073

Audio signal analysis is performed using a software package written in the LabView programming environment [7–9].

The method of analyzing audio signals is as follows.

1) The signal duration is increased by adding zeros to the array of samples. The procedure for adding zeros allows you to increase the frequency resolution of the spectrum.

2) With the help of the fast Fourier transform procedure, the amplitude-frequency spectrum of the signal is calculated.

3) The frequency with the maximum value of the amplitude is being chosen in the spectrum.

2. Results and discussion

The sound of an impact on the pipe was recorded at a sampling rate of 44.1 kHz. The length of recorded audio signals (without adding zeros) is 44,100 samples.

To increase the spectrum frequency resolution, zeros were added to the signal. An example of a signal supplemented with zeros is shown in figure 3.

Figure 4 and 5 show the spectra of signals received from a pipe with a pressure of 0 and 5 bar, respectively.

The amplitude-frequency spectra were calculated for signals with a length of 44100, 88200, 176400 and 352800 samples. The results are summarized in table 1.

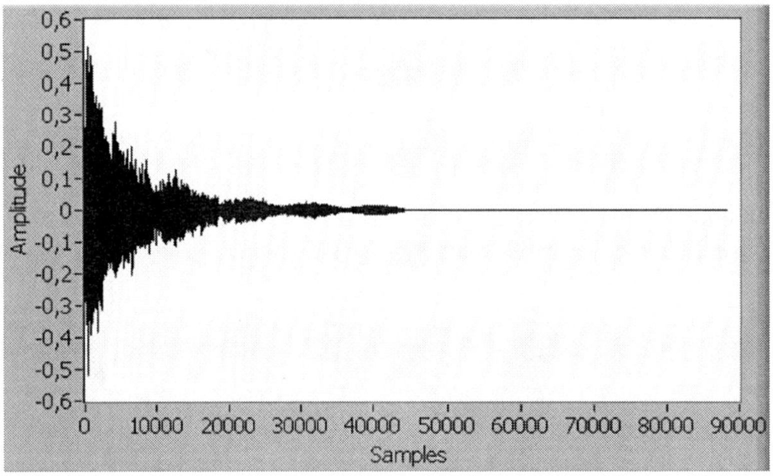

Figure 3. The signal supplemented with zeros.

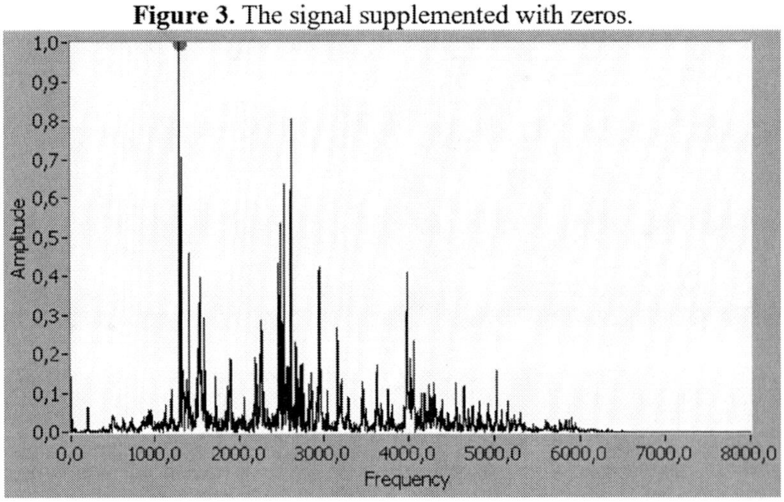

Figure 4. The spectrum of the pipe with internal pressure 0 bar.

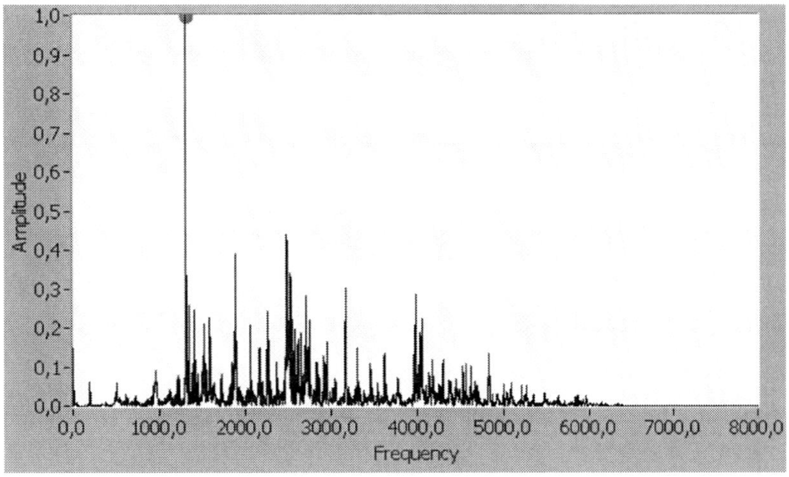

Figure 5. The spectrum of the pipe with internal pressure 5 bar.

Table 1. The results.

		Signal length 44100 samples	Signal length 88200 samples	Signal length 176400 samples	Signal length 352800 samples
Base harmonic frequency, Hz	under a pressure of 0 bar	1296	1296	1296	1296
	under a pressure of 1 bar	1298	1298	1298	1298
	under a pressure of 2 bar	1300	1300	1299.75	1299.875
	under a pressure of 3 bar	1302	1302	1302	1302
	under a pressure of 4 bar	1304	1304	1303.75	1303.75
	under a pressure of 5 bar	1306	1306	1305.75	1305.875
Frequency resolution, Hz		1	0.5	0.25	0.125
Coefficient of determination for linear regression		1	1	0.9992	0.9996
Error in determining the pressure from the vibroacoustic signal, bar		0.5	0.25	0.125	0.063

Table 1 demonstrates that with increasing internal pressure, the frequency of the main harmonic of pipe vibrations increases linearly (the coefficient of determination for linear regression $R^2 = 1$). The pressure changes by 1 bar leads to an increase in the frequency of pipe vibrations by 2 Hz.

The estimation of the error in determining the pressure from the vibroacoustic signal is made using the formula:

$$\delta = \frac{\Delta f}{2},$$

Δf is the frequency resolution, Hz; 2 is the frequency difference of the pipe with a difference in internal pressure of 1 bar, Hz.

The dependency graph of the natural pipe frequency on its internal pressure is shown in figure 6.

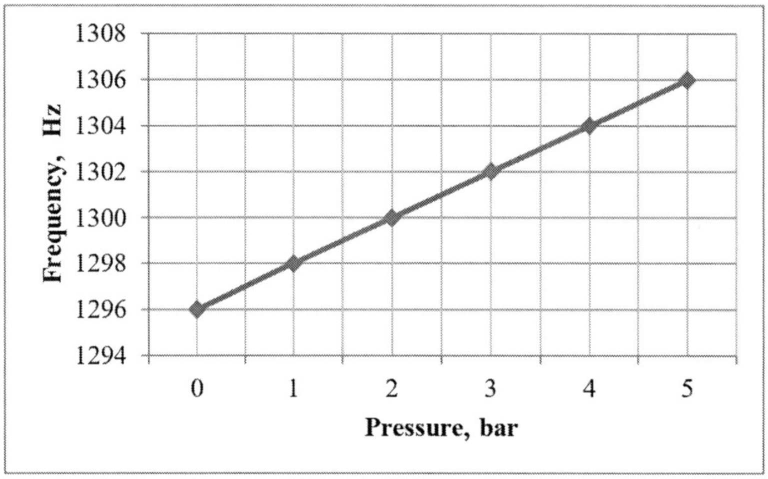

Figure 6. The dependence of the pipe frequency from internal pressure.

3. Conclusion

Studies have confirmed the possibility of monitoring the pressure by changing the natural frequency of equipment vibrations.

It is established experimentally that the main frequency of free-damping vibrations increases linearly with increasing internal pressure of the pipe.

References

[1] Zamaletdinova E Yu and Egorychev A I 2014 Sravnitel'nyj analiz metodov izmereniya davleniya [Comparative Analysis of Pressure Measurement Methods] *Bulletin of Kazan Technological University* **8** 124–7

[2] Bereznev A V 2011 Vliyanie vnutrennego gidrostaticheskogo davleniya na chastoty svobodnyh kolebanij krivolinejnogo uchastka truboprovoda [The Effect of Internal Hydrostatic Pressure on the Frequency of Free Vibrations of a Curved Pipeline Section] *News of Higher Educational Institutions Oil and Gas* **3** 77–80

[3] Efimov A A 2008 Sobstvennye kolebaniya morskogo glubokovodnogo nefteprovoda bol'shogo diametr [Natural Vibrations of a Large Diameter Marine Deep-water Oil Pipeline] *Bulletin of Civil Engineers* **4(17)** 26–9

[4] Sokolov V G, Ogorodnova Yu V, Dmitriev A V and Maslennikov A M 2019 Kolebaniya podzemnyh tonkostennyh magistral'nyh truboprovodov s uchetom vnutrennego davleniya i prodol'noj sily [The Vibrations of Underground Thin-walled Main Pipelines with Regard to Internal Pressure and Longitudinal Power] *Bulletin of Civil Engineers* **5(76)** 105–112

[5] Balakin R A and Timets V M 2012 *Pat. of the Russian Federation* No 2470274 appl. 29.07.2011 publ. 20.12.2012

[6] Nazarychev S A, Vankov Yu V, Politova T O, Fominykh K S and Shlychkov V V 2019 *J. of Phys.: Conf. Series* **1328** 012057

[7] Shvetsov I V, Vankov Yu V and Zagretdinov A R 2018 *IOP Conf. Ser.: Mater. Sci. Eng.* **441** 012051

[8] Zagretdinov A R, Kondratyev A E and Gaponenko S O 2017 Reliability Increasing Solutions for Multilayer Composite Structures Shock-Acoustic Control *Procedia Engineering* **206** 656–61

[9] Vankov Yu, Rumyantsev A, Ziganshin Sh, Politova T, Minyazev R and Zagretdinov A 2020 Assessment of the Condition of Pipelines Using Convolutional Neural Networks *Energies* **13(3)** 618

Assessment of the technical condition of pipelines in polyurethane insulation

I V Shvetsov[1,4], Sh G Ziganshin[2], A R Zagretdinov[2] and D V Zhelnov[3]

[1] Yaroslav-the-Wise Novgorod State University, ul. B. St. Petersburgskaya, 41 173003 Veliky Novgorod, Russia
[2] Kazan State Power Engineering University, ul. Krasnoselskaya, 51 420066 Kazan, Russia
[3] Department of Expertise and Certification of the Union "St. Petersburg Chamber of Commerce and Industry", pr. Lesnoy, 20 194044 St. Petersburg, Russia

[4] E-mail: Igor.Shvetsov@novsu.ru

Abstract. The paper considers issues of improving the reliability of operation of pipelines of heating networks. A description of the created diagnostic complex is given, with the help of which it is possible to determine the technical condition of pipelines in polyurethane foam insulation. Finite-element models were constructed and modal calculations of defective and defect-free pipelines in polyurethane foam insulation were carried out. The analysis of the received acoustic signals was carried out using neural network algorithms. The results of processing acoustic vibrations of pipelines by a neural network showed a high efficiency in identifying pipelines by their state (defective / defect-free).

1. Introduction
Ensuring the required reliability of heating systems is an important task, the solution of which will allow to maintain the accident-free operation of pipelines, reduce the heat losses and the cost of transfer of thermal energy. Creating a pipeline monitoring system can be one of the options for solving this problem. Modern heating systems mainly use pre-isolated pipes with polyurethane foam insulation. For this type of pipeline, there is a mandatory remote control of the watering of polyurethane foam with damage detectors. But this is not enough, as by this way it is impossible to determine exactly the location of the leak and the type of the defect. In this regard, there is a problem of lack of equipment for monitoring and diagnosing pipes in polyurethane foam insulation.

2. Results and discussion
This paper describes the diagnostic complex that has been created, through which it is possible to determine the technical condition of pipelines in polyurethane insulation. Diagnosis is based on an acoustic method of non-destructive control due to a number of features [1, 3, 4]. Information parameters are the frequencies of free vibrations or resonances of forced vibrations [10, 11]. These frequencies are related to the geometric parameters of the products and the speed at which the sound is distributed.

A pipeline in polyurethane insulation with the following geometric characteristics was selected as a the object of the study: the length L = 995 mm, the outer diameter D_1 = 220 mm, the thickness of

Content from this work may be used under the terms of the Creative Commons Attribution 3.0 licence. Any further distribution of this work must maintain attribution to the author(s) and the title of the work, journal citation and DOI.
Published under licence by IOP Publishing Ltd

the wall $\delta_1 = 6$ mm, the thickness of the layer of polyurethane insulation $\delta = 45$ mm and the thickness of the polyethylene sheath $\delta_2 = 5$ mm. The dependence of the change in the acoustic characteristics of the pipeline on the presence of a "hole-type" defect on it was investigated. In order to obtain frequencies of their own fluctuations of defective and non-defective pipelines, it was necessary to carry out their modal analysis. Modal analysis suggests that the system is linear. All types of non-linearity – non-linear behavior of the material, contact boundary conditions, final movements – are ignored. It is assumed that free non-fading fluctuations are made:

$$[M] \cdot \{u''\} + [K] \cdot \{u\} = 0 ,$$

where $\{u\}$ is the vector of nodal displacements for the whole body, $\{u''\}$ is the vector of acceleration of body points, [*K*], [*M*] are "global" matrixes of rigidity and mass for the whole body.

We note that the rigidity matrix of the design can include the effect of pre-loading For the linear system free vibrations will be harmonic:

$$(u) = (\varphi)_i \cdot cos\omega_i t$$

where the natural vector represents the i-form of oscillations; *i* is natural circular frequency (radian per unit of time); *t* is time.

The values of their own circular frequencies (ω) and own frequencies (f) are related by the following ratio:

$$f_i = \frac{\omega_i}{2 \cdot \pi}$$

where i is natural frequency (cycles per unit of time) [5].

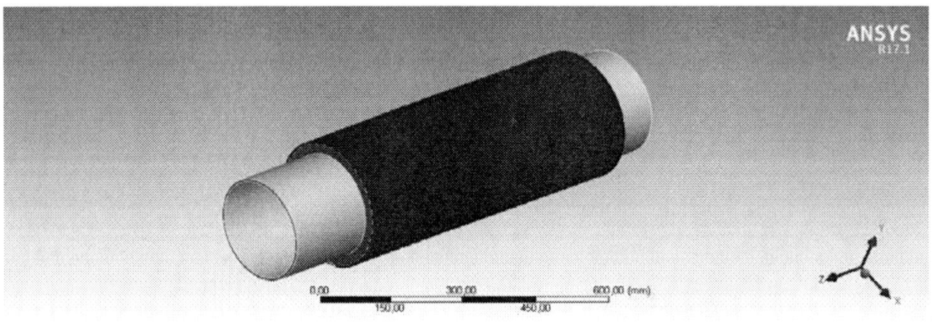

a)

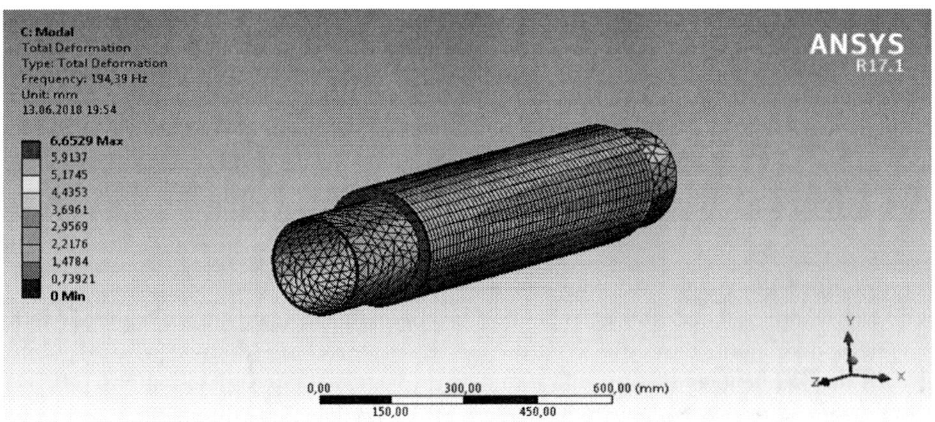

b)

Figure 1. Modeled in the ANSYS software complex, a defective pipeline in polyurethane insulation: (a) general view; (b) Pipeline model broken down into a grid of end elements.

A finite elemental simulation was used in the ANSYS software complex to solve the problems of modal analysis. The modeled pipeline (figure 1) consists of three layers with the following characteristics:

1. Steel St3 with the Poisson coefficient of ν = 0,3, elasticity module and density;
2. Polyurethane with the Poisson coefficient of ν = 0,496, elasticity module and density;
3. Polyethylene with the Poisson coefficient of ν = 0,42, elasticity module and density.

The resulting dependence of the frequencies of own vibrations on modes for defect-free and defective pipelines are shown in figure 2.

As far as can be seen from the graph in figure 2, the noticeable difference in the frequencies of defect-less and defective vibrations is in intervals from the 10th to the 20th and from the 40th to the 80th modes In the future, these areas should be preferred when analyzing acoustic signals.

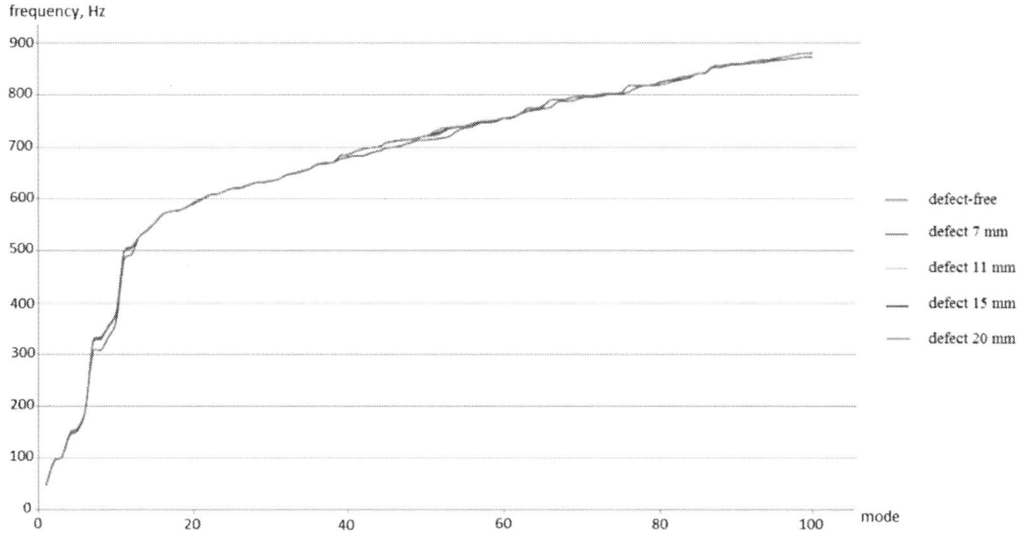

Figure 2. Dependences on the frequency of natural oscillations from modes for defect-less and defective pipelines in polyurethane insulation.

To confirm the possibility of assessing the condition of pipelines in polyurethane insulation by acoustic method, an experimental installation was created, the structural scheme of which is shown in figure 3. The excitation frequency in the pipeline under study was carried out with the same period and force of impact. Reception, primary conversion and transmission of the signal is done using the ADC NI USB-6229. The analog signal from the sensors is converted by the controller and transmitted to a personal computer. Data is processed using a program written on the basis of the LabView software complex.

With the installation described above, it is possible to obtain the frequency of pipeline oscillations that can be judged on their condition, but the problem is the processing of a large amount of data and the unmistakable identification of pipeline defects. Therefore, further processing and analysis of the resulting frequencies were carried out using a neural network (figure 4). The use of neural networks allows you to quickly and efficiently analyze data and reduce error in decision-making [2, 6–9, 12].

Input signals are sent to the neural network, which are the frequencies of pipeline oscillations. During the initialization of the neural network, the weights of swaying frequencies take a randomized value. On the structural scheme of the neural reverse distribution network (figure 4), all the input information multiplied by their weight is the hidden layer:

$$f_{i,input} = (I_1 \cdot W_1) + (I_2 \cdot W_2) + \cdots + (I_i \cdot W_i)$$

where I_1 is an input neuron, and W_1 is weight.

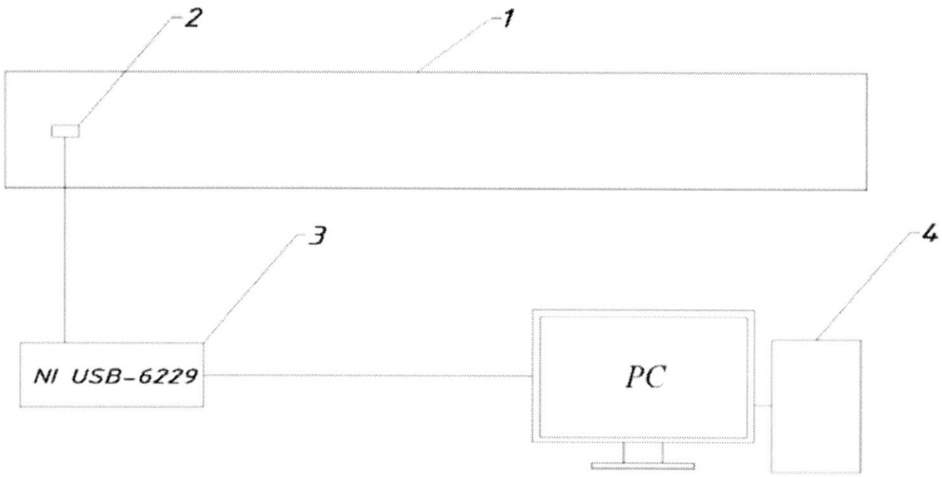

Figure 3. Structural scheme of the experimental installation: 1 – the pipeline being studied; 2 – piezoelectric sensor; 3 – NI USB-6229 DAQ-device; 4 – personal computer.

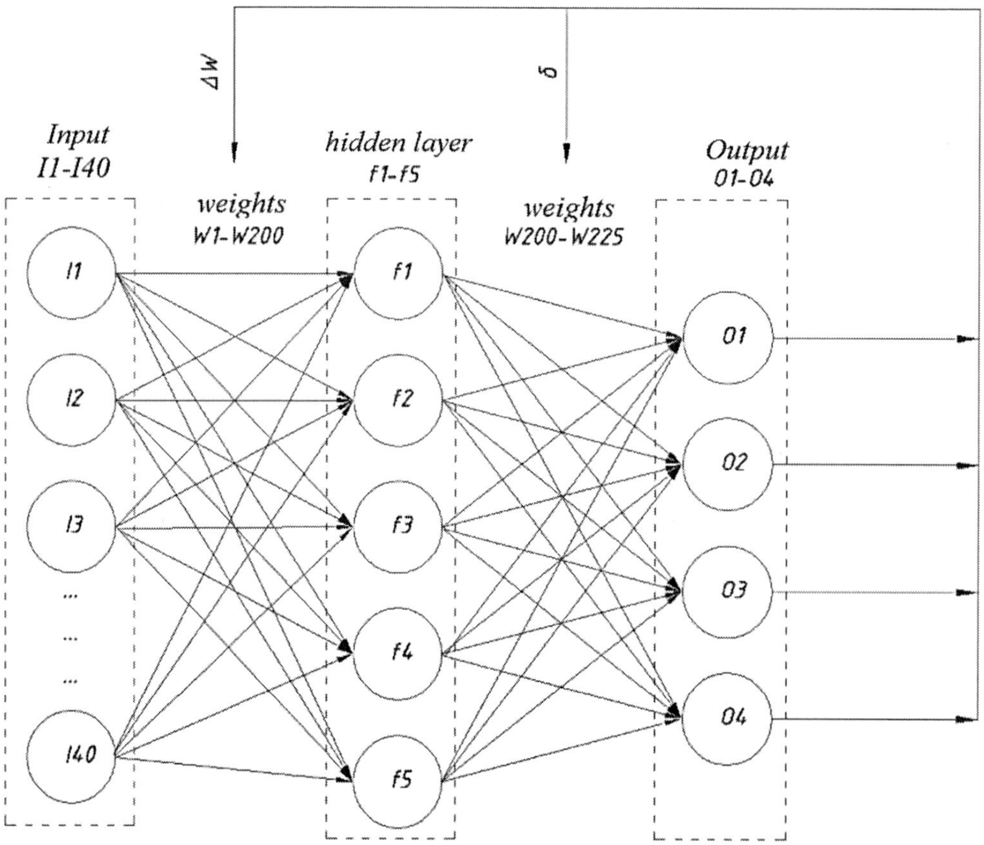

Figure 4. Structural scheme of the neural network of reverse spread.

Since the range of number operations in a neural network is [0; 1] or [–1; 1], the input values ($f_{i,input}$) are converted through the activation function

$$f(x) = \frac{1}{1 + e^{-x}}$$

When the weight data was processed, randomized values were obtained, in subsequent calculations the weight coefficients were recalculated taking into account their gradient, and each subsequent iteration took place with the updated weights. To more accurately find the parameters of output neurons, the minimum error rate at which the calculation ended was 0.001. The number of iterations of data processing is 1085.

3. Conclusion

The results of the processing of acoustic oscillations of pipelines by the neural network showed high efficiency (98%) identification of pipelines by their condition: defective-less or defective. The efficiency of determining the size of the defect was at the level of 75%. To improve the effectiveness of diagnostics, the authors plan to accumulate a larger training base of acoustic signals of pipelines and use of deep convolutional neural networks.

References

[1] Saifullin E R, Zagretdinov A R, Ziganshin S G and Vankov Y V 2018 Control of the rotary equipment disbalance by the spectrum of envelope vibroacoustic signal *J. of Advanced Research in Dynamic and Control Systems* **10** 2242–7

[2] Saifullin E R, Ziganshin S G, Vankov Y V and Zagretdinov A R 2018 Assessment of the technical condition of polyurethane foam thermal insulation pipelines of heating networks using neural network technologies *Int. J. of Eng. and Tech.* (UAE) **7(4)** 241–4

[3] Ziganshin Sh G, Izmailova E V and Maryashev A V 2017 Technique for search of pipeline leakage according to acoustic signals analysis *Int. Conf. on Industrial Engineering, Applications and Manufacturing (*ICIEAM 2017 Proceedings) 8076416

[4] Vankov Yu V, Ziganshin Sh G, Izmailova E V and Serov V V 2015 *IOP Conf. Ser.: Mater. Sci. Eng.* **86** 012035

[5] Bate K and Vilson E 1982 *Chislennye metody analiza i metod konechnyh elementov* [Numerical analysis methods and the finite element method] (Moscow: Strojizdat) p 450

[6] Saifullin E R, Ziganshin Sh G, Vankov Y V and Serov V V 2017 Neural network analysis of vibration signals in the diagnostics of pipelines *J of Fundamental and Applied Sciences* **9**(2S) 1139–51

[7] Srivastava R K, Greff K and Schmidhuber J 2015 Training very deep networks *Proc. of the 29th Annual Conf. on Neural Information Processing Systems pp* 2377–2385

[8] LeCun Y, Kavukcuoglu K and Farabet C 2010 Convolutional networks and applications in vision Circuits and Systems (ISCAS) *Proc. of 2010 IEEE Int. Symposium* 253–6

[9] Minyazev R Sh, Rumyantsev A A, Dyganov S A and Baev A A 2018 X-Ray Image Analysis for the Neural Network-Based Detection of Pathology *Bulletin of the Russian Academy of Sciences: Phys.* **12** 1529–31

[10] Nazarychev S A, Gaponenko S O, Kondratiev A E and Shakurova R Q 2019 *J of Phys.: Conf. Series* **1328** 012055

[11] Gaponenko S O, Kondratiev A E and Zagretdinov A R 2016 Low-frequency Vibro-acoustic Method of Determination of the Location of the Hidden Canals and Pipelines *Procedia Engineering* **150** 2321–6

[12] Vankov Y, Rumyantsev A, Ziganshin S, Politova T, Minyazev R and Zagretdinov A 2020 Assessment of the condition of pipelines using convolutional neural networks *Energies* **13(3)** 618

IOP Conf. Series: Materials Science and Engineering **939** (2020) 012075 doi:10.1088/1757-899X/939/1/012075

Energy performance analysis of the temperature control system of an object with a random principle of thermal perturbations

A B Sulin[1], A L Timofeevskiy[1], T V Ryabova[1], A A Nikitin[1], S S Muraveinikov[1] and S I Arendateleva[2,3]

[1] Saint-Petersburg National Research University of Information Technologies, Mechanics and Optics, ul. Lomonosova, 9 191002 St. Petersburg, Russia
[2] Yaroslav-the-Wise Novgorod State University, ul. B. St. Petersburgskaya, 41 173003 Veliky Novgorod, Russia

[3] E-mail: Svetlana.Arendateleva@novsu.ru

Abstract. The application of simulation modeling to the development of a temperature control system for an object with a random nature of thermal loads is considered. The object under study is a facility with enclosing structures made of fiberglass, inside of which there is heat-generating equipment. Thermostatic control of the object is carried out by a fully recirculating ventilation system that provides even blowing of the inner walls of the object with fresh air. To create a simulation model that is closest to real conditions, based on statistical meteorological data for the city of St. Petersburg, a model of meteorological conditions was created, suggesting every three hours the following parameters: ambient temperature, wind speed, solar radiation intensity, cloudiness. Heat generating equipment located inside the serviced facility does not operate continuously. The turning-on of this equipment occurs at random, the duration of work and the amount of heat dissipation also vary. To take into account this circumstance, the specified parameters are selected by the program in accordance with the Simpson distribution. The article presents graphs of changes in temperature and energy characteristics of the system for the three cold and three warm days of the considered annual period of the system operation. Conclusions are drawn about the feasibility of using the simulation method in the design and development of such systems, on the possible increase in their energy efficiency, as well as on the reduction of capital and operating costs.

1. Introduction

Modern high-tech equipment, which includes components subject to the action of the atmosphere, usually should work in a certain range of environmental parameters, the proper choice and implementation of which depends on the ability of equipment to function normally and ensure its design performance, as well as uptime [1, 2]. Among such requirements, one can single out the requirements for the quality of the surrounding air, its humidity, temperature, mobility and other parameters. In some cases, to ensure the required parameters of the ambient air, it is necessary to develop a temperature control system that is complex and saturated with equipment, which should operate year-round, quickly responding not only to changes in ambient temperature, solar radiation, but also to fluctuations in internal heat generation from the equipment itself. An important factor

Content from this work may be used under the terms of the Creative Commons Attribution 3.0 licence. Any further distribution of this work must maintain attribution to the author(s) and the title of the work, journal citation and DOI.
Published under licence by IOP Publishing Ltd

is to ensure maximum energy efficiency of the temperature control system, the maintenance of which is often very expensive [3].

2. Results and Discussions

The development and calculation of such a system is a labor-intensive process that requires taking into account a significant number of factors affecting its behavior, as well as taking into account their mutual influence [4, 5, 6], since it is the summation and superposition of peak loads on each other that is most often taken for the main design mode of operation of the temperature control system. To study complex systems with a large number of variables, it is advisable to use simulation. Further, using an example of a simulation model of an object's temperature control system with a large number of unsteady external and internal parameters created in the AnyLogic program [7, 8], we will consider the annual cycle of its operation closest to reality. In the framework of this model, the system is considered as a system with lumped parameters. The combined action of various kinds of factors affecting the operation of the temperature control system sometimes poses quite unexpected tasks for it, which cannot always be taken into account with traditional design methods, but it is their consideration that makes it possible to obtain the most balanced and stable operating system.

The object under consideration represents a confined volume, inside of which there is periodically switched on equipment that releases a significant amount of heat (20–30 kW) during operation. The shell of the object consists of fiberglass with a thickness of about 30 mm. The facility is located in the open air in the city of St. Petersburg. Permissible environmental parameters inside the shell for the normal functioning of the equipment: air temperature from +5 to +45 °C, relative humidity not more than 70%, complete absence of drip moisture, which necessitates temperature control. The air thermostatic control system of the internal volume of the object is organized according to the principle of 100% recirculation. The air preparation system that processes it before being fed into the serviced volume allows first draining it to a dew point of about + 5 °C and then heating it to the required temperature. The schematic diagram of the object under consideration is shown in figure 1.

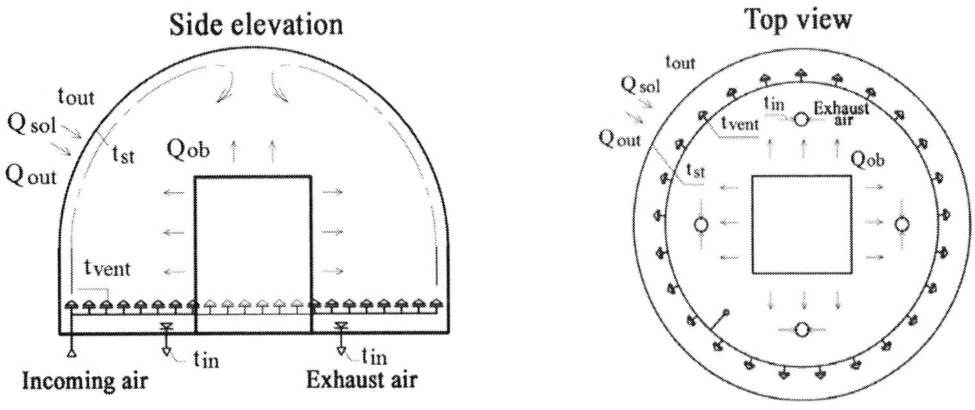

Figure 1. Scheme of the research object. t_{out} is the outdoor temperature, Q_{ob} is heat dissipation from the equipment, Q_{out} is heat gain from the environment, Q_{sol} is heat gain from solar radiation; t_{out}, is the ambient air temperature, tvent is the supply air temperature, t_{in} is the exhaust air temperature, t_{st} is the temperature of the internal surface the walls.

To develop the most accurate model of the impact of the atmosphere on the object, we used statistical climatic data for the city of St. Petersburg, which were recorded daily for a year in increments of 3 hours. From the entire array of meteorological data, the following outdoor parameters were used to develop a simulation model, which have the greatest impact on the temperature control system:

– outdoor temperature – for calculating heat influx from the environment;

– wind speed – for calculating the heat transfer coefficient from fences to the outside air;
– cloud cover – for calculating solar radiation on the wall of an object.

In addition to these data, to calculate the heat input to the object from solar radiation, a parameter was added for the total (direct and scattered) daily average solar radiation flux density for each month of the year, taking into account the duration of daylight hours [9].

As noted earlier, under the shell of the object there is the equipment that during operation emits a time-varying thermal power. Due to the periodic nature of the heat in the serviced volume and the impossibility to determine their magnitude accurately, to calculate the heat balance of the room we used the distribution of its values according to the triangular law (Simpson's law), which is often used when there is insufficient information about the parameters of the object. In this case, to specify a triangular distribution, it is necessary to know three quantities:

– is the minimum value of the parameter (min);
– is the maximum value of the parameter (max);
– the most probable value of the parameter (mode).

In accordance with the Simpson distribution, the following parameters were changed:

– frequency of turning the equipment on: it is accepted that the equipment is turned on from 1 to 2 times a month, most likely 1 time per month;
– duration of operation of the equipment: if it is turned on, the duration ranges from 24 to 48 hours, most likely 24 hours;
– heat from the equipment ranges from 20 to 30 kW, most likely 25 kW.

The algorithm of the temperature control system involves four modes of its functioning: two for the warm season (when the outdoor temperature is more than +15 °C) and two for the cold season (the outside temperature is less than +15 °C). Of the two modes possible in a certain period of the year, one is realized when the equipment inside the facility is turned off, and the other when it is turned on. An enlarged version of the algorithm is presented in figure 2.

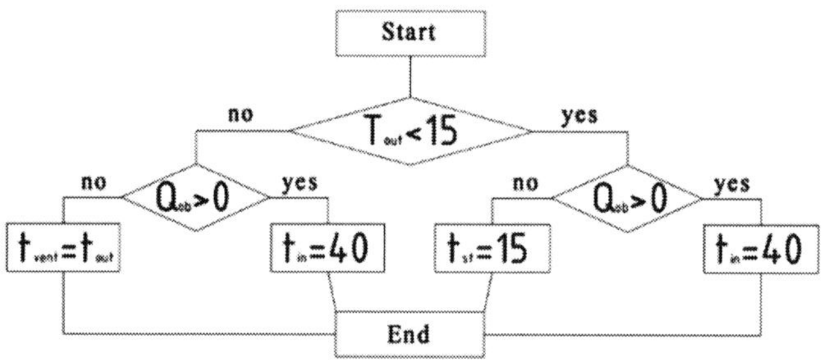

Figure 2. The enlarged algorithm of the temperature control system. T_{out} is the outdoor temperature, is Q_{ob} is the heat dissipation from the equipment, t_{vent} is the supply air temperature, t_{in} is the exhaust air temperature, t_{st} is the temperature of the inner surface of the wall.

Each operating mode is described in more detail below. Due to the impossibility of presenting, within the framework of this article, annual graphs of the temperature and energy characteristics of the system (due to their large size and saturation), graphs for three days of the warm and three days of the cold period of the year are presented to demonstrate the operation of the simulation model.

Modeling the system in the cold season. The two main tasks that the developed algorithm of the temperature control system had to solve in the cold season were to ensure the complete absence of drip moisture in the internal volume of the facility and increase its energy efficiency. Since the air preparation system for the object (not considered in this article) involves drying it to a dew point of +5 °C followed by heating to the required temperature, in order to realize the condition of complete absence of moisture condensation in the object, it is necessary to maintain the temperature of all its surfaces at a level above + 5 °C. The coldest surface of an object, on which droplet moisture from air can form, is the inner surface of its shell; therefore, in the cold season it is necessary to maintain the temperature of this surface above the dew point of the air blowing at it. Air supply to the inner surface of the shell of the object cannot always be ensured with sufficient uniformity [10], therefore, to ensure prevention of moisture condensation, the temperature of the inner surface of the shell of the object was set to +15 °C, i.e., 10 °C above the dew point of the supply air. In addition to the non-uniformity of the blowing of the inner surface of the shell with the supply air, it became possible to level out the influence of such factors as the unevenness of the wind load acting on the object from the outside and the non-isothermal supply air stream propagating along the internal wall of the object.

When the equipment located in the simulation facility was suddenly turned on, the generated heat causes significant heating of the internal volume. This circumstance allows using the excess heat to heat it and not to spend additional energy on heating the supply air [11]. In this case, it is advisable to transfer the temperature control system from the mode of maintaining the wall temperature at +15 °C to the mode of maintaining the maximum allowable air temperature in the internal volume of the object, which, according to the terms of reference, should not be higher than +45 °C. Due to the non-uniformity of the inside temperature field, an air temperature of +40 °C is accepted. The above described system operation algorithm is used at the outdoor temperature below +15 °C.

At higher outdoor temperatures, the discussed below algorithm of the system for the warm period of the year is used. Graphs of temperature changes in the temperature control system of the facility for three days of the cold period of the year, during which the equipment is turned on, are shown in figure 3. The beginning of the experiment was the time point 00 hours 00 minutes 01 second January 01. Further, the countdown of model time was carried out in minutes.

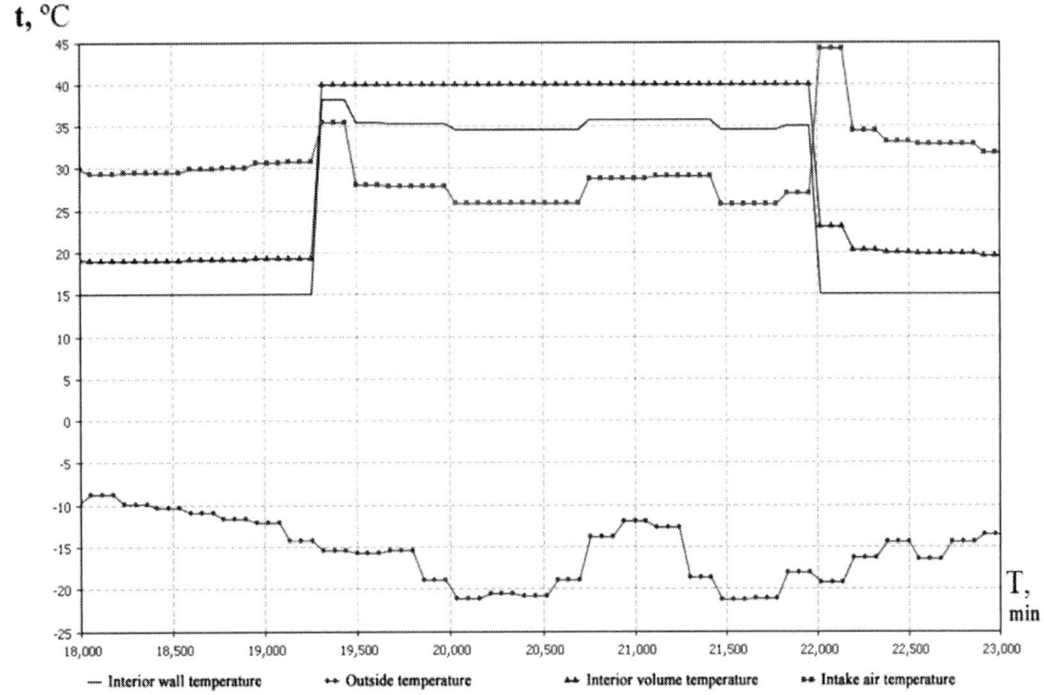

Figure 3. Graph of the change in the main temperatures of the temperature control system for three days of the cold season (December).

The graph of changes in the energy characteristics of the object for the same three days of the cold period of the year is shown in figure 4. The equipment was turned on in the time interval from 19,200 min to 22,000 min.

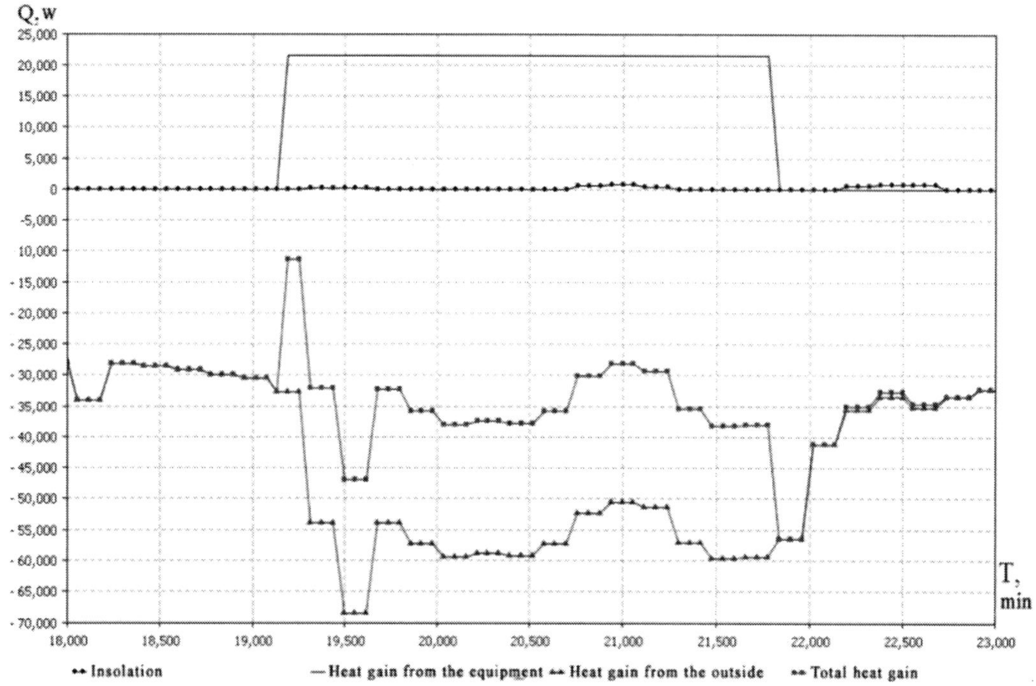

Figure 4. Graph of changes in the energy characteristics of the temperature control system for three days of the cold season (December).

As can be seen from figure 3, at low outdoor temperatures, the supply air temperature and the air temperature inside the serviced volume do not go beyond the required limits (+5 °C ... +45 °C) both when the heat-generating equipment is turned on and off. In the cold season, during operation of the heat-generating equipment, the recirculation air is cooled before it is supplied to the serviced volume by using the external temperature potential and does not entail significant energy costs [12].

Modeling the system in the warm season. With the unconditional need to keep the temperature of the internal volume within the specified limits, the control of the operation of the temperature control system of the facility in the warm season is aimed, inter alia, at minimizing energy costs to maintain the required parameters. Maintaining the temperature of the shell wall, which was emphasized in the algorithm of the system during the cold season, in this case does not matter, since the inclusion of the algorithm of the system for the warm period of the year occurs only if the ambient temperature reaches +15 °C and more. In normal mode, when the heat generating equipment is turned off, the supply air temperature is taken equal to the ambient air temperature. This condition can significantly reduce the heat transfer of the internal volume of the object with the environment, and minimize the amount of artificially obtained cold required for cooling the supply air before its subsequent supply to the serviced volume (the system is considered as a fully recirculation one).

When the heat generating equipment is turned on, the temperature control system in any period of the year maintains a temperature of +40 °C in the internal volume of the facility. This circumstance makes it possible to increase the outflow of excess heat into the environment by maintaining the maximum possible temperature difference outside and inside the object, as well as to minimize the power spent on cooling the recirculated air before its subsequent supply to the serviced volume while maintaining the required initial temperature data. A graph of temperature changes over time for three days of the warm period of the year (August), which also includes the period of activation of the heat-generating equipment, is shown in figure 5. In the period from 311000 minutes to 313100 minutes, the equipment is turned on.

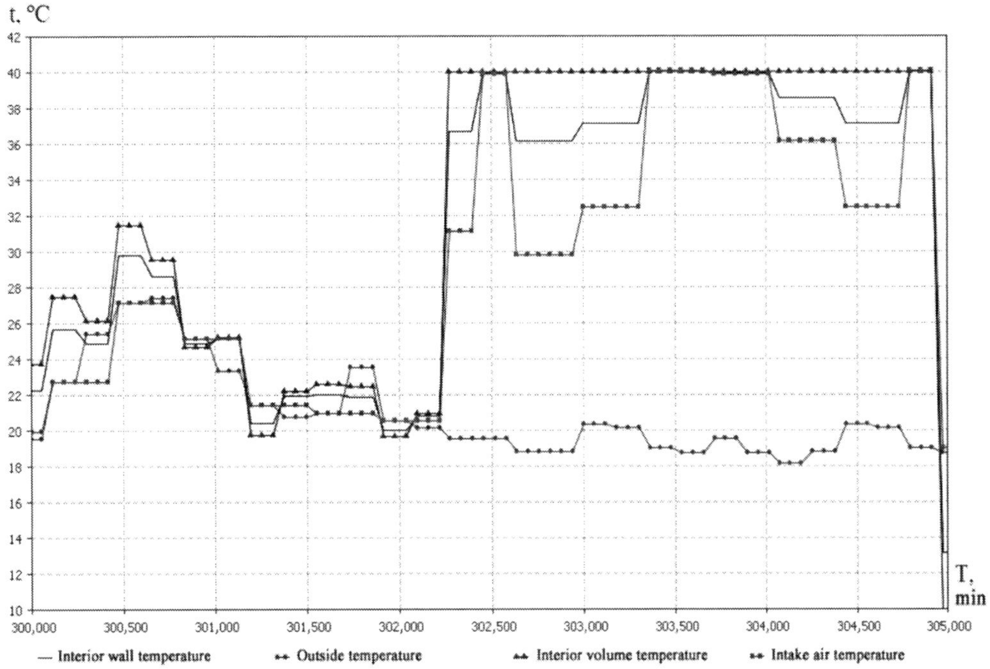

Figure 5. Graph of changes in the main temperatures of the temperature control system for three days of the warm period of the year (August).

The graph of changes in the energy characteristics of the temperature control system for the same three days of the warm period of the year (August) is shown in figure 6.

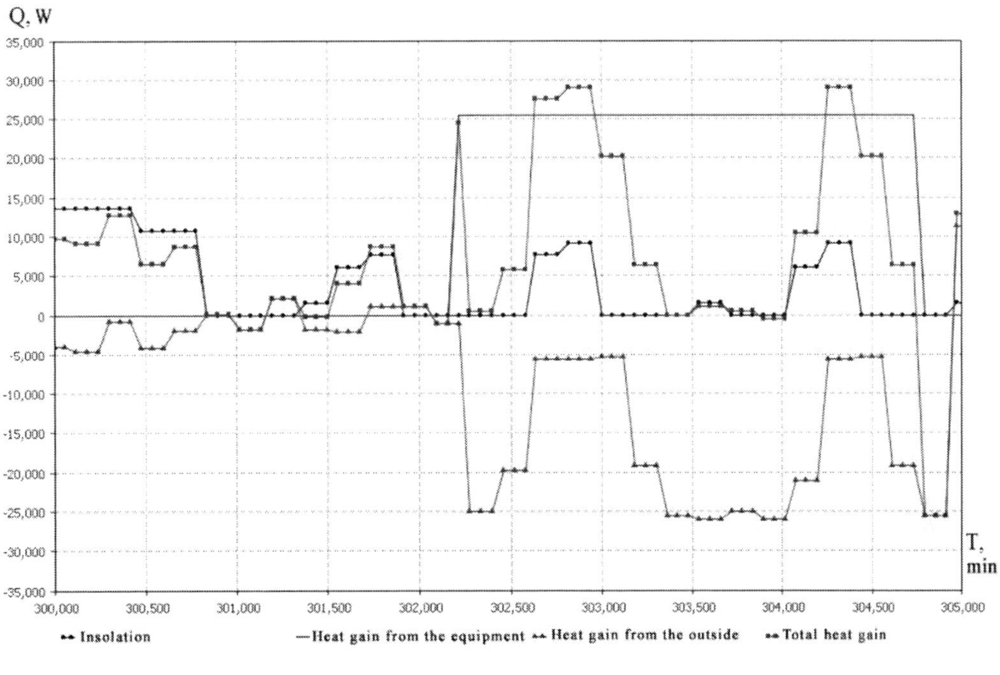

Figure 6. Graph of changes in the basic energy characteristics of the temperature control system for three days of the warm period of the year (August).

3. Conclusion

The graphs for the warm and cold periods of the year clearly show that the simulated system in the entire range of outdoor temperatures, under the influence of solar radiation, as well as internal thermal loads, the occurrence of which is random, provides the required thermal regime inside the object, while maintaining high level energy efficiency [13, 14].

In the cold season, when the heat generating equipment is turned off, the cost of heating the supply air is minimized by maintaining its temperature at the minimum acceptable level, which helps prevent moisture condensation on the inner surface of the shell.

In the warm season, when the heat generating equipment is turned off, the cost of cooling the supply air is minimized by maintaining its temperature at the level of the outdoor temperature. As can be seen from the graph in figure 6, the level of total heat input to the object in this case is maintained at around 13 kW, which is insignificant for such an object.

In the event of a sudden turning on of the heat generating equipment in the cold season, the need for heating the supply air with a temperature control system completely disappears, and, conversely, there is a need for cooling it, which is relatively easy to do by using the low-temperature potential of outdoor air.

When the heat generating equipment is suddenly turned on during the warm season, a controlled increase in air temperature inside the facility to a maximum of +40 °C allows maximizing the outflow of excess heat into the environment and reducing the load on the air conditioning system.

The developed algorithm allows for research and experiments to optimize the operating modes of ventilation and heat exchange equipment, while minimizing capital and operating costs for the functioning of the temperature control system.

References

[1] Bykov V J and Ilyin G N 2010 Mnogokanal'naya cifrovaya sistema termostatirovaniya ehlementov radiometricheskogo bloka. [Multichannel digital temperature control system for radiometric unit elements *Proceedings of the Institute of Applied Astronomy RAN* **21** 288–293 [In Russ.]

[2] Kravchenko A V, Plaksin S V and Sokolovskij I I 2005 Aktivnoe termostatirovanie poluprovodnikovyh SVCH-generatorov [Active temperature control of semiconductor microwave generators] Technology and design in electronic equipment [In Russ.]

[3] Kobeleva S A 2011 Metodicheskie podhody proektirovaniya resurso-i ehnergoehffektivnyh zdanij [Methodological approaches to the design of resource and energy efficient buildings. Construction and reconstruction] *Stroitel'stvo i rekonstrukciya* **5(37)** 18–20 [In Russ.]

[4] Mar'yasin O YU and Ogarkov A A 2017 Imitacionnoe modelirovanie i optimizaciya ehnergopotrebleniya ofisnogo zdaniya [Simulation and energy optimization of an office building] *Simulation modeling. Theory and practice* pp 480–4 [In Russ.]

[5] Shon D *et al.* 2017 CFD modelling of air temperature reduction and airflow induced by the use of chilled wall panels based on the biological principles of zebra stripes *Architectural Science Review* vol 60 **6** 507–515

[6] Bastani M S and Asadi S 2015 Simulating the impact of feedback on energy consumption and emission production in commercial buildings using agent-based approach *AEI 2015: Birth and Life of the Integrated Building* pp. 78–90

[7] Borshchev A 2013 *The big book of simulation modeling: multimethod modeling with AnyLogic* 6 (Chicago: AnyLogic North America) p 614

[8] Borshchev A, Karpov Y and Kharitonov V 2002 Distributed simulation of hybrid systems with AnyLogic and HLA *Future Generation Computer Systems* vol 18 **6** pp 829–839

[9] 2012 Svod pravil SP131.13330.2012: Stroitel'naya klimatologiya. Aktualizirovannaya redakciya [Construction climatology. Updated edition] SNiP 23-01-99 (Moscow: NIISF RAASN)

[10] Lepesh G V, Sproge G A and Odnodvorec Yu V 2015 Imitacionnoe modelirovanie

differencirovannogo obogreva ventiliruemogo pomeshcheniya kompleksom sovremennyh otopitel'nyh priborov [Simulation of differentiated heating of a ventilated room with a complex of modern heating devices] *TTPS* **1(31)** 31–37 [In Russ.]

[11] Szász C 2014 Intelligent building development and LabView-based modelling of a net zero-energy strategy *International Review of Applied Sciences and Engineering* vol 5 **2** 157–166

[12] Hesselbach J *et al.* 2008 Energy efficiency through optimized coordination of production and technical building services *LCE 2008: 15th CIRP Int. Conf. on Life Cycle Engineering: Conf. Proc. CIRP* p 624

[13] Baranskia M, Fütterera J and Müllera D 2017 *Development of a generic model-assisted control algorithm for buildings HVAC systems*

[14] Menassa C C *et al.* 2013 Conceptual framework to optimize building energy consumption by coupling distributed energy simulation and occupancy models *J. of Computing in Civil Eng.* vol 28 **1** 50–62

Development of the design and manufacturing technology of a bactericidal ionizer

A A Tikhonov[1]

[1] Yaroslav-the-Wise Novgorod State University, B. St. Peterburgskaya, 41 173003, Velikiy Novgorod, Russia

E-mail: atnu1@yandex.ru

Abstract. The design and manufacturing technology of a bactericidal ionizer with electrolytic coatings of copper and silver forming a galvanic cell are described. The ionizer is capable of creating solutions containing silver and copper ions. Using an ionizer, one can get preventive solutions that strengthen the immune system; medical preparations; cosmetic compositions; receive disinfectant solutions that can destroy a number of pathogenic bacteria and viruses. The ionizer has a low cost; it is easy to use and surpasses analogues in performance. A manufacturing technology for such an ionizer has been developed.

1. Introduction

At present, in connection with the spread of coronavirus infection worldwide, it is very important to increase immunity, use effective disinfecting solutions at home, drink disinfected water and drinks, and be able to prepare some medications at home. An ionizer allows creating solutions containing copper and silver ions. Copper, being a completely irreplaceable microelement, in addition to antimicrobial action, has immunostimulating, antioxidant and anti-inflammatory properties. Thus, a 99% reduction in the number of initial bacteria occurred in 10 hours at a concentration of copper ions of 0.01 mg/l, in three hours at a concentration of copper ions of 0.1 mg/l, and after one hour at a copper ion content of 1 mg / l [1, 2]. Silver in ionic form has an immunomodulating, bactericidal, bacteriostatic, antiviral, antifungal and antiseptic effect against more than 500 pathogenic microorganisms. One milligram of silver ions in one liter of water for half an hour causes inactivation of the influenza viruses A, B, Miter and Sendai, has a therapeutic effect in the treatment of viral enteritis, and also inhibits the initial development of the AIDS virus. The effect of silver killing bacteria is 1.500 times higher than the action of the same phenol concentration and 3.5 times higher than the action of mercuric chloride [3–5]. In this regard, the authors of the utility model [6] developed plastic utensils with bactericidal properties containing silver nanopowder. When using this glassware, liquids poured into it are saturated with silver ions. However, the concentration of silver may not always be sufficient for a full disinfecting effect. Since the area of contact of silver and the liquid used in the proposed bactericidal dishes is insignificant, even the presence of silver in the form of nanopowder does not completely solve the problem. A significantly higher concentration of silver and copper ions can be obtained using electrochemical ionizers. In particular, the utility model "Ionizer for disinfecting drinking water with silver or copper ions and preparing silver or copper aqueous solutions of a given concentration" is known [7]. The ionizer contains a grid electrometering device, where a current isolation is provided, ensuring the safety of the ionizer, stabilizing the current and

Content from this work may be used under the terms of the Creative Commons Attribution 3.0 licence. Any further distribution of this work must maintain attribution to the author(s) and the title of the work, journal citation and DOI.
Published under licence by IOP Publishing Ltd

reversing the polarity. The ionizer has a timer, a stabilizer of the concentration of ions in a given volume of drinking water and electrodes made of chemically pure silver or copper, made identical in shape and size, parallel to each other. The electrodes are predominantly triangular in shape and perforated with round holes, enclosed in a dielectric cassette of a similar shape and perforation, while the cassette is equipped with a handle made in one piece with the cassette, and has the shape of a "spoon" for the forced separation of silver or copper ions from the electrodes by stirring water throughout the electrolytic process. The ionizer is equipped with an autonomous power source, and an electronic device (tester) is nested in a "spoon" to visually monitor the amount of dissolved silver or copper. Using this ionizer, you can get solutions containing copper and silver ions of any desired concentration. Simpler in design are such ionizers as antimicrobial ionizer [8] and copper-silver ionizer [9]. These ionizers contain a spiral anode and a cathode which are short-circuited to each other by direct electrical contact, made of metals with various electrochemical potentials. They are installed in a vessel of limited capacity. The following electrochemical metal pairs are used as electrodes: copper (anode) and silver or silver-plated copper (cathode), and in the second version, silver (anode) and gold or gilded silver (cathode). The formation of copper and silver ions occurs due to the galvanic cell operation. In this case, electrodes with surface areas of 10.5 cm^2 each are used in a copper-silver ionizer. Tests have shown an increase in resistance to colds and a decrease in the frequency of exacerbation of chronic otolaryngological diseases if using an ionic solution. Herewith, they have no inhibitory effect on the ciliary epithelium. No side effects from the use of the solution were noted [9]. However, this ionizer also has disadvantages. It has low productivity, because small area electrodes are used. In this ionizer it is impossible to obtain solutions with different ratios of copper and silver ions, as well as solutions containing only copper ions or only silver ions. The ionizer is not very convenient to use, since it takes a relatively long time to clean and service it. Thus, the problem of creating an inexpensive, productive, universal, convenient ionizer for home use is quite relevant. In this regard, the goal of this work is to develop the design and manufacturing technology of such an ionizer.

2. Experiment: materials and methodic

The samples had dimensions of 30x20x0.5mm; they were made of low-carbon steel 10kp containing 0.1% carbon. For the preparation of electrolytes, distilled water and chemically pure chemicals were used. The electrodeposition of the coatings was carried out using an AC rectifier with current stabilization and short circuit protection. The setup graph is shown in the experimental part. Electrodeposition of coatings was performed in laboratory galvanic baths with a volume of electrolytes of one liter. The electrolyte was stirred with a magnetic stirrer at a speed of 60 - 80 r/min. The thickness of the coatings and the deposition rate were determined by the weight method (by weight gain) on an analytical balance with an accuracy of 0.1 mg. The wear resistance of the coatings was determined at the stand of the reciprocating motion of the LTI structure. The porosity of the coatings was determined by applying a filter paper impregnated with a solution containing 10 g/l of potassium iron-hydrogen sulfide and 20 g/l of sodium chloride, after 20 minutes exposure. The adhesion strength of coatings was tested by such methods as applying a grid of scratches and brushing with a brass brush.

3. Results and discussion

The proposed ionizer consists of containers of various capacities from 25 ml to several tens of liters and a spoon. Tanks can have a different cylindrical shape with either the same diameters at the top and bottom or expanding up. These containers can be made of high purity copper; silver can be coated on the lower inner part of the container so that the ratio of the internal surfaces of copper and silver coating is from 0.8:1 to 1:1. The second, more economical option, presents the tank made of low-carbon steel 05kp, 08kp or 10kp, and then its entire surface is coated with copper. In the third variant, the container is made of plastic. Then the entire surface of the tank is covered with copper. After that, a silver coating is on the lower inner part of the vessel, so that the ratio of the inner surfaces of copper and silver coating is from 0.8:1 to 1:1. In this case, it must be borne in mind that the liquid is not poured into the tank to the very top. The ionizer is a galvanic cell consisting of copper and silver coatings in contact with each other wetted by an electrolyte. As an electrolyte, a low concentrated solution of citric acid (100 mg/l is enough) in distilled water, mineral or drinking water, physiological saline with citric acid, milk whey, tomato juice containing sodium chloride and other electrolytes depending on the purpose of the solution can be used. In this pair of coatings, a silver coating having a standard potential of + 0.8V is cathodic, and a copper coating having a standard potential of + 0.3V is anodic. In actually used solutions, the coating potentials, of course, will differ from the standard ones. However, in any of the prepared solutions, the potential of copper will be more negative compared to the potential of silver. Accordingly, the copper coating will be the anode and will dissolve faster than the silver coating due to the reaction (1):

$$Cu^{o} - 2e = Cu^{2+} \quad (1)$$

In this case, silver oxide is formed on the surface of the silver coating, which, slowly dissolving, also passes into the solution. The set of processes can be expressed with formulas:

$$Cu + 2Ag + H_2O + 2H^{+} \rightarrow Cu^{2+} + Ag_2O + 2H_2 \quad (2)$$

$$Ag_2O + H_2O = 2Ag^{+} + 2OH^{-} \quad (3)$$

To make a therapeutic and preventive solutions using copper ions in combination with silver ions and cosmetics with the participation of copper and silver, which can stimulate the processes of generating biological tissue without side effects, one can use the recommendations set forth in the patent [9]. In the ionizer with the ratio of the areas of their electrodes Cu/Ag: 1/0.7; 1/0.8; 1/0.9; 1/1 the following ratios of concentrations of Ag/Cu in drinking water were obtained: 1/6500, 1/4900, 1/4800, 1/4500. The authors of this patent found that the ionizer with the above parameters in 400 ml of physiological solution with the addition of citric acid at a liquid temperature of 18-20°C releases 0.4 mg of copper and 0.00008 mg of silver within an hour. The amount of emitted ions is directly proportional to the exposure time. So, when installing the ionizer in a standard container with physiological saline (400 ml) with the addition of citric acid 100 mg/l, at an exposure of 6, 12 and 24 hours, respectively, the following silver concentrations were obtained: 0.0012; 0.002; 0.0048 mg/l and for copper: 6.0; 12.0; 24.0 mg/l.

When using the proposed bactericidal dishes, the exposure time should be significantly reduced, since the anode surface area in the proposed bactericidal dishes (the same volume of 400 ml) is an order of magnitude higher than the anode area (10.5 cm^2) of the ionizer proposed in the patent [9]. For the preparation of preventive solutions simultaneously containing both copper and silver ions, it is sufficient to keep the fluid used for 5-10 minutes in the ionizer. To prepare preventive solutions containing only silver ions, one should pour liquid inside the ionizer so that its upper level does not touch the copper coating. To prepare preventive solutions containing only copper ions, it is necessary to place the bactericidal container inside another non-metallic (glass, ceramic) container with a larger diameter and pour liquid between the outer copper surface of the ionizer and the inner surface of another non-metallic container, as shown in figure 1.

When preparing preventive solutions containing only copper or only silver, the exposure time should be significantly increased to several hours (2–24 hours), since in these cases the dissolution of copper or silver is much slower, not due to the work of the galvanic cell, but due to contact discharge. The reason is that the solubility of copper and silver is very low. Since the solubility of silver in water is 0.04 μg/l, Ag + silver ions form hydrated ions that maintain stability for a long time:

$$Ag(H_2O)^{+} \rightarrow H^{+} + (Ag^{+} + OH^{-}) \quad (4)$$

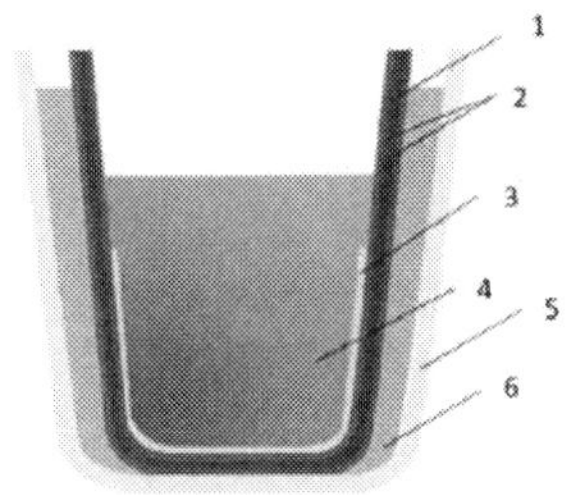

Figure 1. Ionizer (bactericidal container). 1 – steel or plastic glass; 2 – copper coating; 3 – silver coating; 4 – solution saturated with copper and silver ions; 5 – ceramic or glassware; 6 – solution saturated with copper ions.

Another element of the ionizer is a bactericidal spoon. A bactericidal spoon, as well as a bactericidal container, can be made of copper and coated with half of the copper area (taking into account the liquid level) with silver. A bactericidal spoon, as well as a bactericidal container, can be made of low-carbon steel and first totally covered with copper, and then half (taking into account the liquid level) coated with silver, observing the ratio of the surfaces of copper and silver coating from 0.8:1 to 1:1. A spoon can also be used to prepare solutions. A solution simultaneously containing copper and silver ions is prepared if the spoon is immersed in the liquid almost completely, so that both the silver and copper coatings are wetted by the liquid. To prepare a solution containing only copper ions, only part of the copper-coated spoon is immersed in the liquid. To prepare a solution containing only silver ions, only part of a spoon with silver coatings is immersed in a liquid. A bactericidal spoon is easier to make, but it can mainly be used to prepare bactericidal prophylactic solutions, and it is better to use bactericidal dishes for the preparation of therapeutic and cosmetic solutions, since it usually has a larger surface of the coating electrodes. The content of silver and copper in drinking water is regulated by the norms of Sanitary Rules and Norms (hereinafter – SanPiN) 2.1.4.1074-01 "Drinking water. Hygienic requirements for water quality of centralized drinking water supply systems. Quality control". The copper content should not exceed 1 mg/l, silver should not be more than 0.05 mg/l. For packaged water, the content of silver ions should not exceed 0.025 mg/l in accordance with SanPiN 2.1.4.1116-02 "Drinking water. Hygienic requirements for the quality of water packaged in containers. Quality control". Concentrations of silver and copper close to the maximum permissible concentrations do not adversely affect the human body. If the solution is prepared on the basis of drinking water, it is necessary to consider the concentration of these ions in the water of particular region. Prolonged use of solutions with silver and copper ions in amounts significantly exceeding the maximum permissible concentration can have a negative effect, in particular, lead to argyria.

For the manufacture of an ionizer, shaping and coating technologies are used. The most productive methods of forming from copper and steel belong to the field of pressure treatment. To make a spoon, molding is used. Tanks can be made by hood. In small-scale production, for example, for the manufacture of a pilot batch, the shaping of metal containers for the ionizer can be performed by the end rolling process. Forming of spoons and containers made of plastic, most often, is carried out by injection molding with injection molding machines. Next, it is necessary to apply copper and silver coatings. On plastic products, the first layer of copper coating can be applied by vacuum condensation, and then the local silver coating is applied galvanically, because it is the easier way. The second variant of metallization is the deposition of copper and silver plating. This option is more multi-operational, but it allows getting an ionizer with higher performance due to the creation of coatings with large actual areas. First, we consider the technology of metallization of steel products in a galvanic mode. To prepare the surface, degreasing and etching solutions were used; they are presented in table 1.

Table 1. Solutions for preparing the surface of low carbon steel.

Components, g/l and modes	Solution 1	Solution 2
Sodium hydroxide	15–25	
Sulphuric acid		120–150
Sodium carbonate	30–50	
Trisodium phosphate	30–50	
Sodium liquid glass	10–20	
Inhibitor KPI-1		0.5–1.0
Preparation OS-20	0.1–0.5	
Temperature, °C	60–80	18–30
Density of alternating current, A/dm^2	3–8	
Time, min	4–5	3–4

Electrochemical degreasing on alternating current was used. It made possible to use cheaper electrical equipment and increase the load of the degreasing bath, since in this case it is not necessary to use additional electrodes. The etching time may vary depending on the amount of oxides on the surface of the steel. After etching, the operation of decapitation in a solution of sulfuric acid 50–70g/l was performed for 0.5–1 minutes. Then washing was performed in tap and distilled water. After that, galvanic coatings were applied. For electrodeposition of coatings, the installation shown in Figure 1 and electrolytes, the compositions of which are given in tables 2, 4, 5, were used.

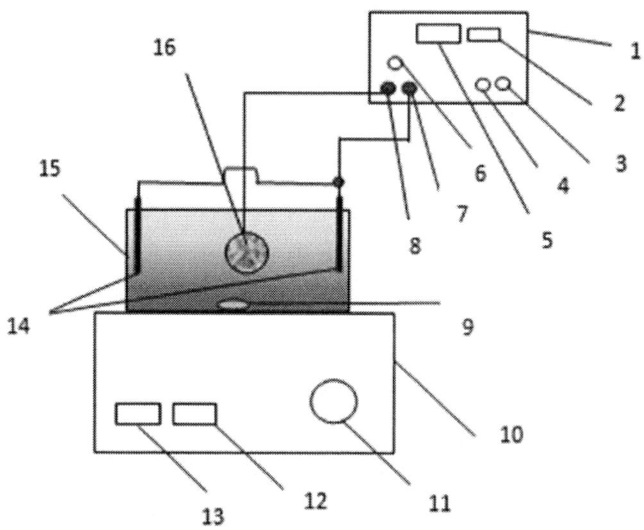

Figure 2. Installation for applying electrolytic coatings. 1 – ACRG (AC rectifier designed by V.M. Goreev); 2 – current regulator; 3 - tumbler switch on of electricity grid; 4 – tumbler switch on of current supply; 5 – digital milliamperemeter; 6 – tumbler switch on of ACRG; 7 – anode terminal "+"; 8 – cathode terminal "–"; 9 – mixing element of the stirrer; 10 – magnetic stirrer; 11 – mixing speed controller; 12 – heating switch on button; 13 – power switch on button; 14 – anodes; 15 – bath with electrolyte; 16 – metallized product.

The thickness of the coatings was determined by the gravimetric method using formula 5.

$$h = \frac{(m_2 - m_1)10000}{\rho \cdot S} \quad (5)$$

IOP Conf. Series: Materials Science and Engineering **939** (2020) 012076 doi:10.1088/1757-899X/939/1/012076

where $\boldsymbol{h}$ is the coating thickness, microns; m_2 is the mass of the coated sample, g; m_1 is the mass of the sample without coating, g; ρ_1 is the density of the coating metal, g/cm^3; $\boldsymbol{S}$ is the area of the covered area, cm^2.

The most common copper plating electrolytes are sulfate electrolytes. However, acidic copper plating electrolytes do not allow obtaining sufficient, necessary adhesion of the copper coating on steel due to contact release of copper. Contact release of copper is not completely suppressed in the presence of special additives introduced into acidic electrolytes. A number of copper plating alkaline electrolytes are devoid of this disadvantage. Pyrophosphate electrolytes are the simplest in composition, stable and harmless compared to other copper plating alkaline electrolytes. For copper plating, two pyrophosphate electrolytes were proposed; their operating modes and compositions are given in tables 2 and 4, preliminary copper plating electrolyte 3 and electrolyte 6 of the main, more high-speed copper plating. To suppress the contact release of copper in the preliminary copper plating electrolyte, the copper concentration was reduced and citric acid, trioxygluaric acid and sodium sulfosalicylate were introduced. To increase the adhesion of the copper coating to steel, the samples were loaded under current and a current boost was performed for 30–45 seconds with a current density of 1.5–2 times higher than the working cathode current density. Then the coating was applied to the required thickness at the working current density. Table 2 also shows the compositions of electrolytes and modes of preliminary nickel plating and tinning.

Table 2. Compositions and modes of electrolytes of preliminary copper plating, nickel plating and tinning.

Components, g/l and modes	3	4	5
Copper sulfate	3–8		
Potassium pyrophosphate	100–150		
Nickel fluoroborate		300–400	
Tin sulfate			40–60
Nickel chloride		10–15	
Sulphuric acid			80–100
Citric acid	15–25		
Boric acid		10–15	
Trioxyglutaric acid	5–7		
Preparation OS-20			2–5
Sodium sulfosalicylate	20–30		
Temperature, °C	18–25	18–50	18–25
pH	8.3–8.7	3–3.5	
$D_к$, A/dm^2 without stirring			1–2
$D_к$, A/dm^2 with stirring	0.2–0.5	5–20	
Process duration, min	15–38	1–3	4–8

To prevent contact release of copper on carbon steel, electrolytic tinning and nickel plating can be used. For these purposes, nickel sulfate electrolyte (Watts electrolyte) is most often used. However, this electrolyte does not have high nickel rate and dissipation ability, and in these parameters it is inferior to a hydrofluoric nickel plating electrolyte. In addition, the fluoroborate nickel plating electrolyte is stable in operation, it allows obtaining light, not hard, elastic coatings; therefore, the nickel fluoroborate electrolyte was used (electrolyte 4 in table 2). The use of this electrolyte significantly reduced the time of preliminary nickel plating. However, the fluoroborate nickel plating electrolyte is inferior in its scattering power to tinning electrolyte and copper plating pyrophosphate electrolyte. For preliminary metallization of steel, tin sulfate electrolyte 5 was also used (table 2). Tin electrolytes are prone to delamination and suspension, so they are not mixed, which is usually done to increase the deposition rate of coatings. To determine the required thickness of coatings deposited using pre-metallization electrolytes, we determined the effect of coating thicknesses on adhesion. Coating of pre-metallization electrolytes with a thickness of 1.5 and 3 μm was applied. Then, all samples were coated with copper with a thickness of 6 μm using pyrophosphate electrolyte of the main copper plating 6. The results are shown in table 3.

Table 3. The effect of thickness and composition of coatings and preliminary metallization on adhesion.

Coatings	Thickness 1.5 μm	Thickness 3 μm
Copper	+/+	+/+
Nickel	+/–	+/+
Tin	+/–	+/+

Table 3 shows the method of measuring adhesion: in the numerator – brushing with a brass brush, in the denominator – scratching (drawing marks). As can be seen, from the data given in table 3, in order to ensure sufficient adhesion, the thickness of the copper coating electrodeposited in the electrolyte prior to copper plating should be 1.5 μm. Nickel and tin coatings should have a thickness of 3 μm. The duration of electrolysis to obtain such thicknesses is given in table 2, depending on the current density used. Covered containers can have a complex shape for coating, for example, a large ratio of the height of the container to the diameter, etc. Electrolytes may not have enough opacity and dispersion for high-quality metallization of such containers. In these cases, it is advisable to carry out the coating with an additional anode, which is installed inside the covered tank, both during preliminary metallization and during the main build-up of the required coating thickness.

After preliminary coating the operation of copper coating extension to the required thickness can be performed in copper plating electrolytes 6 and 7, which is presented in table 4.

Table 4. Electrolytes of the main copper plating.

Electrolyte Components (g/l) and Modes	6	7
Copper sulfate	80–100	
Copper Fluoroborate		200–250
Potassium pyrophosphate	300–350	
Hydrofluoric acid (free)		2–3
Citric acid	15–25	
Boric acid		15–16
Boron nitride	5–10	5–10
Temperature,°C	40–60	40–60
pH	8.3–8.7	1.2–1.7
D_K, A/dm2 with stirring	1.5–2.5	5–20
Process duration, min	36–60	6–18

As can be seen from table 4, pyrophosphate and hydrogen fluoride electrolytes were used for the main copper plating. Pyrophosphate has a significantly higher scattering power. In addition, the use of pyrophosphate electrolytes of preliminary and main copper plating allows abandoning the washing operation in purified water. Products, after preliminary copper plating, can be transferred without washing to the main pyrophosphate copper plating electrolyte. However, pyrophosphate electrolytes, by almost an order of magnitude, have a lower copper plating rate compared with hydrogen fluoride. On the other hand, a hydrogen fluoride electrolyte is more toxic. When choosing the electrolyte of the main copper plating, first of all, it is necessary to take into account the configuration of the bactericidal dishes. Copper electrolytes 6 and 7 contain cubic boron nitride (elbor) MRTU 2-036-1-68 brand LM 1/0. The grain size of the main fraction (95%) is 1 μm and smaller, i.e., the main fraction contains boron nitride nanopowder *BN*. Cubic *BN* in hardness is second only to diamond and penetrating into coatings significantly increases their wear resistance [10]. It is known that nanopowders can adsorb metal cations and improve their delivery to the cathode. Nanopowders, as a rule, increase the adhesion of coatings and reduce their porosity [11]. Thus, the introduction of boron nitride into the electrolyte made it possible to reduce the required thickness of the copper coating. Since the copper coating should be continuous without through pores reaching the base metal of the steel, so that iron ions do not fall into the solutions created with the help of bactericidal dishes. In this regard, the porosity of the coatings was determined, the results are given in table 5.

Table 5. The effect of pre-coating type and the thickness of the main copper coating on the number of through pores, pieces/cm^2.

Pre-coating type	Main copper 6 μm	Main copper 12 μm	Main copper 18 μm
Copper 1.5 μm	19	8	0
Nickel 3 μm	12	5	0
Tin 3 μm	26	14	2

As can be seen from the results obtained with a thickness of 18 μm, almost all coatings become non-porous. The maximum permissible concentration (MPC) of iron in drinking water: from artesian wells, 0.5-30 mg/l, and for schools, hospitals and other similar institutions – 0.3 mg/l. Given that the MPC of iron in drinking water is relatively large, the thickness of the main copper coating of 18 μm will be enough for safe operation of the ionizer. Table 4 gives the time of electrolysis to obtain a coating of this thickness depending on the cathode current density.

After applying the copper coating, a silvering operation was performed. The greatest cathodic polarization during the electrodeposition of silver is possessed by cyanide electrolytes, which so far remain the best in terms of the quality of the coatings obtained, in the stability of the electrolyte, in scattering power, and other parameters. However, cyanide electrolytes are highly toxic due to the presence of free potassium or sodium cyanide in their composition. Among non-cyanide electrolytes, the most close in its characteristics to cyanide electrolytes is ferruginous-heterodistor (hexacyanoferrate II) electrolyte. This is due to the fact that silver is present as a cyanide complex in ferruginous-ferroelectric electrolytes, but the electrolyte does not contain free cyanide. To dissolve the anodes, potassium thiocyanate is introduced into the electrolyte. The compositions of the ferruginous-hydrodynamic electrolyte and the modes of their operation are given in table 6.

Table 6. Ferruginous-heterodistor electrolytes of silver plating.

Electrolyte components (g/l) and electrolysis mode	8	9
Silver (in terms of metal)	1–2	25–30
Potassium ferruginous (free)	40–60	40–80
Potassium rhodonite	80–100	80–100
Potassium carbonate		35–40
Boron nitride	5–10	5–10
Temperature, °C	18–25	18–25
D_K, A/dm^2 without stirring	0.1–0.2	
D_K, A/dm^2 with stirring		0.6–1.2

Electrolyte 8 is a pre-silvering electrolyte. To prevent contact emission of silver on the copper coating, the silver concentration is reduced to 1–2 g/l. To apply a silver coating only on the inner lower surface of bactericidal dishes, one should pour silver electrolyte to the required level, install a silver anode in the cover in the electrolyte, connect the container to the negative pole of the rectifier and conduct electrolysis. The electrolysis time in the pre-silvering bath is 5 minutes. The recommended thickness of the silver coating on the dishes (spoons, forks, etc.) is 24 microns. However, taking into account the fact that the wear resistance of the silver composite coating, determined on the wear stand of the reciprocating movement, increased by 8.6 times, it is possible to reduce the thickness of the coating. Of course, the wear stand does not reproduce the real conditions of the abrasion of the silver coating on the dishes; however, the coating thickness can be reduced at least by half. In the main silver plating electrolyte 9, a coating thickness of 12 μm was electrodeposited at $D_k = 1$ A/dm^2 in 24 minutes. Silver composite coatings were matte, light, even without visible defects and passed the test for adhesion strength.

4. Conclusion

As a result of this work, an ionizer has been developed that surpasses analogues without a current source in the rate of saturation of liquid media with copper and silver ions. A technology has been developed for the manufacture of bactericidal dishes using copper, silver, nickel and tin plating. The influence of the composition and thickness of coatings, preliminary metallization, on the adhesion and porosity of multilayer coatings is established. Two technological schemes for the metallization of bactericidal dishes have been created. The first scheme is using pyrophosphate electrolytes of preliminary and main copper plating, which allows applying copper coatings to containers of a relatively complex shape. The second, more productive, technological scheme with the use of preliminary nickel plating or tinning can be applied for the manufacture of ionization spoons and ionization tanks. At the same time, electrolytic coating of containers with complex configurations using hydrofluoric nickel plating and copper plating electrolytes, as well as tinning, should be carried out using additional anodes. A technology has been developed for producing a composite, wear-resistant electrochemical coating consisting of silver and boron nitride.

References

[1] Gutenev V V, Khasanov M B, Montvila O I and Azhgirevich A I 2001 Bakteritsidnyye svoystva medi i vliyaniye na nikh razlichnykh faktorov [Bactericidal properties of copper and the influence of various factors on them] *Water and ecology: problems and solutions* **3** 21–27

[2] Azhgirevich A I 2011 Primeneniye v khimiko-bakteritsidnykh tekhnologiyakh dezinfektantov, soderzhashchikh med' i tsink [Application of disinfectants containing copper and zinc in chemical and bactericidal technologies] *Ekologiya urbanizirovannykh territoriy* [Ecology of urban areas] **4** 45–51

[3] Bukina Yu A and Sergeyeva Ye A 2012 Antibakterial'nyye svoystva i mekhanizm bakteritsidnogo deystviya nanochastits i ionov serebra [Antibacterial properties and mechanism of bactericidal action of nanoparticles and silver ions] *Bulletin of Kazan Technological University* vol 15 **14** 170–172

[4] Ivanov V N *et al.* 1995 Nekotoryye eksperimental'nyye i klinicheskiye rezul'taty primeneniya kationov serebra v bor'be s lekarstvenno-ustoychivymi mikroorganizmami [Some experimental and clinical results of the use of silver cations in the fight against drug-resistant microorganisms] *Serebro v meditsine, biologii i tekhnike* **4** 53–62

[5] Azhgirevich A I 2019 Aspekty povysheniya bakteritsidnoy aktivnosti serebrosoderzhashchikh dezinfekrantov [Aspects of increasing the bactericidal activity of silver-containing disinfectants] *Problemy bakteritsidnoy ekologii* **6** 144–150

[6] Antonov I M and Antonov A I 2011 Patent RF no 104835 *Bull.* **15**

[7] Abdulmenov F F 2005 Patent RU48530 *Bull.* **30**

[8] Gavrikova Ye I 2012 Patent RU115773 *Bull.* **13**

[9] Rodimin Ye M and Rodimin V Ye 2005 Patent RF 2264220 *Bull.* **32**

[10] Tikhonov A A, Filippov D A, Arendateleva S I and Perepelitsa T I 2019 *IOP Conf. Ser.: Mater. Sci. Eng.* **656** 012051

[11] Ilyin In A, Semenychev In B, Naletov B P, Salakhova R K and Tyurikov E V 2010 Patent RF no 2 389 828 *Bull.* **14**

IOP Conf. Series: Materials Science and Engineering **939** (2020) 012077 doi:10.1088/1757-899X/939/1/012077

Metallization of aluminum alloys with electrodeposition by rubbing

A A Tikhonov[1]

[1] Yaroslav-the-Wise Novgorod State University, ul. B. St. Petersburgskaya, 41 173003 Veliky Novgorod, Russia

[2] E-mail: atnu1@yandex.ru

Abstract. The technology of metallization of duralumin alloy D16 and cast aluminum alloy AL4 with method of electrodeposition by rubbing has been developed, which allows excluding operations of chemical etching and clarification without reducing the adhesion of coatings. In these operations, such high-concentration acids as nitric, sulfuric, hydrofluoric and others can be used. Instead, it is proposed to use cathode-anode processing operations and application of the zinc-nickel-boron nitride alloy in a universal, much more environmentally friendly electrolyte for simultaneous etching and deposition of coatings. A technology for applying a gradient zinc-nickel-boron nitride alloy with electrodeposition by rubbing has been developed. The process of copper plating of aluminum alloys from pyrophosphate electrolytes with electrodeposition by rubbing is studied. Such modes of coating as copper and zinc-nickel-boron nitride ones using anode tools with an area of 1 cm^2 and 3.75 cm^2 were established.

1. Introduction

Aluminum and its alloys are used in many industries because they have a number of advantages: low density, high corrosion resistance, increased electrical conductivity, good thermal conductivity and ductility. Alloys of aluminum with silicon – silumins have very good casting properties. It is, first of all, high fluidity. Duralumin, high-strength and granular (powder) alloys have a high specific strength. However, aluminum and aluminum alloys are not without drawbacks. Duralumin aluminum alloys have low corrosion resistance. Aluminum and its alloys are poorly soldered, have a high coefficient of friction, low hardness and wear resistance and not always sufficient decorative properties. To eliminate the shortcomings of aluminum and aluminum alloys and significantly improve their properties, electrolytic coatings can be used. Coatings allow combining the positive (valuable) properties of aluminum with the properties of the applied coatings. Aluminum and its alloys on the surface have a strongly bonded oxide film Al_2O_3. After complete removal, alumina is rapidly formed again. Therefore, despite the fact that various multi-operational technologies are used in electroplating for coating, it is not possible to achieve high-quality metallization of aluminum and its alloys, especially those working under harsh operating conditions. In this regard, the problem of creating high-quality, reliable, low-operational and inexpensive metallization of aluminum and its alloys is quite relevant [1]. One of the most common methods of metallization of these alloys is galvanic, based on the use of electrolysis. Electrolysis is also used in the electrodeposition by rubbing method. In contrast to the galvanic method, there is no galvanic bath in the method of electrodeposition by rubbing. A small volume of electrolyte

Content from this work may be used under the terms of the Creative Commons Attribution 3.0 licence. Any further distribution of this work must maintain attribution to the author(s) and the title of the work, journal citation and DOI.
Published under licence by IOP Publishing Ltd

is held by a porous material attached to the anode tool. This method is more environmentally friendly, versatile, mobile and high-speed compared to the traditional galvanic method [2–4].

2. Experiment: materials and methodic

Samples were made of duralumin alloy D16 containing copper 3.8–4.9%, magnesium 1.2–1.8%, manganese 0.3–0.9% and silicon 0.5%, and from a cast aluminum alloy AL4 containing silicon 8–10%, manganese 0.25–0.5% and magnesium 0.17–0.3%. The samples were 40×60×1 mm in size.

For electrodeposition with rubbing, anode tools with a working area of 1 cm^2 and 3.75 cm^2 were used. The conductive part of the anode tools was made of 08H18N10T stainless steel containing carbon 0.8%, titanium about 1%, nickel 10% and chromium 18%. To hold the electrolyte and prevent a short circuit, the lower conductive part of the anode tools was wrapped with fabrics. First, a thin terry cloth with electrolyte absorption of 0.07 ml/cm^2 was used, and the second layer was made of a wear-resistant synthetic fabric consisting of polyester fibers (crimpl) with electrolyte absorption – 0.02 ml/cm^2.

To supply electrolytes, the dipping method and the jet method were used. The feed rate of the electrolyte by the jet method was 12–18 ml/cm^2 min. The speed of movement of the anode instruments was maintained from 15 to 20 m/min. The ratio of the cathode (coated) surface to the anode (area of the anode tool) was 2. Electrolytes were prepared using distilled water and chemically pure chemicals. To prepare the surface and for electrodeposition with rubbing an AC rectifier with current stabilization and protection against short circuit was used. The installation and anode instrument circuits are given in [5, 6]. The thickness of the coatings and the deposition rate were determined by the gravimetric method (by weight gain) on an analytical balance with an accuracy of 0.1 mg and using a high-precision micrometer. The adhesion strength of the coatings was checked by the following methods: applying a grid of scratches (grooves), brazing with a brass brush and heat treatment with heating to 190–200 °C, holding for 1 hour and cooling in air. In addition, the adhesion of the zinc-nickel-boron nitride coating was evaluated by the results of diffusion vacuum annealing with a heating temperature of 400 °C.

3. Results and discussion

Considerable attention in this work was given to the development of surface preparation of aluminum alloys to increase the adhesion of the applied coatings, reduce the number of operations and low environmental hazards. Technological schemes were developed: "A" – without immersion preparation of surface, that is, with direct coating and scheme "B" with immersion preparation of surface. To reduce the number of operations of the technological process, the principle of combining operations was used, which can be implemented in the method of electrodeposition by rubbing. The first combination used in both schemes "A" and "B" is the simultaneous execution of machining and electrochemical degreasing. For this, an electrolyte suspension with micropowder of cubic boron nitride with a dispersion of the main fraction of 10 μm was used. Degreasing was performed using anode tools. Samples for all experiments were degreased in the following solution (g/l): sodium carbonate 40–50, trisodium phosphate 40–50, sodium metasilicate 3–5, syntanol (DS-10) 1–2, boron nitride up to 200. The electrolyte was heated to 60 – 70 °C and cathodic degreasing was performed at a current density of 5–10 A/dm^2 for 7 minutes. The quality of degreasing was controlled with the wetting of the sample surface with water.

For direct electrodeposition of the coating on aluminum alloys, a coating based on the zinc-nickel *Zn – Ni* electrolytic alloy was used. This alloy can be galvanized on aluminum, steel and other metals. The alloy is applied to provide reliable, anode "active protection" from corrosion of steel parts. With such protection, the coating is an anode in regards to steel and, dissolving, protects it from corrosion, therefore corrosion products are formed on the coating. Nevertheless, more and more often, protective coatings are required to ensure that active protection is accompanied by maintaining the surface almost unchanged during long-term operation. To some extent, this is characteristic of zinc-nickel alloy coating (12–15% *Ni*). Recent studies have focused on further improving the protective ability of these coatings and addressing some of the drawbacks of *Zn – Ni* coatings [7, 8]. It is known that when applying zinc coatings by selective rubbing onto a steel base of a large area, there is a significant etching of the zinc coating at the edges. This occurs at a time when the electrode tool is far from the edge of the coating and this section has a low cathodic polarization [9]. The *Zn – Ni* alloy has advantages over zinc because of the less corrosion that occurs directly during rubbing. Considering the data of [10], we will compile table 1 of a comparative analysis of acid and alkaline electrolytes used for the galvanic deposition of *Zn–Ni* alloy, it will help us make the right choice of electrolyte for electrodeposition by rubbing.

Table 1. Comparison of acidic and alkaline electrolytes used for *Zn – Ni* coating.

Indicators	Acidic	Alkaline
Current output	More	Less
Coating speed	Higher	Lower
Scattering ability	Lower	Higher
Coating ability – "starting charactersitics"	Higher	Lower
Protective abilities	High	High

As can be seen from table 1, the main advantage of alkaline electrolytes is their high dissipation ability, this property is very important for traditional galvanic coating, it is secondary, not the most important for electrodeposition by rubbing. Moreover, the heightened "starting characteristics" of the electrolyte are especially important for the metallization of aluminum and other poorly coated metals. Acid electrolytes are more promising for direct coating of aluminum alloys with electrodeposition by rubbing. In scheme "A", the technology of combining two operations in one universal electrolyte was used: electrochemical etching and almost immediately applied coating. We used this technology for metallization of molybdenum alloys and achieved good adhesion [11]. The same principle was used in another work [12] for coating aluminum alloys with the galvanic method. In this work, a *Zn – Ni* galvanic alloy of sulfate electrolyte was deposited on aluminum alloys, in which the cathodic treatment and then the anode treatment were first performed. The electrolyte, in addition to zinc sulfate and nickel sulfate, to increase electrical conductivity contains sodium sulfate and boric acid, as a buffer additive stabilizing the pH. The electrolyte also contains sodium fluoride. Sodium fluoride dissolves the oxide film on aluminum; aluminum binds to sodium fluoride into a fluoroaluminate complex. This contributes to obtaining a fine-grained structure of the coating and increases the adhesion of the coating to aluminum alloys. We suggest a technology for producing a gradient zinc-nickel-nitride boron coating due to changes in the concentration of electrodeposit metals and changes in the composition of electrolytes. To do this, we used the electrolytes shown in table 2. The purpose of obtaining such a gradient coating is to increase the adhesion of the applied coating and increase corrosion resistance. To solve this problem, a zinc-nickel-nitride boron coating *Zn – Ni – BN* was used with a gradual increase in the nickel content in the coating. The standard potential of aluminum is 1.66V, zinc – 0.76V, nickel – 0.25V. The potentials of coatings in real operating conditions will differ from standard ones, however, in real operating conditions, the zinc potential will be more negative compared to the potential of nickel. Accordingly, in the *Zn – Ni – BN* gradient coating, the potential will gradually change from more negative to more positive, this should increase the protective ability of such a coating. In the gradient *Zn – Ni – BN* coating, the coefficient of thermal linear expansion (CTLE) will change more smoothly. For aluminum, it is 1.7 times higher than that of nickel and slightly differs from the CTLE of zinc. Thus, in the gradient coating, the CTLE will also change more smoothly, and this should increase the adhesion of the coatings, especially when the temperature changes, for example, during soldering

IOP Conf. Series: Materials Science and Engineering **939** (2020) 012077 doi:10.1088/1757-899X/939/1/012077

Table 2. The composition of the electrolytes to obtain a gradient coating *Zn – Ni – BN*.

Components, g/l	Electrolyte 1	Electrolyte 2	Electrolyte 3
Zinc sulfate	240–300	160–200	
Zinc chloride			40–50
Nickel sulfate	80–100	176–220	
Nickel chloride			232–290
Sodium sulfate	40–100		
Glycine		6–12	30–50
Sodium acetate		20–30	20–30
Boric acid	20–30		
Sodium fluoride	15–20		
Preparation OS-20	0.1–0.5	0.1–0.5	0.1–0.5
Gelatin		0.1–0.3	0.1–0.3
Saccharin			0.5–1.5
Boron nitride	5–10	5–10	5–10
pH	1.9–2.0	2.5–3.0	3.5–4.0

Electrolytes 1–3 contain cubic boron nitride (elbor) MRTU 2-036-1-68 brand LM 1/0. The grain size of the main fraction (95%) is 1 μm and smaller, i.e., the main fraction contains boron nitride nanopowder *BN*. Cubic *BN* is second only to diamond in hardness. In addition, *BN* when electrically rubbing coatings on aluminum alloys activates the surface, which is especially important when applying coatings on aluminum alloys. Also, during electrodeposition by rubbing, *BN* can smooth out microprotrusions, which contributes to the production of smoother, semi-shiny coatings. It is known that nanopowders can adsorb cations of molecules dissociated in the electrolyte, which improve their delivery to the cathode. Hitting the nanoparticles with the deposition surface, they locally destroy the double electric layer (DEL), energetically activate the deposition zone, while at the same time bringing ions adsorbed on the surface of the nanoparticles to a distance less than the thickness of the DEL. Nanopowders, as a rule, increase the adhesion of coatings and reduce their porosity [13]. Electrolytes 1–3 contain the preparation OS-20 which is a mixture of polyethylene glycol esters of synthetic primary higher fatty alcohols. OS-20 is a surfactant and helps to increase the sedimentation stability of nanosuspension. OS-20 eliminates pitting, improves the wettability of the surface of aluminum alloys with electrolyte. Electrolytes 2 and 3 contain gelatin. Gelatin and OS-20 contribute to the production of fine crystalline and shiny coatings. Presumably, they form colloidal ion complexes with metal ions and increase cathodic polarization. Gelatin can be adsorbed on microprotrusions and block their growth; accordingly, it is a shiner. However, it must be borne in mind that gelatin, especially at relatively high concentrations in electrolytes, reduces the ductility of the coatings. Therefore, the gelatin content in electrolytes does not exceed 0.3 g/l.

Saccharin also contributes to the production of shiny coatings and reduces internal stresses, primarily nickel, so it is part of electrolyte 3. Electrolyte 2 contains aminoacetic acid (glycine) and sodium acetate, these substances have good buffering properties. Glycine with nickel can form a complex $[Ni\ (GLy)]+$, which in an acidic environment is a cation and acquires a positive charge. This complex compound has a negative logarithm of instability constants (ion equilibrium constants) $lgK = 5.7$. This complex is concentrated near a negatively charged cathode and it has a positive effect on the quality of the deposited coatings. In the electrolyte, complexes could appear: $[Ni(GLy)_2]$ c $lgK = 10.54$ and $[Ni(GLy)_3]^-$ c $lgK = 14.26$ [14]. The complex compounds of glycine based on zinc $[Zn(GLy)_2]$ are formed at a pH above 6, and at a pH above 4.5 the complex $[Zn(GLy)]^+$ is formed, it has $lgK = 5.3$ and acts as a cation. In [15], data are presented on the low resistance of zinc glycine complexes and, accordingly, the electrodeposition of zinc from such solutions occurred with a slight overvoltage. During the deposition of nickel and zinc from complex cations, there is an increase in the concentration of glycine in the cathode layer and, accordingly, conditions are created for the formation of stronger nickel complexes [$Ni(GLy)_2$] и [$Ni(GLy)_3$]. Complex compounds of nickel with glycine are stronger than complex compounds of zinc and, accordingly, glycine contributes to the enrichment of the alloy with nickel, bringing the polarization curves of nickel deposition to the polarization curves of zinc deposition. The content of nickel and zinc was defined with analytical method. The content of boron nitride was calculated by subtracting the total content of zinc and nickel (percentages are given by weight) from 100%. The results are presented in table 3.

Table 3. The effect of electrolytes on the composition of the coatings.

Composition	Electrolyte 1	Electrolyte 2	Electrolyte 3
Nickel	4.3	11.4	23.1
Zinc	89.5	82.9	72.0
Boron nitride	6.2	5.7	4.9

Thus, when using electrolytes from 1 to 3, the nickel concentration in the coatings increases due to an increase from 0.3 to 3 in the ratio of nickel to zinc in the electrolyte, and also due to an increase in the glycine concentration from 0 to 50 g/l in the electrolyte. In this case, the potential of the first coating layers is closest to the potential of aluminum, and as the coating thickness increases, the potential becomes more electropositive. Moreover, if during the application of electrolytic coatings on steel, we try not to exceed the nickel content above 12–16% in order to maintain the anode, active corrosion protection mechanism, in the galvanic pair "aluminum alloy – coating electrolytic $Zn-Ni$ alloy of any composition", the coatings will be cathodic. Since $Zn-Ni$ alloys with increased nickel content become more chemically and corrosion resistant, there is a need for a gradual increase of nickel in the coating, i.e. in creating a gradient coating with an increased concentration of nickel. However, too high nickel content of the coating can create high internal stresses in the coating. To eliminate lamination and to make transitions in nickel concentration in the coating more smooth, diffusion (homogeneous) annealing in a vacuum furnace can be used.

Thus, the technological scheme "A" will consist of the following operations: combined operation of simultaneous degreasing and grinding, performed by rubbing electrodeposition; rinsing in deionized water; cathodic treatment by rubbing electrodeposition in electrolyte 1 with a current density of 10 A/dm^2 for 20 seconds; anodic treatment by rubbing electrodeposition in electrolyte 1 with a current density of 10A/dm^2 for 20 seconds; electrodeposition by rubbing of $Zn-Ni-BN$ alloy in electrolyte 1 with a cathode current density (D_c) of 30 A/dm^2 for 4 min; electrodeposition by rubbing of $Zn-Ni-BN$ alloy in electrolyte 2 with D_c = 30A/dm^2 for 4 min; electrodeposition by rubbing of $Zn-Ni-BN$ alloy in electrolyte 3 with D_c = 30A/dm^2 for 4 min; rinsing in deionized water; drying; diffusion vacuum annealing with an isothermal holding temperature of 400°C for 3 hours and cooling together with the furnace; quality control.

The coating thickness was determined with the gravimetric method using formula 1.

$$h = \frac{(m_2 - m_1)10000}{(\rho_1 n_1 + \rho_2 n_2 + \rho_3 n_3)\cdot S}, \quad (1)$$

where: h is the coating thickness, microns; m_2 is the mass of the coated sample, g; m_1 is the mass of the sample without coating, g; ρ_1 is the density of zinc in g/cm^3; n_1 is the fraction of zinc in the coating in fractions of a unit; ρ_2 is the density of nickel in g/cm^3; n_2 is the fraction of nickel in the coating in fractions of a unit; S is the area of the covered area in cm^2, ρ_3 is the density of boron nitride in g/cm^3; n_3 is the fraction of boron nitride in the coating in fractions of a unit.

For electrodeposition with rubbing, an anode tool with an area of 3.75 cm^2 with jetting of electrolyte was used. When using the electronic rubbing modes specified in the technological scheme "A", smooth, semi-shiny coatings without visible defects were obtained. When measured with a high-precision micrometer, the thickness varied from 36 to 49.5 μm in different parts of the coated sample. The total average coating thickness calculated by formula 1, obtained under these conditions according to scheme "A", was 47.6 μm. The average coating rate was 4 μm/min.

To implement the technological scheme "B" immersion solutions were prepared; their compositions are given in table 4. The immersion treatment of aluminum alloys was performed by dipping. Processing time in electrolytes 4 and 5 of a duralumin alloy is 8–20 seconds, of cast aluminum alloy is 10–30 seconds. After this, the samples were washed in water, especially thorough washing should be performed after treatment in electrolyte 4 due to the presence of sodium hydroxide in its composition, however, solution 4 is cheaper than solution 5. Differences in the adhesion of copper coatings obtained when immersion solutions are used in surface preparation 4 and 5 were not found.

Table 4. Compositions for applying immersion coatings *Zn* and *Zn – Ni* on aluminum alloys: duralumin D16 and casting AL4.

Components, g/l	Electrolyte 4	Electrolyte 5
Zinc Sulfate	20–30	
Hydrofluoride Zinc		20–40
Hydrobromide Nickel		200–220
Sodium hydroxide	100–150	
Ammonium hydrogen fluoride		30–40
Potassium nitrate tartrate	8–10	
Ferric chloride	1–3	
Sodium nitrite	0.5–1	

In technological scheme B, the process of electrodeposition by rubbing of copper coatings was used. Alkaline copper plating electrolytes were applied, since one of the main disadvantages of the main acid copper plating electrolytes is the inability to obtain a sufficiently high, necessary adhesion of the copper coating on aluminum alloys due to the contact release of copper. Contact copper release is not completely suppressed even in the presence of special additives introduced into acidic electrolytes. Many alkaline copper plating electrolytes do not have this disadvantage. Pyrophosphate electrolytes are the simplest in composition, stable and harmless in comparison with other alkaline electrolytes of copper plating. For electrodeposition by rubbing pyrophosphate electrolytes 6 and 7 were used, their compositions are given in table 5.

Table 5. Compositions of pyrophosphate copper plating electrolytes for electrodeposition by rubbing.

Components, g/l	Electrolyte 6	Electrolyte 7
Copper sulfate	32–40	96–120
Potassium pyrophosphate	300–350	300–350
Citric acid	15–20	
Sodium selenic acid	0–0.3	0–0.3
Additive CPEO in ml/l		15–25
pH	8.3–8.7	8.3–8.7

Pyrophosphate copper electrolyte 6 (table 5) is intended for preliminary copper plating of aluminum alloys. In this electrolyte, it is possible to apply both shiny coatings in the presence of sodium selenic acid concentration of 0.1–0.3 g/l, and matte in the absence of this salt. Copper electrolyte 7 can be used for faster main copper plating. It contains CPEO additive which is a condensation product of oxalic acid and ethylene glycol. The use of the CPEO additive in galvanic electrolytes decreased the porosity of the copper coating, increased the penetrability of complex-shaped products while expanding the range of the working current density [16]. The additive has also been shown to be effective in electrodeposition by rubbing. To determine the modes of electrodeposition by rubbing, the maximum permissible cathodic current density was established for pyrophosphate electrolytes, these data are presented in table 6.

Table 6. The effect of the area of the anode tools ($S_{a\text{-}t}$) and the composition of the electrolytes on the maximum allowable cathodic current densities (A/cm^2) when dipping the anode tools into the electrolyte in the numerator and when jetting the electrolyte in the denominator.

$S_{a\text{-}t}$, cm^2	Electrolyte 6	Electrolyte 7
1	8/16	20/38
3.75	3.2/7	9/19

As can be seen from table 6, in electrolyte 7, the maximum allowable current densities for electrodeposition by rubbing are much higher than in electrolyte 6. This is due to the higher concentration of copper ions in electrolyte 7, compared with the copper content in electrolyte 6. The maximum allowable cathodic current density is also influenced by the area of the anode tools, the smaller the anode tool, the higher the maximum allowable cathodic current densities can be applied. This is due to improved diffusion of copper ions. So, for example, when using an anode tool with an area of 0.02 cm^2, even in electrolyte 6 the maximum permissible cathodic current density of 200 A/dm^2 was reached. The feed rate of the electrolyte in the jet method is an order of magnitude higher than the feed rate of the electrolyte in the dipping method; it allowed increasing the permissible cathodic current densities. For depositing a copper coating with a total thickness of at least 24 μm, you can use the modes with jet feeding of electrolytes presented in table 7.

Table 7. Modes (cathodic current density D_c, A/dm^2 and time) for applying a copper coating with a thickness of 25 ± 1 μm for electrodeposition by rubbing with anode tools with an area of 1cm^2 and 3.75 cm^2.

Modes	S = 1 cm^2		S = 3.75 cm^2	
	Electrolyte 6	Electrolyte 7	Electrolyte 6	Electrolyte 7
D_c, A/ dm^2	16	38	7	19
Time, min	1.5	4	2.5	8

As can be seen from table 7, the total time for electrodeposition by rubbing when using an anode tool with an area of 1 cm^2 is almost two times less than for an anode tool with an area of 3.75 cm^2. This is due to the use of higher D_c due to better diffusion when using smaller anode tools. However, the area of the resulting coating for the anode tool S = 3.75 cm^2 is 3.7 times larger than for the anode tool with S = 1 cm^2.

Thus, the technological scheme "B" consists of the following operations:

– combined operation of simultaneous degreasing and grinding, performed by electrodeposition by rubbing;
– washing in deionized water;
– immersion treatment;
– rinsing in deionized water;
– electrodeposition by rubbing in the electrolyte of preliminary copper plating 6;
– electrodeposition by rubbing in electrolyte 7 of the main copper plating;
– washing in deionized water;

- drying;
- quality control.

Quality control included external inspection, determination of coating thickness and adhesion test. Copper coatings were electrodeposited without visible defects from dull to semi-shiny and shiny, depending on the content of shine-forming additives in the electrolyte. The results of checking the adhesion of zinc-nickel and copper coatings obtained using the preparation scheme A and B are presented in table 8.

Table 8. Adhesion of coatings by a gradient alloy based on zinc (Scheme A) and copper (Scheme B) to the cast aluminum alloy AL4 in the numerator and adhesion to the duralumin aluminum alloy D16 in the denominator.

Schemes	The method of measuring adhesion		
	1	2	3
Scheme A	+/+	+/+	+/+
Scheme Б	+/+	–/+	

Table 8 shows the methods for measuring adhesion: 1 – by brushing with a brass brush, 2 – by scratching (making grooves), 3 – the method of heat treatment (heating). The coating control standard does not recommend a heating method to test the adhesion of copper coatings on aluminum alloys, so this method was used only for zinc-based alloys.

As can be seen from the results obtained, when using technological scheme A, all samples passed the adhesion test. When using scheme B, coatings on a cast aluminum alloy did not pass the adhesion tests by the grooves method. This is due to the fact that the cast aluminum alloy contains more silicon compared to the duralumin alloy. For metallization of cast aluminum alloys, if necessary, to have a top layer of copper, technological scheme A was used and a $Zn - Ni - BN$ coating was applied, and then a copper coating of pyrophosphate electrolytes was applied. This option was tested, and it gave a good adhesion result. As can be seen from table 5, the adhesion of the coatings is higher to the duralumin alloy. Samples of D16 duralumin alloy coated with both a gradient zinc-nickel-boron nitride alloy using technological scheme A and copper using scheme B passed all adhesion tests.

4. Conclusion

In comparison with standard technological schemes two more environmentally friendly technological schemes for the metallization of aluminum alloys by the method of electrodeposition by rubbing with a reduced number of operations have been developed. Scheme B using immersion coatings made it possible to apply high-quality shiny copper coatings to duralumin alloys with high adhesion. In scheme A (direct coating of aluminum alloys) through the use of combined operations of simultaneous grinding and degreasing, etching and deposition of zinc-nickel-boron nitride alloy in one universal electrolyte reduced the number of operations. Technological scheme A allows obtaining high-quality, semi-shiny coatings with high adhesion strength to duralumin and cast aluminum alloys. The applied coating is a gradient zinc-nickel-boron nitride alloy with an increased nickel content as the coating thickness increases from 4.3% to 23.1%.

References

[1] Skopintsev V D, Margolin L N and Farafonov V V 2019 Galvanotekhnika i obrabotka poverkhnost *Electroplating and surface treatment* vol 27 **2** 9–22

[2] Semenychev V V 2017 Galvanotekhnika i obrabotka poverkhnosti *Electroplating and surface treatment* vol 25 **1** 23–27

[3] Demin S A and Vinogradov S S 2018 *Works of VIAM* **9** 43–50 DOI: 10.18577/2307-6046-2018-0-9-43-50

[4] Naletov B P and Tyurikov E V 2009 *Aviation materials and technologies* **2(11)** 25–29
[5] Moskovits M 1993 Galvanotekhnika i obrabotka poverkhnosti *Electroplating and surface treatment* vol 2 **3** 40–45
[6] Tikhonov A A and Sokolova A Yu 2010 *Bulletin of the Novgorod state University Yaroslav The Wise* **60** 102–105
[7] Shekhanov R F, Gridchin S N and Balmasov A V 2019 Galvanotekhnika i obrabotka poverkhnosti. *Electroplating and surface treatment* vol 27 4–8
[8] Kuznetsova I O, Saveljev A V, Nyrkov N P and Shuvalov D A 2019 Vestnik nauki *Bulletin of Science* vol 3 **6(15)** 382–8
[9] Ivshin Ya V and Kadrikov R A 2014 Galvanotekhnika i obrabotka poverkhnosti *Electroplating and surface treatment* vol 22 **3** 40–4
[10] Ralf Kraus 2013 Galvanotekhnika i obrabotka poverkhnosti *Electroplating and surface treatment* vol 21 **4** 24–9
[11] Tikhonov A A, Arendateleva S I and Perepelitsa T I 2019 *IOP Conf. Series: Materials Science and Engineering* **656** 012050
[12] Novikov A N and Bakaeva N V 2004 *Patent RU2237757*
[13] Ilyin In A, Semenychev In B, Naletov B P, Salakhova R K and Tyurikov E V 2010 Patent RF 2 389 828 *Bul.* **14**
[14] Dolgih O V, Sotskaya N V and Shamanaeva E S 2006 Kondensirovannyye sredy i mezhfaznyye granitsy *Condensed media and interfaces* vol 8 **4** 275–282
[15] Berezin N B, Berezina T N, Mezhevich Zh V and Chevela V V 2006 Vestnik Kazanskogo universiteta *Kazan University Herald* vol 16 **5** 267–8
[16] Kolchugin A V, Opolovnikov V R, Priyatkin G M and Vasiljev V V 1997 *Patent RU 2094543*

Electrodeposition by rubbing. The method of calculating the modes and the development of the high-speed copper plating process

A A Tikhonov[1,3], D A Filippov[1], S I Arendateleva[1], V A Popov[1] and T I Perepelitsa[2]

[1] Yaroslav-the-Wise Novgorod State University, ul. B. St. Petersburgskaya, 41 173003 Veliky Novgorod, Russia
[2] SC Rosatom, JSC Atomproekt, ul. Savushkina, 82 197183 St. Petersburg, Russia

[3] E-mail: atnu1@yandex.ru

Abstract. Taking into account the processes carried out during electrodeposition by rubbing, high-speed copper electrolytes have been developed: low-acid and nitrate sulfate. There are proposed modes for anode-tools with area of 1 cm^2, 3.75 cm^2 and 4.43 dm^2, allowing for application of matte, semi-slush and shiny copper coatings with maximum speed up to 16 μm/min. A procedure has been created for calculating the permissible cathode current density and the required rate of electrolyte supply by dipping and jet methods. The effect of the area of anodes – tools on the maximum permissible cathode current density is determined. The effect of cathodic current density in high-speed copper plating electrolytes on current output and the rate of copper coating application has been established.

1. Introduction

The electrodeposition by rubbing method uses electrolysis. For this purpose the part is attached to the negative pole of the rectifier, whereas the anode tool is wrapped with a cloth, impregnated with electrolyte, attached to the positive pole and rubbed onto the covered part of the part. The main difference between the method of electrodeposition by rubbing and the galvanic method is the absence of a galvanic bath, the function of which is performed by a porous material impregnated with an electrolyte. Pollution of wastewater and ambient air are reduced due to the small volumes of used electrolytes.

Electrodeposition by rubbing is a mobile, portable method that allows for coating deposition to various parts, including large ones. Coatings are deposited with electrodeposition by rubbing at a high speed: nickel at a speed of up to 15 μm/min, zinc up to 10 μm/min, cadmium up to 9 μm/min, chromium up to 60 μm/min, iron up to 33 μm/min. With electrodeposition by rubbing it is easier to apply selective (local) coatings [1–7]. The maximum copper plating speed of 25 μm/min was indicated in [8], herewith in 10 minutes of electrodeposition by rubbing, a coating with a thickness of 100 to 250 μm was obtained on the sample; accordingly, when determining the copper plating speed by weight gain, this would be about 17.5 μm/min. Electrodeposition by rubbing has disadvantages. The process of electrodeposition by rubbing, as a rule, has a low level of mechanization and automation. It is required to have a large number of anode tools for various parts. During electrodeposition by rubbing, coatings may be uneven. The coating rate depends on the ratio of the areas of the working

Content from this work may be used under the terms of the Creative Commons Attribution 3.0 licence. Any further distribution of this work must maintain attribution to the author(s) and the title of the work, journal citation and DOI.
Published under licence by IOP Publishing Ltd

surface of the anode tool and the area of the coated part area. There are fewer electrolytes for electrodeposition by rubbing, and electrolyte compositions, as a rule, are not disclosed. There are no methods for calculating the permissible cathodic densities and electrolyte feed rate and there is no data on the influence of the area of the anode tool on the process of electrodeposition by rubbing. In this regard, in order to eliminate some of the above shortcomings, this research was carried out using copper plating, one of the most popular processes. Electrolytic copper coatings can be applied to steel, as well as to titanium, molybdenum, aluminum, copper, zinc and alloys based on them. Electronically rubbed copper coatings can be used to improve solderability, increase electrical conductivity, restore worn parts, protect against corrosion (for example, in multilayer coatings), for local protection of steel parts from cementation, nitration, boriding and other purposes [9–15].

2. Experiment: materials and methodic

For electrodeposition by rubbing, anode tools with a working area of 1 cm^2, 3.75 cm^2 and 443 cm^2 were used. The conductive part of the anode tools was made of 08H18N10T stainless steel containing 0.8% carbon, about 1% titanium, 10% nickel, and an increased chromium content of 18%. It is these last two elements that increase the chemical resistance of steel to the greatest degree. The speed of movement of the anode instruments was maintained from 7 to 9 m/min. The ratio of the cathode (coated) surface to the anode (area of the anode tool) was maintained equal to two. Electrolytes were prepared using distilled water and chemically pure chemicals. To supply electrolytes, dipping and jet methods were used.

Samples were made from unalloyed carbon steel coated with copper in a pyrophosphate electrolyte. For electrodeposition by rubbing the anode tool with an area of 4.43 dm^2, samples with a size of $400\times700\times0.5$ mm were used, and for electron rubbing with the rest of the anode tools – $40\times60\times1$ mm. Before coating, the samples were polished, degreased with Viennese lime (a mixture of calcium oxide 90% and magnesium oxide 10%), decapitated in a 20% sulfuric acid solution, washed first with tap water and then with distilled water.

For electrodeposition by rubbing, AC rectifiers with current stabilization and short circuit protection were used. The installation diagram of the electrodeposition by rubbing and anode tools are presented in [16, 17]. The thickness of the coatings, the deposition rate, and the current efficiency were determined by the gravimetric method (by weight gain) on an analytical balance with an accuracy of 0.1 mg. The adhesion strength of the coatings (adhesion) was tested by the method of applying grooves.

3. Results and discussion

Considerable attention was paid to the development of high-speed copper plating electrolytes. In electrodeposition by rubbing, the slowest stage of the electrochemical process, as well as in a galvanic bath, is the delivery of ions of electrodeposited metals to the surface of the cathode (part). In this regard, the developed by us electrolytes for electrodeposition by rubbing have a high concentration of copper ions. This is justified, because in electrodeposition by rubbing small volumes of electrolytes are used, compared with the volumes of electrolytes used in galvanic baths. In addition, in order to use high cathodic current densities, it is important to use effective buffer additives to prevent alkalization in the cathode region, which can occur as a result of reaction (1) and lead to the copper hydroxides precipitation. In electrodeposition by rubbing, mainly insoluble anodes are used, which are close to the cathode. In this regard, the electrolyte may be acidified due to the formation of hydroxonium. Moreover, if the current efficiency of reaction (3) is equal to 100% or close to 100% at the anode, then, in addition to hydrogen evolution, a metal ion reduction reaction takes place at the cathode (2). On the other hand, reaction (1) proceeds directly on the surface of the cathode, and hydroxonium forms in the anode space and diffuses to the cathode. The final pH in the cathode space will depend on the ratio of the rates of the above processes, on the initial pH of the electrolyte and on the diffusion rate of the ions. Therefore, we need a high buffer capacity of the electrolyte in both acid and alkali.

$$2H^{+} + 2e = H_2 \tag{1}$$

$$Me^{n+} + ne = Me \tag{2}$$

$$OH^{-} + 4e = 2H_2O + O_2 \tag{3}$$

Considerable attention should be paid to the choice of additives: shine-forming, inhibitory, accelerating, wetting. Many of them can significantly increase the rate of deposition of copper coatings, improve the quality of coatings, contribute to the production of shiny coatings [18–23], and this was taken into account when developing electrolytes. It is important that the electrolyte has high electrical conductivity, does not strongly corrode the metal base, does not emit toxic substances during operation, and has a low sensitivity to impurities. The following acidic electrolytes are used for applying copper coatings in electroplating and electrotype: sulfate, fluoroborate, silicofluoride, nitrate, citrate, tartrate, trilonate, chloride, iodide, sulfamide and others. The fastest electrolytes are fluoroborate, silicofluoride and sulfamide ones. However, these electrolytes are more expensive, more harmful and more sensitive to impurities, in comparison with sulfate and nitrate electrolytes. The remaining acidic and alkaline copper plating electrolytes did not attract our attention due to the inability to conduct electrodeposition at high speeds. The most common copper plating electrolytes are sulfate. Copper sulfate electrolytes are simple in composition, stable in operation and have a low cost. It is known that sulfuric acid in sulfate electrolytes of copper plating increases the electrical conductivity of the electrolyte, prevents the hydrolysis of copper salts, however, the solubility of copper sulfate decreases with increasing sulfuric acid. Since reaction (3) proceeds during electrodeposition by rubbing, leading to an increase in acidity at the insoluble anode close to the cathode, an electrolyte with a lower concentration of sulfuric acid and a higher content of copper sulfate can be used accordingly. A lower concentration of sulfuric acid can also be used due to the constant updating of the cathode space with new portions of fresh electrolyte. Such a low-acid sulfate electrolyte is electrolyte 1; its composition is shown in table 1. Glossy coatings can be applied from this electrolyte when a shine-forming BS-1 additive is added to its composition with a concentration of 4 to 6 g/l. BS-1 additive is a stabilized mixture of aqueous solutions of safranin dye and dipropanedisulfonic acid disulfide. In addition to the shining effect, additives of the BS series (BS-1, BS-2) increase the scattering ability, the permissible cathode density. Copper coatings, electrodeposited from electrolytes containing only organic additives, have low ductility. To increase the ductility, sodium chloride with a concentration of 0.03–0.1 g/l is introduced into the electrolytes. In addition, chloride ion in sulfate electrolytes increases the cathode current density [24]. Nitrate electrolytes are of interest for electrodeposition by rubbing because of the high solubility of copper nitrate and a significantly higher limiting current density in comparison with other copper plating electrolytes. However, it is known that during electrodeposition of copper from copper nitrate solutions it is not possible to obtain high-quality coatings. The coatings are obtained friable with dendrides at the edges of the cathode due to the fact that in addition to copper precipitation during the electrolysis on the cathode, reactions of reduction of nitrate ions to nitrite ions and ammonium ions occur. These reactions reduce current efficiency by up to 20% and degrade the quality of coatings. Nevertheless, in the presence of inhibitors that suppress the reaction of reduction of nitrate ions, it is possible to obtain matte and even shiny copper coatings. Inorganic additives (chlorine ions, pyrophosphate and others) increase the current output of nitrate electrolytes to 100%. Organic compounds: acetic acid, citric acid, aminoacetic acid and trilon B, as a rule, increase the current efficiency to a lesser extent. However, citric acid and trilon B contribute to shiny coatings [25, 26]. Electrolyte 2 is a copper plating nitrate electrolyte in which the addition of XX chloride-quinonic and citric acid perform such functions. For the preparation of electrolyte 2, nitric acid copper was used, and for the preparation of electrolyte 1, copper sulfate was used. The range of copper ion content in electrolytes was established with a 20% decrease in copper concentration from the maximum for each specific electrolyte. Most often, it is precisely such a decrease in concentration as a result of copper production during electrodeposition with an insoluble anode rubbing that does not lead to a deterioration in the quality of the coatings, the concentration of copper sulfate from 200 to 250 g/l is given in the classic sulfate electrolyte of copper plating, i.e. with a difference of 20%.

Table 1. Compositions of copper plating electrolytes for electrodeposition with rubbing.

Components g/l	1	2
Copper (ions)	64–82.5	112–140
Sulfuric acid	20–30	
Citric acid		20–40
Sodium Chloride	0.03–0.1	
Shine-forming BS-1 additive	4–6	
Additive XX (ml/l)		30–50

For electrodeposition with rubbing, various feed rates of electrolytes are recommended, as a rule, selected experimentally. In this regard, a methodology was developed for calculating the supply of electrolyte. First of all, it is necessary to determine the feed rate of electrolyte, which compensates for the production of copper ions during electrodeposition with rubbing. To do this, we calculate the allowable decrease in the concentration of copper ions ΔC in each electrolyte, based on the composition of the electrolytes. So for electrolyte 1, the maximum concentration is 82.5 g/l, and the minimum is 64 g/l, respectively, for electrolyte 1 this value is $\Delta C = 0.0185$ g/ml, for electrolyte 2 $\Delta C = 0.028$ g/ml. The mass of electrodeposited copper and m removed from the electrolyte can be calculated by the formula (4).

$$m = K_e \cdot I \cdot \tau \cdot \eta \tag{4}$$

Taking into account formula (4), it is possible to propose formula (5) for calculating the rate of compensatory supply of electrolyte (G_c, ml/cm^2 min), which does not allow the generation of ions of an electrodeposited metal below the allowable by prescription.

$$G_c = \frac{K_e \cdot \eta \cdot I \cdot \tau}{\Delta C \cdot S \cdot \tau} = \frac{K_e \cdot \eta \cdot I}{\Delta C \cdot S} = \frac{K_e \cdot \eta \cdot D_c}{\Delta C} \tag{5}$$

In formulas (4) and (5), m is the mass of electrodeposited copper, K_e is the electrochemical equivalent in g/A min, I is the current strength in A, η is the current output in fractions of a unit, τ is the electrolysis time in minutes, G_c is the electrolyte feed rate, which compensates for the production of ions of the electrodeposited metal, ml/cm^2·min, ΔC is the decrease in the concentration of the electrodeposited metal, g/ml, S is the cathode (coated area) area in cm^2, D_c is the cathodic current density in A/cm^2.

We use the formula (5) to calculate the compensatory feed rate of copper plating electrolytes. To ensure the supply rate of electrolyte with a margin, to a first approximation, we will make calculations by taking the current efficiency as 100% or in fractions of a unit of 1. The calculation results are presented in table 2.

Table 2. Effect of electrolyte composition and cathodic current density (D_c, A/cm^2) on the compensating feed rate of electrolyte G_c in ml/min·cm^2.

Electrolyte	D_c = 0.05 A/cm^2	D_c = 0.2 A/cm^2	D_c = 1.0 A/cm^2	D_c = 2.0 A/cm^2
1	0.053	0.21	1.05	2.1
2	0.035	0.14	0.7	1.4

As can be seen from the data presented in table 2, the higher the cathodic current density is, the faster the generation of copper ions in the cathode space occurs, accordingly, a higher electrolyte feed rate is required. Naturally, the smaller the initial maximum concentration of copper ions in the electrolyte, respectively, the smaller is the allowable interval for decreasing the concentration of copper ions ΔC, the larger should be the supply of electrolyte at the same current density. Therefore, the feed rate of electrolyte 1, ceteris paribus, is higher than electrolyte 2. The easiest way to supply electrolyte is by dipping an anode tool wrapped in cloth or other porous material into the electrolyte. Usually, it is one dipping per minute. In this case, the fabric or other porous material must have sufficient water absorption (electrolyte absorption), high wear resistance and good capillary properties, which ensure the movement of the electrolyte from the non-working zone to the cathode space. To increase the wear resistance and the volume held by the electrolyte fabric, combinations of fabrics were used taking into account their properties. For the manufacture of anode tools, we used a thin terry cloth with electrolyte absorption of 0.07 ml/cm^2 and a synthetic fabric made of polyester fibers (crimp) with electrolyte absorption of 0.02 ml/cm^2. Terry cloth was used for the first layer of fabric touching the conductive part of the electrode-tools, and the most wear-resistant fabric, crimp, was used for the second layer touching the cathode surface. It is possible to increase the electrolyte absorption of the anode tool using holes, recesses, and grooves on the anode tool. In particular, holes with a diameter of 4 mm were drilled on the largest of our anode tools with an area of 4.43 dm^2. It is easier to use the dipping method to provide the required electrolyte feed rate for small anode tools, the smaller the anode tool, the easier it is to provide the required electrolyte feed rate by dipping. This is due to the fact that the ratio of the working area of the anode tool to the area of the fabric used for small anode tools is much larger. So for the manufacture of a large anode tool with an area of 4.43 dm^2, a two-layer fabric of 7.09 dm^2 was required. This was enough to create a two-layer coating on the working surface of the anode tool and fixing the fabric on the conductive part. For the manufacture of the anode instrument with an area of 1 cm^2, 20 cm^2 of a two-layer fabric was required. Thus, the ratio of the tissue area to the working area of the anode tool for the large electrode tool was 1.6, and for the small one was 20. Accordingly, this ensures a better supply of fresh batches of electrolyte to the cathode reaction zone. To calculate the feed rate of the electrolyte with a two-layer fabric on the anode tools, we use the formula (6).

$$G_a = \frac{(W_1+W_2)\cdot S_T}{\tau} \tag{6}$$

In the formula (6) G_a a is the absorption rate of electrolyte supply using tissues, ml/min; W_1 and W_2 – electrolyte absorption of two types of tissue, m/cm^2; S_T is the area of the two-layer fabric used to make the anode tool, cm^2; τ is the time interval between two electrolyte feeds, min. We take one minute, because after a minute a new supply of electrolyte was made by dipping. Using formula (6), we calculate the feed rates of electrolytes by tissues for various tissue combinations for anode instruments with an area of 1 sm^2, 3.75 cm^2 and 443 cm^2. Considering that for the manufacture of an anode tool with an area of 1 sm^2, fabrics with an area of 20 cm^2 were used, for an anode tool with an area of 3.75 cm^2 fabrics with an area of 28 cm^2 should be used. The results of these calculations are presented in table 3.

Table 3. The feed rate of the electrolyte with tissues in ml/min (G_{as}) for anode tools of various sizes, depending on the total electrolyte absorption by tissue combinations (W).

W, ml/cm^2	Anode tool area, cm^2		
	1	3.75	443
0.04 (2 layers of crimp)	0.8	1.12	28.4
0.09 (terry and crimp)	1.8	2.52	63.8

Table 3 shows the rates of supply of electrolytes by tissues in ml/min, however, not all of the electrolyte supplied by the tissue manages to take part in electrochemical cathodic reactions before renewing (dipping) the anode tool in a fresh portion of the electrolyte. To determine the *P* – fraction (part) of the electrolyte participating in the electrolysis, we determined the permissible cathodic current density, which allows obtaining high-quality coatings.

Above this current density, the quality of the coating deteriorated; more often, a burn-out formed. The permissible current density was determined for the anode instrument with a two-layer fabric (crimp and terry cloth). The feed rates of electrolytes were determined: G_{cs} compensating for the anode tool of a certain area according to the formula (7) and G_a the absorption rates of electrolyte supply by the tissue by the formula (6). The ratio of these speeds allows determining the approximate share (*P*) of the supply of electrolyte involved in electrolysis.

$$G_{cs} = G_c \cdot S_a \cdot \tau \tag{7}$$

In formula (7), G_{cs} is the electrolyte supply rate that compensates for the generation of electrodeposited metal for an anode tool of a certain area, ml/min; G_c is the electrolyte supply rate that compensates for the generation of electrodeposit metal, ml/ cm^2 min, S_a is the area of the anode tool, sm^2, τ is the time interval between two electrolyte feeds, min. We take one minute, because after a minute a new supply of electrolyte was made by dipping. The results are presented in tables 4 and 5.

Table 4. The influence of the area of the anode tool on the permissible cathodic current density, the feed rate of the electrolytes and the fraction of their participation in electrolysis when electrically rubbed by anode tools with a two-layer fabric (crimp and terry cloth) for electrolyte 1.

S anode tool	D_c, A/cm^2	G_{cs}, ml/min	G_a, ml/min	*P*, fraction
1	0.75	0.79	1.80	0.439
3.75	0.30	1.2	2.52	0.476
443	0.10	46.5	63.8	0.729

Table 5. The influence of the area of the anode tool on the permissible cathodic current density, the feed rate of the electrolytes and the fraction of their participation in electrolysis when electrically rubbed by anode tools with a two-layer fabric (crimp and terry cloth) for electrolyte 2.

S anode tool	D_c, A/cm^2	G_{cs}, ml/min	G_a, ml/min	*P*, fraction
1	1.0	0.70	1.80	0.389
3.75	0.4	1.06	2.52	0.421
443	0.14	43.4	63.8	0.680

As can be seen from the data presented in tables 4 and 5, the smaller the area of the anode tool, the higher cathodic current densities can be used. When using electrolyte 2, for the anode tool with an area of 1 sm^2 D_c = 1 A/cm^2, and for the anode tool with an area of 443 sm^2 it is only 0.14 A/cm^2, since for large anode tools the concentration polarization is higher. For a large anode instrument, the proportion of used electrolyte feed is about 0.7.

For anode instruments with areas of 1 cm^2 and 3/75 cm^2, the average value of this fraction is 0.43. This is due to the fact that the ratio of the area of the anode tool to the area of the tissue for the anode tool with S_a = 443 cm^2 is small – 1.6, and for other anode tools this figure is much larger. Now, using the formula (8), it is possible to calculate the real feed rates of electrolytes provided by the tissues and, subject to the conditions of the formula (9), the coating will be of high quality without burning and other defects that arise when working with D_c above the limit.

$$G_r = G_{as} \cdot P \tag{8}$$

$$G_r \geq G_{cs} \tag{9}$$

Formula (10) will determine at what permissible current density it is possible to electrically rub the coating using the dipping method to supply the electrolyte. For this, we determine D_c from formula (5) and replacing the compensating feed rate with the real feed rate, we obtain formula (10).

$$D_{cd} = \frac{G_r \cdot \Delta C}{K_e \cdot \eta} \tag{10}$$

In formulas (8, 9, 10), D_{cd} is the cathodic current density when feeding the electrolyte by dipping in A/cm^2, G_r is the actual electrolyte supply rate provided by the tissues in ml/min, G_{as} is the electrolyte supply rate by the tissues is absorption for the anode instruments of a certain area in ml/min, G_{cs} is electrolyte supply rate, which compensates for the production of electrodeposited copper ions for an anode instrument of a certain area in ml/min, P is a fraction (part) of electrolyte that managed to take part in electrolysis, Δ C is an allowable decrease in concentration by copper ions in g/ml, K_e is the electrochemical equivalent of copper in g/A·min, I is the current strength in A, η is the current output in fractions of a unit.

The dipping method can be recommended for a small anode instrument with a working area of up to about 4 cm^2. However, the dipping method does not allow electrodeposition with rubbing processes with high cathode current densities, especially when using a large area anode tool. Firstly, a large area anode tool is more difficult to provide with the necessary volume of electrolyte to make up for the produced copper ions by the dipping method. Secondly, when the anode tool is of smaller size, the mixing of the electrolyte in the cathode space and, accordingly, the diffusion of copper ions proceed more intensively. This leads to an increase in the limiting cathodic current density.

It is possible to significantly improve the supply and control of electrolyte supply using a jet supply of electrolyte, as well as the exchange (supply and withdrawal) of electrolyte using a pump. The last option for supplying electrolyte justifies itself in large-scale production, when it is necessary to apply thick coatings, for example, to restore worn parts. Such installations with supply and removal of electrolyte by a pump are based on metal-cutting machines. Their efficiency is very high due to the obtaining of high coating rates. However, they have a higher cost, require the use of large volumes of electrolytes and do not have mobility and portability.

Thereby, in this research the electrolyte supply was used by the dipping method and the jet-gravity method. To implement the latter, a container with an electrolyte with a volume of five liters was fixed one meter above the anode tool, the anode tool was connected to the container with a PVC polymer tube and, using a set of clamps and a tap, the necessary electrolyte supply was established. Such an electrolyte supply system made it possible to increase the permissible cathodic current density and the deposition rate of copper coatings. It is known that increasing the diffusion rate of ions of an electrodeposited metal due to mixing of the electrolyte decreases concentration polarization and increases the limiting current density and the deposition rate. In the jet method, in comparison with the dipping method, the diffusion rate of copper ions can be increased by increasing the supply of electrolyte. It was found how the feed rate of the electrolyte affects the deposition rate of copper coatings and current efficiency. The results are shown in table 6.

Table 6. The influence of the electrolyte feed rate by the jet method (V_j, ml/min) on the current output (η,%) and the copper coating deposition rate (R, μm/min) by an anode tool with an area of 1 cm^2 when using D_{cj}= 2 A/cm^2.

V_j, ml/min	R, μm / min	η, %
0.7	9.8	22.1
2.1	13.9	31.4
6.3	16.2	36.7
18.9	16.9	38.2

As can be seen from table 6, the coating rate increases with an increase in the supply of electrolyte by 9 times from 0.7 ml/min to 6.3 ml/min, with a further increase in the supply of electrolyte, the copper plating rate is almost unchanged. It is possible to determine the optimal jet velocity of the electrolyte supply by the formula (11).

$$C_j = G_c \cdot S \cdot N_j \tag{11}$$

In formula (11), G_c is the electrolyte feed rate that compensates for the production of ions of the electrodeposit metal ml/ cm^2·min; S is the area of the anode tool in cm^2; N_j – how many times it is necessary to increase the speed of the jet feed in comparison with the compensating feed rate of the electrolyte. In our experiments, N_j for the anode tool with an area of 1 cm^2 is 9–10, for the anode tool with an area of 3.75 cm^2 N_j = 18, and for the anode tool with an area of 443 cm^2 N_j = 32. These values were used to calculate the jet feed rate for other experiments. Then we determined N_d, i.e., how many times D_{cj} – the permissible current density when using the jet method can be larger than D_{cd} – the permissible current density when using the dipping method to supply the electrolyte. We obtained the value N_d = 2.1. Thus, it is possible to approximately calculate D_{cj} by the formula (12).

$$D_{cj} = N_d \cdot D_{cd} \tag{12}$$

After that, the current efficiency and the deposition rate of copper coatings in electrolytes 1 and 2 were determined. The results obtained are presented in table 7.

Table 7. The effect of the cathodic current density (D_{cj}, A/cm^2) on the copper plating speed (in μm/min) and current efficiency (in %) in copper plating electrolytes 1 and 2 when using an anode tool of 1 cm^2. In the numerator, the copper plating speed is given in μm/min, and in the denominator the current efficiency is given in %.

Electrolyte	0.5A/cm^2	1.0A/cm^2	1.5A/cm^2	2.0A/cm^2
1	9.2/83.2	11.6/52.5	12.3/37.2	–
2	8.4/76.0	13.1/59.3	15.9/48.0	16.2/36.7

As can be seen from table 7, with an increase in the cathodic current density, the deposition rate of copper coatings increases, moreover, first up to D_c = 1A/cm^2 more sharply, then slowly, this is due to a significant decrease in current efficiency with increasing cathodic current density. The maximum copper coating deposition rate was obtained in nitrate electrolyte 2 due to a higher initial concentration of copper ions of 140 g/l, while in electrolyte 1 this concentration was 82.5 g/l. In electrolytes 1 and 2, in the absence of shining additives, even pink-red copper coatings were deposited. Gloss depended on the cathodic current density, coating thickness, initial roughness of the sample, the type of electrolyte, and the concentration of the shiner. The degree of gloss in the presence of shiners in the sulfuric acid low acid electrolyte 1 was higher than in nitrate 2, from which semi-gloss coatings were mainly deposited. The degree of gloss decreased with increasing coating thickness and increasing cathodic current density. All coatings passed the adhesion test.

4. Conclusion

As a result of research work, the operating modes and compositions of high-speed copper plating electrolytes are proposed. Nitrate copper plating electrolyte, which allows applying high-quality copper coatings from semi-brilliant to matte with a maximum speed of 16 μm/min by the method of electrodeposition by rubbing. Low acid sulfate electrolyte is inferior to nitrate electrolyte in coating rate. However, the sulfate electrolyte is cheaper, more stable, and more reliable in operation and allows for the application of copper coatings from matte to shiny. A technique has been developed that allows calculating a combination of fabrics for the manufacture of anode tools, feed rates of electrolytes by dipping and jetting, allowable current densities to achieve high coating rates with the rational use of electrolytes. The calculation procedure is confirmed by the experimental results. The influence of the area of the anode tools on the maximum allowable cathodic current density is established. For a small anode tool (up to about 4 cm^2), electrolytes can be applied by dipping; for an anode tool with an area measured by several dm^2, it is better to carry out jetting of electrolyte. It was determined how the cathodic current density in copper plating electrolytes affects the current efficiency and the rate of deposition of copper coatings.

References

[1] Ivshin Ya V and Kadrikov R A 2014 *Electroplating and surface treatment* vol 22 **3** 34–9
[2] Salakhova R K, Naletov B P and Tyurikov E V 2009 *Aviation materials and technologies* **2(11)** 25–9
[3] Semyonychev V V and Salakhova R K 2015 *Physics and chemistry of materials processing* **4** 59–64
[4] Demin S Aand Vinogradov S S 2018 *Works of VIAM* **9** 43–50 DOI: 10.18577/2307-6046-2018-0-9-43-50
[5] Ilyin V A, Semenychev V V, Naletov B P, Tyurikov E V and Salakhova R K 2009 Patent RF No 2 353 713 *Bull* **12**
[6] Ilyin In A, Semenychev In B, Naletov B P, Salakhova R K and Tyurikov E V 2010 Patent RF No 2 389 828 Bull **14**
[7] Batishchev A N 2008 *Works of GOSNITI* vol 102 102–5
[8] Ivshin Ya V, Kaidrikov R A and Zilberg A I 2014 *Bulletin of the Kazan Technological University vol* 17 **13** 321–24
[9] Salakhova R K, Tikhoobrazov A B and Smirnova T B 2019 P*roceedings of VIAM* **12(84)** 80–7 DOI: 10.18577/2307-6046-2019-0-12-80-87
[10] Maiselis A A, Bayrachny B I, Zaitseva V Yu, Trubnikova L V and Maiselis Z A 2014 *Progressive technologies and systems of mechanical engineering* **1(47)** 183–9 (In Ukrain.)
[11] Belov P A, Korotkovsky V I, Petrova V O 2019 *Electronic scientific journal of Kursk state University* **1(21)** 5
[12] Tikhonov A A 2018 Patent RF No 2653515 *Bull* **13**
[13] Tikhonov A A, Arendateleva S I and Perepelitsa T I 2019 *IOP Conf. Series: Mater. Sc. Eng.* **656** 012050
[14] Devyatkina T I, Markova T V, Rogozhin V V and Mikhalenko M G 2016 Patent RF 2 588 702 *Bull* **19**
[15] Kapustin Yu I, Averina Yu M, Nyrkov N P, Shuvalov D A and Moiseeva N A 2018 *Advances in chemistry and chemical technology* **32(14)** 26–9
[16] Moskovits M 1993 *Electroplating and surface treatment* **2(3)** 40–5
[17] Tikhonov A A and Sokolova A Yu 2010 *Bulletin of the NovSU* **60** 102–5
[18] Pasquale M and Gassa Arvia A 2008 *Electrochim. Acta* **53(20)** 5891–5904 DOI: 10.1016/j.electacta.2008.03.073
[19] Jo Y E, Yu D Y & Cho, K S 2020 *J. Appl. Electrochem* **50** 245–253 DOI: 10.1007/s10800-019-01381-4
[20] Qibo Zhang, Yu Xiaolv, Yixin Hua and Wei Xue 2015 *J. Appl. Electrochem* **45** 79–86
[21] Arezoo Dianat *et al.* 2018 *J. Appl. Electrochem* **48** 211–219 DOI: 10.1007/s10800-018-1150-1
[22] Tomin Liu, Rui Vilar, Sónia Eugénio and Joseph Gron 2015 *J. Appl. Electrochem* **45** 87–93
[23] Ivchenko N S and Nemikina O V 2017 *Proc. Univ. Appl. Chem. and Biotech.* vol 7 **3** 39–45 (In Russ.)
[24] Balakai V I, Shpanova K A and Chiryaev M G 2018 *Bulletin of Science* **3(8)** 42–44
[25] Pravda A A, Lukashchuk T S and Larin V I 2016 *Ecology and industry* **3(48)** 68–72
[26] Pravda A A, Radchenkova A P, Larin V I and Lukashchuk T S 2013 *Bulletin of the NovSU* **73-2** 103–107

Dependencies of nonlinear hereditary creep theory for concrete and creep type equations of hardening under plane stress

A M Tokmuratov[1], **B K Dujsenbekov**[1], **A S Varenik**[2] **and K A Varenik**[2,3]

[1] M Auezov South Kazakhstan State University, Tauke khan avenue, 5 160012 Shymkent, Republic of Kazakhstan
[2] Yaroslav-the-Wise Novgorod State University, ul. B. St. Petersburgskaya, 41 173003 Veliky Novgorod, Russia

[3] E-mail: vkirillv89@mail.ru

Abstract. Concrete creep is of great practical importance and is taken into account in calculation and design of structures. The article discusses various options for describing the theory of creep of concrete, which are divided into three main groups: the theory of an elastic-creeping body (hereditary theory of aging), the theory of elastic heredity (the theory of an visco-elastic body), and the theory of aging. The main advantages and disadvantages of their use are demonstrated depending on various conditions and processes. Collisions in the study of concrete creep are clearly visible when considering expressions for the creep measure, as well as creep and relaxation nuclei. Different scientists propose their own modification of expressions that reflect various features of the material, and which are often very different from each other. The authors of the article obtained the full creep equations of structural materials.

1. Introduction

Let us consider some types of dependences of the nonlinear hereditary theory of concrete creep and hardening type creep equations for the case of a plane stress state.

It is known that creep under two-axial stress conditions has some peculiarities. For example, the experimental data of some authors show that compression in the X direction reduces creep deformations in the Y direction in some cases up to 70%. Since thin-walled elements are usually made of concrete with an increased water-cement ratio, this causes more intense creep. There are various, mutually unrelated, proposals for describing the dependences between the creep measures of concrete with two-dimensional stress state c_{2x} on the one hand, and simple compression c_x on the other. The ratios c_{2x}/c_x are associated with the existing normal stresses and the transverse creep coefficient, a formula that relates the creep measures c_x and c_{2x} is proposed:

$$c_x = \frac{1}{1+\sigma_y/\sigma_x}, \tag{1}$$

where σ_y, σ_x are normal stresses in concrete, and $\sigma_y \leq \sigma_x$.

Static processing of experimental data of I. E. Prokopovich allowed one to establish an approximate dependence of the ratio c_{2x}/c_x on the thickness of a flat disk h:

Content from this work may be used under the terms of the Creative Commons Attribution 3.0 licence. Any further distribution of this work must maintain attribution to the author(s) and the title of the work, journal citation and DOI.
Published under licence by IOP Publishing Ltd

$$\frac{c_{2x}}{c_x} = -0.054h^2 + 0.1344h + 0.258, \quad (2)$$

where 2 cm $\leq h \leq$ 10 cm.

It must be assumed that the values of the c_{2x}/c_x ratio are also significantly affected by the stress level and a number of other factors.

Experimental data of the researchers shows that the coefficient of transverse creep deformation of concrete $\nu_2(t, \tau)$ is slightly less than the coefficient of transverse elastic-instantaneous deformation $\nu_1(\tau)$. However, by the smallness of the variables v1 and v2, we can accept the assumptions about their equality:

$$\nu_1(\tau) = \nu_2(t,\tau) = const. \quad (3)$$

As demonstrated by I. E. Prokopovich, hypothesis (3) does not lead to large errors (5–6%).

2. Theories of concrete creep

According to a number of the most common features, the variety of variants of the theory of creep of concrete can now be reduced into three different groups: the theory of an elastic-creeping body (hereditary theory of aging), the theory of elastic heredity (theory of an visco-elastic body), and the theory of aging. The most accurately constructed theory of creep should take into account the partial reversibility of creep strains.

The hereditary theory of aging (Maslov-Arutyunyan's theory of an elastic-creeping body) takes into account the partial reversibility of creep deformations caused by the influence of concrete aging and therefore most correctly describes the process of stress relaxation in concrete structures, provided that the core is well selected in the original integral equation. Mathematically, this is expressed in the fact that the creep measure is taken in the form of:

$$C(t,\tau) = \left(C_1 + \frac{A_1}{\tau}\right)\left[1 - e^{-\gamma(t-\tau)}\right], \quad (4)$$

where: C_1 is the limit value;
A_1, γ is the experimental data.

The theory of elastic heredity does not take into account the presence of an irreversible part of creep deformations and, as a consequence of this, leads to the greatest possible restoration of stresses in the relaxation process. In this case, the creep measure is written as:

$$C(t,\tau) = C_1\left[1 - e^{-\gamma(t-\tau)}\right]. \quad (5)$$

$$C_1 = const, E(\tau) = E_0 = const. \quad (6)$$

Consequently, the theory of elastic heredity is applicable only to old and dried or old and non-drying concrete.

The theory of aging, created specifically for the calculation of concrete and reinforced concrete structures, postulates the complete irreversibility of creep deformation and, consequently, the absence of the process of stress recovery during relaxation. This theory is based on the hypothesis and "parallelism" of curves of simple creep proposed by Whitney C.:

$$C(t,\tau) = \frac{1}{E(0)}[\varphi(t) - \varphi(\tau)], \quad (7)$$

where $\varphi(t)$ and $\varphi(\tau)$ are characteristics of creep;
E (0) is the strain modulus at the initial instant of time.

If, when considering a process of sufficiently long duration, we neglect the features of the development of nonlinear components of creep deformations in the initial section, then we can assume the affine similarity of simple creep curves corresponding to different stress levels $\eta = \frac{\sigma}{R}$ and present the expressions for the creep measure in the nonlinear region as:

$$C(\sigma,t,\tau) = f(\sigma)C(t,\sigma), \tag{8}$$

where $C(t,\tau)$is the measure of creep in a conditionally linear region;
$f(\sigma)$ is a nonlinearity function, which is selected on the basis of experimental data depending on the brand of concrete and other factors.

In determining the effect of concrete creep on the stress-strain state of reinforced concrete structures, the hereditary theory of aging is most accurate. In the field of high stresses, one of the most justified equations having a large experimental justification is the nonlinear creep equation Arutyunyan N. Kh. with elastic instantaneous characteristics, which for the case of a two-axial stress state taking into account equality (3) is written as:

$$\varepsilon_{11}(t) = \frac{\sigma_{11}(t)-\upsilon\sigma_{22}(t)}{E(t)} - \int_{\sigma_1}^{t}[\sigma_{11}(\tau) - \upsilon\sigma_{22}(\tau)]\left\{\frac{\partial}{\partial\tau}\left[\frac{1}{E(\tau)}\right] + f\left[\sigma_i(\tau)\frac{\partial c(t,\tau)}{\partial c}\right]\right\}d\tau;$$

$$\varepsilon_{22}(t) = \frac{\sigma_{22}(t)-\upsilon\sigma_{11}(t)}{E(t)} - \int_{\sigma_1}^{t}[\sigma_{22}(\tau) - \upsilon\sigma_{11}(\tau)]\left\{\frac{\partial}{\partial\tau}\left[\frac{1}{E(\tau)}\right] + f\left[\sigma_i(\tau)\frac{\partial c(t,\tau)}{\partial c}\right]\right\}d\tau;$$

$$\gamma_{12}(t) = \frac{2(1+\upsilon)}{E}\tau_{12}(t) - 2(1+\upsilon)\int_{\tau_1}^{t}\tau_{12}(\tau)\left\{\frac{\partial}{\partial\tau}\left[\frac{1}{E(\tau)}\right] + f\left[\sigma_i(\tau)\frac{\partial c(t,\tau)}{\partial c}\right]\right\}d\tau. \tag{9}$$

If the kernels of an integral equation of type (9) are degenerate, that is, they can be represented by the product of the functions τ and t or the sum of such products, then after twofold differentiation with respect to t of the left and right sides of (9), it becomes possible to extract a second-order differential equation with variable coefficients:

$$\ddot{\sigma}_{11}(t) - \upsilon\ddot{\sigma}_{22}(t) + (\dot{\sigma}_{11} - \upsilon\dot{\sigma}_{22})\left[\gamma - \frac{\dot{E}(t)}{E^2(t)}\right] + \gamma E(t)\varphi_1(t)\dot{\sigma}_{11}(\dot{\sigma}_{11} - \upsilon\sigma_{22})f(\sigma_i) = E(t)(\ddot{\varepsilon}_{11} + \gamma\dot{\varepsilon}_1);$$

$$\ddot{\sigma}_{22}(t) - \upsilon\ddot{\sigma}_{11}(t) + (\dot{\sigma}_{22} - \upsilon\dot{\sigma}_{11})\left[\gamma - \frac{\dot{E}(t)}{E^2(t)}\right] + \gamma E(t)\varphi_1(t)\dot{\sigma}_{22}(\dot{\sigma}_{11} - \upsilon\sigma_{11})f(\sigma_i) = E(t)(\ddot{\varepsilon}_{22} + \gamma\dot{\varepsilon}_{11});$$

$$\ddot{\tau}_{12} + \dot{\tau}_{11} + \left[\gamma - \frac{\dot{E}(t)}{E^2(t)}\right] + \gamma E(t)\varphi_1(t)[\dot{\tau}_{12}]^2 f(\sigma_i) = \frac{E(t)}{2(1+\upsilon)}(\ddot{\gamma}_{12} + \gamma\ddot{\gamma}_{12}). \tag{10}$$

For a case of old concrete with $E(t) = E(0) = const$ the dependences (10) are written as:

$$\dot{\sigma}_{11} - \upsilon\dot{\sigma}_{22} - \dot{\varepsilon}_{11} - E(0) = \gamma[\sigma_{11} + C(\infty)E(0)(\sigma_{11} - \upsilon\sigma_{22})f(\sigma_i) - \varepsilon_{11}E(0)];$$

$$\dot{\sigma}_{22} - \upsilon\dot{\sigma}_{11} - \dot{\varepsilon}_{22} - E(0) = \gamma[\sigma_{22} + C(\infty)E(0)(\sigma_{22} - \upsilon\sigma_{11})f(\sigma_i) - \varepsilon_{22}E(0)]$$

$$\dot{\tau}_{12} - \frac{E(0)\dot{\gamma}_{12}}{2(1+\upsilon)} = \tau_{12}\left[\gamma + \frac{\gamma c(\infty)E(0)f(\sigma_i)}{2(1+\upsilon)}\frac{\gamma}{2(1+\upsilon)}\right]. \tag{11}$$

where C (∞) is the limit creep measure.

It is generally accepted that the theory of N. Kh. Arutyunyan [2] gives good results for cases of slow reduction or increase in stress in concrete. Such modes are characteristic for the operation of building structures. The disadvantage of this theory is that it does not take into account the softening of nonlinearity with increasing time.

The basic equation of nonlinear creep of the theory of aging, proposed by I. I. Ulitsky, is written for a two-axial stress state as:

$$\varepsilon_{11}(t) = \frac{\sigma_{11}(t)-\upsilon\sigma_{22}(0)}{E(0)}[1 + f(\sigma_i)\varphi(t)] + \int_0^t\left\{\frac{d}{d\tau}[\sigma_{11}(\tau) - \upsilon\sigma_{22}(\tau)]\left[\frac{1}{E(t)} + \frac{\varphi(t)-\varphi(\tau)}{E(0)}f(\sigma_i)\right]\right\}\sigma\tau;$$

$$\varepsilon_{22}(t) = \frac{\sigma_{22}(t)-\upsilon\sigma_{11}(0)}{E(0)}[1 + f(\sigma_i)\varphi(t)] + \int_0^t\left\{\frac{d}{d\tau}[\sigma_{22}(\tau) - \upsilon\sigma_{11}(\tau)]\left[\frac{1}{E(t)} + \frac{\varphi(t)-\varphi(\tau)}{E(0)}f(\sigma_i)\right]\right\}\sigma\tau;$$

$$\gamma_{12}(t) = \frac{2(1+\upsilon)}{E(0)}\tau_{12}(0)[1 + f(\sigma_i)\varphi(t)] + \int_0^t 2(1+\upsilon)\left\{\frac{\partial\tau_{12}(\tau)}{\partial\tau}\left[\frac{1}{E(\tau)} + \frac{\varphi(t)-\varphi(\tau)}{E(0)}f(\sigma_i)\right]\right\}d\tau. \tag{12}$$

The kernel of the integral equation (12) in the theory of aging in accordance with (7) depends only on τ, which allows, by differentiating the right and left sides of (12) with respect to t, to reduce them to first order differential equations.

The disadvantages of the theory of aging associated with the hypothesis of "parallelism" and the corresponding complete irreversibility of creep deformations make it unsuitable for describing long-term processes in the presence of repeated or variable short-term and long-term effects. Nevertheless, this theory attracts many authors with the simplicity of solutions, which manifests itself most significantly in the nonlinear version. Its application gives quite acceptable results in solving problems of the theory of creep, especially in the case of considering long single exposures.

A special advantage of expressions (4) and (7) is that they allow one to reduce the solution of the basic integral creep equations (9) and (12) to the solution of linear differential equations of the second or first order. This circumstance significantly facilitates the numerical implementation of the tasks to be solved, which will be illustrated in this paper.

3. Creep equations

Experiments show that in a complex stress state the creep of structural materials is determined by shear stresses and proceeds in general according to the laws of “ordinary” plastic deformation. In this regard, the following main points are adopted:

1) a change in body volume is an elastic deformation, that is:

$$\xi^c = \xi_{11}^c + \xi_{22}^c + \xi_{33}^c = 0, \tag{13}$$

where ξ^c, ξ_{11}^c, ξ_{22}^c, ξ_{33}^c are the creep strain rates;

2) the principal stresses of the creep strain rate tensors and stresses coincide;

3) the shapes of stress deviators and creep strain rates coincide;

4) at each point a simple (or close to simple) loading is realized;

5) the intensity of creep shear strain rates η_i^c is a function of the shear stress intensity τ_i characteristic for a given material at a given temperature.

This dependence is taken as:

$$\eta_i^c = f(\tau_i)\tau_i, \tag{14}$$

where the scalar function $f(\tau_i)$ depends, generally speaking, on scalar parameters associated with the stress and strain state, on time t and temperature T [1].

The form of the function can be found by considering the case of uniaxial tension and comparing (14) with experimental data.

From the similarity of Mohr using condition (13), we can obtain formulas that relate the components of the creep strain rate tensor and stress tensor:

$$\xi_{11} = 0.5 f(\tau_i)(\sigma_{11} - \sigma);$$
$$\eta_{13} = f(\tau_i)\tau_{13}. \tag{15}$$

where σ is the average pressure.

Adding to (15) the rates of nonlinear elastic strains, we obtain the complete creep equations of structural materials, which for the biaxial stress state can be written as:

$$\dot{\varepsilon}_{11} = \frac{1}{E}(\dot{\sigma}_{11} - \upsilon\dot{\sigma}_{22}) + \frac{1}{3} f(\tau_i, t)\left(\sigma_{11} - \frac{1}{2}\sigma_{22}\right);$$
$$\dot{\varepsilon}_{22} = \frac{1}{E}(\dot{\sigma}_{22} - \upsilon\dot{\sigma}_{11}) + \frac{1}{3} f(\tau_i, t)\left(\sigma_{22} - \frac{1}{2}\sigma_{11}\right);$$
$$\dot{\gamma}_{12} = \frac{2(1+\upsilon)\tau_{12}}{E} + f(\tau_i, t)\tau_{12}. \tag{16}$$

For practical calculations, it is convenient to set the function in some analytical form. The inevitable spread of the experimental data makes various analytical approximations of the creep law acceptable.

The most convenient are exponential approximations:

$$f(\tau_i, t) = \frac{3\varepsilon_e}{\sigma_i}\exp\left(\frac{\sigma_i}{\sigma_e}\right) \quad (17)$$

and power approximations:

$$f(\tau_i, t) = 3\sigma_i^{n-1}\frac{\varepsilon_n}{\sigma_n^n}, \quad (18)$$

where ε_e, σ_e, ε_n, σ_n are the temperature functions, which are usually set graphically or in the form of tables;

$\sigma_{i=}\sqrt{3}\tau_i$ – the intensity of normal stresses.

Sometimes when using formula (18), the values of the exponent n turn out to be very large (for example, $n = 20$). In these cases, you can use the following approximation:

$$f(\tau_i, t) = \dot{\varepsilon}_n\left(\frac{\sigma_i}{\dot{\sigma}_n} - 1\right)^n \text{ if } \sigma_i > \dot{\sigma}_n;$$

$$f(\tau_i, t) = 0 \text{ if } \sigma_i \leq \dot{\sigma}_n. \quad (19)$$

The value can be called the creep limit, although the use of (19) does not mean at all that the existence of a physical creep limit is allowed. The value of exponent n in (19) turns out to be significantly less than in formula (18).

References

[1] Tokmuratov A M, Sanzharovskij R S and Dujsenbekov B K 2017 Konechno-raznostnye uravneniya sharnirno-opiraemoj pologoj zhelezobetonnoj obolochki [Finite-difference equations of a hinged-supported shallow reinforced concrete shell] *Vestnik KazGASA* **4(66)** 117–120 [In Russ.]

[2] Arutyunyan N H 1952 *Nekotorye voprosy teorii polzuchesti* [Some questions of creep theory (Moscow –Leningrad: GosTekhIzdat) p 323 [In Russ.]

IOP Conf. Series: Materials Science and Engineering **939** (2020) 012080 doi:10.1088/1757-899X/939/1/012080

Short-term and long-term longitudinal load tests of wooden rods

K A Varenik[1,2], A S Varenik[1], A V Kirillov[1] and M V Shuvalov[1]

[1] Yaroslav-the-Wise Novgorod State University, ul. B. St. Petersburgskaya, 41 173003 Veliky Novgorod, Russia

[2] E-mail: vkirillv89@mail.ru

Abstract. The purpose of the experimental studies was to verify the reliability of the method for solving the stability problem of centrally compressed rods, taking into account the creep of wood. To achieve this goal, it is necessary to have a diagram of deformations of real wood and experimental data on the behavior of rods of the same wood under continuous load. Experimental studies include short-term axial compression tests of small specimens with recording of load-deformation diagrams, as well as short-term and long-term tests of cores of different flexibility for central compression, the description of which is presented in this article. The authors of the paper obtained the values of critical stresses under longitudinal bending for wooden rods with different lengths and flexibility. These results have a high convergence with the values obtained as a result of theoretical calculations.

1. Introduction

In [1] and [2], the authors obtained analytical expressions of the long-term critical force for the longitudinal bending of a wooden rod taking creep into account. To verify the reliability of the proposed method for solving the stability problem of centrally compressed rods, taking into account the creep of wood, experimental studies were carried out.

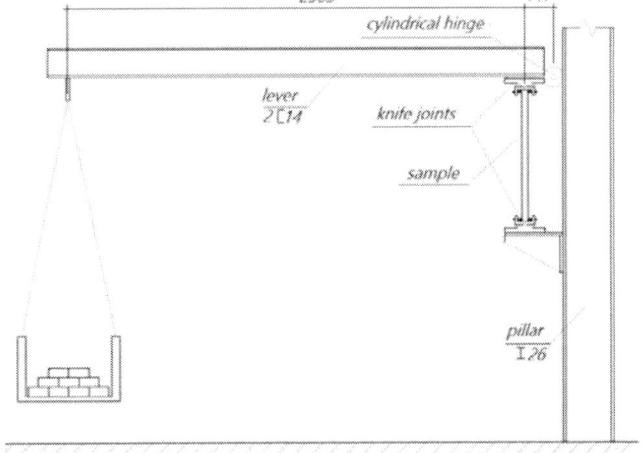

Figure 1. Installation diagram.

Content from this work may be used under the terms of the Creative Commons Attribution 3.0 licence. Any further distribution of this work must maintain attribution to the author(s) and the title of the work, journal citation and DOI.
Published under licence by IOP Publishing Ltd

To test the rods for longitudinal bending, a steel linkage was designed and manufactured. The installation diagram is shown in figure 1. The device of the installation ensures the fulfillment of the theoretical boundary conditions of the test wooden poles and allows for long-term testing with controlled load without the need for additional loading during the creep process.

Calibration of the installation was carried out using a DIN-1 electronic dynamometer. With a force measurement range of 0 ... 100 kN and a relative error limit of force measurement of 0.1 ... 0.5%. For loading weights of 2 and 4 kg were used. Thus, due to the ratio of the shoulders, a 2kg increase in the load created a compressive force on the rod of 31.35 kgf.

2. Experiment description

Sample preparation was carried out in accordance with the requirements of GOST 16483.0-89 "Wood. General requirements for physical and mechanical tests". Timber was harvested from the plantations in the Novgorod region. A pine with a butt-log portion diameter of 34 cm was selected.A beam of 3.5 m length with a section of 200 × 200 mm was cut from the bole. For testing, samples with a section of 30 × 30 mm with lengths of 540 and 630 mm were made out of the beam. In accordance with the accepted lengths, taking into account the end hinges (25 mm), the rods had flexibility of about 65 and 76. 5 samples of each flexibility were tested for short-term compression and 5 for long-term action of a centrally applied load.

The air temperature in the laboratory during short-term and long-term tests was 21-23 °C, relative humidity was 65 ± 2%. The average humidity of the samples was 10 ± 0.5%.

The ends of the samples were fixed in knife hinges. The horizontalness and alignment of the joints were carefully verified using level and theodolite. The position of the hinges during the tests was controlled using dial gauges with a division price of 0.01 mm. For the most accurate load application along the geometrical axis of the rod, test loads were applied (up to 40% of the expected breaking load). In this case, in the middle of the sample length, the deflection value was fixed using a dial gauge. After each test load, the ends of the sample were displaced relative to the hinge knives until a position was reached at which the indicator arrow did not depart from the initial reference.

The loading of the samples was carried out by the own weight of the lever and manually using weights. The dead weight of the lever and the suspended empty cargo platform was about 30% of the breaking load. This load was smoothly transferred to the sample by lowering the lever with a screw jack.

In short-term tests, the loading rate was about 20 kg per second. The sample was being loaded for about 1.5 minutes. The destruction of the samples was brittle, in the form of an instantaneous rupture of the fibers along the sawtooth surface.

The test results of the rods under short-term loading are presented in table 1.

Table 1. Short-term test results for rods.

Sample №	Length l, cm	Section b×h, cm^2	Flexibility λ	Breaking load N_{cr}, kgf	Breaking stresses σ_{cr}, kgf/cm^2
1	56.5	3.00×3.01=9.03	65	3127	346
2	56.5	3.00×3.01=9.03	65	3065	339
3	56.5	3.00×3.00=9.00	65	3284	365
4	56.5	3.00×2.99=8.97	65	3347	373
5	56.5	3.01×2.98=8.97	65	3378	377
6	65.5	3.00×3.01=9.03	76	2698	299
7	65.5	2.98×3.01=8.97	76	2510	280
8	65.5	3.00×3.01=9.03	76	2698	299
9	65.5	2.98×3.01=8.97	76	2573	287
10	65.5	3.00×3.01=9.03	76	2729	302

The average value of breaking stresses for rods with a flexibility of 65 is 360 kgf/cm^2 (coefficient of variation is 5%). The average value of breaking stresses for rods with a flexibility of 76 is 293 kgf/cm^2 (coefficient of variation is 3%).

After testing the rods, 120 mm high prisms were cut from intact areas and tests were carried out to determine the ultimate strength. According to the results of testing 10 samples for compression along the fibers, the average value of the tensile strength was 578 kgf/cm^2 (coefficient of variation is 4%).

The loading of the rods during long-term tests was performed step by step. At the first loading step, which amounted to about 30% of the short-term critical load, the rods were centered according to the method described above. Each further loading step was 62.7 kgf (a weight of 4 kg on the platform), and when approaching destruction it was 31.35 kgf (a weight of 2 kg).

At each step, the position of the supports was monitored and the parameters of the stress-strain state of the wooden post were measured, which corresponds to the developed theoretical model for calculating the post.

Deflections in the middle of the sample were measured by dial gauges, with a division value of 0.01 and 0.001 mm. To measure the relative edge deformations, we used resistive strain gages with a base of 10 mm, a sensitivity of K = 2.19, and a resistance of P = 99.6 + 0.1 Ω. On opposite sides, exactly in the middle of the length of the samples, two sensors were glued — one on each side. To measure the output signals of resistive strain gages, we used the multichannel universal meter-recorder TEREM-4.1 of the scientific and production enterprise Interpribor (Chelyabinsk). The readings from each of the resistive strain gages were taken and automatically recorded in the memory of the device every 4 hours.

Loading to the next stage was carried out if for seven days the deflection of the rod with an accuracy of 0.01 mm remained unchanged. The duration of testing for each sample was about two months.

Figure 2 shows the characteristic curves of the deflection growth over time of one of the samples with flexibility 76 depending on the level of stresses. For convenience, the curves are reduced to a common origin of coordinates.

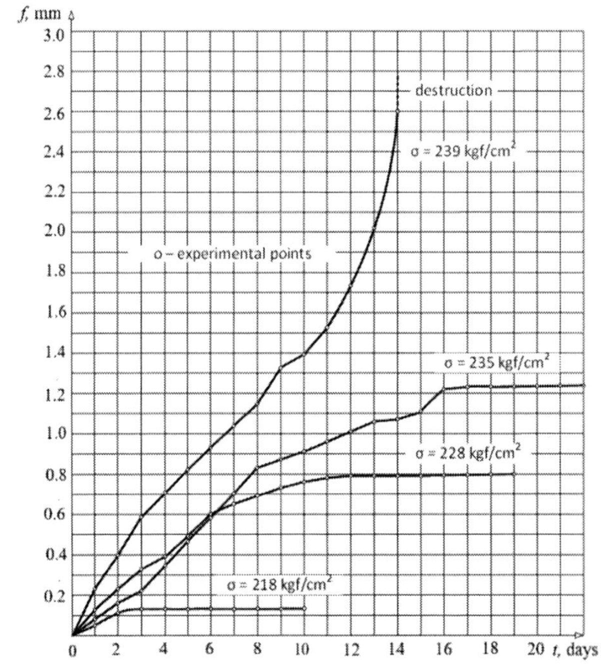

Figure 2. Growth curves of rod deflection over time as a function of stress level.

IOP Conf. Series: Materials Science and Engineering **939** (2020) 012080 doi:10.1088/1757-899X/939/1/012080

The growth curve of the relative edge compression strains of the rod according to the readings of the resistive strain gage at the stage of destructive stresses of 239 kgf/cm² is shown in figure 3.

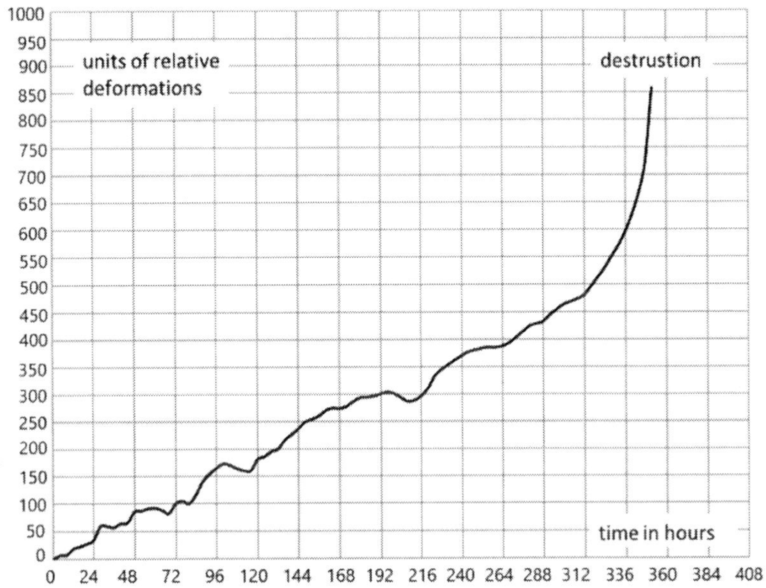

Figure 3. Growth curve of edge relative deformations according to strain gauges.

3. Conclusion

In general, the obtained curves of the development of creep deformations are in good agreement with the data of experiments on creep of wood by foreign researchers [3, 4].

The experiments showed that the exhaustion of the bearing capacity of wooden rods occurred from a loss of stability during longitudinal bending.

The results of long-term longitudinal bending test of the rods are presented in table 2.

Table 2. Long rod test results.

Sample №	Length l, cm	Section b×h, cm²	Flexibility λ	Breaking load N_{cr}, kgf	Breaking stresses σ_{cr}, kgf/cm²
1	56.5	3.00×3.01=9.03	65	2249	249
2	56.5	3.00×3.01=9.03	65	2093	232
3	56.5	3.00×3.01=9.03	65	2312	256
4	56.5	2.99×3.01=9.00	65	2218	246
5	65.5	3.00×3.01=9.03	76	2156	239
6	65.5	3.00×3.01=9,03	76	2218	218
7	65.5	2.98×3.01=8.97	76	1999	223
8	65.5	2.98×3.01=8.97	76	2061	230

The average value of breaking stresses for rods with a flexibility of 65 is 246 kgf/cm² (coefficient of variation is 4%). The average value of breaking stresses for rods with flexibility of 76 is 227 kgf/cm² (coefficient of variation is 4%).

The theoretical and experimental results obtained by the author are presented in the form of stress-flexibility relationships in Figure 4. The results are close to the data of L.P. Drozdova, who conducted extensive experiments on the central compression of the elements of wooden structures under long-term load [5].

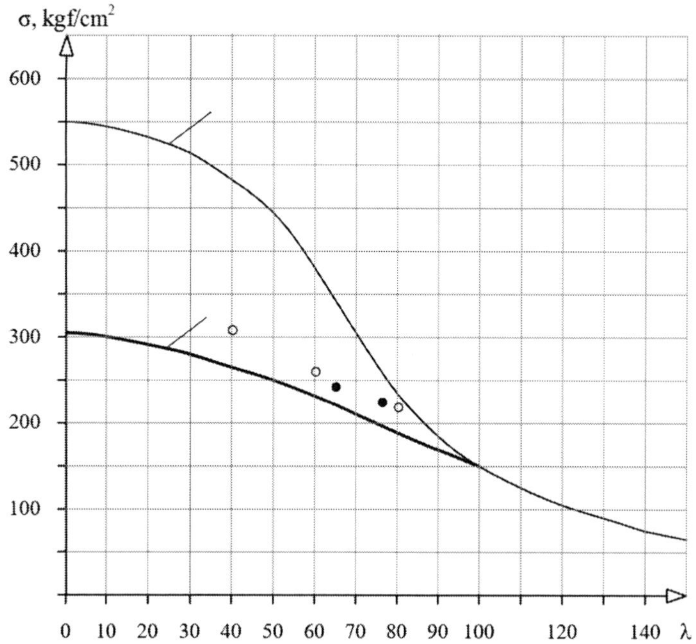

Figure 4. Theoretical and experimental dependences σ-λ for short-term and long-term central compression of wooden rods.

In figure 4: 1 – theoretical curve of short-term stability; 2 – theoretical curve of long-term bearing capacity taking into account the creep of wood; * – experimental points for short-term loading; • – experimental points during prolonged loading; ○ – experimental points under prolonged loading by L.R.Drozdova.

The maximum deviation of the experimental critical stress values during prolonged loading of wooden rods from theoretical data is less than 12%. Thus, we can conclude that the method for solving the problem of stability of centrally compressed rods taking into account the creep of wood proposed by the authors in [1] and [2] is reliable.

References

[1] Varenik K A and Varenik A S 2014 Dlitelnaya nesushchaya sposobnost derevyannyh konstrukcij [Long-term bearing capacity of wooden structures] *Stroitelnaya mekhanika inzhenernyh konstrukcij i sooruzhenij* **2** 23–31[In Russ.]

[2] Varenik K A, Varenik A S and Sanzharovskij R S 2014 Ustojchivost szhatyh derevyannyh konstrukcij s uchetom mgnovennoj nelinejnosti i nelinejnoj polzuchesti [Stability of compressed wooden structures taking into account instantaneous nonlinearity and nonlinear creep] *Nauchnoe obozrenie* **8(2)** 572–575 [In Russ.]

[3] Gressel P 1984 Kriechverhalten von Holz und Holzwerkstoffen – Folgerungen für den Formänderungsnachweis in DIN 1052 *Bauen mit Holz* **4** 216–223 [In Germ.]

[4] Schänzlin J 2010 Modeling the long-term behavior of structural timber for typical serviceclass-II-conditions in South-West Germany *Habilitation. Mitteilung des Instituts für Konstruktion und Entwurf* (Stuttgart: 2nd Institut für Konstruktion und Entwurf) p 185

[5] Drozdova L P 1978 *Ustojchivost celnyh centralno-szhatyh elementov derevyannyh konstrukcij pri dlitelnoj nagruzke* [Stability of integral centrally-compressed elements of wooden structures under prolonged load] Abstract thesis. PhD. tech. sciences (Moscow) p 18 [In Russ.]

Improving stability of self-propelled jib cranes

I A Zaselyaev[1]

[1] Yaroslav-the-Wise Novgorod State University, ul. B. St. Petersburgskaya, 41 173003 Veliky Novgorod, Russia

E-mail: Igor.Zaselyaev@novsu.ru

Abstract The article describes the developed and proposed to use device of outriggers, which provides an additional anchor force during operation of the crane on soil bases, that increases the holding torque of the crane. In the proposed external outrigger of the lifting machine, a screw which is screwed into the ground, and which has a conical tip and an angle of elevation of the helical surface less than the angle of friction of this surface against the soil, serves as a base plate. Upon that, the outrigger is equipped with an auger rotation drive, which, together with the feed drive, is screwed into the ground without chip separation. A calculation example shows that such outriggers can increase load stability, therefore, the lifting capacity or working radius of the crane can increase by 1.79 times and even more opposed to when the crane is working on conventional outriggers.

1. Introduction

Self-propelled jib cranes are among the most common machines used in the construction and installation works, due to their ability to move freely around the construction area. Such machines are also widely used to mechanize production processes in almost all industries. Among self-propelled jib cranes, predominantly automobile and pneumatic wheel cranes are equipped with external (optional) outriggers. When installed on outriggers near the crane, the support contour increases and its stability also increases, which allows for safely lifting heavier loads. The support contour is a contour formed by horizontal projections of straight lines, connecting the vertical axles of the crane supporting elements (of wheels or outriggers). Here it can be noted that the platform on which the crane is installed will usually have a slope relative to a perfectly horizontal plane and will not geometrically coincide with the reference contour. In modern self-propelled jib cranes, various outrigger device schemes (retractable, rotary, folding) are used, in which the outrigger support plates are pressed to the ground surface (soil) by means of mainly hydraulic cylinders and a hydraulic drive system. Outriggers are mounted on the running frame of the crane, as a result of which they unload the axles and pneumatic wheels of the machine when working with loads. To the greatest extent this effect is achieved when the crane is completely hung out on the outriggers, when the wheels get off from the ground surface. In the conditions of a construction or assembly site, self-propelled jib cranes often have to work on soil foundations of various load-bearing capacities. Moreover, cases of capsizing or accident of cranes as a result of loss of their cargo stability when lifting loads due to insufficient bearing capacity of soil bases are not uncommon. Therefore, according to the safety measures developed in Russia for self-propelled jib cranes, they must be installed and moved on soil bases, the bearing capacity of which is 0.3 ... 0.9 MPa, depending on the type and capacity of the crane. The bearing capacity of the soil must be determined before installing a crane on it. If the resistibility of the soil base is insufficient for the operation of the crane on it, the soil must be

Content from this work may be used under the terms of the Creative Commons Attribution 3.0 licence. Any further distribution of this work must maintain attribution to the author(s) and the title of the work, journal citation and DOI.
Published under licence by IOP Publishing Ltd

compacted or underlying devices must be used on it. When using underlying devices in the form of shields or plates, it is recommended to pour sand, gravel, and gravel with a layer of 5 ... 10 cm in order to avoid being crushed in loose or waterlogged soil. When installing the crane on a soil base, its angle of inclination,which is equal to the sum of the angles of inclination of the platform and draft caused by uneven deformation of the soil under the crane, should not exceed the values indicated in the crane passport. All of the above leads to complication of production processes and to in crease of cost of work during which self-propelled jib cranes working on soil bases are used, therefore, there is a need to develop outrigger devices that increase the cargo stability of cranes.

2. Stability analysis of cranes operating on outriggers

A feature of jib cranes is the lifting and moving of cargo in an area that extends beyond the support contour of the crane. The external forces impacting the crane create a tilting moment relative to one of the edges of the support contour (tipping ribs), and the crane's own weight creates a holding moment, respectively. The stability of the crane in working condition with exposed outriggers can be estimated by one of the load stability factors $\boldsymbol{k_1}$, defined as the ratio of the holding moment ($\boldsymbol{M_1}$) created by the weight of the crane without taking into account the additional loads and the slope of the platform on which the crane is installed, to the overturning moment created lifted load (M2).

$$k_1 = M_1/M_2 = G\,a/Q\,b \geq 1.5, \quad (1)$$

where G is the weight of the crane;

a is the distance from the projection point onto the horizontal plane of the coordinate of the center of gravity of the crane to the tipping rib;

Q is the weight of the load being lifted;

b is the distance from the tipping rib to the vertical axis of hoisting.

More precisely, the degree of stability of the crane in working condition is determined by the coefficient of cargo stability, calculated taking into account the additional loads (inertial, centrifugal forces, wind load and others) and the slope of the platform on which the crane is installed, that is, taking into account factors that reduce the holding torque from tipping over of the crane towards cargo.

If we evaluate the effectiveness of the outriggers by calculating the crane's load stability on the supports and without them, we can conclude that the crane stability in the process of lifting the load is provided only by its own weight and the dimensions of the support contour, affecting the position of the tipping rib. The outriggers of modern cranes are designed in such a way that, when lifting a load, the main part of the work to ensure the stability of the crane is done by the outriggers set in the direction of the lifted load, and they also make a tipping rib.

The outriggers placed in the opposite direction from the load being lifted practically do not participate in this work. Obviously, the stability of the crane would be higher if the outriggers set in the direction of the load to be lifted were pressed less into the ground when lifting the load, thereby providing a smaller inclination of the platform on which the crane is installed. Also, the stability of the crane could increase if the outriggers set in the opposite direction from the load being lifted would also actively participate in ensuring the stability of the crane, working as an effective soil anchors, creating an anchor force holding the crane and an additional holding moment that prevents the crane from tilting and tipping over.

3. The device of outriggers, providing increased stability for the crane

It should be noted that among the published inventions and utility models related to outriggers of machines, there are devices that are a helical anchor immersed in the ground by communicating translational and rotational motion to it [1, 2]. The helical anchor in this case is executed as a rod with one or two turns of the helical surface. In this work, we propose a device for the outrigger of a lifting machine for the case of work on soil foundations with low bearing capacity, in which a conical tip screw screwed into the ground and the angle of elevation of the screw surface is smaller than the angle of friction of this surface on the ground can be used as the base plate of the outrigger support. In this case, the outrigger is equipped with an auger rotation drive, which, together with the feed drive, ensures its screwing into the ground without chip separation. With this screwing, the screw should work in the ground not as a screw anchor immersed in the soil loosened by its lower blades, but as, for example, as a self-tapping screw screwed into wood. At the same time, to ensure the movement (feed) of the screw, corresponding to its rotation and screwing, and not drilling, the parameters of the rotation and feed drives should be interconnected by the ratio:

$$V = n\,L, \tag{2}$$

where V is the feed rate of the screw; n is the screw rotation speed; L is the pitch of the screw surface of the screw.

In the proposed device, the supporting surface (the lower plane of the screw surface of the screw) of the outrigger will provide a more effective support on the ground than conventional outriggers of cranes and will have a large total area of support. Such an outrigger, set in the opposite direction from the load to be lifted, interacting with the soil with the upper plane of the screw surface of the screw, can create a significant holding force that prevents the machine from tipping over. Outrigger device is shown in figure 1.

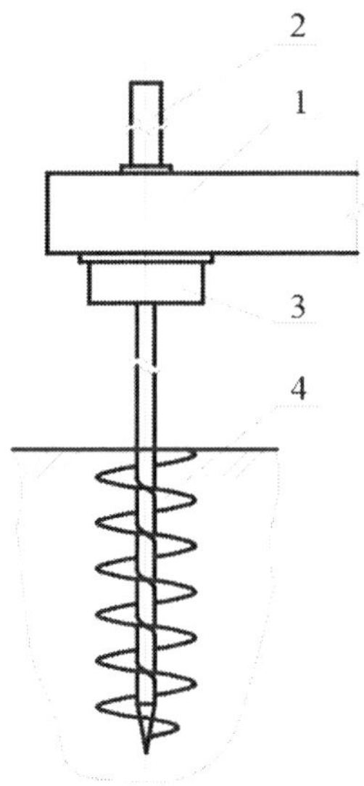

Figure 1. The device of the outrigger of the lifting machine, where: 1 – outrigger; 2 – feed drive; 3 – rotation drive; 4 – screw screwed into the ground.

Comparing the proposed outrigger with the outriggers that are already installed on cranes and are currently being used, and other hoisting machines, it can be noted that the outrigger in the form of a screwed screw, which is, as it were, a support plate for the currently used external support devices, the support will not occur on the ground, but on the buried, more dense layers of soil, characterized by a greater bearing capacity [3]. In this case, the support on the ground of one support will occur not by the area of one support plate, but by the area of all turns of the screw immersed in the ground, from which the total area of support can be very significant. The use of a screw with an angle of elevation of its screw surface less than the angle of friction of this surface against the ground will ensure that the screw is not twisted out of the soil if the drive mechanism for rotating the screw is not braked when pulling the screw out of the ground from the roll of the machine. This also ensures good grip of the screw with the ground. To achieve significant bearing capacity of the screw working as an anchor, the depth of its screwing it into the ground, if one is using the recommendations [4], should be 5 ... 6 diameters of the screw blade. It is also obvious that the screw which is screwed into the soil to the indicated depth will provide significant resistance to the inclination of its axle. By installing a crane for working on outriggers with the help of auger movement drives in the horizontal position of its frame, we can assume that the slope factor of the platform on which the crane is installed will lose its value. At the same time, the geological and geodetic conditions of the site on which the crane is installed can lead to unequal screwing depths of screw outriggers. The proposed outrigger, having a significant holding or anchor force, preventing the rollover of the machine, will significantly increase the load stability of the lifting machine, working on soils with low bearing capacity and this, in particular, will serve to prevent accidents. Such outriggers can be considered as a kind of prefabricated anchor foundations on which the crane is installed during its operation. In cases of operation of the machine with the proposed outriggers on solid foundations that do not allow screw screw to be screwed into the ground, as well as on durable coatings, the proposed supports can be installed on them without screwing and work as usual, resting on the surface of the solid foundation (coating) through a metal gasket which is in such cases put under the tip of the screw.

4. Evaluation of the effectiveness of the proposed outriggers

Let us evaluate the stability of a crane working on the proposed outriggers creating a holding anchor force with a load stability coefficient of k_{1A} using a specific example. The assessment will also be carried out simplified as in the calculation of k_1 according to formula (1), calculating the holding moment M_1 in the numerator of the formula without taking into account additional loads and the slope of the platform on which the crane is installed. In this case, we take into account only the additional holding force F_A, which creates an additional holding moment, for each of the two supports set in the opposite direction from the load being lifted. The factor of providing the proposed outriggers exposed in the direction of the load to be lifted, support occurring not on the ground, but on the buried, denser layers of the soil, as well as the fact that the support on the ground will occur over the area of all turns of the screw immersed in the ground, and not over the area of one base plate, due to the uncertainty of the specific geological conditions of the crane and to simplify the estimated calculations will not be taken into account.

The scheme for calculating the coefficient k_{1A} is shown in figure 2.

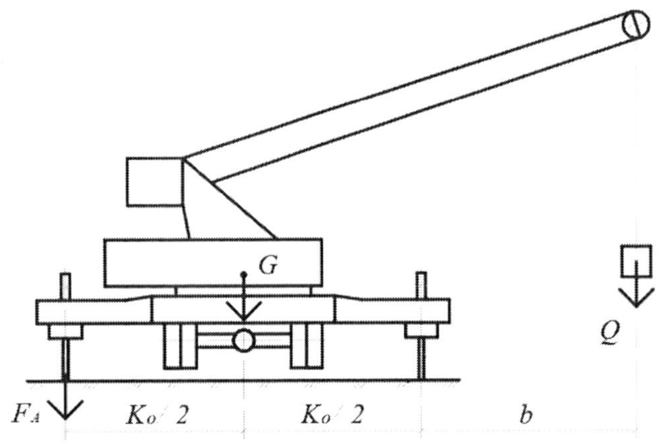

Figure 2. Scheme of a jib crane on outriggers and loads for calculating cargo stability.

Then the formula for calculating k_{1A} will be written as:

$$k_{1A} = M_1/M_2 = (G\,K_o/2 + 2F_A\,K_o)/Q\,b \geq 1.5, \tag{3}$$

where G is the weight of the crane;

K_o is the distance between the outriggers;

F_A is the anchor force created by the outrigger;

Q is the weight of the load being lifted;

b is the distance from the tipping rib to the vertical axle of the load.

We will carry out a calculated assessment for the KS-55715 "Galichanin" crane by calculating k_{1A}, respectively, according to formula (2) for the specified crane, as if working on the proposed outriggers, and k_1 calculations according to formula (1) for the crane working on its outriggers pylons. Calculations are carried out for the case of lifting the load from the side as shown in figure 2. According to the reference data, this crane with a jib of 21.7 m is characterized by: a lifting capacity of 7 tons at a working radius (distance from the axle of rotation of the turning part to the vertical axle of the hook suspension) 6 m, and a lifting capacity of 1 t at a working radius of 18 m; the size of the support contour formed by the outriggers along the axle of the chassis (B_o) and across the axle; of the chassis of the crane (K_o), respectively 4.6 and 5.6 m; weight of the crane is 22.8 tons. To estimate the values of F A in the formula (3), we use the graphic dependences of the anchor force created by screw anchors in [4] in Figure 13 on the diameter of the anchor blade and the type of soil. Having taken the immersion depth of the bearing (lower) anchor blade in soil of 1 m, the blade diameter of 200 mm and the type of soil with low bearing capacity (soft foul clay), we determine the bearing capacity of one screw anchor – 45 kN or about 4.5 tf.

As a result of calculations of k_{1A} according to the formula (2) and k_1 according to the formula (1) for the crane lifting capacity of 7 tons and a working radius of 6 m, the coefficients k_{1A} and k_1 are obtained; they are, respectively, 5.1 and 2.85. For the crane lifting capacity of 1 t and the working radius of 18 m, the values of the coefficients k_{1A} and k_1 are obtained, and they are, respectively, 7.52 and 4.2. The ratios of the coefficient k_{1A} to k_1 for the indicated different values of lifting capacity and the working radius are the same, equal to 1.79.

5. Conclusions

The estimated calculations indicate a very significant increase in the cargo stability of cranes operating on soil foundations with low bearing capacity, if the proposed outriggers are used on them. An increase in the cargo stability coefficient of the crane by 1.79 times also means that with the same value of the cargo stability coefficient as when the crane was working on its outriggers, and when the proposed outriggers were used, it is possible to increase the crane's carrying capacity on the working radiuses indicated in the example. Also, when the crane is working on the proposed supports, an increase in working radiuses is possible at the same load capacities. This will provide the jib crane with great opportunities for installation and lifting works. It can also be noted that when using the proposed outriggers, there is no need for such expensive and time-consuming measures as compaction of the soil or coating of soft soils with shields and slabs. A patent was obtained for the proposed device [5].

Based on the foregoing, it is possible to recommend the general requirements for the devices of the outriggers of cranes or other lifting machines working on soil bases: the outriggers should transfer the load from the weight of the machines with the load not to the ground's surface, but to the buried layers of soil and work at the same time as the soil anchors.

References

[1] Levchenko S A, Mirkin S N and Egorov K D 2007 *The portable support of the earth moving machine* [Vynosnaya opora zemlerojno-transportnoj mashiny] Patent RU 61699 U1, MKI B66S 23/78 [In Russ.]

[2] Scherbakov V S and Bychkov I S 2019 *Remote support of the hoisting machine* [Vynosnaya opora gruzopodjemnoj mashiny] Patent RU 187891 U1, MKI B66S 23/78 [In Russ.]

[3] Verruijt A 2018 *An introduction to soil mechanics* (Springer) p 420

[4] Penchuk V A 1985 *Screw piles and anchors for supports* [Vintovye svai i ankery dlya opor] (Kyiv: Budivelnik) p 96

[5] Zaselyaev I A 2019 *Remote support of the hoisting machine* [Vynosnaya opora gruzopodjemnoj mashiny] Patent RU 2707158 C1, MKI B66S 23/78 [In Russ.]

IOP Conf. Series: Materials Science and Engineering **939** (2020) 012082 doi:10.1088/1757-899X/939/1/012082

Technology for forming micro devices based on gallium nitride

A V Zhelannov[1,3], B I Seleznev[2] and D G Fedorov[1]

[1] JSC «OCB - Planet», ul. Bolshaya Moskovskaya, 13a, room 1n, 173004, Veliky Novgorod, Russia
[2] Yaroslav-the-Wise Novgorod State University, ul Bolshaya St.Petersburgskaya, 41 173003, Veliky Novgorod, Russia

[3] E-mail: ZhelannovAV@okbplaneta.ru

Abstract. The technology of forming micro devices based on gallium nitride, including the main stages of manufacturing, namely, inter-assembly isolation, formation of non-straightening (ohmic) and straightening (Schottky barrier) contacts, surface passivation, formation of inter-electric connections in the form of "air bridges", plate grinding, cutting plate into crystals and sorting out, is considered. A short technological cycle of manufacturing microstructures with a description of the main operations is given. As a result of this work, experimental samples of transistors with a gate length of 0.5 microns were obtained.

1. Introduction

Currently, much attention is paid to such wideband semiconductors as gallium nitride, silicon carbide and diamond as promising materials for the creation of effective devices for power and high-temperature electronics. The technology of devices based on gallium nitride is developing most intensively, thanks to its use for the production of LEDs. At the same time, a unique combination of physical properties, including a large band gap, a high drift velocity of electron saturation, high breakdown voltages, high thermal conductivity, and high chemical and thermal stability, makes it possible to consider nitride compounds not only in optoelectronics, but also for the development of a component base of power SHF electronics [1].

The system of solid solutions based on $Al_xGa_{1-x}N$, as a material for high-temperature, high-voltage, high-frequency and high-current applications, can significantly expand the operational capabilities of semiconductor technology. A unique combination of physical properties allows considering these materials to create a new generation of microelectronic devices. It is with the introduction of such materials that a breakthrough in SHF and high-current electronics is currently associated.

The aim of this work is to develop the basic technological operations of transistors formation based on the AlGaN/GaN structure in order to create a prototype and to study the characteristics of the obtained instrument structures.

2. Methods of Experiment

AlGaN/GaN heterostructures grown by the MOCVD method (chemical vapor deposition using organometallic compounds) on a 2-inch diameter sapphire substrate were used as the starting material for the development of the transistor manufacturing process. The mobility and concentration of charge

Content from this work may be used under the terms of the Creative Commons Attribution 3.0 licence. Any further distribution of this work must maintain attribution to the author(s) and the title of the work, journal citation and DOI.
Published under licence by IOP Publishing Ltd

carriers in the used heterostructures were 1650 cm^2/V s and 1.27 $\times 10^{13}$ cm^{-2}. The structure of the source material is shown in figure 1.

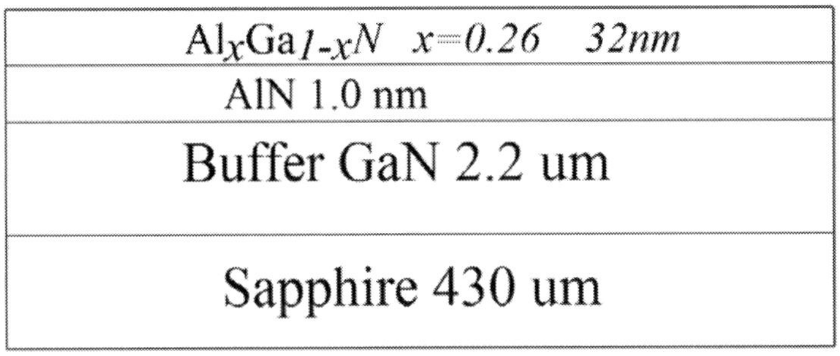

Figure 1. Schematic cross-sectional view of AlGaN/GaN heterostructure.

For the formation of microstructures based on GaN, it is necessary to perform a number of technological operations. The determining operations that make the main contribution to the parameters of microstructures are the creation of reliable inter-instrument insulation and the formation of high-quality metal – semiconductor contacts.

2.1. The formation of inter-instrumental insulation of structures based on gallium nitride

For etching materials of group A^3B^5, both chemical ("liquid") and plasma ("dry") etching methods can be used. "Liquid" methods of material removal involve the use of liquid solutions for chemical etching of a semiconductor areas unprotected with mask. These methods are simple to implement and do not require sophisticated technological equipment. But when working with AlGaN/GaN heterostructures, the "liquid" method is not widely used, because due to the strong chemical bonding in nitride semiconductors it is difficult to choose solutions that allow the chemical etching of the material. GaN photostimulated electrochemical etching is reported; however, this method depends on many parameters and is poorly reproducible [2, 3]. Due to the limited possibilities of liquid etching of nitride semiconductors, the main efforts of the technologists were directed to the development of dry etching methods. To implement the dry etching method in this work, we used a plasma-chemical etching unit with an inductively coupled plasma (ICP) source.

Inter-instrument isolation in the form of a mesa structure is carried out by etching the heteroepitaxial structure below the level of a two-dimensional electron gas formed at the interface between the AlGaN and GaN layers.

The binding energy for GaN is 8.92 eV/atom; therefore, the etching conditions for this material are significantly different from silicon, gallium arsenide and indium phosphide, semiconductors with well-studied etching processes.

To form the mesa areas, etching modes of nitride layers in plasma based on the $Cl_2/BCl_3/Ar$ gas mixture were developed. The development of etching processes consisted in the selection of modes that ensure a sufficiently wide range of etching rates required for various applications [3].

For the heterostructures considered in this work, the mesa areas were etched to a depth of about 200 nm, which ensured reliable inter-instrument isolation and the necessary relief for combining the topological pattern on the plate.

When choosing masking coatings, it is necessary that they have chemical resistance to an aggressive plasma gas atmosphere, have high adhesion to the substrate, and are easily removed after etching without changing the morphology of the semiconductor surface [4].

To carry out the etching process, a mask based on SP15 photoresist prepared for etching in a chlorine-containing medium was used. Such a mask provides reliable protection of the heterostructure, the absence of erosion of the edges of the mask and ease of removal after the etching process.

The structures were etched at an inductively coupled plasma source power of 600 W, a displacement of 100 W, a pressure of 1.2 Pa, and a gas flow ratio of $BCl_3/Cl_2/Ar$ equal to 20/60/10 cm^3/min. The etching time was 25 seconds.

Thus, the formation of inter-instrument insulation requires the following sequence: photolithography of etching windows of a semiconductor, plasma-chemical etching of a semiconductor, removal of a photoresist mask and control of the depth of etching.

2.2. The formation of ohmic contacts to heterostructures based on gallium nitride

The next stage in the formation of the instrument structure is the formation of ohmic contacts to the heterostructure. Ohmic contacts must meet a number of requirements: low contact resistance, good adhesion of metals to the semiconductor, a flat edge of the contact boundary, resistance to electromigration, high thermal stability, preservation of morphology during heat treatment, good reproducibility.

To create an ohmic contact to nitride heterostructures, Ti-based multicomponent contacts are most often used, which during heat treatment form compounds with a low output work. The low resistance of the ohmic contact is usually associated with the formation of nitrogen vacancies due to the interaction of the semiconductor with the contact material, for example Ti.

Contacts are formed by explosive photolithography using a two-layer photoresist system that provides a negative slope of the wall profile to improve the "explosion" and remove metal. This is achieved by reversing the latent positive image. Figure 2 shows the main stages of contact formation.

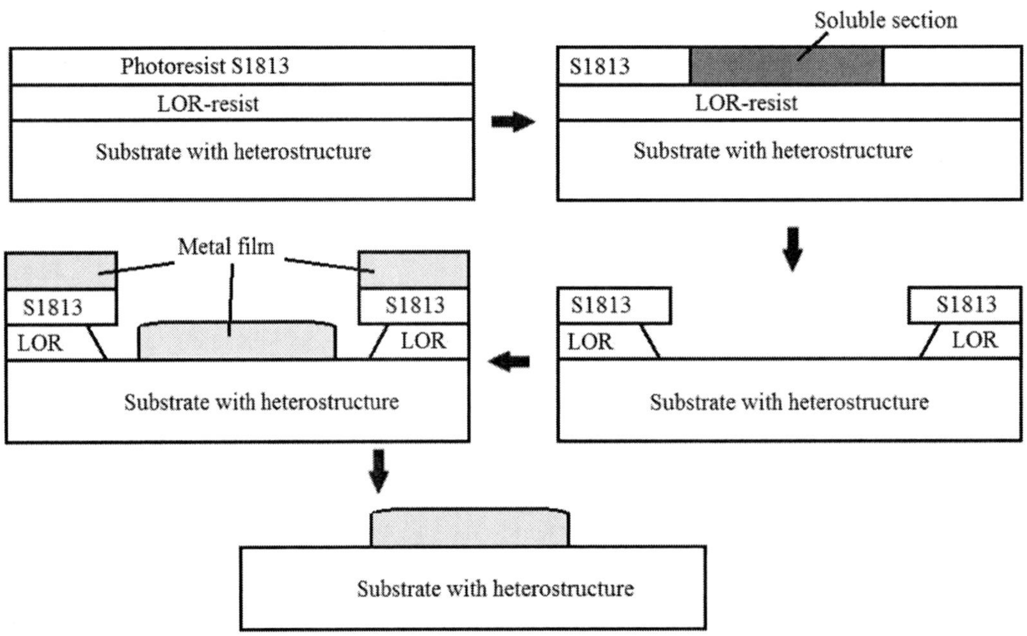

Figure 2. Stages of forming a resistive mask by explosive photolithography.

In this work, the Ti/Al/Ni/Au system was used as metallization of ohmic contacts. Contacts were formed using explosive photolithography and electron beam metal sputtering.

After creating the contact pads, annealing was carried out at a temperature of 720 °C for 45 seconds. Contact resistance was measured with test elements using the long line method [5] at the Cascade Microtech semi-automatic station. The Agilent B1505A semiconductor performance meter was used as a measuring part for monitoring parameters at a direct current. Current-voltage characteristics and the appearance of ohmic contacts are shown in figures 3 and 4.

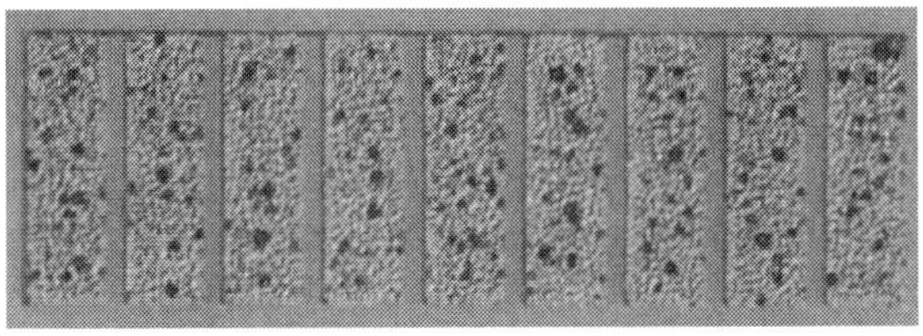

Figure 3. Optical microscopy images of Ti/Al/Ni/Au on the AlGaN/GaN heterostructure.

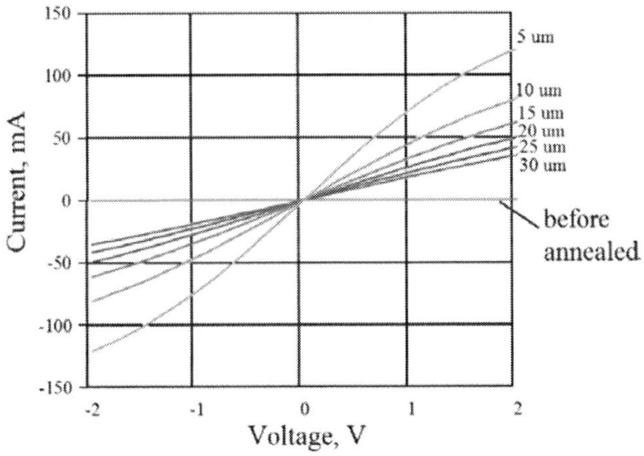

Figure 4. The *I–V* characteristic for Ti/Al/Ti/Au to AlGaN/GaN before and after annealing at 720°C.

Calculation of specific contact resistance (ρ_c), performed by the long-line method, shows ρ_c at the level of 0.6 Ohm · mm.

2.3. Formation of a straightening contact (Schottky barrier) to heterostructures based on gallium nitride

The SHF parameters of modern microwave transistors at heterojunctions are mainly determined by the gate parameters of circuits based on them. Obtaining MIS amplifiers for a certain frequency range is determined by the material of the semiconductor and the geometry of the gate.

The formation of a mushroom-shaped gate consists of the following key operations: electron beam lithography to obtain a profile of the future gate; application of barrier metallization; "explosion" of an electronic resist.

Let us consider the stage of formation of the mushroom-shaped profile in the resist.

After dehydration on the tile at a temperature of 120° C, a multilayer system of electronic resists with intermediate drying on the tile was applied to the plate by centrifugation (figure 5a, 5b). The system consists of layers PMMA950K/PMGI/Copolymer/PMGI/PMMA950K with a total thickness of about 1.3 μm. The sequence of layers and their thicknesses were optimized for separate exposure and controlled manifestation of mushroom-shaped profiles.

The prepared plate is exposed in an electron beam exposure apparatus. The upper areas of the barriers 0.8 μm wide are the first to be exposed, as well as the barrier pads (first exposure). Next, a sequential manifestation of the upper three layers of the resist in the developers was carried out

After that, a second exposure was carried out to form lithography under the gate subfeet with the required nominal size and dose supplement on the sites. Next, the lower two layers of the resist appeared (figure 5c, 5d).

As a result, a mushroom-shaped profile of the developed areas in the resist was formed. After that, stripping in oxygen plasma from the residues of the resist at the bottom of the developed areas is carried out.

After stripping, the necessary gate metallization is applied in oxygen plasma (figure 5e). Gate metallization based on the Ni/Au system was formed on the electron beam spraying unit. The "explosion" of metallization was carried out with dimethylformamide (figure 5f).

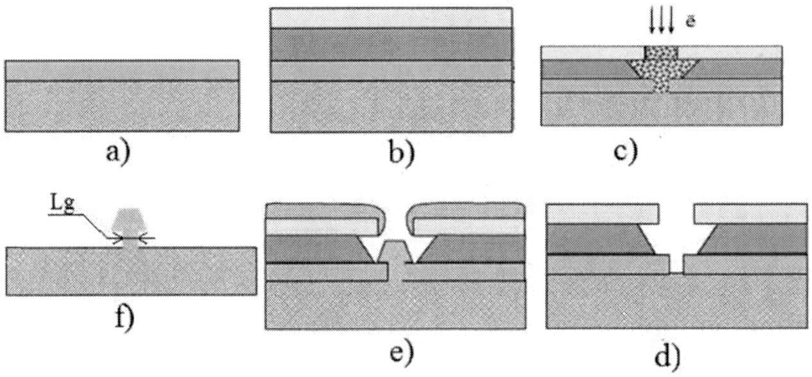

Figure 5. Stages of forming a mask of electronic resists to form a T-shaped gate.

The gate length, L_g, obtained by electronic lithography was 0.5 μm. The gate metallization contacts are a set of single gates with a width of 100 μm and are shown in figure 6.

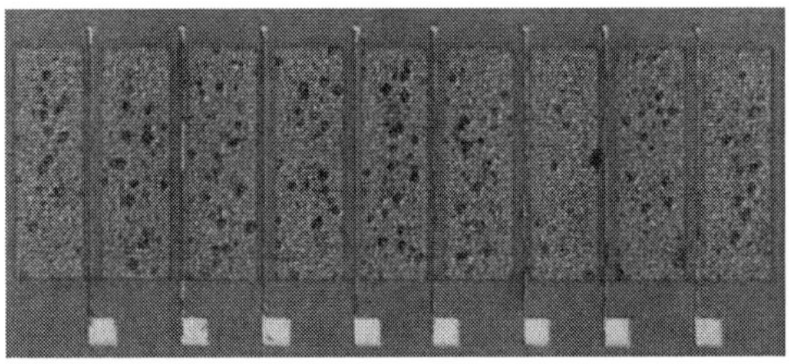

Figure 6. Optical microscopy images of gates.

2.4. Surface passivation, formation of contact pads and inter-electric connections

Surface passivation protects the structures from environmental influences and reduces surface effects. As passivation, a silicon nitride film was used, obtained by plasma-chemical deposition using a gas mixture based on monosilane, argon, and nitrogen.

The contact pads of the drain, source, and gate are formed using explosive photolithography and electron beam sputtering of the Ti/Au system.

The source contacts are connected using air bridge technology. Herewith, it first performs photolithography of the bridge resist, which determines the height of the bridge, and then the pattern in the photoresist is exposed to temperature to form a sloping edge necessary for continuous deposition of the metal layer. Then, a Ti/Au inoculating layer is formed at the electron beam spraying unit. At the next stage, photolithography is carried out to build up galvanic gold. After growing gold, unnecessary areas of the inoculating layer are removed by chemical etching. The final operation is the removal of the under-bridge photoresist.

The image of the finished transistor structures is shown in figure 7.

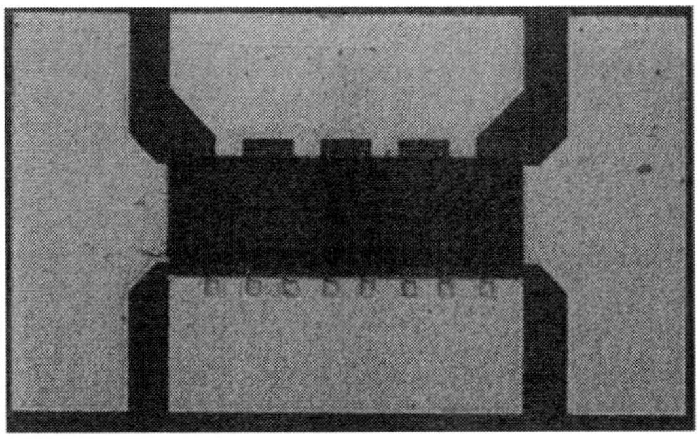

Figure 7. Optical microscopy images of finished transistor structures.

After the formation of transistor structures on the plate, the operation of thinning the sapphire substrate to a thickness of 80 μm was carried out with further laser separation into crystals. Figure 8 shows the image of the transistor crystal obtained by the technology described above.

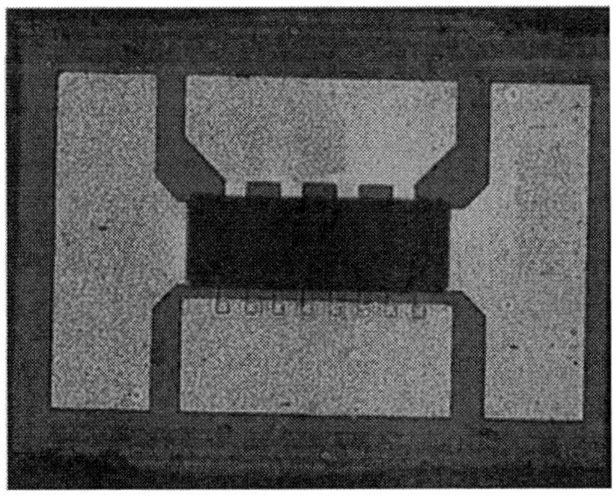

Figure 8. Optical microscopy images of crystal transistor with periphery 8×100 um.

3. Results of measurements

On the obtained transistor crystals, an analysis of the output current-voltage and transition characteristics is carried out. Figure 9 shows the output current-voltage characteristics of transistor crystals.

The maximum drain current is 0.60 A/cm^2, the steepness is 155 mS/mm.

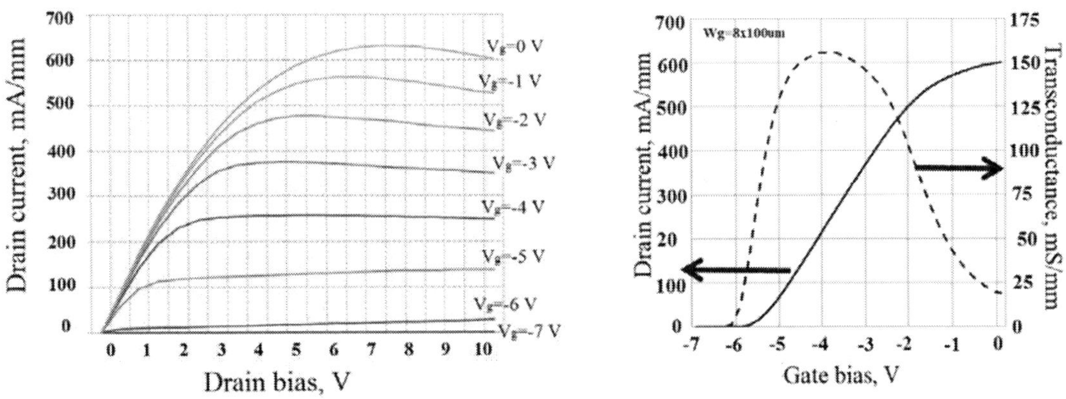

a) b)

Figure 9. The (a) output characteristics and (b) transfer characteristics of an AlN/GaN HEMT with a 0.5 um gate length.

IOP Conf. Series: Materials Science and Engineering **939** (2020) 012082 doi:10.1088/1757-899X/939/1/012082

Another important parameter of transistors is the value of the gate-drain capacitance. The value of capacitance has a key effect on the frequency characteristics of transistors. The dependences of the gate-drain capacitance on the gate voltage for the structures under consideration are shown in figure 10. At a gate voltage of 0 V, the capacitance values are 1.12 pF.

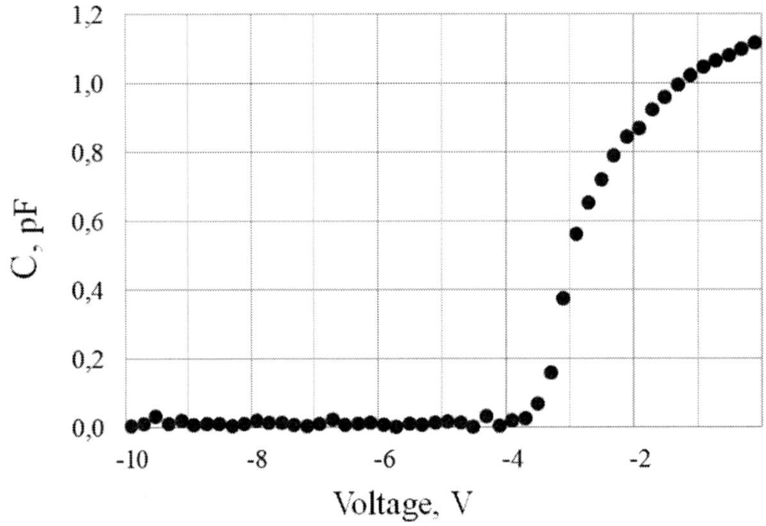

Figure 10. Gate-drain capacitance voltage on the gate.

For SHF measurements, transistor crystals were mounted on Al_2O_3 boards, as shown in figure 11. The board was formed according to standard technology using current-carrying tracks based on V-Cu-V metallization and a through-hole system.

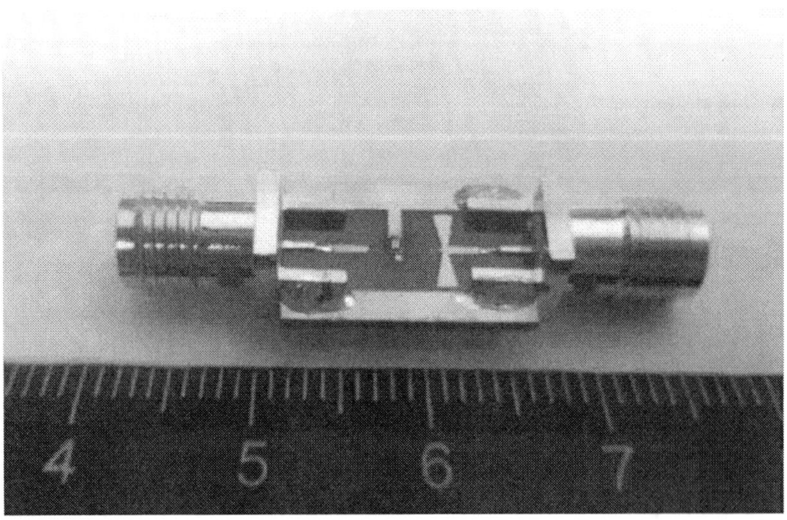

Figure 11. Optical microscopy images of crystal transistor with periphery 8×100 um.

The SHF properties were studied on a 5230A vector analyzer. The measurement results are presented in figure 12.

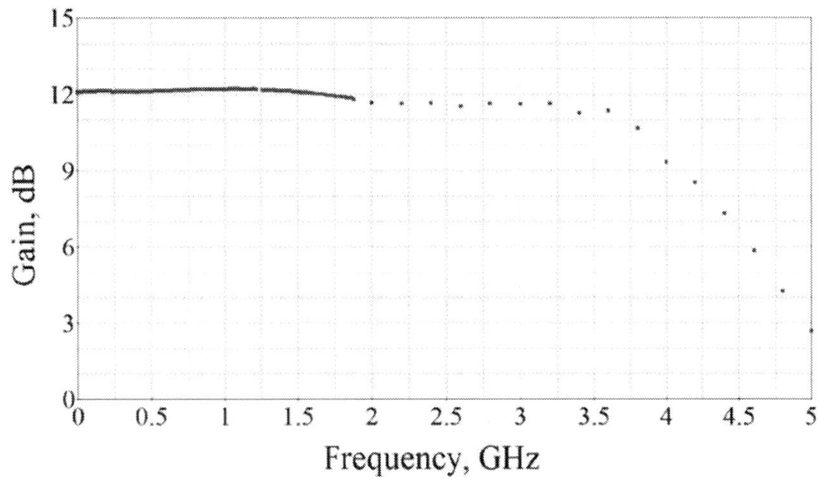

Figure 12. SHF performance of the 0.5-um-gate-length HEMT.

The output power at a frequency of 3.5 GHz was 0.25 W/mm.

4. Conclusion

As a result of the research, the technology of post-growth processing of transistor structures based on gallium nitride was obtained. The key operations of manufacturing gallium nitride-based micro devices are considered, namely: the formation of inter-instrument insulation, the formation of non-straightening (ohmic) and straightening (Schottky barrier) contacts, surface passivation, the formation of inter-electrical connections in the form of "air bridges", grinding of a plate, cutting of a plate into crystals and sorting out. Experimental samples of transistor crystals with fairly high characteristics were obtained.

Acknowledgments

The publication was prepared with the financial support of the Ministry of Science and Higher Education of Russia within the framework of the project part of the state assignment, project No. 3.3572.2017/PCh.

References

[1] Balakirev A 2015 *Jelektronika: NTB* **4** 64–68
[2] Ma L *at al.* 2006 *Proc. of the CSMANTECH Conf.* 105–108
[3] Usikov A S 2016 *Tech. Phys. Lett.* **42(9)** 80–87
[4] Zhelannov A V *at al.* 2017 *Nano- and microsystem technology* **19(7)** 399–405
[5] Protasov D Yu 2014 *Tech. Phys.* **84(9)** 96-99
[6] Desmaris V 2004 *Electrochemical and Solid-State letters* **7(4)** G72–74

AUTHOR INDEX

Institute of Physics
Dirac House, Temple Back
Bristol BS1 6BE UK

ISSN: 1757-8981
ISBN 978-1-7138-4774-8

CURRAN ASSOCIATES INC.
proceedings.com